BAYESIAN STATISTICS 8

BAYESIAN STATISTICS 8

Proceedings of the Eighth Valencia
International Meeting

June 2–6, 2006

Edited by

J. M. Bernardo
M. J. Bayarri
J. O. Berger
A. P. Dawid
D. Heckerman
A. F. M. Smith
and
M. West

OXFORD
UNIVERSITY PRESS

OXFORD
UNIVERSITY PRESS

Great Clarendon Street, Oxford OX2 6DP

Oxford University Press is a department of the University of Oxford.
It furthers the University's objective of excellence in research, scholarship,
and education by publishing worldwide in

Oxford New York

Auckland Cape Town Dar es Salaam Hong Kong Karachi
Kuala Lumpur Madrid Melbourne Mexico City Nairobi
New Delhi Shanghai Taipei Toronto

With offices in

Argentina Austria Brazil Chile Czech Republic France Greece
Guatemala Hungary Italy Japan Poland Portugal Singapore
South Korea Switzerland Thailand Turkey Ukraine Vietnam

Published in the United States
by Oxford University Press Inc., New York

First published 2007

British Library Cataloguing in Publication Data
Data available

Library of Congress Cataloging in Publication Data
Data available

Typeset by editors using TEX
Printed in Great Britain
on acid-free paper by
Biddles Ltd., King's Lynn, Norfolk

ISBN 978–0–19–921465–5

1 3 5 7 9 10 8 6 4 2

PREFACE

The Eighth Valencia International Meeting on Bayesian Statistics was held in Benidorm (Alicante, Spain), 150 kilometres south of Valencia, from June 2nd to June 6th 2006. The meeting was convened with the Eighth World Meeting of the International Society for Bayesian Analysis (ISBA). Valencia/ISBA 8 continued the tradition of this premier conference series – established in 1979 with the first Valencia International Meeting – as the forum for a definitive overview of current concerns and activities in Bayesian statistics.* In this tradition, Valencia/ISBA 8 encompassed a wide range of theoretical and applied research, and also notably highlighted the breadth, vitality and impact of Bayesian thinking in interdisciplinary research.

The Valencia organising committee invited 20 leading experts to present papers, each of which was followed by discussion led by an invited discussant. ISBA selected 32 contributed papers for plenary oral presentation, and a further 326 papers were presented in three poster sessions. The conference was preceded by a day of expository tutorials on Bayesian statistics.

These *Proceedings* contain the 20 invited papers with their discussions, and synopses of 19 contributed papers (of which five were presented orally and 14 as posters) that were selected by a rigorous refereeing process. The full versions of these contributed papers appear in the *Bayesian Analysis* electronic journal.

The papers cover a broad range of topics.

Foundational issues in statistics are addressed by several authors. At the inferential interfaces, **Mira and Baddeley** are interested in the potential for estimating equation methods to provide unifying opportunities in deriving statistical estimators, while **Rousseau** explores theoretical relationships between null and interval testing from both Bayesian and non-Bayesian viewpoints. Objective Bayesian foundations are addressed by **Wallstrom**, who revisits marginalization paradoxes and proposes resolutions in countably additive settings.

Disciplinary interface foundations are investigated in two papers. **Bishop and Lasserre** discuss discriminative and generative approaches to learning, representing interface statistics-machine learning perspectives in classification problems. The foundations of Bayesian statistics intersect the foundations of quantum theory, and **Schäck** provides an overview of recent and emergent

* The *Proceedings* of previous meetings have been published: the first by the University Press, Valencia (1980); the second by North-Holland, Amsterdam (1985); and the third, fourth, fifth, sixth and seventh by Clarendon Press, Oxford (1988, 1992, 1996, 1999, 2003). The editors in each case were the members of the organizing committee.

advances in interpreting and defining quantum mechanical probabilities from a subjective Bayesian viewpoint.

Research in Bayesian non-parametrics percolates through the proceedings. Several papers focus on extending and applying variants of Dirichlet process models and mixtures. **Gelfand, Guindani and Petrone** provide a synthesis of Bayesian spatial modelling with Dirichlet processes to generate a new class of non-parametric spatial models, while **Ghahramani, Griffiths and Sollich** discuss Dirichlet process and related models for binary latent variables applications. Theoretical developments are discussed by **Kokolakis and Kouvaris**, who construct classes of continuous random measures by mixing Dirichlet process building blocks.

Flexible models for Bayesian non-parametric regression and function fitting are the primary focus of two papers. **Clyde and Wolpert** develop a class of Bayesian non-parametric regression models, and associated computational methodology, based on Lévy process priors, while **Mertens** develops semi-parametric, basis-function approaches to regression with a proteomic application.

The growth and development of objective Bayesian methods in the last several years is reflected in a number of papers. Among these, **Sun and Berger** develop and explore classes of objective priors in multilinear/normal contexts, **Bernardo and Pérez** inject objective Bayesian decision theory using information-theoretic loss functions in the normal means problem, while **Cano, Kessler and Salmerón** discuss objective prior specification and testing in the related one-way random effects model.

Theory and methods for model assessment and testing are the primary focus of a number of papers. **Almeida and Mouchart** discuss model specification testing of a parametric null model embedded in an encompassing class of non-parametric alternatives, while **Chakrabarti and Ghosh** develop theoretical studies of cross-validatory Bayes factors and related methods in model selection for prediction. Developments in Bayesian decision theory linked to 'large p' multiple testing problems are addressed by **Müller, Parmigiani and Rice**, motivated by genomic applications. **Peruggia** develops Bayesian analysis for checking a linear hierarchical model by embedding in models with correlated error structures, while **Spitzner** develops theoretical investigation of testing under both objective and smoothness priors in the 'large p' normal means problem.

In time series and forecasting, **Carvalho and West** introduce and develop a synthesis of matrix-variate dynamic models with graphical modelling, illustrated in applications to Bayesian portfolio decision-making in finance.

The growth of structured probabilistic modelling for challenging problems in molecular biology, genetics and genomics is seen in several papers.

Among these, **Brooks, Manolopoulou and Emerson** develop highly structured mixture models, and associated computational methods, in problems of evolutionary molecular genetics with ecological/phylogeographic applications, while **Cowell, Lauritzen and Mortera** discuss Bayesian mixture modelling in DNA forensic analysis. **Merl and Prado** introduce, develop and apply structured Bayesian models of DNA sequence data in challenging problems of detecting natural selection, while **Xing and Sohn** develop hidden Markov Dirichlet process models in studies that aim to jointly model recombination and coalescence events in molecular population genetics.

While computational research is evident in many papers, several authors are concerned primarily with computational questions. The evaluation of marginal likelihood values in Bayesian analysis is addressed in two papers. **Raftery, Newton, Satagopan and Pavel** develop new computational approaches to this problem based on the harmonic mean identity, while **Skilling** introduces a new 'nested sampling' approach inspired by methods in statistical physics. Sequential computation using sequential Monte Carlo methods is broadly reviewed and developed theoretically by **Del Moral, Doucet and Jasra**, while **Holmes and Pintore** explore sequential "relaxation" approaches for iterative development of regression and distributional models. In the MCMC domain, **Möller and Mengersen** provide theoretical development of a method for estimation of posterior expectations of monotone functions using theory of dominating processes.

Biomedical applications of Bayesian methods continue to represent a major area of success and growth of more realistic, complex statistical modelling. **Dukíc and Dignam** develop flexible multiresolution models for survival hazard functions applied to breast cancer recurrence, while **Jirsa, Quinn and Varga** discuss sensitive prior modelling of response curves in radiotherapy studies.

Bayesian research and applications in spatial statistics have expanded substantially over the last decade, and several authors address aspects of this broad field. Practical questions of prior elicitation are developed in **Denham and Mengersen** in spatial/ecological contexts. **Gamerman, Salazar and Reis** overview a range of recent methodological developments and introduce new spatio-temporal systems using classes of dynamic Gaussian process models, while **Ma and Carlin** develop hierarchical spatial models utilizing multivariate conditional autoregressions.

Bayesian methods in social and policy sciences are evident in several papers. **Little and Zheng** discuss and overview Bayesian thinking and methods in finite population survey sampling studies, while **Madrigal** extends influence diagrams to incorporate experimental design intervention nodes motivated by policy decision assessment studies.

The growth and success of complex Bayesian modelling for applications in challenging scientific areas is reflected in several papers. Among these, **Schmidler** introduces novel Bayesian models for geometric shapes, and efficient computational approaches to shape matching, with applications to problems of protein structure alignment and classification, while **Short, Higdon and Kronberg** develop Bayesian process convolution and smoothing spline models in a study of estimation of galactic magnetic fields.

Estimation of multi-regime models, including change-point problems, are considered by several authors. **Girón, Moreno and Casella** revisit change-point problems in regression from an objective Bayesian viewpoint, while **Hutter** discusses MCMC and model selection questions for piecewise constant response functions.

We are most grateful to a number of organizations that provided support for the meeting. These include the *Universitat de València*, the *International Society for Bayesian Analysis*, *Microsoft Corporation*, and the *National Science Foundation of the USA*.

The review and selection of contributed papers for parallel publication in this volume and *Bayesian Analysis* (`http://ba.stat.cmu.edu`) was handled by the editorial staff of the journal. We are most grateful for the efforts of founding editor-in-chief Rob Kass, current editor-in-chief Brad Carlin, managing editor Herbie Lee, system managing editor Pantelis Vlachos, editors Alicia Carriquiry, Phil Dawid, David Dunson, David Heckerman, Michael Jordan, Fabrizio Ruggeri, and Dalene Stangl, and of course numerous anonymous associate editors and referees. The papers accepted and appearing in Bayesian Analysis are represented here by their extended synopses.

We are also most grateful to Mailo Albiach, Josefina Rodríguez, Lizbeth Roman, Vera Tomazella, and Dolores Tortajada for their invaluable assistance on matters administrative, technical and social, and in particular to Dolores Tortajada for preparing the final LaTeX version of these *Proceedings*.

We look forward to the Ninth Valencia International Meeting on Bayesian Statistics, which will again be convened jointly with ISBA and is planned to take place in the early summer of 2010.

M.J. Bayarri
J.O. Berger
J.M. Bernardo
A.P. Dawid
D. Heckerman
A.F.M. Smith
M. West

CONTENTS

I. INVITED PAPERS (with discussion)

II. CONTRIBUTED PAPERS (synopsis)

INVITED PAPERS
(with discussion)

BAYESIAN STATISTICS 8, pp. 3–24.
J. M. Bernardo, M. J. Bayarri, J. O. Berger, A. P. Dawid,
D. Heckerman, A. F. M. Smith and M. West (Eds.)

Generative or Discriminative? Getting the Best of Both Worlds

CHRISTOPHER M. BISHOP
Microsoft Research, UK
cmbishop@microsoft.com

JULIA LASSERRE
Cambridge University. UK
jal62@cam.ac.uk

SUMMARY

For many applications of machine learning the goal is to predict the value of a vector $\boldsymbol{c}$ given the value of a vector $\boldsymbol{x}$ of input features. In a classification problem $\boldsymbol{c}$ represents a discrete class label, whereas in a regression problem it corresponds to one or more continuous variables. From a probabilistic perspective, the goal is to find the conditional distribution $p(\boldsymbol{c}|\boldsymbol{x})$. The most common approach to this problem is to represent the conditional distribution using a parametric model, and then to determine the parameters using a training set consisting of pairs $\{\boldsymbol{x}_n, \boldsymbol{c}_n\}$ of input vectors along with their corresponding target output vectors. The resulting conditional distribution can be used to make predictions of $\boldsymbol{c}$ for new values of $\boldsymbol{x}$. This is known as a discriminative approach, since the conditional distribution discriminates directly between the different values of $\boldsymbol{c}$.

An alternative approach is to find the joint distribution $p(\boldsymbol{x}, \boldsymbol{c})$, expressed for instance as a parametric model, and then subsequently uses this joint distribution to evaluate the conditional $p(\boldsymbol{c}|\boldsymbol{x})$ in order to make predictions of $\boldsymbol{c}$ for new values of $\boldsymbol{x}$. This is known as a generative approach since by sampling from the joint distribution it is possible to generate synthetic examples of the feature vector $\boldsymbol{x}$. In practice, the generalization performance of generative models is often found to be poorer than than of discriminative models due to differences between the model and the true distribution of the data.

When labelled training data is plentiful, discriminative techniques are widely used since they give excellent generalization performance. However, although collection of data is often easy, the process of labelling it can be expensive. Consequently there is increasing interest in generative methods since these can exploit unlabelled data in addition to labelled data.

Although the generalization performance of generative models can often be improved by 'training them discriminatively', they can then no longer make use of unlabelled data. In an attempt to gain the benefit of both generative and discriminative approaches, heuristic procedure have been proposed which interpolate between these two extremes by taking a convex combination of the generative and discriminative objective functions.

Julia Lasserre is funded by the Microsoft Research European PhD Scholarship programme.

Here we discuss a new perspective which says that there is only one correct way to train a given model, and that a 'discriminatively trained' generative model is fundamentally a new model (Minka, 2006). From this viewpoint, generative and discriminative models correspond to specific choices for the prior over parameters. As well as giving a principled interpretation of 'discriminative training', this approach opens the door to very general ways of interpolating between generative and discriminative extremes through alternative choices of prior. We illustrate this framework using both synthetic data and a practical example in the domain of multi-class object recognition. Our results show that, when the supply of labelled training data is limited, the optimum performance corresponds to a balance between the purely generative and the purely discriminative. We conclude by discussing how to use a Bayesian approach to find automatically the appropriate trade-off between the generative and discriminative extremes.

Keywords and Phrases: GENERATIVE, DISCRIMINATIVE, BAYESIAN INFERENCE, SEMI-SUPERVIZED, UNLABELLED DATA, MACHINE LEARNING

1. INTRODUCTION

In many applications of machine learning the goal is to take a vector $\boldsymbol{x}$ of input features and to assign it to one of a number of alternative classes labelled by a vector $\boldsymbol{c}$ (for instance, if we have C classes, then $\boldsymbol{c}$ might be a C-dimensional binary vector in which all elements are zero except the one corresponding to the class).

In the simplest scenario, we are given a training data set $\boldsymbol{X}$ comprising N input vectors $\boldsymbol{X} = \{\boldsymbol{x}_1, \ldots, \boldsymbol{x}_N\}$ together with a set of corresponding labels $\boldsymbol{C} = \{\boldsymbol{c}_1, \ldots, \boldsymbol{c}_N\}$, in which we assume that the input vectors, and their labels, are drawn independently from the same fixed distribution. Our goal is to predict the class $\widehat{\boldsymbol{c}}$ for a new input vector $\widehat{\boldsymbol{x}}$, and so we require the conditional distribution

$$p(\widehat{\boldsymbol{c}}|\widehat{\boldsymbol{x}}, \boldsymbol{X}, \boldsymbol{C}). \tag{1}$$

To determine this distribution we introduce a parametric model governed by a set of parameters $\boldsymbol{\theta}$. In a *discriminative* approach we define the conditional distribution $p(\boldsymbol{c}|\boldsymbol{x}, \boldsymbol{\theta})$, where $\boldsymbol{\theta}$ are the parameters of the model. The likelihood function is then given by

$$L(\boldsymbol{\theta}) = p(\boldsymbol{C}|\boldsymbol{X}, \boldsymbol{\theta}) = \prod_{n=1}^{N} p(\boldsymbol{c}_n|\boldsymbol{x}_n, \boldsymbol{\theta}). \tag{2}$$

The likelihood function can be combined with a prior $p(\boldsymbol{\theta})$, to give a joint distribution

$$p(\boldsymbol{\theta}, \boldsymbol{C}|\boldsymbol{X}) = p(\boldsymbol{\theta})L(\boldsymbol{\theta}) \tag{3}$$

from which we can obtain the posterior distribution by normalizing

$$p(\boldsymbol{\theta}|\boldsymbol{X}, \boldsymbol{C}) = \frac{p(\boldsymbol{\theta})L(\boldsymbol{\theta})}{p(\boldsymbol{C}|\boldsymbol{X})} \tag{4}$$

where

$$p(\boldsymbol{C}|\boldsymbol{X}) = \int p(\boldsymbol{\theta})L(\boldsymbol{\theta})\, d\boldsymbol{\theta}. \tag{5}$$

Predictions for new inputs are then made by marginalizing the predictive distribution with respect to $\boldsymbol{\theta}$ weighted by the posterior distribution

$$p(\widehat{\boldsymbol{c}}|\widehat{\boldsymbol{x}}, \boldsymbol{X}, \boldsymbol{C}) = \int p(\widehat{\boldsymbol{c}}|\widehat{\boldsymbol{x}}, \boldsymbol{\theta}) p(\boldsymbol{\theta}|\boldsymbol{X}, \boldsymbol{C}) \, d\boldsymbol{\theta}. \tag{6}$$

In practice this marginalization, as well as the normalization in (5), are rarely tractable and so approximation, schemes such as variational inference, must be used. If training data is plentiful a point estimate for $\boldsymbol{\theta}$ can be made by maximizing the posterior distribution to give $\boldsymbol{\theta}_{\mathrm{MAP}}$, and the predictive distribution then estimated using

$$p(\widehat{\boldsymbol{c}}|\widehat{\boldsymbol{x}}, \boldsymbol{X}, \boldsymbol{C}) \simeq p(\widehat{\boldsymbol{c}}|\widehat{\boldsymbol{x}}, \boldsymbol{\theta}_{\mathrm{MAP}}). \tag{7}$$

Note that maximizing the posterior distribution (4) is equivalent to maximizing the joint distribution (3) since these differ only by a multiplicative constant. In practice, we typically take the logarithm before maximizing as this gives rise to both analytical and numerical simplifications. If we consider a prior distribution $p(\boldsymbol{\theta})$ which is constant over the region in which the likelihood function is large, then maximizing the posterior distribution is equivalent to maximizing the likelihood. More generally, we make predictions by marginalizing over all values of $\boldsymbol{\theta}$ using either Monte Carlo methods or an appropriate deterministic approximation framework (Bishop, 2006). In all cases, however, the key quantity for model training is the likelihood function $L(\boldsymbol{\theta})$.

Discriminative methods give good predictive performance and have been widely used in many applications. In recent years there has been growing interest in a complementary approach based on *generative* models, which define a joint distribution $p(\boldsymbol{x}, \boldsymbol{c}|\boldsymbol{\theta})$ over both input vectors and class labels (Jebara, 2004). One of the motivations is that in complex problems such as object recognition, where there is huge variability in the range of possible input vectors, it may be difficult or impossible to provide enough labelled training examples, and so there is increasing use of *semi-supervised* learning in which the labelled training examples are augmented with a much larger quantity of unlabelled examples. A discriminative model cannot make use of the unlabelled data, as we shall see, and so in this case we need to consider a generative approach.

The complementary properties of generative and discriminative models have led a number of authors to seek methods which combine their strengths. In particular, there has been much interest in 'discriminative training' of generative models (Bouchard and Triggs, 2004; Holub and Perona, 2005; Yakhnenko, Silvescu, and Honavar, 2005) with a view to improving classification accuracy. This approach has been widely used in speech recognition with great success (Kapadia, 1998) where generative hidden Markov models are trained by optimizing the predictive conditional distribution. As we shall see later, this form of training can lead to improved performance by compensating for model mis-specification, that is differences between the true distribution of the process which generates the data, and the distribution specified by the model. However, as we have noted, discriminative training cannot take advantage of unlabelled data. In particular it has been shown (Ng and Jordan, 2002) that logistic regression (the discriminative counterpart of a Naive Bayes generative model) works better than its generative counterpart, but only for a large number of training data points (large depending on the complexity of the problem), which confirms the need for using unlabelled data.

Recently several authors (Bouchard and Triggs, 2004; Holub and Perona, 2005; Raina, Shen, Ng, and McCallum, 2003) have proposed hybrids of the generative and discriminative approaches in which a model is trained by optimizing a convex combination of the generative and discriminative log likelihood functions. Although the motivation for this procedure was heuristic, it was sometimes found that the best predictive performance was obtained for intermediate regimes in between the discriminative and generative limits.

In this paper we develop a novel viewpoint (Minka, 2005, Bishop, 2006) which says that, for a given model, there is a unique likelihood function and hence there is only one correct way to train it. The 'discriminative training' of a generative model is instead interpreted in terms of standard training of a different model, corresponding to a different choice of distribution. This removes the apparently ad-hoc choice for the training criterion, so that all models are trained according to the principles of statistical inference. Furthermore, by introducing a constraint between the parameters of this model, through the choice of prior, the original generative model can be recovered.

As well as giving a novel interpretation for 'discriminative training' of generative models, this viewpoint opens the door to principled blending of generative and discriminative approaches by introducing priors having a soft constraint amongst the parameters. The strength of this constraint therefore governs the balance between generative and discriminative.

In Section 2 we give a detailed discussion of the new interpretation of discriminative training for generative models, and in Section 3 we illustrate the advantages of blending between generative and discriminative viewpoints using a synthetic example in which the role of unlabelled data and of model mis-specification becomes clear. In Section 4 we show that this approach can be applied to a large scale problem in computer vision concerned with object recognition in images, and finally we draw some conclusions in Section 5.

2. A NEW VIEW OF 'DISCRIMINATIVE TRAINING'

A generative model can be defined by specifying the joint distribution $p(\boldsymbol{x}, \boldsymbol{c}|\boldsymbol{\theta})$ of the input vector $\boldsymbol{x}$ and the class label $\boldsymbol{c}$, conditioned on a set of parameters $\boldsymbol{\theta}$. Typically this is done by defining a prior probability for the classes $p(\boldsymbol{c}|\boldsymbol{\pi})$ along with a class-conditional density for each class $p(\boldsymbol{x}|\boldsymbol{c}, \boldsymbol{\lambda})$, so that

$$p(\boldsymbol{x}, \boldsymbol{c}|\boldsymbol{\theta}) = p(\boldsymbol{c}|\boldsymbol{\pi})p(\boldsymbol{x}|\boldsymbol{c}, \boldsymbol{\lambda}) \tag{8}$$

where $\boldsymbol{\theta} = \{\boldsymbol{\pi}, \boldsymbol{\lambda}\}$. Since the data points are assumed to be independent, the joint distribution is given by

$$L_{\mathrm{G}}(\boldsymbol{\theta}) = p(\boldsymbol{X}, \boldsymbol{C}, \boldsymbol{\theta}) = p(\boldsymbol{\theta}) \prod_{n=1}^{N} p(\boldsymbol{x}_n, \boldsymbol{c}_n|\boldsymbol{\theta}). \tag{9}$$

This can be maximized to determine the most probable (MAP) value of $\boldsymbol{\theta}$. Again, since $p(\boldsymbol{X}, \boldsymbol{C}, \boldsymbol{\theta}) = p(\boldsymbol{\theta}|\boldsymbol{X}, \boldsymbol{C})p(\boldsymbol{X}, \boldsymbol{C})$, this is equivalent to maximizing the posterior distribution $p(\boldsymbol{\theta}|\boldsymbol{X}, \boldsymbol{C})$.

In order to improve the predictive performance of generative models it has been proposed to use 'discriminative training' (Yakhnenko *et al.*, 2005) which involves maximizing

$$L_{\mathrm{D}}(\boldsymbol{\theta}) = p(\boldsymbol{C}, \boldsymbol{\theta}|\boldsymbol{X}) = p(\boldsymbol{\theta}) \prod_{n=1}^{N} p(\boldsymbol{c}_n|\boldsymbol{x}_n, \boldsymbol{\theta}) \tag{10}$$

in which we are conditioning on the input vectors instead of modelling their distribution. Here we have used

$$p(\boldsymbol{c}|\boldsymbol{x},\boldsymbol{\theta}) = \frac{p(\boldsymbol{x},\boldsymbol{c}|\boldsymbol{\theta})}{\sum_{\boldsymbol{c}'} p(\boldsymbol{x},\boldsymbol{c}'|\boldsymbol{\theta})}. \tag{11}$$

Note that (10) is not the joint distribution for the original model defined by (9), and so does not correspond to MAP for this model. The terminology of 'discriminative training' is therefore misleading, since for a given model there is only one correct way to train it. It is not the training method which has changed, but the model itself.

This concept of discriminative training has been taken a stage further (Yakhnenko, Silvescu, and Honavar, 2005) by maximizing a function given by a convex combination of (9) and (10) of the form

$$\alpha \ln L_{\mathrm{D}}(\boldsymbol{\theta}) + (1-\alpha)\ln L_{\mathrm{G}}(\boldsymbol{\theta}) \tag{12}$$

where $0 \leqslant \alpha \leqslant 1$, so as to interpolate between generative ($\alpha = 0$) and discriminative ($\alpha = 1$) approaches. Unfortunately, this criterion was not derived by maximizing the distribution of a well-defined model.

Following (Minka, 2005) we therefore propose an alternative view of discriminative training, which will lead to an elegant framework for blending generative and discriminative approaches. Consider a model which contains an additional independent set of parameters $\widetilde{\boldsymbol{\theta}} = \{\widetilde{\boldsymbol{\pi}}, \widetilde{\boldsymbol{\lambda}}\}$ in addition to the parameters $\boldsymbol{\theta} = \{\boldsymbol{\pi}, \boldsymbol{\lambda}\}$, in which the likelihood function is given by

$$q(\boldsymbol{x},\boldsymbol{c}|\boldsymbol{\theta},\widetilde{\boldsymbol{\theta}}) = p(\boldsymbol{c}|\boldsymbol{x},\boldsymbol{\theta})p(\boldsymbol{x}|\widetilde{\boldsymbol{\theta}}) \tag{13}$$

where

$$p(\boldsymbol{x}|\widetilde{\boldsymbol{\theta}}) = \sum_{\boldsymbol{c}'} p(\boldsymbol{x},\boldsymbol{c}'|\widetilde{\boldsymbol{\theta}}). \tag{14}$$

Here $p(\boldsymbol{c}|\boldsymbol{x},\boldsymbol{\theta})$ is defined by (11), while $p(\boldsymbol{x},\boldsymbol{c}|\widetilde{\boldsymbol{\theta}})$ has independent parameters $\widetilde{\boldsymbol{\theta}}$.

The model is completed by defining a prior $p(\boldsymbol{\theta},\widetilde{\boldsymbol{\theta}})$ over the model parameters, giving a joint distribution of the form

$$q(\boldsymbol{X},\boldsymbol{C},\boldsymbol{\theta},\widetilde{\boldsymbol{\theta}}) = p(\boldsymbol{\theta},\widetilde{\boldsymbol{\theta}}) \prod_{n=1}^{N} p(\boldsymbol{c}_n|\boldsymbol{x}_n,\boldsymbol{\theta})p(\boldsymbol{x}_n|\widetilde{\boldsymbol{\theta}}). \tag{15}$$

Now suppose we consider a special case in which the prior factorizes, so that

$$p(\boldsymbol{\theta},\widetilde{\boldsymbol{\theta}}) = p(\boldsymbol{\theta})p(\widetilde{\boldsymbol{\theta}}). \tag{16}$$

We then determine optimal values for the parameters $\boldsymbol{\theta}$ and $\widetilde{\boldsymbol{\theta}}$ in the usual way by maximizing (15), which now takes the form

$$\left[p(\boldsymbol{\theta}) \prod_{n=1}^{N} p(\boldsymbol{c}_n|\boldsymbol{x}_n,\boldsymbol{\theta})\right] \left[p(\widetilde{\boldsymbol{\theta}}) \prod_{n=1}^{N} p(\boldsymbol{x}_n|\widetilde{\boldsymbol{\theta}})\right]. \tag{17}$$

We see that the resulting value of $\boldsymbol{\theta}$ will be identical to that found by maximizing (11), since it is the same function which is being maximized. Since it is $\boldsymbol{\theta}$ and not $\widetilde{\boldsymbol{\theta}}$ which determines the predictive distribution $p(\mathbf{c}|\boldsymbol{x}, \boldsymbol{\theta})$ we see that this model is equivalent in its predictions to the 'discriminatively trained' generative model. This gives a consistent view of training in which we always maximize the joint distribution, and the distinction between generative and discriminative training lies in the choice of model.

The relationship between the generative model and the discriminative model is illustrated using directed graphs in Fig. 1.

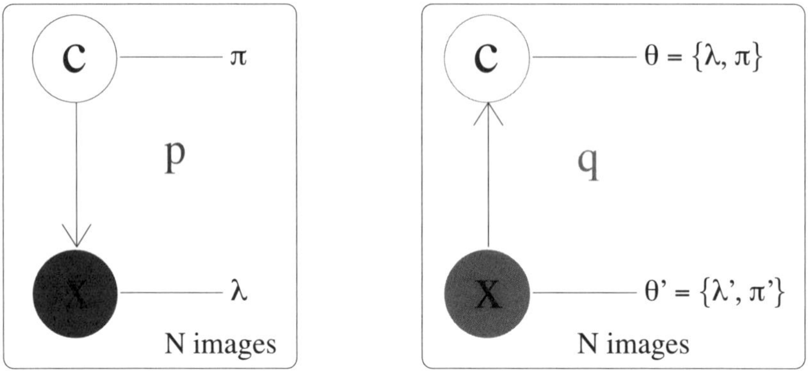

Figure 1: *Probabilistic directed graphs, showing on the left, the original generative model, and on the right the corresponding discriminative model.*

Now suppose instead that we consider a prior which enforces equality between the two sets of parameters

$$p(\boldsymbol{\theta}, \widetilde{\boldsymbol{\theta}}) = p(\boldsymbol{\theta})\delta(\boldsymbol{\theta} - \widetilde{\boldsymbol{\theta}}). \tag{18}$$

Then we can set $\widetilde{\boldsymbol{\theta}} = \boldsymbol{\theta}$ in (13) from which we recover the original generative model $p(\boldsymbol{x}, \mathbf{c}|\boldsymbol{\theta})$. Thus we have a single class of distributions in which the discriminative model corresponds to independence in the prior, and the generative model corresponds to an equality constraint in the prior.

2.1. *Blending Generative and Discriminative*

Clearly we can now blend between the generative and discriminative extremes by considering priors which impose a soft constraint between $\widetilde{\boldsymbol{\theta}}$ and $\boldsymbol{\theta}$. Why should we wish to do this?

First of all, we note that the reason why 'discriminative training' might give better results than direct use of the generative model, is that (15) is more flexible than (9) since it relaxes the implicit constraint $\widetilde{\boldsymbol{\theta}} = \boldsymbol{\theta}$. Of course, if the generative model were a perfect representation of reality (in other words the data really came from the model) then increasing the flexibility of the model would lead to poorer results. Any improvement from the discriminative approach must therefore be the result of a mis-match between the model and the true distribution of the (process which generates the) data. In other words, the benefit of 'discriminative training' is dependent on model mis-specification.

Conversely, the benefit of the generative approach is that it can make use of unlabelled data to augment the labelled training set. Suppose we have a data set comprising a set of inputs $\boldsymbol{X}_{\mathrm{L}}$ for which we have corresponding labels $\boldsymbol{C}_{\mathrm{L}}$, together with a set of inputs $\boldsymbol{X}_{\mathrm{U}}$ for which we have no labels. For the correctly trained generative model, the function which is maximized is given by

$$p(\boldsymbol{\theta}) \prod_{n \in \mathrm{L}} p(\boldsymbol{x}_n, \boldsymbol{c}_n|\boldsymbol{\theta}) \prod_{m \in \mathrm{U}} p(\boldsymbol{x}_m|\boldsymbol{\theta}) \tag{19}$$

where $p(\boldsymbol{x}|\boldsymbol{\theta})$ is defined by

$$p(\boldsymbol{x}|\boldsymbol{\theta}) = \sum_{\boldsymbol{c}'} p(\boldsymbol{x}, \boldsymbol{c}'|\boldsymbol{\theta}). \tag{20}$$

We see that the unlabelled data influences the choice of $\boldsymbol{\theta}$ and hence affects the predictions of the model. By contrast, for the 'discriminatively trained' generative model the function which is now optimized is again the product of the prior and the likelihood function and so takes the form

$$p(\boldsymbol{\theta}) \prod_{n \in \mathrm{L}} p(\boldsymbol{x}_c|\boldsymbol{x}_n, \boldsymbol{\theta}) \tag{21}$$

and we see that the unlabelled data plays no role. Thus, in order to make use of unlabelled data we cannot use a discriminative approach.

Now let us consider how a combination of labelled and unlabelled data can be exploited from the perspective of our new approach defined by (15), for which the joint distribution becomes

$$\begin{aligned} q(\boldsymbol{X}_{\mathrm{L}}, \boldsymbol{C}_{\mathrm{L}}, \boldsymbol{X}_{\mathrm{U}}, \boldsymbol{\theta}, \widetilde{\boldsymbol{\theta}}) &= p(\boldsymbol{\theta}, \widetilde{\boldsymbol{\theta}}) \\ &\left[\prod_{n \in \mathrm{L}} p(\boldsymbol{c}_n|\boldsymbol{x}_n, \boldsymbol{\theta}) p(\boldsymbol{x}_n|\widetilde{\boldsymbol{\theta}})\right] \left[\prod_{m \in \mathrm{U}} p(\boldsymbol{x}_m|\widetilde{\boldsymbol{\theta}})\right]. \end{aligned} \tag{22}$$

We see that the unlabelled data (as well as the labelled data) influences the parameters $\widetilde{\boldsymbol{\theta}}$ which in turn influence $\boldsymbol{\theta}$ via the soft constraint imposed by the prior.

In general, if the model is not a perfect representation of reality, and if we have unlabelled data available, then we would expect the optimal balance to lie neither at the purely generative extreme nor at the purely discriminative extreme.

As a simple example of a prior which interpolates smoothly between the generative and discriminative limits, consider the class of priors of the form

$$p(\boldsymbol{\theta}, \widetilde{\boldsymbol{\theta}}) \propto p(\boldsymbol{\theta}) p(\widetilde{\boldsymbol{\theta}}) \frac{1}{\sigma} \exp\left\{-\frac{1}{2\sigma^2} \|\boldsymbol{\theta} - \widetilde{\boldsymbol{\theta}}\|^2\right\}. \tag{23}$$

If desired, we can relate σ to an α like parameter by defining a map from $(0, 1)$ to $(0, \infty)$, for example using

$$\sigma(\alpha) = \left(\frac{\alpha}{1 - \alpha}\right)^2. \tag{24}$$

For $\alpha \to 0$ we have $\sigma \to 0$, and we obtain a hard constraint of the form (18) which corresponds to the generative model. Conversely for $\alpha \to 1$ we have $\sigma \to \infty$ and we obtain an independence prior of the form (16) which corresponds to the discriminative model.

3. ILLUSTRATION

We now illustrate the new framework for blending between generative and discriminative approaches using an example based on synthetic data. This is chosen to be as simple as possible, and so involves data vectors $\boldsymbol{x}_n$ which live in a two-dimensional Euclidean space for easy visualization, and which belong to one of two classes. Data from each class is generated from a Gaussian distribution as illustrated in Fig. 2.

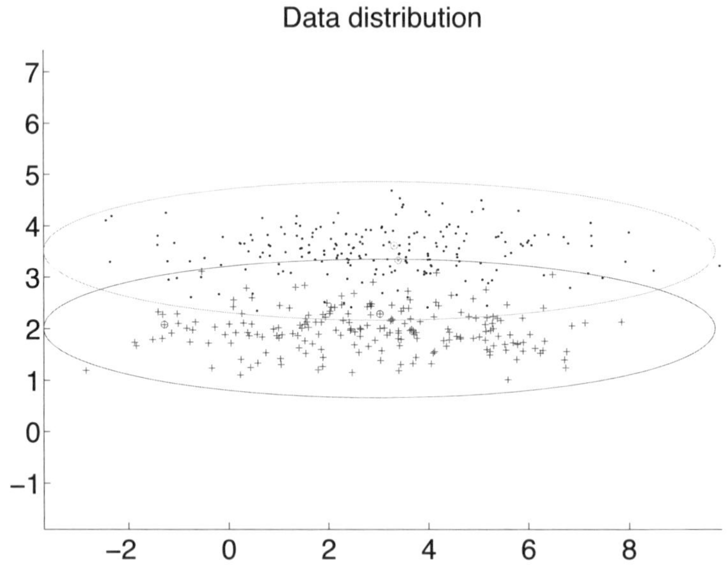

Figure 2: *Synthetic training data, shown as crosses and dots, together with contours of probability density for each of the two classes. Two points from each class are labelled (indicated by circles around the data points).*

Here the scales on the axes are equal, and so we see that the class-conditional densities are elongated in the horizontal direction.

We now consider a continuum of models which interpolate between purely generative and purely discriminative. To define this model we consider the generative limit, and represent each class-conditional density using an isotropic Gaussian distribution. Since this does not capture the horizontally elongated nature of the true class distributions, this represents a form of model mis-specification. The parameters of the model are the means and variances of the Gaussians for each class, along with the class prior probabilities.

We consider a prior of the form (23) in which $\sigma(\alpha)$ is defined by (24). Here we choose $p(\boldsymbol{\theta}, \widetilde{\boldsymbol{\theta}}|\alpha) = p(\boldsymbol{\theta})\ N(\widetilde{\boldsymbol{\theta}}|\boldsymbol{\theta}, \sigma(\alpha))$, where $p(\boldsymbol{\theta})$ is the usual conjugate prior (a Gaussian-gamma prior for the means and variances, and a Dirichlet prior for the class label). The generative model consists of a spherical Gaussian per class, with mean $\boldsymbol{\mu}$ and a diagonal precision matrix $\Delta \boldsymbol{I}$, so that $\boldsymbol{\theta} = \{\boldsymbol{\mu}_k, \Delta_k\}$ and $\widetilde{\boldsymbol{\theta}} = \{\widetilde{\boldsymbol{\mu}}_k, \widetilde{\Delta}_k\}$. Specifically we have chosen

$$p(\boldsymbol{\theta}, \widetilde{\boldsymbol{\theta}}|\alpha) \propto \mathcal{N}(\boldsymbol{\theta}|\boldsymbol{\theta}', \sigma(\alpha)) \prod_k \left[\mathcal{N}({\boldsymbol{\mu}'}_k|\mathbf{0}, (10{\Delta'}_k)^{-1})\mathcal{G}({\Delta'}_k|0.01, 100)\mathcal{G}(\Delta_k|0.01, 100)\right] \quad (25)$$

where $\mathcal{N}(\cdot|\cdot, \cdot)$ denotes a Gaussian distribution and $\mathcal{G}(\cdot|\cdot, \cdot)$ denotes a gamma distribution.

The training data set comprises 200 points from each class, of which just two from each class are labelled, and the test set comprises 200 points all of which are labelled. Experiments are run 10 times with differing random initializations (including the random selection of which subset of training points to label) and the results used to computer a mean and variance over the test set classification, which are shown by 'error bars' in Fig. 3.

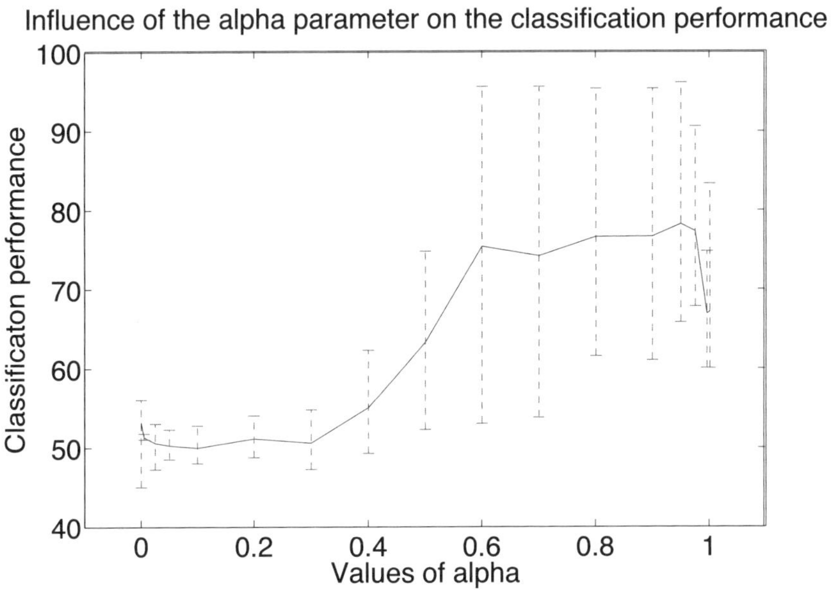

Figure 3: *Plot of the percentage of correctly classified points on the test set versus α for the synthetic data problem.*

We see that the best generalization occurs for values of α intermediate between the generative and discriminative extremes.

To gain insight into this behaviour we can plot the contours of density for each class corresponding to different values of α, as shown in Fig. 4.

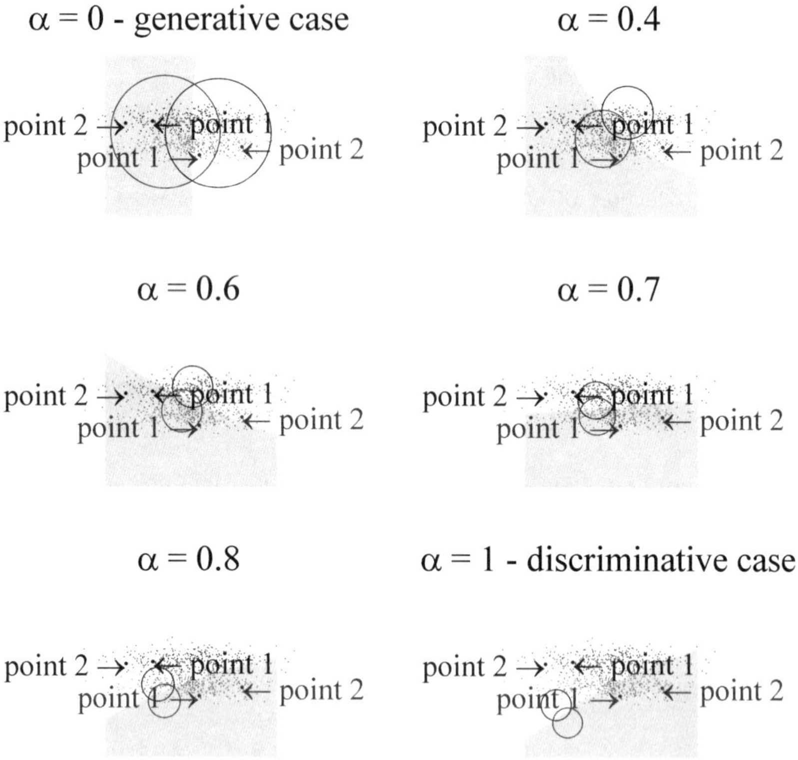

Figure 4: *Results of fitting an isotropic Gaussian model to the synthetic data for various values of α. The top left shows $\alpha = 0$ (generative case) while the bottom right shows $\alpha = 1$ (discriminative case). The gray area corresponds to points that are assigned to the red class.*

We see that a purely generative model is strongly influenced by modelling the density of the data and so gives a decision boundary which is orthogonal to the correct one. Conversely a purely discriminative model attends only to the labelled data points and so misses useful information about the horizontal elongation of the true class-conditional densities which is present in the unlabelled data.

4. OBJECT RECOGNITION

We now apply our approach to a realistic application involving object recognition in static images. This is a widely studied problem which has been tackled using a range of different discriminative and generative models. The long term goal of such research is to achieve near human levels of recognition accuracy across thousands of object classes in the presence of wide variations in location, scale, orientation and lighting, as well as changes due to intra-class variability and occlusion.

4.1. *The Data*

We used eight different classes: airplanes, bikes, cows, faces, horses, leaves, motorbikes, sheep (Bishop, 2006). Together these images exhibit a wide variety of poses, colours, and illumination, as illustrated by the sample images shown in Fig. 5. The goal is to assign images from the test set to one of the eight classes.

Figure 5: *Sample images from the training set.*

4.2. *The Features*

Our features are taken from Winn, Criminisi, and Minka (2005), in which the original RGB images are first converted into the CIE (L, a, b) colour space (Kasson and Plouffe, 1992). All images were re-scaled to 300×200, and raw patches of size 48×48 were extracted on a regular grid of size 24×24 (i.e. every 24th pixel). Each patch is then convolved with 17 filters, and the set of corresponding pixels from each of the filtered images represents a 17-dimensional vector. All these feature vectors are clustered using K-means with $K = 100$. Since this large value of K is computationally costly in later stages of processing, PCA is used to give a 15-dimensional feature vector. Winn *et al.* (2005) use a more powerful technique to reduce the number of features, but since this is a supervised method based on fully labelled training data, we did not re-implement it here. The cluster centers obtained through K-means are called *textons* (Varma and Zisserman, 2005).

The filters are quite standard: the first three filters are obtained by scaling a Gaussian filter, and are applied to each channel of the colour image, which gives $3 \times 3 = 9$ response images. Then a Laplacian filter is applied to the L channel, at four different scales, which gives four more response images. Finally two DoG (difference of Gaussians) filters (one along each direction) are applied to the L channel, at two different scales, giving another four responses.

From these response images, we extract every pixel on a 4×4 grid, and apply K-means to obtain K textons. Now each patch will be represented by a histogram of these textons, i.e. by a K-dimensional vector containing the proportion of each texton. Textons were obtained from 25 training images per class (half of the training set). Note that the texton features are found using only unlabelled data. These vectors are then reduced using PCA to a dimensionality of 15.

4.3. *The Model*

We consider the generative model introduced by Ulusoy and Bishop (2005), which we now briefly describe. Each image is represented by a feature vector $\boldsymbol{x}_n$, where $n = 1, \ldots, N$, and N is the total number of images. Each vector comprises a set of J patch vectors $\boldsymbol{x} = \{\boldsymbol{x}_{nj}\}$ where $j = 1, \ldots, J$. We assume that each patch belongs to one and only one of the classes, or to a separate 'background' class, so that each patch can be characterized by a binary vector $\boldsymbol{\tau}_{nj}$ coded so that all elements of $\boldsymbol{\tau}_{nj}$ are zero except the element corresponding to the class. We use $\boldsymbol{c}_n$ to denote the image label vector for image n with independent components $c_{nk} \in \{0, 1\}$ in which $k = 1, \ldots C$ labels the class.

The overall joint distribution for the model can be represented as a directed graph, as shown in Fig. 6.

We can therefore characterize the model completely in terms of the conditional probabilities $p(\boldsymbol{c})$, $p(\boldsymbol{\tau}|\boldsymbol{c})$ and $p(\boldsymbol{x}|\boldsymbol{\tau})$. This model is most easily explained generatively, that is, we describe the procedure for generating a set of observed feature vectors from the model.

First we choose the overall class of the image according to some prior probability parameters ψ_k where $k = 1, \ldots, C$, and $0 \leqslant \psi_k \leqslant 1$, with $\sum_k \psi_k = 1$, so that

$$p(\boldsymbol{c}) = \prod_{k=1}^{C} \psi_k^{c_k}. \tag{26}$$

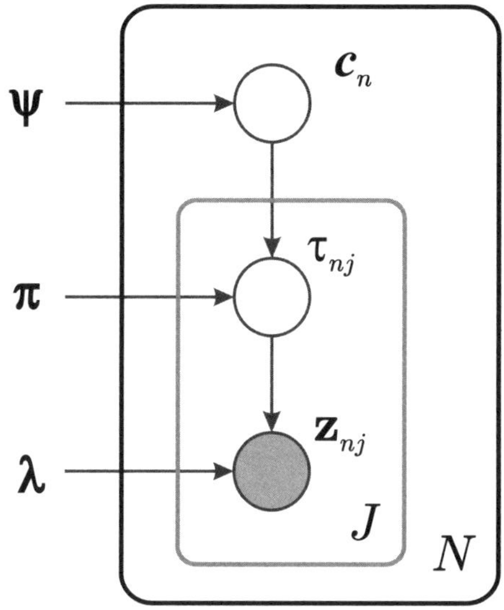

Figure 6: *The generative model for object recognition expressed as a directed acyclic graph, for unlabelled images, in which the boxes denote 'plates' (i.e. independent replicated copies). Only the patch feature vectors $\{\boldsymbol{x}_{nj}\}$ are observed, corresponding to the shaded node. The image class labels $\mathbf{c}_n$ and patch class labels $\boldsymbol{\tau}_{nj}$ are latent variables.*

Given the overall class for the image, each patch is then drawn from either one of the foreground classes or the background ($k = C + 1$) class. The probability of generating a patch from a particular class is governed by a set of parameters π_k, one for each class, such that $\pi_k \geqslant 0$, constrained by the subset of classes actually present in the image. Thus

$$p(\boldsymbol{\tau}_j|\boldsymbol{c}) = \left(\sum_{l=1}^{C+1} c_l \pi_l\right)^{-1} \prod_{k=1}^{C+1} (c_k \pi_k)^{\tau_{jk}}. \tag{27}$$

Note that there is an overall undetermined scale to these parameters, which may be removed by fixing one of them, e.g. $\pi_{C+1} = 1$.

For each class, the distribution of the patch feature vector $\boldsymbol{x}$ is governed by a separate mixture of Gaussians which we denote by

$$p(\boldsymbol{x}|\boldsymbol{\tau}_j) = \prod_{k=1}^{C+1} \phi_k(\boldsymbol{x}_j; \boldsymbol{\lambda}_k)^{\tau_{jk}} \tag{28}$$

where $\boldsymbol{\lambda}_k$ denotes the set of parameters (means, covariances and mixing coefficients) associated with this mixture model.

If we assume N independent images, and for image n we have J patches drawn independently, then the joint distribution of all random variables is

$$\prod_{n=1}^{N}\left[p(\boldsymbol{c}_n)\prod_{j=1}^{J}p(\boldsymbol{x}_{nj}|\boldsymbol{\tau}_{nj})p(\boldsymbol{\tau}_{nj}|\boldsymbol{c}_n)\right]. \tag{29}$$

Here we are assuming that each image has the same number J of patches, though this restriction is easily relaxed if required.

The graph shown in Fig. 6 corresponds to unlabelled images in which only the feature vectors $\{\boldsymbol{x}_{nj}\}$ are observed, with both the image category and the classes of each of the patches being latent variables. It is also possible to consider images which are 'weakly labelled', that is each image is labelled according to the category of object present in the image. This corresponds to the graphical model of Fig. 7 in which the node $\boldsymbol{c}_n$ is shaded.

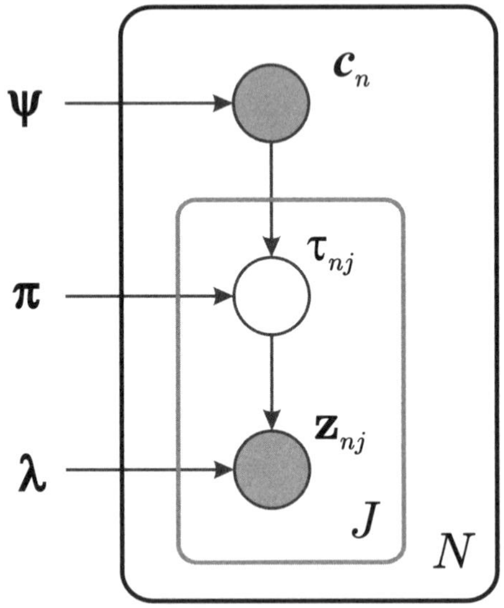

Figure 7: *Graphical model corresponding to Fig. 6 for weakly labelled images.*

Of course, for a given size of data set, better performance is expected if all of the images are 'strongly labelled', that is segmented images in which the region occupied by the object or objects is known so that the patch labels $\boldsymbol{\tau}_{nj}$ become observed variables. The graphical model for a set of strongly labelled images is shown in Fig. 8.

Strong labelling requires hand segmentation of images, and so is a time consuming and expensive process as compared with collection of the images themselves. For a given level of effort it will always be possible to collect many unlabelled or weakly labelled images for the same cost as a single strongly labelled image. Since the variability of natural images and objects is so vast we will always be operating in a regime in which the size of our data sets is statistically small (though they will often be computationally large).

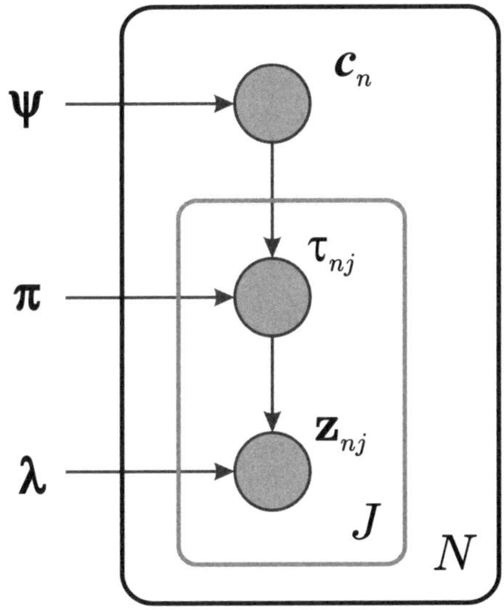

Figure 8: *Graphical models corresponding to Fig. 6 for strongly labelled images.*

For this reason there is great interest in augmenting expensive strongly labelled images with lots of cheap weakly labelled or unlabelled images in order to better characterize the different forms of variability. Although the two stage hierarchical model shown in Fig. 6 appears to be more complicated than in the simple example shown in Fig. 1, it does in fact fall within the same framework. In particular, for labelled images the observed data is $\{\boldsymbol{x}_n, \boldsymbol{c}_n, \boldsymbol{\tau}_{nj}\}$, while for 'unlabelled' images only $\{\boldsymbol{x}_n\}$ are observed. The experiments described here could readily be extended to consider arbitrary combinations of strongly labelled, weakly labelled and unlabelled images if desired.

If we let $\boldsymbol{\theta} = \{\psi_k, \pi_k, \boldsymbol{\lambda}_k\}$ denote the full set of parameters in the model, then we can consider a model of the form (22) in which the prior is given by (23) with $\sigma(\alpha)$ defined by (24), and the terms $p(\boldsymbol{\theta})$ and $p(\widetilde{\boldsymbol{\theta}})$ taken to be constant.

We use conjugate gradients to optimize the parameters. Due to lack of space we do not write down all the derivatives of the log likelihood function required by the conjugate gradient algorithm. However, the correctness of the mathematical derivation of these gradients, as well as their numerical implementation, can easily be verified by comparison against numerical differentiation (Bishop, 1995). The conjugate gradients is the most used technique when it comes to blending generative and discriminative models, thanks to its flexibility. Indeed, because of the discriminative component $p(\boldsymbol{c}_n|\boldsymbol{x}_n, \boldsymbol{\theta})$ which contains a normalizing factor, an algorithm such as EM would require much more work, as nothing is directly tractable anymore. However, a comparison of the two methods is currently being investigated.

4.4. *Results*

We use 50 training images per class (giving 400 training images in total) of which five images per class (a total of 40) were fully labelled, *i.e.*, both the image and the individual patches have class labels. All the other images are left totally unlabelled, i.e. not even the category they belong to is given. Note that this kind of training data is (1) very cheap to get and (2) very unusual for a discriminative model. The test set consists of 100 images per class (giving a total of 800 images), the task is to label each image.

Experiments are run five times with differing random initializations and the results used to compute a mean and variance over the test set classification, which are shown by 'error bars' in Fig. 9.

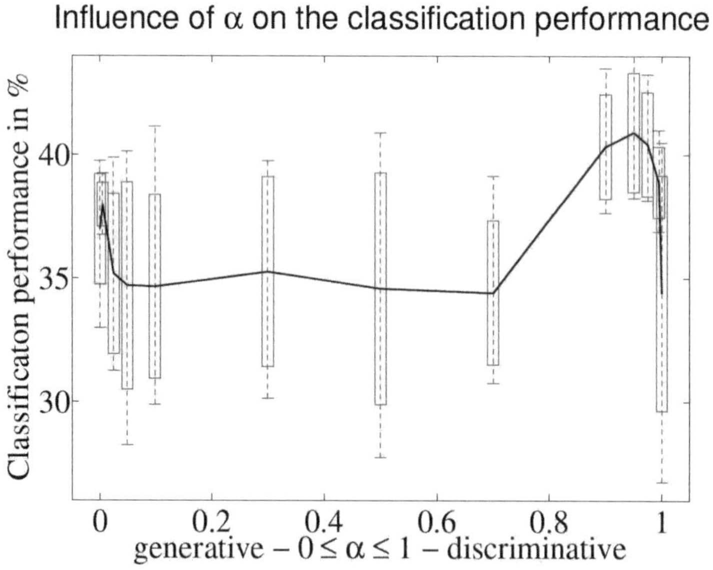

Figure 9: *Influence of the term α on the test set classification performance.*

Note that, since there are eight balanced classes, random guessing would give 12.5% correct on average. Again we see that the best performance is obtained with a blend between generative and discriminative extremes.

5. CONCLUSIONS

In this paper we have shown that 'discriminative training' for generative models can be re-cast in terms of standard training methods applied to a modified model. This new viewpoint opens the door to a wide range of new models which interpolate smoothly between generative and discriminative approaches and which can benefit from the advantages of both. The main drawback of this framework is that the number of parameters in the model is doubled leading to greater computational cost.

Although we have focussed on classification problems, the framework is equally applicable to regression problems in which $\mathbf{c}$ corresponds to a set of continuous variables.

A principled approach to combining generative and discriminative approaches not only gives a more satisfying foundation for the development of new models, but it also brings practical benefits. In particular, the parameter α which governs the trade-off between generative and discriminative is now a hyper-parameter within a well defined probabilistic model which is trained using the (unique) correct likelihood function. In a Bayesian setting the value of this hyper-parameter can therefore be optimized by maximizing the marginal likelihood in which the model parameters have been integrated out, thereby allowing the optimal trade-off between generative and discriminative limits to be determined entirely from the training data without recourse to cross-validation (Bishop, 2006). This extension of the work described here is currently being investigated.

REFERENCES

Bishop, C. M. (1995). *Neural Networks for Pattern Recognition.* Oxford: University Press

Bishop, C. M. (2006). *Pattern Recognition and Machine Learning.* Berlin: Springer-Verlag

Bouchard, G. and Triggs, B. (2004). The trade-off between generative and discriminative classifiers. *IASC 16th International Symposium on Computational Statistics, Prague, Czech Republic*, 721–728.

Holub, A. and Perona, P. (2005). A discriminative framework for modelling object classes. *IEEE Conference on Computer Vision and Pattern Recognition*, San Diego (California), USA. IEEE Computer Society.

Jebara, T. (2004). *Machine Learning: Discriminative and Generative.* Dordrecht: Kluwer

Kapadia, S. (1998). *Discriminative Training of Hidden Markov Models.* PhD Thesis, University of Cambridge, UK.

Kasson, J. M. and Plouffe, W. (1992). An analysis of selected computer interchange color spaces. *ACM Transactions on Graphics* **11**, 373–405.

Minka, T. (2005). Discriminative models, not discriminative training. *Tech. Rep.*, Microsoft Research, Cambridge, UK.

Ng, A. Y. and Jordan, M. I. (2002). On discriminative vs. generative: A comparison of logistic regression and naive Bayes. *Advances in Neural Information Processing Systems* **14**, (T. G. Dietterich, S. Becker, and Z. Ghahramani, eds.) Cambridge, MA: The MIT Press, 841–848.

Raina, R., Shen, Y., Ng, A. Y. and McCallum, A. (2003). Classification with hybrid generative/discriminative models. *Advances in Neural Information Processing Systems* **16** Cambridge, MA: The MIT Press, 545–552.

Ulusoy, I. and Bishop, C. M. (2005). Generative versus discriminative models for object recognition. *Proceedings IEEE International Conference on Computer Vision and Pattern Recognition*, CVPR., San Diego.

Varma, M. and Zisserman, A. (2005). A statistical approach to texture classification from single images. *IJCV* **62**, 61–81.

Winn, J., Criminisi, A. and Minka, T. (2005). Ob ject categorization by learned universal visual dictionary. *IEEE International Conference on Computer Vision*, Beijing, China. IEEE Computer Society.

Yakhnenko, O., Silvescu, A. and Honavar, V. (2005). Discriminatively trained Markov model for sequence classification. *5th IEEE International Conference on Data Mining*, Houston (Texas), USA. IEEE Computer Society.

DISCUSSION

HERBERT K. H. LEE (*University of California, Santa Cruz, USA*)

Let me start by congratulating the authors for this paper. In terms of 'Getting the best of both worlds', this paper can also be seen as crossing between machine learning and statistics, combining useful elements from both. I can only hope that our two communities continue to interact, deepening our connections. The perspectives are often complementary, which is an underlying theme of this discussion.

One of the obstacles to working across disciplines is that a very different vocabulary may be used to describe the same concepts. Statisticians reading this paper may find it helpful to think of supervised learning as classification, and unsupervised learning as clustering. Discriminative learning is thus classification based on a probability model (which it typically is in statistics) while the generative approach is clustering based on a model (such as a mixture model approach).

While machine learning and statistics are really quite similar, there is a key difference in perspective. In machine learning, the main goal is typically to achieve good predictions, and while a probability model may be used, it is not explicitly required. In contrast, most statistical analyses see the probability model as the core of the analysis, with the idea that optimal predictions will arise from accurate selection and fitting of the model. In particular, Bayesian analyses rely crucially on a fully-specified probability model. Thus one of the core points of this paper, that of a unique likelihood function, should seem natural to a Bayesian. Yet it is an important insight in the context of the machine learning literature. Bringing these machine learning algorithms into a coherent probabilistic framework is a big step forward, and one that is not always fully valued. This is an important contribution by these authors and their collaborators.

Uncertainty about the likelihood function can be dealt with by embedding the functions under consideration into a larger family with one or more additional parameters, and this is exactly what has been done here. This follows a strong tradition in statistics, such as Box-Cox transformations and model averaging, but represents a relatively untapped potential in machine learning. In contrast, it is more common in machine learning to use implicit expansions of the model class (or of the fitting algorithm, when the model may not be explicitly stated). Examples include bagging (Breiman, 1996), where individual predictions from over-fit models are averaged over bootstrap samples to reduce over-fitting, and boosting (Freund and Schapire, 1997), where overly-simple models are combined to create an improved ensemble prediction. Such implicit expansion can work well in practice, but it can be difficult to understand or describe the expanded class of models, and hence difficult to leverage related knowledge from the literature.

On a related note, the authors argue that a key benefit of using discriminative training for generative models is that it improves performance when the model is mis-specified, as vividly demonstrated by the example in Section 3. In practice, this is quite useful, as our parametric models are typically only approximations to reality, and the approximations can be quite poor. But this does leave open the possibility of explicit model expansion. A larger parametric family may encompass a model which is close enough to reality. Or taking things even further, one could move to a fully nonparametric approach. Then it becomes less clear what the trade-offs are.

Many highly innovative and creative ideas have arisen in machine learning, and the field of statistics has gained by importing some of these ideas. Statistics, in turn, can offer a substantial literature that can be applied once a machine learning algorithm can be mapped to a probability model. From the model, one can draw from the literature to better understand when the algorithm will work best, when it might perform poorly, what diagnostics may be applicable, and possibly how to further improve the algorithm. The key is connecting the algorithm to a probability model, either finding the model which implicitly underlies the algorithm, or showing that the algorithm approximates a particular probability model. These sorts of connections benefit both fields.

Thus thinking more about Bayesian probability models, some possible further directions for this current work come to mind. It would seem natural to put a prior on α and to treat it as another parameter. At least from the experiments shown so far, it appears that there may be some information about likely best ranges of α, allowing the use of an informative prior, possibly in comparison to a flat prior. In addition to the possibility of marginalizing over α, one could also estimate α to obtain a 'best' fit.

Another natural possible direction would be to look at the full posterior, rather than just getting a point estimate, such as the maximum a posteriori class estimate. Knowing about the uncertainty of the classification can often be useful. It may also be useful to move to a more hierarchical model, particularly for the image classes. It would seem that images of horses and cows would be more similar to each other, and images of bicycles and motorbikes would be similar to each other, but that these two sets would be rather different from each other, and further different from faces or leaves. Working within a hierarchical structure should be straightforward in a fully Bayesian paradigm.

In terms of connections between machine learning and statistics, it seems unfortunate that the machine learning literature takes little notice of related work in the statistics literature. In particular, there has been extensive work on model-based clustering, for example, Fraley and Raftery (2002) and even a poster presented at this conference (Frühwirth-Schnatter and Pamminger, 2006). It would be great if the world of machine learning were more cognizant of the statistical literature.

In summary, this paper presents a practical solution to a problem that is definitely in need of attention, and which has received relatively little attention in the statistical literature. The likelihood-based approach is particularly promising. The authors make a very positive contribution in helping to bridge the gap between machine learning and statistics, and I hope to see more such work in the future.

STEPHEN E. FIENBERG (*Carnegie Mellon University, USA*)

The presentation by Bishop and Laserre (BL) is an especially welcome contribution to the Bayesian statistical world. It attempts to formulate a principled approach to combining discriminative and generative approaches to learning in a setting focused on classification. As someone engaged in both activities (although with far more emphasis going into the generative modeling) and whose appointments are currently in both a department of statistics and in the first machine learning department in the world, I admire the clarity of their initial distinction between the two approaches and at their framing of their strengths and weaknesses. I'd like to offer two observations.

First, in the statistical world we actually spend a lot of time worrying about problems other than classification. Thus many of us spend a large amount of our time worrying about generative models in a much broader context. We focus on producing families of parametric models that span the space of reasonable models for the phenomenon of interest and attempt to imbue the parameters θ, often associated with latent quantitics which in the Bayesian context are also random variables, with interpretations that are part of a generative process, often of a dynamic form. Inevitably, our models are oversimplifications and we learn both from the activity of parameter estimation and model choice and from careful examination and interpretation of the posterior distribution of the parameters as well as from the form of various predictive distributions, e.g., see Blei *et al.* (2003a,2003b), Erosheva (2003), and Erosheva *et al.* (2004). When I teach statistical learning to graduate students in the Machine Learning Department, I emphasize that the world of machine learning would be enriched by taking on at least part of this broader perspective.

My second observation is closely related. While I very much appreciated the BL's goal and the boldness of their effort to formulate a different likelihood function to achieve an integration of the two perspectives, I think the effort would be enhanced by consideration of some of the deeper implications of the subjective Bayesian paradigm. As Bishop noted in his oral response to the discussion, machine learning has been moving full force to adopt formal statistical ideas, and so it is now common to see the incorporation of MAP and model averaging methodologies directly into the classification setting. But as some of the other presentations at this conference make clear, model assessment is a much more complex and subtle activity which we can situate within the subjective Bayesian framework, cf., Draper (1999). In particular, I commend a careful reading of the Lauritzen (2006) discussion of the paper by Chakrabarti and Ghosh (2006), in which he emphasized that we can only come to grips with the model choice problem by considering model comparisons and specifications with respect to our own subjective Bayesian prior distribution (thus taking advantage of the attendant coherence in the sense of de Finetti (1937)) until such time as we need to step outside the formal framework and reformulate our model class and likelihoods. Thus BL's new replicate distribution for θ' possibly should be replaced by a real subjective prior distribution, perhaps of a similar form, and then they could explore the formulation of the generative model without introducing a likelihood that departs from the one that attempts to describe the underlying phenomenon of interest. This, I suspect, would lead to an even clearer formulation that is more fully rooted in both the machine learning and statistical learning worlds.

REPLY TO THE DISCUSSION

We would like thank the discussants for their helpful remarks and useful insights. We would also like to take the opportunity to make some comments on an important issue raised by both discussants, namely the relationship between the fields of machine learning and statistics.

Historically, these fields evolved largely independently for many years. Much of the motivation for early machine learning algorithms, such as the perceptron in the 1960s, and multi-layered neural networks in the 1980s, came from loose analogies with neurobiology. Later, the close connections to statistics became more widely recognized, and these have strongly shaped the subsequent evolution of the field during the 1990s and beyond (Bishop 1995 and Ripley 1996). Today, there are many similarities in the techniques employed by the two fields, although as the discussants note, both the vocabulary and the underlying motivations can differ.

The discussants express frustration at the lack of appreciation of the statistics literature by some machine learning researchers. Naturally, the converse also holds, and indeed this conference has provided examples of algorithms proposed as novel which are in fact well known and widely cited in the machine learning world. To some extent this lack of appreciation is understandable. The literature in either field alone is vast, and the issue of vocabulary mentioned above is a further obstacle to cross-fertilization.

It could be even be argued that the relative independence of the two fields has brought some benefits. For example, the use of large, highly parameterized black-box models trained on large data sets, of the kind which characterized much of the applied work in neural networks in the 1990s, did not fit well with the culture of the statistics community at the time, and met with scepticism from some quarters. Yet these efforts have led to numerous large-scale applications of substantial commercial importance.

Nevertheless, it seems clear that greater cross-fertilization between the two communities would be desirable. Conference such as AI Statistics (Bishop 2003) explicitly seek to achieve this, and several text books also span the divide (Hastie 2001 and Bishop 2006).

Increasingly, the focus in the machine learning community is not just on well-defined probabilistic models, but on fully Bayesian treatments in which distributions over unknown variables are maintained and updated. However, almost any model which has sufficient complexity to be of practical interest will not have a closed-form analytical solution, and hence approximation techniques are essential. For many years the only general purpose approximation framework available was that of Monte Carlo sampling, which is computationally demanding and which does not scale to large problems.

A crucial advance, therefore, has been the development of a wide range of powerful deterministic inference techniques. These include variational inference, expectation propagation, loopy belief propagation, and others. Like Markov chain Monte Carlo, these techniques have their origins in physics. However, they have primarily been developed and applied within the machine learning community. They complement naturally the recent advances in probabilistic graphical models, such as the development of factor graphs. Also, they are computationally much more efficient than Monte Carlo methods, and have permitted, for instance, a fully Bayesian treatment of the player ranking problem in on-line computer games through the TrueSkill^{TM} system, in which millions of players are compared and ranked, with ranking updates and team matching being done in real time (Herbrich 1996). Here the Bayesian treatment, which maintains a distribution over player skill levels, leads to substantially improved ranking accuracy compared to the traditional ELO method , which used a point estimate and which can be viewed as a maximum likelihood technique. This type of application would be inconceivable without the use of deterministic approximation schemes. In our view, these methods represent the single most important advance in practical Bayesian statistics in the last 10 years.

ADDITIONAL REFERENCES IN THE DISCUSSION

Breiman, L. (1996). Bagging predictors. *Machine Learning* **26**, 123–140.

Blei D. M., Jordan, M. I. and Ng, A. (2003a). Hierarchical Bayesian models for applications in information retrieval. *Bayesian Statistics 7* (J. M. Bernardo, M. J. Bayarri, J. O. Berger, A. P. Dawid, D. Heckerman, A. F. M. Smith and M. West, eds.) Oxford: University Press, 25–45.

Blei D. M., Jordan, M. I. and Ng, A. (2003b). Latent Dirichlet allocation. *J. Machine Learning Research* **3**, 993–1022.

Chakrabarti, A. and Ghosh, J. K. (2006). Some aspects of Bayesian model selection for prediction. *In this volume.*

de Finetti, B. (1937). La prévision: ses lois logiques, ses sources subjectives. *Annales de l'Institut Henri Poincaré* **7**, 1–68.

Draper D. (1999). Model uncertainty yes, discrete model averaging maybe. (discussion of 'Bayesian model averaging: a tutorial,' by Hoeting *et al.*). *Statist. Science* **14**, 405–409.

Erosheva, E. A. (2003). Bayesian estimation of the grade of membership model. *Bayesian Statistics 7* (J. M. Bernardo, M. J. Bayarri, J. O. Berger, A. P. Dawid, D. Heckerman, A. F. M. Smith and M. West, eds.) Oxford: University Press, 501–510.

Erosheva, E. A., Fienberg, S. E., and Lafferty, J. (2004). Mixed-membership models of scientific publications. *Proc. National Acad. Sci.* **97**, 11885–11892.

Fraley, C. and Raftery, A. E. (2002). Model-based clustering, discriminant analysis, and density estimation. *J. Amer. Statist. Assoc.* **97**, 611–631.

Freund, Y. and Schapire, R. E. (1997). A decision-theoretic generalization of on-line learning and an application to aoosting. *J. Computer and System Sciences* **55**, 119–139.

Frühwirth-Schnatter, S. and Pamminger C. (2006). Model-based clustering of discrete-valued time deries sata. *Tech. Rep.*, Johannes Kepler Universität Linz, Austria.

Lauritzen, S. (2006). Discussion of Chakrabarti and Ghosh. *In this volume.*

BAYESIAN STATISTICS 8, pp. 25–50.
J. M. Bernardo, M. J. Bayarri, J. O. Berger, A. P. Dawid,
D. Heckerman, A. F. M. Smith and M. West (Eds.)

Assessing the Effect of Genetic Mutation: A Bayesian Framework for Determining Population History from DNA Sequence Data

S. P. Brooks, I. Manolopoulou
University of Cambridge, UK
steve@statslab.cam.ac.uk

B. C. Emerson
University of East Anglia, UK
b.emerson@uea.ac.uk

Summary

In cases where genetic sequence data are collected together with associated physical traits it is natural to want to link patterns observed in the trait values to the underlying genealogy of the individuals. If the traits correspond to specific phenotypes, we may wish to associate specific mutations with changes observed in phenotype distributions, whereas if the traits concern spatial information, we may use the genealogy to look at population movement over time.

In this paper we discuss the standard approach to analyses of this sort and propose a new framework which overcomes a number of shortcomings in the standard approach. In particular, we allow for uncertainty associated with the underlying genealogy to fully propagate through the model to directly interact with the inferences of primary interest, namely the effects of genetic mutations on phenotype and/or the dispersal patterns of populations over time.

Keywords and Phrases: Reversible jump MCMC; posterior model probability; cladogram; cluster analysis; phenotypic analysis; phylogeographic analysis; nested clade analysis.

The authors gratefully acknowledge the financial support of Trinity College, Cambridge and the Leverhulme Trust. The authors would also like to thank Sara Goodacre for her invaluable input to this project.

1. INTRODUCTION

This paper is motivated by the increasing interest in phenotypic and phylogeographic analyses arising from recent developments in genetic research. These methods attempt to explain phenotypic or geographic variations in terms of the underlying genealogy in order to, for example, identify mutations associated with characteristic traits (phenotypic analysis) or infer population movement over time (phylogeographic analysis).

These methods are important for many reasons. Phenotypic analyses can be used to prevent or predict unwanted genetic traits which predispose the associated individual to disease, encourage the development of desired traits and even control specific genes to influence the associated phenotype. Phylogeographic analyses are used to help understand the long-term evolutionary history and behaviour of species such as colonisation and range expansion, as well as investigating the effect of geological events such as glaciation or seismic activity, for example.

Both forms of analysis require similar data and proceed in similar ways. We begin with data from a series of individuals comprising for each individual i a DNA sequence S_i together with an associated (perhaps vector) variable $\boldsymbol{T}_i$ recording the characteristic traits of interest. The latter might be indicators for the presence of disease or the geographic location of the associated individual, for example. The task then is to determine patterns in the traits observed across individuals that can be explained by the underlying genealogy, *e.g.*, do all those with a particular trait have commonalities between their corresponding sequences?

In answering such questions our first task is to determine the underlying genealogy, that is to construct a network (or so-called cladogram) of nodes which correspond to specific sequences and edges which correspond to specific genetic mutations (Cassens *et al.*, 2005, and Posada and Crandall, 2001) . Once we have this network which shows how each individual is related to the rest through a sequence of one or more genetic mutations, we can look to see whether deleting any edge (or collection of edges) creates clusters of sequences whose associated traits are similar within a cluster but distinct between clusters. An alternative motivation for the analysis is to view it as a simple clustering problem in which we wish to allocate individuals to two or more groups in terms of the observed traits but in a way that is consistent with the underlying genealogy,

This approach is probably best explained via a simple example. Suppose we have data from 6 individuals with sequences

$$ATCGA, ATCGA, ATCGA, ATTGA, ACCGA \text{ and } TTCGA$$

and corresponding trait values $1.9, 3.5, 3.1, 4.4, 1.2$ and 2.0, say. We see that the first three individuals have the same sequence and so in our cladogram, they will all correspond to the same node. We thus have four distinct nodes and we connect any two nodes by an edge if those two nodes differ in value at only one point in the sequence. The corresponding cladogram is given in Fig. 1.

Our aim is to draw some conclusion of the type 'a mutation at the ith nucleotide position is associated with a significant change in the values of the observed trait'. In order to do this, we first need to assign a distribution to the trait values (in order to determine 'significance') and then consider each partition in turn that can be obtained by deleting a single edge from the associated cladogram. Here, we have only three edges and so only three potential arrangements of the observations into two clusters. We can thus cluster the traits as $(\{4.1\}, \{1.9, 3.5, 2.1, 1.2, 2.0\})$

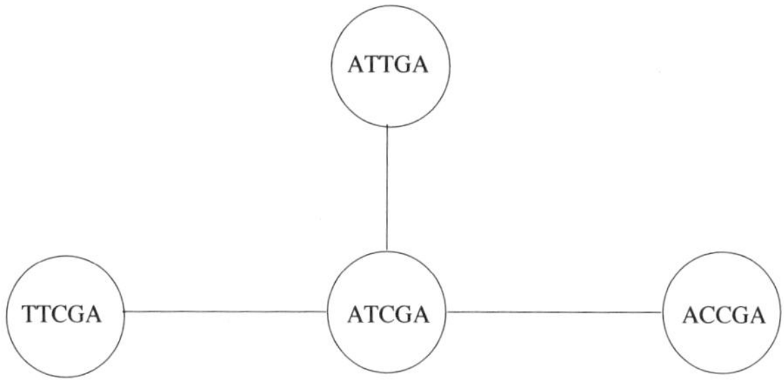

Figure 1: *Basic Cladogram showing the sequences observed and the mutations linking them.*

$(\{1.2\}, \{1.9, 3.5, 2.1, 4.1, 2.0\})$ or $(\{2.0\}, \{1.9, 3.5, 2.1, 4.1, 1.2\})$ by deleting the edge corresponding to a mutation at the third, second and 1firstst nucleotide position respectively. An ad-hoc interpretation might suggest that a mutation at the third nucleotide position causes a change in the associated trait and we will discuss and develop more formal methods for testing this in due course. However, we note here that without the sequence data, if we were to cluster the trait values directly, we might well wish to group 4.1 and 3.5 together, a combination that is not permitted by the underlying genealogy. Thus, by using the sequence data, we reduce the number of possible clusters to those which have a direct interpretation in terms of the mutations required to obtain the sequences observed in the data.

Of course, this is a very simple example. In practice the cladogram is not so easily determined. Problems associated with missing values, homoplasy (multiple mutations at the same nucleotide position) and recombinations (mutations not at a single nucleotide position but obtained by swapping a partial sequence from one node with that from another) mean that the cladogram is non-unique and so the clustering procedure should take account of the associated uncertainty in the underlying genealogy though, in practice, this uncertainty is often ignored.

Current best statistical practice relies on a series of analytic steps in which each stage is conditioned on the outcome of the previous one. In particular, the 'best' network is usually determined using a combination of statistical and ad-hoc procedures and then a form of iterative ANOVA is used to determine the mutations linking significantly distinct clusters. This modular approach is not able to fully propagate uncertainty at each stage into the final analysis and so inferences tend to overestimate the support for any one hypothesis over another and may, potentially, result in substantial inferential bias.

In this paper we describe a Bayesian approach in which different networks are assigned appropriate weights and then network-averaged mixture models are used to determine appropriate clusters. It is worth noting here, the existence of the already large but growing literature on inferring network structures given sequence data. By far the largest literature is concerned with the derivation of phylogenetic trees which incorporate both network and temporal structure by allowing edge lengths to denote times between observed mutations. See, for example, Mau *et al.* (1999), Newton *et al.* (1999), Larget and Simon (1999), Huelsenbeck *et al.* (2002), Altekar *et al.* (2004) and Stamatakis *et al.* (2005). Similarly a large number of software packages have been developed which create the phylogenctic tree from sequence data, *e.g.*, MrBayes, BAMBE and RAxML. However, here we do not require temporal information but do wish instead to infer the state of missing nodes in the tree. Though there is a direct link between the cladogram and the phylogenetic tree, here we work directly with the network structure (or cladogram).

We begin, in Section 2 with a review of the traditional method of undertaking such analyses based upon an analysis of variance built around the so-called nested cladogram. Then, in Section 3, we introduce a Bayesian alternative based around the idea of bivariate normal mixture models to help determine appropriate geographic clusters using the underlying haplotype network to determine cluster membership. In Section 4 we extend the basic analysis to deal with problems associated with missing values and homoplasy both of which occur frequently in data of this sort. We conclude with the results of our analysis in Section 5 and finally a discussion in Section 6 which highlights a variety of potential future research areas.

2. NESTED CLADE ANALYSIS

The main method used for associating physical traits with underlying genetic mutations is the so-called nested clade analysis. The standard nested clade analysis proceeds in three steps: first, the cladogram is constructed; second, a nested cladogram is formed; and, finally, we test for associations between clades and the corresponding traits.

2.1. *The Cladogram*

As we discussed in the previous section, the cladogram is a graph comprising nodes which correspond to specific sequences and edges which correspond to mutations at specific nucleotide positions. For the basic cladogram, we join all nodes which differ at just one nucleotide position by an edge and, if the resulting graph is incomplete we introduce additional nodes which correspond to sequences that are believed either to be extant but were missed in the sampling, or to be extinct. In adding missing nodes we typically appeal to a parsimony argument and assume that cladogram with the smallest number of missing nodes is the correct one in that it most accurately reflects the evolutionary history of the species under study.

In the absence of homoplasy (and, indeed recombination, which we do not deal with here as we focus only on mitochondrial DNA) the cladogram is structurally unique in terms of the observed sequences. Thus, we know exactly which sequences are connected to which and, where there are missing values, we know the number of missing values between any two observed nodes and the locations at which the mutations for those missing sequences occur (though we may not know their order). For example, the cladogram corresponding to the sequences

$$ATCGA, ACCGA, TTCGA, ATTGA, ATTGC$$

is unique and has no missing values (see Fig. 2). However, if the sequence $ATCGA$ were not observed then we know, in the absence of homoplasy, exactly what the missing sequence is and where it lies because of its two neighbours. If ATTGA were missing, we would know it, but we not know what the missing sequence actually was (it could be ATTGA or ATCGG). Note, however, that if $TTCGA$ were missing, we would not know it and this will be true for any leaf node.

The presence of homoplasy may create cycles or loops in the cladogram which means that one or more sequences can join the graph in two or more distinct ways. The presence of homoplasy can also lead to non-uniqueness of missing sequences

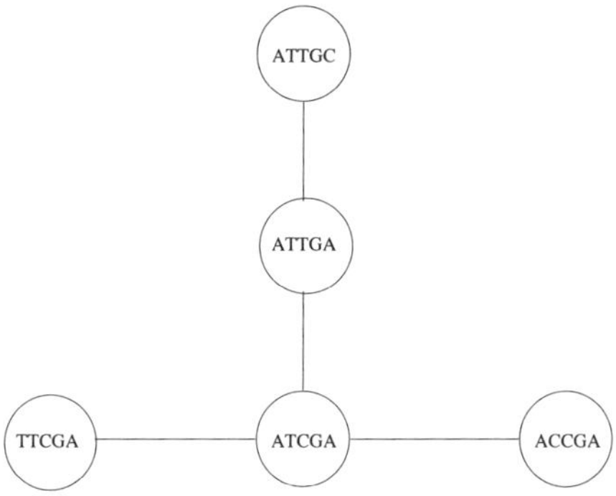

Figure 2: *A simple illustration of a cladogram with five nodes.*

and, indeed, can lead to uncertainty as to the number of missing sequences required to complete the graph.

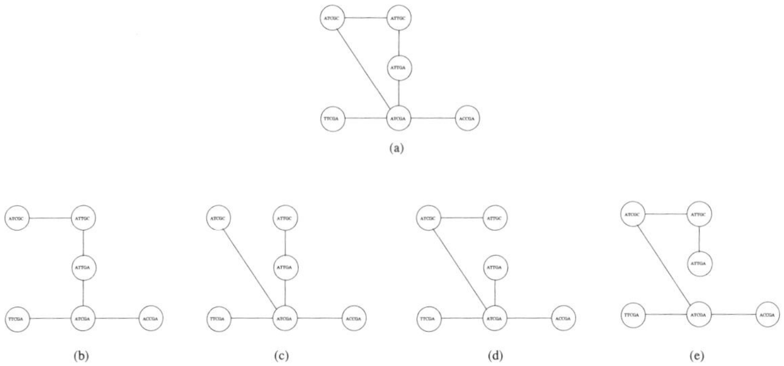

Figure 3: *(a) A 6-node cladogram with homoplasy. Figures (b)-(e) provide the corresponding cladograms with the homoplasy removed.*

For example, suppose we observe a sixth sequence: $ATCGC$. This sequence is just one nucleotide position apart from the sequences $ATCGA$ and $ATTGC$ and so could be joined to either one or both of these nodes (see Fig. 3a). Again, the principle of parsimony is used to justify the assumption that any sequence will not occur via mutation from any other sequence in more than one way. This means that the true cladogram will be a sub-graph of the cyclical graph in Fig. 3(a) formed by deleting one or more edges so that the graph is both complete and contains no loops. In this case, there are four possible graphs, see Fig. 3(b)-(e). Note that if the sequence $ATTGA$ were not observed (as above) then no cycle would exist, there would be no observed homoplasy and the cladogram would be unique.

Again, it is common to use the parsimony principle to assume that both homoplasy and missing values should only be considered if they are required to complete the graph. Templeton *et al.* (1992) develop a sequential method for determining the number of missing values and for minimising homoplasy (and, indeed, recombination events). See also Lloyd and Calder (1991), Clement *et al.* (2000,2002), Crandall and Templeton (1993), Templeton *et al.* (1988) and Felsenstein (1992). We shall see in the following section how similar methods may be incorporated into the Bayesian analysis to account for the uncertainty associated with the underlying cladogram.

2.2. *The Nested Cladogram*

Once we have the cladogram, we next form the so-called nested cladogram. This is done following the guidelines set out in Templeton *et al.* (1987) as follows.

We start with our 0-step clades, which are just the sequences represented as leaves in our cladogram. Given the n-step clades, the $n+1$-step clades are formed by taking the union of all n-step clades which can be joined up by moving one mutational step back from the terminal node of each n-step clade. We continue the process recursively until all the nodes in our cladogram have been nested. As before, the nesting process is probably best explained via an example.

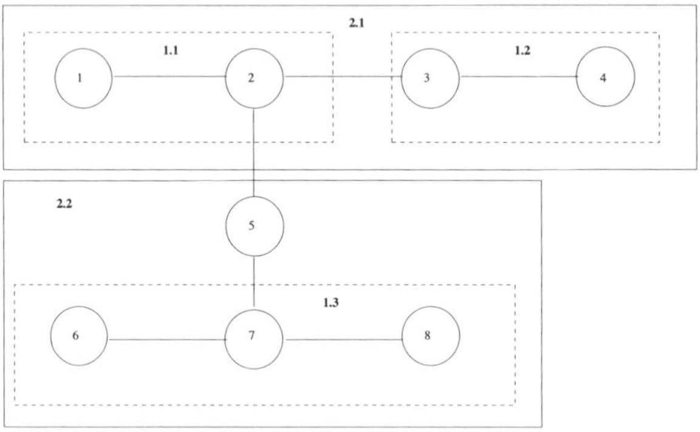

Figure 4: *A simple illustration of a nested cladogram comprising three one-step clades and two two-step clades.*

Fig. 4 provides a simple illustration of a cladogram with iight nodes. Following Templeton's algorithm, we first join the leaves to their neighbours to form the one-step clades. In the case where two leaves are joined to the same neighbour, as for example occurs with nodes 6, 7 and 8, the corresponding clades are joined together. Thus, we obtain 1-2, 3-4 and 6-7-8 as our one-step clades which we arbitrarily label as clades 1.1, 1.2 and 1.3 respectively. At the next stage these three groups are essentially treated as a single node and form the two step clades by combining 1-2 with 3-4 and 6-7-8 with 4. We label the corresponding two-step clades 2.1 and 2.2. At the third stage we would combine the two two-step clades to form a single group containing all nodes in the cladogram.

We can see here the link between the clusters obtained following Templeton's algorithm and those obtained by deleting edges in the graph as discussed in the previous section. However, not all edge-based clusters are obtained using Templeton's method and there are many cases where Templeton's approach leads to ambiguities or inappropriate clusters. Templeton's method might be compared with that of stepwise regression in which we attempt to identify the best model by successively adding and deleting parameters from an initial starting model. It is well recognised that this process does not necessarily lead to the best model and there are many examples where more exhaustive comparison have identified significantly better solutions than these sequential methods. Adhoc comparisons are used within Templeton's algorithm (based upon merging pairs of neighbouring clusters with smallest sample size, for example, Templeton and Sing, 1993) which introduces a somewhat arbitrary element to the clustering process in the final stage of the analysis. We will show how the Bayesian approach based upon creating clusters by deleting sets of edges from the graph can be used to more efficiently and reliably identify potentially significant clusters of sequences as well as allowing for the uncertainty associated with different clustering combinations to be properly propagated to the final stage of the analysis.

2.3. *Nested Clade Analysis*

The final stage of the standard phenotypic and phylogeographic analysis concerns the identification of significant clades, *i.e.*, those clusters of nodes that appear to well explain the variation in the observed traits. We distinguish here between the two forms of analysis as slightly different procedures are used for each.

The Phenotypic Analysis

The phenotypic analysis is based upon a standard (nested) analysis of variance (Templeton *et al.*, 1987). Suppose we have a cladogram with 0-step, one-step, up to k-step clades (where the $(k+1)$-step clade contains all nodes in the graph) then we perform a $(k+1)$-way (M)ANOVA using the trait values to determine which clade levels exhibit significant deviation from the null hypothesis of equality of means. Sums of squares that are deemed significant are further decomposed to gain finer resolution on the mutations associated with differences in the means. For example, if the one-step clades are deemed significant, then the associated sum of squares is sub-divided into independent sums of squares formed within each 2-step clade. For example, with our nested cladogram in Fig. 4, we would decompose the Sum of Squares for the one-step clades into two: the sum within clade 2.1 and that within 2.2. If the former were found to be significant, this suggests that the mutation between nodes 2 and 3 is associated with a significant change in the mean trait level.

This approach suffers from a number drawbacks associated with the risk of the effect of significant mutations carrying through to different clades masking the true patterns in the data. In practice, multiple comparison statistics are used together with Bonferoni-type corrections to ensure the correct overall significant level. We shall see in the next section, how the traditional ANOVA can be replaced by Bayesian mixture modelling which, once again, provides a more comprehensive and robust procedure for identifying mutations associated with changes in phenotype values.

The Phylogeographic Analysis

Phylogeographic analyses are treated slightly differently since the aims of the analysis are typically somewhat different from those of phenotypic analysis. Here we are interested in determining associations between observed genetic mutations and spatial patterns observed in the data. See Ibrahim *et al.* 1996, Templeton 1995, Gomez-Zurita *et al.*, 2000, and Emerson and Hewitt, 2005, for example. There are three major biological factors that can cause a significant spatial association with haplotype variation: (i) restricted gene flow; (ii) past fragmentation; and (iii) range expansion (including colonisation). The standard phylogeographic analysis proceeds in two steps: a permutational contingency test is carried out to determine whether or not haplotypes are distributed randomly at all clade levels; then, if the hypothesis of randomness is rejected then further tests under assumptions of the three different scenarios above are conducted and the results interpreted via an inference key provided originally by Templeton (1998) and based upon a series of simulation studies.

The so-called nested clade phylogeographic analysis (NCPA) works by estimating two distinct geographic measures in the context of the hierarchical design inherent in the nested cladogram: the clade distance $D_c(X)$ for clade X, which represents the geographical range of that clade; and D_n (X), which measures how that particular clade is geographically distributed relative to its closest evolutionary clades (i.e. clades in the same higher-level nesting category as X). Specifically, $D_c(X)$ is the mean distance of clades within clade X from the geographical centre of that clade and $D_n(X)$ is the mean distance of the clades within clade X to the geographical centre of all same-level clades that lie within the higher-level clade that contains X. See Templeton *et al.* (1995), for example. The distribution of these distance measures under the null hypothesis of no geographical association within clade X is obtained by using a Monte Carlo approach, resampling the measures under permutations of the low-level nesting clades. See Templeton and Sing (1993) and Templeton *et al.* (1995), for example.

If the null hypothesis of no association is rejected then the permutation test is repeated under different assumptions as to the underlying nature of the geographic association. In practice, however, the D_c and D_n values are calculated and a prescriptive key provided by Templeton (1998) is followed to determine whether any of the three factors described above can be identified as adequately describing the spatial patterns of observed haplotypes.

In essence, the first stage of the NCPA is equivalent to an ANOVA based upon distance and suffers from similar problems associated with multiple testing and poor resolution in terms of identifying the level of clade at which significant mutations occur. As we shall see in the next section, this inferential process can be easily incorporated into the Bayesian framework that we develop here.

The three-stage approach described here, which represents current best practice for problems of this sort, suffers from problems at each stage that can be resolved by using a Bayesian approach. A further advantage of the Bayesian approach is that all three stages can be combined into a single analysis so that it is no longer necessary to conduct any stage conditioning on the results from the previous stage. In this way, we fully propagate all sources of uncertainty into our final inference and avoid the need to conduct conditional analyses based upon what are often adhoc procedures for picking between different solutions at each stage.

3. THE BAYESIAN ALTERNATIVE

The problem with Nested Clade Analyses is that by conditioning on the best option at every stage of the process, we lose the information concerning the associated uncertainties. In this section we describe a Bayesian approach which combines all three stages into one so that uncertainty in the underlying tree structure is fully propagated throughout the model and is properly accounted for when assigning individuals to clusters.

We will focus on the phylogeographic analysis here though essentially identical methods may be applied to the phenotypic analysis. In line with the classical NCA we will use Euclidean distance between individuals as basis for clustering which is essentially equivalent to adopting a bivariate normal distribution to describe the spatial distribution of individuals within a cluster. Thus we will use a bivariate normal mixture model in which the clustering of individuals arises from the separation of the underlying tree into disjoint parts. For phenotypic analyses alternative distributions to the normal may be more appropriate.

For the moment, we will assume the simplest case in which the data contains no missing values and there is no homoplasy. In this case the underlying tree is unique and can be determined analytically. The only uncertainty, then, arises from the parameters of the bivariate normal parameters and the partition of the tree into disjoint sets to determine component membership. We will assume here that we do not know how many clusters exist in the data and so use a combination of standard and reversible jump MCMC methods to undertake the analysis. We begin with a brief description of the mixture modelling framework.

3.1. *The Mixture Model*

Suppose we observe n individuals so that our data $\mathcal{D}$ comprises the set $\{(S_i, \boldsymbol{T}_i) : i = 1, ..., n\} = \{\mathcal{S}, \mathcal{T}\}$, say, of sequences and associated traits (in this case geographic locations so that $\boldsymbol{T}_i = (x_i, y_i)$, say). Suppose also that within $\mathcal{D}$ we observe $m \leq n$ distinct haplotypes so that our associated cladogram comprises m nodes $v_1, ..., v_m$ and $m-1$ edges $e_1, ..., e_{m-1}$. Under the assumption of no homoplasy or missing values, the cladogram is a complete connected graph with no loops. Thus, any individual i will be associated with exactly one node $v_{r(i)}$, say, which corresponds to the associated sequence S_i for individual i. Clearly, any node v_j may be associated with more than one individual if and only if $m < n$.

To model the locations, we adopt a mixture of k bivariate normals for the variables $\boldsymbol{T}_i$ in which allowable groupings of the n individuals correspond only to partitions of the graph obtained by deleting $k-1$ edges. That is, deleting any set of $k-1$ edges creates a partition of the graph into k disjoint sub-graphs. If we label the subgraphs $j = 1, ..., k$ then the individuals associated with component j of the mixture are those individuals associated with nodes within subgraph j. We index the partitions by $\boldsymbol{e}(k)$ which denotes a set of k edges chosen from the complete set $\mathcal{E} = \{e_1, ..., e_{m-1}\}$ for deletion. Under partition $\boldsymbol{e}(k)$, we allocate individual i to component $z_i(\boldsymbol{e}(k))$ for all $i = 1, ..., n$. Note that any two individuals with the same haplotype will therefore be assigned to the same component. Finally, for any individual i in component j we assume that $\boldsymbol{T}_i \sim \mathcal{N}_2(\boldsymbol{\mu}_j, \Sigma_j)$ with associated density $f(\boldsymbol{T}_i|\boldsymbol{\mu}_j, \Sigma_j)$, say.

Thus, we obtain a joint probability distribution for the data given the number of components k, the partition $\boldsymbol{e}(k)$ and the set $\mathcal{M}_k$ of component means $\mathcal{V}_k$ of component covariance matrices:

$$f(\mathcal{T}|k, \boldsymbol{e}(k), \mathcal{M}_k, \mathcal{V}_k) = \prod_{i=1}^{n} f(\boldsymbol{T}_i|\boldsymbol{\mu}_{z_i(\boldsymbol{e}(k))}, \Sigma_{z_i(\boldsymbol{e}(k))}).$$

Note that since at this stage we are modelling only the geographic locations, we have here only a probability distribution for the traits $\mathcal{T}$. We shall derive a similar distribution for the sequences $\mathcal{S}$ in the next section.

We shall assume, for the moment, that the value of k is fixed and adopt a uniform prior over all valid partitions, together with independent bivariate normal and inverse Wishart priors for the component mean vectors and covariance matrices respectively. Thus

$$\pi(\boldsymbol{e}(k), \mathcal{M}_k, \mathcal{V}_k|\mathcal{T}, k) \propto f(\mathcal{T}|k, \boldsymbol{e}(k), \mathcal{M}_k, \mathcal{V}_k)p(\boldsymbol{e}(k)) \prod_{i=1}^{k} p(\boldsymbol{\mu}_i, \Sigma_i). \tag{1}$$

3.2. *The MCMC Algorithm*

In order to explore the posterior distribution given in equation (1) above we use a standard MCMC algorithm comprising a series of Gibbs and Metropolis Hastings updates for the different parameters. Here, we take the following priors for the $\boldsymbol{\mu}_i$ and Σ_i:

$$\begin{aligned} \Sigma_j &\sim \text{InvWishart}(t, \Psi); \text{ and} \\ \boldsymbol{\mu}_j|\Sigma &\sim \mathcal{N}_2\left(\mathbf{0}, \frac{1}{\tau}\Sigma_j\right). \end{aligned}$$

Thus, we obtain the following posterior conditional distributions for $j = 1, ..., k$:

$$\begin{aligned} \boldsymbol{\mu}_j|\Sigma_j, \boldsymbol{e}(k), \mathcal{T}, k &\sim \mathcal{N}_2\left(\frac{n_j \boldsymbol{T}_j}{n_j+\tau}, \frac{1}{n_j+\tau}\Sigma_j\right); \text{ and} \\ \Sigma_j|\boldsymbol{\mu}_j, \boldsymbol{e}(k), \mathcal{T}, k &\sim \text{InvWishart}\Big(n+t, \Psi + \sum_{i:z_i(\boldsymbol{e}(k))=j} \boldsymbol{T}_i \boldsymbol{T}_i' \\ &\qquad - \Big(n_j \boldsymbol{T}_j \boldsymbol{T}_j' + \frac{n_j\tau}{n_j+\tau}(\boldsymbol{T}_j - \boldsymbol{\mu}_j)(\boldsymbol{T}_j - \boldsymbol{\mu}_j')\Big)\Big), \end{aligned}$$

where n_j denotes the number of individuals assigned to component k (dropping the dependence on $\boldsymbol{e}(k)$ for notational convenience) and $\boldsymbol{T}_j$ denotes the vector of all $\boldsymbol{T}$ vectors associated with individuals in cluster j, *i.e.*,

$$\boldsymbol{T}_j = \frac{1}{n_j} \sum_{i:z_i(\boldsymbol{e}(k))=j} \boldsymbol{T}_i.$$

Thus, the MCMC algorithm for drawing from the posterior distribution given in equation (1) proceeds by updating the component means and variances using Gibbs updates, in which new values are drawn from the posterior conditional distributions given above, followed by a move which updates the edge set $\boldsymbol{e}(k)$. Obviously, $\boldsymbol{e}(k)$ is a discrete random variable and so can be updated directly using a Gibbs update. However, for datasets with large numbers of distinct sequences (i.e., large m), a Gibbs update for the edge set can prove extremely computationally expensive due to the fact that the posterior probabilities associated with each of the '$m-1$ choose k' potential edge sets must be calculated. A far more efficient update is obtained by using a Metropolis Hastings procedure in which only the posterior probabilities associated with the existing and proposed new edge set need be calculated.

In practice we found that updating only the edge set led to extremely poor mixing because of the strong posterior association between the edge set and the corresponding component means and covariance. Thus, we propose updating the edge set as follows. Select an edge within the edge set with equal probability and propose to replace it uniformly at random with any other edge within the graph that is not already in the edge set, *i.e.*, from the set $\mathcal{E} \setminus \boldsymbol{e}(k)$. We then propose new values for all means in $\mathcal{M}_k$ and covariance matrices in $\mathcal{V}_k$ from the corresponding posterior conditionals under the proposed new edge set. This new edge set, the associated cluster means and covariances are then accepted with the usual Metropolis Hastings acceptance ratio.

This, then, completes the updates required to generate a series of realisations from the posterior distribution given in equation (1) from which we can obtain empirical estimates for any posterior statistics of interest.

3.3. *Label Switching*

As is common with mixture modelling, our model suffers from the so-called label-switching problem caused by symmetry in the joint probability distribution of the data given the model parameters. This has the practical consequence that the collection of nodes labelled in one iteration of the MCMC algorithm as group j (with associated mean and covariance μ_j and Σ_j) may may be labelled as group $l \neq j$ in the next iteration.

The standard approach to overcoming such difficulties is to introduce an essentially arbitrary identifiability constraint such as ordering the components in terms of the associated component parameters e.g., the mean. This sort of approach was unsatisfactory in the range of examples we considered due, mainly, to the multi-dimensional nature of the mixture distributions and the tendency for there to be considerable overlap between the marginal distributions associated with the bivariate normal components. Ordering based upon distance of the mean vector from a fixed point as well as ordering based upon the properties of the tree (e.g., start at the left of the tree and label the components as you visit them) all failed to produce sensible results.

Thus, we adopt the relabelling algorithm of Stephens (2000) to draw inference about component parameters from the sample of observations obtained via the MCMC algorithm described earlier. Suppose we have a sample of T observations. For any observation t we require a permutation ν_t of the the associated component labels that provides us with a consistent labelling for all $t = 1, ..., T$. Stephens suggests the following iterative algorithm. Given suitable permutations for iterations $1, ..., t$ choose a permutation ν_{t+1} for iteration $t + 1$ that maximises

$$\sum_{i=1}^{n} \log \frac{1}{t} \left(\sum_{j=1}^{t} \mathbb{I}_{c_i^{(j)} = c_i^{(t)}} \right).$$

This, then, provides a series of permutations that assign labels to the different components that leads to 'as many nodes as possible belonging to their favourite group so far'.

From a practical perspective, this approach proved extremely reliable with only minimal additional computational complexity. Indeed the labelling process can even be introduced as an additional (deterministic) step within the MCMC algorithm itself.

3.4. *How Many Clusters?*

Finally, we consider the case where the number of clusters in the data is unknown *a priori*. In this case, we must first extend the posterior distribution in equation (1) to allow for uncertainty in the value of k, which is easily achieved by specifying an appropriate prior $p(k)$ so that

$$\pi(\boldsymbol{e}(k), \mathcal{M}_k, \mathcal{V}_k, k|\mathcal{T}) \propto \pi(\boldsymbol{e}(k), \mathcal{M}_k, \mathcal{V}_k, k|\mathcal{T}, k)p(k). \quad (2)$$

Here, we assume k follows a uniform distribution on the discrete values $0, ..., k_{\max}$ though other priors may be also be appropriate. For example, we may wish to specify a prior which weights each value of k by the reciprocal of the number of possible edge sets available when k edges can be deleted. In any case, appropriate re-weighting of the posterior model probabilities under the uniform prior for k can provide the required the probabilities associated with any other and so we stick with the uniform here. Note here that the priors for $\boldsymbol{e}(k)$, $\mathcal{M}_k$ and $\mathcal{V}_k$ must be normalised since they will be functions of k. For the fixed-k case this was not necessary. The normalisation constants are all easily obtained.

As we have extended our posterior distribution, we must now extend our MCMC algorithm to allow for the additional move required to update the parameter k. As this necessitates changing the number of parameters in the model (i.e., adding or deleting component means and covariances) this requires a reversible jump MCMC (Richardson and Green, 1997) update.

We might begin each model move by first deciding to increase k by splitting an existing cluster or to decrease k by merging two adjacent clusters (where adjacency is determined in terms of the tree). In practice this means that we add or delete an edge to the edge set with probabilities P_{add} and P_{delete} respectively. If we propose to move beyond the $[0, k_{\max}]$ range we automatically reject the proposal. A simpler way is to simply select an edge and if it is already in the current edge set we propose to delete it and if it is not, then we propose to add it. This simplifies the acceptance ratio, but leads to fairly high probabilities for a split move. In practice, we found that this caused us no computational difficulties, but for other problems it may be better to specify fixed split/merge probabilities to gain better control over the algorithm's performance.

Suppose we propose to add a new edge to the edge set, moving from k to $k+1$, then this automatically splits an existing cluster. We therefore need to propose values for the normal parameters associated with the two new clusters obtained by splitting the original. We do this by drawing values from the associated posterior conditionals, conditioning on the proposed new edge set and associated assignments of nodes (individuals) to each cluster. Similarly, when merging two clusters, we draw new normal parameter values from the posterior conditional with individuals from the two original clusters are assigned to the proposed new one. In this way, the Jacobian term required for the reversible jump acceptance ratio reduces to 1. In fact, the entire acceptance ratio therefore reduces to a ratio of appropriate normalisation constants and is easily derived.

Combining this step with the Gibbs and Metropolis Hastings steps described in Section 3.2 we obtain an MCMC algorithm capable of obtaining draws from the posterior distribution described in equation (2).

Recall that at the beginning of this section we made the assumption that there was (i) no missing values and (ii) no homoplasy. In practice these two assumptions will rarely be valid. Thus, in practice we will require a number of extensions to our method in order to deal with these two common situations.

4. EXTENDING THE BASIC FRAMEWORK

Here we extend our analysis to allow both for missing values and homoplasy and therefore need to deal with uncertainties in the underlying tree. Note the distinction here with the previous section which essentially assumes that the underlying tree is fixed and focuses only on modelling the traits. In order to assess tree probabilities, we must derive a form of model for the observed sequences $\mathcal{S}$ from which the associated model parameters can be inferred. This requires estimation of the mutation probabilities from the mutations observed to occur in the data which, in turn requires the derivation of a root node which determines the direction of the mutations in the tree.

4.1. *Finding the Root*

Here, we discuss methods for determining the root of the tree, that is the 'oldest' haplotype from which the other observed haplotypes can be assumed to have been derived. We can make various assumptions about the properties of the root. For example, we would normally expect the root to be towards the 'centre' of the cladogram and that it might often be surrounded by missing nodes representing haplotypes that might now be extinct. It might also be assumed that the root node would tend to be of higher degree than other nodes, though this can be masked by population dynamics such as changing productivity rates. Such information can be used to determine an appropriate prior for the root node. In the absence of suitable data on what constitutes a 'central" node, we will assume a flat prior for the root over all nodes (both observed and missing) in the graph.

4.2. *Estimating Mutation Rates*

Having determined the root, the graph now becomes directed and it is possible to estimate the mutation rates. We begin by parameterising the mutation probabilities as follows.

We assume that the nucleotide frequencies are at equilibrium and that the mutation process is time-reversible. We use a Generalised Time-Homogeneous Time-Reversible Markov Process model (Tavaré, 1986) for the mutation rates (which are assumed to be independent and identical across all sites) generated by a Q-matrix:

$$Q = \begin{pmatrix} \cdot & \alpha\pi_G & \beta\pi_C & \gamma\pi_T \\ \alpha\pi_A & \cdot & \delta\pi_C & \epsilon\pi_T \\ \beta\pi_A & \delta\pi_G & \cdot & \zeta\pi_T \\ \gamma\pi_A & \epsilon\pi_G & \zeta\pi_C & \cdot \end{pmatrix}$$

where the π_i's ($i = A,\ G,\ C,\ T$) represent the equilibrium probabilities of the nucleotides (we assume that the gene frequencies are stationary), and $\alpha, \ldots, \zeta$ the relative transition and transversion probabilities, so that we have 9 degrees of freedom. For simplicity we refer to $(\pi_A,\ \pi_G,\ \pi_C,\ \pi_T)$ as $(\pi_1,\ \pi_2,\ \pi_3,\ \pi_4)$ respectively. This chain is time-reversible since $\pi_i q_{ij} = \pi_j q_{ji},\quad i, j = 1, \ldots, 4$. Here we assume that the generator matrix is identical and independent across all sites, though it would be possible to extend the methods discussed her to relax this assumption.

The Q-matrix above is normalised to obtain the following matrix of mutation probabilities:

$$P = \begin{pmatrix} 0 & c_1\alpha\pi_2 & c_1\beta\pi_3 & c_1\gamma\pi_4 \\ c_2\alpha\pi_1 & 0 & c_2\delta\pi_3 & c_2\epsilon\pi_4 \\ c_3\beta\pi_1 & c_3\delta\pi_2 & 0 & c_3\zeta\pi_4 \\ c_4\gamma\pi_1 & c_4\epsilon\pi_2 & c_4\zeta\pi_3 & 0 \end{pmatrix}$$

where the c_i are simple normalisation constants chosen so that the entries in each row sum to 1.

Since the π_i sum to 1, we adopt a Dirichlet prior for the vector of nucleotide frequencies, $\boldsymbol{\pi}$. Here, we assume that $\boldsymbol{\pi} \sim Dir(B_1, B_2, B_3, B_4)$ where the $B_i \sim \mathcal{N}(1, \sigma_B^2)$. Alternatively, we might adopt the Jeffreys prior in which the B_i are fixed to be equal for $i = 1, ..., 4$. Again, we observed no clear sensitivity to the choice of prior in the primary posterior summary statistics of interest. Under the hierarchical Dirichlet prior, it is simple to show that the posterior conditional for $\boldsymbol{\pi}$ given the other parameters is itself a Dirichlet with parameters $B_i + b_i$, where b_i denotes the number of observed genes of type i within the observed sequences.

For the transition/transversion rates, we assume independent normal priors chosen on the basis of expert information as to the relative probabilities of transitions and transversions so that

$$\alpha, \zeta \sim \mathcal{N}(\mu_s, \sigma_s^2) \text{ and } \beta, \gamma, \delta, \epsilon \sim \mathcal{N}(c\mu_s, \sigma_s^2)$$

for some appropriate values of c, μ_s and σ_s^2. Note also, that the value of c might itself have a prior distribution if the likely transition/transversion ratio is unknown. For the moment, we assume that the prior parameters are fixed and, if $\boldsymbol{\theta}$ denotes the set of transition/transversion rates, we have prior $p(\boldsymbol{\theta}|c, \mu_s, \sigma_s^2)$.

4.3. *Assessing Tree Probabilities*

Now that we have the mutation probabilities, we can assess the probabilities associated with any given tree. In particular, we can use these to determine the state of missing nodes and, more importantly, break loops by deleting appropriate edges.

The basic tree probability is simply the product over all edges of the mutation probability associated with that edge. These probabilities are normalised by the sum over all trees consistent with the data and here we will make the simplifying assumption that the minimum number of missing nodes will be added to ensure completeness of the tree. Suppose that our tree contains η edges, $\mathcal{E} = \{e_1, ..., e_\eta\}$ with probability of mutation $p(e_j)$ for edge e_j derived from the mutation probability matrix given in Section 4.2 and the directions indicated by the root node identified in Section 4.1. Then the probability of the tree with edges $\mathcal{E}$ is given by

$$p(\mathcal{E}|r, \phi, \boldsymbol{\theta}, \eta) \propto \prod_{i=1}^{\eta} p(e_i).$$

Often the state of any missing node (i.e., the sequence with which it is associated) will be uniquely determined but in the presence of homoplasy or where strings of more than one missing nodes occur in the cladogram, the true state of the missing node will be unknown *a priori* and can be estimated. Of course, once the set of edges $\mathcal{E}$ is known, the states of the missing nodes are uniquely determined and so they are automatically estimated as a by-product of the estimation of the non-deterministic edges that we discuss below.

Homoplasy has the effect in our network of creating two or more different edges representing a mutation on the same restriction site. In some cases, loops may be formed and these need to be eradicated in order to form a permissible cladogram. In practice, loops are removed by deleting one or more edges (as few as possible) in much the same way as edges are removed in order to form clusters for the trait variables. The mutation probabilities can be used to weight edges according to the their relative probabilities and so the clustering algorithm described in the previous section can be averaged over the trees obtained by breaking loops in the full tree using the associated probabilities to weight the inference.

An additional complication is that when we wish to undertake a phenotypic analysis, if any edge in the edge set corresponds to a mutation which happens elsewhere in the graph (this is only possible if homoplasy occurs) then all edges corresponding to that mutation must be added to the edge set. Since we are not attempting to make causal associations in the same way for the phylogeographic analyses, this problem can be ignored in this case.

4.4. *The MCMC Algorithm*

The posterior distribution from the previous section is amended so that the distribution of the edge set depends upon the set of edges $\mathcal{E}$ within the tree. We decompose this set of edges into two: the set $\mathcal{E}_S$ of edges in the full graph that can be derived with certainty from the observed sequence data $\mathcal{S}$, *i.e.*, edges only between observed nodes that are not part of a loop within the cladogram; and $\mathcal{E}_0$ which denotes the set of uncertain edges i.e., edges connected to missing nodes and subsets of the edges in the original tree that form loops. Thus, $\mathcal{E}_0 \cup \mathcal{E}_S$ forms a connected tree without loops. Then we amend our posterior in equation (2) to obtain

$$\begin{aligned} &\pi(\boldsymbol{e}(k), \mathcal{M}_k, \mathcal{V}_k, \mathcal{E}_0, \boldsymbol{\theta}, \boldsymbol{\pi}, \boldsymbol{B}, r, k|\mathcal{D}, \mathcal{E}_S) \\ &\quad\propto\quad f(\mathcal{T}|k, \boldsymbol{e}(k), \mathcal{M}_k, \mathcal{V}_k) p(\boldsymbol{e}(k)|\mathcal{E}) p(\mathcal{E}|r, \phi, \boldsymbol{\theta}) \\ &\qquad\times\ p(\boldsymbol{\theta}) p(\boldsymbol{\pi}|\boldsymbol{B}) p(\boldsymbol{B}) p(r) \prod\nolimits_{i=1}^{k} p(\boldsymbol{\mu}_i, \Sigma_i) p(k) \end{aligned} \tag{3}$$

The associated MCMC algorithm therefore involves updating the original parameters in much the same way as described in Section 3.2. The additional parameters introduced in this section are updated as follows:

Updating r: Here, we propose a new root randomly from the observed nodes in the graph and accept the move with the standard Metropolis Hastings acceptance probability. An alternative is to propose only neighbouring nodes to the current root. The latter tends to perform better in general but it is possible to get stuck in local modes of the distribution from which it is hard for the chain to escape.

Updating $\boldsymbol{\theta}$: The transition/transversion probabilities are updated using a random walk Metropolis Hastings update.

Updating $\boldsymbol{\pi}$: The nucleotide frequencies are updated via a Gibbs step, since their posterior conditional is Dirichlet.

Updating $\boldsymbol{B}$: The hyperprior parameters for the Dirichlet prior on the nucleotide frequencies are updated using a random walk Metropolis Hastings proposal on the log scale.

Updating $\mathcal{G}_0$: Here, we retain a list of edges involved in loops within the graph and at each iteration we draw uniformly at random a collection of edges that both removes all loops within the graph and which provides a complete tree. These edges are then accepted using the usual Metropolis Hastings acceptance ratio. If we wish to update the state of the missing nodes and, indeed, their associated trait value, these can be sampled directly from their corresponding posterior conditionals.

Now that we have the MCMC methodology required to sample from the posterior of interest, we may begin to analyse data and interpret the results.

5. EXAMPLE: A PHYLOGEOGRAPHIC ANALYSIS

Here, we consider a specific data set and conduct a phylogeographic analysis using the methodology described above. We begin with a description of the data before providing the results and associated interpretation and discussion.

5.1. *The Data*

Here, we focus on data from the geologically young and well-characterised island of La Palma from within the Canary Islands archipelago to generate phylogeographic predictions for *Brachyderes rugatus rugatus*, a flightless curculionid beetle species occurring throughout the island in the forests of *Pinus canariensis.*

The data comprises a sample of 135 beetles from 18 locations across the island and for each we record 570 base pairs of sequence data for the mitochondrial DNA cytochrome oxidae II gene. The data are summarised in Fig. 5 which superimposes the sampling locations on a map of La Palma together with forest density. At each location the number of distinct haplotypes observed is recorded, with 69 distinct haplotypes observed in all. See Emerson *et al.* (2000, 2006), for example.

Geological studies of the island provide us with a fairly complete understanding of the island's geological history. The northern part of the island is mainly older volcanic terrain with the southern part comprising a ridge of more recent volcanic origin. It is a reasonable assumption that the *Brachyderes* beetle population, with their limited mobility, would have been strongly influenced by La Palma's volcanic and erosional history and there is also strong evidence that the population was seeded by immigration from the nearby island of Tenerife to the East. Thus we might expect the oldest haplotypes to be concentrated in the northern part of the island and to observe evidence of a more recent range expansion to the southern tip. If we were to cluster the haplotypes geographically we should therefore find that those to the North should be more central to the haplotype network, and those to the south should be placed towards the tips of the network.

5.2. *The Analysis*

Here, we apply the methodology described in the previous sections to the beetle data.

Table 1: *Posterior probabilities for different numbers of clusters.*

No. Clusters	2	3	4	5	6
Post. Model Prob.	0.02	0.13	0.24	0.36	0.25

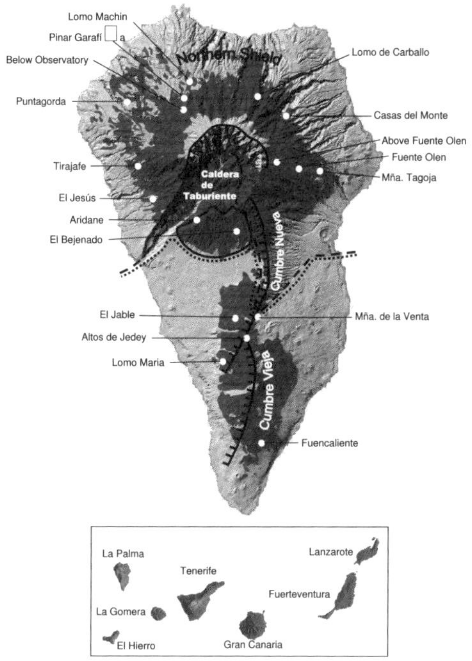

Figure 5: *Map of La Palma with sampling locations and tree density superimposed.*

Table 1 provides the posterior model probabilities associated with models containing between two and six clusters. Clearly, the 5-cluster model is *a posteriori* most probable, but there is little to choose between many of the models under consideration.

Fig. 6 provides the maximum *a posteriori* (MAP) estimate of the underlying cladogram, together with an indication of the clusters to which each node belongs. We can see that the assignment of nodes to specific clusters is certain *a posteriori* for many nodes, except a few towards the top of the cladogram which are most likely associated with the first (upper left) cluster but also have non-zero probability of being allocated to the second one. Perhaps most intriguingly there are no nodes that are assigned to the last (bottom left) cluster with absolute certainty with most nodes being assigned with considerable probability to the fourth cluster if they have any probability of being assigned to the last. At first sight, this may seem strange especially when you consider that in general we would expect posterior certainty for the last cluster to increase as we move closer to the leaf node towards the bottom left-hand corner of the cladogram. The reason for this is that there is considerable uncertainty as the underlying cladogram structure in this region with relatively substantial probability that the dotted edge given in Fig. 6 might replace one of the two dashed lines in the graph. The uncertainty here arises from two distinct ways to delete a loop in the original graph and this clearly highlights the value of incorporating the uncertainty of the underlying graph into the analysis rather than simply conditioning on the most likely graph determined from a separate analysis.

The *a posteriori* most probable root node under the flat prior is the single second node of degree 5, though there is considerable uncertainty as to the exact location with the second most likely root node attracting 80% of the posterior probability associated with the first. The second cluster is the *a posteriori* most probable to contain the root node.

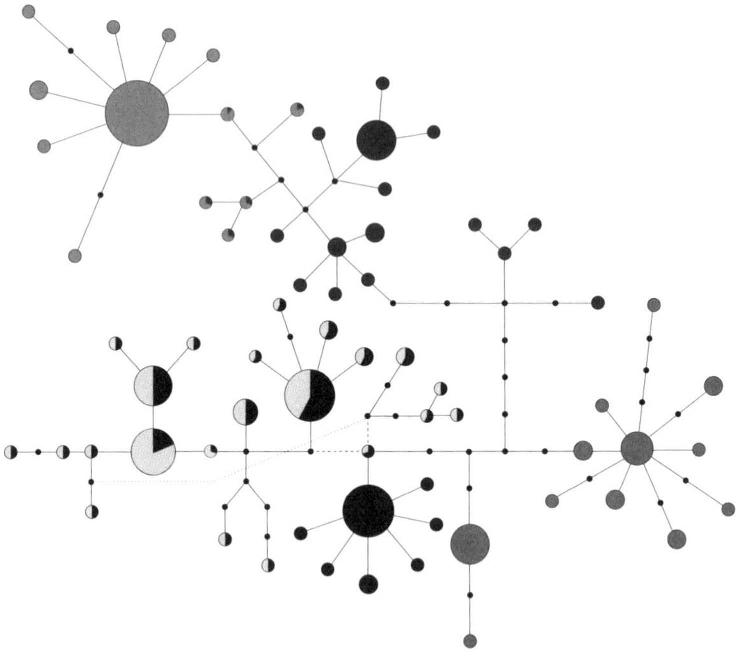

Figure 6: *The MAP estimate of the underlying cladogram. Each node is represented by a pie chart indicating the proportion of times it is assigned to each of the five clusters in the MAP model. The dotted line indicates the next most probable edge that could be added and the two dashed lines indicate the two edges either of which would be deleted in order to remove the corresponding loop in the original graph.*

To get a feel for the spatial structure of the haplotype dispersal, Fig. 7 provides a plot of the contours of the bivariate normal distributions (evaluated at the posterior means of the mean and covariance parameters) associated with each component of the mixture. Here, we use the same colour scheme as for the cladogram in Fig. 7, for comparison. It is clear the the identification of the three clusters at the North of the island could not have been possible without the additional information provided by the underlying cladogram. The individual locations are marked on the figure as black circles, with the largest circle corresponding to the location most likely *a posteriori* to contain the root node under a flat prior. This is located to the East of the island in the so-called northern shield at Mna Tagoja, though the neighbouring locations around Fuente Olen also have substantial posterior probability of containing the root node.

We can see a clear distinction between clusters in terms of the locations, spread and associated covariance structures. With some clusters, such as the red cluster, being roughly circular and with only local spread, whereas others, such as the grey cluster, has much greater spread in one direction (North-South) than the other. These latter shapes are a result of clusters containing haplotypes spread over a wide area and typically in a specific direction. This type of pattern is consistent with range expansion in the population, whereas the more symmetric and focused clusters would be more indicative of a geographically stable population.

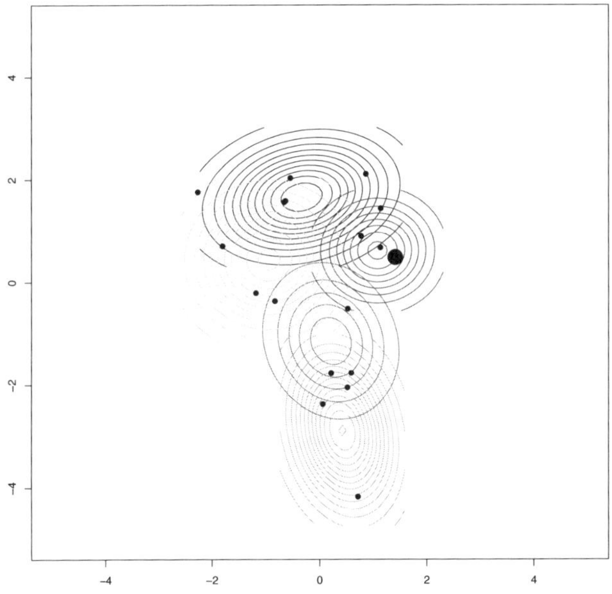

Figure 7: *Bivariate normal contour plots for the five components under the MAP model and evaluated at the posterior means. The circles indicate sampling locations, with the larger circle indicating the location most likely to include the root node in the cladogram under a flat prior.*

6. DISCUSSION

The results provided in the previous section consistently point to the red cluster as being the population source as it both most likely to contain the root node and exhibits a suitably stable spatial structure. The MAP cladogram suggests a comparatively recent range expansion South East to the Cumbre Nueva terrain running vertically down the southern half of the island from the centre. We infer the recency of the expansion from the lack of missing nodes (indicating both missed samples and haplotype extinctions) between the two clusters in the cladogram. The largest green node is clearly a source for local expansion due to its high degree.

Less recently, the red cluster appears to have been the source for two major expansions one much further to the South resulting in the grey cluster and one to the North West resulting in first the blue and then yellow clusters. This, again, is inferred from the structure of the cladogram in Fig. 6 and the spatial structure apparent in Fig. 7.

These conclusions are consistent with complementary evidence and previous analyses of similar data. For example, La Palma is at the western end of the Canary Islands and so any colonisation of the island is most likely to have occurred along the Eastern coast. Previous studies of these data by Emerson *et al.* (2006) drew very similar conclusions in that the island was colonised in the mid-east and saw three range expansions, two to the South and one to the West. Indeed the MAP cladogram in Fig. 6 is fairly similar to their cladogram C, though quite distinct structurally from the other two potential cladograms that they discuss. Emerson *et al.* (2006) hypothesised that the observed range expansions followed similar expansions in the host species *Pinus canariensis* and it would be interesting to conduct a complementary phylogeographic study to determine the extent of any such association. The only major discrepancy between our results and those of Emerson *et al* (2006) concern the colonisation of the Northern Shield. Our analysis suggests that the Northern part of the island was colonised first by a movement to the North West and then South West down the coast, essentially moving in an anti-clockwise direction. Emerson *et al.* (2006) on the other hand suggest that the colonisation occurred first to the West and then to the North East, moving in a clockwise direction.

Our analysis has several advantages over the classical nested clade analysis. Perhaps the most important is that when it comes to making inference about the underlying dynamics of the population over time, our approach is able to provide quantitative measures of statistics of interest. For example, we are able to associate probabilities with any one location as being the source population for later colonisation of the island. The second advantage is that by combining all three stages of the analysis we are able to make sure that, for example, information from the geographical location of the individuals is allowed to feedback and inform the associated genealogy represented by the cladogram. It is clear that certain cladograms may lead only to nonsensical clusterings, but these can be avoided in the integrated analysis. This did not cause any problems for these particular data but, for others, the ability to incorporate all information into any one component of the analysis is bound to be a distinct advantage.

It is worth pointing out here that there are several ways in which this work can and should be further extended. In constructing our original graph we allowed only the minimum number of missing nodes necessary to complete the graph. It would be interesting to relax this assumption so that an even greater range of cladograms is permissible. In addition, it would be useful to allow for recombinations with the population. These won't have occurred in these particular data, because we are using mitochondrial DNA but other datasets may well exhibit both recombinations and homoplasy. Adding the possibility of recombinations significantly adds to the complexity of the graph construction, and various assumptions (based primarily on parsimony) will be required to make the problem tractable. Finally, it would be very useful to determine a quantitative measure along the lines of Templeton's inference key so that we could actually associate (posterior) probabilities with different clusters being associated with different forms of dispersal over evolutionary timescales.

Each of these tasks are the focus of current work by the authors.

REFERENCES

Altekar, G., Dwarkadas, S., Huelsenbeck, J. P., Ronquist, F. (2004). Parallel Metropolis coupled Markov chain Monte Carlo for Bayesian phylogenetic inference. *Bioinformatics* **20**, 407–415.

Cassens, I., Mardulyn, P. and Milinkovitch, M. C. (2005). Evaluating intraspecific 'network' construction methods using simulated sequence data: Do existing algorithms outperform the global maximum parsimony approach? *Systematic Biology* **54**, 363–372.

Clement, M., Posada, D., and Crandall, K. A. (2000). TCS: A computer program to estimate gene genealogies. *Molecular Ecology* **9**, 1657–1659.

Clement, M., Snell, Q., Walker, P., Posada, D., and Crandall, K. (2002). TCS: estimating gene genealogies. *International Workshop on High Performance Computational Biology.*

Crandall, K. A. and Templeton, A. R. (1993). Empirical tests of some predictions from coalescent theory with applications to intraspecific phylogeny reconstruction. *Genetics* **134**, 959–969.

Emerson, B. C., Forgie, S., Goodacre, S. and Oromi, P. (2006). Testing phylogeographic predictions on an active volcano island: *Brachyderes rugatus* (Coleoptera: Curculionidae) on La Palma (Canary Islands). *Molecular Ecology* **15**, 449–458.

Emerson, B. C. and Hewitt, G. M. (2005). Phylogeography. *Current Biology* **15**, 369–371.

Emerson, B. C., Oromi, P. and Hewitt, G. M. (2000). Colonization and diversification of the species *Brachyderes Rugatus* (Coleoptera) on the Canary Islands: Evidence from mitochondrial DNA COII gene sequences. *Evolution* **54**, 911–923.

Felsenstein, J. (1992). Phylogenies from restriction sites: A maximum-likelihood approach. *Evolution* **46**, 159–173.

Gomez-Zurita, J., Petitpierre, E. and Juan, C. (2000). Nested cladistic analysis, phylogeography and speciation in the *Timarcha goettingensis* complex (Coleoptera, Chrysomelidae). *Molecular Ecology* **9**, 557–570.

Huelsenbeck, J. P., Larget, B., Miller, R. E. and Ronquist, F. (2002). Potential applications and pitfalls of Bayesian inference of phylogeny. *Systematic Biology* **51**, 673–688.

Ibrahim, K. M., Nichols, R. A. and Hewitt, G. M. (1996). Spatial patterns of genetic variation generated by different forms of dispersal during range expansion. *Heredity* **77**, 282–291.

Larget, B., and Simon, D. (1999). Markov Chain Monte Carlo Algorithms for the Bayesian analysis of phylogenetic trees. *Molecular Biology and Evolution* **16**, 750–759.

Lloyd, D. G., and Calder, V. L. (1991). Multi-residue gaps, a class of molecular characters with exceptional reliability for phylogenetic analyses *J. Evolutionary Biology.* **4**, 9–21.

Mau, B., Newton, M., and Larget, B. (1999) Bayesian phylogenetic inference via Markov chain Monte Carlo methods. *Biometrics* **55**, 1-12.

Newton, M. A., Mau, B., and Larget, B. (1999). Markov chain Monte Carlo for the Bayesian analysis of evolutionary trees from aligned molecular sequences. *Statistics in Molecular Biology and Genetics.* (F. Seillier-Moseiwitch, ed.) IMS Lecture Notes-Monograph Series, **33**, 143–162.

Posada, D. and Crandall, K. A. (2001). Intraspecific gene genealogies: trees grafting into networks. *Trends in Ecology and Evolution*, **16**, 37-45.

Richardson, S. and Green, P. (1997). On Bayesian analysis of mixtures with an unknown number of components. *J. Roy. Statist. Soc. B* **59**, 731–792.

Stamatakis, A., Ludwig, T. and Meier, H. (2005). RAxML-III: a fast program for maximum likelihood-based inference of large phylogenetic trees. *Bioinformatics* **21**, 456–463.

Stephens, M. (2000). Dealing with label-switching in mixture models. *J. Roy. Statist. Soc. B* **62**, 795–809.

Tavaré, S. (1986). *Some Probabilistic and Statistical Problems in the Analysis of DNA sequences.* Lectures on Mathematics in the Life Sciences 17. American Mathematical Society.

Templeton, A. R. (1995). A cladistic analysis of phenotypic associations with haplotypes inferred from restriction endonuclease mapping or DNA sequencing. V. Analysis of case/control sampling designs: Alzheimer's disease and apoprotein E locus. *Genetics* **140**, 403–409.

Templeton, A. R. (1998) Nested clade analysis of phylogeographic data: testing hypotheses about gene flow and population history. *Molecular Ecology* **7**, 381-397.

Templeton, A. R., Boerwinkle, E., and Sing, C. F. (1987). A cladistic analysis of phenotypic associations with haplotypes inferred from restriction endonuclease mapping. I. Basic theory and an analysis of alcohol dehydrogenase activity in *Drosophila. Genetics* **117**, 343–351.

Templeton, A. R., Crandall, K. A., and Sing, C. F. (1992). A cladistic analysis of phenotypic associations with haplotypes inferred from restriction endonuclease mapping. III. Cladogram estimation.*Genetics* **132**, 619-633.

Templeton, A. R., Routman, E. and Phillips, C (1995). Separating population structure from population history: a cladistic analysis of the geographical distribution of mitochondrial DNA haplotypes in the Tiger Salamander, *Ambystoma tigrinum. Genetics* **140**, 767–782.

Templeton, A. R., and Sing, C. F. (1993). A cladistic analysis of phenotypic associations with haplotypes inferred from restriction endonuclease mapping. IV. Nested analyses with cladogram uncertainty and recombination. *Genetics* **134**, 659–669.

Templeton, A. R., Sing, C. F., Kessling, A. and Humphries, S. (1988). A cladistic analysis of phenotypic associations with haplotypes inferred from restriction endonuclease mapping. II. The analysis of natural populations. *Genetics* **120**, 1145–1154.

DISCUSSION

BANI K. MALLICK (*Texas A&M University, USA*)

I congratulate the authors on their very fine paper. It contains novel methodology emerged from a very interesting application. Process of grouping a collection of objects into subsets or clusters such that elements within each cluster are more closely related to one another than objects assigned to different clusters. This clustering has been done using some type of similarity measure (usually measuring the Euclidean distance between the data). In this particular problem there are additional information like DNA sequence data which makes it a constrained clustering problem. Cladograms are trees or graphs that are helpful in relating individuals from a population(s) given their sequence data. Partitioning such trees may allow us to associate specific genetic mutations with differences within the studied population. The authors have proposed novel Bayesian framework for creating disjoint partitions of cladograms.

There are two components of the data: S: Sequences, T: Traits. A Bayesian model will specify a joint distribution for $p(S, T)$. It has been done conveniently by individually modeling $p(T|S)$ and $P(S)$. Mixture model has been used for $p(T|S)$ as well as extra uncertainties have been added by modeling S. I have some comments about the general methodology.

Hierarchical clustering? Hierarchical clustering develops a tree (dendrogram) whose leaves are the data points and whose internal nodes represent nested clusters of various sizes. The nested cladogram method by Templeton *et al.* (1987) looks like very similar to hierarchical clustering method. Due to a nice structure in the data through the cladogram is hierarchal clustering is more scientifically interpretable than the random clustering method proposed here?

Bayesian hierarchical clustering is of recent interest (Heard *et al.*, 2006; Heller and Ghahramani, 2005). In this problemt to calculate the normalizing constant a determinant of the matrix

$$\phi^{-1} + S_j + \frac{\tau . n_j}{(\tau + n_j)} \bar{T}_j \bar{T}_j{}'$$

has to be calculated where S_j is the sample variance-covariance matrix of the jth cluster. This could be time consuming specially when the dimension of the trait is very high. In a hierarchal clustering framework it is simple to find an iterative relationship between S (or $\bar{T}\bar{T}'$) and their children so you do not need to recalculate these quantities at each iteration. This way it could lead to considerable savings in computation time specially when the dimension is high.

Specification of prior distributions. As it is an applied problem, I was expecting more subjective priors in this analysis. Also prior sensitivity needs to be verified. How would the clustering structures change if you change the prior for k from uniform to other distributions which will penalize too many clusters? If there are no strong subjective priors available then you can exploit model-based clustering methods using maximum likelihood estimation via EM algorithm (McLachlan and Peel, 2000). That way you can avoid the problem of specifying hyper priors and related prior sensitivities.

*Mixture model vs DPP based clustering.*An alternative way to produce clusters consist of using Dirichlet process prior (DPP) based models. An advantage of this method is the number of clusters has been determined automatically. Is it possible to use DPP based method in this setup?

Clustering of graphs. Graph cutting, partitioning or clustering as it is variously known is the decomposition of a graph into roughly equal sized pieces while minimizing the number of edges between those pieces (Chung, 1997). Looks like the method developed in this paper is closely related to clustering of graphs. The main difference is in this problem the cladogram (the dependence structure) has been provided or when there is some uncertainty in the tree structure still biologically interpretable mutation models are helpful to rebuild the structure. In graph clustering problem the dependence structure is completely unknown, so is it possible to model the dependence structure as well as the clustering simultaneously? This will be very useful to develop gene regulatory networks. To understand the nature of the cellular function, it is necessary to study the behavior of the genes in a holistic rather than individual way. Gene networks can be developed based on gene expression data where the dependence structure among the genes are completely unknown. With very large number of genes, it is believed that the whole gene network can be partitioned into small sub-networks. This is same as clustering the network model in small pieces. Graph cutting is a popular tool (Chung, 1997) where you adaptively create the graph as well as cluster it in sub-graphs. Recently it has been shown that the graph cutting algorithm is equivalent to kernel k-means clustering algorithm (Dhillon *et al.*, 2003). Existing Graph cutting techniques fail to provide uncertainty measures. In a recent paper Ray, Mallick, Dougherty (2006) developed a Bayesian graphical model with DPP priors to obtain the posterior distribution of clusters. The method developed by Brooks et al. is a novel way to produce clusters using the the set of edges and it will be interesting to see how this can be extended in the complex situations to identify the clusters within an adaptive graph model.

PAUL FEARNHEAD (*Lancaster University, UK*)

I would like to congratulate the authors for demonstrating how Bayesian methods can be used to coherently address challenging problems in statistical genetics; and the advantages they have over more classical approaches which split the analysis into a number of stages and that essentially assume that the inferences within each preceeding stage are exact.

The work implicitly assumes no recombination, so that there is a unique genealogical tree that describes the historical relationship of the sample; and this is appropriate for the phylogeographic analysis considered as mitochondrial DNA does not recombine. However, I wonder if you have considered how to extend your method to situations where there is recombination? As most DNA data comes from recombining regions of the genome, extending the method to this case is of particular importance, especially for possible applications relating to mapping disease genes.

In these cases there is no single genealogical tree, but (potentially) different genealogical trees for each site in the DNA sequence; this information can be described by something called the Ancestral Recombination Graph (ARG). Applying phylogenetic methods to this case is inappropriate - while a tree could be inferred, it no longer has the natural interpretation that there is in the no recombination case, and ignoring recombination has shown to lead to biases in various situations (Sheirup and Hein 2000). In theory the Bayesian approach can be adapted to this case, by replacing mixing over possible trees with mixing over ARGs, but in practice this is likely to be computationally infeasible due to the dimension of the space of ARGs. On a related point, there is a substantial literature within the field of population genetics that is relevant to this work. In particular there is a natural population genetic model for the underlying tree (or ARG) for the data based around a stochastic process called the coalescent (Kingman 1982). There are numerous methods for inference under coalescent models (see for example Stephens and Donnelly 2000, Drummond *et al.* 2002 and references therein), some of which can include geographic information (Griffiths and Bahlo 1998, Beerli and Felsenstein 1999). While these are primarily only for non-recombining DNA, there are a few methods for the recombining case (e.g. Fearnhead and Donnelly 2001). There are a number of advantages that these coalescent-based methods have over the more phylogenetic approach described in the paper; for example they would enable trees to be inferred without assuming parsimony, better estimates of the root of the tree, and inferences about times on the tree (for example to date possible migration events). Though these are likely to come at a higher computational cost.

Also of relevance are various accurate approximate methods that have been devised in population genetics. The most generic of which is the Product of Approximate Conditional Likelihood method of Li and Stephens (2003). See also Wilson and McVean (2006) for an application. This framework may give computationally feasible way of extending the work in this paper to the recombining DNA case. There is also related work by Zöllner and Pritchard (2005) on methods for mapping disease genes; and by Pritchard *et al.* (2000) and Falush *et al.* (2003) on detecting population structure from genetic data.

REPLY TO THE DISCUSSION

We are very grateful to Bani Mallick for his thoughtful discussion of our paper. He makes a number of interesting observations to which we respond below.

Hierarchical clustering. Calculation of the normalisation constant with our bivariate normal clusters is a trivial exercise in the context of phylogeographic analyses. However, for phenotypic analyses higher dimensional distributions may well be appropriate in which case techniques such as those described may well be very useful. That being said, it is not clear that the normal assumption will necessarily be valid. In many cases, phenotypic traits may well be binary, noting the presence/absence of disease for example, and so entirely different distributions would need to be used.

Priors. We use vague priors in this analysis to demonstrate the ability of the method to extract information from the data that is consistent with what is already known. This paper represents a very early foray into this area and we are keen to demonstrate the utility of our method at this early stage. To a large degree, the results of our analysis are consistent with expert opinion. As a demonstration of the method's value the message would be less clear if we adopted more informative priors only to gain a posterior which then agreed with them. A more statistically rigorous defense of our choice of vague priors is based on the observation that the knowledge we already have about the model parameters (and, indeed, model probabilities) is predominantly based on previous analyses of these same data and so would be inappropriate for use as a prior. Of course, there is some independent information from studies on related species or on nearby islands, but it's very difficult to disentangle these from knowledge based upon earlier analyses of these data. Current research undertaking analyses of new data adopts a more rigorous approach to this issue using reference priors on occasion and carefully sourcing information for more informative priors where both possible and appropriate.

We were intrigued by the suggestion that in the absence of informative prior information, we should resort to a classical analysis as a means to 'avoid the problem of specifying hyper-priors and related prior sensitivities'. Ignoring any philosophical objections, it is worth noting that the EM algorithm would necessarily have to ignore the uncertainty in the underlying tree, converging to that with the highest likelihood. Therefore, such an algorithm would offer little advance over traditional methods. The Bayesian method, on the other hand, can incorporate this uncertainty and allow it to propogate through to the clustering component of the analysis.

Mixtures vs DPP. It may well be possible to produce an alternative scheme based upon DPP models. This is not something that we have explored but may well be worth investigating in due course. Our proposed method determines the number of clusters automatically too, so it is unclear to us the extent to which DPP-based models would provide an improvement as opposed to a simple alternative at this stage. It may, nonetheless, be worth exploring in more detail.

Clustering graphs. We had not thought to link our work with the literature on partitioning graphs. The similarity measures in the latter case are functions of the graph structure itself, whereas in our case, they are functions of trait values associated with the relevant nodes. It is certainly possible that some cross-fertilisation of ideas might benefit one or other of the two areas and we will explore this as we continue our research.

Paul Fearnhead also kindly contributes to the discussion by providing a tutorial overview of a few related ideas together with additional references that augment our own. In particular, he discusses the problem in which recombinations may occur. This is not appropriate for the data we discuss here but may be appropriate for other datasets and readers may well find these additional references useful.

ADDITIONAL REFERENCES IN THE DISCUSSION

Bahlo, M. and Griffiths, R. C. (1998). Inference from gene trees in a subdivided population. *Theor. Pop. Biol.* **57**, 79–95.

Beerli, P. and Felsenstein, J. (1999). Maximum-Likelihood estimation of migration rates and effective population numbers in two populations using a coalescent approach. *Genetics* **152** 763–773.

Chung, F. R. K. (1997). *Spectral Graph Theory*, CBMS Lecture Notes, AMS Publication.

Dhillon, I. S., Guan, Y., and Kulis, B. (2004). Kernel k-means, spectral clustering and normalized cuts. *Proc. 10th ACM SIGKDD Int'l Conf on Knowledge Discovery and Data Mining (KDD)*, 551–556.

Drummond, A. J., Nicholls, G. K., Rodrigo, A. G. and Solomon W. (2002). Estimating mutation parameters, population history and genealogy simultaneously from temporally spaced sequence data. *Genetics* **161**, 1307–1320.

Falush, D., Stephens, M. and Pritchard, J. K. (2003). Inference of population structure using multilocus genotype data: Linked loci and correlated allele frequencies. *Genetics* **164**, 1567–1587.

Fearnhead, P. and Donnelly, P. (2001). Estimating recombination rates from population genetic data. *Genetics* **159**, 1299–1318.

Green, P. and Richardson, S. (2001). Modelling heterogeneity with and without Dirichlet process. *Scandinavian J. Statist.* **28**, 355–375.

Heard, N., Holmes, C. and Stephens, D. (2006). A quantitative study of gene regulation involved in the immune response of anopheline mosquitoes: an application of Bayesian hierarchical clustering of curves. *J. Amer. Statist. Assoc.* **473**, 18–29.

Heller, K. and Ghahramani, Z. (2005). Bayesian hierarchical clustering. *Proceedings of the 22nd International conference on machine learning.*

Kingman, J. F. C. (1982). The coalescent. *Stoch. Proc. and App.* **13**, 235–248.

Li, N. and Stephens, M. (2003). Modelling LD, and identifying recombination hotspots from SNP data. *Genetics* **165**, 2213–2233.

McLachlan, G. and Peel, D. (2000). *Finite Mixture Models*, New York: Wiley

Pritchard, J.K̃., Stephens, M. and Donnelly, P. (2000). Inference of population structure using multilocus genotype data. *Genetics* **155**, 945–959.

Ray, S. and Mallick, B. (2006). Functional clustering by Bayesian wavelet methods. , *J. Roy. Statist. Soc. B* **68**, 305–320.

Schierup, M. H. and Hein, J. (2000). Consequences of recombination on traditional phylogenetic analysis. *Genetics* **156**, 879–891.

Stephens, M. and Donnelly, P. (2000). Inference in molecular population genetics (with discussion). *J. Roy. Statist. Soc. B* **62**, 605–655.

Ray, S., Mallick, B. and Dougherty, E. (2006). Bayesian graph cutting. *Tech. Rep.*, Texas A&M University.

Wilson, D. J. and McVean, G. A. T. (2006). Estimating diversifying selection and functional constraint in the presence of recombination. *Genetics* **172**, 1411–1425.

Zöllner, S. and Pritchard, J. K. (2005). Coalescent-based association mapping and fine mapping of complex trait loci. *Genetics* **169**, 1071–1092.

BAYESIAN STATISTICS 8, pp. 51–90.
J. M. Bernardo, M. J. Bayarri, J. O. Berger, A. P. Dawid,
D. Heckerman, A. F. M. Smith and M. West (Eds.)

Some Aspects of Bayesian Model Selection for Prediction

ARIJIT CHAKRABARTI
Indian Statistical Institute, India
arc@isical.ac.in

JAYANTA K. GHOSH
Purdue University, Indiana, USA and *Indian Statistical Institute, India*
ghosh@stat.purdue.edu and jayanta@isical.ac.in

SUMMARY

We consider cross-validatory Bayes factors for prediction in both the $\mathcal{M}$-open and $\mathcal{M}$-closed cases and explore how much of a sample is to be used for estimating parameters of competing models and how much for model selection. The dimension p of the model may be fixed or go to infinity. We also re-examine critically the relation between classical cross-validation and AIC. The latter is shown to be Parametric Empirical Bayes in prediction problems and its performance is studied in all subsets model selection.

Keywords and Phrases: CROSS-VALIDATION, CROSS-VALIDATORY BAYES FACTOR, AIC, PARAMETRIC EMPIRICAL BAYES

1. INTRODUCTION

This paper is an attempt to explore different aspects of a deep question raised by Professor Pericchi at a Bayesian conference last year. How much of the sample should be used to estimate the parameters of competing models and how much for model check and model selection?

A major goal for us is to study this problem in $\mathcal{M}$-open examples as described in Bernardo and Smith (1994, pp. 404) and Key *et al.* (1999). The true model is not in the model space and one considers the predictive distribution with a logarithmic score. They introduce criteria based on cross-validation, e.g.,

$$\frac{1}{n}\sum_{i=1}^{n} \log \frac{p_2(x_j|x_{(n-1)}(j))}{p_1(x_j|x_{(n-1)}(j))}, \tag{1}$$

Arijit Chakrabarti is an Assistant Professor at the Indian Statistical Institute, Kolkata. Jayanta K. Ghosh is a Professor of Statistics at Purdue University, West Lafayette, Indiana, USA and Professor Emeritus at the Indian Statistical Institute, Kolkata.

where p_1, p_2 correspond to the two models of interest, none of which is true and $x_{n-1}(j)$ is the set of $(n-1)$ observations with the j-th deleted. In (1), $(\frac{n-1}{n})$ proportion of observations is used for estimation and $(\frac{1}{n})$ for validation of a chosen model. Is this always a good assignment of observations for estimation and validation? In addition to (1), we consider the more general Bayesian criterion

$$\log \frac{p(x_s|x_{s^c}, M_2)}{p(x_s|x_{s^c}, M_1)}, \tag{2}$$

and its average over $[\frac{n}{k}]$ disjoint subsets $\{x_s = (x_i, i \in s)\}$ of a given size k from the whole sample, where

$$p(x_s|x_{s^c}, M_2) = \int f(x_s|\theta)\pi(\theta|x_{s^c})\, d\theta, \tag{3}$$

$\{x_{s^c} = (x_i, i \notin s)\}$, $[.]$ denotes the greatest integer less than or equal to a real number. The criterion (2) was considered by Gelfand and Dey (1994) who drew upon earlier work of Geisser (1975). We do not consider the closely related IBF's of Berger and Pericchi (1995) because they are already quite well studied in the basic paper of Berger and Pericchi just mentioned and Ghosh and Samanta (2002). Moreover, in Mukhopadhyay *et al.* (2005), it was shown that relating (1) and IBF's isn't fruitful.

Our second goal arises from the first. We provide a critical review of what is known in similar problems in the classical case, and try to extrapolate from there to Bayesian analysis. We believe that the basic issues relating to cross-validation are very similar but even in the classical case, not much is really known.

A review of what is known about cross-validation in Hastie, Tibshirani and Friedman (2001) and Gu (2002), including a thumb rule for sample allocation in the first book is presented in Section 2. We do not discuss an interesting inequality presented in Vapnik (1995).

Cross-validation inevitably leads to AIC and theorems of Stone (1977) and Shao (1993). These are discussed in Sections 2 and 3. Our search indicates that there has not been much theoretical study of cross-validation. Even the famous result of Stone (1977) relating cross-validation to AIC seems overinterpreted in the sense that what Stone claimed and proved is much weaker than what is popularly believed. In Section 3 we provide a counter-example to show that AIC and cross-validation are not equivalent. However, we also show that cross-validation is actually equivalent to or better than AIC under conditions similar to those of Stone.

Since our main interest is in Bayesian model selection rules in the context of prediction, should we include or exclude AIC? Following Mukhopadhyay (2000) we show AIC can be interpreted as a PEB model selection rule both in the nested and in the all subsets case (under orthogonality). In the non-nested case, an additional Binomial penalty, as in George and Foster (2000) makes AIC conservative in a precise sense as in Bogdan, Doerge and Ghosh (2004). This modification of AIC, called mAIC henceforth, is also shown to have the same PEB interpretation in the all subsets model selection case, under orthogonality.

Criteria like (2) (or their averages) are considered in detail in Section 4 and their asymptotic properties are explored. In this relatively simple scenario, we can exhibit examples that require very different ratios of k and $(n-k)$ from that suggested in the thumb rule of Hastie, Tibshirani and Friedman (2001). In each case there is a natural explanation. We consider not only the cross-validatory fixed dimensional Bayes rules of Gelfand and Dey (1994) but also high-dimensional examples of Chakrabarti and Ghosh (2005). The high-dimensional problems are considered in the simpler setting of $\mathcal{M}$-closed model selection, and instead of the averages, criterion (2) itself is only studied for any given x_s of size k. We mention here that our asymptotics in this paper is informal in the sense that common regularity conditions will not be spelt out in detail. Some often used tools like Laplace approximation and well known results on m.l.e. will be used, with the understanding that appropriate conditions need to be invoked whenever necessary so that these become valid, though the details of the conditions will often be skipped.

In Section 5, we explore whether a reasonable choice of k can be found from simulation for the examples in Section 4. What is reasonable is decided in the light of the theoretical results in Section 4, i.e, the values of k which would provide good discrimination between models as approximation to the true distribution.

In Section 6, we explore the performance of usual AIC and mAIC with a Binomial penalty as in George and Foster (2000).

Our main findings are summarized in Section 7.

2. CROSS-VALIDATION IN CLASSICAL STATISTICS

This section is based mainly on Chapter 7 of Hastie, Tibshirani and Friedman (2001) (to be referred to as HTF for the rest of this section) and partly on Gu (2002). Model selection using least squares or (Bayes) shrinkage estimation is included. They point out that cross-validation avoids the double use of the same data for both estimation and validation. Double use would lead to underestimation of a risk, whereas cross-validation avoids double use and is believed to provide an estimate of risk that is usually conservative.

HTF suggest dividing sample size n into m disjoint sets of k observations and m different cross-validations with one set of observations being used for cross-validation of a model and the remaining $n-k = (m-1)k$ observations being used for estimating the model parameters. This procedure is referred to as 'leave k out cross-validation'. The special case 'leave one-out' is still popular but is supposed to underestimate the variability of estimates of model parameters owing to large overlap of the different $(n-k)$ observations used for estimation. The criterion (2) or its average over $[\frac{n}{k}]$ disjoint sets is based on similar ideas.

A popular choice is $k = 10$. The reason for its popularity is both smaller computational burden than 'leave one-out' as well as better handling of variability. This particular choice, *i.e.*, $k = 10$ is mentioned by HTF (pp. 214). In their thumb rule, HTF(pp. 196) suggest a break up of 50% for estimation, 25% for validation and a further 25% for testing the final selected model. In our problem of assignment to validation and estimation, this suggests a ratio $k/(n-k) = \frac{1}{3}$. It seems plausible that this thumb rule is proposed as a possibility rather than a recommendation. Our discussion below, based on the assumption that this is a recommendation, aims to show why such a detailed specification will rarely be possible, nor would one expect a single rule to do well in all cases.

There are some exact and approximate calculations to show how the total frequentist risk is decomposed in various kinds of bias as well as a variance. Also there is an approximation to the cross-validation estimate that avoids the repeated calculations.

We haven't seen any results on consistency of cross-validation estimates in HTF. Results have however been proved on consistency of cross-validation and generalized cross-validation in the case of penalized least squares estimates for nonparametric regression. See for example Theorems 3.1 and 3.3 in Gu (2002) who attributes equivalent results to Li (1986) and partly to Craven and Wahba (1979). Consistency is also considered by Jun Shao (1993). For normal linear models, he shows under certain conditions that 'leave-one-out' cross-validation is inconsistent in the sense of failing to choose the true model with smallest dimension with probability tending to 1, whereas 'leave k out' cross validation with $k/n \to 1$ achieves consistency in this regard. These results are shown when the dimension of the model is fixed. Our results in Section 4 have strong similarities with his conclusions.

There is a good discussion of the various factors that are involved in a good choice of k and $(n-k)$ but our Bayesian ideas, to be presented in Section 4, seem to be simpler and intuitive.

Finally, we do not find any discussion of why the thumb rule of $k/(n-k) \sim \frac{1}{3}$ is proposed. We offer some speculations. One possibility is that the rule is based on theoretical results on relative magnitude of different components of the risk function like $(\text{bias})^2$ and variance in the worst possible scenarios. The rather precise results for minimax procedures that are needed, not just minimax rates of convergence, do not seem to be known (except for nonparametric regression). The other possibility is that the rule is based on the authors' empirical experience. Also it seems likely that nonparametric or high-dimensional examples underlie this recommendation. On the other hand, our simple insights (vide results in Section 4) are based mostly on parametric models used as approximations in an $\mathcal{M}$-open case and the high-dimensional results are based on the $\mathcal{M}$-closed case. More fundamentally, we use a different notion of either choosing the best approximation to truth (in Kullback–Leibler divergence) or apply the principle of parsimony in case both models provide equally good approximation.

3. MODEL SELECTION BY CLASSICAL CROSS-VALIDATION AND AIC

Consider the variable selection problem in the regression setup

$$\mathbf{Y} = \mathbf{X}\boldsymbol{\beta} + \epsilon, \tag{4}$$

where $\epsilon \sim N(\mathbf{0}, \sigma^2 \mathbf{I}_{\mathbf{n}\times\mathbf{n}})$, where σ^2 is known, $\mathbf{Y}$ is an $(n \times 1)$ vector, $\mathbf{X}$ is an $(n \times p)$ nonrandom matrix and $\boldsymbol{\beta}$ is a $(p \times 1)$ vector. We begin by showing that AIC can be viewed as a Parametric Empirical Bayes (PEB) rule for prediction. We prove this for only the orthogonal case (with $\mathbf{X}'\mathbf{X}$ the identity matrix of order p) for only two nested models but the same calculations can be repeated for any two models in a given class of nested models without orthogonality when the true model is at least as complex as each model in the model class. There is no conflict with the $\mathcal{M}$-open scenario.

Given two nested orthogonal models, we can reduce the two models, via canonical reduction, to the following form. Under the simpler model M_1,

$$\mathbf{Z} = \begin{pmatrix} \boldsymbol{\beta}^{(1)} \\ \mathbf{0} \end{pmatrix} + \epsilon \tag{5}$$

with the order of $\boldsymbol{\beta}^{(1)}$ as p_1. Under the more complex model M_2, we have

$$\mathbf{Z} = \begin{pmatrix} \boldsymbol{\beta}_1 \\ \boldsymbol{\beta}_2 \\ \mathbf{0} \end{pmatrix} + \epsilon \tag{6}$$

where $\boldsymbol{\beta}' = (\boldsymbol{\beta}_1, \boldsymbol{\beta}_2)'$ is of order $p_2 > p_1$, where $p_2 \le p$. The models will be used in two ways.

The statistician knows or believes M_2 is true and $\boldsymbol{\beta} \sim N(0, \tau^2 I_{p_2 \times p_2})$. He has the option of using M_1 or M_2 and the corresponding least squares estimates for prediction of a new replicate $\mathbf{Z}^{new}$. The restriction to least squares estimates is a constraint but the option of using M_1 even though M_2 is true allows more flexibility and hence less posterior risk than being forced to use M_2.

Without loss of generality, consider only two models in the action space to be used for prediction and under M_2 the $\boldsymbol{\beta}$ is an arbitrary element in $\mathbf{R}^p$ and under M_1, $\boldsymbol{\beta}$ is identically equal to zero. If M_1 is chosen, we take the action $\hat{\boldsymbol{\beta}} = \mathbf{0}$ with utility loss

$$L(\boldsymbol{\beta}, \mathbf{0}) = \sum_{i=1}^{p} \beta_i^2 - 2p \log(1-w) I_1$$

and if M_2 is chosen, the least squares estimate $\hat{\boldsymbol{\beta}}$ is chosen with utility loss,

$$L(\boldsymbol{\beta}, \hat{\boldsymbol{\beta}}) = \sum_{i=1}^{p} (\hat{\beta}_i - \beta_i)^2 - 2p \log(w) I_1$$

where $I_1 = 1$ if we have all subsets model selection scenario and $I_1 = 0$ if we have a nested model scenario. We use the normal prior as mentioned before. The first sum of squares in the utility losses above, can be shown, upto a constant, as the expected prediction loss in predicting a new replicate, given the current observation $\mathbf{Y}$. Note that $\hat{\boldsymbol{\beta}}$ is a sufficient statistic for $\boldsymbol{\beta}$ and $E(\beta_i|\hat{\boldsymbol{\beta}}) = \frac{\hat{\beta}_i \tau^2}{\sigma^2+\tau^2}$ and $Var(\beta_i|\hat{\boldsymbol{\beta}}) = \frac{\sigma^2\tau^2}{\sigma^2+\tau^2}$. The posterior risk for M_1 is

$$\sum_{i=1}^{p} E(\beta_i^2|\hat{\boldsymbol{\beta}}) - 2p\log(1-w) = \sum_{i=1}^{p} \hat{\beta}_i^2 \left(\frac{\tau^2}{\sigma^2+\tau^2}\right)^2 + p\frac{\sigma^2\tau^2}{\sigma^2+\tau^2} - 2p\log(1-w) \quad (7)$$

and that for M_2 is

$$\sum_{i=1}^{p} E((\hat{\beta}_i - \beta_i)^2|\hat{\boldsymbol{\beta}}) - 2p\log(w) = \sum_{i=1}^{p} \hat{\beta}_i^2 \left(\frac{\sigma^2}{\sigma^2+\tau^2}\right)^2 + p\frac{\sigma^2\tau^2}{\sigma^2+\tau^2} - 2p\log(w). \quad (8)$$

Taking the difference of the r.h.s. of (8) and (7), we get,

$$\sum_{i=1}^{p} \hat{\beta}_i^2 \left\{\frac{\sigma^4 - \tau^4}{(\sigma^2+\tau^2)^2}\right\} - 2p\log\left(\frac{w}{1-w}\right).$$

The PEB estimate of $\sigma^2 + \tau^2$ is $\frac{\sum_{i=1}^{p} \hat{\beta}_i{}^2}{p}$. Hence the difference becomes

$$\left\{-\sum_{i=1}^{p} \hat{\beta}_i^2 + 2p\sigma^2\right\} - 2p\log\left(\frac{w}{1-w}\right),$$

which is our new AIC criterion mAIC. Same steps as in (7) and (8) above, but without the terms $-2p\log(1-w)$ and $-2p\log(w)$ yield the same result for AIC also.

Consider now the case of two non-nested models. If the two models M_1 and M_2 are not nested, then in addition to, say, p_3 non-zero β's in M_2 but not in M_1, there will be, say, p_4 β's in M_1 but not in M_2. Assuming these β's are i.i.d. $N(0, \tau')$, τ' estimated from these p_4 β's, the difference in posterior risk for these β's (using squared error loss) + difference in twice the log-likelihoods $= -2p_4\sigma^2$. Adding these equations for the two sets of β's, total difference in posterior risk for M_2 and $M_1 = -$(difference in twice log likelihood $+ 2(p_4 - p_3)\sigma^2$). This shows that AIC can have a PEB interpretation in the non-nested case also. Using same steps as these, but applying a modification of the usual squared error loss by subtracting $-2(\text{modelsize})\log(w/1 - w)$, one can show that difference in posterior risk for M2 and M1 = - (difference in twice log likelihood $+ 2(p_4 - p_3)\sigma^2$ + difference in Binomial penalty). This shows that mAIC has a PEB interpretation in the all subsets model selection case.

The effect of the Binomial penalty is to help parsimonious model selection. To some extent, this effect can be seen in the following proposition.

Proposition 1. Suppose $M_1 \subset M_2$ and M_1 is true so that both M_1 and M_2 are true. Then given α we can choose w sufficiently small so that the probability of preferring M_2 to M_1 (under M_1) is less than α.

Proof. The proof follows from the argument in the appendix (pp. 899) of Bogdan *et al.* (2004) □

We will now discuss the connection of AIC and cross-validation in the context of variable selection in linear regression. It is usually believed, vide Stone (1977), Shao (1993, 1997) that leave-one-out cross-validation and AIC are equivalent procedures asymptotically, in that the difference between the criteria for a given model tends to zero in probability. But this heuristics crucially depends, among other things, on the assumption that the model under consideration is actually a correct model. We investigate this matter further on a more general framework and derive some interesting new facts. We start by stating and discussing two key assumptions which will be used later in the section and also in Section 4.

Suppose we have observations $X_1, \ldots, X_n$ which are assumed to be i.i.d. with density f. Because of the DeFinetti–Hewitt–Savage representation theorem for exchangeable random variables, the assumption of independence and identical distributions may not be unrealistic. We do not always assume f is in our parametric model spaces. This causes important differences in the asymptotics. In view of this, we list below, the two assumptions.

Assumption A. Give a model M_j specifying parametric densities $f_{j,\boldsymbol{\theta}_j}, \boldsymbol{\theta}_j \in \boldsymbol{\Theta}_j$, where $\boldsymbol{\Theta}_j$ is a subset of a p_j-dimensional Euclidean space, there exists an interior point $\boldsymbol{\theta}_{j0}$ such that

$$K(f, f_{j,\boldsymbol{\theta}_{j0}}) = \inf_{\boldsymbol{\theta}_j \in \boldsymbol{\Theta}_j} K(f, f_{j,\boldsymbol{\theta}_j}), \tag{9}$$

where $K(.,.)$ is the well-known Kullback–Leibler divergence between two densities. Moreover the family $\{f_{j,\boldsymbol{\theta}_j}, \boldsymbol{\theta}_j \in \boldsymbol{\Theta}_j\}$ satisfies standard regularity conditions and

$$\frac{\partial}{\partial \boldsymbol{\theta}_{jl}} E_f \left(\log f_{j,\boldsymbol{\theta}_j}\right) |_{\boldsymbol{\theta}_j = \boldsymbol{\theta}_{j0}} = E_f \left(\frac{\partial}{\partial \boldsymbol{\theta}_{jl}} \log f_{j,\boldsymbol{\theta}_j} |_{\boldsymbol{\theta}_j = \boldsymbol{\theta}_{j0}} \right) = 0 \tag{10}$$

and

$$-E_f\left(\frac{\partial^2}{\partial\boldsymbol{\theta}_{jl}\partial\boldsymbol{\theta}_{jl'}}\log f_{j,\boldsymbol{\theta}_j}|_{\boldsymbol{\theta}_j=\boldsymbol{\theta}_{j0}}\right) \tag{11}$$

is the (l,l') th entry of a positive definite information matrix $[((I_{ll'}))]$.

Remark 1. By definition of $\boldsymbol{\theta}_{j0}$ and calculus, $\frac{\partial}{\partial\boldsymbol{\theta}_{jl}}E_f(\log f_{j,\boldsymbol{\theta}_j})|_{\boldsymbol{\theta}_j=\boldsymbol{\theta}_{j0}} = 0$. By (10) we get also an interchange of differentiation and integration. A similar remark and partial justification applies to (11).

Assumption B. In addition to (9), (10) and (11), f belongs to the model space of densities under M_j and

$$E_f\left(\frac{\partial^2}{\partial\boldsymbol{\theta}_{jl}\partial\boldsymbol{\theta}_{jl'}}\log f_{j,\boldsymbol{\theta}_j}\right)|_{\boldsymbol{\theta}_j=\boldsymbol{\theta}_{j0}} =$$

$$-E_f\left[\left(\frac{\partial}{\partial\boldsymbol{\theta}_{jl}}\log f_{j,\boldsymbol{\theta}_j}\right)\left(\frac{\partial}{\partial\boldsymbol{\theta}_{jl'}}\log f_{j,\boldsymbol{\theta}_j}\right)|_{\boldsymbol{\theta}_j=\boldsymbol{\theta}_{j0}}\right]. \tag{12}$$

Remark 2. Assumption B ensures that some key heuristic 'expectations' and the crucial assumption of the model under consideration being in the model space, as in Stone (1977) are satisfied.

We consider p possible models $M_1 \subset M_2 \subset \cdots \subset M_p$, in the normal linear regression problem as described in the beginning of this section. Under model M_j, $\beta_{j+k} = 0$ for $k \geq 1$. We will indicate when Stone's basic argument can be used with or without employing Assumption B and how the equivalence of AIC and cross-validation breaks down in a simple example where assumption B is not satisfied.

The AIC criterion function is

$$AIC(j) = L(\hat{\boldsymbol{\beta}}_j) - j - \left(-\frac{1}{2}\sum_{i=1}^{n}(Y_i - X_i\hat{\boldsymbol{\beta}}_j)^2 + constant\right) - j, \tag{13}$$

where $L(\hat{\boldsymbol{\beta}}_j)$ is the maximized log-likelihood of the whole data under model M_j, and $\mathbf{X}_i$ is the ith row of the $\mathbf{X}$ matrix. The CV criterion is obtained by minimizing over $1 \leq j \leq n$ the quantity

$$CV(j) = \frac{1}{n}\sum_{i=1}^{n}(Y_i - \hat{Y}_{(-i,j)})^2$$

where $\hat{Y}_{(-i,j)}$ is the estimate of Y_i under model M_j, leaving out observation i. But minimizing $CV(j)$ is equivalent to maximizing over $1 \leq j \leq n$ the criterion

$$CV'(j) = -\frac{1}{2}\sum_{i=1}^{n}(Y_i - \hat{Y}_{(-i,j)})^2 + \text{ same } constant \text{ as in (13)}$$

which is the same as $-\frac{1}{2}\sum_{i=1}^{n}(Y_i - \mathbf{X}_i\hat{\boldsymbol{\beta}}_{(-i,j)})^2 +$ constant in (13), where $\hat{\boldsymbol{\beta}}_{-i,j}$ is the estimate of $\boldsymbol{\beta}$ under model M_j leaving out observation i. So $CV'(j)$ can be written as $\sum_{i=1}^{n} l(Y_i|\hat{\boldsymbol{\beta}}_{(-i,j)})$ and similarly

$$AIC(j) = \sum_{i=1}^{n} l(Y_i|\hat{\boldsymbol{\beta}}_j) - j,$$

$l(Y_i|\boldsymbol{\beta})$ being the log-likelihood of Y_i evaluated at parameter value $\boldsymbol{\beta}$.

Remark 3. In the usual regression model, the Y_i's are not i.i.d. But if we repeat the $\mathbf{Y}$ vector r times for a given design matrix X, *i.e.*, we have $\mathbf{Y}_1,\ldots,\mathbf{Y}_r$, then indeed the $\mathbf{Y}_i$'s become i.i.d. In that case, as the definition of $AIC(j)$ we will be $AIC(j) = \sum_{i=1}^{r} l(\mathbf{Y}_i|\hat{\boldsymbol{\beta}}_j) - j$ and $CV'(j)$ will be $CV'(j) = \sum_{i=1}^{r} l(\mathbf{Y}_i|\hat{\boldsymbol{\beta}}_{(-i,j)})$, where $\hat{\boldsymbol{\beta}}_{-i,j}$ is the estimate of $\boldsymbol{\beta}$ under model M_j leaving the vector $\mathbf{Y}_i$ out. Stone's result hold true in the regression problem in the orthogonal design matrix setup, as in Proposition 3, provided a whole vector $\mathbf{Y}_i$ is kept out at a time for cross-validation. But if the cross-validation is done leaving one co-ordinate of $\mathbf{Y}_1,\ldots,\mathbf{Y}_r$ out at a time, the result does not hold, as shown in Example 1.

Remark 4. Consider the nested linear model setup. Let M_{j_0} be true. Then M_{j_0+k} is also true for $k \geq 1$ and for each of these models $AIC(j) - CV'(j)$ will converge to zero asymptotically provided conditions (i)–(iv) of Stone(1977) can be rigorously justified. Those conditions are as follows. Stone represents $CV'(j)$ as

$$CV'(j) = L(\hat{\boldsymbol{\beta}}_j) + \sum_{i=1}^{n} l'(y_i|\hat{\boldsymbol{\beta}}_{(-i,j)})^T[L''_{ji}]^{-1}l'(y_i|\hat{\boldsymbol{\beta}}_j + a_i(\hat{\boldsymbol{\beta}}_{(-i,j)} - \hat{\boldsymbol{\beta}}_j))$$

where $0 \leq |a_i| \leq 1$, and $[L''_{ji}]$ is the matrix whose k-th row is

$$\left\{\frac{\partial^2}{\partial\boldsymbol{\beta}_1\partial\boldsymbol{\beta}_k} l(\mathbf{Y}|\hat{\boldsymbol{\beta}}_j + b_{ik}(\hat{\boldsymbol{\beta}}_{(-i,j)} - \hat{\boldsymbol{\beta}}_j)),\ldots,\frac{\partial^2}{\partial\boldsymbol{\beta}_{p_j}\partial\boldsymbol{\beta}_k} l(\mathbf{Y}|\hat{\boldsymbol{\beta}}_j + b_{ik}(\hat{\boldsymbol{\beta}}_{(-i,j)} - \hat{\boldsymbol{\beta}}_j))\right\},$$

where $\mathbf{Y} = \{Y_1,\ldots,Y_n\}$, and $0 \leq |b_{ik}| \leq 1$, for $1 \leq k \leq p_j$, where p_j is the number of non-zero parameters under model j. Letting $\boldsymbol{\beta}_{j0}$ be the point in $\boldsymbol{\Theta}_j$ (the parameter space under model M_j) such that $E^P\{l(Y|\boldsymbol{\beta})\}$ is maximized, where P is the true distribution of each of $Y_1,\ldots,Y_n$ (Y_i's may be vectors), Stone argues that $CV'(j) - AIC(j)$ converges to zero asymptotically provided model M_j is true and (i) $\hat{\boldsymbol{\beta}}_j \xrightarrow{P} \boldsymbol{\beta}_{j0}$, (ii) $\hat{\boldsymbol{\beta}}_{(-i,j)} \xrightarrow{P} \boldsymbol{\beta}_{j0}$, (iii) $n^{-1}[L''_{ji}] \xrightarrow{P} E\{l''(Y|\boldsymbol{\beta}_{j0})\} = L_2$, and (iv) $n^{-1}\sum_{i=1}^{n} l'(Y_i|\hat{\boldsymbol{\beta}}_j + a_i(\hat{\boldsymbol{\beta}}_{(-i,j)} - \hat{\boldsymbol{\beta}}_j))l'(Y_i|\hat{\boldsymbol{\beta}}_{(-i,j)}) \xrightarrow{P} L_1 = E\{l'(Y|\boldsymbol{\beta}_{j0})l'(Y|\boldsymbol{\beta}_{j0})^T\}$. Using properties (i)-(iv) Stone argues that $CV'(j) \approx L(\hat{\boldsymbol{\beta}}_j) + trace(L_2^{-1}L_1)$ and that $trace(L_2^{-1}L_1) = -j$ provided M_j is a true model. Also note that under our Assumption A, existence of $\boldsymbol{\beta}_{j0}$ and (i) are assured, while (ii)–(iv) are expected to happen.

The following example shows that this phenomenon of equivalence of AIC and CV will not hold in general.

Example 1. Consider the model

$$\mathbf{Y} = \begin{pmatrix} \beta_1 & + & \beta_2 \\ \beta_1 & - & \beta_2 \end{pmatrix} + \epsilon \tag{14}$$

where $\epsilon \sim N(\mathbf{0}, I_{2\times 2})$, and assume that M_2 is true and $\beta_2 \neq 0$. Also let $\mathbf{Y}$ is repeated r times, *i.e.*, we have i.i.d. observations $\mathbf{Y}_1, \ldots, \mathbf{Y}_r$, but the cross-validation is done leaving one coordinate of the $\mathbf{Y}$'s at a time. First note that $AIC(1) = L(\hat\beta_1) - 1$ and $CV'(1) = L(\hat\beta_1) - \frac{1}{2r}\sum_{l=1}^{2r} l'(y_l|\hat\beta_{(-l,1)})l'(y_l|\hat\beta_1 + a_l(\hat\beta_{(-l,1)} - \hat\beta_1))$, where $y_1, \ldots, y_{2r}$ is a new indexing of Y_{11}, Y_{12}, Y_{21},... where $y_1 = Y_{11}$, $y_2 = Y_{12}$, $y_3 = Y_{21}$ etc. Here $\hat\beta_1$ is the least squares estimate of β_1 under model M_1. So,

$$CV'(1) = L(\hat\beta_1) - \frac{1}{2r}\sum_{l=1}^{2r}(y_l - \hat\beta_{(-l,1)})\{(y_l - \hat\beta_1) - a_l(\hat\beta_{(-l,1)} - \hat\beta_1)\}.$$

But $\hat\beta_{(-l,1)} = \sum_{k=1}^{2r}(y_k - y_l)/(2r-1)$ whence $y_l - \hat\beta_{(-l,1)} = (y_l - \hat\beta_1)(2r)/(2r-1)$. Also, $\hat\beta_{(-l,1)} - \hat\beta_1 = (\hat\beta_1 - y_l)/(2r-1)$. After some algebra, one has,

$$CV'(1) = L(\hat\beta_1) - \frac{1}{2r}\sum_{l=1}^{2r}(y_l - \hat\beta_1)^2\left(\frac{2r}{2r-1}\right)\left(1 + \frac{a_l}{2r-1}\right).$$

Now,

$$\frac{1}{2r}\sum_{l=1}^{2r}(y_l - \hat\beta_1)^2\left(\frac{2r}{2r-1}\right)\left(1 + \frac{a_l}{2r-1}\right)$$

$$= \frac{1}{2r}\sum_{l=1}^{2r}(y_l - \hat\beta_1)^2 + \left(\frac{1}{2r}\sum_{l=1}^{2r}(y_l - \hat\beta_1)^2\right)O_p\left(\frac{1}{2r-1}\right),$$

since $0 < |a_l| \leq 1$. Now note that $\frac{1}{2r}\sum_{l=1}^{2r}(y_l - \hat\beta_1)^2 = \frac{1}{2r}\sum_{l=1}^{2r}(y_l - \beta_1)^2 - (\beta_1 - \hat\beta_1)^2$, where β_1 is the true value of β_1. Now as M_2 is true, $y_l \sim N(\beta_1 + \beta_2, 1)$ if l is odd and $y_l \sim N(\beta_1 - \beta_2, 1)$ if l is even. Using WLLN and this fact, it follows that $\frac{1}{2r}\sum_{l=1}^{2r}(y_l - \beta_1)^2 \xrightarrow{P} 1 + \beta_2^2 > 1$, as $r \to \infty$, noting that $\beta_2 \neq 0$. Also, it is easy to show that $(\hat\beta_1 - \beta_1)^2 \xrightarrow{P} 0$. So

$$CV'(1) - AIC(1) = 1 - \frac{1}{2r}\sum_{l=1}^{2r}(y_l - \hat\beta_1)^2\left(\frac{2r}{2r-1}\right)\left(1 + \frac{a_l}{2r-1}\right) \xrightarrow{P} -\beta_2^2 < 0.$$

Also $AIC(2) - AIC(1) = L(\hat{\beta}_1, \hat{\beta}_2) - L(\hat{\beta}_1) - 1$ and $CV(2) - CV(1) \approx L(\hat{\beta}_1, \hat{\beta}_2) - L(\hat{\beta}_1) - 1 + \beta_2^2$, asymptotically, where $(\hat{\beta}_1, \hat{\beta}_2)$ is the estimate of (β_1, β_2) under M_2. So if AIC chooses M_2, CV will also choose M_2. On the other hand, if CV chooses M_2, AIC may or may not choose M_2. So CV in this case is seen to handle the model selection problem better than AIC.

The following proposition generalizes the argument of Example 1, the proof of which will be skipped because of space.

Proposition 2. Let us consider the normal linear regression model with an orthogonal design matrix, with norm of each column being equal to 1. Let $\mathbf{Y}_1, \ldots, \mathbf{Y}_r$ be the observation vectors, each observation being a $k \times 1$ vector. Let M_{j_0} be the true model such that $\beta_{j_0} \neq 0$ and $\beta_{j_0+k} = 0$ for $k \geq 1$. Also the cross-validation is done with leaving one component of $\mathbf{Y}_i$'s at a time. Then, for $j < j_0$, $AIC(j) - CV'(j)$ converges in probability to $\sum_{i=1}^{k}(X_{i,j}X'_{i,j})(X_i\boldsymbol{\beta}_0 - X_{i,j}\boldsymbol{\beta}_j)^2$, where X_i is the i-th row of $\mathbf{X}$, $X_{i,j}$ is the sub-vector of X_i consisting of the first j elements of X_i, $\boldsymbol{\beta}_0$ is the true $\boldsymbol{\beta}$ and $\boldsymbol{\beta}_j$ consists of the first j elements of $\boldsymbol{\beta}_0$.

Remark 5. Observe that the proposition shows the difference between these two criteria always converges to a non-negative quantity. By properly choosing the design matrix and the true $\boldsymbol{\beta}_0$, it can be made strictly positive (and, for that matter, a large positive value) in many examples.

Remark 6. Example 1 and Proposition 1 highlight the fact that the popular notion of AIC and cross-validation being equivalent is far from being true. However, the following result shows that if cross-validation is done by leaving out one whole vector at a time, the AIC and CV are equivalent in the orthogonal design normal model, in that the difference between the criteria for a given model converges to zero in probability, no matter whether the model is true or not. The proof can be obtained essentially following Stone (1977), except that for the case when model M_j under consideration is not the true model, one can directly argue using normality and orthogonality of the columns of the design matrix that $trace(L_2^{-1}L_1) = j$

Proposition 3. In the normal linear model setup, for a fixed orthogonal design matrix $\mathbf{X}$, suppose we observe $\mathbf{Y}_1, \ldots, \mathbf{Y}_r$ and M_{j_0} is the true model where $\beta_{j0} \neq 0$ and $\beta_{j_0+k} = 0$ for $k \geq 1$. Then $AIC(j) - CV'(j)$ will converge to zero in probability as $r \to \infty$, for all models M_j, $1 \leq j \leq n$, provided cross-validation is done by keeping a whole vector $\mathbf{Y}_i$ at a time.

Remark 7. Assumption B holds for models $\{M_{j_0+k} : 0 \leq k \leq p - j_0\}$, but it does not hold for $\{M_{j_0-k} : 1 \leq k \leq j_0 - 1\}$. Yet the equivalence is seen to be true, since in this setup the existence of $\boldsymbol{\beta}_{j_0-k}$ is ensured and conditions (i)–(iv) are easily seen to be satisfied. A direct argument using orthogonality shows that $trace(L_2^{-1}L_1) = j_0 - k$. The reason we got the counterexamples is that condition (iv) can't be applied, since the y_i's are not i.i.d in those cases. But this type of situation is quite likely to happen in the regression setups.

4. CROSS-VALIDATORY BAYES FACTORS

In this section we will study asymptotic properties of cross-validatory Bayes factors and their averages described in the introduction. This study will be done both in the $\mathcal{M}$-open and $\mathcal{M}$-closed scenarios under assumptions A and B respectively of Section 3. A major goal will be to understand what should be the choice of the proportions of the sample to estimate the parameters of the competing models and model selection. Most of this will be done when the parametric dimension of the models under consideration is fixed in large samples. However, in a few cases we also let the model dimension grow with sample size. The fixed and high-dimensional examples are treated separately in subsections 4.1 and 4.2 respectively.

4.1. *Fixed Dimensional Problems*

Consider first cross-validatory predictive distribution under a given model M_j with p_j parameters, *i.e.*, we will consider

$$f_j(\mathbf{X}_{(k)}|\mathbf{X}_{(-k)}, M_j) = \int_{\Theta_j} f_j(\mathbf{X}_{(k)}|\boldsymbol{\theta}_j)\pi_j(\boldsymbol{\theta}_j|\mathbf{X}_{(-k)})\, d\boldsymbol{\theta}_j, \tag{15}$$

where $\mathbf{X}_{(k)} = \{X_i : i \in \{i_1, \ldots, i_k\}\}$ and $\mathbf{X}_{(-k)} = \{X_i : i \notin \{i_1, \ldots, i_k\}\}$, for a fixed choice of k indices $\{i_1, \ldots, i_k\}$ from $\{1, \ldots, n\}$. We start with the following useful proposition.

Proposition 1. Assume that

$$l_{jn}(\boldsymbol{\theta}_j) = \log f_j(\mathbf{X}_{(n)}|\boldsymbol{\theta}_j), \quad l_{j,(n-k)}(\boldsymbol{\theta}_j) = \log f_j(\mathbf{X}_{(n-k)}|\boldsymbol{\theta}_j),$$

have sharp global maxima $\hat{\boldsymbol{\theta}}_{jn}$ and $\hat{\boldsymbol{\theta}}_{j,(n-k)}$ in the interior of Θ_j. Also assume that $\pi_j(\boldsymbol{\theta}_j)$ is a sufficiently well-behaved positive density. Further let the conditions in assumption A of Section 3 be satisfied by $f_j(.)$. Then as $n \to \infty$ and $n - k \to \infty$,

$$\begin{aligned}\log f_j(\mathbf{X}_{(k)}|\mathbf{X}_{(-k)}, M_j) &= (l_{jn}(\hat{\boldsymbol{\theta}}_{jn}) \\ &- l_{j,(n-k)}(\hat{\boldsymbol{\theta}}_{j,(n-k)})) + \frac{p_j}{2}\log\left(1 - \frac{k}{n}\right) + o_p(1). \end{aligned}\tag{16}$$

Proof. Note that

$$\log f_j(\mathbf{X}_{(k)}|\mathbf{X}_{(-k)}, M_j) = \log \frac{\int_{\Theta_j} e^{l_{jn}(\boldsymbol{\theta}_j)}\pi_j(\boldsymbol{\theta}_j)\, d\boldsymbol{\theta}_j}{\int_{\Theta_j} e^{l_{j,(n-k)}(\boldsymbol{\theta}_j)}\pi_j(\boldsymbol{\theta}_j)\, d\boldsymbol{\theta}_j}.$$

Using Laplace approximation in the numerator and denominator of the quantity inside the logarithm above, as in Gelfand and Dey (1994), one can write the above quantity as

$$\begin{aligned}\Bigg[(l_{j,n}(\hat{\boldsymbol{\theta}}_{j,n}) - \frac{p_j}{2}\log n) &+ \log \pi_j(\hat{\boldsymbol{\theta}}_{j,n}) + \frac{p_j}{2}\log 2\pi - \frac{1}{2}\log\left(\left|-\frac{l''_{jn}(\hat{\boldsymbol{\theta}}_{j,n})}{n}\right|\right)\Bigg] \\ &- \Bigg[\left(l_{j,(n-k)}(\hat{\boldsymbol{\theta}}_{j,(n-k)}) - \frac{p_j}{2}\log(n-k)\right) \\ &+ \log \pi_j(\hat{\boldsymbol{\theta}}_{j,(n-k)}) + \frac{p_j}{2}\log 2\pi \\ &- \frac{1}{2}\log\left(\left|-\frac{l''_{j,(n-k)}(\hat{\boldsymbol{\theta}}_{j,(n-k)})}{n}\right|\right)\Bigg] + o_p(1),\end{aligned}\tag{17}$$

where l''_{jn} and $l''_{j,(n-k)}$ denote the Hessian matrices of $l_{jn}(\boldsymbol{\theta})$ and $l_{j,(n-k)}(\boldsymbol{\theta})$. Now, using the facts that $\pi_j(\boldsymbol{\theta}_{j0}) > 0$ by our assumption on $\pi_j(.)$, and under assumption A, $\hat{\boldsymbol{\theta}}_{j,n} \to \boldsymbol{\theta}_{j0}$ and $\hat{\boldsymbol{\theta}}_{j,(n-k)} \to \boldsymbol{\theta}_{j0}$ in probability as $n \to \infty$ and $n-k \to \infty$ and $I(\boldsymbol{\theta}_{j0})$ is positive definite, (16) follows. □

Remark 1. Using Proposition 1, it is easy to study the behaviour of cross-validatory Bayes factor as a model selection criterion, by comparing the cross-validatory predictive distributions under different models. In Proposition 2, we study the behaviour of this model selection criterion for any fixed choice of indices $\{i_1, \ldots, i_k\}$ in $\mathbf{X}_{(k)}$. But the result holds for any other choice of $\{i_1, \ldots, i_k\}$ of k indices from $(1, \ldots, n)$.

Proposition 2. Consider two models M_1 and M_2. Let all the conditions of Proposition 1 be satisfied for the likelihoods and priors under both models M_1 and M_2. Assume that n and k vary in such a way that $k \to \infty$ and $n-k \to \infty$. Then the following hold.

(a)
$$\begin{aligned}&\log f_2(\mathbf{X}_{(k)}|\mathbf{X}_{(-k)}, M_2) - \log f_1(\mathbf{X}_{(k)}|\mathbf{X}_{(-k)}, M_1) = \\ &(l_{2,n}(\hat{\boldsymbol{\theta}}_{2,n}) - l_{1,n}(\hat{\boldsymbol{\theta}}_{1,n})) - (l_{2,(n-k)}(\hat{\boldsymbol{\theta}}_{2,(n-k)}) - l_{1,(n-k)}(\hat{\boldsymbol{\theta}}_{1,(n-k)})) \\ &+\tfrac{p_2-p_1}{2}\log(1-\tfrac{k}{n}) + o_p(1).\end{aligned} \tag{18}$$

(b) Let $M_1 \subset M_2$ and $\boldsymbol{\theta}_{10} \neq \boldsymbol{\theta}_{20}$. Then criterion in the l.h.s. of (18) chooses model M_2 with probability tending to 1 if $k \to \infty$.
(c) Let $M_1 \subset M_2$ and $\boldsymbol{\theta}_{10} = \boldsymbol{\theta}_{20} = \boldsymbol{\theta}_0$, say. A sufficient condition for the choice of model M_1 by criterion on the l.h.s. of (18) is $k/n \to 1$.

Proof. (a) Follows directly by application of Proposition 1 separately for model M_1 and M_2.
(b) Using part (a), one has, using Taylor expansion of each log-likelihood

$$\begin{aligned}&\log f_2(\mathbf{X}_{(k)}|\mathbf{X}_{(-k)}, M_2) - \log f_1(\mathbf{X}_{(k)}|\mathbf{X}_{(-k)}, M_1) = \\ &(l_{2,n}(\boldsymbol{\theta}_{20}) - l_{1,n}(\boldsymbol{\theta}_{10})) - (l_{2,(n-k)}(\boldsymbol{\theta}_{20}) - l_{1,(n-k)}(\boldsymbol{\theta}_{10})) \\ &+\tfrac{1}{2}[(\boldsymbol{\theta}_{20} - \hat{\boldsymbol{\theta}}_{2,n})' l''_{2,n}(\psi_{2,n})(\boldsymbol{\theta}_{20} - \hat{\boldsymbol{\theta}}_{2,n}) - (\boldsymbol{\theta}_{10} - \hat{\boldsymbol{\theta}}_{1,n})' l''_{1,n}(\psi_{1,n})(\boldsymbol{\theta}_{10} - \hat{\boldsymbol{\theta}}_{1,n})] \\ &-\tfrac{1}{2}[(\boldsymbol{\theta}_{20} - \hat{\boldsymbol{\theta}}_{2,(n-k)})' l''_{2,(n-k)}(\psi_{2,(n-k)})(\boldsymbol{\theta}_{20} - \hat{\boldsymbol{\theta}}_{2,(n-k)}) \\ &-(\boldsymbol{\theta}_{10} - \hat{\boldsymbol{\theta}}_{1,(n-k)})' l''_{1,(n-k)}(\psi_{1,(n-k)})(\boldsymbol{\theta}_{10} - \hat{\boldsymbol{\theta}}_{1,(n-k)})] \\ &+\tfrac{p_2-p_1}{2}\log(1-\tfrac{k}{n}) + o_p(1),\end{aligned} \tag{19}$$

where ψ_{jn} is a point between the line segment joining $\boldsymbol{\theta}_{j0}$ and $\hat{\boldsymbol{\theta}}_{j,n}$, $j = 1, 2$ and $\psi_{j,(n-k)}$ is a point between the line segment joining $\boldsymbol{\theta}_{j0}$ and $\hat{\boldsymbol{\theta}}_{j,(n-k)}$, $j = 1, 2$. It follows from assumption A by an easy argument that each of the quadratic forms in the r.h.s. of (19) is $O_p(1)$ as $n \to \infty$ and $n-k \to \infty$. So, one can write the r.h.s of (19) as $(l_{2,n}(\boldsymbol{\theta}_{20}) - l_{1,n}(\boldsymbol{\theta}_{10})) - (l_{2,(n-k)}(\boldsymbol{\theta}_{20}) - l_{1,(n-k)}(\boldsymbol{\theta}_{10})) + O_p(1) + \frac{p_2-p_1}{2}\log(1-\frac{k}{n})$. Now note that $M_1 \subset M_2$ implies that

$$\inf_{\boldsymbol{\theta}_1 \in \boldsymbol{\Theta}_1} K(f, f_{1,\boldsymbol{\theta}_1}) > \inf_{\boldsymbol{\theta}_2 \in \boldsymbol{\Theta}_2} K(f, f_{2,\boldsymbol{\theta}_2})$$

i.e., $K(f, f_{1,\boldsymbol{\theta}_{10}}) > K(f, f_{2,\boldsymbol{\theta}_{20}})$. By an easy argument it follows that $(l_{2,n}(\boldsymbol{\theta}_{20}) - l_{1,n}(\boldsymbol{\theta}_{10})) - (l_{2,(n-k)}(\boldsymbol{\theta}_{20}) - l_{1,(n-k)}(\boldsymbol{\theta}_{10})) = k\{[K(f, f_{1,\boldsymbol{\theta}_{10}}) - K(f, f_{2,\boldsymbol{\theta}_{20}})] + O_p(\frac{1}{\sqrt{k}})\}$ as $k \to \infty$. If

$$\limsup_{k\to\infty, n\to\infty} \frac{k}{n} < 1,$$

the conclusion is immediate. If, on the other hand, $k \sim n$, as $n \to \infty$, $k \to \infty$ and $n - k \to \infty$, observing that $\frac{n-k}{n} \geq \frac{1}{n}$, it follows that

$$e^{kc}\left(\frac{n-k}{n}\right)^{\frac{p_2-p_1}{2}} \geq \frac{e^{kc}}{n^{\frac{p_2-p_1}{2}}} \to \infty$$

for any $c > 0$. Using this fact the result follows in this case also.
(c) In this case $l_{2n}(\boldsymbol{\theta}_{20}) = l_{1n}(\boldsymbol{\theta}_{10})$ and $l_{2,(n-k)}(\boldsymbol{\theta}_{20}) = l_{1,(n-k)}(\boldsymbol{\theta}_{10})$, so the the r.h.s. of (19) becomes $\frac{p_2-p_1}{2}\log(1-\frac{k}{n}) + O_p(1)$. The conclusion follows immediately. □

Remark 2. Since part (c) of Proposition 2 is proved under assumption A, the conclusion will also hold under assumption B. So, this covers both the $\mathcal{M}$-open and $\mathcal{M}$-closed scenarios.

Remark 3. Parts (b) and (c) of Proposition 2 indicate the usefulness of criterion in the l.h.s. of (18) as a model selection tool in the nested model scenario. In part (b), it is shown that the model with a smaller divergence from the truth is chosen with probability tending to 1 under fairly mild conditions on the rate of growth of k as $n \to \infty$. On the other hand, in part (c), it is shown that if two models are equally divergent from the truth, then the more parsimonious model will be chosen if $k/n \to 1$ as $n \to \infty$ and $k \to \infty$. It can be shown in some simple cases of normal means problems that in this situation, $k/n \to 0$ will imply that the criterion will tend to zero in probability asymptotically, so it may not be able to discriminate between the two models. We expect this phenomenon to hold in general also, but it needs a rigorous argument, which we have not been able to come up with as yet. But our simulation outputs seem to support our belief, as commented in Section 5.1.

Remark 4. For part (b), consider now a non-nested scenario and $\boldsymbol{\theta}_{10} \neq \boldsymbol{\theta}_{20}$. Exactly same calculation as in the proof goes through, but in this case one can have $K(f, f_{1,\boldsymbol{\theta}_{10}}) > K(f, f_{2,\boldsymbol{\theta}_{20}})$, $K(f, f_{1,\boldsymbol{\theta}_{10}}) < K(f, f_{2,\boldsymbol{\theta}_{20}})$ or $K(f, f_{1,\boldsymbol{\theta}_{10}}) = K(f, f_{2,\boldsymbol{\theta}_{20}})$. Hence the r.h.s. of (18) becomes, as before, as $k \to \infty$,

$$k\left\{[K(f, f_{1,\boldsymbol{\theta}_{10}}) - K(f, f_{2,\boldsymbol{\theta}_{20}})] + O_p\left(\frac{1}{\sqrt{k}}\right)\right\} + O_p(1) + \frac{p_2-p_1}{2}\log\left(1-\frac{k}{n}\right)$$

as $k \to \infty$. If $K(f, f_{1,\boldsymbol{\theta}_{10}}) > K(f, f_{2,\boldsymbol{\theta}_{20}})$, it follows by exactly same argument as before that M_2 will be chosen with probability tending to 1 asymptotically whatever may be values of p_2 and p_1. In the case where $K(f, f_{1,\boldsymbol{\theta}_{10}}) < K(f, f_{2,\boldsymbol{\theta}_{20}})$, using same argument it can be shown that M_1 will be chosen with probability tending to 1 asymptotically, whatever may be the values of p_1 and p_2. However, when $K(f, f_{1,\boldsymbol{\theta}_{10}}) = K(f, f_{2,\boldsymbol{\theta}_{20}})$, it is desirable to have the model with smaller dimension chosen. Note that the r.h.s of (19) now becomes $O_p(\sqrt{k}) + \frac{p_2-p_1}{2}\log(1-\frac{k}{n})$, the term involving the difference of the (minimum) Kullback–Leibler divergences being zero. It can be shown, using the central limit theorem, that in case $p_2 > p_1$, for no choice of k, M_1 will ever be chosen with probability tending to 1, but it will be chosen with probability bounded away from zero. Similarly if $p_2 < p_1$, M_2 will also not be chosen with probability tending to 1 for any choice of k. Situations as in part (c) also could happen in the non-nested case, but those are not interesting cases and will not be discussed here.

Remark 5. As mentioned before, we consider also averages of the logarithm of cross-validatory Bayes factor (as in the l.h.s. of (18)) for mutually disjoint sets of $\mathbf{X}_{(k)}$'s; *i.e.*, we consider the average of $[\frac{n}{k}]$ quantities like the l.h.s of (18). Going back to Proposition 2(a), now we have an $o_p(1)$ term in the r.h.s. of (18) for each choice of $\mathbf{X}_{(k)}$. If k and n are of the same order of magnitude, one can straightaway say that

$$\begin{aligned}\frac{1}{[\frac{n}{k}]}\sum_S \log\frac{f_2(\mathbf{X}_{(k)}|\mathbf{X}_{(-k)},M_2)}{f_1(\mathbf{X}_{(k)}|\mathbf{X}_{(-k)},M_1)} = \frac{1}{[\frac{n}{k}]}\sum_S[\{l_{2,n}(\hat{\boldsymbol{\theta}}_{2,n}) - l_{1,n}(\hat{\boldsymbol{\theta}}_{1,n})\}-\\ \{l_{2,n-k}(\hat{\boldsymbol{\theta}}_{2,n-k}) - l_{1,n-k}(\hat{\boldsymbol{\theta}}_{1,n-k})\}] + \tfrac{p_2-p_1}{2}\log(1-\tfrac{k}{n}) + o_p(1),\end{aligned} \tag{20}$$

where S is the collection of $[\frac{n}{k}]$ indices of size k given by $(1,\ldots,k)$, $(k+1,\ldots,2k)$ and so on. If $n/k\to\infty$, appealing to arguments for dealing with remainders in Laplace approximation, it follows that if a moderate deviation result as the following

$$P_{\boldsymbol{\theta}_{j0}}\left\{|\hat{\boldsymbol{\theta}}_{j,n-k} - \boldsymbol{\theta}_{j0}| > c\frac{\sqrt{\log(n-k)}}{\sqrt{n-k}}\right\} = O\left(\frac{1}{(n-k)^{\frac{s-2}{2}}}\right), \tag{21}$$

holds for a suitably large integer $s\geq 3$ and a constant $c\geq\sqrt{s-2+\delta}>0$ for some $\delta>0$, for $j=1,2$, as $(n-k)\to\infty$, the above representation (20) will still hold. (Such a result will hold in this situation using Corollary 17.12 of Bhattacharya and Rao (1976, pp. 178–179) and standard representation of maximum likelihood estimators.) This is so, since under (21), it follows that

$$P_{\boldsymbol{\theta}_{j0}}\left\{\max|\hat{\boldsymbol{\theta}}_{j,(n-k)} - \boldsymbol{\theta}_{j0}| > c\frac{\sqrt{\log(n-k)}}{\sqrt{n-k}}\right\} = O\left(\frac{[\frac{n}{k}]}{(n-k)^{\frac{s-2}{2}}}\right)\to 0,$$

if s is large for $j=1,2$ where the maximum is taken over $\hat{\boldsymbol{\theta}}_{j,(n-k)}$ based on $[\frac{n}{k}]$ $\mathbf{X}_{(-k)}$'s corresponding to disjoint $\mathbf{X}_{(k)}$'s. This ensures that the $o_p(1)$ term in (20) is $o_p(1)$ uniformly over disjoint choices of $\mathbf{X}_{(k)}$'s.

We now state and briefly indicate the proof of Proposition 3. We assume that the conditions of Proposition 1 are satisfied for each choice of $\mathbf{X}_{(k)}$ and also keep in mind that the moderate deviation inequality will hold in this case.

Proposition 3. For any two models M_1 an M_2, if model selection is done using criterion in the l.h.s. of (20), and if $K(f,f_{2,\boldsymbol{\theta}_{20}})\neq K(f,f_{1,\boldsymbol{\theta}_{10}})$, model M_2 or M_1 will be chosen with probability tending to 1 depending on which model is closer to the truth, provided $k\to\infty$, $n-k\to\infty$ and $n\to\infty$.

Proof. If $k\sim n$, the proof follows exactly as before , as in Proposition 2(b) and Remark 4. Now consider the case where $\frac{k}{n}\to 0$ or $\frac{n}{k}\to\infty$. One can write the r.h.s. of (20), upto a $o_p(1)$ term as

$$\begin{aligned}\left\{l_{2,n}(\hat{\boldsymbol{\theta}}_{2,n}) - l_{1,n}(\hat{\boldsymbol{\theta}}_{1,n})\right\} \quad - \quad & \frac{1}{[\frac{n}{k}]}\sum_S\left\{l_{2,(n-k)}(\hat{\boldsymbol{\theta}}_{2,(n-k)}) - l_{1,(n-k)}(\hat{\boldsymbol{\theta}}_{1,(n-k)})\right\}\\ + \quad & \frac{p_2-p_1}{2}\log\left(1-\frac{k}{n}\right).\end{aligned}$$

Using Taylor expansions of the log-likelihoods around $\boldsymbol{\theta}_{10}$ and $\boldsymbol{\theta}_{20}$, whichever is appropriate, and using the moderate deviation result on the quadratic forms in these expansions, one can write the above as

$$\begin{aligned}\Big\{l_{2,n}(\boldsymbol{\theta}_{20}) \quad &- \quad l_{1,n}(\boldsymbol{\theta}_{10})\Big\} - \frac{1}{[\frac{n}{k}]}\sum_{S}\Big\{l_{2,(n-k)}(\boldsymbol{\theta}_{20}) \\ &- \quad l_{1,(n-k)}(\boldsymbol{\theta}_{10})\Big\} + \frac{p_2-p_1}{2}\log\Big(1-\frac{k}{n}\Big) + O_p(1).\end{aligned}$$

This again equals

$$\frac{1}{[\frac{n}{k}]}\sum_{S}\{\log f_2(\mathbf{X}_{(k)}|\boldsymbol{\theta}_{20}) - \log f_1(\mathbf{X}_{(k)}|\boldsymbol{\theta}_{10})\} + \frac{p_2-p_1}{2}\log\left(1-\frac{k}{n}\right) + O_p(1).$$

Now, expectation of the first quantity above if $k[K(f, f_{1,\boldsymbol{\theta}_{10}}) - K(f, f_{2,\boldsymbol{\theta}_{20}})]$ and the variance is $O(k^2/n)$. Hence the first term is

$$k\left[(K(f, f_{1,\boldsymbol{\theta}_{10}}) - K(f, f_{2,\boldsymbol{\theta}_{20}})) + O_p\Big(\frac{1}{\sqrt{n}}\Big)\right].$$

So it follows that the model closer to the truth will be chosen with probability tending to 1, by invoking arguments from Proposition 2. □

Remark 6. In the case of nested models, under assumptions A or B, if $M_1 \subset M_2$ and also $\boldsymbol{\theta}_{10} = \boldsymbol{\theta}_{20}$, then as commented before, this criterion will be consistent if $k/n \to 1$, i.e, it will choose model M_1. In the case of non-nested models, if both models are false but equidistant from the truth, then this criterion will be inconsistent in the sense of choosing a larger dimensional model with probability bounded away from zero for any choice of $k \to \infty$.

4.2. *High Dimensional Problems*

We made some investigation in the high-dimensional normal linear model setup described below. One has observations

$$x_{ij} = \mu_i + \epsilon_{ij}; \tag{22}$$

$i = 1, \ldots, p$; $j = 1, \ldots, r$ and ϵ_{ij} are i.i.d. $N(0, 1)$. One is interested in deciding whether $M_1 : \mu_i = 0\ \forall i = 1, \ldots, p$ is true or $M_2 : \mu \in \mathbf{R}^p$ is true. We assume that the prior under M_2 is the well-known Zellner–Siow multivariate Cauchy prior given below

$$\pi(\mu) = \int_0^\infty \frac{t^{\frac{p}{2}}}{(2\pi)^{\frac{p}{2}}} e^{-\frac{t}{2}\mu'\mu} \frac{1}{\sqrt{2\pi}} e^{-\frac{t}{2}} t^{-\frac{1}{2}}\, dt.$$

In Proposition 4, it is shown that a sufficient condition for the cross-validatory Bayes factor to be consistent under M_2 is that $\liminf_{r\to\infty, k\to\infty} \frac{k}{r} > 0$. Letting $\mathbf{X}_j = (x_{1j}, \ldots, x_{pj})' \sim N(\mu, I_{p\times p})$ where $\mu = (\mu_1, \ldots, \mu_p)'$, one has r i.i.d. observations $(\mathbf{X}_1, \ldots, \mathbf{X}_r)$. So here r takes the role of n. In this setup we do not consider an average of criterion in the l.h.s of (18), as already mentioned in the introduction, but consider the behaviour of the cross- validatory Bayes factor when

$\mathbf{X}_{(k)} = (\mathbf{X}_1, \ldots, \mathbf{X}_k)$ and $\mathbf{X}_{(-k)} = (\mathbf{X}_{(k+1)}, \ldots, \mathbf{X}_r)$. Denote by $BF_{21}^{(r)}$ the Bayes factor of model M_2 to model M_1 based on all observations and by $BF_{21}^{(r-k)}$ the Bayes factor model of M_2 to model M_1 based on $\mathbf{X}_{(-k)} = (\mathbf{X}_{(k+1)}, \ldots, \mathbf{X}_r)$

Proposition 4. Under M_2, a sufficient condition for the cross-validatory Bayes factor to be consistent is that

$$\liminf_{k\to\infty, r\to\infty} \frac{k}{r} > 0,$$

in that, under M_2 if this condition is satisfied, then

$$\frac{{BF_{21}}^{(r)}}{{BF_{21}}^{(r-k)}} \to \infty$$

in probability asymptotically, provided $\lim_{p\to\infty} \frac{1}{p}\sum_{i=1}^{p} \mu_i^2 = \tau^2 > 0$.

Proof. It can be shown, by a simple algebra (see Mukhopadhyay (2000)), that

$$BF_{21}^{r} = \int_{t=0}^{\infty} e^{\frac{p}{2}[rc_p(1-(1+\frac{r}{t})^{-1}) + \log(1+\frac{r}{t})^{-1}]}\pi(t)\,dt,$$

where $\pi(t) = \frac{1}{\sqrt{2\pi}}e^{-\frac{t}{2}}t^{-\frac{1}{2}}$ and $c_p = \frac{1}{p}\sum_{i=1}^{p}\bar{y}_i^2$ and $\bar{y}_i = \frac{1}{r}\sum_{j=1}^{r} x_{ij}$. Letting $x = 1-(1+\frac{r}{t})^{-1}$, one has

$$BF_{21}^{(r)} = \int_{x=0}^{1} e^{\frac{p}{2}[f(x)]}\tilde{\pi}(dx),$$

where

$$f(x) = rc_p x + \log(1-x),\ \tilde{\pi}(dx) = \frac{1}{\sqrt{2\pi}}e^{-\frac{r(1-x)}{2x})}\left[\frac{r(1-x)}{x}\right]^{-\frac{1}{2}}\frac{r}{x^2}\,dx.$$

It can be similarly shown that $BF_{21}^{(r-k)} = \int_{x=0}^{1} e^{\frac{p}{2}[f_1(x)]}\tilde{\pi_1}(dx)$, where $f_1(x) = (r-k)c'_p x + \log(1-x)$, where $c'_p = \frac{1}{p}\sum_{i=1}^{p}\bar{y}^2_{i,(n-k)}$ and $\bar{y}_{i,(n-k)} = \frac{1}{(r-k)}\sum_{j=k+1}^{r} x_{ij}$. $\tilde{\pi_1}(dx)$ is obtained by replacing r by $r-k$ in the definition of $\tilde{\pi}(dx)$.

Now note that $E(c_p) = \frac{1}{r} + \frac{1}{p}\sum_{i=1}^{p}{\mu_i}^2$, $Var(C_p) = \frac{2}{r^2 p} + \frac{4}{rp}\left(\frac{\sum_{i=1}^{p}{\mu_i}^2}{p}\right)$. Also the expectation and variance of c'_p will be obtained by replacing r by $r-k$ in the corresponding expressions for c_p. Hence

$$\frac{\frac{p}{2}rx(c_p - \tau^2)}{\frac{p}{2}rx\tau^2} = \frac{c_p}{\tau^2} - 1 = o_p(1)$$

and also

$$\frac{\frac{p}{2}(r-k)x(c'_p - \tau^2)}{\frac{p}{2}(r-k)x\tau^2} = \frac{c'_p}{\tau^2} - 1 = o_p(1)$$

as $r \to \infty$ and $r - k \to \infty$. Note that the $o_p(1)$ terms above are uniformly so in $x \in (0,1)$. So, we can write

$$\frac{BF_{21}{}^{(r)}}{BF_{21}{}^{(r-k)}} = \frac{\int_0^1 e^{[\frac{p}{2} r x \tau^2 (1+o_p(1)) + \frac{p}{2}\log(1-x)]} \tilde{\pi}(dx)}{\int_0^1 e^{[\frac{p}{2}(r-k) x \tau^2 (1+o_p(1)) + \frac{p}{2}\log(1-x)]} \tilde{\pi_1}(dx)}. \tag{23}$$

Now fix $0 < \epsilon < 1$ and $\eta > 0$. Let $\frac{1}{\epsilon} = 1 + \Delta$. Fix any positive δ less than $1 - \epsilon(1 + \frac{\Delta}{2})$. One can find $r_0 = r_0(\epsilon, \eta)$ and $p = p_0(\epsilon, \eta)$ such that the $(1 + o_p(1))$ quantity in the numerator of the r.h.s of (23) is, for all $x \in (0,1)$, greater than $1 - \delta$ with probability greater than $1 - \eta/2$ for all $r \geq r_0$ and $p \geq p_0$. It follows that, for r and p like this, for all $x \in \left(\frac{\epsilon(1+\frac{\Delta}{2})}{(1-\delta)}, 1\right)$, the term $r x \tau^2 (1 + o_p(1))$ is greater than $r\epsilon\tau^2 + r\epsilon\tau^2 \frac{\Delta}{2}$ with probability greater than $(1 - \eta/2)$. Now by choosing a large enough r satisfying, $r \geq r_0$, we can make $r\epsilon\tau^2 \frac{\Delta}{2} + \log(1-x) > 0$, for all $x \in \left(\frac{\epsilon(1+\frac{\Delta}{2})}{(1-\delta)}, \frac{\epsilon(1+\frac{\Delta}{2})}{(1-\delta)} + \frac{\gamma}{2}\right)$ where $\gamma = 1 - \frac{\epsilon(1+\frac{\Delta}{2})}{(1-\delta)}$. Let us denote for notational similarity $\frac{\epsilon(1+\frac{\Delta}{2})}{(1-\delta)} = A$ and $\frac{\epsilon(1+\frac{\Delta}{2})}{(1-\delta)} + \frac{\gamma}{2} = B$. So, the numerator in the r.h.s of (23) is greater than $e^{\frac{p}{2} r\epsilon\tau^2} \int_A^B \tilde{\pi}(dx)$. Now fix $\delta' > 0$, arbitrarily small. Note that by choosing p and r sufficiently large, the $(1 + o_p(1))$ term in the denominator of the r.h.s. of (23) is less than $(1 + \delta')$ with probability greater than $1 - \frac{\eta}{2}$. But this implies, by noting that $(1-x)^{\frac{p}{2}} \leq 1$, $\int_0^1 \tilde{\pi}(dx) = 1$, that $BF_{21}{}^{(r-k)} \leq e^{\frac{p}{2}(r-k)\tau^2(1+\delta')}$ with probability greater than $1 - \frac{\eta}{2}$ for all large enough p and r. Hence, by Bonferroni's inequality, we have, for large enough p and r, with probability greater than $(1 - \eta)$,

$$\frac{BF_{21}{}^{(r)}}{BF_{21}{}^{(r-k)}} > e^{\frac{p}{2} r\tau^2[\epsilon - (1-\frac{k}{r})(1+\delta')]} \times \int_A^B \tilde{\pi}(dx). \tag{24}$$

After some algebra, the right hand side of (24) can be shown to be greater than

$$C(A,B) e^{[\frac{p}{2} r\tau^2 \{\epsilon - (1-\frac{k}{r})(1+\delta')\} - \frac{(1-A)}{A}\frac{1}{2r} + \frac{1}{2}\log r]},$$

where $C(A,B)$ is strictly positive quantity. If quantity in the exponent goes to infinity as $r \to \infty$, so does the cross-validatory Bayes factor. A sufficient condition for this to happen is $\{\epsilon - (1-\frac{k}{r})(1+\delta')\} > 0$ as $k \to \infty$ and $r \to \infty$, in other words $\frac{k}{r} > (1 - \frac{\epsilon}{(1+\delta')})$ as $k \to \infty$ and $r \to \infty$. If $\liminf_{k\to\infty, r\to\infty} \frac{k}{r} > 0$, this condition will be satisfied, if to start with, ϵ and δ' are chosen such that $(1 - \frac{\epsilon}{(1+\delta')}) < \liminf_{k\to\infty, r\to\infty} \frac{k}{r}$. Hence the proof. □

We now present a heuristic argument to show that under M_1, if $\frac{k}{r}$ is in $(0,1]$ in the limit asymptotically, the criterion will choose M_1 with probability tending to 1 asymptotically. We also show that under M_1, if $k/r \to 0$ with $\frac{pk}{r} \to 0$, then the cross-validatory Bayes factor will converge to 1 in probability, in which case it may not be able to discriminate between the two models. Arguments for both these cases are based on the further assumption that $p/r \to 0$ and $p/(r-k) \to 0$.

First note that we can write

$$\frac{{BF_{21}}^{(r)}}{{BF_{21}}^{(r-k)}} = \frac{\int_0^1 e^{\frac{p}{2}[r(c_p-\frac{1}{r})x+x+\log(1-x)]} e^{-\frac{r(1-x)}{2x}} [\frac{r(1-x)}{x}]^{-\frac{1}{2}} \frac{r}{x^2}\, dx}{\int_0^1 e^{\frac{p}{2}[(r-k)(c_p{}'-\frac{1}{r-k})x+x+\log(1-x)]} e^{-\frac{(r-k)(1-x)}{2x}} [\frac{(r-k)(1-x)}{x}]^{-\frac{1}{2}} \frac{r-k}{x^2}\, dx}.$$

Note that $r(c_p - \frac{1}{r}) = O_p(\frac{1}{\sqrt{p}})$ and also $(r-k)(c_p{}' - \frac{1}{r-k}) = O_p(\frac{1}{\sqrt{p}})$ under M_1. Now note that the contribution for the integral in the numerator and denominator in the range $(0, \delta)$ for any fixed $\delta < 1$ is negligible. So, we concentrate on the integrand around $x = 1$. Note that the dominant term of the integrand in the numerator in this neighbourhood is

$$e^{\frac{p}{2}[r(c_p-\frac{1}{r})x+x+\log(1-x)]-\frac{r}{2x}} \frac{1}{\sqrt{1-x}}.$$

The maximum of the integrand will be attained around $x = 1$. The expression above can be rewritten as

$$e^{\frac{p}{2}[r(c_p-\frac{1}{r})x+x]+\frac{p-1}{2}\log(1-x)-\frac{r}{2x}}.$$

The maximum of this quantity is approximately same as the maximum of $\exp[\frac{p}{2}x - \frac{r}{2x} + \frac{p}{2}\log(1-x)]$, since $r(c_p - \frac{1}{r}) = O_p(\frac{1}{\sqrt{p}})$, whereby the maximum is attained at the point satisfying $px^3 = r(1-x)$. Since $p \to \infty$, if $x \approx 1$ is a solution, then $px^3 \to \infty$ and we need $r(1-x) \to \infty$ also. Also, for any solution near $x = 1$, $p/r = (1-x)/x^3 \to 0$. Hence, taking $\hat{x} = 1 - \frac{p}{r}$, we approximately have the above equation satisfied. In the denominator we take as a solution $\hat{x} = 1 - \frac{p}{r-k}$. By doing a Laplace approximation to evaluate the integrals in the numerator and denominator, one can easily see that the ratio of the two terms as

$$\frac{{BF_{21}}^{(r)}}{{BF_{21}}^{(r-k)}} \sim \sqrt{1-\frac{k}{r}}\, \frac{e^{[O_p(\sqrt{p})+\frac{p}{2}(1-\frac{p}{r})+\frac{p}{2}\log\frac{p}{r}+\frac{r}{2}-\frac{r}{2(1-\frac{p}{r})}-\frac{1}{2}\log\frac{p}{r}]}}{e^{[O_p(\sqrt{p})+\frac{p}{2}(1-\frac{p}{r-k})+\frac{p}{2}\log\frac{p}{r-k}+\frac{r-k}{2}-\frac{r-k}{2(1-\frac{p}{r-k})}-\frac{1}{2}\log\frac{p}{r-k}]}}. \tag{25}$$

In the case $0 < \frac{k}{r} \leq 1$, after some algebra one can show that the quantity in the r.h.s of (25) is less than

$$\sqrt{1-\frac{k}{r}} \times e^{[O_p(\sqrt{p})-cp]}$$

for all large r, for some constant $c > 0$. But this goes to zero in probability as $p \to \infty$, whence M_1 will be chosen with probability tending to 1.

However, if $k/r \to 0$, one can show, after some algebra, that the $O_p(\sqrt{p})$ term is actually $O_p(\sqrt{\frac{pk}{r}})$, and that the cross-validatory Bayes factor is asymptotically of the same order as

$$\sqrt{1-\frac{k}{r}} \times e^{[O_p(2\sqrt{pk/r})-\frac{kp}{r}]}.$$

So the ratio converges to 1 in probability, if, $r \to \infty$, $r-k \to \infty$, $k/r \to 0$, $p/r \to 0$ and $p/r-k \to 0$ in such a way that $\frac{pk}{r} \to 0$, *i.e.*, if $k = o(\frac{r}{p})$. Hence the criterion may not be able to discriminate between M_1 and M_2 in such a case. However for given large r, k and p, in some cases the Bayes Factor is expected to be greater than one , while in some cases it is expected to be less than 1, depending on the sign of the quantity in the exponent. But its magnitude is expected to be very close to 1 asymptotically.

5. SIMULATIONS FOR CROSS-VALIDATORY BAYES FACTORS

In this section, we present simulation results on the performance of the average of logarithm of cross-validatory Bayes factor studied in Section 4. The simulations are done for both the $\mathcal{M}$-open and $\mathcal{M}$-closed scenario, for low dimensional examples and for the $\mathcal{M}$-closed scenario in some high-dimensional cases. In all the simulations, we consider, as in Section 4.2, the normal linear model setup

$$x_{ij} = \mu_i + \epsilon_{ij}; \tag{26}$$

$i = 1, \ldots, p$; $j = 1, \ldots, r$ and ϵ_{ij} are i.i.d. $N(0, 1)$. For a fixed j, letting $\mathbf{X}_j = (x_{1j}, \ldots, x_{pj})' \sim N(\mu, I_{p\times p})$ where $\mu = (\mu_1, \ldots, \mu_p)'$, one has r i.i.d. observations $(\mathbf{X}_1, \ldots, \mathbf{X}_r)$. For calculating the cross-validatory Bayes factor, we consider different choices of $k = 10l$, where l is an integer satisfying $l \in [1, (\frac{r}{10})0.95]$. For each k, we divide the r observations into $[\frac{r}{k}]$ mutually exclusive groups, and consider the average of the logarithm of the Bayes factor

$$\frac{\int_{\Theta_2} f_2(\mathbf{X}_{(k)}|\boldsymbol{\theta}_2)\pi_2(\boldsymbol{\theta}_2|\mathbf{X}_{(-k)})\, d\boldsymbol{\theta}_2}{\int_{\Theta_1} f_1(\mathbf{X}_{(k)}|\boldsymbol{\theta}_1)\pi_1(\boldsymbol{\theta}_1|\mathbf{X}_{(-k)})\, d\boldsymbol{\theta}_1}, \tag{27}$$

calculated for each $\mathbf{X}_{(k)}$ formed with one such group. More specifically, we divide $(\mathbf{X}_1,\ldots,\mathbf{X}_r)$ into the following groups $(\mathbf{X}_1, \ldots, \mathbf{X}_k)$, $(\mathbf{X}_{k+1}, \ldots, \mathbf{X}_{2k})$ and so on. We describe in Sections 5.1 and 5.2 respectively, the low-dimensional and high-dimensional examples. For both types of problems, we use the Zellner–Siow prior defined in Section 4.2 for the parameters under both M_1 and M_2.

5.1. *Simulations with Small p*

In these simulations, we consider the case where $p = 7$. First we describe our findings in the $\mathcal{M}$-open scenario. In the $\mathcal{M}$-open non-nested case, our simulations indicate that the model which is closer to the true model (in the Kullback–Leibler sense) is chosen for all k with the criterion steadily becoming more favourable for its choice for larger values of k. A typical example of this phenomenon is presented in Fig. 1, where M_2 is more divergent from the truth than M_1. This result agrees with the theoretical conclusion in Proposition 3 in Section 4.1. (Note that in Fig. 1 through Fig. 8, the horizontal axis is in the scale of $\frac{k}{10}$.)

For the nested $\mathcal{M}$-open case with $M_1 \subset M_2$, where both models are equally divergent from the true model, it is observed that the smaller dimensional model is being chosen for all $k's$ used in our simulation, with the criterion steadily decreasing for larger k's, this second fact being expected from the comment after Proposition 3. A typical example is presented in Fig. 2. One must note from the plot that the criterion. although negative for all k's, is very close to zero when k is very small compared to r, hence unable to discriminate. We investigated this fact further and found examples in this situation where the criterion is extremely close to zero when $k/r \approx 0$. This comment closely agrees with Remark 3 in Section 4.1, where it was mentioned that we expect the criterion to actually converge to zero if $k/r \to 0$.

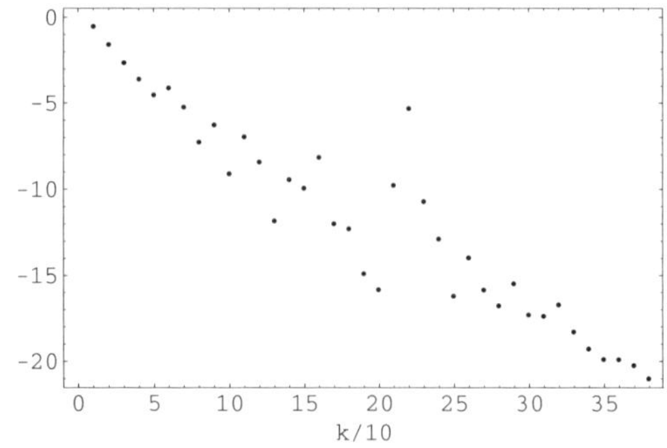

Figure 1: *Average value of logarithm of cross-validatory Bayes factor, when true* $\boldsymbol{\mu} = (0.8, 0.9, 0.5, 0.5, 0.9, 0.7, 0)$, $M_2 : \mu_1 = \mu_2 = \mu_3 = \mu_7 = 0$, *other* μ_i*'s arbitrary and* $M_1 : \mu_4 = \mu_5 = \mu_6 = \mu_7 = 0$, *other* μ_i'*s arbitrary.*

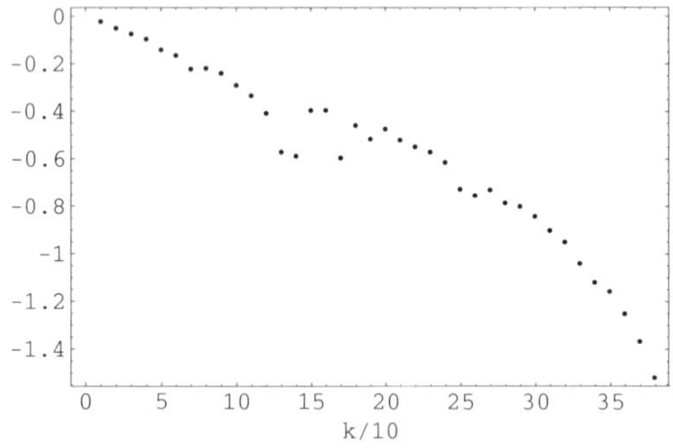

Figure 2: *Average value of logarithm of cross-validatory Bayes factor, when true* $\boldsymbol{\mu} = (0.5, 0.6, 0.05, 0.02, 0.01, 0, 0)$, $M_2 : \mu_5 = \mu_7 = 0$, *other* μ_i*'s arbitrary and* $M_1 : \mu_5 = \mu_6 = \mu_7 = 0$, *other* μ_i'*s arbitrary.*

We add here that for the nested $\mathcal{M}$-open case, when one model is more divergent from the true model than the others, the model with smaller divergence is seen to be chosen in our simulations. In Fig. 3, we present such a plot. Consider now the $\mathcal{M}$-closed scenario. In this case we consider some examples where the true model has non-zero μ_i's for $i = 1, 2, 3, 4$ and $\mu_i = 0$ for $i > 4$. We separately consider the cases where only one of the competing model contains the true model in its model space and when both models have this property. For the first case, we consider, as described below Fig. 4 and 5, $M_1 \subset M_2$, and M_2 is the true model, whereby divergence of the true model from M_1 is larger. It is observed that M_2 was chosen for a sample size of $r = 400$ if the (minimum) Kullback–Leibler divergence for model M_1 is sizeable, as in Fig. 4. But if the (minimum) Kullback–Leibler divergence is small, for $r = 400$, M_1 is chosen, as indicated in Fig. 5. We investigated this further and realized that we need to increase the sample size in this case to have better discriminating power. For the $\mathcal{M}$-closed cases where both models have the true model in their model spaces, *i.e.*, when the divergence from the true model is zero for both models, M_1 is chosen for all k's in our simulations, with the criterion

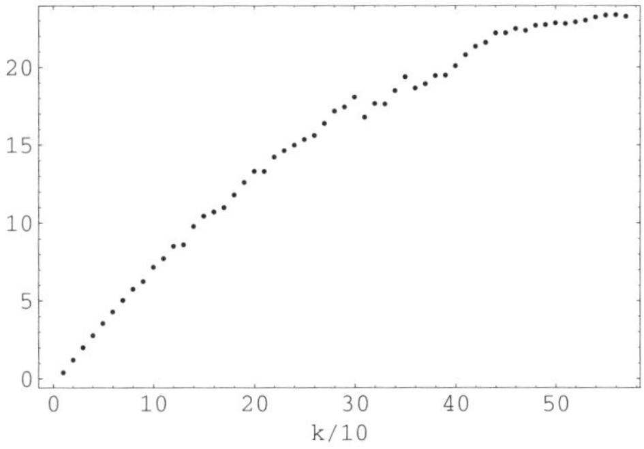

Figure 3: *Average value of logarithm of cross-validatory Bayes factor, when true* $\boldsymbol{\mu} = (0.08, 0.09, 0.5, 0.2, 0.3, 0, 0)$, $M_2 : \mu_4 = \mu_6 = \mu_7 = 0$, *other* μ_i*'s arbitrary and* $M_1 : \mu_4 = \mu_5 = \mu_6 = \mu_7 = 0$, *other* μ_i'*s arbitrary.*

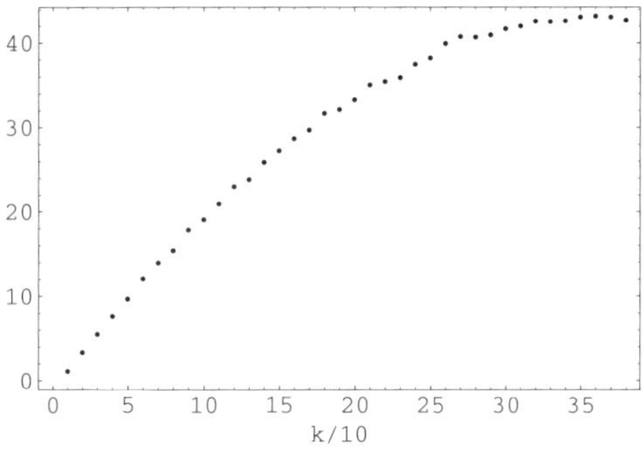

Figure 4: *Average value of logarithm of cross-validatory Bayes factor, when true* $\boldsymbol{\mu} = (0.8, 0.9, 0.5, 0.5, 0, 0, 0)$, $M_2 : \mu_6 = \mu_7 = 0$, *other* μ_i*'s arbitrary and* $M_1 : \mu_4 = \mu_5 = \mu_6 = \mu_7 = 0$, *other* μ_i'*s arbitrary.*

decreasing steadily for large k's, this second phenomenon, again, is expected vide the comment after Proposition 3. Fig. 6 below shows one such simulation output. But here also one must note that for $k/r \approx 0$, the criterion is extremely close to zero.

5.2. *Simulations with Large p*

In this situation, we only consider the $\mathcal{M}$-closed cases. Two models considered are $M_1 : \mu_i = 0$, for $i = 1, \ldots, p$ vs $M_2 : \boldsymbol{\mu} \in \mathbf{R}^p$. When M_2 is true, it is observed that for k/r approximately between 0.6 and 0.7, the criterion reaches its maximum and then tends to fall off for larger k's while it steadily increases as k is increased from being very small. The value of the criterion is pretty close to zero for $\frac{k}{r}$ very small. This seems to indicate that the criterion might choose M_1 with non-zero probability if $k/r \to 0$. A typical simulation result is shown in Fig. 7.

Under M_1, it is generally observed in our simulations that M_2 is chosen for small values of k, while for large values of k, M_1 is seen to be chosen. The simulations are

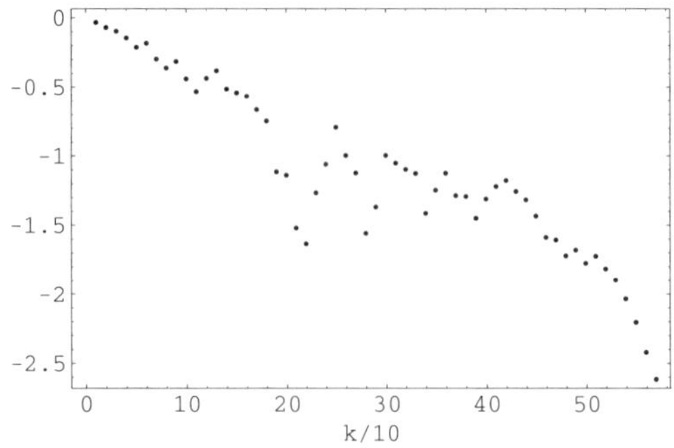

Figure 5: *Average value of logarithm of cross-validatory Bayes factor, when true* $\boldsymbol{\mu} = (0.1, 0.08, 0.06, 0.04, 0, 0, 0)$, $M_2 : \mu_6 = \mu_7 = 0$, *other* μ_i*'s arbitrary and* $M_1 : \mu_4 = \mu_5 = \mu_6 = \mu_7 = 0$, *other* μ_i'*s arbitrary.*

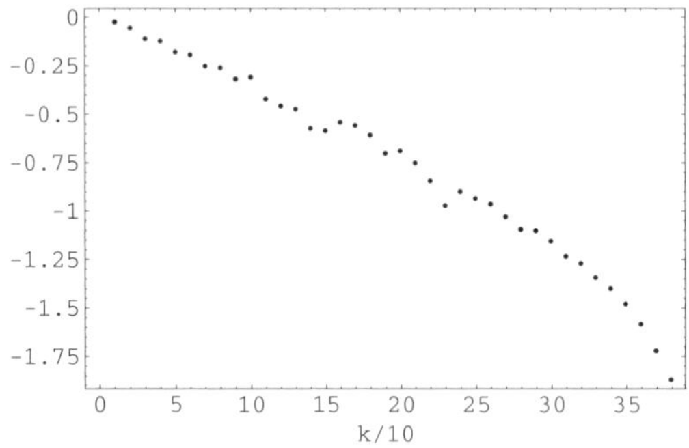

Figure 6: *Average value of logarithm of cross-validatory Bayes factor, when true* $\boldsymbol{\mu} = (0.8, 0.9, 0.5, 0.5, 0, 0, 0)$, $M_2 : \mu_7 = 0$, *other* μ_i'*s arbitrary. and* $M_1 : \mu_6 = \mu_7 = 0$, *other* μ_i'*s arbitrary.*

consistent with the heuristic argument in Section 4 that indicate (a) if $0 < k/r \leq 1$ then with probability tending to 1 the criterion will choose M_1 (under M_1), and (b) if $k/r \to 0$ in such a way that $k = o(r/p)$ then the cross-validatory Bayes factor will converge to 1 in probability. Fig. 8 illustrates this phenomenon.

6. SIMULATIONS ON AIC AND MODIFIED AIC

In this section, we report the findings of our simulation study on the role of an additional penalty for AIC (motivated by George and Foster, 2000) in all subsets model selection. This new criterion has been defined and theoretically investigated in Section 3. As in Section 3, we consider the following normal linear model setup

$$\mathbf{Y} = \mathbf{X}\boldsymbol{\beta} + \epsilon, \tag{28}$$

where $\epsilon \sim N(\mathbf{0}, \mathbf{I}_{n \times n})$, $\mathbf{Y}$ is an $(n \times 1)$ vector, $\mathbf{X}$ is an $(n \times p)$ matrix and $\boldsymbol{\beta}$ is a $(p \times 1)$ vector.

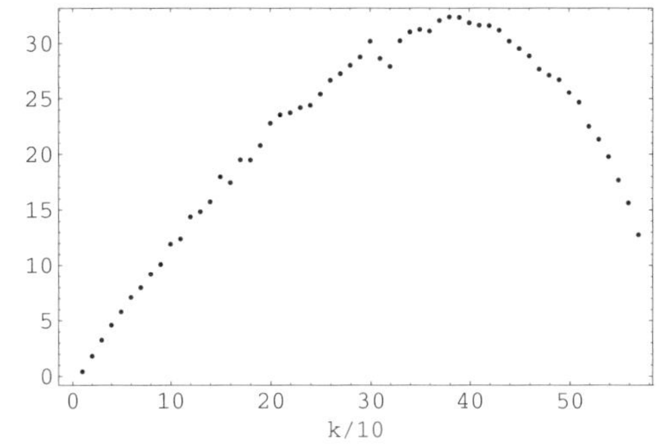

Figure 7: *Logarithm of cross-validatory Bayes factor, when M_2 is true and $p = 40$.*

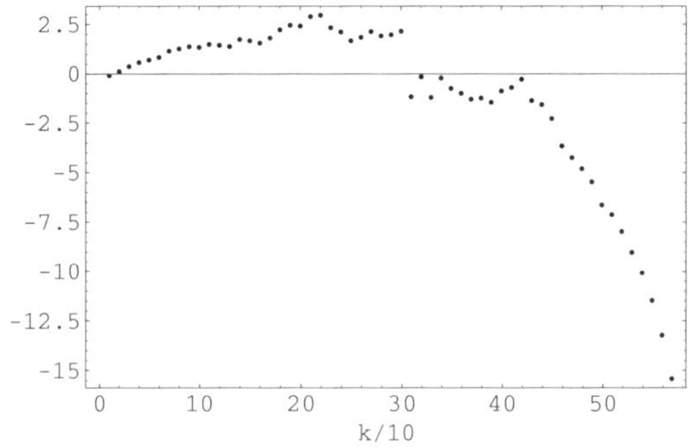

Figure 8: *Logarithm of cross-validatory Bayes factor, when M_1 is true and $p = 30$.*

In our simulations we choose $p = 10$ and $n = 50$. We first simulate a non-orthogonal $\mathbf{X}$ matrix, with $(\mathbf{X}_j, \mathbf{X}_{j+1})$ exchangeable for $j = 1, 2, 3, 4, 5$, where $\mathbf{X}_j$ is the j-th column of $\mathbf{X}$. This particular $\mathbf{X}$ matrix is used for all our calculations. We then fix a value, say $\boldsymbol{\beta}_0$, of $\boldsymbol{\beta}$, the true parameter value. Given this $\boldsymbol{\beta}_0$, 100 random vectors $\mathbf{Y}$ are simulated from $N(\mathbf{X}\boldsymbol{\beta}_0, I_{n\times n})$. For each such $\mathbf{Y}$, a model is selected among $\{2^{10} - 1 = 1023\}$ models using model selection by AIC, mAIC and the Oracle method (which chooses a model by minimizing over all models M, the quantity $||\mathbf{X}\boldsymbol{\beta}_0 - \mathbf{X}\hat{\boldsymbol{\beta}}_M||^2$, where $\hat{\boldsymbol{\beta}}_M$ is the least squares estimate under model M). For each such simulation step, we consider for each of these three methods, the loss $||\mathbf{X}\boldsymbol{\beta}_0 - \mathbf{X}\hat{\boldsymbol{\beta}}^s_{M_s}||^2$ in estmating $\mathbf{X}\boldsymbol{\beta}_0$ by $\mathbf{X}\hat{\boldsymbol{\beta}}^s_{M_s}$, where $\hat{\boldsymbol{\beta}}^s_{M_s}$ is the least squares estimate of β under model M_s chosen by that method in simulation s, where $1 \le s \le 100$. (Note that, this loss is, upto a constant, equal to the expected loss in predicting a future observation vector $\mathbf{Y}^{\text{new}}$ by $\mathbf{X}\hat{\boldsymbol{\beta}}^s_{M_s}$, given the realized value of $\mathbf{Y_s}$ in that particular simulation.) Note that here and above M_s is not a fixed model, but a random one depending on Y_s, chosen by the particular method under study. Finally an average is taken over these quantities over all s between 1 and 100. We repeated this whole procedure for different choices of $\boldsymbol{\beta}_0$ for the initially chosen design matrix $\mathbf{X}$ and the important findings are summarized in the following discussion and the simulation outputs presented in Table 1.

Table 1: *Simulation results for AIC, mAIC and Oracle methods.*

Beta	AIC	mAIC $w = 0.2$	mAIC $w = 0$	Oracle
{1,1,0,0,2,2,.5,.5,0,0}	(8.3,6.6)	(8.8,6.0)	(8.9,7.1)	(5.9,6.0)
{1,1,0,0,.1,.1,.5,.5,0,0}	(8.1,5.1)	(7.7,4.2)	(9.0,5.9)	(4.9,5.4)
{1,1,0,0,2,2,0,0,3,3}	(8.0,6.5)	(6.7,6.1)	(8.5,6.9)	(6.3,6.0)
{.05,.05,0,0,0,0,.02,.02,0,0}	(5.9,1.8)	(4.1,1.1)	(7.5,2.9)	(0.4,1.9)
{1,0,.5,0,.2,0,.1,0,.05,0}	(7.7,3.6)	(6.6,2.5)	(8.6,4.7)	(3.2,4.3)
{1,1,0,0,0,0,0,0,0,0}	(6.8,0.3)	(3.7,2.2)	(8.1,4.0)	(2.4,2.0)
{.5,.5,0,0,0,0,0,0,0,0}	(7.3,3.1)	(5.4,2.2)	(8.9,4.2)	(2.7,2.0)
{.1,.1,0,0,0,0,0,0,0,0}	(8.2,2.3)	(5.4,1.1)	(9.4,3.3)	(0.9,2.0)
{0,0,0,0,0,0,0,0,0,0}	(6.2,1.8)	(3.8,1.0)	(8.1,3.0)	(0.02,1)
{.05,.05,.02,.02,1,1,.02,.02,.01,.01}	(7.2,3.5)	(4.3,2.3)	(8.6,4.4)	(2.6,4.1)
{1,1,.5,.5,.2,.2,.1,.1,.5,.5}	(11.7,7.7)	(14.3,6.9)	(11.1,8.1)	(8.8,9.1)
{1,1,.5,.5,.2,.2,.1,.1,.05,.05}	(9.6,6.1)	(11.3,4.9)	(9.3,6.9)	(6.0,7.7)

In Table 1, we present the average losses and the average size of the model selected, for all the three methods (for $w = 0.2$ and $w = 0.6$ separately for mAIC) for 12 different choices of the true $\boldsymbol{\beta}_0$. For a given method and a given true $\boldsymbol{\beta}_0$, the first quantity inside the parantheses indicates the average loss in 100 simulations, whereas the second one indicates the average size of the models chosen by that method in 100 simulations.

It appears from our simulations that on the whole, based on average loss, mAIC with a small value of w does a substantially better job than mAIC with a larger w and AIC. Even in some cases with small value of w, where mAIC does not do as good as AIC or mAIC with a larger value of w, it is clear that the percentage reduction in average loss using the other methods is at best very moderate. If the statistician has some prior idea about the true value of $\boldsymbol{\beta}$, we feel that it might be helpful in deciding what sort of w to use in mAIC to get good results. The following paragraph tries to address this point based on our simulations.

It is generally observed that if the actual number of non-zero parameters is small or moderate, mAIC with small w performs better than AIC and mAIC with large w. Also, it appears that if the number of non-zero parameters is large, but the parameter values are not large, *i.e.*, when the signal is not a strong one, the same phenomenon occurs. On the other hand, if the number of non-zero parameters is large with a strong signal, then mAIC with large w does somewhat better than AIC and mAIC with a small w. AIC seems to do moderately better than mAIC with a large w in case of sparse signals and mAIC with small w when the number of non-zero parameters is large and the signal is strong. On the whole, mAIC with small w outperforms AIC in the case of sparse signals and non-sparse weak signals. From the average model size chosen by mAIC with small w in our simulations, it appears that the choice of a model with small dimension is facilitated by its use when more than one nested model is true or close to the true model.

7. CONCLUDING REMARKS

We first consider the cross-validatory Bayes factor (2) with fixed dimension p. If the minimum Kullback–Leibler divergences of M_1 and M_2 from the true density are not equal, then under regularity conditions, the larger the k, the better the discriminating power of the Bayes factor, subject to $n - k \to \infty$ and $k \to \infty$. The leading term is k times the difference in the (minimum) Kullback–Leibler divergences from the truth. Ideally, for best discriminatory power, $k \to \infty$, $n - k \to \infty$ in such a way that $k/(n-k) \to \infty$, but as long as $k \to \infty$, even if $k/n \to 0$, the model with bigger divergence from the truth will be selected. The simulations in Section 5.1 (Figs. 1, 3 and 4) bear this out. Under additional conditions, as in Proposition 3 in Section 4.1, similar results hold for the various sorts of averages over the set x_s. In other words, the basic intuition behind (1) and (2) is justified. As in the above case, there is considerable flexibility in the choice of k.

The case where the models are nested and both models are equally divergent, the leading term is zero. In this case the basic intuition behind (1) and (2) is no longer valid. For consistency in the sense parsimony, *i.e.*, for the Bayes factor to choose M_1, a sufficient condition is that $k/n \to 1$ in the limit. Simulations in Figs. 2, 6 in Section 5.1 bear this out. They also seem to indicate that the criterion might not be able to discrminate between the models well if in such a case $k/n \approx 0$.

If the dimension $p \to \infty$ in our specific nested linear model with normal error and the more complex model is true, then the true model is chosen chosen if $n \to \infty$, $n - k \to \infty$ and $k \to \infty$ in such a way that k/n is bounded away from zero in the limit. The simlation output in Fig. 7 in Section 5.1 bear this out. It is also evident from the plot that the criterion is very close to zero if $\frac{k}{n} \approx 0$, even though the difference in (minimum) Kullback–Leibler divergences from the truth for two models is non-zero. This is where, the high dimensional problem is different from the fixed dimensionl problem, since in a similar situation in the fixed dimensional problem, we proved that $k \to \infty$ suffices, no matter how slowly it goes to infinity. A heuristic argument shows that if the lower dimensional model is true, *i.e.*, if the (minimum) Kullback–Leibler divergence is same for both the models, the cross-validatory Bayes factor will converge to 1 in probability, *i.e.*, will not be able to discriminate between M_2 and M_1 if k/n is small compared with $1/p$. Also, it is shown heuristically that if $0 < k/n \leq 1$, asymptotically, then M_1 will be chosen asymptotically if M_1 is true.

We have also compared the AIC and the modified AIC in a predictive setting, with all subsets allowed in the models. There is some evidence from the simulations that the modified AIC with a relatively small w of 0.2 may be preferred on the grounds of parsimony. It would be interesting to have similar studies for what seems like a major generalization DIC (Deviance Information Criterion) of Spiegelhalter *et al.* (2002).

There are several interesting problems that we have not addressed. Suppose we choose all $\binom{n}{k}$ BF's and take an arithmetic or geometric mean of them. It would be interesting to have an asymptotic theory for these means and some comparison with the computationally more attractive and theoretically simpler averages based on $[n/k]$ disjoint x_s's.

Finally, our theoretical studies for the high dimensional case and simulation studies for both high and low dimensional cases have been done entirely for normal distributions. We do not know what effect other distributions would have on our conclusions. We have been told that the model selection rule of Spiegelhalter *et al.* (2002) seems to do better for exponential families than for mixtures. This may be due to the well-known difficulty of analysing mixtures through Bayesian or likelihood based methods.

A natural way to handle the $\mathcal{M}$-open case is through Bayesian nonparametrics. That's a promising direction for further research. We are aware that some work has been done by Walker but couldn't locate it.

REFERENCES

Berger, J. O. and Pericchi, L. R. (1995). The intrinsic Bayes Factor for linear models. *Bayesian Statistics 5* (J. M. Bernardo, J. O. Berger, A. P. Dawid and A. F. M. Smith, eds.) Oxford: University Press, 23–42 (with discussion).

Bernardo, J. M. and Smith, A. F. M. (1994). *Bayesian Theory.* Chichester: Wiley.

Bhattacharya, R. N. and Rao, R. R. (1976). *Normal Approximation and Asymptotic Expansions.* New York: Wiley.

Bogdan, M., Ghosh, J. K. and Doerge, R. W. (2004). Modifying the Schwarz Bayesian information criterion to locate multiple interacting quantitative trait loci. *Genetics* **167**, 989–999.

Chakrabarti, A. and Ghosh, J. K. (2006). A generalization of BIC for the general exponential family. *J. Statist. Planning and Inference* **136**, 2847–2872.

Craven, P. and Wahba, H. (1979). Smoothing noisy data with spline functions: Estimating the correct degree of smoothing by the method of generalized cross-validation. *Num. Math.* **31**, 377–403.

Geisser, S. (1975). The predictive sample reuse method with applications. *J. Amer. Statist. Assoc.* **70**, 320–328.

Gelfand, A. and Dey, D. (1994). Bayesian model choice: asymptotics and exact calculations. *J. Roy. Statist. Soc. B* **56**, 501–514.

George, E. I. and Foster, D. P. (2000). Calibration and empirical Bayes variable selection. *Biometrika* **87**, 731–747.

Ghosh, J. K and Samanta, T. (2002). Nonsubjective Bayes testing – An overview. *J. Statist. Planning and Inference* **103**, 205-223.

Gu, C. (2002). *Smoothing Spline Anova Models.* New York: Springer-Verlag

HastieT., Tibshirani, R. and Friedman, J. (2001). *The Elements of Statistical Learning.* New York: Springer-Verlag.

Key, J. T., Pericchi, L. R. and Smith, A. F. M. (1999). Bayesian model choice: what and why? *Bayesian Statistics 6* (J. M. Bernardo, J. O. Berger, A. P. Dawid and A. F. M. Smith, eds.) Oxford: University Press, 343–370 (with discussion).

Li, K. C. (1986). Asymptotic optimality of C_L and generalized cross-validation in the ridge regression with application to spline smoothing. *Ann. Statist.* **14**, 1101-1112.

Mukhopadhyay, N. D. (2000). Bayesian model selection for high dimensional models with prediction error loss and $0-1$ loss. Thesis submiitted to Purdue University, Department of Statistics.

Mukhopadhyay, N. D., Ghosh, J. K. and Berger, J. O. (2005). Some Bayesian predictive approaches to model selection. *Statist. Probab. Lett.* **73**, 369–379.

Shao, J. (1993). Linear model selection by cross-validation. *J. Amer. Statist. Assoc.* **88**, 486–494.

Spiegelhalter, D. J., Best, N. G., Carlin, B. P and van der Linde, A. (2002). Bayesian measures of model complexity and fit. *J. Roy. Statist. Soc. B* **64**, 583–649 (with discussion).

Stone, M. (1977). An asymptotic equivalence of choice of model by cross-validation and Akaike's criterion. *J. Roy. Statist. Soc. B* **39**, 44–47.

Vapnik, V., N. (1998). *Statistical Learning Theory.* New York: Wiley.

DISCUSSION

STEFFEN L. LAURITZEN (*University of Oxford, UK*)

Discussion of model selection can be confusing because 'model' has many meanings in the statistical literature, and indeed in the article under discussion. Statisticians tend to use this word in ways different from any other scientist. It seems therefore worthwile to dwell on this variation in terminology before we proceed. Examples of use of the term are

- There is a *true model* P_0 which typically has a frequentist interpretation so its probability statements refer to a conceptual infinite ensemble of repeated observations.
- A *Fisherian model* states that this true model P_0 belongs to a specified family of probability distributions $\mathcal{P} = \{P_\theta : \theta \in \Theta\}$ but is otherwise unknown.
- A *Bayesian model* adds a prior distribution over $\mathcal{P}$ to the Fisherian model, quantifying the prior uncertainty associated with the unknown P_0.
- A *Bayesian subjective model* expresses individual beliefs for observables through a personal probability distribution P.

These and other concepts of model are different but related and much conflict and confusion is due to mixing up the concepts. To focus the discussion of the present article we choose to set a *Bayesian scene.*

Fundamental model. The fundamental model for our focus begins with a sequence $X = (X_1, X_2, \ldots)$ of *observables* and a probability distribution P which is *Your Bayesian model* for X, *i.e.*, a Bayesian subjective model as just described above.

This P is based on Your insight into the problem considered, it is specifying the settings for Your *reasoning and ideal inference* and it is *not the object of automatic selection* of any kind. Your *predictive distribution* is then

$$p(y_1, \ldots, y_k \mid x_1, \ldots, x_n) = \\ P(X_{n+1} = y_1, \ldots, X_{n+k} = y_k \mid X_1 = x_1, \ldots, X_n = x_n).$$

In principle this is the end of the prediction story in a Bayesian setup and no further arguments, principles, methods, or selection procedures are needed. However, some elaboration is necessary to transfer this fundamental chain of reasoning into practical use.

deFinetti–Hewitt–Savage Theorem. For simplicity we shall henceforth assume that P is *exchangeable:*

$$p(x_1, \ldots, x_n) = p(x_{\pi(1)}, \ldots, x_{\pi(n)}), \; \forall \pi \in S(n), n = 1, 2, \ldots$$

corresponding most closely to the setting in the article under discussion. A fundamental theorem (deFinetti 1931, Hewitt and Savage 1955) now says that, under exchangeability and very general regularity conditions, we have

(i) $p(x_1, \ldots, x_n) = \int_{\mathcal{P}(\mathcal{X})} \prod_{i=1}^n q(x_i) \mu(dQ)$ for some $\mu \in \mathcal{P}(\mathcal{P}(\mathcal{X}))$;

(ii) $\lim_{n \to \infty} \frac{1}{n} \sum_{i=1}^n \delta_{X_i} = Q_0 \in \mathcal{P}(\mathcal{X})$;

(iii) $\lim_{n\to\infty} p(y_1, \ldots, y_k \,|\, X_1, \ldots, X_n) = \prod_{i=1}^k q_0(y_i)$;

(iv) $Q_0 \sim \mu$.

The random distribution Q_0 is referred to as the *'true' distribution.* This Q_0 is unobservable, but has a *frequency interpretation* from (ii), and is also the *limiting predictive distribution* from (iii). The measure μ is the *prior distribution* for Q_0. In this way, the theorem creates a bridge between the Fisherian and Bayesian models. We omit details concerning the precise form of limits and necessary technical conditions.

Model selection. The problem is that the subjective probability distribution P is too complex and therefore unavailable. Following Bernardo and Smith (1994), Chapter 6, we wish to replace P by an approximation P_j, where

$$p_j(x_1, \ldots, x_n) = \int_{\Theta_j} \prod_{i=1}^n p_j(x_i \,|\, \theta_j)\mu_j(d\theta_j), j \in \mathcal{J}$$

where $\mathcal{M}_j \sim (\Theta_j, \mu_j), j \in \mathcal{J}$ are possible models. *Bayesian model selection* refers to the process of selecting one of P_j as an approximation to P, using observed data. Generally this is done using the *Bayes factor*

$$\mathrm{BF}_{jl}^n = \frac{p_j(x_1, \ldots, x_n)}{p_l(x_1, \ldots, x_n)} \cdot$$

The Bayes factor between the models M_j and M_l is in fact equal to the *likelihood ratio* between the models and another name is only needed to emphasize that the likelihood ratio is calculated for a non-Fisherian model. Again, this settles the model selection problem in principle.

Model selection for prediction. Following Bernardo and Smith (1994), model selection is best discussed in the context of a well-defined decision problem. Predicting a single observation with logarithmic scoring rule, the *risk associated with using the model* $\mathcal{M}_j$ is

$$R_j(\boldsymbol{x}_n) = \int_{y\in\mathcal{X}} \log p_j(y \,|\, \boldsymbol{x}_n) P(dy \,|\, \boldsymbol{x}_n),$$

where $\boldsymbol{x}_n = (x_1, \ldots, x_n)$, and models should be chosen to yield a small risk. *Cross-validation* is approximating this integral by (possibly by Monte Carlo) to

$$\mathrm{CV}_{n-1}^1(j) = \frac{1}{n}\sum_{i=1}^n \log p_j(x_i \,|\, \boldsymbol{x}_{n-1}(i)),$$

where $\boldsymbol{x}_{n-1}(i)$ is obtained from $\boldsymbol{x}_n$ by leaving out x_i (Bernardo and Smith 1994). The justification for this approximation exploits that for large n, $\boldsymbol{x}_n \approx \boldsymbol{x}_{n-1}(i)$ and

$$p(y \,|\, \boldsymbol{x}_n) \approx q_0(y) \approx \frac{1}{n}\sum_{i=1}^n \delta_{x_i},$$

i.e., a type of 'Bayesian bootstrap'. It is not clear to me that this approximation is valid and I wonder (or even doubt) whether it can be given a formal justification?

The criterion can be generalized to predicting k observations and possibly using non-Bayesian predictors $\tilde{p}_j(\boldsymbol{y}_k; \boldsymbol{x}_n)$ as

$$\mathrm{CV}^k_{n-k}(j) = \mathrm{ave}\{\log \tilde{p}_j(\boldsymbol{x}_a; \boldsymbol{x}_b)\},$$

where $a \cup b = \{1, \ldots, n\}$, $|a| = k$, and $|b| = n - k$, leading to cross-validatory Bayes factors. The article under discussion addresses (among other things) *how to choose* k, partly through introducing a new selection criterion, *mAIC*.

Alternative criteria. It seems illuminating to consider the family of criteria

$$\mathrm{CV}_w = \sum_{a \cup b = \{1, \ldots, n\}} w(a, b) \log \tilde{p}_j(\boldsymbol{x}_a; \boldsymbol{x}_b),$$

where w is a system of weights. The previously defined criteria CV^k_{n-k} occur when w is uniform on subsets where $|a| = k$ and $|b| = n - k$. In the following we shall argue that there are other interesting candidates for these weights. Respecting exchangeability suggests we should insist on

$$w(a, b) = r(|a|, |b|)$$

so weights only depend on size.

Prequential validation. Dawid (1984) suggests validating prediction schemes through their performance in a sequential setting. Using again a logarithmic score, this is calculated as

$$\mathrm{PV}_n(j) = \sum_{i=1}^{n} \log \tilde{p}_j(x_i; \boldsymbol{x}_{i-1}).$$

This criterion has a number of desirable properties. However, in general *it does not respect exchangeability* in the sense just described and depends on the particular order in which the x_i appear.

Prequential cross-validation. It seems tempting to symmetrize PV by combining over all permutations

$$\mathrm{PCV}_n(j) = \frac{1}{n!} \sum_{\pi \in S(n)} \sum_{i=1}^{n} \log \tilde{p}_j(x^\pi_i; \boldsymbol{x}^\pi_{i-1}),$$

with $x^\pi = (x_{\pi(1)}, \ldots, x_{\pi(n)})$. A Monte Carlo approximation to this can generally be computed easily. However, PCV simplifies in important special cases as we shall see below.

Plug-in schemes. Consider first the special case where the prediction scheme has the form

$$\tilde{p}_j(\boldsymbol{y}_k; \boldsymbol{x}_{n-k}) = \prod_{l=1}^{k} \tilde{p}_j(y_l; \boldsymbol{x}_{n-k}).$$

This is for example the case for *plug-in* schemes

$$\tilde{p}_j(\boldsymbol{y}_k; \boldsymbol{x}_{n-k}) = \prod_{l=1}^{k} p_j\{y_l \,|\, \hat{\theta}_j(\boldsymbol{x}_{n-k})\},$$

where $\hat{\theta}_j(\boldsymbol{x}_{n-k})$ is some estimator of $\theta_j \in \Theta_j$. A simple combinatorial argument shows that

$$\mathrm{PCV}_n(j) = \sum_{k=1}^{n-1} \mathrm{CV}_{n-k}^k = \mathrm{CV}_w(j), \quad \text{for} \quad w(a,b) \propto \binom{|b|}{|a|}^{-1}. \tag{29}$$

In other words, this leads to *uniform weights* on cross-validative criteria based on 'leave k out' validation for different k, rather than actually choosing k, as suggested in the article under discussion, and I wonder whether there is any specific reason to choose a fixed k in the cross-validatory criteria?

Note that the quantity in (29) can be approximately calculated by a simple Monte Carlo procedure. Just choose k uniformly from $\{1, \ldots, n-1\}$ and subsequently choose a uniform among subsets of size k. See also Gneiting and Raftery (2005) for similar arguments.

Bayesian predictors. As noted by Dawid (1984), any fully Bayesian predictor satisfies

$$\sum_{i=1}^{n} \log p_j(x_i \,|\, \boldsymbol{x}_{n-i}) = \log p_j(x_1, \ldots, x_n)$$

so that then

$$\mathrm{PCV}_n(j) = \mathrm{PV}_n(j) = \log p_j(x_1, \ldots, x_n)$$

and in particular

$$\mathrm{PCV}_n(j) - \mathrm{PCV}_n(l) = \log \mathrm{BF}_{jl}^n, \tag{30}$$

i.e., the usual *Bayes factor* appears again, which seems particularly appealing.

Approximate calculation of Bayes factors. If we turn things around we now get a simple approximation to the Bayes factor:

$$\log \mathrm{BF}_{jl}^n \approx \mathrm{PCV}_n(j) - \mathrm{PCV}_n(l),$$

where the expression on the right-hand side can be calculated with Monte Carlo methods.

In addition, since the relation (30) is exact for coherent Bayesian procedures, it is tempting to conclude that the Monte Carlo variance can be used as a *measure of incoherence* of the prediction method.

In any case, it seems to this discussant that the arguments are not convincingly strong for introducing other criteria than the classical Bayes factor, or likelihood ratio, and approximations to it.

DAVID DRAPER (*University of California at Santa Cruz, USA*)

I would like to expand the scope of discussion of this interesting and valuable paper by placing it in a wider context. The authors focus on cross-validatory Bayes factors for prediction as the sole approach to Bayesian model selection they examine, but would they not agree that the topic of Bayesian model specification is broader than that? There appear to be two different fundamental questions at the heart of Bayesian model selection:

(1) Is model M_1 better than M_2? (This tells us how to discard models in the specification search; asking this pairwise question recursively is sufficient to find the best model in a class of model structures of any finite size.)

(2) Is M_1 good enough? (This provides a stopping rule for the specification search.)

It would seem almost self-evident that it is not possible to answer either of these questions without being explicit about the purpose to which the final model chosen will be put; for example, if you are asked 'Is M_1 good enough?' the only possible answer is to reply with the follow-up question 'Good enough for what?' This means (*e.g.*, Draper, 1996, Key *et al.*, 1999) that model specification is really a decision problem, and must be approached via maximization of expected utility after specifying a utility structure that is appropriate to the purpose of the model selection exercise. In some cases (e.g., Draper and Krnjajić, 2006) a predictive log scoring approach like the one taken by Chakrabarti and Ghosh can be given an explicit utility justification, but in other cases involving Bayesian model selection the utility structure implied by this approach is not appropriate. An example is given by Draper and Fouskakis (2000); also see Fouskakis and Draper, 2002), in which the problem of variable selection in generalized linear models with a data collection cost constraint is solved by specifying a cost-benefit utility structure that selects predictor variables only if they predict well enough given how much it costs to collect their data values; the model selection approach taken in the paper discussed here will fail in the problem studied by Draper and Fouskakis, because lack of parsimony—the only implicit or explicit cost accounted for in the methods examined by Chakrabarti and Ghosh–is the wrong measure of cost in the Draper–Fouskakis problem. Thus utility is the key in Bayesian model specification, not cross-validatory Bayes factors (or indeed any other kind of Bayes factors, unless their implied utility structure is appropriate to the task at hand). It may seem reasonable to focus on questions such as 'Which model is the most probable, given the data?' (which Bayes factors based on quantities of the form $p(M_i|\text{data})$ address) or 'Which model makes the best predictions, given the data?' (which model selection criteria based on the predictive log score address), but these are not always the right questions to ask in Bayesian model specification.

LUIS R. PERICCHI (*Universidad de Puerto Rico at Rio Piedras, Puerto Rico*)

Is there an optimal training sample size in Bayesian hypothesis testing and model selection?

This is a main question that the authors set themselves to explore. This article by Professor Ghosh and collaborators deserves careful study and assimilation. They are certainly among the serious thinkers in the areas of Bayesian hypothesis testing and model selection.

This contribution is an initial reaction to their work. I follow the author's unconventional notation of calling k the validatory sample size and $n-k$ the training sample size.

I motivate here a crucial point: *there is a (one to one) correspondence between training sample size and the precision of the (implicit or intrinsic) prior.* To do this we need an alternative method to Laplace's asymptotics, that is also suitable for finite training or validatory samples: the alternative is what has been called the

Intrinsic Prior method, which follows. Let's call AVE a generic average,

$$AV_{ij} = AVE_l \left[\frac{f_i(x^l_{(k)}|x^l_{(-k)})}{f_j(x^l_{(k)}|x^{(l)}_{(-k)})} \right],$$

of $l : 1, \ldots, L$ training samples of size $n - k$. The authors confine themselves to the logarithm average, but other averages, notably the arithmetic average has been found important (Berger and Pericchi, 1996). Notice the fundamental relationship

$$f_i(x^l_{(k)}|x^l_{(-k)}) = \frac{f_i^N(x)}{f_i^N(x^l_{(-k)})},$$

so that

$$AV_{ij} = \frac{f_i^N(x)}{f_j^N(x)} \times AVE_l \frac{f_j^N(x^l_{(-k)})}{f_i^N(x^l_{(-k)})},$$

where the superscript N means that an objective prior (typically non-integrable in the overall parameter space) has been employed. Take the arithmetic average for instance. Under some conditions we have,

$$AV_{ij} \to \frac{f_i^N(x)}{f_j^N(x)} \times E^{f_S(X(-k)|\theta_S)} \left[\frac{f_j^N(x^l_{(-k)})}{f_i^N(x^l_{(-k)})} \right],$$

where $f_S(X(-k)|\theta_S)$ denotes the sampling model. When the expectation can be calculated (under model M_i) then the simplest solution of the Intrinsic Prior equations is (Berger and Pericchi, 1996):

$$\pi^I(\theta_i) = \pi^N(\theta_i) \times E^{f_S(X(-k)|\theta_S)} \left[\frac{f_j^N(x^l_{(-k)})}{f_i^N(x^l_{(-k)})} \right],$$

$$\pi_j^I(\theta_j) = \pi_j^N(\theta_j).$$

Let us see a simple instance. Consider the example of a Normal mean μ and known variance σ_0, and $H_0 : \mu = \mu_0$ VS $H_1 : \mu \neq \mu_0$. This situation, for Arithmetic Average, is covered by Theorem 1 of Berger and Pericchi (1996) so that the Intrinsic Prior is fully proper, and in fact it is given by:

$$\pi^I(\mu) = \mathrm{N}\left(\mu \,\middle|\, \mu_0\,, \frac{2\sigma_0^2}{(n-k)}\right).$$

(The geometric average, leads to a Normal prior mean μ_0 and variance $\sigma_0^2/(n-k)$ but multiplied by a (modest) constant, equal to $\exp[-0.5]$.) Observing $\pi^I(\mu)$ a first natural assumption is that $(n-k)$ is one, or small and fixed (with the sample size) to be a sensible prior, at least in absence of substantive additional information, like that the data had been produced by a careful designed experimental procedure, as suggested by Ghosh and Samantha in their discussion of Berger and Pericchi (2001). Other considerations are also possible, like the interest on reducing the variance of the Bayes Factor.

If we follow the arithmetic average (the geometric would lead to somewhat similar conclusions) the Bayes Factor based on Intrinsic Priors is equal to:

$$B_{10}^{I}(n-k) = \frac{N(\bar{y}|\mu_0, \sigma_0^2(1/n + 2/(n-k)))}{N(\bar{y}|\mu_0, \sigma_0^2/n)} .$$

This expression imposes very specific conditions over k in order that the the Bayes Factor is consistent under H_0, for example (as the authors suggest) that k cannot be fixed with n. Stronger conclusions can be extracted from the expression of B_{10}^{I}. These conditions and related discussions, under more general settings, will be published elsewhere.

AKI VEHTARI (*Helsinki University of Technology, Finland*)

Congratulations to the authors for an interesting paper. The authors consider model selection for prediction. My comments concern the different predictive assumptions made in the criteria used.

The criterion (1) estimates the difference in predictive performance when trying to predict just the next single observation given the observed data. I think this is natural choice, since in real applications after obtaining the next observation we could update the posterior or in many cases the model is used for many predictions without updating the posterior. The latter case happens, for example, if in regression or classification we observe future values of x but not y (or y is observed with long delay).

The average of criterion (2) estimates the difference in predictive performance when trying to predict jointly k future observations given the observed data with an effective sample size $n-k$. In some applications predicting several observations jointly is useful, for example, in time series prediction for predicting the next k time steps, where k is defined by the problem. Using the chain rule, the joint prediction of the next k observations corresponds to predicting the observations sequentially one at a time and updating the posterior after each observation. In some applications, model could be used like this but then the selection of k defines how many times the posterior updating will be done. In criterion (2) each k corresponds to a different prediction problem, and considering k just as a parameter in a model selection method ignores how the model is actually going to be used.

The predictive performance of the model depends on how much data was available for updating the posterior distribution. The learning curve can be used to describe how much additional data would improve the predictive performance. In the case of flexible models (like non-parametric models) the learning curve has a considerable difference in predictive performance with sample sizes n and $n-k$ with large k. If n is large and the learning curve is relatively flat between the values $n-k$ and n, then there is no large difference in predictive performance given sample size of $n-k$ or n, and thus more data can be used as a proxy for future observations. When considering asymptotic results the learning curve is flat when $n \to \infty$. The models used in simulations in the paper are quite simple and the sample size compared to the number of parameters in the models is relatively large, and thus it is reasonable to assume that the learning curve is flat between $n-k$ and n for moderate values of k. In the case of flexible complex models and small or moderate n, the learning curve is not flat and thus asymptotic results and simulations results in the paper are not necessarily helpful for deciding which value of k to use.

Dependence on k is further increased by the way authors compute the average of the criterion (2) over $[\frac{n}{k}]$ groups. Each group contains k observations and thus each group contribute k terms (via chain rule). With $[\frac{n}{k}]$ groups a total number of terms is n. Taking the average over groups means that the n terms are divided by $[\frac{n}{k}]$ instead of more natural choice of n, which would give the average contribution of a single observation. The averaging used by the authors could explain the near linear behaviour of the criterion with moderate values of k before other above-mentioned issues start to have larger effect.

It is possible to consider the prediction problem stated by criterion (1) but still due to computational reasons leave k samples out at time giving a criterion

$$\frac{1}{n}\sum_{j=1}^{n}\log\frac{p_2(x_j|x_{s(j)^c}, M_2)}{p_2(x_j|x_{s(j)^c}, M_1)},$$

where $s(j)$ is group of k observations which includes the j-th observation. This criterion still assumes that we are interested predicting just the single next observation. In this case too, predictions are conditioned on sample size $n-k$, and above discussion concerning the learning curve is valid. For this criterion it is possible to make a first order estimate of the learning curve and compute a corrected estimate (see, *e.g.*, Vehtari and Lampinen, 2003).

Comparisons were made to AIC, which is based on maximum likelihood, and assumes the $\mathcal{M}$-closed case. I think it would be more relevant to make comparisons to NIC (Murata *et al.*, 1994), which may be computed using maximum a posteriori and assumes the $\mathcal{M}$-open case. Note that, similar equation was also given by Takeuchi (1976), and Stone (1977, eq. 4.5). See also review of NIC by Ripley (1996).

REPLY TO THE DISCUSSION

We completely agree with the comments of Prof. Draper about the need to specify the purpose of model selection and choose an appropriate utility function. However, our goal is not to propose new model selection criteria with appropriate utilities. We look at criteria already in use and ask how in this context we may answer Prof. Pericchi's question. We also appreciate Prof. Pericchi's exact, rather than asymptotic, arguments that seems to support some of our general conclusions.

Dr. Vehtari raises interesting points which we study below after clarifying that the question we try to answer seems to us somewhat different from his points. We are not trying to estimate the integral

$$I_1 = \int \log\frac{p_2(y|\mathbf{x})}{p_1(y|\mathbf{x})} p(y|\mathbf{x})\, d\, y,$$

as in Bernardo and Smith (1994, pp. 404), where $\mathbf{x} = (x_1, \ldots, x_n)$ is the observed sample, y is a future observation, p is the posterior predictive density under the true model and p_2 and p_1 are those under the competing models M_2 and M_1 respectively. Indeed only (1) is a natural estimate of I_1 as $n \to \infty$. The average of our general Cross-Validatory Bayes Factor (CVBF) of order k (see (2)) shown in the l.h.s. of (19), may be considered to be an estimate of the analogous integral

$$I_k = \int \log\frac{p_2(y_1, \ldots, y_k|\mathbf{x})}{p_1(y_1, \ldots, y_k|\mathbf{x})} p(y_1, \ldots, y_k|\mathbf{x})\, d\, y_1 \ldots d\, y_k,$$

not of I_1 as defined before. In the context of I_k's or more directly in terms of the average Cross-Validatory Bayes Factor, vide equation (20), we try to decide which k to choose. Our answer in the paper is based on the discriminatory power of average CVBF's, not quite on their quality as estimates of I_1 or I_k, though we agree with Dr. Vehtari that both are related. Another focus has been on the simple Figs. 1-8 of Section 5, which seem to contain all the information in our propositions and suggest which range of values of k are suitable for leaving out.

It is indeed true, as suggested by Dr. Vehtari, that for estimation of the integral I_1, the average of leave $k = 1$ out CVBF is the most natural one and for average of CVBF with arbitrary k, an additional divisor of k is needed. We examine these renormalized average CVBF's, henceforth called CVBF$_k$'s through figures analogous to those in the paper.

Since I_1 is not easy to calculate (because it is typically not possible to find a closed form expression of $\log[p_2(y|\mathbf{x})/p_1(y|\mathbf{x})]$ and it is also random because of its dependence on $x_1, \ldots, x_n$, we present below two different approaches to see how CVBF$_k$ performs in the case when the difference between the (minimum) Kullback–Leibler divergence of the two competing models from the truth is nonzero. We first offer an approximate calculation based on the fact that

$$I_1 \approx \min_{\boldsymbol{\theta}_1 \in \boldsymbol{\Theta}_1} K(f, f_{1,\boldsymbol{\theta}_1}) - \min_{\boldsymbol{\theta}_2 \in \boldsymbol{\Theta}_2} K(f, f_{2,\boldsymbol{\theta}_2}).$$

A plot of $(\text{CVBF}_k - I_1)^2$ for different k's is shown below in Fig. 9 . We get a curve with a minimum at a very large and very small k. If we do not approximate I_1, but calculate it by Monte Carlo, we get a similar picture (see Fig. 10) below. For both these figures, $n = r = 100$ is used, and the same true model and setup as in Fig. 1 of Section 5.1 are used.

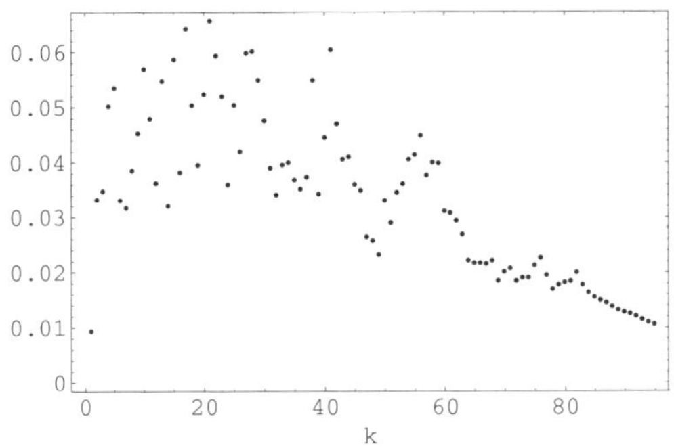

Figure 9: *Plot of squared difference of CVBF$_k$'s from* $K(f, f_{1,\boldsymbol{\theta}_{10}}) - K(f, f_{2,\boldsymbol{\theta}_{20}})$, *when true* $\boldsymbol{\mu} = (0.8, 0.9, 0.5, 0.5, 0.9, 0.7, 0)$, $M_2 : \mu_1 = \mu_2 = \mu_3 = \mu_7 = 0$, *other* μ_i'*s arbitrary and* $M_1 : \mu_4 = \mu_5 = \mu_6 = \mu_7 = 0$, *other* μ_i*'s arbitrary.*

We have also redrawn some of our figures with our original criterion replaced by CVBF$_k$'s. It is observed that when the two models have different (minimum) Kullback–Leibler divergence from the truth, the minimum (or maximum) is attained at values of k which are neither too large or too small. Figures 11 and 12 below

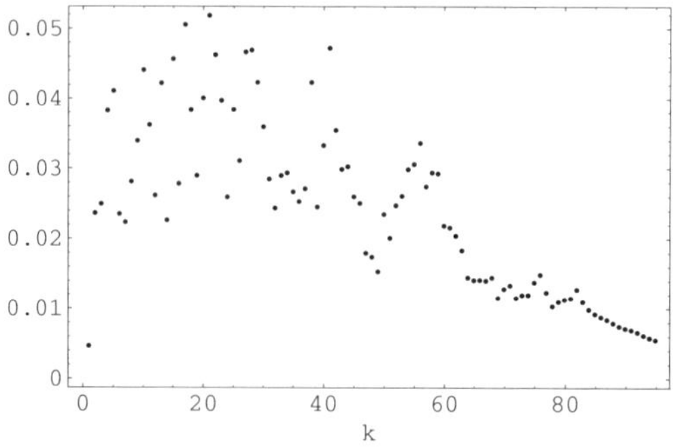

Figure 10: *Plot of squared difference of $CVBF_k$'s from I_1, when true* $\boldsymbol{\mu} = (0.8, 0.9, 0.5, 0.5, 0.9, 0.7, 0)$, $M_2 : \mu_1 = \mu_2 = \mu_3 = \mu_7 = 0$, *other* μ_i*'s arbitrary, and* $M_1 : \mu_4 = \mu_5 = \mu_6 = \mu_7 = 0$, *other* μ_i'*s arbitrary.*

present two such simulations, with $n = r = 100$ and the same setup as in Fig. 1 of our paper. In Fig. 11, $K(f, f_{1,\boldsymbol{\theta}_{10}}) - K(f, f_{2,\boldsymbol{\theta}_{20}}) = -0.075$, while that in Fig. 12 is 1.5.

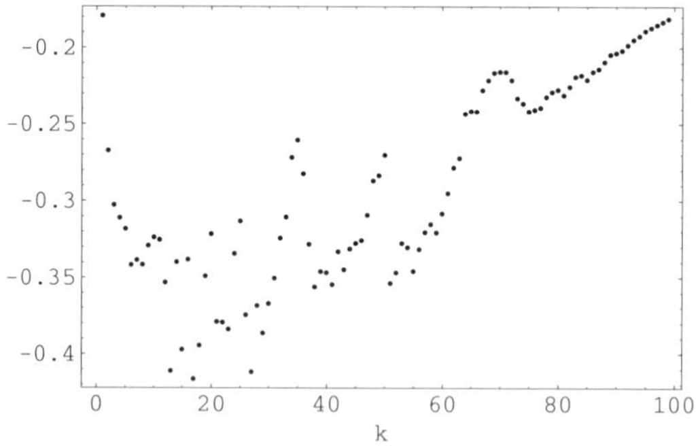

Figure 11: *Plot of $CVBF_k$'s for different k's, when true* $\boldsymbol{\mu} = (0.8, 0.9, 0.5, 0.5, 0.9, 0.7, 0)$, $M_2 : \mu_1 = \mu_2 = \mu_3 = \mu_7 = 0$, *other* μ_i*'s arbitrary, and* $M_1 : \mu_4 = \mu_5 = \mu_6 = \mu_7 = 0$, *other* μ_i'*s arbitrary.*

When both models are equally divergent, we don't see much point in estimating the integral I_1, which is approximately zero. In this case parsimony requires that we choose the simpler model, *i.e.*, the criterion should be negative if M_1 is the simpler model (as in Figs. 13 and 14 below). We remarked in our paper that 'the criterion might not be able to discriminate between the models if $k/n \approx 0$'. This fact is further substantiated by Figs. 13 and 14 below, when for $k/n \approx 0$, our original criterion is seen to choose both the smaller dimensional model and the larger model in a nested $\mathcal{M}$-open scenario same as in Fig. 2 of the paper. In Fig. 13, $n = r = 100$ is used while in Fig. 14, $n = r = 1000$ is used. Also, in all these simulations, when k/n is large, the parsimonious model is selected as desired. By investigating this phenomenon theoretically, we proved that for $k = 1$ in some situations our criterion will in fact choose the larger model with probability bounded away from zero asymptotically. (But we must comment here that in many simulations, as in

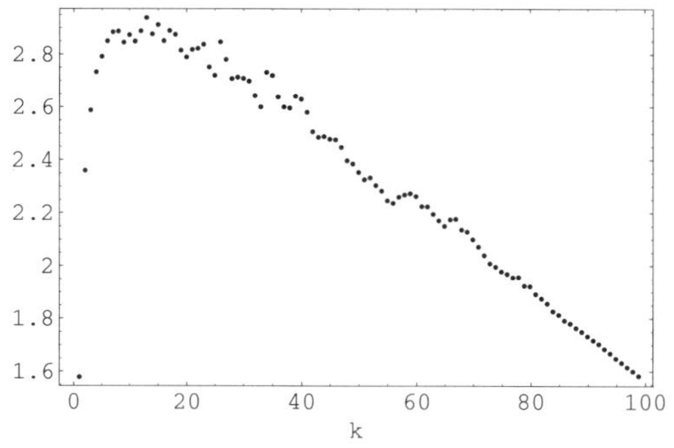

Figure 12: *Plot of $CVBF_k$'s for different k's, when true* $\boldsymbol{\mu} = (1, 0.9, 0.5, 0.5, 0.9, 2, 0)$, $M_2 : \mu_1 = \mu_2 = \mu_3 = \mu_7 = 0$, *other* μ_i'*s arbitrary and* $M_1 : \mu_4 = \mu_5 = \mu_6 = \mu_7 = 0$, *other* μ_i'*s arbitrary.*

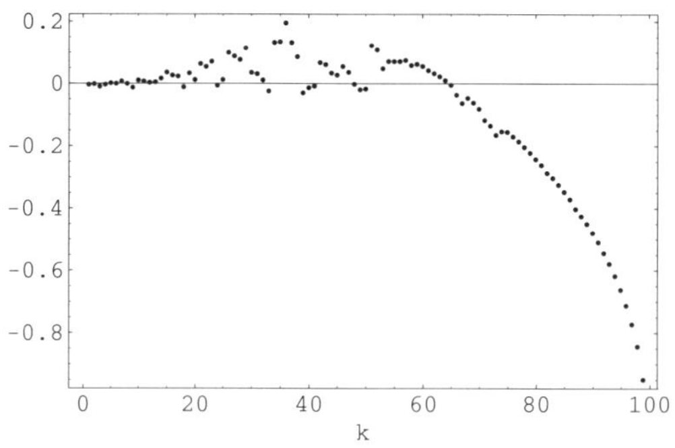

Figure 13: *Average value of logarithm of cross-validatory Bayes factor, when true* $\boldsymbol{\mu} = (0.5, 0.6, 0.05, 0.02, 0.01, 0, 0)$, $M_2 : \mu_5 = \mu_7 = 0$, *other* μ_i'*s arbitrary and* $M_1 : \mu_5 = \mu_6 = \mu_7 = 0$, *other* μ_i'*s arbitrary.*

Fig. 2, we observed that the smaller model is chosen for all k's as should be the case, but phenomena as in Figs. 13 and 14 are also observed frequently.)

Taking into account the above discussion as well as the fact that we will not know in advance whether the (minimum) Kullback–Leibler divergences are same or different, a very large k seems desirable, in fixed-dimensional problems.

Professor Lauritzen has proposed interesting new Bayes Factors by starting with the prequential predictive set-up of Dawid and then imposing exchangeability by considering all permutations of the data. One of the Bayes Factors contains a sum over k, which seems undesirable in view of our discussion above in the nested case. Professor Lauritzen also makes an interesting comment that the variance of a Cross-Validatory Bayes Factor may be a measure of its incoherence.

Professor Lauritzen asks whether the CVBF_k with $k = 1$ is consistent for the integral in Bernardo and Smith (1994) (for predicting a single y based all $x_1, \ldots, x_n$ and logarithmic utility). Yes, it consistent, in that both these quantities converge to the same limit, namely the difference of the minimum Kullback–Leibler divergence of the two models from the true density. But if that is the goal there seems to be other reasonable estimates also as our answer to Dr. Vehtari shows.

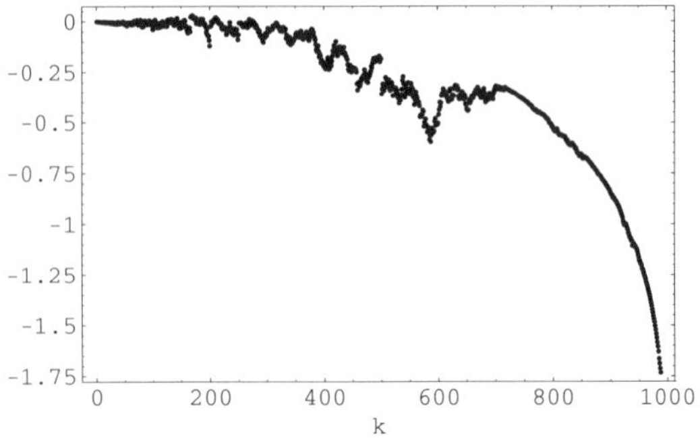

Figure 14: *Average value of logarithm of cross-validatory Bayes factor, when true* $\boldsymbol{\mu} = (0.5, 0.6, 0.05, 0.02, 0.01, 0, 0)$, $M_2 : \mu_5 = \mu_7 = 0$, *other* μ_i*'s arbitrary, and* $M_1 : \mu_5 = \mu_6 = \mu_7 = 0$, *other* μ_i'*s arbitrary.*

Our goal was different. As explained in our reply to Prof. Draper and Dr. Vehtari, we look at the variety of Cross Validatory Bayes Factors that have been proposed and examine them in, say, the $\mathcal{M}$-open case, and explore which choice of k seems to give good discriminatory power between two models. We would like to choose the model nearer to f, if both models are not equally divergent. In the equally divergent case, we would like to choose the more parsimonious model. This was our interpretation of Prof. Pericchi's question, quoted in the paper.

Prof. Lauritzen also suggests that the usual Bayes Factor or an average of the Cross Validatory Bayes Factors over k is to be preferred. The answer to this question is that our work, theoretical and numerical, shows that the (cross-validatory) Bayes Factors are not equally good. One should choose k suitably, not average over k. About the suggestion regarding the usual Bayes Factor, it must be pointed out at the very beginning that it is not well defined if any of the priors is improper since an improper prior is defined up to an arbitrary constant. In this case conditioning with respect to some observations produces a proper posterior which is used as a surrogate prior for the CVBF's. Also it is believed (see O'Hagan (1995) that the CVBF's are more robust than the usual BF with respect to choice of prior, even in the case of proper priors, at least for small or moderate n. In the $\mathcal{M}$-open case the usual Bayes Factor has no special normative status, at least as far as we know. We believe Bernardo and Smith (1994) were the first to address this problem, but further work is needed. We made some simulations in both the low-dimensional $\mathcal{M}$-open and $\mathcal{M}$-closed and high-dimensional $\mathcal{M}$-closed cases in our setup of Sections 5.1 and 5.2, comparing the discriminatory power of the usual Bayes Factor and the (average) Cross Validatory Bayes Factors studied in our paper. In terms of magnitude, the usual Bayes factor seems to be comparable with the criterion studied in this paper for large k, in both the high-dimensional and low-dimensional problems when the difference in the (minimum) Kullback–Leibler divergence from the true density for the two models is non-zero and this difference is not too small. In one such typical simulation, in an $\mathcal{M}$-open nested case, the usual (log) Bayes Factor is 16.39, which is close to our criterion value for $k/n = 0.8$, which eventually gets as large as 17.5 for larger k's. If the difference in (minimum) Kullback–Leibler divergence is zero, the usual (log) Bayes factor seems to show a larger magnitude in favour of the model which should be ideally chosen. In a typical example of this kind, in an $\mathcal{M}$-open nested case, the usual (log) Bayes Factor is -2.13 while the smallest value of the criterion studied in this paper is -1.2. In terms of choosing a model, on the whole, the usual Bayes Factor is seen to be choosing the model which should be chosen in terms of being closer to the truth or being a simpler model, if another model is equally close to the truth. But we also observed that in some simulations in the nested case with the difference in (minimum) Kullback–Leibler divergence for the two competing models very small, the usual Bayes Factor chooses the simpler model when in fact the more complex model is true and the simpler model is not true or the more complex model is closer to the truth, while the criterion studied in this paper chooses the model which should ideally be chosen in such situations. This last phenomenon was observed both in the low and high dimensional problems with moderate/large sample sizes. Incidentally, it must mentioned that neither the CVBF's or the usual Bayes Factor seem to be the obvious Bayesian procedure in the $\mathcal{M}$-open case.

In this connection it might be noted that it has been a moot question whether the Berger–Pericchi IBF can be interpreted in the sort of predictive framework discussed by Bernardo and Smith (1994) in the $\mathcal{M}$-open case. It has been argued elsewhere, (see Mukhopadhyay, Berger and Ghosh (2005)) that the differernce between IBF and the presumed Bernardo and Smith (1994) type criterion can actually be quite large. We believe this is possibly because the IBF is not normalized like the Cross-Validatorty Bayes Factor in equation (1) to be an estimate of the Bernardo and Smith (1994) integral.

Other discussants have informed us that they are intrigued by our observation that even asymptotically AIC and cross-validation are not equivalent in general, and, in fact, cross-validation has an asymptotically higher chance than AIC of selecting the true model in many cases. However, we like to point out that difference between the two methods, though not tending to zero, isn't substantial. Moreover the superiority of CV in selecting the true model may be compensated by inferior prediction.

Our results on classical Cross-Validation and AIC, e.g., that AIC can be regarded as Parametric Empirical Bayes or that mAIC with an additional binomial penalty seems to do better than the usual AIC in our simulations, are part of our general discussions on cross-validation and predictive Bayes rules but have no direct bearing on choice of k for the Cross-Validatory Bayes Factors.

Finally, we have benefitted from the discussion of cross-validation in Hastie *et al.* (2001) even though we don't agree with all their recommendations even in the classical setting.

ADDITIONAL REFERENCES IN THE DISCUSSION

Berger J. O. and Pericchi, L. R. (1996). The intrinsic Bayes factor for model selection and prediction. *J. Amer. Statist. Assoc.* **91**, 109–122.

Berger, J. O. and Pericchi, L. R. (2001). Objective Bayesian methods for model Selection: Introduction and comparison. *IMS Lecture Notes-Monograph Series* **8**, 135–207 (with discussion).

Dawid, A. P. (1984). Statistical theory. The prequential approach. *J. Roy. Statist. Soc. A* **147**, 277–305.

de Finetti, B. (1931). Funzione caratteristica di un fenomeno aleatorio. *Atti della R. Academia Nazionale dei Lincei, Serie 6. Memorie, Classe di Scienze Fisiche, Mathematice e Naturale* **4**, 251–299.

Draper, D. (1996). Utility, sensitivity analysis, and cross-validation in Bayesian model-checking. *Statistica Sinica* **6**, 760–767 (discussion of 'Posterior predictive assessment of model fitness via realized discrepancies,' by A. Gelman, X.-L. Meng, and H. Stern).

Draper, D. and Fouskakis, D. (2000). A case study of stochastic optimization in health policy: problem formulation and preliminary results. *Journal of Global Optimization* **18**, 399–416.

Draper, D. and Krnjajić, M. (2006). Bayesian model specification. *Submitted.*

Fouskakis, D. and Draper, D. (2002). Stochastic optimization: A review. *Internat. Statist. Rev.* **70**, 315–349.

Gneiting, T. and Raftery, A. E. (2005). Strictly proper scoring rules, prediction, and estimation. *Tech. Rep.*, University of Washington, USA..

Hewitt, E. and Savage, L. J. (1955). Symmetric measures on Cartesian products. *Trans. Amer. Math. Soc.* **80**, 470–501.

Murata, N., Yoshizawa, S. and Amari, S.-I. (1994). Network Information Criterion – Determining the number of hidden units for an Artificial Neural Network model. *IEEE Trans. Neural Networks* 6, 865–872.

O'Hagan, A. (1995). Fractional Bayes Factors and model comparison. *J. Roy. Statist. Soc. B* **57**, 99–138.

Ripley, B. D. (1996). *Pattern Recognition and Neural Networks.* Cambridge: University Press.

Takeuchi, K. (1976). Distribution of informal statistics and criterion for model fitting (in Japanese). *Suri-Kagaku* 153, 12–18.

Vehtari, A. and Lampinen, J, (2003). Expected utility estimation via cross-validation. *Bayesian Statistics 7* (J. M. Bernardo, M. J. Bayarri, J. O. Berger, A. P. Dawid, D. Heckerman, A. F. M. Smith and M. West, eds.) Oxford: University Press, 701–710.

BAYESIAN STATISTICS 8, pp. 91–114.
J. M. Bernardo, M. J. Bayarri, J. O. Berger, A. P. Dawid,
D. Heckerman, A. F. M. Smith and M. West (Eds.)

Nonparametric Function Estimation Using Overcomplete Dictionaries

MERLISE A. CLYDE and ROBERT L. WOLPERT
Duke University, U.S.A.
clyde@stat.duke.edu rlw@stat.duke.edu

SUMMARY

We consider the problem of estimating an unknown function based on noisy data using nonparametric regression. One approach to this estimation problem is to represent the function in a series expansion using a linear combination of basis functions. Overcomplete dictionaries provide a larger, but redundant collection of generating elements than a basis, however, coefficients in the expansion are no longer unique. Despite the non-uniqueness, this has the potential to lead to sparser representations by using fewer non-zero coefficients. Compound Poisson random fields and their generalization to Lévy random fields are ideally suited for construction of priors on functions using these overcomplete representations for the general nonparametric regression problem, and provide a natural limiting generalization of priors for the finite dimensional version of the regression problem. While expressions for posterior modes or posterior distributions of quantities of interest are not available in closed form, the prior construction using Lévy random fields permits tractable posterior simulation via a reversible jump Markov chain Monte Carlo algorithm. Efficient computation is possible because updates based on adding/deleting or updating single dictionary elements bypass the need to invert large matrices. Furthermore, because dictionary elements are only computed as needed, memory requirements scale linearly with the sample size. In comparison with other methods, the Lévy random field priors provide excellent performance in terms of both mean squared error and coverage for out-of-sample predictions.

Keywords and Phrases: GAUSSIAN RANDOM FIELD; INFINITELY DIVISIBLE; KERNEL REGRESSION; LÉVY RANDOM FIELD; NONPARAMETRIC REGRESSION; RELEVANCE VECTOR MACHINE; REVERSIBLE JUMP MARKOV CHAIN MONTE CARLO; SPATIAL-TEMPORAL MODELS; SPLINES; SUPPORT VECTOR MACHINE; WAVELETS.

Merlise Clyde is Associate Professor of Statistics at Duke University, Durham, North Carolina, USA. Robert Wolpert is Professor of Statistics at Duke University. The authors would like to thank Jen-hwa Chu, Leanna House, and Chong Tu for their contributions. This material is based upon work supported by the National Science Foundation under Grant Number DMS-0342172, DMS-0422400 and DMS-0406115. Any opinions, findings, and conclusions or recommendations expressed in this material are those of the author(s) and do not necessarily reflect the views of the National Science Foundation.

1. INTRODUCTION

The canonical setup for the nonparametric regression problem consists of having n measurements $\boldsymbol{Y} = (Y_1, \ldots, Y_n)^T$ of an unknown real valued function $f(\boldsymbol{x})$ defined on some space $\boldsymbol{\mathcal{X}}$,

$$Y_i = f(\boldsymbol{x}_i) + \epsilon_i \tag{1}$$

observed at points $\boldsymbol{x}_i \in \boldsymbol{\mathcal{X}}$. In the regression formulation the errors, ϵ_i, will typically represent white noise, $\epsilon_i \overset{iid}{\sim} \mathrm{N}(0, \sigma^2)$, but the nonparametric model may be extended to other exponential family models where $g(\mathrm{E}[Y_i]) = f(\boldsymbol{x}_i)$ for some link function g, as in generalized additive models. The function $f(\cdot)$ will often be regarded as an element of some separable Hilbert space $\mathcal{H}$ of real-valued functions on a compact space $\boldsymbol{\mathcal{X}}$. For Bayesian inference regarding the unknown mean function f, we must first place a prior distribution on f. If we are to model f nonparametrically, then we *should* place a prior distribution over the infinite dimensional space $\mathcal{H}$ of possible functions. However, in practice it is common to place a prior on the finite dimensional vector $\boldsymbol{f}_n \equiv (f(\boldsymbol{x}_1), \ldots, f(\boldsymbol{x}_n))^T$ for $\boldsymbol{f}_n \in \mathbb{R}^n$, for example, by expressing $\boldsymbol{f}_n$ at the observed points $\boldsymbol{x}_i$ in terms of a finite dimensional basis and placing prior distributions only on the coefficients or coordinates of $\boldsymbol{f}_n$ with respect to the basis. While a class of priors in the finite dimensional version may lead to reasonable behaviour of posteriors with modest sample sizes, one would hope that the finite dimensional prior remains sensible in the infinite dimensional limit and as the sample size n increases. In this paper, we promote the use of Lévy random field priors for stochastic expansions of f and show how Lévy random fields provide a natural limiting extension of certain finite dimensional prior distributions. We begin in Section 2 by reviewing some of the popular choices of priors in the the finite dimensional version of the problem. In Section 3, we present priors for stochastic expansions of f using Lévy random fields and show how these priors arise as natural limits of certain prior distributions on finite dimensional spaces. The connection between Lévy random fields and Poisson random fields provides the key to tractable computation using (reversible jump) Markov chain Monte Carlo sampling for the stochastic expansions. In Section 4 we describe the resulting hierarchical model and discuss prior specifications. In Section 5 we discuss how the Lévy random field priors lead to penalized likelihoods and contrast these expressions with other model selection criteria. We highlight some of our applications of Lévy random fields in Section 6. For many problems, Lévy random fields provide an attractive alternative to Gaussian random field priors. We conclude by discussing some areas for future research.

2. PRIOR DISTRIBUTIONS ON FUNCTIONS

When it comes to placing prior distributions over nonparametric functions of explanatory variable(s) $\boldsymbol{x}$, Gaussian process (or random field) priors are perhaps the most accessible. If the function f has a Gaussian Process (GP) prior with mean μ and covariance function $\boldsymbol{\Sigma}(\cdot, \cdot; \boldsymbol{\theta})$ (a positive definite function on $\boldsymbol{\mathcal{X}} \times \boldsymbol{\mathcal{X}}$),

$$f(\cdot) \sim \mathrm{GP}\left(\mu, \boldsymbol{\Sigma}(\cdot, \cdot; \boldsymbol{\theta})\right) \tag{2}$$

then this implies that any finite dimensional vector $(f(\boldsymbol{x}_1), \ldots, f(\boldsymbol{x}_n))^T$ has an n dimensional multivariate normal distribution with mean μ and $n \times n$ covariance matrix, $\text{Cov}[f(\boldsymbol{x}_i), f(\boldsymbol{x}_j)] = \boldsymbol{\Sigma}(\boldsymbol{x}_i, \boldsymbol{x}_j; \boldsymbol{\theta})$. The hyperparameter vector $\boldsymbol{\theta}$ controls various features of the covariance function, and hence the process. While widely used in spatial-temporal modelling (Cressie, 1993, Chapter 3) and smoothing splines (Wahba, 1990), these priors are well suited for the general nonparametric regression and classification problems (O'Hagan, 1978; Neal, 1999). For Gaussian error models, Gaussian random field priors are particularly appealing as the unknown function f may be integrated out analytically, leaving the marginal likelihood of the (typically) much lower dimensional parameters $\boldsymbol{\theta}$ and σ^2. While sampling from the posterior distribution of the parameters $\boldsymbol{\theta}$ and σ^2 is generally straightforward using MCMC algorithms, implementation typically require repeated inversion of $n \times n$ matrices within the MCMC loop, limiting the applicability to modest n.

One may avoid matrix inversions by working with the Karhunen–Loève (Karhunen, 1946; Loève, 1955) expansion of f,

$$f(\boldsymbol{x}) = \sum_j \psi_j(\boldsymbol{x})\beta_j \tag{3}$$

where ψ_j are the orthonormal eigenfunctions of the integral operator based on the covariance function and β_j are the coordinates of f. For the mean zero Gaussian process, the β_j are independent normal random variables with mean 0 and variance equal to the eigenvalues λ_j. In practice, one may obtain a finite dimensional approximation to f by using the eigenfunctions corresponding to the n largest eigenvalues and setting the remaining $\beta_j, j > n$ to zero. However, starting with a covariance function and determining the corresponding orthonormal eigenfunctions is a challenging task (see Xia and Gelfand, 2005, for discussion and alternative solutions).

The connection between covariance functions in Gaussian processes and kernels in a reproducing Kernel Hilbert space $\mathcal{H}$ has led to a tremendous resurgence of interest in kernel representations of f (see http://kernel-machines.org) in both the statistics and machine learning communities. A Hilbert space associated with a positive definite function $k(\cdot,\cdot)$ on $\boldsymbol{\mathcal{X}} \times \boldsymbol{\mathcal{X}}$, such as the covariance function $\boldsymbol{\Sigma}$, may be constructed as the collection of all finite linear combinations of the form $\sum k(\cdot, \boldsymbol{x}_j)$ and their limits (under the norm induced by the inner product $\langle k(\cdot, \boldsymbol{x}_i), k(\cdot, \boldsymbol{x}_j)\rangle = k(\boldsymbol{x}_i, \boldsymbol{x}_j)$ (Wahba, 2002)). The representer theorem (Kimeldorf and Wahba, 1971) leads to a finite dimensional representation of f:

$$f_\lambda(\cdot) = \sum_{j=1}^{n} k(\cdot, \boldsymbol{x}_j)\beta_j \tag{4}$$

as the solution to the optimization problem of finding $f \in \boldsymbol{\mathcal{H}}$ to minimize the penalized loss functional

$$\sum_i L\left(y_i, f(\boldsymbol{x}_i)\right) + \lambda ||f||_k^2 \tag{5}$$

where L is typically (although not necessarily) convex in f. The equivalence between the penalized loss (5) and negative log posterior under a Gaussian prior on β leads to a maximum *a posteriori* (MAP) estimate of f using kernels at the observed data points $\boldsymbol{x}_i$ and provides a Bayesian interpretation of support vector machines (SVM) (Law and Kwok, 2001; Sollich, 2002). These representations, however, use as many basis functions as observations.

Starting with the finite dimensional representation based on the MAP estimate, Tipping (2001) and Chakraborty *et al.* (2004) provide extensions of the SVM model by replacing the Gaussian prior on the β_j with independent scale mixtures of normals,

$$\beta_j \mid \lambda_j \stackrel{ind}{\sim} \mathrm{N}(0, 1/\lambda_j) \tag{6}$$

where λ_j is given a Gamma prior distribution. As with the LASSO (Tibshirani, 1996), which is based on a scale mixture of normals corresponding to a double exponential prior, the posterior *modes* for a subset of the β_j may be zero, resulting in a MAP estimate of f based on fewer than the n basis functions used in the original Bayesian SVM solution with Gaussian priors.

As suggested by the above expansions, an alternative to using a Gaussian process prior is to expand f directly in terms of a countable collection of basis functions $\phi_j(\cdot), j \in \mathcal{J}$

$$f(\boldsymbol{x}_i) = \sum_{j \in \mathcal{J}} \psi_j(\boldsymbol{x}_i)\beta_j \tag{7}$$

where the $\{\beta_j, j \in \mathcal{J}\}$ are the unique (but unknown) coefficients of f with respect to the basis $\{\psi_j(\cdot), j \in \mathcal{J}\}$ for $\mathcal{H}$. This includes basis functions determined by kernels $g(\cdot, \boldsymbol{x}_j; \boldsymbol{\theta})$, piecewise polynomials, Fourier series, splines, wavelets, etc. One advantage of this approach is that there is no need to restrict the generator to be a positive (Mercer) kernel (Tipping, 2001) allowing for a more flexible representations. A problem with many classical bases, such as a Fourier basis or spline basis, is that they are *non-local*, meaning that many basis functions may contribute to values of the decomposition at 'spatially' distant points. Typically, this will lead to decompositions with many non-zero coefficients and is inefficient in the sense that there may be significant "cancellation' of coefficients in order to determine the value of the function at a given point. Wavelet bases and bases generated from kernels with compact support, on the other hand, are 'local', leading to more adaptive and parsimonious representations. Even so, orthonormal wavelet bases have a disadvantage in that the positions and the scales of the basis functions are subject to dyadic constraints and may require potentially more basis functions than with a non-orthonormal wavelet basis.

Recent developments using overcomplete representations through frames and dictionaries where the number of functions $|\mathcal{J}|$ in the expansion of $\boldsymbol{f}$ is potentially greater than n (in the finite dimensional case) show great promise (Wolfe *et al.*, 2004; Johnstone and Silverman, 2005). While inherently redundant, overcomplete representations have been advocated due to their increased flexibility and adaptation over orthonormal bases, particularly for finding sparse representations of f. Examples of overcomplete dictionaries include unions of bases, Gabor frames, non-decimated or translational invariant wavelets and wavelet packets. Because of the redundancy of overcomplete representations, coefficients β_j in the expansion (7) using all dictionary elements are not unique. This lack of uniqueness *a priori* is advantageous, permitting more parsimonious representations to be extracted from the dictionary than those obtained using any single basis as shown by Wolfe *et al.* (2004) .

Working with Gabor frames, Wolfe *et al.* (2004) used the variable selection priors popularized by the Stochastic Search Variable Selection (SSVS) model of George and McCulloch (1997). Dictionary elements are identified by introducing an indicator variable γ_j, such that

$$\beta_j \mid \lambda_j, \gamma_j \stackrel{ind}{\sim} \mathrm{N}(0, \sigma^2\gamma_j/\lambda_j). \tag{8}$$

When γ_j is zero, the distribution of β_j is degenerate at 0, so that function ψ_j is not included in the expansion. A hierarchical prior for the indicators variables $\gamma_j, j \in \mathcal{J}$ completes the specification; by default the γ_j are taken as independent Bernoulli random variables with $\Pr(\gamma_j = 1) = \pi \in (0, 1)$. Bayesian variable selection has been used successfully to identify sparse solutions in the finite dimensional regression formulation and provides the canonical model for soft and hard thresholding of wavelet coefficients (Abramovich *et al.*, 1998; Clyde *et al.*, 1998; Clyde and George, 2000; Johnstone and Silverman, 2005; Vidakovic, 1999) and models for automatic curve fitting using splines or piecewise polynomials (Smith and Kohn, 1996; Denison *et al.*, 1998a,b) (see Clyde and George (2004) for an overview and additional references).

2.1. *Limiting Version of SSVS*

As the sample size n increases, it is common to consider a richer collection of potential dictionary elements, and an important challenge is to characterize the limiting behaviour of the SSVS prior. This is particularly relevant in overcomplete frames in $\mathbb{R}^n$ for fixed sample size n, as the number of dictionary elements increases. If we let $J \equiv \sum_{j \in \mathcal{J}} \gamma_j$ denote the number of functions in the expansion with nonzero coefficients, then the independent Bernoulli priors on the γ_j lead to a $\text{Bi}(|\mathcal{J}|, \pi)$ distribution for J. As $|\mathcal{J}| \to \infty$ and $\pi \to 0$, with $\pi\,|\mathcal{J}|$ converging to a constant ν_+, the number of dictionary elements in the expansion will have a limiting Poisson distribution with mean ν_+, and the number of terms in the expansion will be finite almost surely. Assuming independent prior distributions for β_j (as in Wolfe *et al.*, 2004), the resulting limiting prior distribution on f is a *compound Poisson random field*, which is a special case of a Lévy random field. The examples in Wolfe *et al.* (2004) used a finite (but large) dictionary. Note, however, that the collection of dictionary elements may not even be countable, for example, in the case of free knot splines (DiMatteo *et al.*, 2001), continuous wavelets (Vidakovic, 1999, Chapter 3) or kernels evaluated at points other than the observed data points. Lévy random fields as limits of sequences of compound Poisson fields provide a natural choice for priors on expansions of functions using overcomplete dictionaries. While proper prior distributions on β_j are required in trans-dimensional problems in order for marginal likelihoods to be well determined (Clyde and George, 2004), we will see that the Lévy random field priors lead to improper measures for the coefficients and infinite ν_+ in the limiting version. This provides a new avenue for objective Bayesian analysis in nonparametric modeling.

3. LÉVY RANDOM FIELDS AND STOCHASTIC EXPANSIONS

For illustration, we will consider overcomplete dictionaries created by a generating function g,

$$\phi_{\boldsymbol{\omega}}(\boldsymbol{x}) \equiv g(\boldsymbol{x}, \boldsymbol{\omega}) \qquad \boldsymbol{\omega} \in \boldsymbol{\Omega} \tag{9}$$

a Borel measurable function $g : \mathcal{X} \times \boldsymbol{\Omega} \to \mathbb{R}$ where $\boldsymbol{\Omega}$ is a complete separable metric space and for fixed $\boldsymbol{\omega}$, $g(\cdot, \boldsymbol{\omega})$ is an element in $\mathcal{H}$. Examples of generating functions include density functions (either normalized or un-normalized) from location-scale families of distributions. In the one dimensional setting, the choice

$$g(x, \boldsymbol{\omega}) = \exp\{-\lambda |x - \chi|^p\} \tag{10}$$

includes the un-normalized Gaussian ($p = 2$) and Laplace densities ($p = 1$), where $\boldsymbol{\omega}^T \equiv (\chi, \lambda)^T$ and $\chi \in \mathbb{R}$ and $\lambda \in \mathbb{R}^+$ (p may also be included in $\boldsymbol{\omega}$ as well). Of course, there is no need to restrict attention to location families or symmetric generators. For example, the un-normalized exponential density or other asymmetric densities may be more appropriate for modeling pollution dissipation over time.

Shifts and scales of a wavelet function ψ,

$$g(x, \boldsymbol{\omega}) = \lambda^{1/2} \psi\Big(\lambda(x - \chi)\Big) \tag{11}$$

provide other generating functions with desirable features for functions exhibiting non-stationarity, where the choice of wavelet may be based on a particular smoothness class of the function. The original Haar wavelet with $\boldsymbol{\mathcal{X}} = [0, 1]$

$$g(x, \boldsymbol{\omega}) \equiv \mathbf{1}_{\{0 < \lambda(x-\chi) \le 1\}} \tag{12}$$

is the simplest. Many popular wavelets with compact support do not have explicit closed forms, but fast numerical calculation of the continuous wavelet at any point may be obtained using filters and the Daubechies–Lagarias (Vidakovic, 1999, p. 89).

3.1. *Lévy Random Fields*

With the goal of extracting potentially sparse representations for f from an over-complete dictionary, we consider expansions of the form

$$f(x) \equiv \sum_{0 \le j < J} g(\boldsymbol{x}, \boldsymbol{\omega}_j)\, \beta_j \tag{13}$$

for a random number $J \le \infty$ of randomly drawn pairs $\beta_j \in \mathbb{R}$, $\boldsymbol{\omega}_j \in \boldsymbol{\Omega}$. Note that the sum is equivalent to an integral with respect to a random *signed* Borel measure $\mathcal{L}(d\boldsymbol{\omega}) \equiv \sum \beta_j \delta_{\boldsymbol{\omega}_j}(d\boldsymbol{\omega})$ on $\boldsymbol{\Omega}$ with random support points $\boldsymbol{\omega}_j$ and 'jumps' at $\boldsymbol{\omega}_j$ of size β_j. This leads to the equivalent stochastic integral representation:

$$f(x) = \int_{\boldsymbol{\Omega}} g(x, \boldsymbol{\omega}) \mathcal{L}(d\boldsymbol{\omega}) \tag{14}$$

for the random prior measure $\mathcal{L}$. Our goal is to deduce the random measure $\mathcal{L}$ based on information in the data, a form of inverse problem (Wolpert *et al.*, 2003).

As suggested by the limiting version of the SSVS priors and other priors in trans-dimensional problems, an intuitive construction of such random measures begins by choosing any positive number $\nu_+ > 0$ and assigning J a Poisson distribution, $J \sim \text{Po}(\nu_+)$. Then, conditionally on J, accord the $(\beta_j, \boldsymbol{\omega}_j) \in \mathbb{R} \times \boldsymbol{\Omega}$ independent identical distributions, $(\beta_j, \boldsymbol{\omega}_j) \sim \pi(d\beta, d\boldsymbol{\omega})$, where π is a probability distribution on $\mathbb{R} \times \boldsymbol{\Omega}$. In that case, the random measure $\mathcal{L}$ assigns independent infinitely-divisible (ID) random variables,

$$\mathcal{L}(A_i) \equiv \sum_{0 \le j < J} \mathbf{1}_{A_i}(\boldsymbol{\omega}_j)\, \beta_j \tag{15}$$

to disjoint Borel sets $A_i \subset \boldsymbol{\Omega}$. The random variables $\mathcal{L}(A_i)$ for a collection of Borel sets A_i is an example of a Lévy random field with 'Lévy measure' $\nu(d\beta, d\boldsymbol{\omega}) = \nu_+ \pi(d\beta, d\boldsymbol{\omega})$, the product of the Poisson rate ν_+ for J and the distribution $\pi(d\beta, d\boldsymbol{\omega})$ for $\{(\beta_j, \boldsymbol{\omega}_j)\}$. Here π is proper distribution and $\nu(\mathbb{R} \times \boldsymbol{\Omega})$ is finite (by construction).

More generally, Khinchine and Lévy (1936) showed that any ID random variable and indeed any ID random measure (Rajput and Rosiński, 1989, Prop. 2.1) has characteristic function

$$\mathrm{E}\left[e^{it\mathcal{L}(A)}\right] = \exp\left\{it\delta(A) - \tfrac{1}{2}t^2\Sigma(A) + \int\!\!\int_{\mathbb{R}\times A}\left(e^{it\beta} - 1 - ith(\beta)\right)\nu(d\beta, d\boldsymbol{\omega})\right\} \tag{16}$$

determined uniquely by the characteristic triplet of sigma-finite measures (δ, Σ, ν) consisting of a signed measure $\delta(d\boldsymbol{\omega})$ and a positive measure $\Sigma(d\boldsymbol{\omega})$ on $\boldsymbol{\Omega}$, and a positive measure $\nu(d\beta, d\boldsymbol{\omega})$ on $\mathbb{R}\times\boldsymbol{\Omega}$ that for each compact $K \subset \boldsymbol{\Omega}$ satisfies

$$\int\!\!\int_{\mathbb{R}\times K}(1\wedge\beta^2)\nu(d\beta, d\boldsymbol{\omega}) < \infty \tag{17}$$

and $\nu(\{0\}, \boldsymbol{\Omega}) = 0$. The function $h(\beta) \equiv \beta\mathbf{1}_{[-1,1]}(\beta)$ is known as the 'compensator' and is required to make the last integrand in (16) bounded and $O(\beta^2)$ for β in a neighborhood of zero. When the Lévy measure satisfies the more restrictive condition

$$\int\!\!\int_{\mathbb{R}\times K}\left(1\wedge|\beta|\right)\nu(d\beta, d\boldsymbol{\omega}) < \infty \tag{18}$$

for each compact $K \subset \boldsymbol{\Omega}$, we may take the compensator $h(\beta)$ to be zero. From the characteristic function, $\mathcal{L}(A)$ may be recognized as the sum of two independent parts: a Gaussian component (with mean and covariance determined by the first two terms in (16)) and a 'pure jump' component with Lévy measure ν. We will restrict attention to random measures $\mathcal{L}$ with $\delta \equiv 0$ and $\Sigma \equiv 0$ (no Gaussian component).

A random measure $\mathcal{L}$ satisfying (16) (without Gaussian component) induces a *random field*, a linear mapping from functions $\phi : \boldsymbol{\Omega} \to \mathbb{R}$ to random variables

$$\mathcal{L}[\phi] \equiv \int_{\boldsymbol{\Omega}} \phi(\boldsymbol{\omega})\mathcal{L}(d\boldsymbol{\omega}) \tag{19}$$

with characteristic function

$$\mathrm{E}\left[e^{it\mathcal{L}[\phi]}\right] = \exp\left\{\int\!\!\int_{\mathbb{R}\times\boldsymbol{\Omega}}\left(e^{it\phi(\boldsymbol{\omega})\beta} - 1 - it\phi(\boldsymbol{\omega})\,h(\beta)\right)\nu(d\beta, d\boldsymbol{\omega})\right\}. \tag{20}$$

When the bound (18) holds, the class of functions ϕ and spaces $\boldsymbol{\Omega}$ for which (20), and hence the random field $\mathcal{L}[\phi]$, is well defined includes all bounded measurable compactly supported functions. More generally, Rajput and Rosiński (1989) show that the space Φ of functions that are integrable with respect to an ID random measure $\mathcal{L}(d\boldsymbol{\omega})$ with no Gaussian component are certain *Musielak–Orlicz* modular spaces.

For expansions of f, let $\mathcal{G}$ denote the linear space of measurable functions $g : \boldsymbol{\mathcal{X}} \times \boldsymbol{\Omega} \to \mathbb{R}$ for which $g(\boldsymbol{x}, \cdot) \in \Phi$ for each $\boldsymbol{x} \in \boldsymbol{\mathcal{X}}$. For any generator function $g \in \mathcal{G}$, we can construct a random function $f : \boldsymbol{\mathcal{X}} \to \mathbb{R}$ by $f(\boldsymbol{x}) \equiv \mathcal{L}[g(\boldsymbol{x}, \cdot)]$; in particular this holds for the generators in Sec. 3 (see Tu *et al.*, 2006 for more details on conditions and examples). Moments of $f(x) = \mathcal{L}[g(x, \cdot)]$, when they exist, are easy to compute from the characteristic function (20) for any $\mathcal{L}[\phi]$:

$$\mathrm{E}\Big[f(x)\Big] = \int\!\!\int_{\mathbb{R}\times\Omega} g(x, \boldsymbol{\omega})\Big[\beta - h(\beta)\Big]\nu(d\beta, d\boldsymbol{\omega}) \tag{21}$$

$$\mathrm{Cov}\Big[f(x_1), f(x_2)\Big] = \int\!\!\int_{\mathbb{R}\times\Omega} g(x_1, \boldsymbol{\omega})\,g(x_2, \boldsymbol{\omega})\,\beta^2\,\nu(d\beta, d\boldsymbol{\omega}). \tag{22}$$

3.2. *Poisson Representation*

Tu *et al.* (2006) give an equivalent construction of the Lévy random fields from a sequence of compound Poisson random fields, which is key to tractable posterior inference. Let $N(d\beta, d\boldsymbol{\omega}) \sim \text{Po}(\nu)$ denote a Poisson random measure on $(\mathbb{R} \times \Omega)$ that assigns independent random variables with $\text{Po}(\nu(C_i))$ distributions to disjoint Borel sets $C_i \subset (\mathbb{R} \times \Omega)$ and denote the *compensated* or centered Poisson measure $\tilde{N}(d\beta, d\boldsymbol{\omega}) \equiv N(d\beta, d\boldsymbol{\omega}) - \nu(d\beta, d\boldsymbol{\omega})$ (Sato, 1999, page 38). Then equivalently

$$\mathcal{L}[\phi] \stackrel{d}{=} \int\!\!\int_{\mathbb{R}\times\Omega} [\beta - h(\beta)]\, \phi(\boldsymbol{\omega}) N(d\beta, d\boldsymbol{\omega}) + \int\!\!\int_{\mathbb{R}\times\Omega} h(\beta)\, \phi(\boldsymbol{\omega}) \tilde{N}(d\beta, d\boldsymbol{\omega}) \qquad (23)$$

which simplifies to

$$\mathcal{L}[\phi] \stackrel{d}{=} \int\!\!\int_{\mathbb{R}\times\Omega} \beta\phi(\boldsymbol{\omega}) N(d\beta, d\boldsymbol{\omega}) \qquad (24)$$

in the case ν satisfies (18).

In the case of infinite Lévy measure, the number of points $N(\mathbb{R}, \boldsymbol{\Omega})$ will be infinite almost surely. While the integrals $\mathcal{L}[g]$ will be well behaved, stochastic expansions with an infinite number of terms are not practical for posterior simulation of the distribution of f. One solution is to truncate jumps less than a certain threshold in absolute value, since we are primarily interested in sparse solutions based on the jumps with largest absolute magnitudes. When the integrability condition (18) holds, we can always approximate $\mathcal{L}$ and $\mathcal{L}[\phi]$ by choosing some small $\epsilon > 0$, and replacing $\mathbb{R}$ in (24) by $[-\epsilon, \epsilon]^c$. Thus,

$$\mathcal{L}_\epsilon[\phi] \equiv \int\!\!\int_{[-\epsilon,\epsilon]^c\times\boldsymbol{\Omega}} \phi(\boldsymbol{\omega})\beta\, N(d\beta, d\boldsymbol{\omega}) = \int\!\!\int_{\mathbb{R}\times\boldsymbol{\Omega}} \phi(\boldsymbol{\omega})\beta\, N_\epsilon(d\beta, d\boldsymbol{\omega}) \qquad (25)$$

where N_ϵ is a Poisson measure on $\mathbb{R} \times \boldsymbol{\Omega}$ with intensity measure

$$\nu_\epsilon(d\beta, d\boldsymbol{\omega}) \equiv \nu(d\beta, d\boldsymbol{\omega})\mathbf{1}_{\{|\beta|>\epsilon\}}. \qquad (26)$$

Consequently, the Lévy random field $\mathcal{L}[\phi]$ may be approximated by a sequence of compound Poisson random fields, $\mathcal{L}_\epsilon[\phi]$ where $\mathcal{L}_\epsilon[\phi]$ converges in distribution to $\mathcal{L}[\phi]$ as $\epsilon \to 0$. Similar approximations are possible in the case the Lévy measure satisfies the more general bound (17), however, one may need to include an ϵ dependent deterministic adjustment due to the compensator (Tu *et al.*, 2006).

Truncating the support is not the only way to construct suitable approximating sequences of finite Lévy measures $\nu_\epsilon(d\beta, d\boldsymbol{\omega})$ for which the integrals $\mathcal{L}_\epsilon[\phi]$ converge in distribution to $\mathcal{L}[\phi]$. The Lévy measure $\nu(d\beta, d\boldsymbol{\omega}) = \alpha(d\boldsymbol{\omega})\beta^{-1}e^{-\beta\tau}\mathbf{1}_{\mathbb{R}_+}(\beta)d\beta$ for the Gamma $\text{Ga}(\alpha(d\boldsymbol{\omega}), \tau)$ random field, for example, may be approximated by $\nu_\epsilon(d\beta, d\boldsymbol{\omega}) \equiv \gamma(d\boldsymbol{\omega})\beta^{\epsilon-1}e^{-\beta/\tau}\mathbf{1}_{\mathbb{R}_+}(\beta)d\beta$, a finite measure (if $\gamma(\boldsymbol{\Omega}) < \infty$ and $\epsilon > 0$) with full support. Truncation has an advantage, in that it maintains the exact conditional distribution for all included mass points $(\beta_j, \boldsymbol{\omega}_j)$, and merely sets to zero the smallest coefficients. This focus on the large magnitude coefficients is desirable, as we are interested in finding sparse expansions.

3.3. *Examples of Lévy Measures*

Examples of Lévy random fields with infinite Lévy measures include the symmetric α-stable (SαS) family for $0 < \alpha < 2$ (including the Cauchy process with $\alpha = 1$), with Lévy measure

$$\nu(d\beta, d\boldsymbol{\omega}) = c_\alpha \gamma(d\boldsymbol{\omega})|\beta|^{-1-\alpha}\, d\beta \qquad (27)$$

for some constant $c_\alpha > 0$, giving $\mathcal{L}[A] \sim \mathrm{St}(\alpha, 0, \gamma(A), 0)$. For $1 \le \alpha < 2$, the construction of $\mathcal{L}[\phi]$ requires compensation as in (23). Interestingly, the Lévy measure for the symmetric stable may be represented as a mixture of normals

$$\nu(d\beta, d\boldsymbol{\omega}) \propto \gamma(d\boldsymbol{\omega}) \int_0^\infty \xi^{1/2} \exp\left(-\frac{1}{2}\beta^2 \xi\right) \xi^{\alpha/2-1}\, d\xi \tag{28}$$

where the mixing distribution on ξ is a limiting version of a $\mathrm{Ga}(\alpha/2, b)$ distribution with $b = 0$. This suggests a way of extending the independent normal priors in SVMs such that the limiting infinite distribution on f is an integral with respect to a Stable random field. The prior used by Tipp (2001), in fact corresponds to taking $\alpha = b = 0$, which is not equivalent to using a Lévy measure for an α-stable random field, and does not lead to a proper posterior.

The Gamma random field is another important example used in constructing non-negative functions, with Lévy measure

$$\nu(d\beta, d\boldsymbol{\omega}) = \gamma(d\boldsymbol{\omega})\beta^{-1} e^{-\beta/\tau} \mathbf{1}_{\{\beta>0\}}\, d\beta \tag{29}$$

for some σ-finite measure $\gamma(d\boldsymbol{\omega})$ on $\boldsymbol{\Omega}$, giving $\mathcal{L}[A] \sim \mathrm{Ga}(\gamma(A), 1/\tau)$ for $A \subset \boldsymbol{\Omega}$ of finite γ measure. A symmetric analogue of the Gamma random field (29) has Lévy measure

$$\nu(d\beta, d\boldsymbol{\omega}) = |\beta|^{-1} e^{-|\beta|/\tau}\, d\beta\, \gamma(d\boldsymbol{\omega}) \tag{30}$$

on all of $\mathbb{R} \times \Omega$, leading to random variables $\mathcal{L}(A)$ distributed as the difference of two independent $\mathrm{Ga}(\gamma(A), 1/\tau)$ variables, with characteristic functions

$$\mathrm{E}[e^{it\mathcal{L}(A)}] = (1 + t^2\tau^2)^{-\gamma(A)}.$$

Both the standard positive and this symmetric Gamma random measures satisfy the bound (18), thus no compensation is required and we may use (25) to approximate (24).

The means $\mathrm{E}[f(x)]$ are available directly from (21); they vanish for the symmetric Gamma random field, $\mathrm{E}[f(x)] = 0$, or for any other Lévy random field with a symmetric (in $\pm\beta$) Lévy measure that satisfies (18). Covariances (when they exist) are available from (22). Nearly all of the commonly used isotropic geostatistical covariance functions (see Chilès and Minières, 1999, Chapter 2.5) may be achieved by the choice of a suitable generating kernel $g(x, \cdot)$ and Lévy measure $\nu(d\beta, d\boldsymbol{\omega})$. The Gaussian generating kernel $|\,(x)$ of (10) with $p = 2$ and the symmetric Gamma Lévy measure ν from (30) with $\gamma(d\boldsymbol{\omega}) = \gamma\pi(d\lambda)\, d\chi$ for some $\alpha > 0$ and $\pi(d\lambda)$ a point mass at $\lambda_0 > 0$ lead to

$$\mathrm{Cov}\Big[f(x_1), f(x_2)\Big] = 2\alpha\tau^2 \sqrt{\pi/\lambda_0}\, e^{-\lambda_0 (x_1 - x_2)^2/4},$$

the isotropic Gaussian covariance function, while the choice $\pi(d\lambda) = \mathrm{Ga}(a, b)$ for some $a > \frac{1}{2}$ and $b > 0$ leads to the generalized Cauchy model (Yaglom, 1987, p. 365):

$$\mathrm{Cov}\Big[f(x_1), f(x_2)\Big] = 2\alpha\tau^2 \sqrt{\pi b}\, \frac{\Gamma(a - 1/2)}{\Gamma(a)} \left[1 + \frac{(x_1 - x_2)^2}{4b}\right]^{1/2-a}.$$

For the Laplace generating kernel (10) with $p = 1$ with the same Lévy measure ν, the covariance is

$$\begin{aligned}\mathrm{Cov}\Big[f(x_1), f(x_2)\Big] &= 2\alpha\tau^2 \int_{\mathbb{R}+} \frac{1}{\lambda} e^{-\lambda|x_1-x_2|}\Big(1+\lambda|x_1-x_2|\Big)\pi(\lambda)d\lambda \\ &= \frac{2\alpha\tau^2}{a-1}\,\frac{b+a|x_1-x_2|}{[1+|x_1-x_2|/b]^a}.\end{aligned}$$

Other examples of kernels and resulting covariance functions may be found in Tu *et al.* (2006).

4. HIERARCHICAL MODEL

Capitalizing on the (approximate) compound Poisson representation of $\mathcal{L}$ based on the truncation of small jumps, we may state the likelihood and prior of f in hierarchical fashion

$$Y_i \mid f(\boldsymbol{x}_i) \overset{ind}{\sim} \mathrm{N}(f(\boldsymbol{x}_i), \sigma^2) \qquad f(\boldsymbol{x}_i) = \sum_{0\le j<J} g(\boldsymbol{x}_i, \boldsymbol{\omega}_j)\beta_j \tag{31}$$

$$J \sim \mathrm{Po}(\nu_+) \quad \text{where } \nu_+ \equiv \nu_\epsilon(\mathbb{R}, \boldsymbol{\Omega}) \tag{32}$$

$$(\beta_j, \boldsymbol{\omega}_j) \mid J \overset{iid}{\sim} \pi(d\beta_j, d\boldsymbol{\omega}_j) \equiv \frac{\nu_\epsilon(d\beta_j, d\boldsymbol{\omega}_j)}{\nu_\epsilon(\mathbb{R}, \boldsymbol{\omega})} \quad \text{for } j = 1, \ldots, J \tag{33}$$

where J is the random number of terms in the stochastic expansion, $(\beta_1, \ldots, \beta_J)$ represents the unknown coefficients and $(\boldsymbol{\omega}_1, \ldots, \boldsymbol{\omega}_J)$ represents the collection of generator specific parameters. A prior distribution on σ^2 completes the prior specification for the first stage of the hierarchy. While we have stated the model in terms of the Gaussian error model (1), other distributions for Y, of course, can replace the normal assumption without any loss of generality.

We may also place a prior distribution on parameters in the Lévy measure. With either the Gamma (29) or Stable (27) random fields and default stationary measure $\gamma(d\boldsymbol{\omega}) \equiv \gamma\pi(d\lambda)\,d\chi$ for some distribution $\pi(d\lambda)$, we may place a Gamma prior on γ, which leads to J having a Negative Binomial distribution. This provides robustness to a fixed choice of γ by providing over-dispersion. Hyperparameters in the Negative-Binomial may be elicited by specifying various quantiles, for example fixing a probability that there are no components (just a constant mean) and specifying the 95th quantile for J. This has given reasonable behaviour in a wide variety of problems. The prior distribution $\pi(d\lambda)$ needs more careful specification to keep the kernel generating functions from having support that is too narrow. We have used a $\mathrm{Ga}(a_\lambda, b_\lambda)$ prior for λ, with selection of the hyperparameters based on subjective information about the problem.

While expressions for posterior modes or posterior distributions of quantities of interest do not exist in closed form, the prior construction using Lévy random fields permits tractable posterior simulation via a reversible jump Markov chain Monte Carlo algorithm (Green, 1995). Efficient computation is possible because updates to f based on adding/deleting or updating single dictionary elements bypass the need to invert large matrices. Furthermore, because dictionary elements $g(\boldsymbol{x}, \boldsymbol{\omega})$ are only computed as needed, memory requirements scale linearly with the sample size. For generating functions with compact support, further improvements in computational speed are possible.

5. MODEL COMPLEXITY

Given observations $\boldsymbol{Y} \equiv (Y_1, \ldots, Y_n)^T$, the log of the posterior distribution

$$\log\left[\pi\left(\{\beta_j, \boldsymbol{\omega}_j\}_{j=1}^J,\, J,\, \boldsymbol{\theta} \mid \boldsymbol{Y}\right)\right]$$

where $\boldsymbol{\theta}$ represents the fixed dimensional parameters (σ^2, hyperparameters in the Lévy measure, etc.), takes the form of a penalized or regularized likelihood

$$\text{constant} \; - \frac{1}{2\sigma^2}\sum_{i=1}^{n}(Y_i - f(\boldsymbol{x}_i))^2 - \log(J!) + \sum_{j=1}^{J}\log(\nu_\epsilon(d\beta_j, d\boldsymbol{\omega}_j)). \qquad (34)$$

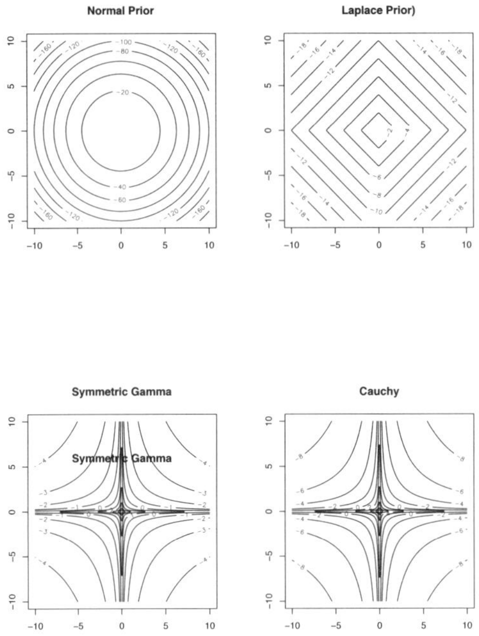

Figure 1: *Log contours for the joint distribution of two β coefficients in the expansion of f under independent normal, independent double exponential, the Gamma random eld prior which is proportional to (30), and the Cauchy random field prior, which is proportional to (27).*

Model complexity is penalized directly through the $\log(J!)$ term, as in ℓ_0 penalties which penalizes the number of coefficients in the expansion. As in other Bayesian model selection examples, model complexity is also indirectly penalized through the choice of prior on the regression coefficients, in this case through the Lévy measure ν. Fig. 1 contrasts the contours of the log prior for two coefficients $(\beta_j, \beta_{j'})$ under the independent normal priors (spherical contours), the independent double exponential prior (diamond) and the priors based on the Lévy measures for the symmetric Gamma and Cauchy random fields. The approximate Lévy measure in the symmetric Gamma and Cauchy random field models may be seen as inducing a sparsity penalty for the addition of generator functions to the function $f(x)$, similar (or stronger in fact) to the L_1 penalties of the LASSO (Tibshirani, 1996). The shape of the joint prior makes it much harder to keep redundant components in the model than with the LASSO.

The Gamma and Cauchy processes are both examples of infinite Lévy measures, and in order to restrict the expansion to a finite number of terms, we must restrict $|\beta| > \epsilon$. This may be related to the idea of practical significance in the non-conjugate version of the SSVS algorithm George and McCulloch (1993); Chipman *et al.* (1997) where the prior distribution on β is a mixture of two normal distributions; one fairly dispersed and the other concentrated around zero. The variance in the concentrated distribution is selected to reflect values of the coefficient that for practical purposes suggest that the variable could be dropped from the model. The choice of ϵ in the Lévy random field framework may be guided by this idea of practical significance for estimating f in the presence of noise, in that dictionary elements with coefficients larger than ϵ in absolute value will be retained. In the limit as $\epsilon \to 0$, the distribution for any β taken on its own is actually improper, with an infinite spike at zero, however, the prior distribution on f with infinite J and infinite measure ν is well defined based on (20).

6. APPLICATIONS

We illustrate the Lévy random field priors in several examples, using both simulated and real applications to highlight the versatility of the priors.

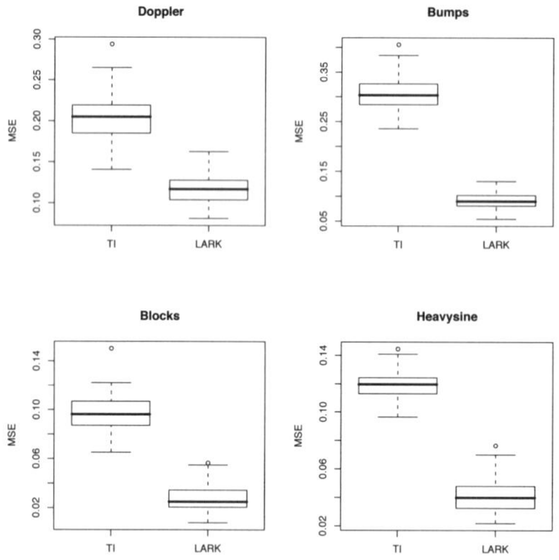

Figure 2: *Average MSE over 100 simulations using the Empirical Bayes estimator with translation-invariant wavelet transform (TI) (Johnstone and Silverman, 2005) and Lévy Adaptive Regression Kernels (LARK) using the ϵ-truncated symmetric Gamma Lévy random field prior (Tu et al., 2006).*

Example 1 (Wavelets). In Tu *et al.* (2006) we compare the performance of the Lévy random field priors to estimators based on translational invariant wavelets (Johnstone and Silverman, 2005), another overcomplete representation, using simulated data and several of the now standard wavelet test functions: Blocks, Bumps, Doppler and Heavysine (Donoho and Johnstone, 1994). Fig. 2 illustrates the significant improvements in mean squared error for estimating the true function using the symmetric Gamma Lévy random field priors.

Because parameters in the generators g are allowed to vary with location, the Lévy random field priors lead to adaptive estimation. Like translation-invariant wavelets, the prior representation leads to an overcomplete representation, however, the posterior has a much sparser representation than with the non-decimated wavelets. We have illustrated the methods using generating functions based on kernels, thus the reduction in MSE may be due to choice of prior, generator, or both. The methodology is easily extended to continuous wavelets using the Daubechies–Lagarias algorithm for evaluating wavelets at arbitrary locations/scales. While preliminary results of Chu *et al.* (2006) using continuous wavelets with the compound Poisson priors with finite J and a normal prior on β as suggested in Abramovich *et al.* (2000) provide improvements over the translational invariant wavelets (using the same wavelet generators), we expect additional reductions may be possible using the heavier-tailed symmetric Gamma or Cauchy random field prior distributions with the overcomplete wavelet dictionary.

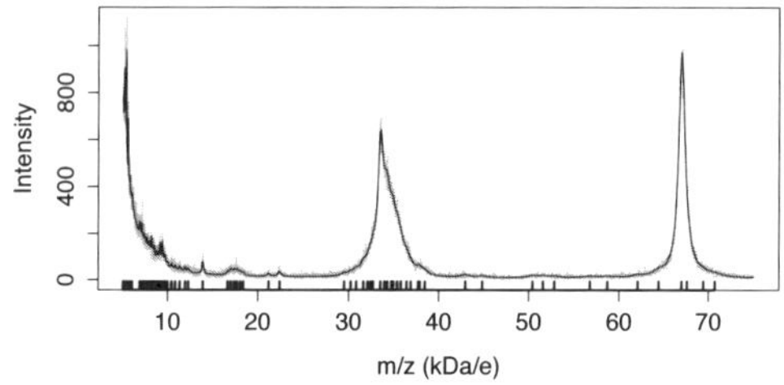

Figure 3: *A raw spectrum and estimates of mean intensity from the highest posterior draw; the rug plot at the bottom indicates locations of the latent proteins (peaks).*

Example 2 (Mass Spectroscopy). In Clyde *et al.* (2006) we use Gamma random field priors to construct models for the latent relative abundance of proteins as a function of their mass/charge (or equivalently time of flight) using data from Matrix Assisted Laser Desorption/Ionization Time of Flight mass spectroscopy. Normalized Gaussian kernels with time varying scale parameters provide a natural choice of generating functions to capture the variation in time of flight of proteins of a given mass/charge. Combined with the positive Gamma measure ν in (29), f will be non-negative. Unlike wavelets or spline models, the parameters in the adaptive kernel model have interesting biological interpretations: J is the number of unknown proteins in the sample, β_j is the unknown concentration for a protein with expected time of flight τ_j, and the resolution parameter ρ_j governs the peak widths λ_j (here we take $\boldsymbol{\omega}_j \equiv (\tau_j, \rho_j)$). This interpretability is a key feature of the Lévy random field models, as it allows us to incorporate subjective prior information regarding resolution and time of flight (a transformation of the mass/charge). In this example, we drop the normality assumption and use a Gamma distribution for Y_i without any additional computational complexity. Fig. 3 illustrates a draw from the RJ-MCMC output, where the rug plot on the x-axis indicates locations of peaks by their time of flight.

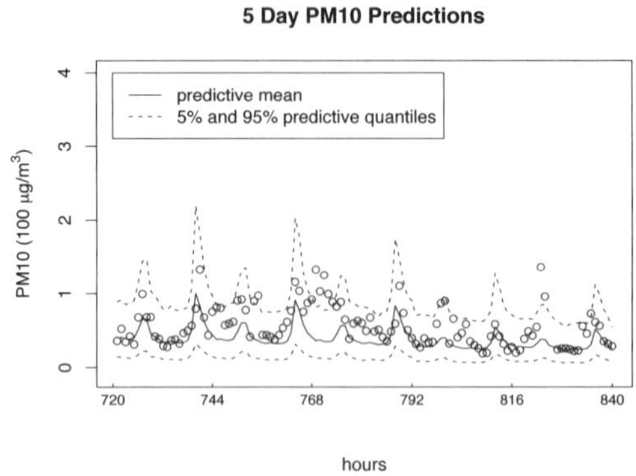

Figure 4: *Five day forecast of PM10 predictions with 90% credible intervals (pointwise).*

Example 3 (Non-Stationary Spatial–Temporal Models). The third area of application concerns development of non-stationary temporal (as well as spatial and spatial-temporal models) for concentrations of one or more criteria pollutants. As expected pollution concentrations are inherently non-negative, the Lévy random field prior based on a Gamma random field ensure that the expected functions are non-negative, and is a natural alternative to the commonly adopted Gaussian random field priors in spatial-temporal models. The models here may be seen as natural extensions of the work of Wolpert and Ickstadt (1998), who developed conjugate Poisson-Gamma spatial temporal models. Unlike Gaussian process convolution models (Higdon *et al.*, 1999; Lee *et al.*, 2002; Xia and Gelfand, 2005), there is no need to evaluate the kernels on a regular lattice, thus the Lévy random field models have the potential to be more parsimonious, as in the case of critically sampled wavelets versus continuous wavelets. The spatial-temporal locations of jumps in the Lévy random field may be interpreted as point sources of pollution, with dispersal over time and space controlled by additional parameters in the kernels. Hierarchical models for parameters in the Lévy measure allow incorporation of meteorological variables which influence both the dispersal parameters and expected concentrations. An extension of the models described here is to utilize marked random fields that allow common jumps (shared impulses) between two or more pollutants or that represent periodic (repetitive) events over time (for example, peaks in concentration due to morning/evening commutes which occur regularly).

Fig. 4 illustrates out-of-sample predictions for a time series of particulate matter from Phoenix, AZ. The model incorporates a periodic process that repeats daily and automatically captures features of the morning/evening traffic, plus a non-periodic process. Unlike in-fill predictions, in the out-of-sample forecasts the non-periodic events are driven by the prior process and are more sensitive to hyperparameter specifications. In comparison with standard methods, the Lévy random field priors provide excellent performance in terms of both mean squared error (RMSE =

$0.28\mu g/m^3$) and 91.5% coverage for nominal 90% credible intervals for out-of-sample model predictions. Additional details and more examples of the multi-pollutant and spatial-temporal models may be found in Tu *et al.* (2006).

7. SUMMARY

In this paper, we have tried to illustrate the potential of Bayesian nonparametric modelling using Lévy random field priors for function estimation. The model is based on a stochastic expansion of functions in an overcomplete dictionary, which may be formulated as a stochastic integration problem with a (signed) random measure. The unknown function may be approximated as a finite sum of generating functions at arbitrary locations where the number of components is a free parameter. The generator parameters are location-specific and thus are adaptively updated given the data. The adaptability of the generators is especially useful for modeling 'spatially' inhomogeneous functions. Unlike many wavelet based methods, there is no requirement that the data are equally spaced. The RJ-MCMC algorithm developed for fitting the models provides an automatic search mechanism for finding sparse representations of a function.

REFERENCES

Abramovich, F., Sapatinas, T. and Silverman, B. W. (1998). Wavelet thresholding via a Bayesian approach. *J. Roy. Statist. Soc. B* **60**, 725–749.

Abramovich, F., Sapatinas, T. and Silverman, B. W. (2000). Stochastic expansions in an overcomplete wavelet dictionary. *Probability Theory and Related Fields* **117**, 133–144.

Chakraborty, S., Ghosh, M. and Mallick, B. K. (2004). Bayesian nonlinear regression for large p small n problem. *Tech. Rep.*, 2004-01, University of Florida, USA.

Chilès, J.-P. and Minières, P. D. (1999). *Geostatistics: Modeling Spatial Uncertainty.* New York: Wiley.

Chipman, H. A., Kolaczyk, E. D. and McCulloch, R. E. (1997). Adaptive Bayesian wavelet shrinkage. *J. Amer. Statist. Assoc.* **92**, 1413–1421.

Chu, J., Clyde, M. A. and Liang, F. (2006). Bayesian function estimation using an overcomplete continuous wavelet dictionary. Discussion Paper 06–11, ISDS, Duke University.

Clyde, M. and George, E. I. (2000). Flexible empirical Bayes estimation for wavelets. *J. Roy. Statist. Soc. B* **62**, 681–698.

Clyde, M. and George, E. I. (2004). Model uncertainty. *Statist. Science* **19**, 81–94.

Clyde, M., Parmigiani, G. and Vidakovic, B. (1998). Multiple shrinkage and subset selection in wavelets. *Biometrika* **85**, 391–401.

Clyde, M. A., House, L. L. and Wolpert, R. L. (2006). Nonparametric models for proteomic peak identification and quantification. *Bayesian Inference for Gene Expression and Proteomics* (K. A. Do, P. Müller and M. Vannucci, eds.). Cambridge: University Press.

Cressie, N. A. C. (1993). *Statistics for Spatial Data.* New York: Wiley.

Denison, D. G. T., Mallick, B. K. and Smith, A. F. M. (1998a). Automatic Bayesian curve fitting. *J. Roy. Statist. Soc. B* **60**, 333–350.

Denison, D. G. T., Mallick, B. K. and Smith, A. F. M. (1998b). Bayesian MARS. *Statist. Computing* **8**, 337–346.

DiMatteo, I., Genovese, C. R. and Kass, R. E. (2001). Bayesian curve-fitting with free-knot splines. *Biometrika* **88**, 1055–1071.

Donoho, D. L. and Johnstone, I. M. (1994). Ideal spatial adaptation by wavelet shrinkage. *Biometrika* **81**, 425–455.

George, E. I. and McCulloch, R. E. (1993). Variable selection via Gibbs sampling. *J. Amer. Statist. Assoc.* **88**, 881–889.

George, E. I. and McCulloch, R. E. (1997). Approaches for Bayesian variable selection. *Statistica Sinica* **7**, 339–374.

Green, P. J. (1995). Reversible jump Markov chain Monte Carlo computation and Bayesian model determination. *Biometrika* **82**, 711–732.

Higdon, D., Swall, J. and Kern, J. (1999). Non-stationary spatial modeling. *Bayesian Statistics 6* (J. M. Bernardo, J. O. Berger, A. P. Dawid and A. F. M. Smith, eds.) Oxford: University Press, 761–768.

Johnstone, I. M. and Silverman, B. W. (2005). Empirical Bayes selection of wavelet thresholds. *Ann. Statist.* **33**, 1700–1752.

Karhunen, K. (1946). Zur Spektraltheorie Stochasticher Prozesse. *Ann. Acad. Sci. Fennicae* **34**, 1–7.

Khinchine, A. Y. and Lévy, P. (1936). Sur les lois stables. *C. R. Acad. Sci. Paris* **202**, 374–376.

Kimeldorf, G. S. and Wahba, G. (1971). Some results on Tchebycheffian spline functions. *J. Math. Anal. Applications* **33**, 82–95.

Law, M. H. and Kwok, J. T. (2001). Bayesian support vector regression. *Proc. 8th International Workshop on Artificial Intelligence and Statistics (AISTATS).* Florida: Key West, 239–244.

Lee, H. K. H., Holloman, C. H., Calder, C. A. and Higdon, D. M. (2002). Flexible Gaussian processes via convolution. *Tech. Rep.*, 09–09, ISDS, Duke University, USA.

Loève, M. M. (1955). *Probability Theory.* Princeton: University Press.

Neal, R. M. (1999). Regression and classification using Gaussian process priors. *Bayesian Statistics 6* (J. M. Bernardo, J. O. Berger, A. P. Dawid and A. F. M. Smith, eds.) Oxford: University Press, 475–501.

O'Hagan, T. (1978). Curve fitting and optimal design for prediction (with discussion). *J. Roy. Statist. Soc. B* **40**, 1–42.

Rajput, B. S. and Rosiński, J. (1989). Spectral representations of infinitely divisible processes. *Probab. Theory Rel.* **82**, 451–487.

Sato, K.-i. (1999). *Lévy Processes and Infinitely Divisible Distributions.* Cambridge: University Press.

Smith, M. and Kohn, R. (1996). Nonparametric regression using Bayesian variable selection. *J. Econometrics* **75**, 317–343.

Sollich, P. (2002). Bayesian methods for support vector machines: Evidence and predictive class probabilities. *Machine Learning* **46**, 21–52.

Tibshirani, R. (1996). Regression shrinkage and selection via the lasso. *J. Roy. Statist. Soc. B* **58**, 267–288.

Tipping, M. E. (2001). Sparse Bayesian learning and the Relevence Vector Machine. *J. Mach. Learn. Res.* **1**, 211–244.

Tu, C. (2006). *Nonparametric Modelling using Lévy Process Priors with Applications for Function Estimation, Time Series Modeling and Spatio-Temporal Modeling.* PhD Thesis, ISDS, Duke University, USA.

Tu, C., Clyde, M. A. and Wolpert, R. L. (2006). Lévy adaptive regression kernels. *Tech. Rep.*, 06–08, ISDS, Duke University, USA.

Vidakovic, B. (1999). *Statistical Modeling by Wavelets.* New York: Wiley.

Wahba, G. (1990). *Spline models for observational data.* Philadelphia, PA: SIAM.

Wahba, G. (2002). Soft and hard classification by reproducing kernel Hilbert space methods. *Proc. National Academy of Sciences* 99, 524–530.

Wolfe, P. J., Godsill, S. J. and Ng, W.-J. (2004). Bayesian variable selection and regularisation for time-frequency surface estimation. *J. Roy. Statist. Soc. B* **66**, 575–589.

Wolpert, R. L. and Ickstadt, K. (1998). Poisson/gamma random field models for spatial statistics. *Biometrika* **85**, 251–267.

Wolpert, R. L., Ickstadt, K. and Hansen, M. B. (2003). A nonparametric Bayesian approach to inverse problems (with discussion). *Bayesian Statistics 7* (J. M. Bernardo, M. J. Bayarri, J. O. Berger, A. P. Dawid, D. Heckerman, A. F. M. Smith and M. West, eds.) Oxford: University Press, 403–418.

Xia, G. and Gelfand, A. (2005). Stationary process approximation for the analysis of large spatial datasets. *Tech. Rep.*, 05-24, ISDS, Duke University, USA.

Yaglom, A. M. (1987). *Correlation Theory of Stationary and Related Random Functions: Basic Results* **1**. Berlin: Springer-Verlag.

DISCUSSION

BRANI VIDAKOVIC (*Georgia Institute of Technology, USA*)

The authors consider the nonparametric regression problem in which the estimation of a function of interest is done in the atomic decomposition domains by Bayesian modeling. It can be shown that many of currently used models for regularizing functions by shrinkage of coefficients in their atomic decompositions are either special cases or approximations of the Clyde–Wolpert model. Examples include SSVS in orthogonal regression, wavelet thresholding, pursuit methods. The authors construct the prior on the parameters of atoms using Lévy Random Field which leads to tractable posterior simulations. The reversible-jump MCMC method used for efficient computation is well suited because updates based on adding/deleting or updating a single dictionary element at the time. The dictionary elements are only computed when needed which leads to calculational and memory storage efficiency.

Introduction and overview. The authors are to be congratulated for an excellent contribution that is unifying for several state-of-art Bayesian statistical models in the 'atomic domains.' It can be shown that many of current models for regularizing functions by shrinkage in the atomic domains are either special cases or approximations of the Clyde–Wolpert model. Examples include SSVS in orthogonal regression, Bayesian wavelet shrinkage of various flavors, Bayesian pursuit methods.

To put the ideas of Clyde and Wolpert in the proper setting, we consider a paradigmatic nonparametric regression model. The observations $Y_i, i = 1, \ldots, n$ have two components: a sampled unknown function f as the systematic part and the zero mean random errors as the stochastic part. The errors are assumed to be iid normal with a constant variance, although more general distributions can be considered.

$$Y_i = f(x_i) + \epsilon_i, \ i = 1, \ldots, n; \quad \epsilon_i \sim N(0, \sigma^2).$$

Assume that the function of interest has atomic decomposition

$$f(x_i) = \sum_{0 \leq j \leq J} g(x_i, \omega_j)\beta_j = \int g(x_i, \omega)\mathcal{L}(d\omega),$$

where $g(x_i, \omega)$ are ω-indexed atoms evaluated at observations x_i and where $\mathcal{L}(d\omega) = \sum \beta_j \delta_{\omega_j}(d\omega)$ is a random signed Borel measure.

The atoms are generated by a function g that could be modulated trigonometric function, wavelet, or a custom made function. For example, if the generator is a wavelet ψ, then the atoms are given by

$$g(x_i, \omega_j) = \sqrt{\lambda_j}\psi(\lambda_j(x_i - \chi_j)), \quad \omega_j = (\chi_j, \lambda_j).$$

where λ_j is a scale and χ_j is a location parameter. The model in (35) is completed by adopting priors on J, β and ω.

How the measure $\mathcal{L}$ was constructed? A rudimentary idea can be traced back to Abramovich *et al.* (2000) who proposed $J \sim Poi(\nu_+)$ and normal prior on β_j. However, their model is not constructive, that is, does not lead to an effective Bayesian simulation. Effective models that use Gaussian random fields are proposed Godsill and his Cambrige team; see Wolfe *et. al.* (2004).

Clyde and Wolpert propose Compound Poisson Representation of $\mathcal{L}$ which is an approximation to Lèvy Measure $\nu(d\beta, d\omega) = \nu_+\pi(d\beta, d\omega)$. This approximation proved to be a constructive. Examples of specific random fields are given in equations (28)-(30).

Additional strength of the paper is wide applicability of the model. The authors provide several examples: (i) wavelet-based function estimation which is described in more detail in Chu *et al.* (2006); (ii) proteomic peak identification that uses the library of normalized Gaussians in peak modeling (research described in Clyde *et al.* (2006); and (iii) non-stationary time series exemplified on measurements of particulate matter from Phoenix, AZ. Here the model is based on Lévy process generated by Gamma random field, as described in Tu (2006).

In the next section we discuss possible construction of calculationally efficient overcomplete dictionaries that potentially can be modeled by compound Poisson representations of $\mathcal{L}$.

Joy and sorrow of redundancy. Albert Einstein once said that models should be simple enough but not simpler. The right measure of complexity versus parsimony in modeling is fundamental philosophical issue that directs the development of modeling methodologies. This trade off is in influenced by the 'vocabulary of models.' An object has parsimonious representation in a rich vocabulary and vice versa, a complex representation in a minimal dictionary.

Figure 5: *Three fundamental colors: red, green and blue (here in gray).*

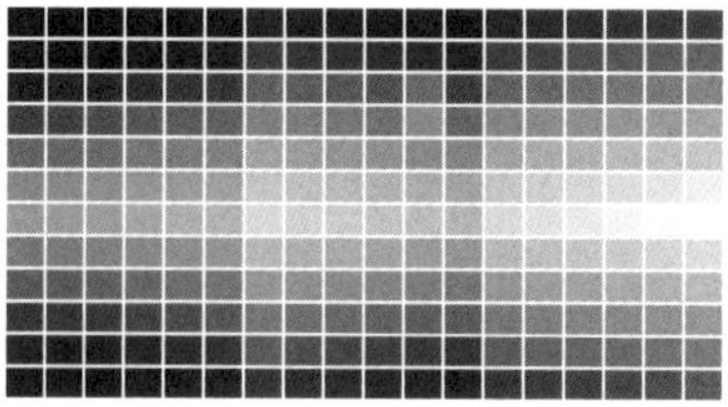

Figure 6: *Pallete with shades*

It is instructive to consider an analogy from chromatic theory. A color painting is to be made and this can be done by one of the two proposed strategies. The first strategy is to paint from a palette with only three basic colors: red, green and blue (RGB, Fig. 5), while the second one is paint from a rich palette consisting of many pre-mixed colors of various shades, Fig. **??**.

The painting can be made using any of the palettes but the redundant palette makes modeling task easier and faster.

There are two camps among atomic modelers: (i) *criticalists*: the researchers who strive to produce representations from the (critically) minimal dictionaries, and (ii) *redundantists*, who enjoy benefits of redundancy for more informative and simpler modeling leading to most compact representations. The orthonormal bases are an example of minimal dictionaries: excluding any element from the distionary will make it incomplete, that is, for some objects the representations are not possible.

Here we list several modeling properties in the light of size of dictionaries. Atomic representations with overcomplete dictionaries (OD): (i) enable deeper sparsity, (ii) can contain more meaningful (in the sense of the phenomena they describe) atoms, and (iii) atoms could have nice properties that are impossible in critical dictionaries (symmetry of compactly supported atoms in the wavelet context, shift invariance, more directional information).

On the other hand, critically minimal dictionaries (i) produce unique representations and are mathematically elegant, (ii) are often computationally (and statistically) more convenient, and (iii) have some modeling properties not shared by the overcomplete models. For example, energy preservation is a key property needed in modeling the scaling, long range dependence and (multi)fractality, and controlling the energy distribution in overcomplete models is difficult if possible at all.

We narrow our discussion to construction of dictionaries that are redundant but share some desirable properties of the minimal bases. The dictionaries would be parameterizable in a simple way and thus amenable to Bayesian prior modeling.

It is desirable that the function generating the dictionary is well localized in the time and frequency domains. For most generators this is an impossible requirement. For example, for wavelet bases, the Fourier dual of Haar wavelet is a sinc function which has poor time localization (decays as $O(1/x)$.)

A theoretical result showing the necessity of overcompleteness if locality is important is celebrated Balian–Low Theorem (Balian, 1981; Low, 1985). It states that for most minimal atomic expansions, e.g., orthonormal bases or tight frames, simultaneous time/frequency locality of atoms is impossible.

Theorem 1 *(Balian–Low Theorem). Let g be an L_2 function (window), a and b positive constants, and $(m, n) \in Z \times Z$.*

If the Gabor system of functions (Gabor, 1946) $\{g_{m,n}(x) = e^{2\pi aimx} g(x - bn)\}$ is a basis, then $ab = 1$, and

$$\left(\int_R |tg(t)|^2 dt \right) \cdot \left(\int_R |\omega \hat{g}(\omega)|^2 d\omega \right) = \infty.$$

For example, it is impossible to form a basis by modulating Gaussian window, $g(x) = e^{-x^2}$.

The solution to Balian–Low obstacle is the redundancy of dictionaries. The examples of standardly used OD are numerous: continuous wavelets, Gabor frames, non-decimated wavelets, wavelet packets, SLEX bases, complex wavelets, to list a few. The drawback for all OD, as we pointed out, is the computational complexity of representing and manipulating the models.

Can we have versatile OD libraries of atoms that retain calculational simplicity of o.n. bases? The answer is positive, and the wavelet packet tables are a nice example. We provide several proposals in the wavelet context and suggest a possibility of Bayesian approaches. The underlying idea is to mix parameterized orthonormal bases.

It is well known that multiresolution analysis by wavelets is fully described by a single filter $\boldsymbol{h}$, called wavelet filter. In fact, the scaling function (informally, the father wavelet) ϕ and its Fourier transform are connected with *transfer function* m_0, and in the sequel, the transfer function uniquely generates the filter $\boldsymbol{h}$. Opposite direction is also possible: given the wavelet filter $\boldsymbol{h}$ one can reconstruct the corresponding scaling function. Schematically,

$$\phi(x) \leftrightarrow \Phi(\omega) \leftrightarrow m(\omega) \leftrightarrow \boldsymbol{h} = (h_0, h_1, \ldots).$$

For example, for the Haar scaling function $\phi(x) = \mathbf{1}(0 \le x \le 1)$ the Fourier transform is $\Phi(\omega) = e^{-i\omega/2}\text{sinc}(\omega/2)$, the transfer function is $m_0(\omega) = \frac{1+e^{i\omega}}{2}$, producing $\boldsymbol{h} = \{1/\sqrt{2}\ 1/\sqrt{2}\}$. For details see Vidakovic (1999, p. 60).

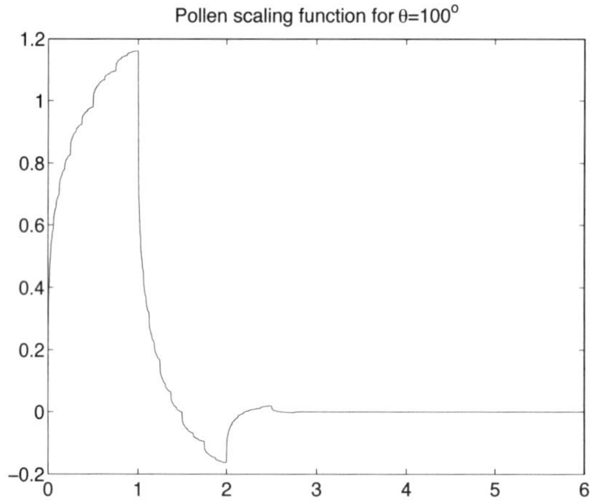

Figure 7: *Scaling function from Pollen family for* $\varphi = 5\pi/9$.

There are many examples of wavelet filters that can be parameterized. Pollen's family (Pollen, 1989) is often used because of its simplicity. An example of four tap Pollen dictionary indexed by a single parameter is provided below. For any value φ from $[0, 2\pi]$ the resulting filter

$$\begin{aligned} h_0 &= (1 + \cos(\varphi) - \sin(\varphi))/2^{3/2} \\ h_1 &= (1 + \cos(\varphi) + \sin(\varphi))/2^{3/2} \\ h_2 &= (1 - \cos(\varphi) + \sin(\varphi))/2^{3/2} \\ h_3 &= (1 - \cos(\varphi) - \sin(\varphi))/2^{3/2} \end{aligned}$$

generates an orthogonal multiresolution analysis, i.e., corresponds to an orthogonal compactly supported wavelet basis. For example, $\varphi = 0$ corresponds to Haar wavelet while $\varphi = \pi/6$ corresponds to a Daubechies 4 tap wavelet. Thus the atom $g(x, \omega, \varphi) = \sqrt{\lambda}\phi_\varphi(\lambda x - \chi)$ is parameterized not only by scale λ and location χ, but also by the *shape* parameter φ. Fig. 7 shows Pollen scaling function from Shi *et al.* (2005) where it was used the context of wavelet-based classification; the best basis from the Pollen library was selected by entropy minimization had $\varphi = 5\pi/9$. Minent bases (as opposed to maxent priors) are closest to Kahrunen-Loève bases in the sense of 'energy packing' and thus produce parsimonious representations.

It would be interesting to let the data (signal) to select φ in Bayesian fashion.

Another interesting wavelet that can produce a versatile library of atoms is the GT wavelet. The following result holds

Theorem 2 *For any* $|b| \geq 1$, $h_0 = -\frac{\sqrt{2}}{2b}$, $h_1 = \frac{\sqrt{2}}{2}$, $h_{2k} = \frac{(b^2-1)\sqrt{2}}{2b^{k+1}}$, $h_{2k+1} = 0$, $k = \pm 1, \pm 2, \ldots$ *is a wavelet filter.*

The GT wavelet has exponential decay and various properties depending on b. For example, if $b = -3$, the scaling function has finite second moment, (Fig. 9). while for $b = \frac{1+\sqrt{5}}{2}$ the scaling function is orthogonal not only to its integer shifts, but also to its 1/2 shifts, $\langle \phi(x), \phi(x + 1/2) \rangle = 0$, Fig. 8. If the parameter b is taken into account when modeling, the posterior of b may lead to efficient data justified atoms.

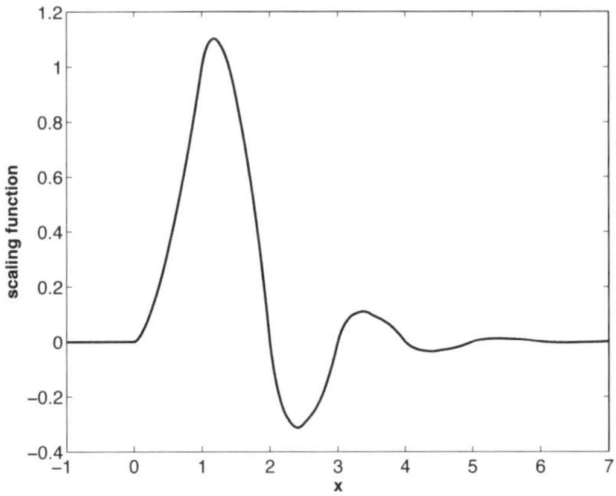

Figure 8: $b = -3$ *has finite* $\phi'(x)$.

There is also possibility to form an overcomplete dictionary by manipulating wavelet-filter equations. Consider the system of wavelet-filter equations generating a Daubechies minimum phase 4 tap filter. If the standard zero-moment condition $0h_0 - 1h_1 + 2h_2 - 3h_3 = 0$ in the system

$$
\begin{array}{l}
h_0 + h_1 + h_2 + h_3 = \sqrt{2} \\
h_0^2 + h_1^2 + h_2^2 + h_3^2 = 1 \\
h_0 h_2 + h_1 h_3 = 0 \\
h_0 - h_1 + h_2 - h_3 = 0; \quad 0h_0 - 1h_1 + 2h_2 - 3h_3 = 0
\end{array}
$$

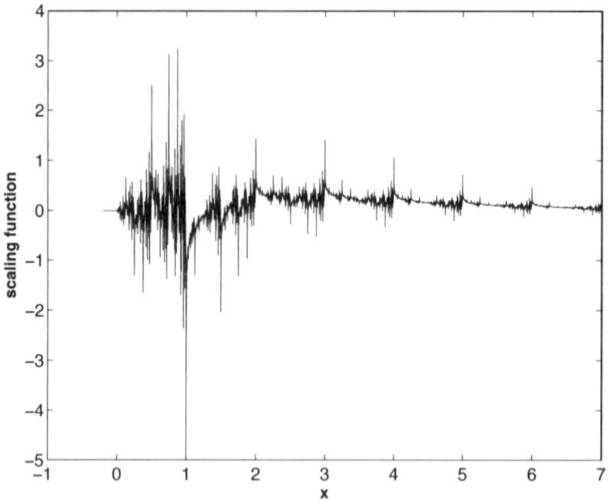

Figure 9: $b = \frac{1+\sqrt{5}}{2}$ *gives* $\langle \phi(x), \phi(x+1/2) \rangle = 0$.

is replaced by $h_3 - f(1,c)h_2 + f(2,c)h_1 - f(3,c)h_0 = 0, \quad f(x,c) = e^{-x^2/c}, \ c > 0,$ then each c would correspond to a valid wavelet basis. In the modeling approach a prior could be be placed on c.

We conclude our discussion with the sketch of construction of a custom-made library, via a parameterized wavelet basis with the scaling function matching the features of the data. The steps of the algorithm for such construction are

(i) Select an arbitrary function g appropriate for the modeling problem. In nonparametric regressions the rationale is to select the scaling function that 'looks like the data.'

(ii) Find Fourier transform of g, $G(\omega) = \hat{g}(\omega)$.

(iii) Normalize G to construct the Fourier transform of a scaling function $\Phi(\omega)$ as

$$\frac{G(\omega)}{\sqrt{\sum_{\ell=-\infty}^{\infty} |G(\omega + 2\ell\pi)|^2}}.$$

(iv) Fourier-invert the $\Phi(\omega)$ to obtain scaling function $\phi(t)$. From $\Phi(\omega)$ find the transfer function m_0 and the wavelet filter $\boldsymbol{h}$.

(v) Project the filter $\boldsymbol{h}$ on the space of finite wavelet filters if compact support of atoms is desired and parameterize the wavelet.

Conclusions. The paper by Clyde and Wolpert is a milestone contribution in Bayesian modeling in atomic domains. It unifies several well known atomic estimation strategies and is practicable. In our discussion we focus on the possibility to improve performance by customizing and mixing the dictionaries.

REPLY TO THE DISCUSSION

We would like to begin by thanking Prof. Vidakovic for his enlightening comments on atomic decompositions and extensions of the models that we have presented here. Prof. Vidakovic traces features of our construction back to Abramovich *et al.* (2000). Indeed, their stochastic expansion may be viewed as a special case of our Lévy random field approach, with a Lévy measure constructed as the product of three terms: a proper Gaussian measure on the coefficient parameter β (conditional on the scale parameter λ); a uniform measure on location $\chi \in [0, 1]$; and a potentially improper marginal measure on λ. Even through the coefficients in their expansion have proper conditional distributions, the joint measure may be infinite. As with the infinite Lévy measures presented here, this will lead to a decomposition with an infinite number of jumps or support points in the measure $\mathcal{L}$. Using a finite approximation to the Lévy measure (for example, by truncating the scale parameter's support), one can pursue posterior simulation and inference as described in more detail in Tu *et al.* (2006).

Starting from the stochastic approximations of Abramovich *et al.* (2000), Chu *et al.* (2006) developed posterior inference using continuous wavelet dictionaries. Using the same test functions that were considered in our Example 1, they also found that the overcomplete expansions provided improved MSE performance and more sparse representations when compared to the nondecimated wavelets of Johnstone and Silverman (2005). Comparisons between the results of Chu *et al.* (2006) and Tu *et al.* (2006) suggest that the LARK method leads to improved MSE–however, it remains uncertain whether this is due to the different choices of prior distributions or of generating functions. In applications we have used separable measures for the scaling parameters and coefficients, leading to independent prior distributions; more general non-separable measures (hence non-stationary priors) are worth further investigation.

Combining the heavy tailed priors based on the Lévy random measures (symmetric gamma or stable families) with wavelet generating functions, particularly the parametrized filters and wavelets that Prof. Vidakovic suggests, is an extension worth pursuing. With parametric families of generating functions (such as that of Pollen, 1989, cited by Prof. Vidakovic), one could place a prior distribution on the indexing parameter and provide posterior inference for functions by either selecting an 'optimal' family or averaging over families, a form of 'model averaging'. One could also incorporate the index parameter into the support of the Lévy measure, allowing additional mixing of the dictionaries; in this case each atom used in the expansion may come from any family. Larger dictionaries could of course be created by mixing over wavelets and kernels. While certainly leading to more flexible representations, the increased computational complexity may become overwhelming. Parametrized generating functions adapted to features of the data offer a more promising avenue for exploration.

While the computational challenges that arise with the Lévy random field priors and wavelet generating functions are straightforward to overcome, there remains the interesting theoretical question of which function space contains the limiting infinite stochastic expansions for infinite Lévy measures. Abramovich *et al.* (2000) prove that the stochastic expansion will remain in the Besov space of the generating wavelet, in their Gaussian context, but for more general Lévy measures and wavelets this remains an open question.

ADDITIONAL REFERENCES IN THE DISCUSSION

Abramovich, F., Sapatinas, T. and Silverman, B. W. (2000). Stochastic expansions in an overcomplete wavelet dictionary. *Probability Theory and Related Fields* **117**, 133–144.

Balian, R. (1981). A strong uncertainty principle in signal theory or in quantum mechanics. *C. R. Acad. Sci.* **292**, 1357–1362.

Gabor, D. (1946). Theory of communication. *J. IEE London* **93**, 429–457.

Johnstone, I. M. and Silverman, B. W. (2005). Empirical Bayes selection of wavelet thresholds. *Ann. Statist.* **33**, 1700–1752.

Low, F. (1985). Passion for Physics. *Essays in Honor of Geoffrey Chew.* (C. DeTar *et al.*, eds.) Singapore: World Scientific, 17.

Pollen, D. (1989). Parametrization of compactly supported wavelets. *Tech. Rep.*, Aware Inc, USA.

Shi, B., Moloney, K. P., Pan, Y., Leonard, V. K., Vidakovic, B., Jacko, J. and Sainfort, F. (2006). Classification of high frequency pupillary responses using Schur monotone descriptors in multiscale domains. *J. Statist. Computation and Simulation* **76**, 431–446.

BAYESIAN STATISTICS 8, pp. 115–148.
J. M. Bernardo, M. J. Bayarri, J. O. Berger, A. P. Dawid,
D. Heckerman, A. F. M. Smith and M. West (Eds.)

Sequential Monte Carlo for Bayesian Computation

PIERRE DEL MORAL
Université de Nice, France
delmoral@math.unice.fr

ARNAUD DOUCET
Univ. British Columbia, Canada
arnaud@cs.ubc.ca

AJAY JASRA
Imperial College London, UK
ajay.jasra@imperial.ac.uk

SUMMARY

Sequential Monte Carlo (SMC) methods are a class of importance sampling and resampling techniques designed to simulate from a sequence of probability distributions. These approaches have become very popular over the last few years to solve sequential Bayesian inference problems (*e.g.*, Doucet *et al.*, 2001). However, in comparison to Markov chain Monte Carlo (MCMC), the application of SMC remains limited when, in fact, such methods are also appropriate in such contexts (e.g. Chopin, 2002); Del Moral et al. 2006). In this paper, we present a simple unifying framework which allows us to extend both the SMC methodology and its range of applications. Additionally, reinterpreting SMC algorithms as an approximation of nonlinear MCMC kernels, we present alternative SMC and iterative self-interacting approximation (Del Moral and Miclo 2004, 2006) schemes. We demonstrate the performance of the SMC methodology on static and sequential Bayesian inference problems.

Keywords and Phrases: IMPORTANCE SAMPLING; NONLINEAR MARKOV CHAIN MONTE CARLO; PROBIT REGRESSION; SEQUENTIAL MONTE CARLO; STOCHASTIC VOLATILITY

1. INTRODUCTION

Consider a sequence of probability measures $\{\pi_n\}_{n\in\mathbb{T}}$ where $\mathbb{T} = \{1, \ldots, P\}$. The distribution $\pi_n(d\mathbf{x}_n)$ is defined on a measurable space $(E_n, \mathcal{E}_n)$. For ease of presentation, we will assume that each $\pi_n(d\mathbf{x}_n)$ admits a density $\pi_n(\mathbf{x}_n)$ with respect to

Pierre Del Moral is Professor of Mathematics at the Université Nice Sophia Antipolis, Arnaud Doucet is Associate Professor of Computer Science and Statistics at the University of British Columbia and Ajay Jasra is Research Fellow in the Department of Mathematics at Imperial College London.

a σ-finite dominating measure denoted $d\mathbf{x}_n$ and that this density is only known up to a normalizing constant

$$\pi_n(\mathbf{x}_n) = \frac{\gamma_n(\mathbf{x}_n)}{Z_n}$$

where $\gamma_n : E_n \to \mathbb{R}^+$ is known pointwise, but Z_n might be unknown. We will refer to n as the time index; this variable is simply a counter and need not have any relation with 'real time'. We also denote by S_n the support of π_n, i.e. $S_n = \{\mathbf{x}_n \in E_n : \pi_n(\mathbf{x}_n) > 0\}$.

In this paper, we focus upon sampling from the distributions $\{\pi_n\}_{n\in\mathbb{T}}$ and estimating their normalizing constants $\{Z_n\}_{n\in\mathbb{T}}$ *sequentially*; i.e. first sampling from π_1 and estimating Z_1, then sampling from π_2 and estimating Z_2 and so on. Many computational problems in Bayesian statistics, computer science, physics and applied mathematics can be formulated as sampling from a sequence of probability distributions and estimating their normalizing constants; see for example Del Moral (2004), Iba (2001) or Liu (2001).

1.1. *Motivating Examples*

We now list a few motivating examples.

Optimal filtering for nonlinear non-Gaussian state-space models. Consider an unobserved Markov process $\{X_n\}_{n\geq 1}$ on space $(\mathsf{X}^{\mathbb{N}}, \mathcal{X}^{\mathbb{N}}, \mathbb{P}_\mu)$ where $\mathbb{P}_\mu$ has initial distribution μ and transition density f. The observations $\{Y_n\}_{n\geq 1}$ are assumed to be conditionally independent given $\{X_n\}_{n\geq 1}$ and $Y_n|\,(X_n = x) \sim g(\cdot|\,x)$. In this case we define $E_n = \mathsf{X}^n$, $\mathbf{x}_n = x_{1:n}$ ($x_{1:n} \triangleq (x_1, \ldots, x_n)$) and

$$\gamma_n(\mathbf{x}_n) = \mu(x_1)\, g(y_1|\,x_1) \left\{ \prod_{k=2}^{n} f(x_k|\,x_{k-1})\, g(y_k|\,x_k) \right\}. \tag{1}$$

This model is appropriate to describe a vast number of practical problems and has been the main application of SMC methods (Doucet *et al.*, 2001). It should be noted that MCMC is not appropriate in such contexts. This is because running P MCMC algorithms, either sequentially (and not using the previous samples in an efficient way) or in parallel is too computationally expensive for large P. Moreover, one often has real-time constraints and thus, in this case, MCMC is not a viable alternative to SMC.

Tempering/annealing. Suppose we are given the problem of simulating from $\pi(\mathbf{x}) \propto \gamma(\mathbf{x})$ defined on E and estimating its normalizing constant $Z = \int_E \gamma(\mathbf{x})\, d\mathbf{x}$. If π is a high-dimensional, non-standard distribution then, to improve the exploration ability of an algorithm, it is attractive to consider an inhomogeneous sequence of P distributions to move 'smoothly' from a tractable distribution $\pi_1 = \mu_1$ to the target distribution $\pi_P = \pi$. In this case we have $E_n = E\ \forall n \in \mathbb{T}$ and, for example, we could select a geometric path (Gelman and Meng 1996; Neal 2001)

$$\gamma_n(\mathbf{x}_n) = [\gamma(\mathbf{x}_n)]^{\zeta_n} [\mu_1(\mathbf{x}_n)]^{1-\zeta_n}$$

with $0 \leq \zeta_1 < \cdots < \zeta_P = 1$. Alternatively, to maximize $\pi(\mathbf{x})$, we could consider $\gamma_n(\mathbf{x}_n) = [\gamma(\mathbf{x}_n)]^{\zeta_n}$ where $\{\zeta_n\}$ is such that $0 < \zeta_1 < \cdots < \zeta_P$ and $1 << \zeta_P$ to ensure that $\pi_P(\mathbf{x})$ is concentrated around the set of global maxima of $\pi(\mathbf{x})$. We will demonstrate that it is possible to perform this task using SMC whereas, typically, one samples from these distributions using either an MCMC kernel of invariant distribution $\pi^*(\mathbf{x}_{1:P}) \propto \gamma_1(\mathbf{x}_1) \times \cdots \times \gamma_P(\mathbf{x}_P)$ (parallel tempering; see Jasra *et al.* (2005b) for a review) or an inhomogeneous sequence of MCMC kernels (simulated annealing).

Optimal filtering for partially observed point processes. Consider a marked point process $\{c_n, \varepsilon_n\}_{n\geq 1}$ on the real line where c_n is the arrival time of the nth point ($c_n > c_{n-1}$) and ε_n its associated real-valued mark. We assume the marks $\{\varepsilon_n\}$ (resp. the interarrival times $T_n = c_n - c_{n-1}$, $T_1 = c_1 > 0$) are i.i.d. of density f_ε (resp. f_T). We denote by $y_{1:m_t}$ the observations available up to time t and the associated likelihood $g\left(y_{1:m_t} | \{c_n, \varepsilon_n\}_{n\geq 1}\right) = g\left(y_{1:m_t} | c_{1:k_t}, \varepsilon_{1:k_t}\right)$ where $k_t = \arg\max\{i : c_i < t\}$. We are interested in the sequence of posterior distributions at times $\{d_n\}_{n\geq 1}$ where $d_n > d_{n-1}$. In this case, we have $\mathbf{x}_n = \left(k_{d_n}, c_{1:k_{d_n}}, \varepsilon_{1:k_{d_n}}\right)$ and

$$\pi_n(\mathbf{x}_n) \propto g\left(y_{1:m_{d_n}} \mid k_{d_n}, c_{1:d_n}, \varepsilon_{1:k_{d_n}}\right) \left(\prod_{k=1}^{k_{d_n}} f_\varepsilon(\varepsilon_k) f_T(c_k - c_{k-1})\right) p_{d_n}(k_{d_n} \mid c_{1:d_n}),$$

where $c_0 = 0$ by convention. These target distributions are all defined on the same space $E_n = E = \biguplus_{k=0}^{\infty}\{k\} \times A_k \times \mathbb{R}^k$ where $A_k = \{c_{1:k} : 0 < c_1 < \cdots < c_k < \infty\}$ but the support S_n of $\pi_n(\mathbf{x}_n)$ is restricted to $\biguplus_{k=0}^{\infty}\{k\} \times A_{k,d_n} \times \mathbb{R}^k$ where $A_{k,d_n} = \{c_{1:k} : 0 < c_1 < \cdots < c_k < d_n\}$, i.e. $S_{n-1} \subset S_n$. This is a sequential, trans-dimensional Bayesian inference problem (see also Del Moral *et al.*, 2006).

1.2. *Sequential Monte Carlo and Structure of the Article*

SMC methods are a set of simulation-based methods developed to solve the problems listed above, and many more. At a given time n, the basic idea is to obtain a large collection of N weighted random samples $\left\{W_n^{(i)}, \mathbf{X}_n^{(i)}\right\}$ ($i = 1, \ldots, N$, $W_n^{(i)} > 0$; $\sum_{i=1}^N W_n^{(i)} = 1$), $\left\{\mathbf{X}_n^{(i)}\right\}$ being named particles, whose empirical distribution converges asymptotically ($N \to \infty$) to π_n; i.e. for any π_n-integrable function $\varphi : E_n \to \mathbb{R}$

$$\sum_{i=1}^{N} W_n^{(i)} \varphi\left(\mathbf{X}_n^{(i)}\right) \longrightarrow \int_{E_n} \varphi(\mathbf{x}_n)\, \pi_n(\mathbf{x}_n)\, d\mathbf{x}_n \text{ almost surely.}$$

Throughout we will denote $\int_{E_n} \varphi(\mathbf{x}_n)\, \pi_n(\mathbf{x}_n)\, d\mathbf{x}_n$ by $\mathbb{E}_{\pi_n}\left(\varphi(\mathbf{X}_n)\right)$. These particles are carried forward over time using a combination of sequential Importance Sampling (IS) and resampling ideas. Broadly speaking, when an approximation $\left\{W_{n-1}^{(i)}, \mathbf{X}_{n-1}^{(i)}\right\}$ of π_{n-1} is available, we seek to move the particles at time n so that they approximate π_n (we will assume that this is not too dissimilar to π_{n-1}), that is, to obtain $\left\{\mathbf{X}_n^{(i)}\right\}$. However, since the $\left\{\mathbf{X}_n^{(i)}\right\}$ are not distributed according to π_n, it is necessary to reweight them with respect to π_n, through IS, to obtain $\left\{W_n^{(i)}\right\}$. In addition, if the variance of the weights is too high, measured through the effective sample size (ESS) (Liu, 2001), then particles with low weights are eliminated and particles with high weights are multiplied to focus the computational efforts in 'promising' parts of the space. The resampled particles are approximately distributed according to π_n; this approximation improves as $N \to \infty$.

In comparison to MCMC, SMC methods are currently limited, both in terms of their application and framework. In terms of the former, Resample-Move (Chopin, 2002; Gilks and Berzuini, 2001) is an SMC algorithm which may be used in the same context as MCMC but is not, presumably due to the limited exposure, of applied statisticians, to this algorithm. In terms of the latter, only simple moves have been previously applied to propagate particles, which has serious consequences on the performance of such algorithms. We present here a simple generic mechanism relying on auxiliary variables that allows us to extend the SMC methodology in a principled manner. Moreover, we also reinterpret SMC algorithms as particle approximations of nonlinear and nonhomogeneous MCMC algorithms (Del Moral, 2004). This allows us to introduce alternative SMC and iterative self-interacting approximation (Del Moral and Miclo, 2004; 2006) schemes. We do not present any theoretical results here but a survey of precise convergence for SMC algorithms can be found in Del Moral (2004) whereas the self-interacting algorithms can be studied using the techniques developed in Del Moral and Miclo (2004; 2006) and Andrieu *et al.* (2006).

The rest of the paper is organized as follows. Firstly, in Section 2, we review the limitations of the current SMC methodology, present some extensions and describe a generic algorithm to sample from any sequence of distributions $\{\pi_n\}_{n\in\mathbb{T}}$ and estimate $\{Z_n\}_{n\in\mathbb{T}}$ defined in the introduction. Secondly, in Section 3, we reinterpret SMC as an approximation to nonlinear MCMC and discuss an alternative self-interacting approximation. Finally, in Section 4, we present three original applications of our methodology: sequential Bayesian inference for bearings-only tracking (*e.g.*, Gilks and Berzuini, 2001); Bayesian probit regression (*e.g.*, Albert and Chib, 1993) and sequential Bayesian inference for stochastic volatility models (Roberts *et al.*, 2004).

2. SEQUENTIAL MONTE CARLO METHODOLOGY

2.1. *Sequential Importance Sampling*

At time $n-1$, we are interested in estimating π_{n-1} and Z_{n-1}. Let us introduce an importance distribution η_{n-1}. IS is based upon the following identities

$$\begin{aligned}\pi_{n-1}(\mathbf{x}_{n-1}) &= Z_{n-1}^{-1} w_{n-1}(\mathbf{x}_{n-1})\,\eta_{n-1}(\mathbf{x}_{n-1}),\\ Z_{n-1} &= \int_{E_{n-1}} w_{n-1}(\mathbf{x}_{n-1})\eta_{n-1}(\mathbf{x}_{n-1})d\mathbf{x}_{n-1},\end{aligned} \tag{2}$$

where the unnormalized importance weight function is equal to

$$w_{n-1}(\mathbf{x}_{n-1}) = \frac{\gamma_{n-1}(\mathbf{x}_{n-1})}{\eta_{n-1}(\mathbf{x}_{n-1})}. \tag{3}$$

By sampling N particles $\left\{\mathbf{X}_{n-1}^{(i)}\right\}$ $(i = 1, \ldots, N)$ from η_{n-1} and substituting the empirical measure

$$\eta_{n-1}^N(d\mathbf{x}_{n-1}) = \frac{1}{N}\sum_{i=1}^{N} \delta_{\mathbf{X}_{n-1}^{(i)}}(d\mathbf{x}_{n-1})$$

(where δ_x is Dirac measure) to η_{n-1} into (2) we obtain an approximation of π_{n-1} and Z_{n-1} given by

$$\pi_{n-1}^N(d\mathbf{x}_{n-1}) = \sum_{i=1}^N W_{n-1}^{(i)} \delta_{X_{n-1}^{(i)}}(d\mathbf{x}_{n-1}), \tag{4}$$

$$Z_{n-1}^N \quad = \tfrac{1}{N} \textstyle\sum_{i=1}^N w_{n-1}\left(\mathbf{X}_{n-1}^{(i)}\right), \tag{5}$$

where

$$W_{n-1}^{(i)} = \frac{w_{n-1}\left(\mathbf{X}_{n-1}^{(i)}\right)}{\sum_{j=1}^N w_{n-1}\left(\mathbf{X}_{n-1}^{(j)}\right)}.$$

We now seek to estimate π_n and Z_n. To achieve this we propose to build the importance distribution η_n based upon the current importance distribution η_{n-1} of the particles $\left\{\mathbf{X}_{n-1}^{(i)}\right\}$. We simulate each new particle $\mathbf{X}_n^{(i)}$ according to a Markov kernel $K_n : E_{n-1} \to \mathcal{P}(E_n)$ (where $\mathcal{P}(E_n)$ is the class of probability measures on E_n), i.e. $\mathbf{X}_n^{(i)} \sim K_n\left(\mathbf{x}_{n-1}^{(i)}, \cdot\right)$ so that

$$\eta_n(\mathbf{x}_n) = \eta_{n-1}K_n(\mathbf{x}_n) = \int \eta_{n-1}(d\mathbf{x}_{n-1})\, K_n(\mathbf{x}_{n-1}, \mathbf{x}_n). \tag{6}$$

2.2. *Selection of Transition Kernels*

It is clear that the optimal importance distribution, in the sense of minimizing the variance of (3), is $\eta_n(\mathbf{x}_n) = \pi_n(\mathbf{x}_n)$. Therefore, the optimal transition kernel is simply $K_n(\mathbf{x}_{n-1}, \mathbf{x}_n) = \pi_n(\mathbf{x}_n)$. This choice is typically impossible to use (except perhaps at time 1) and we have to formulate sub-optimal choices. We first review conditionally optimal moves and then discuss some alternatives.

2.2.1. *Conditionally Optimal Moves*

Suppose that we are interested in moving from $\mathbf{x}_{n-1} = (\mathbf{u}_{n-1}, \mathbf{v}_{n-1}) \in E_{n-1} = U_{n-1} \times V_{n-1}$ to $\mathbf{x}_n = (\mathbf{u}_{n-1}, \mathbf{v}_n) \in E_n = U_{n-1} \times V_n$ ($V_n \neq \emptyset$). We adopt the following kernel

$$K_n(\mathbf{x}_{n-1}, \mathbf{x}_n) = \mathbb{I}_{\mathbf{u}_{n-1}}(\mathbf{u}_n) q_n(\mathbf{x}_{n-1}, \mathbf{v}_n)$$

where $q_n(\mathbf{x}_{n-1}, \mathbf{v}_n)$ is a probability density of moving from $\mathbf{x}_{n-1}$ to $\mathbf{v}_n$. Consequently, we have

$$\eta_n(\mathbf{x}_n) = \int_{V_{n-1}} \eta_{n-1}(\mathbf{u}_n, d\mathbf{v}_{n-1})\, q_n((\mathbf{u}_n, \mathbf{v}_{n-1}), \mathbf{v}_n).$$

In order to select $q_n(\mathbf{x}_{n-1}, \mathbf{v}_n)$, a sensible strategy consists of using the distribution minimizing the variance of $w_n(\mathbf{x}_n)$ conditional on $\mathbf{u}_{n-1}$. One can easily check that the optimal distribution for this criterion is given by a Gibbs move

$$q_n^{opt}(\mathbf{x}_{n-1}, \mathbf{v}_n) = \pi_n(\mathbf{v}_n | \mathbf{u}_{n-1})$$

and the associated importance weight satisfies (even if $V_n = \emptyset$)

$$w_n(\mathbf{x}_n) = \frac{\gamma_n(\mathbf{u}_{n-1})}{\eta_{n-1}(\mathbf{u}_{n-1})}. \tag{7}$$

Contrary to the Gibbs sampler, the SMC framework not only requires being able to sample from the full conditional distribution $\pi_n(\mathbf{v}_n | \mathbf{u}_{n-1})$ but also being able to evaluate $\gamma_n(\mathbf{u}_{n-1})$ and $\eta_{n-1}(\mathbf{u}_{n-1})$.

In cases where it is possible to sample from $\pi_n(\mathbf{v}_n|\mathbf{u}_{n-1})$ but impossible to compute $\gamma_n(\mathbf{u}_{n-1})$ and/or $\eta_{n-1}(\mathbf{u}_{n-1})$, we can use an attractive property of IS: we do not need to compute exactly (7), we can use an unbiased estimate of it. We have the identity

$$\gamma_n(\mathbf{u}_{n-1}) = \widehat{\gamma}_n(\mathbf{u}_{n-1}) \int \frac{\gamma_n(\mathbf{u}_{n-1}, \mathbf{v}_n)}{\widehat{\gamma}_n(\mathbf{u}_{n-1}, \mathbf{v}_n)} \widehat{\pi}_n(\mathbf{v}_n|\mathbf{u}_{n-1})\, d\mathbf{v}_n \tag{8}$$

where $\widehat{\gamma}_n(\mathbf{u}_{n-1}, \mathbf{v}_n)$ is selected as an approximation of $\gamma_n(\mathbf{u}_{n-1}, \mathbf{v}_n)$ such that $\int \widehat{\gamma}_n(\mathbf{u}_{n-1}, \mathbf{v}_n)\, d\mathbf{v}_n$ can be computed analytically and it is easy to sample from its associated full conditional $\widehat{\pi}_n(\mathbf{v}_n|\mathbf{u}_{n-1})$. We can calculate an unbiased estimate of $\gamma_n(\mathbf{u}_{n-1})$ using samples from $\widehat{\pi}_n(\mathbf{v}_n|\mathbf{u}_{n-1})$. We also have

$$\frac{1}{\eta_{n-1}(\mathbf{u}_{n-1})} = \frac{1}{\widehat{\eta}_{n-1}(\mathbf{u}_{n-1})} \int \frac{\widehat{\eta}_{n-1}(\mathbf{u}_{n-1}, \mathbf{v}_{n-1})}{\eta_{n-1}(\mathbf{u}_{n-1}, \mathbf{v}_{n-1})} \eta_{n-1}(\mathbf{v}_{n-1}|\mathbf{u}_{n-1})\, d\mathbf{v}_{n-1} \tag{9}$$

where $\widehat{\eta}_{n-1}(\mathbf{u}_{n-1}, \mathbf{v}_{n-1})$ is selected as an approximation of $\eta_{n-1}(\mathbf{u}_{n-1}, \mathbf{v}_{n-1})$ such that $\int \widehat{\eta}_{n-1}(\mathbf{u}_{n-1}, \mathbf{v}_{n-1})\, d\mathbf{v}_{n-1}$ can be computed analytically. So if we can sample from $\eta_{n-1}(\mathbf{v}_{n-1}|\mathbf{u}_{n-1})$, we can calculate an unbiased estimate of (9). This idea has a limited range of applications as in complex cases we do not necessarily have a closed-form expression for $\eta_{n-1}(\mathbf{x}_{n-1})$. However, if one has resampled particles at time $k \leq n-1$, then one has (approximately) $\eta_{n-1}(\mathbf{x}_{n-1}) = \pi_k K_{k+1} K_{k+2} \cdots K_{n-1}(\mathbf{x}_{n-1})$.

2.2.2. *Approximate Gibbs Moves*

In the previous subsection, we have seen that conditionally optimal moves correspond to Gibbs moves. However, in many applications the full conditional distribution $\pi_n(\mathbf{v}_n|\mathbf{u}_{n-1})$ cannot be sampled from. Even if it is possible to sample from it, one might not be able to get a closed-form expression for $\gamma_n(\mathbf{u}_{n-1})$ and we need an approximation $\widehat{\pi}_n(\mathbf{v}_n|\mathbf{u}_{n-1})$ of $\pi_n(\mathbf{v}_n|\mathbf{u}_{n-1})$ to compute an unbiased estimate of it with low variance. Alternatively, we can simply use the following transition kernel

$$K_n(\mathbf{x}_{n-1}, \mathbf{x}_n) = \mathbb{I}_{\mathbf{u}_{n-1}}(\mathbf{u}_n)\, \widehat{\pi}_n(\mathbf{v}_n|\mathbf{u}_{n-1}) \tag{10}$$

and the associated importance weight is given by

$$w_n(\mathbf{x}_n) = \frac{\gamma_n(\mathbf{u}_{n-1}, \mathbf{v}_n)}{\eta_{n-1}(\mathbf{u}_{n-1})\, \widehat{\pi}_n(\mathbf{v}_n|\mathbf{u}_{n-1})}. \tag{11}$$

Proceeding this way, we bypass the estimation of $\gamma_n(\mathbf{u}_{n-1})$ which appeared in (7). However, we still need to compute $\eta_{n-1}(\mathbf{u}_{n-1})$ or to obtain an unbiased estimate of its inverse. Unfortunately, this task is very complex except when $\mathbf{u}_{n-1} = \mathbf{x}_{n-1}$ (*i.e.*, $V_{n-1} = \emptyset$) in which case we can rewrite (11) as

$$w_n(\mathbf{x}_n) = w_{n-1}(\mathbf{x}_{n-1}) \frac{\gamma_n(\mathbf{x}_{n-1}, \mathbf{v}_n)}{\gamma_n(\mathbf{x}_{n-1})\, \widehat{\pi}_n(\mathbf{v}_n|\mathbf{x}_{n-1})}. \tag{12}$$

This strategy is clearly limited as it can only be used when $E_n = E_{n-1} \times V_n$.

2.2.3. *MCMC and Adaptive moves*

To move from $\mathbf{x}_{n-1} = (\mathbf{u}_{n-1}, \mathbf{v}_{n-1})$ to $\mathbf{x}_n = (\mathbf{u}_{n-1}, \mathbf{v}_n)$ (via K_n), we can adopt an MCMC kernel of invariant distribution $\pi_n(\mathbf{v}_n | \mathbf{u}_{n-1})$. Unlike standard MCMC, there are no (additional) complicated mathematical conditions required to ensure that the usage of adaptive kernels leads to convergence. This is because SMC relies upon IS methodology, that is, we correct for sampling from the wrong distribution via the importance weight. In particular, this allows us to use transition kernels which at time n depends on π_{n-1}, i.e. the 'theoretical' transition kernel is of the form $K_{n,\pi_{n-1}}(\mathbf{x}_{n-1}, \mathbf{x}_n)$ and is approximated practically by $K_{n,\pi_{n-1}^N}(\mathbf{x}_{n-1}, \mathbf{x}_n)$. This was proposed and justified theoretically in Crisan and Doucet (2000). An appealing application is described in Chopin (2002) where the variance of $\widehat{\pi}_{n-1}^N$ is used to scale the proposal distribution of an independent MH step of invariant distribution π_n. In Jasra *et al.* (2005a), one fits a Gaussian mixture model to the particles so as to design efficient trans-dimensional moves in the spirit of Green (2003).

A severe drawback of the strategies mentioned above, is the ability to implement them. This is because we cannot always compute the resulting marginal importance distribution $\eta_n(\mathbf{x}_n)$ given by (6) and, hence, the importance weight $w_n(\mathbf{x}_n)$. In Section 2.3 we discuss how we may solve this problem.

2.2.4. *Mixture of Moves*

For complex MCMC problems, one typically uses a combination of MH steps where the parameter components are updated by sub-blocks. Similarly, to sample from high dimensional distributions, a practical SMC sampler will update the components of $\mathbf{x}_n$ via sub-blocks; a mixture of transition kernels can be used at each time n. Let us assume $K_n(\mathbf{x}_{n-1}, \mathbf{x}_n)$ is of the form

$$K_n(\mathbf{x}_{n-1}, \mathbf{x}_n) = \sum_{m=1}^{M} \alpha_{n,m}(\mathbf{x}_{n-1}) K_{n,m}(\mathbf{x}_{n-1}, \mathbf{x}_n) \tag{13}$$

where

$$\alpha_{n,m}(\mathbf{x}_{n-1}) \geq 0, \qquad \sum_{m=1}^{M} \alpha_{n,m}(\mathbf{x}_{n-1}) = 1,$$

and $\{K_{n,m}\}$ is a collection of transition kernels. Unfortunately, the direct calculation of the importance weight (6) associated to (13) will be impossible in most cases as $\eta_{n-1}K_{n,m}(\mathbf{x}_n)$ does not admit a closed-form expression. Moreover, even if this were the case, (13) would be expensive to compute pointwise if M is large.

2.2.5. *Summary*

IS, the basis of SMC methods, allows us to consider complex moves including adaptive kernels or non-reversible trans-dimensional moves. In this respect, it is much more flexible than MCMC. However, the major limitation of IS is that it requires the ability to compute the associated importance weights or unbiased estimates of them. In all but simple situations, this is impossible and this severely restricts the application of this methodology. In the following section, we describe a simple auxiliary variable method that allows us to deal with this problem.

2.3. *Auxiliary Backward Markov Kernels*

A simple solution would consist of approximating the importance distribution $\eta_n(\mathbf{x}_n)$ via

$$\eta_{n-1}^N K_n(\mathbf{x}_n) = \frac{1}{N}\sum_{i=1}^N K_n\left(\mathbf{X}_{n-1}^{(i)}, \mathbf{x}_n\right).$$

This approach suffers from two major problems. First, the computational complexity of the resulting algorithm would be in $O(N^2)$ which is prohibitive. Second, it is impossible to compute $K_n(\mathbf{x}_{n-1}, \mathbf{x}_n)$ pointwise in important scenarios, e.g. when K_n is an Metropolis–Hastings (MH) kernel of invariant distribution π_n.

We present a simple auxiliary variable idea to deal with this problem (Del Moral *et al.*, 2006). For each forward kernel $K_n : E_{n-1} \to \mathcal{P}(E_n)$, we associate a backward (in time) Markov transition kernel $L_{n-1} : E_n \to \mathcal{P}(E_{n-1})$ and define a new sequence of target distributions $\{\widetilde{\pi}_n(\mathbf{x}_{1:n})\}$ on $E_{1:n} \triangleq E_1 \times \cdots \times E_n$ through

$$\widetilde{\pi}_n(\mathbf{x}_{1:n}) = \frac{\widetilde{\gamma}_n(\mathbf{x}_{1:n})}{Z_n}$$

where

$$\widetilde{\gamma}_n(\mathbf{x}_{1:n}) = \gamma_n(\mathbf{x}_n)\prod_{k=1}^{n-1} L_k(\mathbf{x}_{k+1}, \mathbf{x}_k).$$

By construction, $\widetilde{\pi}_n(\mathbf{x}_{1:n})$ admits $\pi_n(\mathbf{x}_n)$ as a marginal and Z_n as a normalizing constant. We approximate $\widetilde{\pi}_n(\mathbf{x}_{1:n})$ using IS by using the joint importance distribution

$$\eta_n(\mathbf{x}_{1:n}) = \eta_1(\mathbf{x}_1)\prod_{k=2}^{n} K_k(\mathbf{x}_{k-1}, \mathbf{x}_k).$$

The associated importance weight satisfies

$$\begin{aligned} w_n(\mathbf{x}_{1:n}) &= \tfrac{\widetilde{\gamma}_n(\mathbf{x}_{1:n})}{\eta_n(\mathbf{x}_{1:n})} \\ &= w_{n-1}(\mathbf{x}_{1:n-1})\,\widetilde{w}_n(\mathbf{x}_{n-1}, \mathbf{x}_n), \end{aligned} \tag{14}$$

where the incremental importance weight $\widetilde{w}_n(\mathbf{x}_{n-1}, \mathbf{x}_n)$ is given by

$$\widetilde{w}_n(\mathbf{x}_{n-1}, \mathbf{x}_n) = \frac{\gamma_n(\mathbf{x}_n) L_{n-1}(\mathbf{x}_n, \mathbf{x}_{n-1})}{\gamma_{n-1}(\mathbf{x}_{n-1}) K_n(\mathbf{x}_{n-1}, \mathbf{x}_n)}.$$

Given that this Radon-Nikodym derivative is well-defined, the method will produce asymptotically $(N \to \infty)$ consistent estimates of $\mathbb{E}_{\widetilde{\pi}_n}\left(\varphi(\mathbf{X}_{1:n})\right)$ and Z_n. However, the performance of the algorithm will be dependent upon the choice of the kernels $\{L_k\}$.

2.3.1. *Optimal Backward Kernels*

Del Moral et al. (2006) establish that the backward kernels which minimize the variance of the importance weights, $w_n(\mathbf{x}_{1:n})$, are given by

$$L_k^{opt}(\mathbf{x}_{k+1}, \mathbf{x}_k) = \frac{\eta_k(\mathbf{x}_k) K_{k+1}(\mathbf{x}_k, \mathbf{x}_{k+1})}{\eta_{k+1}(\mathbf{x}_{k+1})} \tag{15}$$

for $k = 1, ..., n-1$. This can be verified easily by noting that

$$\eta_n(\mathbf{x}_{1:n}) = \eta_n(\mathbf{x}_n)\prod_{k=1}^{n-1} L_k^{opt}(\mathbf{x}_{k+1}, \mathbf{x}_k).$$

It is typically impossible, in practice, to use these optimal backward kernels as they rely on marginal distributions which do not admit any closed-form expression. However, this suggests that we should select them as an approximation to (15). The key point is that, even if they are different from (15), the algorithm will still provide asymptotically consistent estimates.

Compared to a 'theoretical' algorithm computing the weights (3), the price to pay for avoiding to compute $\eta_n(\mathbf{x}_n)$ (i.e. not using $L_k^{opt}(\mathbf{x}_{k+1}, \mathbf{x}_k)$) is that the variance of the Monte Carlo estimates based upon (14) will be larger. For example, even if we set $\pi_n(\mathbf{x}_n) = \pi(\mathbf{x}_n)$ and $K_n(\mathbf{x}_{n-1}, \mathbf{x}_n) = K(\mathbf{x}_{n-1}, \mathbf{x}_n)$ is an ergodic MCMC kernel of invariant distribution π then the variance of $w_n(\mathbf{x}_{1:n})$ will fail to stabilize (or become infinite in some cases) over time for any backward kernel $L_k(\mathbf{x}_{k+1}, \mathbf{x}_k) \neq L_k^{opt}(\mathbf{x}_{k+1}, \mathbf{x}_k)$ whereas the variance of (3) will decrease towards zero. The resampling step in SMC will deal with this problem by resetting the weights when their variance is too high.

At time n, the backward kernels $\{L_k(\mathbf{x}_{k+1}, \mathbf{x}_k)\}$ for $k = 1, ..., n-2$ have already been selected and we are interested in some approximations of $L_{n-1}^{opt}(\mathbf{x}_n, \mathbf{x}_{n-1})$ controlling the evolution of the variance of $w_n(\mathbf{x}_{1:n})$.

2.3.2. *Suboptimal Backward Kernels*

• *Substituting π_{n-1} for η_{n-1}.* Equation (15) suggests that a sensible sub-optimal strategy consists of substituting π_{n-1} for η_{n-1} to obtain

$$L_{n-1}(\mathbf{x}_n, \mathbf{x}_{n-1}) = \frac{\pi_{n-1}(\mathbf{x}_{n-1}) K_n(\mathbf{x}_{n-1}, \mathbf{x}_n)}{\pi_{n-1} K_n(\mathbf{x}_n)} \tag{16}$$

which yields

$$\widetilde{w}_n(\mathbf{x}_{n-1}, \mathbf{x}_n) = \frac{\gamma_n(\mathbf{x}_n)}{\int \gamma_{n-1}(d\mathbf{x}_{n-1}) K_n(\mathbf{x}_{n-1}, \mathbf{x}_n)}. \tag{17}$$

It is often more convenient to use (17) than (15) as $\{\gamma_n\}$ is known analytically, whilst $\{\eta_n\}$ is not. It should be noted that if particles have been resampled at time $n-1$, then η_{n-1} is indeed approximately equal to π_{n-1} and thus (15) is equal to (16).

• *Gibbs and Approximate Gibbs Moves.* Consider the conditionally optimal move described earlier where

$$K_n(\mathbf{x}_{n-1}, \mathbf{x}_n) = \mathbb{I}_{\mathbf{u}_{n-1}}(\mathbf{u}_n)\, \pi_n(\mathbf{v}_n | \mathbf{u}_{n-1}). \tag{18}$$

In this case (16) and (17) are given by

$$L_{n-1}(\mathbf{x}_n, \mathbf{x}_{n-1}) = \mathbb{I}_{\mathbf{u}_n}(\mathbf{u}_{n-1})\, \pi_{n-1}(\mathbf{v}_{n-1} | \mathbf{u}_{n-1}),$$

$$\widetilde{w}_n(\mathbf{x}_{n-1}, \mathbf{x}_n) = \frac{\gamma_n(\mathbf{u}_{n-1})}{\gamma_{n-1}(\mathbf{u}_{n-1})}.$$

An unbiased estimate of $\widetilde{w}_n(\mathbf{x}_{n-1}, \mathbf{x}_n)$ can also be computed using the techniques described in Section 2.2.1. When it is impossible to sample from $\pi_n(\mathbf{v}_n | \mathbf{u}_{n-1})$ and/or compute $\widetilde{w}_n(\mathbf{x}_{n-1}, \mathbf{x}_n)$, we may be able to construct an approximation $\widehat{\pi}_n(\mathbf{v}_n | \mathbf{u}_{n-1})$ of $\pi_n(\mathbf{v}_n | \mathbf{u}_{n-1})$ to sample the particles and another approximation $\widehat{\pi}_{n-1}(\mathbf{v}_{n-1} | \mathbf{u}_{n-1})$ of $\pi_{n-1}(\mathbf{v}_{n-1} | \mathbf{u}_{n-1})$ to obtain

$$L_{n-1}(\mathbf{x}_n, \mathbf{x}_{n-1}) = \mathbb{I}_{\mathbf{u}_n}(\mathbf{u}_{n-1})\, \widehat{\pi}_{n-1}(\mathbf{v}_{n-1} | \mathbf{u}_{n-1}), \tag{19}$$

$$\widetilde{w}_n(\mathbf{x}_{n-1}, \mathbf{x}_n) = \frac{\gamma_n(\mathbf{u}_{n-1}, \mathbf{v}_n)\,\widehat{\pi}_{n-1}(\mathbf{v}_{n-1}|\,\mathbf{u}_{n-1})}{\gamma_{n-1}(\mathbf{u}_{n-1}, \mathbf{v}_{n-1})\,\widehat{\pi}_n(\mathbf{v}_n|\,\mathbf{u}_{n-1})}. \tag{20}$$

• *MCMC Kernels.* A generic alternative approximation of (16) can also be made when K_n is an MCMC kernel of invariant distribution π_n. This has been proposed explicitly in Jarzynski (1997), Neal (2001), and implicitly in all papers introducing MCMC moves within SMC, (*e.g.*, Chopin, 2002; Gilks and Berzuini, 2001). It is given by

$$L_{n-1}(\mathbf{x}_n, \mathbf{x}_{n-1}) = \frac{\pi_n(\mathbf{x}_{n-1})\,K_n(\mathbf{x}_{n-1}, \mathbf{x}_n)}{\pi_n(\mathbf{x}_n)} \tag{21}$$

and will be a good approximation of (16) if $\pi_{n-1} \approx \pi_n$; note that (21) is the reversal Markov kernel associated with K_n. In this case, the incremental weight satisfies

$$\widetilde{w}_n(\mathbf{x}_{n-1}, \mathbf{x}_n) = \frac{\gamma_n(\mathbf{x}_{n-1})}{\gamma_{n-1}(\mathbf{x}_{n-1})}. \tag{22}$$

This expression (22) is remarkable as it is easy to compute and valid *irrespective* of the MCMC kernel adopted. It is also counter-intuitive: if $K_n(\mathbf{x}_{n-1}, \mathbf{x}_n)$ is mixing quickly so that, approximately, $\mathbf{X}_n^{(i)} \sim \pi_n$ then the particles would still be weighted. The use of resampling helps to mitigate this problem; see Del Moral *et al.* (2006, Section 3.5) for a detailed discussion.

Contrary to (16), this approach does not apply in scenarios where $E_{n-1} = E_n$ but $S_{n-1} \subset S_n$ as discussed in Section 1 (optimal filtering for partially observed processes). Indeed, in this case

$$L_{n-1}(\mathbf{x}_n, \mathbf{x}_{n-1}) = \frac{\pi_n(\mathbf{x}_{n-1})\,K_n(\mathbf{x}_{n-1}, \mathbf{x}_n)}{\int_{S_{n-1}} \pi_n(\mathbf{x}_{n-1})\,K_n(\mathbf{x}_{n-1}, \mathbf{x}_n)\,d\mathbf{x}_{n-1}} \tag{23}$$

but the denominator of this expression is different from $\pi_n(\mathbf{x}_n)$ as the integration is over S_{n-1} and not S_n.

2.3.3. *Mixture of Markov Kernels*

When the transition kernel is given by a mixture of M moves as in (13), one should select $L_{n-1}(\mathbf{x}_n, \mathbf{x}_{n-1})$ as a mixture

$$L_{n-1}(\mathbf{x}_n, \mathbf{x}_{n-1}) = \sum_{m=1}^{M} \beta_{n-1,m}(\mathbf{x}_n)\,L_{n-1,m}(\mathbf{x}_n, \mathbf{x}_{n-1}) \tag{24}$$

where $\beta_{n-1,m}(\mathbf{x}_n) \geq 0$, $\sum_{m=1}^{M} \beta_{n-1,m}(\mathbf{x}_n) = 1$ and $\{L_{n-1,m}\}$ is a collection of backward transition kernels. Using (15), it is indeed easy to show that the optimal backward kernel corresponds to

$$\begin{aligned}
\beta_{n-1,m}^{opt}(\mathbf{x}_n) &\propto \int \alpha_{n,m}(\mathbf{x}_{n-1})\,\eta_{n-1}(\mathbf{x}_{n-1})\,K_n(\mathbf{x}_{n-1}, \mathbf{x}_n)\,d\mathbf{x}_{n-1},\\
L_{n-1,m}^{opt}(\mathbf{x}_n, \mathbf{x}_{n-1}) &= \frac{\alpha_{n,m}(\mathbf{x}_{n-1})\eta_{n-1}(\mathbf{x}_{n-1})K_n(\mathbf{x}_{n-1},\mathbf{x}_{nn})}{\int \alpha_{n,m}(\mathbf{x}_{n-1})\eta_{n-1}(\mathbf{x}_{n-1})K_n(\mathbf{x}_{n-1},\mathbf{x}_n)d\mathbf{x}_{n-1}}.
\end{aligned}$$

Various approximations to $\beta_{n-1,m}^{opt}(\mathbf{x}_n)$ and $L_{n-1,m}^{opt}(\mathbf{x}_n, \mathbf{x}_{n-1})$ have to be made in practice.

Moreover, to avoid computing a sum of M terms, we can introduce a discrete latent variable $M_n \in \mathcal{M}$, $\mathcal{M} = \{1, \ldots, M\}$ such that $\mathbb{P}(M_n = m) = \alpha_{n,m}(\mathbf{x}_{n-1})$ and perform IS on the extended space. This yields an incremental importance weight equal to

$$\widetilde{w}_n(\mathbf{x}_{n-1}, \mathbf{x}_n, m_n) = \frac{\gamma_n(\mathbf{x}_n)\,\beta_{n-1,m_n}(\mathbf{x}_n)\,L_{n-1,m_n}(\mathbf{x}_n, \mathbf{x}_{n-1})}{\gamma_{n-1}(\mathbf{x}_{n-1})\,\alpha_{n,m_n}(\mathbf{x}_{n-1})\,K_{n,m_n}(\mathbf{x}_{n-1}, \mathbf{x}_n)}.$$

2.4. *A Generic SMC Algorithm*

We now describe a generic SMC algorithm to approximate the sequence of targets $\{\pi_n\}$ based on kernel K_n; the extension to mixture of moves being straightforward. The particle representation is resampled using an (unbiased) systematic resampling scheme whenever the ESS at time n given by $\left[\sum_{i=1}^N (W_n^{(i)})^2\right]^{-1}$ is below a prespecified threshold, say $N/2$ (Liu, 2001).

- **At time** $n = 1$. *Sample* $\mathbf{X}_1^{(i)} \sim \eta_1$ *and compute* $W_1^{(i)} \propto w_1\left(\mathbf{X}_1^{(i)}\right)$.

  If ESS $<$ *Threshold, resample the particle representation* $\left\{W_1^{(i)}, \mathbf{X}_1^{(i)}\right\}$.

- **At time** n; $n \geq 2$. *Sample* $\mathbf{X}_n^{(i)} \sim K_n\left(\mathbf{X}_{n-1}^{(i)}, \cdot\right)$ *and compute* $W_n^{(i)} \propto W_{n-1}^{(i)}\widetilde{w}_n\left(\mathbf{X}_{n-1}^{(i)}, \mathbf{X}_n^{(i)}\right)$.

  If ESS $<$ *Threshold, resample the particle representation* $\left\{W_n^{(i)}, \mathbf{X}_n^{(i)}\right\}$.

The target π_n is approximated through

$$\pi_n^N(d\mathbf{x}_n) = \sum_{i=1}^N W_n^{(i)}\delta_{\mathbf{X}_n^{(i)}}(d\mathbf{x}_n).$$

In addition, the approximation $\left\{W_{n-1}^{(i)}, \mathbf{X}_n^{(i)}\right\}$ of $\pi_{n-1}(\mathbf{x}_{n-1})\,K_n(\mathbf{x}_{n-1}, \mathbf{x}_n)$ obtained after the sampling step allows us to approximate

$$\frac{Z_n}{Z_{n-1}} = \frac{\int \gamma_n(\mathbf{x}_n)\,d\mathbf{x}_n}{\int \gamma_{n-1}(\mathbf{x}_{n-1})\,d\mathbf{x}_{n-1}} \text{ by } \widehat{\frac{Z_n}{Z_{n-1}}} = \sum_{i=1}^N W_{n-1}^{(i)}\widetilde{w}_n\left(\mathbf{X}_{n-1}^{(i)}, \mathbf{X}_n^{(i)}\right). \tag{25}$$

Alternatively, it is possible to use path sampling (Gelman and Meng, 1998) to compute this ratio.

3. NONLINEAR MCMC, SMC AND SELF-INTERACTING APPROXIMATIONS

For standard Markov chains, the transition kernel, say Q_n, is a linear operator in the space of probability measures, *i.e.*, we have $\mathbf{X}_n \sim Q_n(\mathbf{X}_{n-1}, \cdot)$ and the distribution μ_n of $\mathbf{X}_n$ satisfies $\mu_n = \mu_{n-1}Q_n$. Nonlinear Markov chains are such that $\mathbf{X}_n \sim Q_{\mu_{n-1},n}(\mathbf{X}_{n-1}, \cdot)$, i.e. the transition of $\mathbf{X}_n$ depends not only on $\mathbf{X}_{n-1}$ but also on μ_{n-1} and we have

$$\mu_n = \mu_{n-1}Q_{n,\mu_{n-1}}. \tag{26}$$

In a similar fashion to MCMC, it is possible to design nonlinear Markov chain kernels admitting a fixed target π (Del Moral and Doucet 2003). Such a procedure is attractive as one can design nonlinear kernels with theoretically better mixing properties than linear kernels. Unfortunately, it is often impossible to simulate exactly such nonlinear Markov chains as we do not have a closed-form expression for μ_{n-1}. We now describe a general collection of nonlinear kernels and how to produce approximations of them.

3.1. *Nonlinear MCMC Kernels to Simulate from a Sequence of Distributions*

Suppose that we can construct a collection of nonlinear Markov kernels such that

$$\widetilde{\pi}_n = \widetilde{\pi}_{n-1} Q_{n,\widetilde{\pi}_{n-1}}$$

where $\{\widetilde{\pi}_n\}$ is the sequence of auxiliary target distributions (on $(E_{1:n}, \mathcal{E}_{1:n})$) associated to $\{\pi_n\}$ and $Q_{n,\mu} : \mathcal{P}(E_{1:n-1}) \times E_{n-1} \to \mathcal{P}(E_{1:n})$. The simplest transition kernel is given by

$$Q_{n,\mu}\left(\mathbf{x}_{1:n-1}, \mathbf{x}'_{1:n}\right) = \Psi_n\left(\mu \times K_n\right)\left(\mathbf{x}'_{1:n}\right) \tag{27}$$

where $\Psi_n : \mathcal{P}(E_{1:n}) \to \mathcal{P}(E_{1:n})$

$$\Psi_n(\nu)\left(d\mathbf{x}'_{1:n}\right) = \frac{\nu\left(d\mathbf{x}'_{1:n}\right)\widetilde{w}_n\left(\mathbf{x}'_{n-1}, \mathbf{x}'_n\right)}{\int \nu\left(d\mathbf{x}_{1:n}\right)\widetilde{w}_n\left(\mathbf{x}_{n-1}, \mathbf{x}_n\right)}.$$

is a Boltzmann–Gibbs distribution.

If $\widetilde{w}_n\left(\mathbf{x}_{n-1}, \mathbf{x}_n\right) \leq C_n$ for any $(\mathbf{x}_{n-1}, \mathbf{x}_n)$, we can also consider an alternative kernel given by

$$\begin{aligned} Q_{n,\mu}(\mathbf{x}_{1:n-1}, d\mathbf{x}'_{1:n}) &= K_n(\mathbf{x}_{n-1}, d\mathbf{x}'_n)\frac{\widetilde{w}_n(\mathbf{x}_{n-1}, \mathbf{x}'_n)}{C_n}\delta_{\mathbf{x}_{1:n-1}}\left(d\mathbf{x}'_{1:n-1}\right) \\ &+ \left(1 - \int_{E_{1:n}} K_n(\mathbf{x}_{n-1}, d\mathbf{x}'_n)\frac{\widetilde{w}_n(\mathbf{x}_{n-1}, \mathbf{x}'_n)}{C_n}\delta_{\mathbf{x}_{1:n-1}}\left(d\mathbf{x}'_{1:n-1}\right)\right) \\ &\times \Psi_n(\mu \times K_n)(d\mathbf{x}'_{1:n}). \end{aligned} \tag{28}$$

This algorithm can be interpreted as a nonlinear version of the MH algorithm. Given $\mathbf{x}_{1:n-1}$ we sample $\mathbf{x}'_n \sim K_n(\mathbf{x}_{n-1}, \cdot)$ and with probability $\widetilde{w}_n(\mathbf{x}_{n-1}, \mathbf{x}'_n)/C_n$ we let $\mathbf{x}'_{1:n} = (\mathbf{x}_{1:n-1}, \mathbf{x}'_n)$, otherwise we sample a new $\mathbf{x}'_{1:n}$ from the Boltzmann–Gibbs distribution.

3.2. *SMC and Self-Interacting Approximations*

In order to simulate the nonlinear kernel, we need to approximate (26) given here by (27) or (28). The SMC algorithm described in Section 2 can be interpreted as a simple Monte Carlo implementation of (27). Whenever $\widetilde{w}_n\left(\mathbf{x}_{n-1}, \mathbf{x}_n\right) \leq C_n$, it is also possible to approximate (28) instead. Under regularity assumptions, it can be shown that this alternative Monte Carlo approximation has a lower asymptotic variance than (27) if multinomial resampling is used to sample from the Boltzmann–Gibbs distribution (Chapter 9 of Del Moral, 2004).

In cases where one does not have real-time constraints and the number P of target distributions $\{\pi_n\}$ is fixed it is possible to develop an alternative iterative approach. The idea consists of initializing the algorithm with some Monte Carlo estimates $\{\widetilde{\pi}_n^{N_0}\}$ of the targets consisting of empirical measures (that is $\frac{1}{N_0}\sum_{i=1}^{N_0}\delta_{X_{n,1:n}^{(i)}}$) of N_0 samples. For the sake of simplicity, we assume it is possible to sample exactly from $\widetilde{\pi}_1 = \pi_1$. Then the algorithm proceeds as follows at iteration i; the first iteration being indexed by $i = N_0 + 1$.

- **At time** $n = 1$. *Sample* $\mathbf{X}_{1,1}^{(i)} \sim \widetilde{\pi}_1$ *and set* $\widetilde{\pi}_1^i = \left(1 - \frac{1}{i}\right)\widetilde{\pi}_1^{i-1} + \frac{1}{i}\delta_{\mathbf{X}_{1,1}^{(i)}}$.
- **At time** n; $n = 2, ..., P$. *Sample* $\mathbf{X}_{n,1:n}^{(i)} \sim Q_{n,\widetilde{\pi}_{n-1}^i}\left(\mathbf{X}_{n-1,1:n-1}^{(i)}, \cdot\right)$ *and set* $\widetilde{\pi}_n^i = \left(1 - \frac{1}{i}\right)\widetilde{\pi}_n^{i-1} + \frac{1}{i}\delta_{\mathbf{X}_{n,1:n}^{(i)}}$.

In practice, we are interested only in $\{\pi_n\}$ and not $\{\widetilde{\pi}_n\}$ so we only need to store at time n the samples $\left\{\mathbf{X}_{n,n-1:n}^{(i)}\right\}$ asymptotically distributed according to $\pi_n(x_n) L_{n-1}(x_n, x_{n-1})$. We note that such stochastic processes, described above, are *self-interacting*; see Del Moral and Miclo (2004; 2006), Andrieu *et al.* (2006) and Brockwell and Doucet (2006) in the context of Monte Carlo simulation.

4. APPLICATIONS

4.1. *Block Sampling for Optimal Filtering*

4.1.1. *SMC Sampler*

We consider the class of nonlinear non-Gaussian state-space models discussed in Section 1. In this case the sequence of target distribution defined on $E_n = \mathsf{X}^n$ is given by (1). In the context where one has real-time constraints, we need to design a transition kernel K_n which updates only a fixed number of components of $\mathbf{x}_n$ to maintain a computational complexity independent of n.

The standard approach consists of moving from $\mathbf{x}_{n-1} = \mathbf{u}_{n-1}$ to $\mathbf{x}_n = (\mathbf{x}_{n-1}, x_n) = (\mathbf{u}_{n-1}, \mathbf{v}_n)$ using

$$\pi_n(\mathbf{v}_n | \mathbf{u}_{n-1}) = p(x_n | y_n, x_{n-1}) \propto f(x_n | x_{n-1}) g(y_n | x_n).$$

This distribution is often referred to (abusively) as the optimal importance distribution in the literature, e.g. Doucet et al. (2001); this should be understood as optimal *conditional upon* $\mathbf{x}_{n-1}$. In this case we can rewrite (7) as

$$w_n(\mathbf{x}_n) = w_{n-1}(\mathbf{x}_{n-1}) p(y_n | x_{n-1}) \propto w_{n-1}(\mathbf{x}_{n-1}) \frac{p(\mathbf{x}_{n-1} | y_{1:n})}{p(\mathbf{x}_{n-1} | y_{1:n-1})}. \tag{29}$$

If one can sample from $p(x_n | y_n, x_{n-1})$ but cannot compute (29) in closed-form then we can obtain an unbiased estimate of it using an easy to sample distribution approximating it

$$\widehat{\pi}_n(\mathbf{v}_n | \mathbf{u}_{n-1}) = \widehat{p}(x_n | y_n, x_{n-1}) = \frac{\widehat{f}(x_n | x_{n-1}) \widehat{g}(y_n | x_n)}{\int \widehat{f}(x_n | x_{n-1}) \widehat{g}(y_n | x_n) dx_n}$$

and the identity

$$\begin{aligned} p(y_n | x_{n-1}) &= \int \widehat{f}(x_n | x_{n-1}) \widehat{g}(y_n | x_n) dx_n \\ &\times \int \frac{f(x_n | x_{n-1}) g(y_n | x_n)}{\widehat{f}(x_n | x_{n-1}) \widehat{g}(y_n | x_n)} \widehat{p}(x_n | y_n, x_{n-1}) dx_n. \end{aligned}$$

An alternative consists of moving using $\widehat{p}(x_n|\, y_n, x_{n-1})$ –see (10)– and computing the weights using (12)

$$w_n(\mathbf{x}_n) = w_{n-1}(\mathbf{x}_{n-1}) \frac{f(x_n|\, x_{n-1})\, g(y_n|\, x_n)}{\widehat{p}(x_n|\, y_n, x_{n-1})}.$$

We want to emphasize that such sampling strategies can perform poorly even if one can sample from $p(x_n|\, y_n, x_{n-1})$ and compute exactly the associated importance weight. Indeed, in situations where the discrepancy between $p(\mathbf{x}_{n-1}|\, y_{1:n-1})$ and $p(\mathbf{x}_{n-1}|\, y_{1:n})$ is high, then the weights (29) will have a large variance. An alternative strategy consists not only of sampling X_n at time n but also of updating the block of variables $X_{n-R+1:n-1}$ where $R > 1$. In this case we seek to move from $\mathbf{x}_{n-1} = (\mathbf{u}_{n-1}, \mathbf{v}_{n-1}) = (x_{1:n-R}, x_{n-R+1:n-1})$ to $\mathbf{x}_n = (\mathbf{u}_{n-1}, \mathbf{v}_n) = (x_{1:n-R}, x'_{n-R+1:n})$ and the conditionally optimal distribution is given by

$$\pi_n(\mathbf{v}_n|\, \mathbf{u}_{n-1}) = p\left(x'_{n-R+1:n}\middle|\, y_{n-R+1:n}, x_{n-R}\right).$$

Although attractive, this strategy is difficult to apply, as sampling from $p(x'_{n-R+1:n}|\, y_{n-R+1:n}, x_{n-R})$ becomes more difficult as R increases. Moreover, it requires the ability to compute or obtain unbiased estimates of both $p(y_{n-R+1:n}|\, x_{n-R})$ and $1/\eta_{n-1}(x_{1:n-R})$ to calculate (7). If we use an approximation $\widehat{\pi}_n(\mathbf{v}_n|\, \mathbf{u}_{n-1})$ of $\pi_n(\mathbf{v}_n|\, \mathbf{u}_{n-1})$ to move the particles, it remains difficult to compute (11) as we still require an unbiased estimate of $1/\eta_{n-1}(x_{1:n-R})$. The discussion of Section 2.3.2 indicates that, alternatively, we can simply weight the particles sampled using $\widehat{\pi}_n(\mathbf{v}_n|\, \mathbf{u}_{n-1})$ by (20); this only requires us being able to derive an approximation of $\pi_{n-1}(\mathbf{v}_{n-1}|\, \mathbf{u}_{n-1})$.

4.1.2. *Model and Simulation Details*

We now present numerical results for a bearings-only-tracking example (Gilks and Berzuini, 2001). The target is modelled using a standard constant velocity model

$$X_n = \begin{pmatrix} 1 & 1 & 0 & 0 \\ 0 & 1 & 0 & 0 \\ 0 & 0 & 1 & 1 \\ 0 & 0 & 0 & 1 \end{pmatrix} X_{n-1} + V_n,$$

with V_n i.i.d. $\mathcal{N}_4(0, \Sigma)$ ($\mathcal{N}_r(a, b)$ is the r-dimensional normal distribution with mean a and covariance b) and

$$\Sigma = 5 \begin{pmatrix} 1/3 & 1/2 & 0 & 0 \\ 1/2 & 1 & 0 & 0 \\ 0 & 0 & 1/3 & 1/2 \\ 0 & 0 & 1/2 & 1 \end{pmatrix}.$$

The state vector $X_n = \left(X_n^1, X_n^2, X_n^3, X_n^4\right)^T$ is such that X_n^1 (resp. X_n^3) corresponds to the horizontal (resp. vertical) position of the target whereas X_n^2 (resp. X_n^4) corresponds to the horizontal (resp. vertical) velocity. One only receives observations of the bearings of the target

$$Y_n = \tan^{-1}\left(\frac{X_n^3}{X_n^1}\right) + W_n$$

where W_n is i.i.d. $\mathcal{N}\left(0, 10^{-4}\right)$; thus, the observations are almost noiseless. This is representative of real-world tracking scenarios.

We build an approximation $\widehat{\pi}_n\left(\mathbf{v}_n \mid \mathbf{u}_{n-1}\right)$ (respectively $\widehat{\pi}_{n-1}\left(\mathbf{v}_{n-1} \mid \mathbf{u}_{n-1}\right)$) of $\pi_n\left(\mathbf{v}_n \mid \mathbf{u}_{n-1}\right)$ (respectively $\widehat{\pi}_{n-1}\left(\mathbf{v}_{n-1} \mid \mathbf{u}_{n-1}\right)$) using the forward–backward sampling formula for a linear Gaussian approximation of the model based on the Extended Kalman Filter (EKF); see Doucet et al. (2006) for details. We compare

• The block sampling SMC algorithms denoted SMC(R) for $R = 1, 2, 5$ and 10 which are using the EKF proposal.

• Two Resample-Move algorithms as described by Gilks and Berzuini (2001), where the SMC(1) is used followed by: (i) one at a time MH moves using an approximation of $p\left(x_k \mid y_k, x_{k-1}, x_{k+1}\right)$ as a proposal (RML(10)) over a lag $L = 10$; and (ii) using the EKF proposal for $L = 10$ (RMFL(10)). The acceptance probabilities of those moves were in all cases between (0.5,0.6).

Systematic resampling is performed whenever the ESS goes below $N/2$. The results are displayed in Table 1.

Table 1: *Average number of resampling steps for 100 simulations, 100 time instances per simulations using* $N = 1000$ *particles.*

Filter	# Time Resampled
SMC(1)	44.6
RML(10)	45.2
RMFL(10)	43.3
SMC(2)	34.9
SMC(5)	4.6
SMC(10)	1.3

The standard algorithms, namely, SMC(1), RML(10) and RMFL(10), need to resample very often as the ESS drop below $N/2$; see the second column of Table 1. In particular, the Resample-Move algorithms resample as much as SMC(1) despite their computational complexity being similar to SMC(10); this is because MCMC steps are only introduced after an SMC(1) step has been performed. Conversely, as R increases, the number of resampling steps required by SMC(R) methods decreases dramatically. Consequently, the number of unique particles $\left\{X_1^{(i)}\right\}$ approximating the final target $p\left(x_1 \mid y_{1:100}\right)$ remains large whereas it is close to 1 for standard methods.

4.2. *Binary Probit Regression*

Our second application, related to the tempering example in Section 1, is the Bayesian binary regression model in (for example) Albert and Chib (1993). The analysis of binary data via generalized linear models often occurs in applied Bayesian statistics and the most commonly used technique to draw inference is the auxiliary variable Gibbs sampler (Albert and Chib 1993). It is well known (e.g. Holmes and Held 2006) that such a simulation method can perform poorly, due to the strong posterior dependency between the regression and auxiliary variables. In this example we illustrate that SMC samplers can provide significant improvements over the auxiliary variable Gibbs sampler with little extra coding effort and comparable CPU times. Further, we demonstrate that the SMC algorithm based on (18) can greatly improve the performance of Resample-Move (Chopin, 2002; Gilks and Berzuini, 2001) based on (21).

4.2.1. *Model*

The model assumes that we observe binary data $Y_1, \ldots, Y_u$, with associated r-dimensional covariates $X_1, \ldots, X_u$ and that the Y_i, $i = 1, \ldots, u$ are i.i.d.:

$$Y_i|\beta \quad \sim \mathcal{B}(\Phi(x_i'\beta))$$

where $\mathcal{B}$ is the Bernoulli distribution, β is a r-dimensional vector and Φ is the standard normal CDF. We denote by x the $u \times r$ design matrix (we do not consider models with an intercept).

Albert and Chib (1993) introduced an auxiliary variable Z_i to facilitate application of the Gibbs sampler. That is, we have:

$$\begin{aligned} Y_i|Z_i &= \begin{cases} 1 & \text{if } Z_i > 0 \\ 0 & \text{otherwise} \end{cases} \\ Z_i &= x_i'\beta + \epsilon_i \\ \epsilon_i &\sim \mathcal{N}(0,1). \end{aligned}$$

In addition, we assume $\beta \sim \mathcal{N}_r(b, v)$. Standard manipulations establish that the marginal posterior $\pi(\beta|y_{1:u}, x_{1:u})$ concides with that of the original model.

4.2.2. *Performance of the MCMC Algorithm*

To illustrate that MCMC-based inference for binary probit regression does not always perform well, we consider the following example. We simulated 200 data points, with $r = 20$ covariates. We set the priors as $b = 0$ and $v = \text{diag}(100)$. Recall that the Gibbs sampler of Albert and Chib (1993) generates from full conditionals:

$$\begin{aligned} \beta|\cdots &\sim \mathcal{N}_r(B, V) \\ B &= V(v^{-1}b + x'z) \\ V &= (v^{-1} + x'x)^{-1} \\ \pi(z_i|\cdots) &\sim \begin{cases} \phi(z_i; x_i'\beta, 1)\mathbb{I}_{\{z_i>0\}}(z_i) & \text{if } y_i = 1 \\ \phi(z_i; x_i'\beta, 1)\mathbb{I}_{\{z_i\leq 0\}}(z_i) & \text{otherwise} \end{cases} \end{aligned}$$

where $|\cdots$ denotes conditioning on all other random variables in the model and $\phi(\cdot)$ is the normal density. It should be noted that there are more advanced MCMC methods for these class of models (*e.g.*, Holmes and Held, 2006), but we only consider the method of Albert and Chib (1993) as it forms a building block of the SMC sampler below. We ran the MCMC sampler for 100 000 iterations, thinning the samples to every 100. The CPU time was approximately 421 seconds.

In Fig. 1 (top row) we can observe two of the traces of the twenty sampled regression coefficients. These plots indicate very slow mixing, due to the clear autocorrelations and the thinning of the Markov chain. Whilst we might run the sampler for an excessive period of time (that is, enough to substantially reduce the autocorrelations of the samples), it is preferable to construct an alternative simulation procedure. This is to ensure that we are representing all of the regions of high posterior probability that may not occur using this MCMC sampler.

4.2.3. *SMC Sampler*

We now develop an SMC approach to draw inference from the binary logistic model. We consider a sequence of densities induced by the following error at time n:

$$\epsilon_i \sim \mathcal{N}(0, \zeta_n), \qquad \zeta_1 > \cdots > \zeta_P = 1.$$

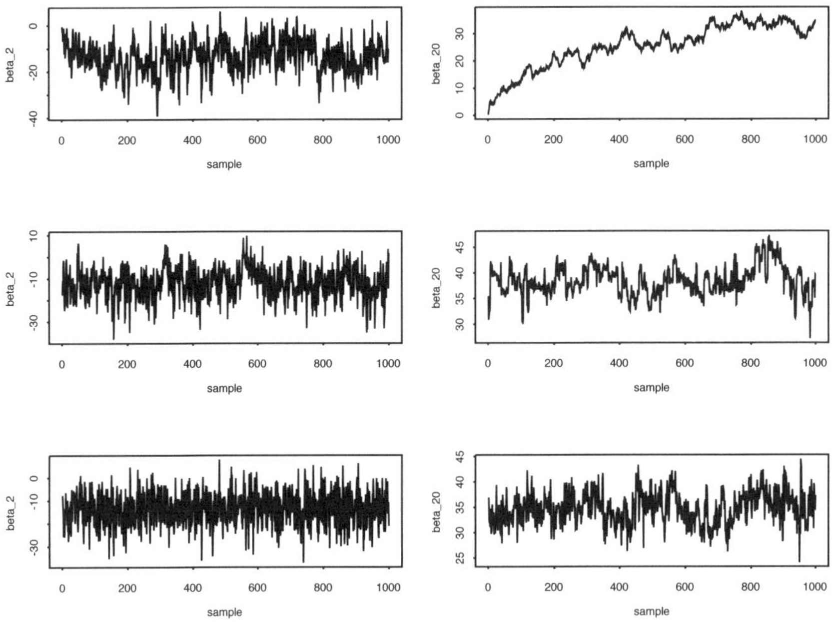

Figure 1: *Sampled coefficients from the binary regression example. For the MCMC (top row), we ran the Gibbs sampler of Albert and Chib (1993) for 100 000 iterations and stored every 100th (CPU time 421 sec). For the reversal SMC (middle row) we ran 1000 particles for 200 time steps (CPU 681 sec). For the Gibbs SMC (bottom row) we did the same except the CPU was 677.*

To sample the particles, we adopt the MCMC kernel above, associated to the density at time n. At time n we sample new $z_{1:u}, \beta$ from:

$$K_n((z_{1:u}, \beta), (z'_{1:u}, \beta')) = \pi_n(z'_{1:u}|\beta, y_{1:u}, x_{1:u})\mathbb{I}_\beta(\beta').$$

We then sample β from the full conditional (since this kernel admits π_n as an invariant measure we can adopt backward kernel (21) and so the incremental weight is 1). For the corresponding backward kernel, L_{n-1}, we consider two options (21) and (18). Since (18) is closer to the optimal kernel, we would expect that the performance under the second kernel to be better than the first (in terms of weight degeneracy).

4.2.4. *Performance of SMC Sampler*

We ran the two SMC samplers above for 50, 100 and 200 time points. We sampled 1000 particles and resampled upon the basis of the ESS dropping to $N/2$ using systematic resampling. The initial importance distribution was a multivariate normal centered at a point simulated from an MCMC sampler and the full conditional density for $z_{1:u}$. We found that this performed noticeably better than using the prior for β.

It should be noted that we did not have to store N, $u-$dimensional vectors. This is possible due to the fact that we can simulate from $\pi_n(z_{1:u}|\cdots)$ and that the incremental weights can be either computed at time n for time $n+1$ and are independent of $z_{1:u}$.

As in Del Moral et al. (2006), we adopted a piecewise linear cooling scheme that had, for 50 time points, $1/\zeta_n$ increase uniformly to 0.05 for the first 10 time points, then uniformly to 0.3 for the next 20 and then uniformly to 1. All other time specifications had the same cooling schedule, in time proportion.

In Figures 1–4 and Table 2 we can observe our results. Figures 2–4 and Table 2 provide a comparison of the performance for the two backward kernels. As expected, (18) provides substantial improvements over the reversal kernel (21) with significantly lower weight degeneracy and thus fewer resampling steps. This is manifested in Fig. 1 with slighly less dependence (of the samples) for the Gibbs kernel. The CPU times of the two SMC samplers are comparable to MCMC (Table 2 final column) which shows that SMC can markedly improve upon MCMC for similar computational cost (and programming effort).

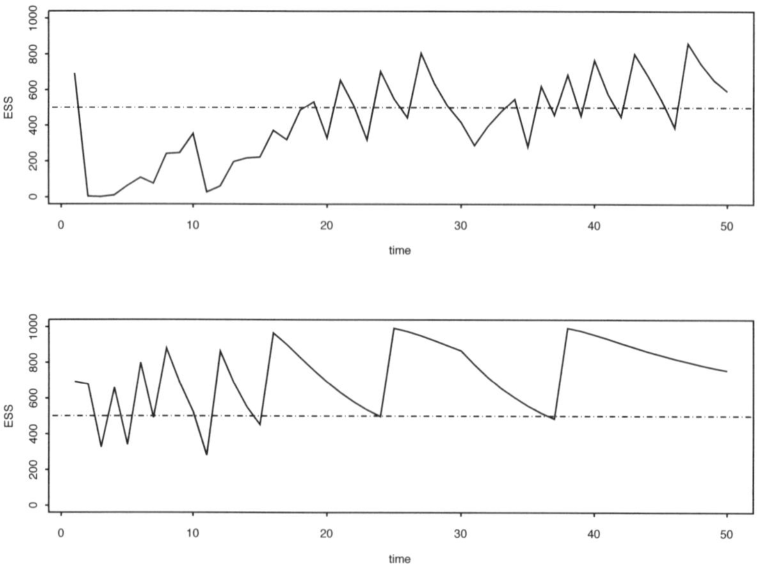

Figure 2: *ESS plots from the binary regression example; 50 time points. The top graph is for reversal kernel (18). We sampled 1000 particles and resampled when the ESS dropped below 500 particles.*

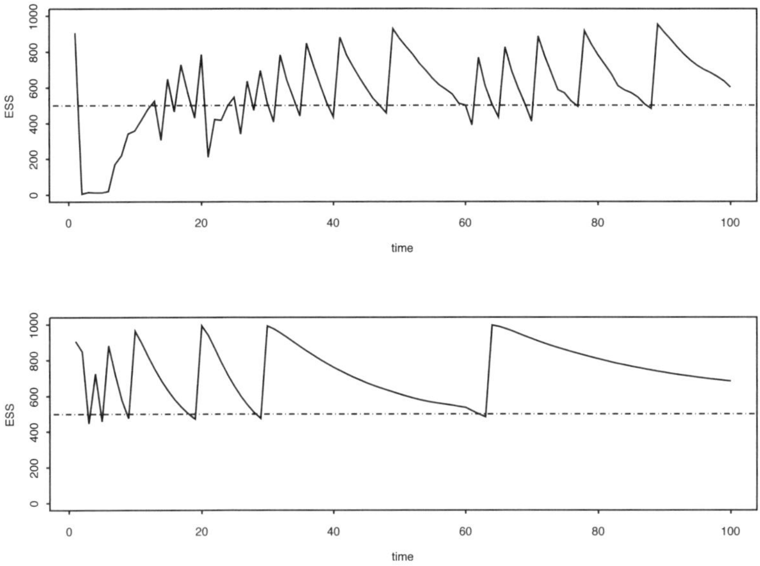

Figure 3: *ESS plots from the binary regression example; 100 time points. The top graph is for reversal kernel (18). We sampled 1000 particles and resampled when the ESS dropped below 500 particles.*

4.2.5. *Summary*

In this example we have established that SMC samplers are an alternative to MCMC for a binary regression example. This was only at a slight increase in CPU time and programming effort. As a result, we may be able to investigate more challenging problems, especially since we have not utilized all of the SMC strategies (e.g. adaptive methods, in Section 2.2).

We also saw that the adoption of the Gibbs backward kernel (18) provided significant improvements over Resample-Move. This is of interest when the full conditionals are not available, but good approximations of them are. In this case it would be of interest to see if similar results hold, that is, in comparison with the reversal kernel (21). We note that this is not meaningless in the context of artifical distributions, where the rate of resampling may be controlled by ensuring $\pi_{n-1} \approx \pi_n$. This is because we will obtain better performance for the Gibbs kernel for shorter time specifications (and particle number) and hence lower CPU time.

4.3. *Filtering for Partially Observed Processes*

In the following example we consider SMC samplers applied to filtering for partially observed processes. In particular, we extend the approach of Del Moral et al. (2006) for cases with $S_{n-1} \subset S_n$, that is, a sequence of densities with nested supports.

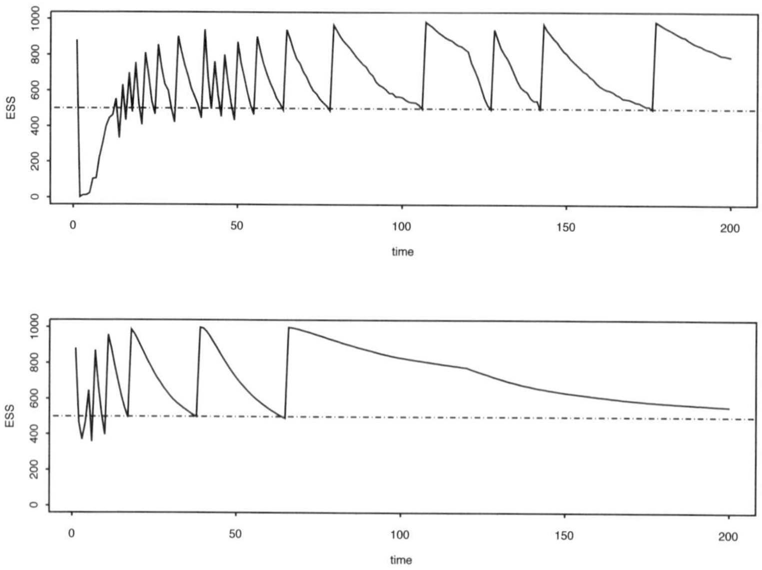

Figure 4: *ESS plots from the binary regression example; 200 time points. The top graph is for reversal kernel (18). We sampled 1000 particles and resampled when the ESS dropped below 500 particles.*

4.3.1. *Model*

We focus upon the Bayesian Ornstein–Uhlenbeck stochastic volatility model (Barndoff-Nielsen and Shepard 2001) found in Roberts et al. (2004). That is, the log-return of an asset X_t at time $t \in [0, T]$ is modelled via the stochastic differential equation (SDE):

$$dX_t = \sigma_t^{1/2} dW_t$$

where $\{W_t\}_{t\in[0,T]}$ is a standard Wiener process. The volatility σ_t is assumed to satisfy the following (Ornstein-Uhlenbeck equation) SDE:

$$d\sigma_t = -\mu\sigma_t dt + dZ_t \tag{30}$$

where $\{Z_t\}_{t\in[0,T]}$ is assumed to be a pure jump Lévy process; see Applebaum (2004) for a nice introduction.

It is well known (Barndoff-Nielsen and Shephard 2001; Applebaum 2004) that for any self-decomposable random variable, there exists a unique Lévy process that satisfies (30); we assume that σ_t has a Gamma marginal, $\mathcal{G}a(\nu, \theta)$. In this case Z_t is a compound Poisson process:

$$Z_t = \sum_{j=1}^{K_t} \varepsilon_j$$

Table 2: *Results from Binary regression example. The first entry is for the reversal (i.e. the first column row entry is the reversal kernel for 50 time points). The CPU time is in seconds.*

Time points	50	100	200
CPU Time	115.33	251.70	681.33
CPU Time	118.93	263.61	677.65
# Times Resampled	29	29	28
# Times Resampled	7	6	8

where K_t is a Poisson process of rate $\nu\mu$ and the ε_j are i.i.d. according to $\mathcal{E}x(\theta)$ (where $\mathcal{E}x$ is the exponential distribution). Denote the jump times of the compound Poisson process as $0 < c_1 < \cdots < c_{k_t} < t$.

Since $X_t \sim \mathcal{N}(0, \sigma_t^*)$, where $\sigma_t^* = \int_0^t \sigma_s ds$ is the integrated volatility, it is easily seen that $Y_{t_i} \sim \mathcal{N}(0, \sigma_i^*)$ with $Y_{t_i} = X_{t_i} - X_{t_{i-1}}$, $0 < t_1 < \cdots < t_u = T$ are regularly spaced observation times and $\sigma_i^* = \sigma_{t_i}^* - \sigma_{t_{i-1}}^*$. Additionally, the integrated volatility is:

$$\sigma_t^* = \frac{1}{\mu}\Big(\sum_{j=1}^{K_t}[1 - \exp\{-\mu(t - c_j)\}]\varepsilon_j - \sigma_0[\exp\{-\mu t\} - 1]\Big).$$

The likelihood at time t is

$$g(y_{t_1:m_t}|\{\sigma_t^*\}) = \prod_{i=1}^{n} \phi(y_{t_i}; \sigma_i^*)\mathbb{I}_{\{t_i<t\}}(t_i)$$

with $\phi(\cdot; a)$ the density of normal distribution of mean zero and variance a and $m_t = \max\{t_i : t_i \leq t\}$. The priors are exactly as Roberts et al. (2004):

$$\begin{aligned}
\sigma_0|\theta, \nu \quad &\sim \mathcal{G}a(\nu, \theta),\ \nu \sim \mathcal{G}a(\alpha_\nu, \beta_\nu),\\
\mu \quad &\sim \mathcal{G}a(\alpha_\mu, \beta_\mu),\ \ \theta \sim \mathcal{G}a(\alpha_\theta, \beta_\theta)
\end{aligned}$$

where $\mathcal{G}a(a, b)$ is the Gamma distribution of mean a/b. We take the density, at time t, of the stochastic process, with respect to (the product of) Lebesgue and counting measures:

$$\begin{aligned}
p_t(c_{1:k_t}, \varepsilon_{1:k_t}, k_t) \quad &= \tfrac{k_t!}{n^{k_t}}\mathbb{I}_{\{0<c_1<\cdots<c_{k_t}<t\}}(c_{1:k_t})\theta^{k_t}\exp\{-\theta\textstyle\sum_{j=1}^{k_t}\varepsilon_j\}\\
&\times\tfrac{(t\mu\nu)^{k_t}}{k_t!}\exp\{-t\mu\nu\}.
\end{aligned}$$

4.3.2. *Simulation Details*

We are thus interested in simulating from a sequence of densities, which at time n (of the sampler) and corresponding $d_n \in (0, T]$ (of the stochastic process) is defined as:

$$\begin{aligned}
\pi_n(c_{1:k_{d_n}}, \varepsilon_{1:k_{d_n}}, k_{d_n}, \sigma_0, \nu, \mu, \theta|y_{t_1:m_{d_n}}) \quad &\propto g(y_{t_1:m_{d_n}}|\{\sigma_{d_n}^*\})\pi(\sigma_0, \nu, \mu, \theta)\\
&\times p_{d_n}(c_{1:k_{d_n}}, \varepsilon_{1:k_{d_n}}, k_{d_n}).
\end{aligned}$$

As in example 2 of Del Moral *et al.* (2006) this is a sequence of densities on trans-dimensional, nested spaces. However, the problem is significantly more difficult as

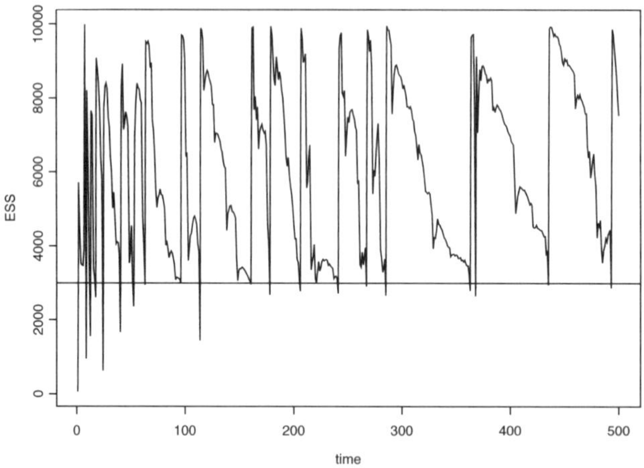

Figure 5: *ESS plot for simulated data from the stochastic volatility example. We ran 10 000 particles with resampling threshold (−−) 3000 particles.*

the full conditional densities are not available in closed form. To simulate this sequence, we adopted the following technique.

If $k_{d_n} = 0$ we select a birth move which is the same as Roberts et al. (2004). Otherwise, we extend the space by adopting a random walk kernel:

$$q((c_{k_{d_{n-1}}-1}, c_{k_{d_{n-1}}}), c_{k_{d_n}}) \quad \propto \exp\{-\lambda|c_{k_{d_n}} - c_{k_{d_{n-1}}}|\}\mathbb{I}_{(c_{k_{d_{n-1}}-1},n)}(c_{k_{d_n}}).$$

The backward kernel is identical if $c_{k_{d_n}} \in (0, d_{n-1})$ otherwise it is uniform. The incremental weight is then much like a Hastings ratio, but standard manipulations establish that it has finite supremum norm, which means that it has finite variance. However, we found that the ESS could drop, when very informative observations arrive and thus we used the following idea: If the ESS drops, we return to the original particles at time $n-1$ and we perform an SMC sampler which heats up to a very simple (related) density and then make the space extension, much like the tempered transitions method of Neal (1996). We then use SMC to return to the density we were interested in sampling from.

After this step we perform an MCMC step (the centered algorithm of Roberts *et al.*, 2004) which leaves π_n invariant allowing with probability 1/2 a Dirac step to reduce the CPU time spent on updating the particles.

4.3.3. *Illustration*

For illustration purposes we simulated $u = 500$ data points from the prior and ran 10 000 particles with systematic resampling (threshold 3000 particles). The priors were $\alpha_\nu = 1.0$, $\beta_\nu = 0.5$, $\alpha_\mu = 1.0$, $\beta_\mu = 1.0$, $\alpha_\theta = 1.0$, $\beta_\theta = 0.1$. We defined the target densities at the observation times $1, 2, \ldots, 500$ and set $\lambda = 10$.

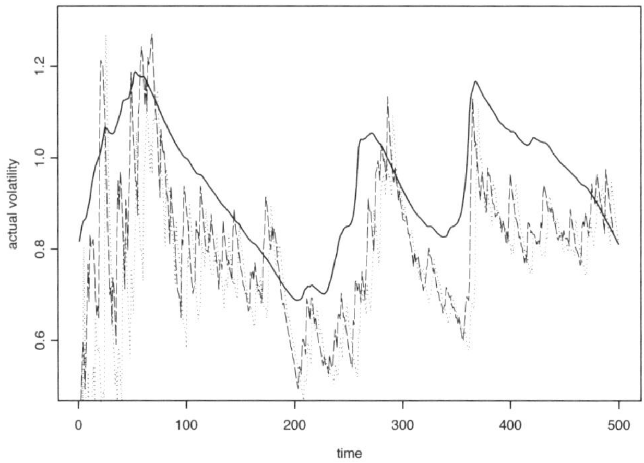

Figure 6: *Actual volatility for simulated data from the stochastic volatility example. We plotted the actual volatility for the final density (full line) filtered (esimated at each timepoint, dot) and smoothed (estimated at each timepoint, lag 5, dash).*

If the ESS drops we perform the algorithm with respect to:

$$\pi_n^{\zeta}(c_{1:k_{d_n}}, \varepsilon_{1:k_{d_n}}, k_{d_n}, \sigma_0, \nu, \mu, \theta | y_{t_1:m_{d_n}}) \propto g(y_{t_1:m_{d_n}} | \{\sigma_{d_n}^*\})^{\zeta} \pi(\sigma_0, \nu, \mu, \theta) \times p_{d_n}(c_{1:k_{d_n}}, \varepsilon_{1:k_{d_n}}, k_{d_n})$$

for some temperatures $\{\zeta\}$. We used a uniform heating/cooling schedule to $\zeta = 0.005$ and 100 densities and performed this if the ESS dropped to 5% of the particle number.

We can see in Fig. 5 that we are able to extend the state-space in an efficient manner and then estimate (Fig. 6) the filtered and smoothed actual volatility σ_i^* which, to our knowledge, has not ever been performed for such complex models. It should be noted that we only had to apply the procedure above, for when the ESS drops, seven times; which illustrates that our original incremental weight does not have extremely high variance. For this example, the MCMC moves can operate upon the entire state-space, which we recommend, unless a faster mixing MCMC sampler is constructed. That is, the computational complexity is dependent upon u (the number of data points). Additionally, due to the required, extra, SMC sampler, this approach is not useful for high frequency data, but is more appropriate for daily returns type data.

5. CONCLUSION

It is well-known that SMC algorithms can solve, numerically, sequential Bayesian inference problems for nonlinear, non-Gaussian state-space models (Doucet et al. 2001). We have demonstrated (in addition to the work of Chopin, 2002; Del Moral *et al.*, 2006; Gilks and Berzuini, 2001; Neal, 2001) that SMC methods are not limited to this class of applications and can be used to solve, efficiently, a wide variety of problems arising in Bayesian statistics.

We remark that, as for MCMC, SMC methods are not black-boxes and require some expertise to achieve good performance. Nevertheless, contrary to MCMC, as SMC is essentially based upon IS, its validity does not rely on ergodic properties of any Markov chain. Consequently, the type of strategies that may be applied by the user is far richer, that is, time-adaptive proposals and even non-Markov transition kernels can be used without any theoretical difficulties. Such schemes are presented in Jasra et al. (2005a) for trans-dimensional problems.

We also believe that it is fruitful to interpret SMC as a particle approximation of nonlinear MCMC kernels. This provides us with alternative SMC and iterative self-interacting approximation schemes as well as opening the avenue for new nonlinear algorithms. The key to these procedures is being able to design nonlinear MCMC kernels admitting fixed target distributions; see Andrieu et al. (2006) and Brockwell and Doucet (2006) for such algorithms.

ACKNOWLEDGEMENTS

The second author would like to thank Mark Briers for the simulations of the block sampling example, Adam Johansen for his comments and Gareth W. Peters. The third author would like to thank Chris Holmes for funding and Dave Stephens for discussions on the examples.

REFERENCES

Albert J. H. and Chib S. (1993). Bayesian analysis of binary and polychotomous response data. *J. Amer. Statist. Assoc.* **88**, 669–679.

Andrieu, C., Jasra, A., Doucet, A. and Del Moral, P. (2006). Non-linear Markov chain Monte Carlo via self interacting approximations. *Tech. Rep.*, University of Bristol.

Applebaum D. (2004) *Lévy Processes and Stochastic Calculus*, Cambridge: University Press.

Barndoff Nielsen, O. E. and Shephard, N. (2001). Non-Gaussian Ornstein–Uhlenbeck-based models and some of their uses in financial economics *J. Roy. Statist. Soc. B* **63** ,167-241, (with discussion). .

Brockwell, A. E. and Doucet, A. (2006). Sequentially interacting Markov chain Monte Carlo for Bayesian computation, *Tech. Rep.*, Carnegie Mellon University, USA.

Chopin, N., (2002). A sequential particle filter method for static models. *Biometrika* **89**, 539-552.

Crisan, D. and Doucet, A. (2000). Convergence of sequential Monte Carlo methods. *Tech. Rep.*, CUED/F-INFENG/TR381, Cambridge University, UK.

Del Moral, P. (2004). *Feynman–Kac Formulae: Genealogical and Interacting Particle Systems with Applications.* New York: Springer-Verlag.

Del Moral, P. and Doucet, A. (2003). On a class of genealogical and interacting Metropolis models. *Séminaire de Probabilités XXXVII,* (Azéma, J., Emery, M., Ledoux, M. and Yor, M., eds.) *Lecture Notes in Mathematics* **1832,**Berlin: Springer-Verlag, 415–446.

Del Moral, P., Doucet, A. and Jasra, A. (2006). Sequential Monte Carlo samplers. *J. Roy. Statist. Soc. B* **68**, 411–436.

Del Moral, P. and Miclo, L. (2004). On convergence of chains with occupational self-interactions. *Proc. Roy. Soc. A* **460**, 325–346.

Del Moral, P. and Miclo, L. (2006). Self interacting Markov chains. *Stoch. Analysis Appl.*, **3** 615–660.

Doucet, A., Briers, M. and Sénécal, S. (2006). Efficient block sampling strategies for sequential Monte Carlo. *J. Comp. Graphical Statist.* **15**, 693–711.

Doucet, A., de Freitas, J. F. G. and Gordon, N. J. (eds.) (2001). *Sequential Monte Carlo Methods in Practice.* New York: Springer-Verlag.

Gelman, A. and Meng, X. L. (1998). Simulating normalizing constants: From importance sampling to bridge sampling to path sampling. *Stat. Sci.* **13**, 163–185.

Gilks, W.R. and Berzuini, C. (2001). Following a moving target – Monte Carlo inference for dynamic Bayesian models. *J. Roy. Statist. Soc. B* **63**, 127–146.

Green, P.J. (2003). Trans-dimensional Markov chain Monte Carlo. in *Highly Structured Stochastic Systems*, Oxford: University Press

Holmes, C. C. and Held, L (2006). Bayesian auxiliary variable models for binary and multinomial regression *Bayesian Analysis* **1**, 145–168,

Iba, Y. (2001). Population Monte Carlo algorithms. *Trans. Jap. Soc. Artif. Intell.* **16**, 279–286

Jarzynski, C. (1997). Nonequilibrium equality for free energy differences. *Phys. Rev. Let.* **78**, 2690–2693.

Jasra, A., Doucet, A., Stephens, D. A. and Holmes, C.C. (2005a). Interacting sequential Monte Carlo samplers for trans-dimensional simulation. *Tech. Rep.*, Imperial College London, UK.

Jasra, A., Stephens, D. A. and Holmes, C.C. (2005b). On population-based simulation for static inference. *Tech. Rep.*, Imperial College London, UK.

Liu, J. S. (2001). *Monte Carlo Strategies in Scientific Computing.* New York: Springer-Verlag.

Neal, R. (1996). Sampling from multimodal distributions via tempered transitions. *Statist. Computing* **6**, 353–366.

Neal, R. (2001). Annealed importance sampling. *Statist. Computing* **11**, 125–139.

Roberts, G. O., Papaspiliopoulos, O. and Dellaportas, P. (2004). Bayesian inference for non-Gaussian Ornstein–Uhlenbeck stochastic volatility processes, *J. Roy. Statist. Soc. B* **66**, 369–393.

DISCUSSION

HEDIBERT FREITAS LOPES (*University of Chicago, USA*)

I would like to start by congratulating the authors on this very important contribution to a growing literature on sequential Monte Carlo (SMC) methods. The authors run the extra mile and show how SMC methods, when combined with well established MCMC tools, can be used to entertain not only optimal filtering problems or, as the authors refer to, sequential Bayesian inference problems, but also for posterior inference in general parametric statistical models.

The paper, similarly to several others in this area, builds upon the simple and neat idea behind sampling importance resampling (SIR) algorithms, in which particles from a (sequence of) importance density are appropriately reweighed in order to obtain (sequences of) approximate draws from a target distribution (see, Smith and Gelfand, 1990, and Gordon, Salmond and Smith, 1995, for a few seminal references).

However, unlike in filtering problems where the nature of the state-space evolution suggests updating schemes for the set of particles, there is no natural rule for sequentially, optimally reweighing the particles that would also be practically feasible. Here lies the main contribution of the paper, where the authors introduce the idea of *auxiliary backward Markov kernels.* SMC methods can then be seen as an alternative to MCMC methods when dealing with sequences of probability distributions that avoids, and to a certain extent eliminates, convergence issues, one of the most cumbersome practical problems in the MCMC literature. Additionally, parallelization of the computation comes as a natural by-product of SMC methods.

I would like to hear (or read) what the authors have to say about a few points that I believe will be in the SMC agenda for the next few years: (i) choice of Markov kernels K for (increasingly) high dimensional state vectorsx_n, such as in modern highly structure stochastic systems; (ii) situations where both fixed parameters and state variables are present, how this distinction helps or makes it difficult to propose kernels? (iii) How to accurately compute and use the variance of weights sequentially when resampling, in principle, makes it more difficult to derive general limiting results? (iv) How does the previous issue relate to the problem of sample impoverishment?

In summary, I humbly anticipate that the next few years will witness a great interaction between *MCMCers* and *SMCers*, both theoretically and empirically. Thanks the authors for writing such an interesting paper.

DAVID R. BICKEL (*Pioneer Hi-Bred Intl., Johnston, Iowa, USA*)

The authors added promising innovations in sequential Monte Carlo methodology to the arsenal of the Bayesian community. Most notably, their backward-kernel framework obviates the evaluation of the importance density function, enabling greater flexibility in the choice of algorithms. They also set their work on posterior inference in a more general context by citing results of the observation that particle filters approximate the path integrals studied in theoretical physics (Del Moral 2004).

My first question concerns another recent advance in SMC, the use of the mixture transition kernel

$$\overline{K}_n\left(\mathbf{x}_{n-1}, \mathbf{x}_n\right) = \sum_{m=1}^{M} \overline{\alpha}_{n,m} \kappa_m \left(\mathbf{x}_{n-1}, \mathbf{x}_n\right),$$

where $\overline{\alpha}_{n,m}$ equals the sum of normalized weights over all particle values that were drawn from the mth mixture component at time $n-1$, and $\kappa_m\left(\mathbf{x}_{n-1}, \mathbf{x}_n\right)$ is an element of the set of M predetermined mixture components (Douc et al. 2007). For example, if the possible transition kernels correspond to Metropolis–Hastings random walk kernels of M different scales chosen by the statistician, then the mixture automatically adapts to the ones most appropriate for the target distribution (Douc et al. 2007). Is there a class of static inference problems for which the backward-kernel approach is better suited, or is it too early to predict which method may perform better in a particular situation?

In his discussion, Hedibert Lopes suggested some opportunities for further SMC research. What areas of mathematical and applied work seem most worthwhile?

I thank the authors for their highly interesting and informative paper.

NICOLAS CHOPIN (*University of Bristol, UK*)

As the authors know already, I am quite enthusiastic about the general SMC framework they developed, and more generally the idea that SMC can outperform MCMC in a variety of 'complex' Bayesian problems. By 'complex', I refer informally to typical difficulties such as: polymodality, large dimensionality (of the parameter, of the observations, or both), strong correlations between components of the considered posterior distribution, etc.

My discussion focuses on the probit example, and particularly the artificial sequence (γ_n) specifically chosen in this application, which I find both intriguing and exciting. As the authors have certainly realised, all the distributions γ_n are equal to the posterior distribution of interest, up to scaling factor ς_n. For a sequence of Gaussian distributions, rescaling with factor ς_n is the same thing as tempering with exponent ς_n^{-2}; therefore, for standard, well-behaved Bayesian problems both approaches can be considered as roughly equivalent. But rescaling is more convenient here for a number of reasons. First, this means that all the Gibbs steps performed within the SMC algorithm are identical to the Gibbs update implemented in the corresponding MCMC algorithm. Thus a clear case is made that, even if update kernels have the same ergodicity properties in both implementations, the SMC implementation provides better estimates (i.e. with smaller variance) than the MCMC one, for the same computational cost.

Second, the fact that all π_n are equivalent, up to appropriate scaling, means that one can combine results from all or part of the iterations, rather than retaining only the output of the last iteration. I wonder if the authors have some guidance on how this can be done in an optimal way. Third, and more generally, I am excited about the idea of defining a sequence of artificial *models* in order to derive the sequence (γ_n). This should make it easier to derive kernels K_n (typically Gibbs like) that allows for efficient MCMC step within the SMC algorithm. I think this is a very promising line of research. My only concern is that some models involving latent variables may be more difficult to handle, because, in contrast with this probit example, N simulations of the vector of latent variables would need to be carried forward across iterations, which seems memory demanding.

YANAN FAN, DAVID S. LESLIE and MATTHEW P. WAND
(*University of New South Wales, Australia and University of Bristol, UK*)

We would like to congratulate the authors on their efforts in presenting a unified approached to the use of sequential Monte Carlo (SMC) samplers in Bayesian computation. In this, and the companion publication (Del Moral et al. 2006), the authors illustrate the use of SMC as an alternative to Markov chain Monte Carlo (MCMC).

These methods have several advantages over traditional MCMC methods. Firstly, unlike MCMC, SMC methods do not face the sometimes contentious issue of diagnosing convergence of a Markov chain. Secondly, in problems where mixing is chronically slow, this method appear to offer a more efficient alternative, see Sisson *et al* (2006) for example. Finally, as the authors point out, adaptive proposals for transition kernels can easily be applied since the validity of SMC does not rely on ergodic properties of any Markov chain. This last property may give the SMC approach more scope for improving algorithm efficiency than MCMC.

In reference to the binary probit regression model presented in Section 4.2, the authors chose to use a multivariate normal distribution as the initial importance distribution, with parameter value given by simulated estimates from an MCMC sampler. An alternative, more efficient strategy may be to estimate the parameters of the multivariate normal distribution by fitting the frequentist binary probit regression model. One can obtain maximum likelihood and the associated variance-covariance estimates for this, and many other, models using standard statistical software packages. We are currently designing a SMC method to fit a general design generalized linear mixed model (GLMM) (Zhao et al. 2006) and find this approach to work well.

The authors adopt an MCMC kernel, and update the coefficients of the covariates β from its full conditional distribution. If one cannot sample directly from the full conditional distributions, a Metropolis–Hastings kernel may be used. The choice of scaling parameters in such kernels can greatly influence the performance of the sampler, and it is not clear if optimal scaling techniques developed in the MCMC literature are immediately applicable here. In our current work to use these techniques for GLMMs we have found that using the variance–covariance estimate from the frequentist model as a guide for scaling the MH proposal variance works well. In general, can the authors offer any guidance on the properties of optimal MCMC kernels for use with SMC samplers?

PAUL FEARNHEAD (*Lancaster University, UK*)

I would like to congratulate the authors on describing an exciting development in Sequential Monte Carlo (SMC) methods: extending both the range of applications and the flexibility of the algorithm. It will be interesting to see how popular and useful these ideas will be in years to come.

The ideas in the paper introduce more choice into the design of SMC algorithms, and it is thus necessary to gain practical insights into how to design efficient algorithms. I would thus like to ask for some further details on the examples.

Firstly I would have liked to see more details about the bearings-only tracking example in Section 4.1. For example, how informative were the priors used, and what did a typical path of the target look like? The choice of prior can have a substantial impact on the efficiency of the standard SMC algorithms that you compare with, with them being very inefficient for relatively diffuse priors (there can be simple ways round this by sampling your first set of particles conditional on the first or first two observations). While the efficiency of EKF approximations to the model can depend on the amount of non-linearity in the observation equation, which in turn depends on the path of the target (with the non-linearity being extreme if the target passes close to the observer). The EKF has a reputation for being unstable on the bearings-only tracking problem (e.g. Gordon *et al.* 1993), so it is somewhat suprising that this block sampler works so well on this problem. (Perhaps the EKF approximation works well because of the informative initial conditions, *i.e.*, a known position and velocity for each particle?)

Also, there are better ways of implementing a SMC algorithm than those that you compare to. In particular, the proposal distribution can be chosen to take into account the information in the most recent observation (Carpenter *et al.*, 1999), which is of particular importance here as the observations are very accurate. How much more efficient is using information from 10 observations as compared to this simpler scheme which uses information from a single observation?

Finally, a comment on your last example (Section 4.3). This is a very challenging problem, and your results are very encouraging, but I have a slight concern about your method whereby if the ESS of the particles drops significantly at an iteration you go back and re-propose the particles from a different proposal. The text reads as though you throw away the particles you initially proposed. If so, does this not introduce a bias into the algorithm (as you will have fewer particles in areas where their importance sampling weight will be large, as these particles will tend to get discarded at such a step)? For batch problems, a simple (and closely related) alternative would be to just let the amount of CPU time/number of particles generated vary with iteration of your algorithm, spending more effort on iterations where the ESS would otherwise drop significantly, and less on iterations where ESS is more robust. For example, this could be done by generating enough particles until the ESS rises above some threshold.

STEVEN L. SCOTT (*University of Southern California, USA*)

Congratulations to Del Moral, Doucet, and Jasra for an informative paper summarizing the current state of the art in sequential Monte Carlo (SMC) methods. SMC has become the tool of choice for Bayesian analysis of nonlinear dynamic models with continuous state spaces. Particularly welcome in the paper are ideas drawn from the MCMC literature. Auxiliary variables and kernel based proposals broaden the SMC toolkit for an audience already comfortable with MCMC.

One goal of the article is to demonstrate SMC as an alternative to MCMC for non-dynamic problems. I wish to make two points regarding the probit regression example from Section 4.2, the only example in Section 4 that is a non-dynamic problem. The first is a clarification about the poor performance of Albert and Chib's method, which would not be so widely used if its typical performance were as bad as suggested. When Albert and Chib's method mixes slowly, its poor performance can usually be attributed to a 'nearly-separating' hyperplane, where most observations have success probabilities very close to 0 or 1. The authors' simulation appears to involve such a phenomenon. If $x_{20} \approx x_2 \approx 1.0$ then β_{20} adds about 30 units to the linear predictor, while β_2 subtracts about 15, leaving the latent probit about 15 standard deviations from zero. There is no information in the paper about the values of the covariates, but a value of 1 is reasonable if the covariates were simulated from standard Gaussian or uniform distributions.

The second point is that the comparison between the authors' SMC algorithm and Albert and Chib is not quite fair because SMC uses a 'trick' withheld from its competitor. SMC achieves its efficiency by manipulating the variance of the latent data. Several authors have used this device in a computationally trivial improvement to Albert and Chib's method. The improvement assumes the latent variables have variance ζ, where Albert and Chib assume $\zeta = 1$ for identifiability. To introduce this improvement in the MCMC algorithm one need only modify the variance of the truncated normal distribution for z and replace the draw from $p(\beta|z)$ with a draw from $p(\beta, \zeta|z)$. Thus computing times for the improved and standard algorithms are virturally identical. Identifiability is restored during post processing by dividing each β by the $\sqrt{\zeta}$ from the same Gibbs iteration. Mathematical results due to Meng and van Dyk (1999), Liu and Wu (1999), and Lavine (2003) guarantee that the post processed β's have the desired stationary distribution and mix at least as rapidly as those from the identified model. van Dyk and Meng (2001) illustrate the effects of increased mixing on a variety of examples from the canon of models amenable to data augmentation, including probit regression. Liu and Wu (1999) compare the performance of the improved and standard Albert and Chib algorithms and find substantial mixing improvements in the presence of nearly separating hyperplanes similar to the simulation described by the current paper. The authors carefully acknowledge other Monte Carlo samplers for probit regression but limit their comparison to Albert and Chib (1993) because of its prevalence in applied work. However, given the nature of this particular SMC algorithm and the minimal burden imposed by the method described above, it would seem more appropriate to compare SMC to 'improved' rather than 'standard' Albert and Chib.

REPLY TO THE DISCUSSION

Firstly we thank the discussants for a set of both stimulating and useful comments. We hope that our work and the comments of the discussants will encourage future research in this developing field of research.

Convergence diagnosis. As noted in the paper and outlined by Lopes and Fan, Leslie and Wand, SMC methods do not rely upon the ergodicity properties of any Markov kernel and as such do not require convergence checks as for MCMC. However, we feel that we should point out that the performance of the algorithm needs to be monitored closely. For example, the ESS needs to be tracked (to ensure the algorithm does not 'crash' to a single particle) and it is advisable to observe the sampled parameters, running the algorithm a few times to check consistent answers; see also Chopin (2004) for more advice.

Optimal MCMC kernels. Fan, Leslie and Wand and Lopes ask about constructing optimal MCMC kernels. We discuss potential strategies that may add to the ideas of the discussants. We clarify that, on observation of the expression of the asymptotic variance in the central limit theorem (Del Moral *et al.*, 2006, Proposition 2), it can be deduced that the faster the kernel mixes the smaller the variance, the better algorithm (in this sense). One way to construct optimal MCMC kernels, i.e. close to iid sampling from π_n, might be to adopt the following adaptive strategy. We will assume that $\pi_{n-1} \approx \pi_n$, (which can be achieved, by construction, in problems where MCMC will typically be used) thus we can seek to use the particles at time $n-1$ to adapt an MCMC kernel at time n. Two strategies that might be used are:

(i) To adopt Robbins Monro type procedures so that the kernel is optimal in some sense. For example, via the optimal scaling of Roberts et al. (1997) for random walk Metropolis. That is, optimal scaling can be useful as it can attempt to improve the efficiency of the MCMC kernel.

(ii) Attempting to approximate the posterior at time n using the particles at time $n-1$; e.g. by using a mixture approximation in the MH independence sampler, as proposed by Andrieu and Moulines (2006) for adaptive MCMC.

To reply to Lopes' point (ii), the problem he mentions is quite difficult. In the context of state-space models, the introduction of MCMC steps of fixed computational complexity at each time step is not sufficient to reduce the accumulation of errors (i.e. does not make the dynamic model ergodic). Two algorithms combining SMC and stochastic approximation procedures have been proposed to solve this problem; see Andrieu, Doucet and Tadić (2005) and Poyadjis, Doucet and Singh (2005).

The backward kernel. In response to Bickel's comment on the use of the D-kernel procedure of Douc *et al.* (2006) against the backward kernel procedure. In the context that the discussant mentions, it would not be possible (except for toy problems) to use the D-kernel procedure; it is typically impossible to compute pointwise a MH kernel and thus the importance weights.

Variance of the weights. Lopes, in point (iii) and (iv) asks about the calculation of the variance of the weights. Typically the variance of particle algorithms is estimated (even in the presence of resampling) via the coefficient of variation (*e.g.*, Liu, 2001) of the unnormalized weight (suppressing the time index):

$$C_v = \frac{\sum_{i=1}^N (W^{(i)} - \frac{1}{N}\sum_{j=1}^N W^{(j)})^2}{(N-1)[\frac{1}{N}\sum_{i=1}^N W^{(i)}]^2}.$$

This is related, to the easier to interpret ESS, estimated as:

$$\text{ESS} = \left(\sum_{i=1}^{N} \left[\frac{W^{(i)}}{\sum_{j=1}^{N} W^{(j)}} \right]^2 \right)^{-1}$$

which is a proxy for the number of independent samples. These are simply indicators of the variance, and the latter quantity is used as a criterion to judge the degeneracy of the algorithm; that is, a low ESS suggests that the algorithm does not contain many diverse samples and that resampling should be performed. Note, the ESS can be misleading; see the discussion in Chopin (2002). In theory, the resampling step can make the analysis of the algorithm more challenging. The mathematical techniques are based upon measure-valued processes; see Del Moral (2004) and the references therein.

The tracking example. In response to Fearnhead's comment about our prior. We selected a reasonably informative prior; i.e. a Gaussian of mean equal to the true initial values and covariance equal to the identity matrix. However, we believe that performance would not degrade drastically if a more diffuse (but not very vague) prior was used. Indeed, in the block sampling strategy described in Section 4.1, the particles at the time origin would be eventually sampled according to an approximation of $p(x_1 | y_{1:R})$ at time R. In the scenarios considered here, the EKF diverged for a small percentage of the simulated paths of length 100. However, this divergence never occurred on a path of length $R \leq 10$ which partly explains why it was possible to use it to build efficient importance sampling densities. In scenarios with an extreme nonlinearity, we do agree with Fearnhead that it is unlikely that such an importance sampling distribution would work well. An importance sampling distribution based on the Unscented Kalman filter might prove more robust.

The probit example. We begin firstly, by stating the objectives of this example explicitly and then to address the points of the discussants one-by-one. The intention of presenting the simulations were two-fold:

- To illustrate, in a static setting, that even when MCMC kernels fail to mix quickly, that the combination of:

  (i) a large population of samples;

  (ii) tempered densities;

  (iii) interaction of the particles;

  can significantly improve upon MCMC for similar coding effort and CPU time.

- That the usage of the backward kernel (17) can substantially improve upon the reversal kernel (20) in terms of variance of the importance weights.

In response to Chopin's point on scaling. It seems that, given the scale parameters, using the idea of reweighting the tempered densities could be used, in order to use the samples targeting those densities other than that of interest. However, in this case, it is clear that this is not a sensible strategy for those densities far away, in some sense (*e.g.*, high variance models, as will be required to induce the tempering effect in the algorithm), as importance weights will have high variance. In the case that the scale parameters are to be determined, the problem of obtaining optimal parameters (in terms of minimizing the variance of importance weights, functionals of interest, etc.) is very difficult as noted in Section 6 of Del Moral *et al.* (2006); some practical strategies are outlined there.

Chopin's point on artificial models is quite insightful and he has identified a particular extension of the sequence of densities idea. We note that such procedures have appeared previously, for example in (Hodgson, 1999). Such references may provide further ideas in developing artificial models in different contexts and hence new sequences of densities to improve the exploration ability of the sampler.

We hope that we have clarified, with our second point, to Fan, Leslie and Wand, that indeed MH kernels could be used, but we were more interested in exploring a different part of the methodology. As Fan, Leslie and Wand note, the initial importance distribution could be improved, and they suggest a better approach than we adopted; however, from a practical point of view it can be much easier to implement our strategy (that is, the code has already been written for the MCMC steps).

In response to Scott's thorough comments on our probit example. The first point, which is accurate is not so relevant, in terms of what we set out to achieve with this example. One objective of SMC samplers, in static contexts, is somehow to try to remove some of the difficulties of considering how and why MCMC kernels do not mix. In essence, for many statistical problems, *e.g.*, stochastic volatility modelling (Kim *et al.*, 1998; Roberts etalc 2004) it can be very difficult to design specific MCMC samplers, such that we have identified the difficulties of the 'vanilla' sampler and then dealt with them. As a result, SMC methods are an attempt to produce *generic* methodology, which can improve over standard MCMC methods without needing too much target specific design (although clearly, this can improve the simulations), but also more freedom in sampler design. We hope that this was demonstrated in our example.

The second point of Scott on the comparison with MCMC is not quite accurate. As noted above, but perhaps not clearly enough in the paper, we wanted to demonstrate that slowly mixing kernels can be improved upon using sequences of densities, resampling and a population of samples. As a result, the 'trick' comment does not take into account that the population and resampling steps allow the algorithm to consider a vast amount of information simultaneously to improve the exploration of the target. Whilst, if we wanted a realistic comparison we could have used a superior MCMC method; however, there can be examples where more advanced MCMC techniques do not work well; see for instance Neal (1996) for examples with both tempered transitions and simulated tempering. More simply put, if the algorithm outlined by Scott mixes poorly, the intention is to use SMC to improve upon it. In Scott's example, we might achieve this via using sequences of densities with pseudo prior distributions $\{p_n(\zeta)\}$ converging close to Dirac measure on the set $\{1\}$.

The Stochastic volatility example. We respond to the comment of Fearnhead concerning the bias of our scheme involving re-proposing the particles. We begin by clarifying exactly what we do:

- If the ESS drops below 5% of the particle number, after a transition at time n, we return to the particles (and weights) before the transition.
- We change (increase) the sequence of densities so that we perform SMC samplers on increasingly more simple densities. Then we extend the state-space

and use SMC samplers to return the particles to the density of which we were interested in sampling.

In view of the above comments, the bias of this procedure can be thought of as similar to ordinary, dynamic resampling SMC techniques. That is, resampling upon the basis of the ESS. Whilst we do not have a theoretical justification for this method (however, we anticipate that we can use the methods of Douc and Moulines (2006) as in dynamic resampling) it provides a simple way to deal with the problem of consecutive densities with regions of high probability in different parts of the state-space, for a fixed $O(N)$ complexity. It is not clear then, that we would have fewer particles where the importance weight is large: firstly, the weight is now different from the first scheme, secondly we would expect that our particle approximation of π_n to be fairly accurate, given the tempering procedure adopted; there is no restriction upon the regions of the space that the particles are allowed to visit.

In response to the suggested idea of adding particles to ensure that the ESS does not drop too far. The first difficulty is when the consecutive densities are so different that it may take a large number of particles (e.g. $10N$) to ensure that our algorithm does not degenerate; this may not be feasible in complex problems due to storage costs. The second drawback that we feel that this procedure has, is that it is not explicitly trying to solve the problem at hand; in effect it is a brute force approach which may not work for feasible computational costs.

ADDITIONAL REFERENCES IN THE DISCUSSION

Andrieu, C., Doucet, A. and Tadić, V. (2005). Online simulation-based methods for parameter estimation in nonlinear non-Gaussian state-space models, *Proc. Control and Decision Conference.*

Andrieu, C. and Moulines, E. (2006). On the ergodicity properties of some adaptive MCMC algorithms, *Ann. Appl. Prob.* **16**, 1462–1505.

Carpenter, J. R., Clifford P., and Fearnhead P. (1999). An improved particle filter for nonlinear problems. *IEE Proc. Radar, Sonar and Navigation* **146**, 2–7.

Douc, R. and Moulines, E. (2006). Limit theorems for weighted samples with applications to sequential Monte Carlo Methods. *Tech. Rep.*, Ecole Polytechnique Palaiseau, France.

Douc, R., Guillin, A., Marin, J. M. and Robert, C. P. (2007). Convergence of adaptive sampling schemes. *Ann. Statist.* (in press).

Gordon, N., Salmond D. and Smith A. F. M. (1993). Novel approach to nonlinear/non-Gaussian Bayesian state estimation. *IEE Proc. Radar, Sonar and Navigation***140**, 107–113.

Hodgson, M. E. A. (1999). A Bayesian restoration of an ion channel. *J. Roy. Statist. Soc. B* **61**, 95–114.

Kim, S., Shephard, N. and Chib, S. (1998). Stochastic volatility: Likelihood inference and comparison with ARCH models, *Rev. Econ. Studies* **65**, 361–393.

Lavine, M. (2003). A Marginal ergodic theorem. *Bayesian Statistics 7* (J. M. Bernardo, M. J. Bayarri, J. O. Berger, A. P. Dawid, D. Heckerman, A. F. M. Smith and M. West, eds.) Oxford: University Press, 577–586.

Liu, J. S. and Wu, Y.-N. (1999). Parameter expansion for data augmentation. *J. Amer. Statist. Assoc.* **94**, 1264–1274.

Meng, X.-L. and van Dyk, D. A. (1999). Seeking efficient data augmentation schemes via conditional and marginal augmentation. *Biometrika* **86**, 301–320

Poyadjis, G., Doucet, A. and Singh, S. S. (2005). Maximum likelihood parameter estimation using particle methods, *Proc. Joint Statistical Meeting, USA.*

Roberts, G. O., Gelman, A. and Gilks, W. (1997). Weak convergence and optimal scaling of random walk Metropolis algorithms, *Ann. Appl. Prob.* **7**, 110–120.

Sisson, S. A., Fan, Y. and Tanaka, M. M. (2006). Sequential Monte Carlo without likelihoods. *Tech. Rep.*, University of New South Wales, Australia.

van Dyk, D. A. and Meng, X.-L. (2001). The art of data augmentation. *J. Comp. Graphical Statist.* **10**, 1–111 (with discussion).

Zhao, Y., Staudenmayer, J., Coull, B. A. and Wand, M. P. (2006). General design Bayesian generalized linear mixed models. *Statist. Science* **21**, 35–51.

BAYESIAN STATISTICS 8, pp. 149–174.
J. M. Bernardo, M. J. Bayarri, J. O. Berger, A. P. Dawid,
D. Heckerman, A. F. M. Smith and M. West (Eds.)

Dynamic Gaussian Process Priors, with Applications to the Analysis of Space-time Data

DANI GAMERMAN, ESTHER SALAZAR and EDNA A. REIS
Universidade Federal do Rio de Janeiro, Brazil
{dani,esalazar,edna}@dme.ufrj.br

SUMMARY

This paper describes the class of dynamic Gaussian processes. These are obtained as straightforward extensions of Gaussian processes when the time dimension is also considered or of dynamic models when the space dimension is also considered. Properties of dynamic Gaussian processes are derived and illustrative examples are provided. They provide useful components to a number of different models designed to analyze space-time data. Their adequacy and limitations are also discussed. Illustrations to simulated data examples show the potential usefulness of these structures.

Keywords and Phrases: FACTOR ANALYSIS; ISOTROPY; POINT PROCESS DATA; SEPARABILITY; SPATIAL DEFORMATIONS; SPATIO-TEMPORAL REGRESSION; STATIONARITY.

1. INTRODUCTION

1.1. *Spatial Structures*

Gaussian processes are by now established as a primary tool to embark when modelling spatial data. They are defined as the process $x(\cdot)$ in a region $D \in \mathbb{R}^d$ ($d = 2$, typically) such that for any $n \geq 1$ and locations $s_1, ..., s_n$, the vector $x(s_1), ..., x(s_n)$ has multivariate normal distribution with vector mean m and variance matrix Σ. Usual simplifying assumptions are

- stationarity implying that $m = \mu 1_n$ and $\Sigma = \sigma^2 R$, where the correlations r_{ij} in R satisfy $r_{ij} = \rho(s_i - s_j)$ for some suitably defined correlation function;

Dani Gamerman is Professor of Statistics and Esther Salazar and Edna A. Reis are PhD students at the Department of Statistical Methods of Universidade Federal do Rio de Janeiro. The authors have benefited from research grants from CNPq, FAPERJ and CAPES, respectively. They are also thankful to Helio Migon, Hedibert Lopes, Marina Paez, Marco Ferreira and Alexandra Schmidt for useful comments based on continued discussion on the subject.

- isotropy implying that the correlation function ρ_θ depends only on the distance between locations and possibly unknowns θ.

Under these assumptions, the notation

$$x(\cdot) \mid \Psi \sim GP(\mu, \sigma^2, \rho_\theta) \tag{1}$$

will be used in the sequel to denote isotropic, stationary Gaussian process, where the hyperparameter is $\Psi = (\mu, \sigma^2, \theta)$. Correlation functions are typically positive and decay monotonically (usually from 1) towards 0 as distance between locations gets larger. Note that these functions depend only on a single argument, the distance between locations.

Gaussian processes provide a flexible framework to describe smooth surfaces over $\mathbb{R}^d$. Smoothness is guaranteed by the large values of the correlation between nearby locations. They can be seen as a non-parametric tool as no specific form for the spatial variation is imposed deterministically. Simplifying assumptions allow for easy estimation due to the small dimension of the hyperparameter. The simplicity comes at the price of restrictions that may not be adequate for the situation under study. Departures from stationarity and isotropy will also be considered a few times in the application section.

The typical set-up for the use of Gaussian processes is the context of spatially correlated data, generically denoted by y. It will be assumed that all spatial correlation present in the data is captured by an underlying process x. Therefore, the observations are conditionally independent (given x) leading to the likelihood for x given by

$$l(x) = \prod_{s \in S_o} p(y(s) \mid x(s)),$$

where S_o is the set of locations where y is observed.

1.2. *Temporal Structures*

Once again, firmly established models with non-parametric flavor are used. Dynamic models have proved a flexible tool to handle temporal correlation (West and Harrison, 1997). They can be described via a p-dimensional process $x(\cdot)$ defined over time according to a temporal difference equation

$$x(t') = G(t', t)x(t) + w(t', t), \qquad \text{with } w(t', t) \sim N(0, W(t', t)), \quad (t' > t) \tag{2}$$

where the transition matrix G conveys the deterministic part of the evolution and the system disturbance w is simply a stochastic component accounting for increased uncertainty (controlled by W) over the temporal evolution. The model is completed with an initial specification for x at say $t = 0$. For temporally equidistant points, the model notation can be simplified to $x(t) = G_t x(t-1) + w_t$. When $G = I_p$, the identity matrix of order p, the model is the random walk

$$x(t) = x(t-1) + w_t. \tag{3}$$

The transition matrix and system disturbance in (2) must satisfy

$$G(t'', t')G(t', t) = G(t'', t), \tag{4}$$
$$w(\cdot, t'', t') + G(t'' - t')w(\cdot, t', t) = w(\cdot, t'', t), \tag{5}$$

for all $t'' \geq t' \geq t$. Unlike Gaussian processes, dynamic models are typically non-stationary but stationary processes may be also be obtained in special cases.

The non-parametric nature of these models are easier to understand with the usual choice of a random walk (3). In this case, the process x is simply undergoing a random walk without imposition of any specific temporal variation and as such is capable of (locally) tracking any smooth trajectory over time. The degree of smoothness is governed by the variances $W(t', t)$. A usual choice is $W(t', t) = (t' - t)W$, obtained from the discretization of an underlying Brownian motion (Revuz and Yor, 1999).

The typical set-up for their use is the context of temporally correlated data, generically denoted by y. It will be assumed that all temporal correlation present in the data is captured by an underlying process x. Therefore, the observations are conditionally independent (given x) leading to the likelihood for x given by

$$l(x) = \prod_{t \in T_o} p(y(t) \mid x(t)),$$

where T_o is the set of times where y is observed.

1.3. *Spatio-Temporal Structures*

Gaussian processes and dynamic models must be extended and combined in order to handle spatio-temporal dependence. This extension is intended to provide a model for a sequence of surfaces $x(s, t)$ and gives rise to the so-called dynamic Gaussian process (DGP, in short). This process may need to be time-continuous in contexts where data is obtained continuously over time. In this case, stochastic differential equations must be used to define DGPs. This is the route used by Brown et al. (2000) and Brix and Diggle (2001) to name a few authors. The resulting inference is more complex than an alternative, simplified version tackling only the case of discrete time. The latter gives rise to spatio-temporal difference equations such as (2) that are easily tractable albeit computationally intensive. These equations are applicable to the context of temporally discrete data collection. In some contexts considered below, they also provide suitable approximations to continuous time data collection.

The idea here is to retain the non-parametric flavor of Gaussian processes in this extension. As a result, surfaces assuming dependence on time and space of no specific form emerge. The approach here contrasts with the work of Cressie and Huang (1999) and Gneiting (2002). Their effort was directed towards obtaining specific parametric forms, imposed on the correlation structure of spatio-temporal processes. The approach here starts by consideration of the underlying latent process and spatio-temporal correlation functions are obtained later, as a consequence.

Hopefully, the latter can be at least used as a preliminary step in the identification of spatio-temporal variations described by the former. Thus, they can be reconciled as complementary steps. The approach here can be seen as a model-based exploratory analysis that may indicate the usefulness of a specific lower-dimensional parametric form for the spatio-temporal dependence.

The typical set-up now is the context of spatio-temporally correlated data, generically denoted by y. It will be assumed that all spatio-temporal correlation present in the data is captured by an underlying process x. Therefore, the observations are conditionally independent (given x) leading to the likelihood for x given by

$$l(x) = \prod_{s,t} p(y(s,t) \,|\, x(s,t)),$$

where the product is over all pairs (s,t) at which observations are made.

Another important issue of spatio-temporal models is associated with their highly dimensional structure. This is the subject of concern because their use require a considerable computational demand. Wikle and Cressie (1999) suggest a basis decomposition for handling models at a reduced dimension. This is a useful strategy that can be adapted to the structures described in this paper. Other forms of dimension reduction involving factor models are described below.

DGPs have been used mostly as a model for the context $y = x + e$ where e is a spatio-temporal white noise, ie, they are used as a model for the mean surface of a spatio-temporal data collection process. There are other situations that are potentially relevant to spatio-temporal data analysis and that can make use of DGPs. The main aim of this paper is to describe them and argue for the usefulness of DGPs in those contexts. Therefore, issues associated with the possible separability of time and space (see definition in Section 2) can be directly considered but are not the main aim. It will be argued that the structure presented here allows for different forms of spatio-temporal dependence both separable and non-separable. The resulting inference about DGPs may also enable a more detailed examination of some of these issues by inspection of the information contained in the posterior.

1.4. *Outline of the Paper*

The next section deals with possible specifications for the underlying process $x(s,t)$. Characteristics of these processes are described and their potential usefulness discussed. This paper concentrates on the simplest versions of spatio-temporal process and only considers models for continuous space. Similar models were entertained for discrete or areal spatial data by Vivar and Ferreira (2005).

This paper focuses on characterization of DGPs and description of their potential uses in different contexts. Thus it will not delve into implementation or computational details. It will also not address the important issue of hyperparameter prior specification or use of reference analysis. All these points must be sought at the corresponding references alluded to in the text.

Section 3 describes a few situations where the DGPs can be applied. All of them involve different contexts of spatio-temporal models. They include regression models, factor analysis, point processes and anisotropic models. They are by no means an exhaustive list of possible uses of DGPs. Hopefully, this paper will instigate researchers into finding other areas of application of DGPs.

Section 4 draws on some concluding remarks.

2. BASIC DEFINITIONS AND PROPERTIES

Some results must be presented before considering possible forms for DGPs due to their possible multivariate nature. A natural multivariate extension of Gaussian processes is given by the definition below.

Definition 1 (Multivariate Gaussian process). *A p-dimensional process* $x(\cdot) = (x_1(\cdot), ..., x_p(\cdot))$ *defined in a region* $D \in \mathbb{R}^d$ *(*$d = 2$*, typically) is a stationary,* *isotropic Gaussian process if for any $n \geq 1$ and locations $s_1, ..., s_n$, the vector $(x'(s_1), ..., x'(s_n))$ has multivariate normal distribution. The case $p = 1$ is given in (1).*

The above definition is too general for practical use. Important special cases are:

- when the scalar processes $x_k(\cdot) \sim GP(\mu_k, \sigma_k^2, \rho_{\theta_k})$ are independent, for $k = 1, ..., p$, each one having its own mean, variance and correlation function. This is denoted as

$$x(\cdot) \sim \prod_{k=1}^{p} GP(\mu_k, \sigma_k^2, \rho_{\theta_k}).$$

  Thus, $Cov(x_k(s_i), x_l(s_j)) = \sigma_k^2 \rho_{\theta_k}(|s_i - s_j|)$, if $k = l$ and 0, otherwise;

- when the vector process $x(\cdot) \sim GP(\mu, \Sigma, \rho_\theta)$ where $\Sigma = (\sigma_{kl})$ is a valid variance matrix and ρ_θ is a (common across processes) correlation function. Thus, $Cov(x_k(s_i), x_l(s_j)) = \sigma_{kl}\rho_\theta(|s_i - s_j|)$;

- when the vector process x is obtained via a linear transformation $x(\cdot) = Az(\cdot)$ where $A = (a_{kl})$ is a p-dimensional square matrix and $z(\cdot) = (z_1(\cdot), ..., z_p(\cdot))$. If $z(\cdot) \sim \prod_{k=1}^{p} GP(\mu_k, \sigma_k^2, \rho_{\theta_k})$ then $Cov(x_k(s_i), x_l(s_j)) = \sum_{m=1}^{p} a_{km}a_{lm}\sigma_m^2$ $\rho_{\theta_m}(|s_i - s_j|)$.

Suitable contexts for application of all special cases above can be found. The independent case was deemed appropriate for the spatio-temporal factor models of Lopes, Salazar and Gamerman (2006) despite its simple covariance structure. The common correlation case was applied in the spatio-temporal regression models of Paez *et al.* (2006). The last case is derived from coregionalization considerations and is used by Gelfand et al. (2004). It generalizes the independent case due to the presence of non-zero off-diagonal elements of A.

Combinations of the multivariate Gaussian processes above with the dynamic models in (2) give rise to many possible definitions of dynamic Gaussian processes. A straightforward version is given below

Definition 2 (Dynamic Gaussian process). *A process $x(\cdot, \cdot)$ is said to follow a dynamic Gaussian process if it can be described by a (temporal) difference equation*

$$x(\cdot, t') = G(t', t)x(\cdot, t) + w(\cdot, t', t) \quad , \quad w(\cdot, t', t) \sim GP \qquad (6)$$

where the multivariate Gaussian process disturbances $w(\cdot, t', t)$ are zero mean and time-independent. The law of the process is completed with a Gaussian process specification for $x(\cdot, 0)$.

In the common correlations across processes and identically distributed equidistant cases, the disturbances satisfy $w(\cdot, t+1, t) \sim GP(0, \Sigma^w, \rho^w)$, for all t and $x(\cdot, 0) \sim GP(\mu, \Sigma^0, \rho^0)$, for some constant mean μ.

Once again, the use of multivariate processes leads to a class that is too large for practical purposes. Below, a few special cases are outlined and their correlation structure is described. Special points of interest are the determinations of separability and stationarity.

A spatio-temporal process x is said to be separable if $Cov(x(s_1,t_1),x(s_2,t_2))$ can be decomposed into $K_s(s_1,s_2)K_t(t_1,t_2)$ or $K_s(s_1,s_2)+K_t(t_1,t_2)$ for suitably defined functions K_s and K_t. A Gaussian spatio-temporal process x is stationary[1] if $E(x(s,t)) = \mu$ and $Cov(x(s_1,t_1),x(s_2,t_2)) = C(h,u)$ where $h = s_2 - s_1$ and $u = t_2 - t_1$. If the process is stationary and separable then $K_s(s_1,s_2) = C_s(s_2-s_1)$ and $K_t(t_1,t_2) = C_t(t_2-t_1)$ where C_s and C_t are a spatial and a temporal covariance functions respectively. These functions can be derived from the correlation functions used in (1) if isotropy is also assumed.

In practice, it is unlikely that most processes are stationary and separable. Stationarity is usually imposed for convenience, to enable estimation of the covariance function from the data (Cressie and Huang, 1999). Many recent papers focus on finding suitable representations of non-separable and/or non-stationary forms (Gelfand et al., 2004). Separability is clearly another simplifying assumption and many authors consider it unrealistic for practical purposes. They suggest concentrating on non-separable forms. The examples below illustrate this discussion.

Example 1 (Random walk DGPs). One special case of interest is provided by $G = I_p$ in which case the process is simply the spatial extension of the purely temporal random walk (3). In the equidistant case, it is given by

$$x(\cdot,t) = x(\cdot,t-1) + w(\cdot) \quad , \quad w(\cdot) \sim GP. \tag{7}$$

As a result, the marginal temporal process $x(s,\cdot)$ at a fixed location s obtained by fixing the location is non-stationary, for all s, as its variance increases without bounds. In fact, the process is non-stationary in space and time as shown below. As previously mentioned, this kind of model may not be appropriate for some applied situations but may prove a useful benchmark for approaching inference in situations where little is known about the underlying process. They lead to many separable and non-separable forms for the covariance structure of x. In the common correlation case, $Cov(x(s_1,t_1),x(s_2,t_2)) = \Sigma^0\rho^0(|s_2-s_1|) + min\{t_2,t_1\}\Sigma^w\rho^w(|s_2-s_1|)$. If the correlation functions ρ^0 of the initial prior and ρ^w of the disturbances coincide, the process is separable but remains non-stationary. Otherwise, the covariance structure is a time-varying mixture of the covariance structures of the prior and of the disturbance.

This mixture of correlation functions is an important feature of the DGPs above. This can be more clearly understood and more thoroughly investigated in the case of the class of DGPs below.

Example 2 (Stationary DGPs). Consider the model

$$x(\cdot,t') = \rho_t(t',t)x(\cdot,t) + w(\cdot,t',t) \quad , \quad w(\cdot,t',t) \sim GP\{0,[1-\rho_t^2(t',t)]\Sigma^w,\rho^w\}.$$

ρ_t is a valid correlation function that must satisfy $\rho_t(t''-t')\rho_t(t'-t) = \rho_t(t''-t)$, for $t'' \geq t' \geq t$, due to (4). Thus, this temporal correlation function must be exponential

[1] Only Gaussian processes are considered in this paper; for non-Gaussian processes, the condition only ensures 2nd order stationarity.

and $\rho_t(u) = \exp\{-u/\theta_t\}$, for $u \geq 0$ where θ_t is a constant. The model is completed with $x(\cdot, 0) \sim GP(0, \Sigma^0, \rho^0)$.

Simple calculations show that the model is valid (*i.e.*, conditions (4) and (5) are satisfied). Thus

$$Cov(x(s_1, t_1), x(s_2, t_2)) = \exp\{-\frac{u}{\theta_t}\}\{\exp\{-\frac{t}{\theta_t}\}\Sigma^0\rho^0(h) + [1 - \exp\{-\frac{t}{\theta_t}\}]\Sigma^w\rho^w(h)\},$$

where $h = |\, s_2 - s_1 \,|$, $t = min\{t_2, t_1\}$ and $u = |\, t_2 - t_1 \,|$.

Note that as $t \to \infty$, the influence of the prior process vanishes and the separable form with $C_s(h) = \Sigma^w\rho^w(h)$ and $C_t(u) = \exp\{-u/\theta_t\}$ is obtained for $Cov(x(s_1, t_1), x(s_2, t_2))$. If $\Sigma^0 = \Sigma^w = \Sigma$ then the marginal processes $x(s, \cdot)$ are temporally stationary for any $s \in D$ in the sense that $x(s, t) \sim N(0, \Sigma)$, for all $t \geq 0$.

For illustrative purposes, consider temporally stationary univariate processes ($p = 1$) where $\Sigma^0 = \Sigma^w$ is scalar and denoted by σ. Figures 1 and 2 show the variation of the covariance function against the spatial and temporal lags for ρ^0 and ρ^w given by commonly used correlation functions. A variety of non-separable forms emerges even for this simple setting.

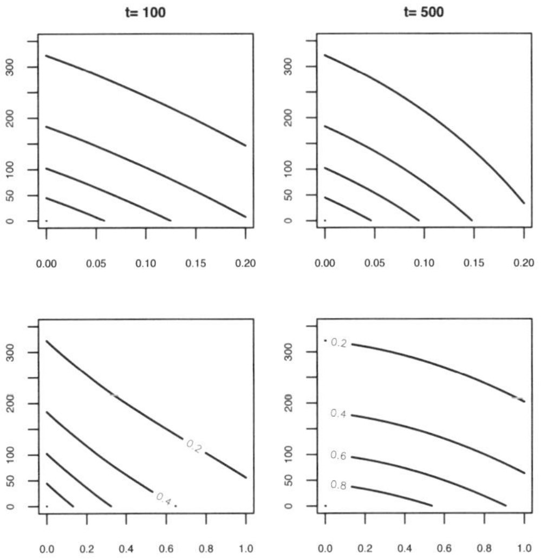

Figure 1: *Covariance contour plots for exponential prior correlation ($\rho(u) = \exp\{-u/\theta\}$). The disturbances correlation functions were: top line–spherical ($\rho(u) = [1 - 1.5(u/\theta) + 0.5(u/\theta)^3]I(u < \theta)$); bottom line–Matérn ($\rho(u) = 2^{1-\kappa}\Gamma^{-1}(\kappa)(u/\theta)^\kappa K_\kappa(u/\theta)$, where K_z is the Bessel function of third kind of order z). The values used were $\sigma = 1$, $\theta = 3$ and $\kappa = 5$.*

DGPs that are stationary in space and time can be obtained by taking $x(\cdot, 0) \sim GP(0, \Sigma^0, \rho^0)$ and $w(\cdot, t', t) \sim \sqrt{1 - \rho_t^2(t', t)}x(\cdot, 0)$, for all $t' > t \geq 0$. The resulting $x(\cdot, t)$ processes are stationary around 0 and also lead to separable covariance structures. A non-zero mean μ are accommodated by replacing $x(\cdot, t)$ by $x(\cdot, t) - \mu$.

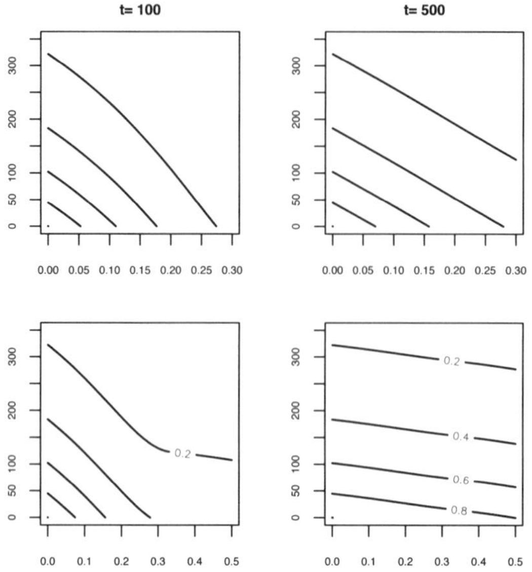

Figure 2: *Covariance contour plots for spherical prior correlation. The disturbances correlation functions were: top line–exponential; bottom line–Matérn. Same values of $\theta = 3$ and $\kappa = 5$ were used.*

Multivariate versions with different correlation functions for each univariate DGP can also be constructed. This opens up the possibility of separate modelling of the mean μ possibly allowing for a temporal structure. This is particularly useful when a non-spatial temporal trend is present.

Finally, there should be mention of space-time interacting versions of DGP that have previously been considered for the analysis of spatio-temporal data. They are based on a kernel convolution that allows for the dependence of the process at future times on values of previous times not only at the same location but also at surrounding locations (see for example Wikle and Cressie (1999) and Brown et al. (2000)). Definition 2 above is recovered in the special case of a trivial kernel concentrated at 0. These general processes may provide more appropriate representation of spatio-temporal processes especially for situations where contagion between locations is involved. However, they depend on suitable choices of kernels and their bandwidths (or scale) and are harder to analyze. Fuentes and Smith (2001) also consider spatial kernel convolutions with time-varying parameters to handle spatio-temporal variation. Once again, basically the same comments apply here.

In any case, inference about x is based on Bayes theorem that combines the information carried by the prior with data information, that may indicate otherwise. These models are reasonably flexible and are capable of adaptation towards any direction the data may point. But they still retain the smoothness restrictions of the prior thus leading to an inference that compromises between these two pieces of information. The complexity and high-dimensionality of the models prevents analytic inference and MCMC methods (Gamerman and Lopes, 2006) are used.

Details of these methods are not presented here but references are provided in each situation.

The focus of this paper is to describe the framework that serves as a common basis for a variety of different areas of application. The applications of the next section are briefly presented to illustrate these areas of spatio-temporal data modelling, where use of DGPs seems fruitful.

3. APPLICATIONS

This section describes some possible uses of DGPs as prior for some components of a model for spatio-temporal data. Initially, spatio-temporal regression models are presented where DGPs arise almost naturally as suitable prior representation for regression coefficients. Then, factor analysis in the context of spatio-temporal data is introduced. The ideas are then applied to point processes data following the seminal work of Brix and Diggle (2001). Finally, the context of regions of interest where spatial stationarity is not reasonable is handled through spatial deformations (Sampson and Guttorp, 1992) are given.

3.1. *Spatio-Temporal Regression*

Gelfand, Banerjee and Gamerman (2005) introduced a general class of multivariate regression models taking into account spatio-temporal heterogeneity in the effect of the explanatory variables. Thus, their model generalizes the space-varying regression models of Gelfand et al. (2003) and dynamic regression models of West and Harrison (1997). It has a q-dimensional response $y(s,t)$ explained by p covariates $z_1, ..., z_p$ in the linear form

$$y(s,t) = \beta_0(s,t) + \beta_1(s,t)z_1(s,t) + ...\beta_p z_p(s,t) + e(s,t), \text{ with } e(s,t) \sim N(0,\sigma^2).$$

Each q-dimensional regression coefficient $\beta_j(s,t)$ is modelled according to a DGP, for $j = 0, 1, ..., p$ as given in (6). Overall, a $q \times (p+1)$ matrix of regression coefficient processes is available at any given time. A more encompassing approach is to decompose each regression coefficient according to

$$\beta_j(s,t) = \mu_j(t) + \gamma_j(s,t) \tag{8}$$

where $\mu_j(\cdot)$ are temporal processes and γ_j are zero-mean DGPs, $j = 0, ..., p$, as mentioned in Example 2. This decomposition separates the purely temporal trend from the spatio-temporal variation and is sometimes useful both computationally and in terms of reporting results. Both components may evolve, for example, according to random walks as described in (3) and (7) respectively. Obviously, the overall process is easily recovered by adding these components.

The above model was applied to environmental data by Gelfand, Banerjee and Gamerman (2005) where the effect of precipitation on temperature was studied. They used the linear transformation of independent Gaussian processes and showed the relevance of both spatial and temporal variations of the both intercept and the regression coefficient. Fig. 3 shows the spatio-temporal variation of the effect of temperature. It reveals a more extreme spatial variation in the months of more extreme weather. The same idea can be extended to other models with (linear) regression terms such as generalized linear models and survival data analysis.

The discussion of stationarity and separability is somewhat irrelevant in these contexts. Even if separable and stationary forms are used for the regression coefficients and white noise for the observation errors, the final covariance structure will

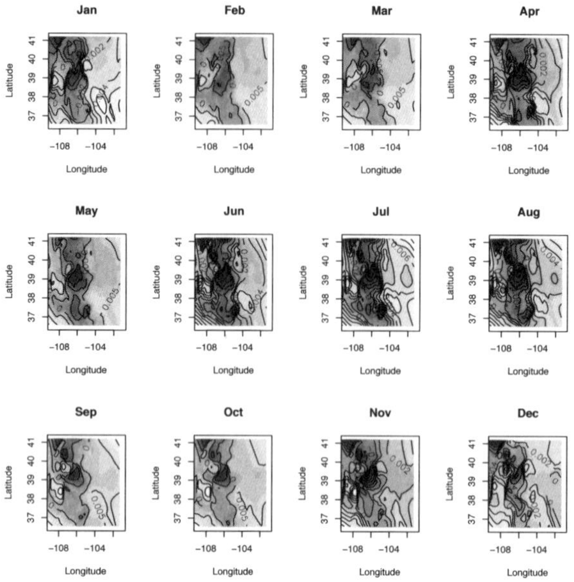

Figure 3: *Posterior mean of the spatio-temporal variation γ of the regression coefficient of precipitation over temperature for a region of the State of Colorado, USA (reprinted from Gelfand, Banerjee and Gamerman, 2005)*

be a complicated function of these elements and the covariates. Thus, it will rarely be stationary or separable.

3.2. *Factor Analysis*

The idea behind factor analysis is basically to reduce dimensionality of multivariate data by identification of a smaller number of factors that are responsible for most of (or ideally all) the correlation between observations. The standard setting is

$$y = \beta f + e \tag{9}$$

where the y is a n-dimensional observation, f is a m-dimensional factor, β is a $n \times m$ matrix of factor loadings and e is a white noise carrying only the idiosyncrasies of the individual observations. When the number of factors m is substantially smaller than n, a considerable reduction in the explanation of the variability is obtained. Usually, some identifiability constraints are imposed on elements of β. References of the Bayesian approach for factor analysis include Geweke and Zhou (1996) and Lopes and West (2004).

The basic context of static factor analysis has been extended in several directions. In the temporal domain, the model (9) has been extended to accommodate observations of multivariate time series and becomes

$$y(t) = \beta f(t) + e(t), \tag{10}$$

where the temporal dependence is captured by the time-varying factors $f(t)$ while the observation errors $e(t)$ remain independent. Aguilar and West (2000) and Pitt and Shephard (1999) have independently used this structure to model financial data and investigate the possibility of a few common factors relating the individual series. Both specify the temporal dependence of the factors through a autoregressive structure on their variances in stochastic volatility models. Other possibilities include autoregressive forms for the factors such as

$$f(t) = \Phi f(t-1) + v(t), \text{ where } v(t) \sim N(0, \Lambda). \tag{11}$$

The loadings associated with the factors, however, remained unchanged.

The possibility of time-varying loadings was investigated by Lopes and Migon (2002) and Lopes and Carvalho (2007). Their model extend (10) by evolving the factor loadings according to the autoregression $\beta(t) = A + B\beta(t-1) + w(t)$. The above evolution was only applied to the unknown components of the loadings and thus identifiability constraints were retained. They were able to show the importance of allowing for time variation of the loadings with some of them remaining relevant only part of the time span of the data.

Lopes, Salazar and Gamerman (2006) consider a factor analysis to determine the relevant aspects of spatial variation in georeferenced data. In their context, the observation vector at any given time consists of a collection of observations on the same variable at different locations. The factors are designed to identify common traits as before but these traits are based on the spatial configuration of the data. This feature is incorporated into the model by loadings that are affected by the location. Calder (2007) and Sansó and Schmidt (2004) use a deterministic dependence of β on the space via smoothing kernels. Lopes, Salazar and Gamerman (2006) allow a more flexible form for spatial variation of the loadings via a stochastic dependence of the loadings on space. Since the factors are independent by construction, it is natural to assume independence between loadings associated with different factors. Thus, the matrix of loadings have independent columns. This discussion leads to and is summarized by the model (10) with the columns β_j $(j = 1, ..., m)$ having independent normal prior distributions with covariance matrix determined by underlying Gaussian processes. Identifiability constraints are no longer required because the Gaussian process prior already imposes (stochastic) constraints on the loadings.

The spatial and temporal variations of the factor loadings can now be easily combined with the use of DGPs. Salazar (2006) and Salazar, Lopes and Gamerman (2006) introduced this idea as a general framework for the analysis of spatio-temporal data. Their model allows for dynamic factors evolving as in (11) and dynamic loadings evolving as described in Section 3.1 with independent priors for each factor loadings.

The independence between factors forces them to explore different aspects of the dependence between observations. Therefore, they are designed to handle a variety of forms of spatial dependence. The resulting model combines them and, as such, is capable of accommodating spatial non-stationarity because of this combination of factors and their respective loadings.

Example 3 (Simulated factor model). A two factor model was simulated over a regular grid in the square $[0, 1] \times [0, 1]$ considering observations made at $N = 30$ locations randomly chosen and displayed in Fig. 4 and over $T = 80$ times. The factors evolve according to (11). Each column of the factor loadings matrix $\beta_j(t)$ follows (8) with random walk evolutions (3) and (7) for μ_j and γ_j, respectively. The spatio-temporal processes γ_j are initialized with $\gamma_j(\cdot, 0) \sim GP(0, \tau_{j0}^2, \rho_{\theta_{j0}})$, for $j = 1, 2$. Data was generated with exponential correlation functions, $\Phi = \text{diag}(0.95, 0.7)$, $\Lambda = \text{diag}(0.49, 0.81)$, $\sigma_i^2 \in [0.02, 0.93]$ $(i = 1, \ldots, 30)$, $W_{\mu_1} = 0.64$ $W_{\mu_2} = 0.25$, $\theta_{10} = 0.15$, $\theta_{20} = 0.4$, $\tau_{10}^2 = 0.75$, $\tau_{20}^2 = 0.9$, $\theta_1 = 0.4$, $\theta_2 = 0.1$, $\tau_1^2 = 1$, $\tau_2^2 = 0.49$.

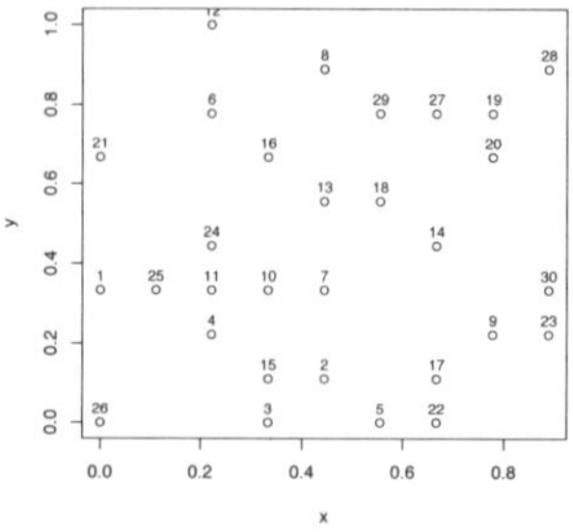

Figure 4: *Locations of generated data in example of two-factor model.*

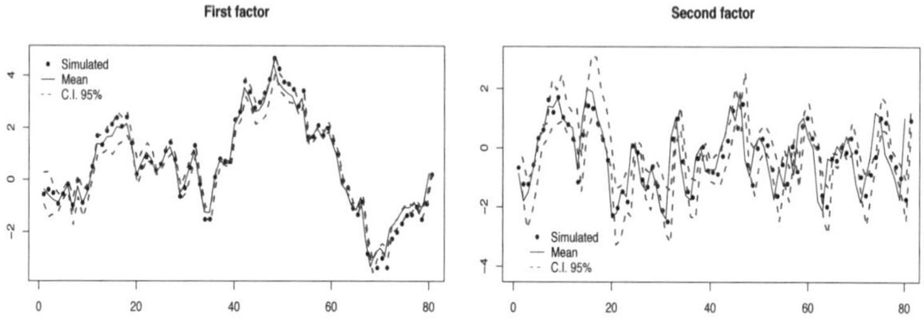

Figure 5: *Posterior means of both first and second factor (solid lines) compared against simulated values (points). Dashed lines correspond to the 95% credible interval limits.*

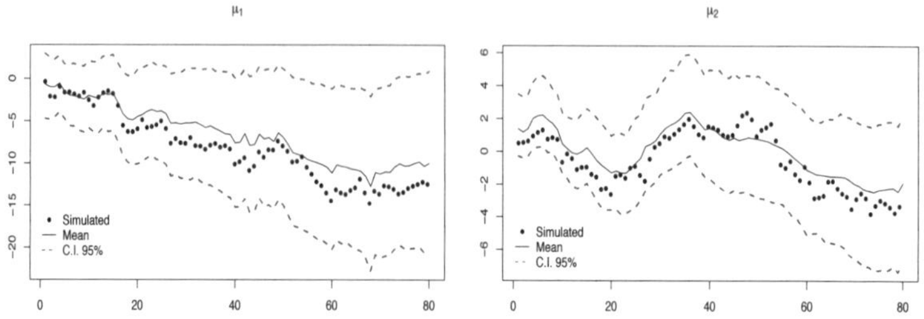

Figure 6: *Posterior means of $\mu_j(t)$ for $j = 1, 2$ and $t = 1, \ldots, 80$ (solid lines) compared against simulated values (points). Dashed lines correspond to the 95% credible interval limits.*

Vague normal priors for the elements of Φ, independent $IG(0.1, 0.1)$ priors for the elements of Λ and for σ_i^2's and priors for the parameters of Gaussian processes based on the reference analysis of Berger, de Oliveira and Sansó (2001) were assumed. The prior hyperparameters θ_{10}, θ_{20}, τ_{10}^2 and τ_{20}^2 were assumed known.

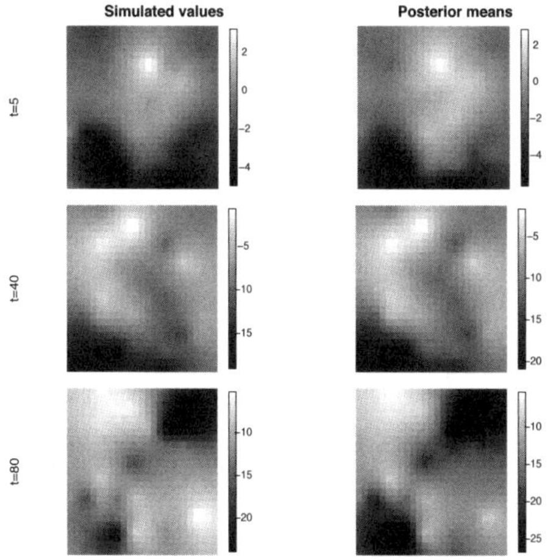

Figure 7: *Loadings of the first factor: simulated values (left) and posterior means (right). The first, second and third rows refer to* $t = 5, 40, 80$*, respectively. Posterior means for loadings at unobserved locations were obtained by Bayesian interpolation.*

Table 1 displays posterior summaries for hyperparameters and Figures 5–8 show the posterior means for factors and temporal components of $\beta_j(t)$ respectively. Good recovery of the factors was observed showing that these structures can be applied to the analysis of spatio-temporal data. It was assumed in the example that the number of factor was known. Uncertainty about the number of factors can be easily incorporated (see Lopes and West, 2000; Lopes, Salazar and Gamerman, 2006).

Table 1: *Posterior summaries of hyperparameters.*

	True	Mean	S.D.	2.5%	50%	97.5%
ϕ_1	0.95	0.927	0.037	0.849	0.930	0.992
ϕ_2	0.7	0.614	0.089	0.442	0.615	0.786
λ_1	0.49	0.459	0.085	0.327	0.450	0.647
λ_2	0.81	0.913	0.42	0.313	0.940	1.670
W_{μ_1}	0.64	0.901	0.380	0.207	0.880	1.710
W_{μ_2}	0.25	0.395	0.260	0.107	0.319	1.070
θ_1	0.4	0.563	0.120	0.378	0.541	0.873
θ_2	0.1	0.121	0.020	0.085	0.120	0.167
τ_1^2	1	1.41	0.26	0.977	1.38	1.98
τ_2^2	0.49	0.528	0.29	0.247	0.387	1.200

3.3. *Point Processes*

One common data collection process is given by the occurrence of events in a given location and the interest lies in determining the pattern associated with the data generation. The basic set-up for this process is to assume an underlying non-homogeneous

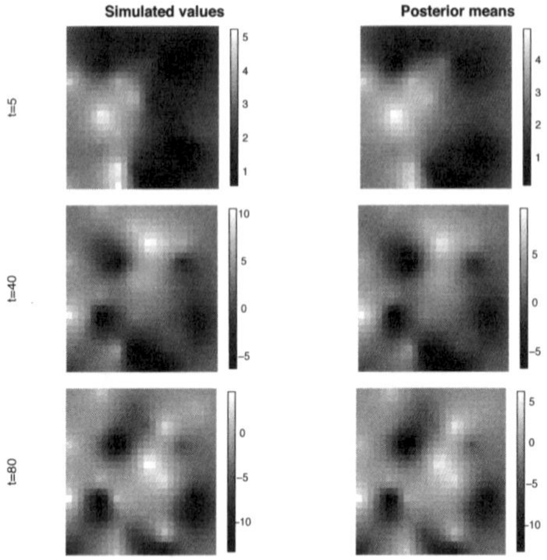

Figure 8: *Loadings of the second factor: simulated values (left) and posterior means (right). The first, second and third rows refer to $t = 5, 40, 80$, respectively. Posterior means for loadings at unobserved locations were obtained by Bayesian interpolation.*

Poisson process with intensity λ leading to the likelihood

$$\prod_{i=1}^{n} \lambda(s_i, t_i) \exp\left\{ - \int_{D\times[0,T]} \lambda(s,t)d(s,t) \right\},$$

where n is the observed number of events, $s_1, ..., s_n$ is the set of their locations, $t_1, ..., t_n$ is the set of their respective times of occurrence, D is the observation region in space and $[0, T]$ is the observation region in time. The statistical theory behind point processes is well established by now (Diggle, 2003) but is mostly concerned with and applied to situations where the temporal dimension is not present.

An usual approach to inference about the intensity λ is through Cox processes (Cox, 1955) where the intensity is assumed to be stochastic. Since λ is positive, one common approach in recent years is to assume log $\lambda = x$ where x is a Gaussian process. This gives rise to the log-Gaussian Cox processes studied by Møller, Syversveen and Waagepetersen (1998).

Brix and Diggle (2001) approached this problem via discretization of space while time remained continuous. They described their process via stochastic differential equations (that are basically a differential version of the difference equations (6)) and studied some of their properties. Inference is a difficult task in this context and they resort to moment estimators.

A natural approach to this problem is to consider directly the DGPs in (6) as a prior for the log intensity x. This route was adopted by Reis (2006). This provides a direct generalization of the work of Paez and Diggle (2006) and Gamerman (1992) that used dynamic processes to model point processes aggregated over space. In practice, some space and time discretizations were performed to enable inference. These can be obtained as a temporal generalization of the Bayesian approach of Christensen and Waagepetersen (2002). Details can be found in Reis (2006).

Example 4 (Simulated Cox processes). Point process data was simulated over a rectangular region in space and over $[0, T]$ with the log intensity x following

$$x(\cdot, t) - \mu = \eta[x(\cdot, t-1) - \mu] + w_t(\cdot), \qquad \text{where } w_t(\cdot) \sim GP(0, (1-\eta^2)\sigma^2, \rho_\theta),$$

where ρ_θ is the exponential correlation function. The model is initialized with $x(\cdot, 0) \sim GP(0, \sigma^2, \rho_\theta)$. From Example 2, this is clearly a stationary process is space and time.

Inference is performed approximately by discretizing space and time. The observation region is divided into N subregions, as in Christensen and Waagepetersen (2002) and Brix and Diggle (2001), and the time span is split into T intervals. Data was generated according to $T = 10$, $\mu = 1$, $\sigma^2 = 1$, $\theta = 1.25$ and $\eta = 0.67$. The point pattern generated can be observed in Fig. 9 for a few selected time points.

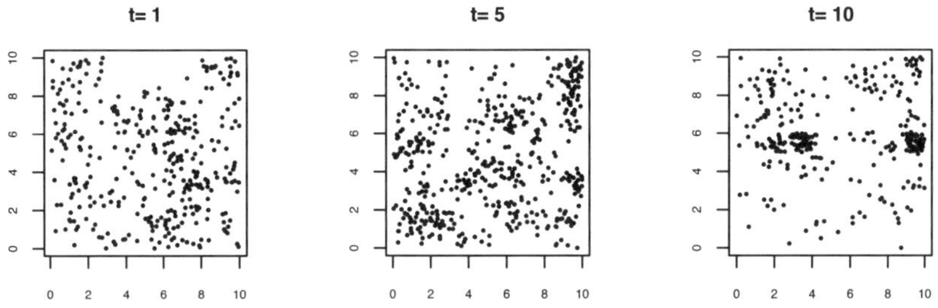

Figure 9: *Locations of generated occurrences for $t = 1, 5$ and 10.*

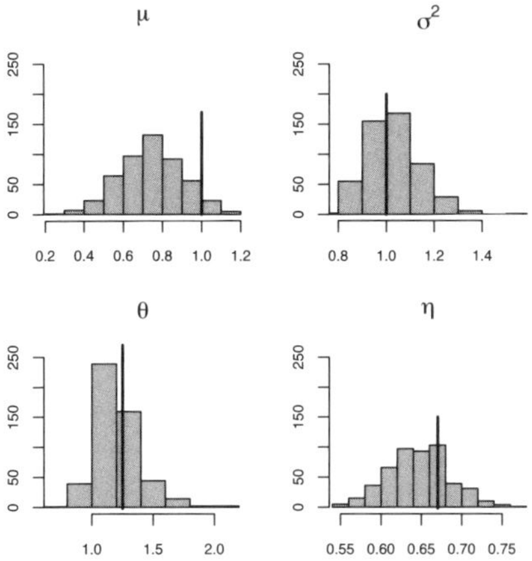

Figure 10: *Histograms of hyperparameters $\mu = 1$, $\sigma^2 = 1$, $\theta = 1.25$ and $\eta = 0.67$ based on 500 posterior samples obtained at every 50 iterations after a burn in of 25,000 iterations.*

Fig. 10 shows the resulting posterior sample of the hyperparameters where good recovery of generated values is achieved. Fig. 11 shows the recovered (mean posterior) maps of the intensity for a number of times and again seems to retrieve the

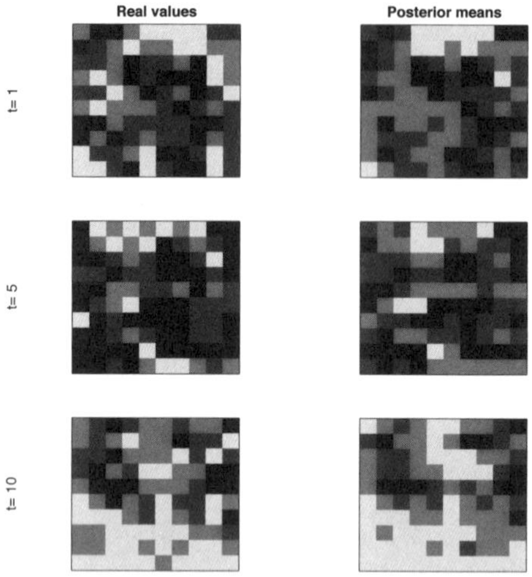

Figure 11: *Map of the posterior mean of log intensity compared against actual values. This value was calculated by integrating the intensity rate over each subregion.*

spatial pattern. The accuracy of the estimation procedure can be evaluated through Figure 12. It summarizes inference results for the log intensity. A good agreement is observed, with results becoming more precise with increasing intensity and a slight upward bias for low intensity regions. The results suggest that these models may be well recognized from the data.

3.4. *Spatial Deformations*

Isotropic models are attractive for their simplicity but this comes at the cost of strong and sometimes unrealistic impositions on the spatial structure. There are many model-based approaches to tackle anisotropy. Higdon, Swall and Kern (1999) introduced the idea of spatially-varying convolutions while Sampson and Guttorp (1992) suggested non-parametric estimation of a deformed space where isotropy could be assumed.

Schmidt and O'Hagan (2003) and Damian, Sampson and Guttorp (2001) proposed similar Bayesian approaches to inference on deformed spaces using Gaussian processes in a hierarchical structure. In the first level, the observations follow a Gaussian process with correlation function depending on latent locations $d(s)$. These locations lie on a deformed space where isotropy can be assumed for the observations. In the second level, the latent locations themselves follow a Gaussian process centered around the geographical locations. The reasoning being that the region of study where observations are taking place is regular enough to become isotropic after some deformation.

Chen and Conley (2001) present an approach to inference with economic panel data where observations are related according to some metric based on their economic characteristics rather than their geographical location. A number of possibilities are considered for them, including fixed and random distances and non-parametric estimation of the mean and variance effects of these distances between

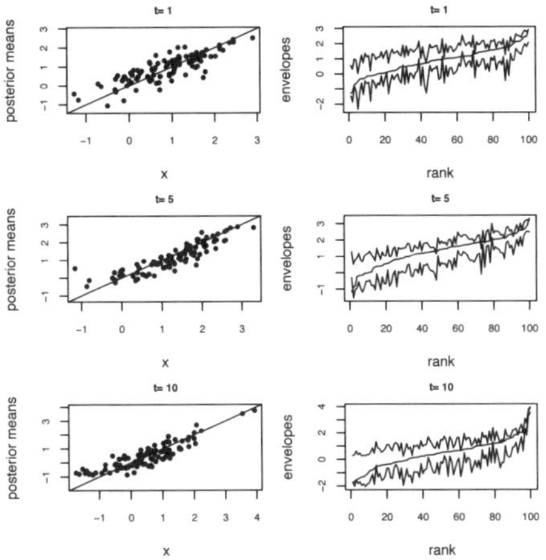

Figure 12: *Summary of inference for the log intensity. The first column shows the plot of posterior mean against actual value of intensity for all* $N = 100$ *subregions. This value was calculated by integrating the intensity rate over all subregions. The second column shows the intensities of each subregion ordered on a line by their actual values along with their respective 90% credibility intervals. The first, second and third rows refer to* $t = 1, 5$ *and* 10*, respectively.*

observation units.

In environmental contexts, deformations are generally caused by feature associated with topography and sometimes climate. In economic contexts, they are caused by shifting structures of the economic environment. It seems more likely to have a less stable spatial (deformation) pattern in studies where measures are associated with human activities. In those cases, it is likely that suitable representation in terms of deformed spaces may need to me placed in a dynamic setting where they are allowed to change.

The above discussion shows that there is scope for consideration of a time-varying pattern in the deformations and one candidate for performing this task is provided by DGPs. This is the subject of ongoing research still in its infancy. We hope to be able to report on its developments in the near future.

4. CONCLUDING REMARKS

This paper was intended to show a host of opportunities for using dynamic Gaussian processes when modelling different aspects of data arising in the spatio-temporal domain. The examples range from established applications to comments of speculative nature that still need to be tested in other areas of applications. They also visited areas that are currently being developed.

The most attractive feature of these processes is that they do not impose any specific form to the temporal and spatial variations. Thus, they can provide useful starting point for identification of underlying data patterns.

The approach is reasonably flexible and only imposes mild smoothness constraints over space and time. More flexibility can be considered with replacement of gaussianity by fatter tail distributions. For example, scale mixtures of Gaussian processes may prove useful to accommodate processes with more aberrant fluctuations over space and time.

The approach can be easily implemented albeit time consuming. The number of parameters is reasonably large and grows with the dimension of the data. Despite their large dimensionality, they typically involve highly sparse matrices because of (near) independence between units far apart in space or time. Substantial gains can be gained from this, making inference computationally viable. Computational issues were not addressed here but further work is certainly required.

REFERENCES

Aguilar, O. and West, M. (2000). Bayesian dynamic factor models and variance matrix discounting for portfolio allocation. *J. Business Econ. Statist.* **10**, 338-357.

Berger, J. O., de Oliveira, V. and Sansó, B. (2001). Objective Bayesian analysis of spatially correlated data. *J. Amer. Statist. Assoc.* **96**, 1361–1374.

Brix, A. and Diggle, P. J. (2001). Spatiotemporal prediction for log-Gaussian Cox processes. *J. Roy. Statist. Soc. B* **63**, 823–841.

Brown, P. E., Kåresen, K. F., Roberts, G. O. and Tonellato, S. (2000). Blur-generated non-separable space-time models. *J. Roy. Statist. Soc. B* **62**, 847–860.

Calder, C. A. (2007). Dynamic factor process convolution models for multivariate space-time data with application to air quality assessment. *Environmental and Ecological Statistics* (to appear).

Chen, X. and Conley, T. G. (2001). A new semiparametric spatial model for panel time series. *J. Econometrics* **105**, 59–83.

Christensen, O. F. and Waagepetersen, R. (2002). Bayesian prediction of spatial count data using generalised linear mixed models. *Biometrics* **58**, 280–286.

Cox, D. R. (1955). Some statistical methods concerned with series of events (with discussion). *J. Roy. Statist. Soc. B* **17**, 129–164.

Cressie, N. and Huang, H.-C. (1999). Classes of non-separable, spatio-temporal stationary covariance functions. *J. Amer. Statist. Assoc.* **94**, 1330–1340.

Damian, D., Sampson, P. and Guttorp, P. (2001). Bayesian estimation of semi-parametric non-stationary spatial covariance structures. *Environmetrics* **12**, 161–178.

Diggle, P. J. (2003). *Statistical Analysis of Spatial Point Patterns.* (2nd. ed.) London: Edward Arnold

Fuentes, M. and Smith, R. L. (2001). A new class of non-stationary spatial models. *Tech. Rep.*, North Carolina State University, USA.

Gamerman, D. (1992). A dynamic approach to the statistical analysis of point processes. *Biometrika* **79**, 39–50.

Gamerman, D. and Lopes, H. F. (2006). *Monte Carlo Markov Chain: Stochastic Simulation for Bayesian Inference* (2nd ed.) London: Chapman and Hall.

Gelfand, A. E., Banerjee, S. and Gamerman, D. (2005). Spatial process modelling for univariate and multivariate dynamic spatial data. *Environmetrics* **16**, 465–479.

Gelfand, A. E., Kim, H., Sirmans, C. F. and Banerjee, S. (2003). Spatial modelling with spatially varying coefficient models. *J. Amer. Statist. Assoc.* **98**, 387–396.

Gelfand, A. E., Schmidt, A. M., Banerjee, S. and Sirmans, C. F. (2004). Nonstationary multivariate process modelling through spatially varying coregionalization. *Test* **13**, 263–312, (with discussion).

Geweke, J. F. and Zhou, G. (1996). Measuring the pricing error of the arbitrage pricing theory. *Rev. Financial Studies* **9**, 557–587.

Gneiting, T. (2002). Non-separable, stationary covariance functions for space-time data. *J. Amer. Statist. Assoc.* **97**, 590–600.

Higdon, D., Swall, J. and Kern, J. (1999). Non-stationary spatial modelling *Bayesian Statistics 6* (J. M. Bernardo, J. O. Berger, A. P. Dawid and A. F. M. Smith, eds.) Oxford: University Press, 761–768.

Lopes, H. F. and Carvalho, C. M. (2007). Factor stochastic volatility with time varying loadings and Markov switching regimes. *J. Statist. Planning and Inference* **136** (to appear).

Lopes, H. F. and Migon, H. S. (2002). Comovements and contagion in emergent markets: stock indexes volatilities. *Case Studies in Bayesian Statistics VI* (C. Gatsonis, R. E. Kass, A. Carriquiry, A. Gelman, D. Higdon, D. K. Pauler and I. Verdinelli, eds.) New York: Springer-Verlag, 285–300,

Lopes, H. F., Salazar, E. and Gamerman, D. (2006). Spatial dynamic factor analysis. *Tech. Rep.*, Universidade Federal do Rio de Janeiro.

Lopes, H. F. and West, M. (2004). Bayesian model assessment in factor analysis. *Statistica Sinica* **14**, 41–67.

Møller, J., Syversveen, A. and Waagepetersen, R. (1998). Log-Gaussian Cox processes. *Scandinavian J. Statist.* **25**, 451-482.

Paez, M. S. and Diggle, P. J. (2006). Cox processes in time for point patterns and their aggregations. *Tech. Rep.*, Universidade Federal do Rio de Janeiro, Brazil.

Paez, M. S., Gamerman, D., Landim, F. M. P. F. and Salazar, E. (2006). Spatially-varying dynamic coefficient models. *Tech. Rep.*, Universidade Federal do Rio de Janeiro, Brazil.

Pitt, M. and Shephard, N. (1999). Time varying covariances: a factor stochastic volatility approach. *Bayesian Statistics 6* (J. M. Bernardo, J. O. Berger, A. P. Dawid and A. F. M. Smith, eds.) Oxford: University Press, 547–570 (with discussion).

Reis, E. A. (2006). *Dynamic Bayesian Modelling of the Temporal Variation on the Log-Gaussian Cox Spatial Point Process. PhD Qualifying Exam.* Universidade Federal do Rio de Janeiro, Brazil (in Portuguese).

Revuz, D. and Yor, M. (1999). *Continuous Martingales and Brownian Motion* (3rd edn.) Berlin: Springer-Verlag

Salazar, E. (2006). *Dynamic spatial factor analysis. PhD Qualifying Exam.* Universidade Federal do Rio de Janeiro (in Portuguese).

Salazar, E., Lopes, H. F. and Gamerman, D. (2006). Time-varying factor loadings in spatial dynamic factor analysis. *Tech. Rep.*, Universidade Federal do Rio de Janeiro, Brazil.

Sampson, P. and Guttorp, P. (1992). Nonparametric estimation of nonstationary spatial covariance structure. *J. Amer. Statist. Assoc.* **87**, 108–119.

Sansó, B. and Schmidt, A. M. (2004). Spatio-temporal models based on discrete convolutions. *Tech. Rep.*, Universidade Federal do Rio de Janeiro, Brazil.

Schmidt, A. M. and O'Hagan, A. (2003). Bayesian inference for nonstationary spatial covariance structures via spatial deformations *J. Roy. Statist. Soc. B* **65**, 743–758.

Vivar, J. C. and Ferreira, M. A. R. (2005). Spatio-temporal models for areal data. *Tech. Rep.*, Universidade Federal do Rio de Janeiro, Brazil.

West, M. and Harrison, J. (1997). *Bayesian Forecasting and Dynamic Models* (2nd edn.) Berlin: Springer-Verlag.

Wikle, C. and Cressie, N. (1999). A dimension-reduced approach to space-time Kalman filtering. *Biometrika* **86**, 815–829.

DISCUSSION

MONTSERRAT FUENTES (*North Carolina State University, USA*)

I would like to congratulate the authors for their very important contributions to the field of spatial temporal modeling. This paper and the authors' other publications in dynamic models have helped advance the modeling tools available for space-time data. I fully agree with the authors that dynamic modeling is becoming a central concept in spatial statistics and a natural framework to handle complex spatial temporal structures.

In the paper the authors describe the use of dynamic Gaussian processes (DGPs) in four different contexts. They include regression models, factor analysis, point processes and anisotropic models. In my discussion, I present some of the limitations, alternative approaches, advantages and disadvantages of using DGPs in each one of the four areas of applications presented in the paper. I also suggest some more potential applications of DGPs in the field of spatial statistics.

Regression models. In a space-time regression context the use of DGPs, in principle, could overcome the problem of modeling complex space-time dependency structures, avoiding then, the issue of having to propose general space-time parametric models for the covariance function.

A common modeling framework for a space-time process, y, would be:

$$y(s,t) = \mu(s,t) + \epsilon(s,t)$$

where $y(s,t)$ has a spatial-temporal large scale trend, $\mu(s,t)$, that is a function of some covariates $z_1, \ldots, z_p$, with unknown coefficients β_i for $i = 1, \ldots, p$:

$$\mu(s,t) = \sum_{i=1}^{p} \beta_i(s,t) z_i(s,t).$$

The coefficients β_i for $i = 1, \ldots, p$, vary in space and time and are DGPs. We assume $y(s,t)$ has zero-mean uncorrelated errors $\epsilon(s,t)$. However, this assumption might not be realistic. Each one of coefficients β_i for $i = 1, \ldots, p$, has a dynamic spatiotemporal process prior with a purely temporal component, γ_{it}, and a spatiotemporal component $\gamma_i(s,t)$:

$$\beta_i(s,t) = \gamma_{it} + \gamma_i(s,t)$$

with

$$\gamma_i(s,t) = \gamma_i(s,t-1) + \eta_i(s,t),$$

where η_i is generally assumed to be an uncorrelated error term.

However, simple DGPs with only white noise error components might not capture complex space-time structures (e.g., Fuentes *et al.*, 2005). Therefore, I disagree with the authors' claim that the discussion of separability and stationary in this context becomes irrelevant. In my experience, a hybrid approach using DGPs with space-time (maybe nonstationary and nonseparable) correlated parametric covariance functions may provide an optimal modeling framework. Thus, general parametric covariance models might still be needed and very helpful to model space-time

---

Fuentes is an Associate Professor of Statistics at North Carolina State University, Raleigh, NC, USA.

data even when we are using DGPs. However, the presence of DGPs should simplify the structure of these covariance functions. For instance, in the previous modeling framework, the η_i terms or the ϵ error term (but not both) could be modeled using parametric space-time nonstationary and nonseparable covariance functions (see Fuentes *et al.*, 2005).

Factor analysis. DGPs are a powerful modeling tool for multivariate data. Reich and Fuentes (2006) have used DGPs for a Bayesian latent factor analysis in the context of multi-pollutants. Some of the ideas presented by Reich and Fuentes (RF) (2006) expand what is introduced in the paper that is discussed.

RF analyze daily average ambient concentrations of PM_{10}, $PM_{2.5}$, and 17 sizes of fine particles with diameters less than 0.40 μm measured at a single monitoring station in downtown Fresno, CA, in 2002. Let y_{dt} be the observed average daily concentration for diameter d at day t, $d = 1, ..., D$ and $t = 1, ..., T$. The vectors of observations for each diameter are standardized to have mean zero and unit variance. The dynamic Bayesian factor analysis model assumes the mean of y_{dt} is a linear combination of $J \leq D$ independent latent time series, i.e.,

$$y_{dt} = \theta_{dt} + \epsilon_{dt}, \quad \text{and} \quad \theta_{dt} = \mu_d + \sum_{j=1}^{J} w_{dj} f_{jt},$$

where θ_{dt} is the true concentration for diameter d at time t, μ_d is the intercept for diameter d, w_{dj} is the loading of the jth factor for diameter d, f_{jt} is the value of the jth latent factor at time t, and $\epsilon_{dt} \sim N(0, \sigma_d^2)$, independent across d and t.

RF model the latent factors $\boldsymbol{f}_j = (f_{j1}, ..., f_{jT})'$ as independent, stationary time series with mean zero and lag-h covariance functions $\rho_j(h)$. In dynamic factor analysis, vague priors are typically selected for the loadings. However, in our setting, the model can be improved by exploiting the natural ordering of the diameters. Let $\mathbf{w}_j = (w_{1j}, ..., w_{Dj})$, the vector of loadings for the jth factor, have prior mean zero and $\text{cov}(w_{d_1 j}, w_{d_2 j}) = \gamma_j(|d_1 - d_2|)$. This prior is used to borrow strength across adjacent diameters.

The induced prior covariance of two true concentrations $\theta_{d_1 t}$ and $\theta_{d_2 t+h}$ is

$$\text{Cov}\,(\theta_{d_1 t}, \theta_{d_2 t+h}) = \sum_{j=1}^{J} \gamma_j(|d_1 - d_2|)\rho_j(h).$$

That is, the covariance between a pair of true concentrations is the sum of the products of the autocovariance functions for time and diameter of the J latent time series. At this level of generality, the factor analysis model results in a non-separable (between diameter and time) covariance function. The latent time series and loading vectors are taken to be independent AR(1) processes. However, more general time series models could be used. Also, temperature, humidity, carbon monoxide level, wind speed, wind direction, and an indicator of weekday are included as predictors of the factors. That is,

$$f_{jt} \sim N(\rho_j f_{jt-1} + x_t'\mathbf{b}_j, \tau_j^2) \text{ and } w_{jd} \sim N(\rho_j^* w_{jd-1}, \delta_j^2), \tag{12}$$

where ρ_j, and $\rho_j^* \in (-1, 1)$, x_t is the vector of explanatory variables on day t, and $\mathbf{b}_j$ are the corresponding regression coefficients for factor j.

Restrictions are necessary to ensure that the factors and loadings are identified. To identify the scale, RF fix the conditional variances of the factors to be one, that is $\tau_j^2 \equiv 1$ for all j. Following Aguilar and West (2000), for the first factor, RF constrain the loading for the smallest diameter w_{11} to be one. For the second factor,

RF set the loading for the smallest diameter w_{21} to zero and, to make identification as strong as possible, RF restrict the loading for the largest diameter w_{2P} to be one. The third loading vector has $w_{31} = w_{3P} = 0$ and $w_{32} = 1$, and so on.

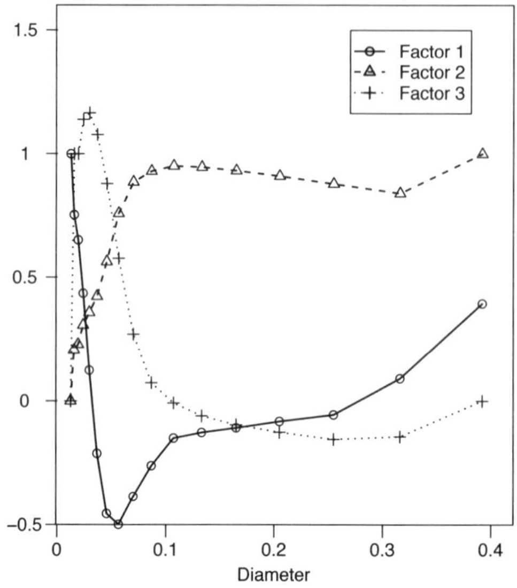

Figure 13: *Posterior medians of the factor loadings of the three-factor model for the ambient ultrafine concentrations (Reich and Fuentes, 2006).*

Figure 13 plots the posterior medians of the loadings. The loadings vary smoothly from one diameter to the next, in part due to the prior for the loadings encourages borrowing strength across nearby diameters. The three factors divide the 17 diameters into three categories: diameters less than 0.02 μm (factor 1), diameters between 0.02 and 0.10 μm (factor 3), and diameters greater than 0.10 μm (factor 2). The breakpoint at 0.1 μm is in agreement with the definition of an ultrafine particle.

The main objective of this research is to explain the association between the multi-pollutants and mortality. Including all $D = 17$ diameters as predictors of mortality leads to substantial multicollinearity and misleading estimates. Clearly, some form of dimension reduction is needed. The factor analysis model represents the ambient concentrations as a linear combinations of the latent time series, $\boldsymbol{f}_1, ..., \boldsymbol{f}_J$. To circumvent multicollinearity, the latent factors are used as predictors of mortality. This results in supervised factor analysis, in that the loadings and latent factors are chosen not only to provide a reasonable fit to the observed ambient concentrations, but also to help explain the health outcome.

However, one of the main challenges of this latent factor analysis approach is how to determine the final number of factors to keep in the regression model. Some of the new ideas presented at the Valencia-8 conference by Vannucci using Dirichlet process mixture models for Bayesian variable selection could be a very good approach to handle this problem. Alternatively, Reversible Jump MCMC (RJMCMC) (Green, 1995) could be used as the authors suggest. However, due to the computational cost of RJMCMC and some of the issues with its convergence, perhaps the stochastic search variable selection (SSVS) (George and McCulloch, 1996) or some of other Bayesian variable selection approaches could be a good solution. In the SSVS the priors for the latent binary variable can be also hard to choose.

Point processes. The intensity function of heterogeneous Poisson processes can be modeled using DGPs as the authors suggest. This is a common approach in environmental epidemiology, in which the mortality counts for some spatial units (e.g., counties or zip codes) due to an environmental stressor of interest are modeled using a heterogeneous Poisson process (e.g., Dominici *et al.*, 2002, Fuentes *et al.*, 2006, Reich and Fuentes, 2006, and many others). DGPs are used to model the long-term trend of the intensity function of the Poisson process, which is explained by spatial covariates. For instance, in the context of environmental epidemiology, these covariates could be weather variables. The covariates (e.g., temperature) are aggregated at the spatial unit of interest and then modeled using DGPs.

However, what is commonly ignored in this context is the link between the original continuous spatial Gaussian processes (temperature) and the obtained discrete spatial processes (e.g., aggregated temperature values for each county or zip code). A more formal study of this association is needed to justify the use of the corresponding DGP model in this context.

Anisotropic models. The issue of anisotropy (directionally) is important in spatial modeling. However, it is more important the problem of lack of stationarity (spatial structure that changes with location), because it is more difficult to model nonstationary processes. The deformation approach is mainly used to model lack of stationarity, not only lack of isotropy. As it was first introduced by Sampson and Guttorp (1992), this approach uses replications over time of a spatial process, to find a transformation of the spatial geographic domain into a deformed space where stationarity is reached. The authors suggest allowing this transformation to change over time. This is a very ambitious proposition, because of the lack of identifiability problem, and also because it is difficult and computationally expensive to find the transformation, even when the observations over time are used only as replications to help to identify a transformation invariant over time. So, it will become much harder when this transformation is allowed to change over time.

It seems that a convolution or mixture approach of local stationary proccess to model nonstationarity, as proposed by Fuentes (2001) and Fuentes and Smith (2001), will provide a more natural framework to use DGPs. More specifically, we could represent the process locally as a stationary isotropic random field with some parameters that describe the local spatial structure. These parameters are DGPs, and are allowed to vary across space to reflect the lack of stationarity of the process. Thus, we write the nonstationary process Z observed on a region D as a convolution of local stationary processes (Fuentes, 2001, 2002, Fuentes and Smith, 2001):

$$Z(\mathbf{x}) = \int_D K(\mathbf{x} - \mathbf{s}) Z_{\boldsymbol{\theta}(\mathbf{s})}(\mathbf{x}) d\mathbf{s}, \tag{13}$$

where $\mathbf{x}$ represents the spatial location (or the spatial-temporal coordinate for a space-time process), K is a kernel function and $Z_{\boldsymbol{\theta}(\mathbf{s})}(\mathbf{x})$, $\mathbf{x} \in D$ is a family of independent stationary processes indexed by $\boldsymbol{\theta}$.

The covariance $C(\mathbf{s_1}, \mathbf{s_2}; \boldsymbol{\theta})$ of Z is a convolution of the local stationary covariances $C_{\boldsymbol{\theta}(\mathbf{s})}(\mathbf{s_1} - \mathbf{s_2})$,

$$C(\mathbf{s_1}, \mathbf{s_2}; \boldsymbol{\theta}) = \int_D K(\mathbf{s_1} - \mathbf{s})K(\mathbf{s_2} - \mathbf{s})C_{\boldsymbol{\theta}(\mathbf{s})}(\mathbf{s_1} - \mathbf{s_2})d\mathbf{s}.$$

The parameter function $\boldsymbol{\theta}(\mathbf{s})$ for $\mathbf{s} \in \mathbf{D}$, measures the lack of stationarity of the process Z. If would be natural to treat $\boldsymbol{\theta}(\mathbf{s})$ as a stochastic Gaussian process (DGP). We consider a hierarchical Bayesian approach to model and take into the account the spatial structure of the parameter $\boldsymbol{\theta}(\mathbf{s})$ in the prediction of Z.

Stage 1:

The process Z is as a convolution of local stationary processes, as in (13). The kernel K has a bandwidth h. Thus, the distribution of Z given $\boldsymbol{\theta}$ and h is Gaussian:

$$[Z|\boldsymbol{\theta}, h] \text{ is Gaussian.}$$

Stage 2:

We model the parameter function $\boldsymbol{\theta}$ as a spatial process,

$$\boldsymbol{\theta}(\mathbf{s}) = \mu_{\boldsymbol{\theta}}(\mathbf{s}) + \epsilon_{\boldsymbol{\theta}}(\mathbf{s}),$$

where $\mu_{\boldsymbol{\theta}}(\mathbf{s})$ represents the spatial trend (large scale structure), and the process $\epsilon_{\boldsymbol{\theta}}(\mathbf{s})$ represents some zero-mean noise. We use a DGP prior (as in Section 1).

A DGP prior is used to explain the space-time structure of the covariance parameters. DGP priors are the natural choice in parametric covariance modeling using convolution approaches. Fig. 14 presents the posterior of the sill parameter (variance of the process) of a Matérn covariance function that is modeled as a DGP. By allowing the covariance parameters to change with location we capture the lack of stationary of the underlying spatial process of interest, in this case the sulfur dioxide concentrations.

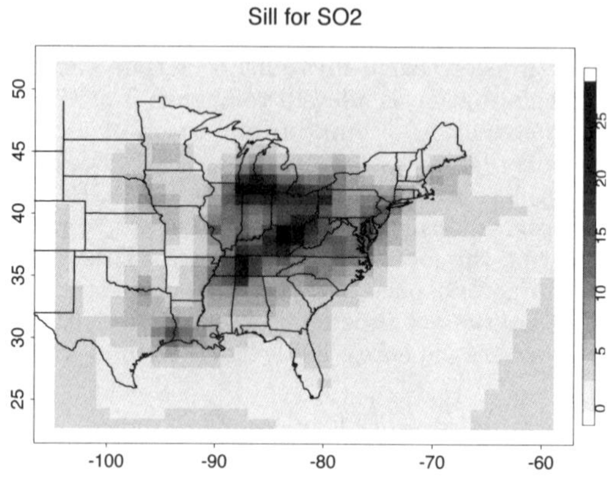

Figure 14: *Map of the modes of the posterior distributions for the sill parameter (that has a DGP prior) of the Matérn covariance for* SO_2 *concentrations* (mg/m^3)*, for the week starting July 11, 1995 (from Fuentes and Smith, 2001).*

Other areas of applications. DGPs can be used to model the covariance parameters of nonstationary processes (as in Section 4 of this Discussion). Similarly, DGPs could be used as process priors for the covariance parameters of nonseparable models.

In terms of modeling multivariate space-time processes where space is discrete, DGP priors can be used to model the conditional covariance and cross-covariance function and/or the weight matrix in multivariate conditional autoregressive models (MCAR).

In the context of multivariate space-time modeling with a spatial continuous structure, DGPs can be used in the linear models of corregionalization (LMC), in which we write the multivariate process Z as a linear combination of underlying space-time processes (W), i.e. $Z = AW$. The authors suggest DGPs to model the underlying space-time processes W, but DGPs with some constraints could be also used to model the weight function A.

REPLY TO THE DISCUSSION

We would like to thank Montse Fuentes for a lively discussion, which reflects her knowledge and expertise in spatial and spatio-temporal modeling. We will use this space merely to clarify a few outstanding issues. We also follow the paper and discussion in dividing the text according to areas of application: factor analysis, point processes and deformations and convolutions.

Factor analysis. Factor analysis is a technique designed to extract latent features and can be used in a number of areas of statistics. It may even be used in the context of spatial applications as mentioned in the discussion. Our approach has a distinctive and apparently novel feature of using factor analysis to specifically model the spatial effect (of georeferenced data locations). Identifiability is ensured via the prior imposition of Gaussian processes and no restrictions were required, unlike standard factor analysis.

The discussion mentions the inclusion of regression effects in the evolution of the factors. This is one of the many possibilities for factor evolution. Lopes, Salazar and Gamerman (2006) discuss some other time series possibilities to account for trend and seasonality in addition to the independent AR(1) factor evolutions.

The important issue of selection of the number m of factors is addresses in Lopes, Salazar and Gamerman (2006). They performed it through formal incorporation of uncertainty about m into the model, as we believe any Bayesian analysis should. They implemented their approach through RJMCMC and did not report any significant computational difficulty.

Note that none of the factor models mentioned in the discussion or outlined in this rejoinder include dynamic Gaussian processes (DGP) in their formulation as our paper did (for the factor loadings). We expect that DGP can help identify time-varying patterns in the spatial influence the latent factors have.

Point processes. We would like to retain the discussion of point process in its original context of continuous space. Discretization is only performed as a last resort and at the final stage of the inference process. Even then, the analysis should always be taking the underlying continuity into consideration. This means that the presence of continuous covariates should imply model aggregation only when using them in a discretized context.

Spatial deformations and convolutions. The scepticism of the discussant about the use of time-varying deformations is justified given the well documented difficulty in the estimation of deformations even in the simpler, static setting. However, this may provide useful information about shifting spatial patterns. For that reason, it is a project worth pursuing. Hopefully, results on this direction will be reported in the near future.

Nevertheless, the use of convolutions of Gaussian processs is very promising and can be applied in many different forms. An important practical aspect is the determination of which convolutions are more appropriate to given spatial settings. For more flexible treatment of spatio-temporal settings, DGP seem to provide a natural starting point for extensions of the lowest stage of the hierarchical models presented in the discussion.

Finally, the other areas of application listed in the discussion are only reinforcing the main point of the paper. They show the relevance of DGP as a powerful tool for handling model building for spatial-temporal data.

ADDITIONAL REFERENCES IN THE DISCUSSION

Aguilar O, West M (2000). Bayesian dynamic factor models and portfolio allocation. *J. Business and Economic Statistics*, **18**, 338–357.

Dominici, F., Daniels, M., Zeger, S.L., and Samet, J. M. (2002). Air pollution and mortality: estimating regional and national dose-response relationships. *J. Amer. Statist. Assoc.* **97**, 100–111.

Fuentes, M. (2001). A High frequency kriging for nonstationary environmental processes. *Environmetrics,* **12**, 1–15.

Fuentes, M. (2002). Periodogram and other spectral methods for nonstationary spatial processes. *Biometrika* **89**, 197–210.

Fuentes, M., Song, H. R, Ghosh, S. and Holland, D. and Davis, J. (2006). Spatial association between speciated fine particles and mortality. *Biometrics* **62**, 855–863.

Fuentes, M.,Chen, L., Davis, J. and Lackmann, G. (2005). A new class of nonseparable and nonstationary covariance models for wind fields. *Environmetrics,* **16**, 449–464.

Fuentes, M. and Smith, R. L. (2001). A new class of nonstationary spatial models. *Tech. Rep.*, North Carolina State University, USA.

George, E .I., McCulloch, R. E. (1996). Stochastic search variable selection. *Markov Chain Monte Carlo in Practice* (W. R. Gilks, S. Richardson, and D. J. Spiegelhalter, eds.) London: Chapman and Hall, 203–214.

Green, P. J. (1995). Reversible jump Markov Chain Monte Carlo computation and Bayesian model determination. *Biometrika* **82**, 711–732.

Lopes, H. F., Salazar, E. and Gamerman, D. (2006). Spatial dynamic factor analysis. *Tech. Rep.*, Universidade Federal do Rio de Janeiro, Brazil.

Reich, B. and Fuentes, M. (2006). A Bayesian analysis of the effects of particualte matter using a human exposure simulator. *Tech. Rep.*, Institute of Statistics, North Carolina State University, USA.

Sampson, P. D. and Guttorp, P. (1992). Nonparametric estimation of nonstationary spatial covariance structure. *J. Amer. Statist. Assoc.* **87**, 108–119.

BAYESIAN STATISTICS 8, pp. 175–200.
J. M. Bernardo, M. J. Bayarri, J. O. Berger, A. P. Dawid,
D. Heckerman, A. F. M. Smith and M. West (Eds.)

Bayesian Nonparametric Modelling for Spatial Data Using Dirichlet Processes

ALAN E. GELFAND
Duke University, USA
alan@stat.duke.edu, USA

MICHELE GUINDANI
MD Anderson Cancer Center, USA
mguindani@mdanderson.org

SONIA PETRONE
Università Bocconi, Milano, Italy
sonia.petrone@unibocconi.it

SUMMARY

Modelling for spatially referenced data is receiving increased attention in the statistics and the more general scientific literature with applications in, e.g., environmental, ecological and health sciences. Bayesian nonparametric modelling for unknown population distributions, i.e., placing distributions on a space of distributions is also enjoying a resurgence of interest thanks to their amenability to MCMC model fitting. Indeed, both areas benefit from the wide availability of high speed computation. Until very recently, there was no literature attempting to merge them. The contribution of this paper is to provide an overview of this recent effort including some new advances. The nonparametric specifications that underlie this work are generalizations of Dirichlet process mixture models. We attempt to interrelate these various choices either as generalizations or suitable limits. We also offer data analytic comparison among these specifications as well as with customary Gaussian process alternatives.

Keywords and Phrases: GAUSSIAN PROCESS, GENERALIZED STICK-BREAKING PROCESS, LOCAL SURFACE SELECTION, NONSTATIONARY PROCESS, SPATIAL RANDOM EFFECTS

1. INTRODUCTION

Point-referenced spatial data is collected in a wide range of contexts, with applications, among others, in environmental, ecological, and health sciences. Modelling for such data introduces a spatial process specification either for the data directly or

---

The authors thank Jason Duan and Athanasios Kottas for valuable contributions in the development of this manuscript.

for a set of spatial random effects associated with the mean structure for the data, perhaps on a transformed scale. In virtually all of this work the spatial process specification is parametric. In fact, it is almost always a Gaussian process (GP), which is often assumed to be stationary.

Within a Bayesian framework the resulting model specification can be viewed as hierarchical (see, e.g., Banerjee, Carlin and Gelfand, 2004). The parameters of the spatial process are unknown, and so they are assigned prior distributions resulting in a random GP. Fitting, using Markov chain Monte Carlo (MCMC) methods, is by now fairly straightforward. See, e.g., Agarwal and Gelfand (2005) and further references therein. Bayesian inference is attractive in analyzing point-referenced spatial data particularly with regard to assessing uncertainty. The alternative likelihood analysis will impose arguably inappropriate assumptions to achieve asymptotic inference, and hence to obtain asymptotic variability (Stein, 1999).

Spatially varying kernel convolution ideas as in Higdon, Swall and Kern (1999) as well as a local stationarity approach as in Fuentes and Smith (2001) remove the stationarity assumption but are still within the setting of GPs. Zidek and colleagues (see, *e.g.*, Le and Zidek, 1992) introduce nonstationarity through a Wishart model for a random covariance matrix with mean based upon a stationary covariance function. However, this construction sacrifices the notion of a spatial process and, given the covariance matrix, the spatial effects are still Gaussian.

There is a rich literature on nonparametric modeling for an expected spatial surface, much of it drawing from the nonparametric regression literature. See, e.g., Stein (1999) and further references therein. Our interest is in nonparametric modeling for the stochastic mechanism producing the spatial dependence structure. In this regard, the literature is very limited. The *nonparametric* variogram fitting approaches, e.g., Shapiro and Botha (1991) and Barry and Ver Hoef (1996) do not fully specify the process; they are nonparametric only in the second moment structure. Arguably, the most significant nonparametric specification of the covariance function is the "deformation" approach of Sampson and Guttorp (1992). The observed locations in the actual (geographic) space are viewed as a nonlinear transformation of locations in a conceptual (deformed) space where the process is assumed stationary and, in fact, isotropic. This approach has been pursued in a Bayesian context by both Damian et al. (2001) and Schmidt and O'Hagan (2003) but again confined to a GP for the likelihood.

Regarding spatiotemporal data, while space is usually taken to be continuous, a primary distinction is whether time is taken to be continuous or discrete. In particular, with continuous time, the space-time dependence structure is provided by a covariance function which may be separable (Mardia and Goodall, 1993) or nonseparable (Stein, 2005 and references therein). With discrete time, the dependence structure is usually conceived as temporal evolution of a spatial process described in the form of a dynamic model. (See, e.g., Gelfand, Banerjee and Gamerman, 2005a.) Such evolution can yield a time series at each spatial location, as with weather or pollution measurements at monitoring stations. It can also occur with cross-sectional data, as with real estate transactions over a region across time. All of this literature is fully parametric.

The goal of this paper is to review and extend recent nonparametric modeling approaches for spatial and spatiotemporal data through the use of the Dirichlet process (DP) (Ferguson, 1973, 1974). DPs have been used to provide random univariate and multivariate distributions, requiring only a baseline or centering distribution and a precision parameter. Using the DP, we describe a probability law for the stochastic process, $\{Y(s) : s \in D\}$. We refer to such a model as a spatial Dirichlet process (SDP) (Gelfand et al., 2005b). As with GPs, we provide this specification through finite dimensional distributions, that is, a random distribution for $(Y(s_1), ..., Y(s_n))$ where n and the set of s_i are arbitrary. The resulting process is nonstationary and the resulting joint distributions are not normal. With regard to the set of random univariate distributions, $\{F(Y(s)) : s \in D\}$ we can achieve a dependent Dirichlet process (DDP) (MacEachern, 2000). More precisely, the $F(Y(s))$ are dependent and, as $s \to s_0$, the realized $F(Y(s))$ converge to the realized $F(Y(s_0))$. This enables us to pool information from nearby spatial locations to better estimate say $F(Y(s_0))$ yielding fully model-based nonparametric spatial prediction analogous to parametric kriging using predictive distributions.

For a fully nonparametric modeling approach, replication of some form will be required; with only a single observation from a multivariate distribution, a nonparametric model is not viable.[1] Replicates from a spatial process typically arise over time. Depending upon the data collection, an assumption of independence for temporal replicates may be inappropriate; fortunately, we can directly embed our methodology within a dynamic model, retaining temporal dependence.

Our contribution is best described as random effects modelling. Just as DP priors are often used to extend random effects specifications beyond usual Gaussian random effects, here we broaden Gaussian spatial process specifications. In this regard, we recall the finite-dimensional Dirichlet process priors (Ishwaran and Zarepour, 2002b references therein), which we denote as DP_K priors, and which converge, under suitable conditions, to Dirichlet process priors. Spatial analogues are easily formulated (we call them SDP_K priors) with similar limiting behavior and thus provide another extension of GP's. Recent work by Duan, Guindani and Gelfand (2005) extends the SDP to a generalized SDP (GSDP) by introducing 'local surface selection'. In essence, for a usual multivariate DP, vectors are drawn from a probability weighted countable collection of vectors. A generalization would allow for individual components of the drawn vector to come from different choices of the countable collection of vectors. Here, we can also formulate a finite-dimensional version which we refer to as a $GSDP_K$. The cartoon presented as Fig. 1 attempts to capture the modelling world that we are focusing on. G denotes "Gaussian" and GP a Gaussian process. The arrows suggest either that the model being pointed to is an extension or generalization of the model it emanates from.

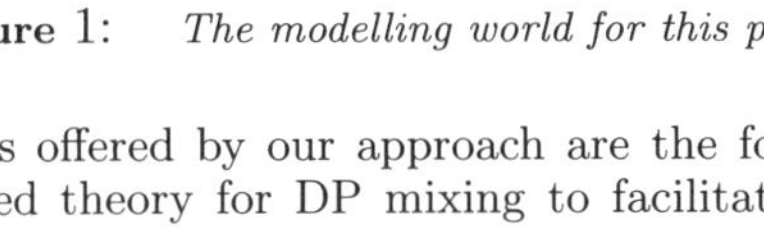

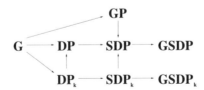

Figure 1: *The modelling world for this paper.*

Possible advantages offered by our approach are the following. We can draw upon the well-developed theory for DP mixing to facilitate interpretation of our

[1]The deformation setting (Sampson and Guttorp, 1992) requires replications to obtain the sample covariance estimate, a *nonparametric* estimate of the process covariance matrix at the observed sites.

analysis. We can implement the required simulation-based model fitting for posterior inference and spatial prediction easily, availing ourselves of established strategies for DP mixture models. See, *e.g.*, Neal (2000) and, more recently, Jain and Neal (2004). We can infer about the random distribution that is operating at any given location or set of locations in the region. An alternative line of nonparametric development using Lévy processes is summarized in Clyde and Wolpert (2006).

The models presented here are in terms of random process realizations over a geographic space. However, we can also envisage random curves or surfaces over a covariate space, i.e., a space of covariate levels. Covariate space might be dose levels with interest in response to treatment, ocean depths with interest in temperature or salinity, or time yielding, say, hormone levels across menstrual cycles. Hence, our models are directly applicable to so-called functional data analysis where we consider replicates of random curves or surfaces as process realizations. Lastly, we can imagine spatial functional data analysis where there is a conceptual random curve at each spatial location. The curves are only observed (with error) at a finite set of locations and a finite number of points at each location. For instance there is an unknown temperature vs depth curve at every location in an ocean and we may gather data at a set of locations and a set of depths at these locations. We would expect that curves would be more similar for locations close to each other, less so when the locations are distant from each other.

The format for the paper is as follows. In Section 2 we review the DP and DP_K specifications. In Section 3 we bring in the spatial aspect with Section 4 providing two examples to compare the GP, SDP, and SDP_K. Section 5 develops the generalized local surface selection versions, the GSDP and $GSDP_K$ leaving Section 6 to provide a comparison among them and the SDP. Section 7 concludes with a summary and some issues that require further examination.

2. A REVIEW

Hierarchical models are frequently built through random effects which typically are specified to be i.i.d. from a mean 0 Gaussian 'population' model. The random effects need not be univariate, e.g., random slopes and intercepts led to random lines and more generally to random curves for instance in growth curve modelling. See, e.g., Diggle, Liang and Zeger (1994) for a parametric treatment with the recent work of Scaccia and Green (2003) offering a nonparametric view. With the computational revolution of the 1990s (Gibbs sampling and MCMC), handling hierarchical models with Gaussian random effects became routine (arguably the most widely used specification in the BUGS software. Of course, concern regarding the Gaussian assumption led to more flexible specifications through mixture modelling including t distributions, Dirichlet processes, and finite mixtures within which we distinguish the finite dimensional Dirichlet process models (what we refer to as DP_K models). We now review this modelling. In the process, we clarify the arrows in Fig. 1 involving G, DP and DP_K.

There has been a growing literature on the use of DP priors primarily because they are easy to specify, attractive to interpret, and ideally suited for model fitting within an MCMC framework. In particular, we recall the stickbreaking representation of the DP (Sethuraman, 1994). Let $\theta_1^*, \theta_2^*, \ldots$ be i.i.d. random elements independently and identically distributed according to the law G_0. G_0 can be a distribution over a general probability space, allowing the θ^*'s to be random objects such as vectors, a stochastic process of random variables (Section 3), or a distribution (Rodriguez *et al.*, 2006). Let $q_1, q_2, \ldots$ be random variables independent of the θ^*'s and i.i.d. among themselves with common distribution Beta$(1, \alpha)$. If we set $p_1 = q_1$, $p_2 = q_2\,(1 - q_1), \ldots, p_k = q_k \prod_{j=1}^{k-1}(1 - q_j), \ldots$ the random probability measure defined by

$$G(\cdot) = \sum_{k=1}^{\infty} p_k\, \delta_{\theta_k^*}(\cdot)\,, \tag{1}$$

is distributed according to a DP.

A wide range of random probability measures can be defined by means of a stick-breaking construction, where the q's are drawn independently from an arbitrary distribution Q on $[0, 1]$ (see Hjort , 2000). For example, the beta-two parameter process (Ishwaran and Zarepour, 2000) is defined by choosing $q_k \sim Beta(a, b)$. If $q_k \sim Beta(1 - a, b + ka)$, $k = 1, 2, \ldots$, for some $a \in [0, 1)$ and $b \in (-a, \infty)$ we obtain the two-parameter Poisson–Dirichlet process (Pitman and Yor, 1997). Ishwaran and James (2001) discuss the general case, $q_k \sim Beta(a_k, b_k)$. More generally, we can consider the discrete random probability measure

$$G_K(\cdot) = \sum_{k=1}^{K} p_k\, \delta_{\theta_k^*}(\cdot), \tag{2}$$

where K is an integer random variable (allowed to be infinite), θ_k^* are i.i.d. from some base distribution G_0 (not necessarily nonatomic) and, independently, the weights p_k can have any distribution on the simplex $\{\mathbf{p} : \sum_{i=1}^{K} p_k = 1, p_k \geq 0\}$ (see Ongaro and Cattaneo, 2004). In these constructions the stick-breaking is *one-dimensional*; the probability p_k is for the selection of the entire random quantity, θ_k^*.

If K is finite, and $(p_1, \ldots, p_K) \sim Dir(\alpha_{1,K}, \ldots, \alpha_{K,K})$, we obtain the class of *finite dimensional Dirichlet priors*, discussed by Ishwaran and Zarepour (2002a). We refer to this class of priors as DP_K priors and denote them by $G_K \sim DP_K(\alpha, G_0)$. In fact, such priors have often been considered as a general approximation of models based on 1, e.g., it has been a common choice to set $\alpha_{i,K} = \lambda_K$ (see Richardson and Green, 1997). However, an important result from Ishwaran and Zarepour (2002a) clarifies when finite sums such as 2 converge to a DP. In particular, let $G_K \sim DP_K(\alpha, G_0)$ and $E_{G_k}(h(x)) = \int h(x) G_K(dx)$ denote a random functional of G_K, where h is a non-negative continuous function with compact support. Then, if $\alpha_{k,K} = \lambda_K$, where $K\,\lambda_K \to \infty$, then $E_{G_k}(h(x)) \xrightarrow{p} E_{G_0}(h(x))$, that is the choice of a uniform Dirichlet prior leads to a limiting parametric model. Instead, if $\sum_{k=1}^{K} \alpha_{k,K} \to \alpha > 0$ and $\max \alpha_{1,K}, \ldots, \alpha_{K,K} \to 0$ as $K \to \infty$, the limit distribution of G_K is really nonparametric, since $E_{G_k}(h(x)) \xrightarrow{\mathcal{D}} (h(x))$, where G is the usual Dirichlet process with finite measure αG_0. The result follows directly from 0 Kingman (1975) and the properties of the Poisson–Dirichlet distribution. In particular, the former includes the common case of $\alpha_{k,K} = \alpha/K$ for some $\alpha > 0$, whose weak convergence to the DP had already been proved by Muliere and Secchi (1995).

In some situations, there could be an interest in indexing the random probability distribution of the observables according to the values of underlying covariates. Following MacEachern (see MacEachern,2000), we refer to these models as dependent random probability measures (see also De Lorio *et al.*, 2004; Griffin and Steel, 2004) noting that there is a previous literature in this vein summarized in Petrone and Raftery (1997). In this setting, the weights and point masses of the random probability measure 2 are indexed by a vector of covariates $\mathbf{z}$, that is $G_K(\mathbf{z}, \cdot) = \sum_{k=1}^{K} p_k(\mathbf{z})\, \delta_{\theta_k^*(\mathbf{z})}(\cdot)$.

3. RANDOM DISTRIBUTIONS FOR SPATIAL LOCATIONS: THE *SDP* AND SDP_K MODELS

We assume data from a random field $\{Y(\mathbf{s}), \mathbf{s} \in D\}$, $D \in \mathbb{R}^d$, such that

$$Y(\mathbf{s}) = \mu(\mathbf{s}) + \theta(\mathbf{s}) + \varepsilon(\mathbf{s}), \tag{3}$$

where the mean structure $\mu(\mathbf{s})$ is, say $\mathbf{x}(\mathbf{s})^T\boldsymbol{\beta}$ with $\mathbf{x}(\mathbf{s})$ a p-dimensional vector comprised of covariates at location $\mathbf{s}$. $\boldsymbol{\beta}$ is a $p \times 1$ vector of regression coefficients. The residual term is partitioned into two pieces. The first, $\theta(\mathbf{s})$, accounts for spatial variability that is not captured by the mean. The second term, $\varepsilon(\mathbf{s})$, is intended to capture variability of a non spatial nature; it can be interpreted either as a pure error term or as a term incorporating measurement errors or microscale variability. When $\theta(\mathbf{s})$ is a mean 0 Gaussian process and $\varepsilon(\mathbf{s})$ is Gaussian white noise, the process $Y(\mathbf{s})$ is Gaussian and stationary. A mixture of Gaussian processes would permit greater flexibility (see, e.g., Brown *et al.*, 2003) as would the Gaussian/logGaussian process described in Palacios and Steel (2007).

Gelfand, Kottas and MacEachern (2005b) generalize in a different way, replacing the Gaussian specification with a DP. They define a spatial Dirichlet process (SDP) by considering the base measure G_0 itself to be a mean zero stationary Gaussian process defined over D. Hence, recalling 1, the θ_k^*'s are realizations of a random field, i.e. surfaces over D, $\theta_k^* = \{\theta_k^*(\mathbf{s}), \mathbf{s} \in D\}$, $k = 1, \ldots$. However, the random weights do not depend on the locations and the spatial dependence is introduced only through the underlying base measure.

More generally, we could extend the random probability measures in 2 so that K is finite, $(p_1, \ldots, p_K) \sim Dir(\alpha_1, \ldots, \alpha_K)$ and θ_k^* are surfaces, which are realizations of a specified random field G_0, thus defining a finite dimensional spatial Dirichlet prior (SDP_K). The above results from Ishwaran and Zarepour (2002a) show that, for certain choices of the α parameters, the SDP_K process provides a finite-approximation to the SDP process. Let G denote a random distribution obtained according to either 1 or 2. In both cases, we need only consider finite dimensional distributions of G at locations, say $(\mathbf{s}_1, \ldots, \mathbf{s}_n)$, denoted by $G^{(n)}$. Then, $G^{(n)} \sim DP(\alpha G_0^{(n)})$ or $G^{(n)} \sim DP_K(\alpha G_0^{(n)})$ where $G_0^{(n)}$ is the associated n-variate finite dimensional (e.g., multivariate normal) distribution of the process G_0.

Then, if $\{\theta(\mathbf{s}), \mathbf{s} \in D\}$ is a random field, such that $\theta(\cdot)|G \sim G$, where $G \sim SDP(\alpha, G_0)$ or $G \sim SDP_K(\alpha, G_0)$, we obtain

$$\begin{aligned} E(\theta(\mathbf{s})|G) &= \textstyle\sum_{k=1}^K p_k\, \theta_k^*(\mathbf{s}), \\ Var(\theta(\mathbf{s})|G) &= \textstyle\sum_{k=1}^K p_k\, (\theta_k^*(\mathbf{s}))^2 - \left\{\sum_{k=1}^K p_k\, \theta_k^*(\mathbf{s})\right\}^2, \end{aligned}$$

and for any two locations $\mathbf{s}_i, \mathbf{s}_j \in D$,

$$Cov(\theta(\mathbf{s}_i), \theta(\mathbf{s}_j)|G) = \sum_{k=1}^K p_k\, \theta_k^*(\mathbf{s}_i)\, \theta_k^*(\mathbf{s}_j) - \left\{\sum_{k=1}^K p_k\, \theta_k^*(\mathbf{s}_i)\right\} \left\{\sum_{k=1}^K p_k \theta_k^*(\mathbf{s}_j)\right\}, \tag{4}$$

where $K = \infty$ if G is a SDP. Hence, for any given G, the process $\theta(\cdot)$ has heterogenous variance and is nonstationary. However, G is "centered" around G_0 which is typically stationary.

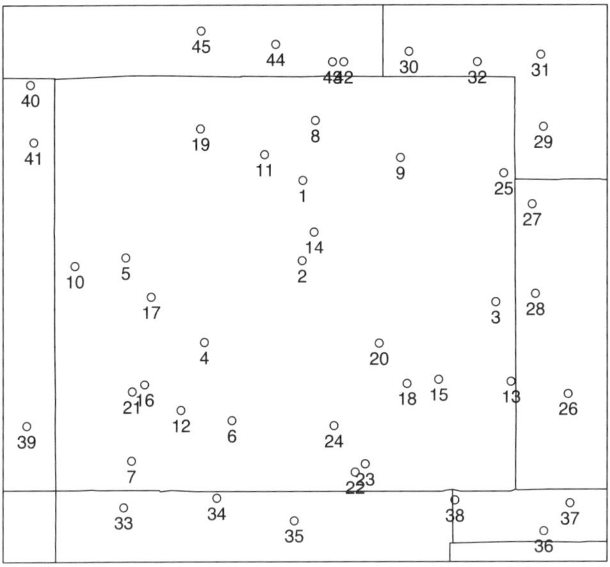

Figure 2: *A region comprising the state of Colorado where we have collected data used in Section 4.*

Finally, consider data $\mathbf{Y}_t = (Y_t(\mathbf{s}_1), \ldots, Y_t(\mathbf{s}_n))^T$, for replications $t = 1, \ldots, T$, and associate with $\mathbf{Y}_t$ a vector $\boldsymbol{\theta}_t = (\theta_t(\mathbf{s}_1), \ldots, \theta_t(\mathbf{s}_n))^T$, such that $\boldsymbol{\theta}_t | G^{(n)} \overset{i.i.d.}{\sim} G^{(n)}$, $t = 1, \ldots, T$. Then, the following semiparametric hierarchical model arises:

$$\begin{array}{rcl} \mathbf{Y}_1, \mathbf{Y}_2, \ldots, \mathbf{Y}_T | \, \boldsymbol{\theta}_1, \boldsymbol{\theta}_2, \ldots, \boldsymbol{\theta}_T, \boldsymbol{\beta}, \tau^2 & \sim & \prod_{t=1}^{T} N(\mathbf{X}_t^T \boldsymbol{\beta} + \boldsymbol{\theta}_t, \, \tau^2 I_n) \\ \boldsymbol{\theta}_1, \boldsymbol{\theta}_2, \ldots, \boldsymbol{\theta}_T | \, G^{(n)} & \overset{i.i.d.}{\sim} & G^{(n)} \end{array} \qquad (5)$$

where X_t is a suitable design matrix for replicate t. The model is completed by choosing an appropriate specification for the unknown G as above together with a suitable choice of prior for the other parameters of the model.

4. A DATA EXAMPLE

Here, we compare the behavior of the GP, the SDP and the SDP_K using temperature and precipitation data collected at 45 weather stations monthly over 40 years (1958-1997) in a region encompassing the state of Colorado (in the US) (see Fig. 2). We assume independence across years by restricting ourselves to average monthly temperature for the single month of July (though embedding (5) within a dynamic model could also be straightforwardly done)

We assume a spatial random effect model as in 3, where $Y_t(s)$ is the average July temperature in year t and $\mu_t(\mathbf{s}) = \beta_0 + \beta_1^T X_t(s)$, with $X_t(s)$ indicating associated

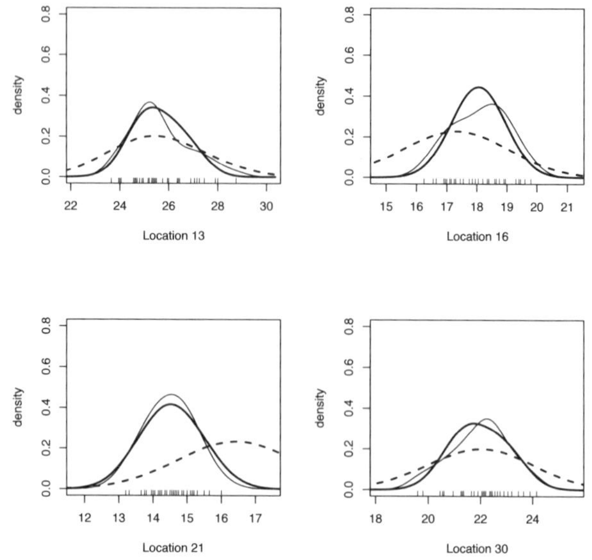

Figure 3: *Posterior predictive densities $Y_{new}(s)|data$ for the SDP(thick line $-$) and GP (thick dotted line $= \; -$). The lighter dotted line $(-)$ is the estimated density from the* 40 *replicates in the Colorado dataset (real data).*

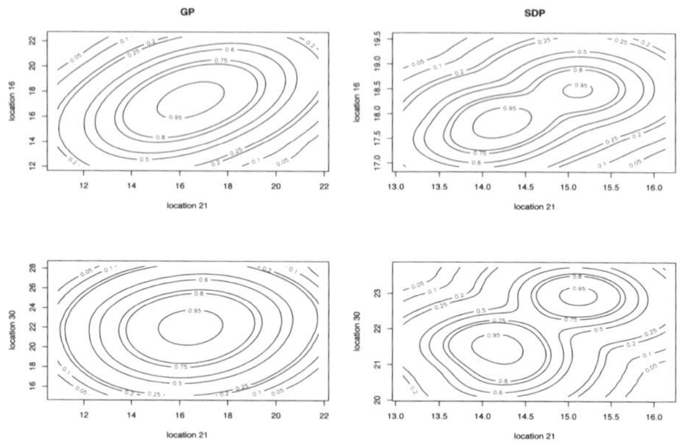

Figure 4: *Bivariate contour plots of the posterior distributions of the mean in the SDP and GP models for locations* 16 *and* 21 *as well as* 21 *and* 30 *and the* 40 *replicates in the Colorado dataset (real data). See Section 4.*

precipitation. The correlation function for the spatial effect is assumed exponential, $\rho(\mathbf{s}-\mathbf{s}') = \sigma^2 \exp\{-\phi\,||\mathbf{s}-\mathbf{s}'||\}$. In all other respects the prior distributions and the hyperparameters in 5 are chosen as in Gelfand *et al.* (2005b), except that here ϕ is given a Gamma prior and is sampled with a Metropolis–Hastings step. Also, in order to facilitate comparison with the SDP_K we fix $\alpha = 10$ (trials with random α in the SDP yielded similar results. For the SDP, this choice corresponds to a number of clusters *a priori* equal to 16 (see Antoniak, 1974), a weak restriction on the number of clusters. Also, for $K = 10$ and $\alpha_{i,K} = \alpha/K$, in the SDP_K the distribution of the p_k is uniform Dirichlet.

With regard to the choice of the GP model for comparison, following (3), we are concerned with the residual $\theta_t(s) + \epsilon_t(s), t = 1, 2, ..., T$. The two extreme cases correspond to $\alpha \to \infty$ (where the $\theta_t(s)$ are all distinct) and $\alpha \to 0$ (where $\theta_t(s) = \theta(s)$). From this perspective, the SDP (and the SDP_K) falls in between. Structurally, the differences are clear. $\theta_t(s) + \epsilon_t(s)$ has dependence within a replication but independence across replications while $\theta(s) + \epsilon_t(s)$ has dependence both within and across replications (see Sethuraman and Tiwari, 1982). With regard to comparison, according to context, a case can be made for each of the extremes. For instance, suppose the spatial locations are associated with apartment buildings, and the replications are the selling prices of condominium units within the building. Each unit should receive a common building level spatial effect (see Gelfand *et al.*, 2006). However, in, *e.g.*, Duan and Gelfand (2006) the $\alpha = \infty$ case was chosen as the 'fair' comparison with the SDP, arguing that, in both cases, 'dependence within a replication with independence across replications' is retained. (Of course, for the former, the mixing distribution is 'known' while for the latter it is not.)

The simple GP ($\alpha \to 0$) is unable to capture the variability of multimodal data. In particular, using $d^2 = \sum_{s,t} [Y_t(s) - E(Y_{new}(s)|\text{data})]^2$, a predictive sum of squared deviations, we find d^2 is roughly 950 for the SDP, SDP_K and 860 for GP ($\alpha \to \infty$), while it is 1600 for the GP ($\alpha \to 0$). Also, as long as the number of mixing components is small relative to K, little difference is seen between the SDP_K and SDP models, in accordance with the theoretical results in Ishwaran and Zarepour (2002a). so, we restrict comparison to the SDP and the GP corresponding to $\alpha \to \infty$.

In Fig. 3, we show the marginal predictive distributions at sites 13, 16, 21, and 30, noting that the SDP captures the estimated density better than the GP at these locations. Next, sites 16 and 21 are close to each other while sites 21 and 30 are far apart. In both cases, the bivariate predictive distribution under the SDP suggests the presence of two clusters, while the GP suggests unimodality (see Fig. 4).

5. GENERALIZING THE SDP AND SDP_K

Here we motivate and formulate the generalized spatial Dirichlet process models.

5.1. *Motivating the Generalized DP*

The DP in (2) selects a θ_k^* with probability p_k. Though clustering can be encouraged say with small α, there is always the chance to draw a new, distinct θ^*. For example, in the problem of species sampling it is attractive to have a mechanism that enables new species types (Pitman, 1996) and a similar perspective would apply in developing say, image classifications.

Though this is possible with usual finite mixture models by allowing the number of mixture components to be random, it is more elegantly handled through the DP.

Still, the DP can be viewed as inefficient in this regard. For instance, assume a species is *defined* through a vector of morphological and genetic traits. In the DP setting, a new species would be characterized by a clearly distinct vector. However, suppose a new species is a hybrid of two existing species. The present DP can not distinguish this case from any other distinct vector. However, a DP that allows for different components of the θ_t's to be drawn from different θ_k^*'s would. Evidently, if some new species arise as such hybrids, the latter process could introduce far fewer clusters yielding a much simpler story for speciation.

If the θ_t's are say $r \times 1$, i.e., $\theta_t^T = (\theta_{t1}, \theta_{t2}, ..., \theta_{tr})$, process specification requires joint probabilities of the form $p_{i_1,i_2,...i_r} = Pr(\theta_{t1} = \theta^*_{i_1,1}, \theta_{t2} = \theta^*_{i_2,2}, ..., \theta_{tr} = \theta^*_{i_r,r})$ where the $i_l \in \{1, 2, ..., K\}$. Such specification will be referred to as r-dimensional stick-breaking and is discussed in some detail in Section 5.3.

Our goal is a bit more ambitious. In the spatial setting, we want local surface selection among the process realizations that define the SDP or SDP_k. Thus, we need to provide such selection for any number of and choice of locations. Moreover, we seek to do this in a spatially structured way. The closer two locations are the more likely they are to select the same surface; when sufficiently far apart, surface selection will be essentially independent. Rather than attempting to sample the p's above, it will prove easier to simulate indicator variables that produce these p's.

5.2. *The Basics of the GSDP*

Though the SDP realizes a countable collection of surfaces it can not capture the situation in which different surfaces can be selected at different locations. For example, in brain imaging, with regard to neurological activity level, researchers imagine healthy brain images (surfaces) as well as diseased or impaired brain images; however, for an actual image, only a portion of the brain is diseased and it is appropriate to envision surface selection according to where the brain is diseased.

To formalize the GSDP, we start by considering a base random field G_0, which, for convenience, we take to be stationary and Gaussian, and indicate with $\theta_l^* = \{\theta_l^*(s), s \in D\}$ a realization from G_0, i.e., a surface over D. Then we define a random probability measure G on the space of surfaces over D as that measure whose finite dimensional distributions almost surely have the following representation: for any set of locations $(s_1, \ldots, s_n) \in D$, and any collection of sets $\{A_1, \ldots, A_n\}$ in $\mathcal{B}(\mathbb{R})$,

$$pr\{\theta(s_1) \in A_1, \ldots, \theta(s_n) \in A_n\} = \sum_{i_1=1}^{K} \cdots \sum_{i_n=1}^{K} p_{i_1,\ldots,i_n} \, \delta_{\theta^*_{i_1}(s_1)}(A_1) \ldots \delta_{\theta^*_{i_n}(s_n)}(A_n), \tag{6}$$

where the θ_j^*'s are independent and identically distributed as $G_0^{(n)}$), i_j is an abbreviation for $i(s_j)$, $j = 1, 2, \ldots, n$, and the weights $\{p_{i_1,\ldots,i_n}\}$, conditionally on the locations, have a distribution defined on the simplex $\mathbb{P} = \{p_{i_1,\ldots,i_n} \geq 0 : \sum_{i_1=1}^K \cdots \sum_{i_n=1}^K p_{i_i,\ldots,i_n} = 1\}$ and independent of that for the $\theta's$.

Evidently, (6) allows the possibility to choose different surfaces at different locations. However, the weights need to satisfy a consistency condition in order to properly define a random process for $\theta(\cdot)$. Specifically, for any set of locations $(s_1, \ldots, s_n)$, $n \in \mathbb{N}$ and for all $\ell \in \{1, \ldots, n\}$, we need

$$p_{i_1,\ldots,i_{\ell-1},i_{\ell+1},\ldots,i_n} = p_{i_1,\ldots,i_{\ell-1},\cdot,i_{\ell+1},\ldots,i_n} \equiv \sum_{j=1}^{K} p_{i_1,\ldots,i_{\ell-1},j,i_{\ell+1},\ldots,i_n}. \tag{7}$$

So, although the suppression of the $s_1, s_2, ..., s_n$ in (7) may disguise it, the collection of probabilities is really a process. Hence, the weights must satisfy a continuity property (essentially, Kolmogorov consistency of the finite dimensional laws); for locations s and s_0, as $s \to s_0$, $p_{i_1,i_2} = pr\{\theta(s) = \theta^*_{i_1}(s), \theta(s_0) = \theta^*_{i_2}(s_0)\}$, tends to the marginal probability $p_{i_2} = pr\{\theta(s_0) = \theta^*_{i_2}(s_0)\}$ when $i_1 = i_2$, and to 0 otherwise.

Extension to three or more locations is clear and we refer this property as almost sure continuity of the weights (suggested by the almost sure continuity of the paths of a univariate spatial process, as defined in Kent (1989) or Banerjee *et al.* (2003). Suppose we also assume the random field G_0 to be almost surely continuous (a univariate spatial process $\theta(s)$, $s \in D$ is said to be almost surely continuous at a point s_0 if $\theta(s) \to \theta(s_0)$ with probability one as $||s - s_0|| \to 0$). Then, for the random field given by (6) if the set of weights $\{p_{i_1,\ldots,i_n}\}$ and the base random field G_0 are almost surely continuous, $\theta(s)$ converges weakly to $\theta(s_0)$ with probability one as $||s - s_0|| \to 0$ for all $s_0 \in D$.

Conditionally on the realized distribution G, the process has first and second moments given by (and comparable with (4))

$$\begin{aligned} E\{\theta(s)|G\} &= \textstyle\sum_{l=1}^K p_l(s)\,\theta^*_l(s) \\ var\{\theta(s)|G\} &= \textstyle\sum_{l=1}^K p_l(s)\,\theta^{*^2}_l(s) - \left\{ \sum_{l=1}^K p_l\,\theta^*_l(s) \right\}^2, \end{aligned}$$

and, for a pair of sites s_i, s_j,

$$\begin{aligned} cov\{\theta(s_i), \theta(s_j)|G\} &= \textstyle\sum_{l=1}^K \sum_{m=1}^\infty p_{l,m}(s_i, s_j)\,\theta^*_l(s_i)\,\theta^*_m(s_j) + \\ &\quad - \left\{ \textstyle\sum_{l=1}^K p_l(s_i)\,\theta^*_l(s_i) \right\} \left\{ \textstyle\sum_{m=1}^K p_m(s_j)\,\theta^*_m(s_j) \right\}. \end{aligned} \tag{8}$$

(8) shows that with almost surely continuous realizations from the base process and of the weights, the GSDP is mean square continuous.

As with the SDP, the process $\theta(\mathbf{s})$ has heterogenous variance and is nonstationary. However, marginalizing over G clarifies the difference. Suppose G_0 is a mean zero stationary Gaussian process with finite variance σ^2 and correlation function $\rho_\phi(s_i - s_j)$. Then, $E\{\theta(s)\} = 0$ and $var\{\theta(s)\} = \sigma^2$ as before, but now

$$cov\{Y(s_i), Y(s_j)\} = \sigma^2 \rho_\phi(s_i - s_j) \sum_{l=1}^K E\{p_{ll}(s_i, s_j)\}. \tag{9}$$

Notice that $\sum_{l=1}^K E\{p_{ll}(s_i, s_j)\} < 1$ so, marginally, the association structure is diminished by the amount of mass that the process (6) is expected to place on the not equally indexed θ^*'s. Moreover, from (9), the process $\theta(s)$ is centered around a stationary process only when $E\{p_{ll}(s_i, s_j)\}$ is a function of $s_i - s_j$ for all s_i and s_j.

5.3. *Multi-dimensional Stick-breaking*

We now turn to the specification of $p_{i_1,\ldots,i_n}$ for any choice of n and $s_1, ..., s_n$. We propose a multivariate stick-breaking construction which we detail in the bivariate setting. Sethuraman's univariate stick-breaking construction has weights p_l defined above (1). Denote the random events $\{\theta = \theta^*_l\}$ by Θ^1_l (with their complements Θ^0_l) and interpret the sequence of weights $\{p_1, p_2, \ldots\}$ as arising from $q_1 = pr\{\Theta^1_l\}$, $q_l = pr\{\Theta^1_l|\Theta^0_m, m < l\} = pr\{Y = \theta^*_l|Y \neq \theta^*_m, m < l\}$, $l = 1, 2, \ldots$. At each location let

$\Theta_l^u(s)$, $u = 0, 1$, be such that $\Theta_l^1(s) = \{\theta(s) = \theta_l^*(s)\}$ and $\Theta_l^0(s) = \{\theta(s) \neq \theta_l^*(s)\}$. Then, for any two locations s_i, s_j, we can consider the probabilities $q_{1,u,v}(s_i, s_j) = pr\{\Theta_1^u(s_i), \Theta_1^v(s_j)\}$, $q_{l,u,v}(s_i, s_j) = pr\{\Theta_l^u(s_i), \Theta_l^v(s_j)|\Theta_m^0(s_i), \Theta_m^0(s_j), m < l\}$, $l \geq 2$, $u, v \in \{0, 1\}$. Note that $q_{l,1,1}(s_i, s_j) + q_{l,1,0}(s_i, s_j) = q_{l,1,+}(s_i, s_j) = q_l(s_i)$. Similarly, $q_{l,+,1}(s_i, s_j) = q_l(s_j)$. Then, we can define

$$\begin{aligned} p_{lm} &= pr\{\theta(s_i) = \theta_l^*(s_i), \theta(s_j) = \theta_m^*(s_j)\} \\ &= pr\{\Theta_l^1(s_i), \Theta_m^1(s_j), \Theta_k^0(s_i), k < l, \Theta_r^0(s_j), r < m\} \\ &= \begin{cases} \prod_{k=1}^{l-1} q_{k,0,0}\, q_{l,1,0} \prod_{r=l+1}^{m-1} (1 - q_r)\, q_m & \text{if } l < m \\ \prod_{r=1}^{m-1} q_{r,0,0}\, q_{m,0,1} \prod_{k=m+1}^{l-1} (1 - q_k)\, q_l & \text{if } m < l \\ \prod_{r=1}^{l-1} q_{r,00}\, q_{l,11} & \text{if } l = m. \end{cases} \end{aligned} \tag{10}$$

Inspection of expression (10) reveals that the weights are determined through a partition of the unit square. For n locations the construction is on the unit n-dimensional hypercube and requires the specification of probabilities $q_{l,u_1,\ldots,u_n}$, $u_j \in \{0, 1\}, j = 1, 2, \ldots, n$, where u_j is an abbreviation for $u(s_j)$, at any set of locations $(s_1, \ldots, s_n)$. This entails defining a spatial process which, conditionally on the locations, has values on the simplex $\mathbb{Q} = \{q_{l,u_1,\ldots,u_n} \geq 0 : \sum_{u_1,\ldots,u_n=0}^{1} q_{l,u_1,\ldots,u_n} = 1\}$, and also satisfies consistency conditions of the type (7), that is

$$q_{l,u_1,\ldots,u_{k-1},u_{k+1},\ldots,u_n} = q_{l,u_1,\ldots,u_{k-1},\cdot,u_{k+1},\ldots,u_n} \equiv \sum_{u_k=0}^{1} q_{l,u_1,\ldots,u_{k-1},u_k,u_{k+1},\ldots,u_n}.$$

We specify the weight process through latent spatial processes.

5.4. *A Fully-Specified GSDP*

To generalize an SDP we can specify the foregoing components consistently if we allow a latent process to determine surface selection, that is, the stick-breaking components $q_{l,u_1,\ldots,u_n}(s_1, \ldots, s_n)$ arise through probabilities associated with the events $\Theta_l^{u_j}(s_j)$. In particular, consider the process $\{\delta_{\Theta_l^1(s)}, s \in D, l = 1, 2, \ldots,\}$, such that at any $l = 1, 2, \ldots$, $\delta_{\Theta_l^1(s)} = 1$ if $\Theta_l^1(s)$ occurs, $\delta_{\Theta_l^1(s)} = 0$ if $\Theta_l^1(s)$ does not occur. In turn, suppose $\Theta_l^1(s)$ occurs if and only if $Z_l(s) \in A_l(s)$. Then, we can work with the equivalent stochastic process defined by $\delta^*_{A_l(s)} = 1$ if $Z_l(s) \in A_l(s)$, $\delta^*_{A_l(s)} = 0$ if $Z_l(s) \notin A_l(s)$ where $\{Z_l(s), s \in D, l = 1, 2, \ldots\}$ is a latent random field. Furthermore, we can write

$$\begin{aligned} q_{l,u_1,\ldots,u_n} \quad & (s_1, \ldots, s_n) = \\ &= pr\{\delta_{\Theta_l^1(s_1)} = u_1, \ldots, \delta_{\Theta_l^1(s_n)} = u_n |\, \delta_{\Theta_i^1(s_j)} = 0, i < l, j = 1, \ldots, n\} \\ &= pr\{\delta^*_{A_l(s_1)} = u_1, \ldots, \delta^*_{A_l(s_n)} = u_n |\, \delta^*_{A_i(s_j)} = 0, i < l, j = 1, \ldots, n\}. \end{aligned}$$

It is easy to see that such a characterization guarantees that 7 is true.

We employ Gausssian thresholding to provide binary outcomes, i.e., $A_l(s) = \{Z_l(s) \geq 0\}$ as in Albert and Chib (1993). This is computationally convenient and, as a model for second stage random effects, there will be little posterior sensitivity to this choice. Suppose $\{Z_l(s), s \in D, l = 1, 2, \ldots\}$ is a countable collection of independent stationary Gaussian random fields with unknown (hence random) mean $\mu_l(s)$, variance 1, and correlation function $\rho_Z(\cdot, \eta)$. It follows that

$$q_{l,u_1,\ldots,u_n}(s_1, \ldots, s_n) = pr\{\delta^*_{\{Z_l(s_1)\geq 0\}} = u_1, \ldots, \delta^*_{\{Z_l(s_n)\geq 0\}} = u_n | \, \mu_l(s_1), \ldots, \mu_l(s_n)\},$$

because of the independence of the processes $\{Z_l(s)\}$ over the index l. For example, for $n = 2$, we get $q_{l,0,1} = pr\{Z_l(s_1) < 0, Z_l(s_2) \geq 0 | \mu_l(s_1), \mu_l(s_2)\}$. If the $\mu_l(s)$ surfaces are independent, $l = 1, 2, ...$, then so are the $q_{l,u_1,\ldots,u_n}(s_1, \ldots, s_n)$'s.

Since $Z_l(s)$ is assumed to be Gaussian, at any location s we obtain

$$q_{l,1}(s) = pr\{Z_l(s) \geq 0\} = 1 - \Phi\{-\mu_l(s)\} = \Phi\{\mu_l(s)\}. \tag{11}$$

If $\mu_l(s) = \mu_l$, for all s, with $\Phi(\mu_l)$ independent Beta$(1, \nu)$ then, marginally, the $\theta(s)$ follow a DP where the marginal weights are same for each s but the marginal distributions are not the same since $\theta^*_l(s) \neq \theta^*_l(s')$. More generally, if the $\mu_l(s)$ are such that the $\Phi\{\mu_l(s)\}$ are independent Beta$(1, \nu)$, $l = 1, 2, \ldots$, now the marginal distribution of $\theta(s)$ is a DP with probabilities that vary with location. Spatially varying weights have recently been considered by Griffin and Steel (2004), who work in the framework of dependent Dirichlet processes.

5.5. *Generalized Finite-Dimensional Spatial Dirichlet Process ($GSDP_K$)*

The above GSDP is, in fact, fitted with a finite sum approximation. But now we consider a much different construction when K is finite in 6. If the random weights are Dirichlet distributed for any choice of n and $s_1, \ldots, s_n$, i.e., $p_{i_1\ldots,i_n} \sim Dir(\{\alpha_{i_1,\ldots,i_n}(s_1, \ldots, s_n)\}$, we say that G is a generalized finite dimensional spatial Dirichlet process, in symbols, $G \sim GSDP_K(\boldsymbol{\alpha}, G_0)$, where $\boldsymbol{\alpha}$ denotes the measure that specifies the $\alpha_{i_1,\ldots,i_n}(s_1, \ldots, s_n)$. In particular, a $GSDP_K$ is a SDP_K process if, for any choice of n and $(s_1, \ldots, s_n)$, $\alpha_{i_1,\ldots,j_n}(s_1, \ldots, s_n) = 0$ unless $i_1 = \cdots = i_n$. We want to allow the possibility of choosing different surfaces at different locations, but, again, as $s \to s'$ we want $\theta(s) \to \theta(s')$ in distribution.

Suppose we want the measure $\alpha(\cdot)$ to have uniform marginals, i.e., $\alpha_i(s) = \alpha/K$, for some real constant α. Again, we illustrate in the case of two locations (s_1, s_2). Let $\alpha_{i+}(s_1) = \sum_{j=1}^K \alpha_{i,j}(s_1, s_2)$ and $\alpha_{+j}(s_2) = \sum_{i=1}^K \alpha_{i,j}(s_1, s_2)$. To obtain uniform marginals, we have to set $\alpha_{i+}(s_1) = \alpha_{+j}(s_2) = \alpha/K$ for any $i, j = 1, \ldots, K$. We can achieve this with $\alpha_{i,i} = \alpha/K$ and $\alpha_{i,j} = 0$ for $i \neq j$ (the SDP_K). The choice $\alpha_{i,j} = \alpha/K^2, i, j = 1, \ldots, K$ achieves independent surface selection across locations. To take into account spatial dependence we take up ideas in Petrone, Guindani and Gelfand (2006). Let $\alpha_{i,j} = \alpha \, E(p_{i,j}) = \alpha \, a_{i,j}$, $i, j = 1, \ldots, K$, where the $a_{i,j}$'s are defined so that $a_i = a_{i+} = \sum_{j=1}^K a_{i,j} = 1/K$. Now, let $H(\cdot, \cdot; \tau)$ be a distribution function on $[0, 1]^2$, with uniform marginals. In other words, let $\boldsymbol{\mathcal{U}} = (\mathcal{U}_1, \mathcal{U}_2)^T$ be a random vector such that $(\mathcal{U}_1, \mathcal{U}_2) \sim H(\cdot, \cdot; \tau)$ and $\mathcal{U}_r \sim U(0, 1)$, $r = 1, 2$. Then, for given K, we can partition the unit interval in the K intervals $(\frac{i-1}{K}, \frac{i}{K}]$, $i = 1, \ldots, K$, and correspondingly consider the induced partition of the unit square made of sets

$$Q_{i,j} = \left(\frac{i-1}{K}, \frac{i}{K}\right] \times \left(\frac{j-1}{K}, \frac{j}{K}\right], \quad i, j = 1, \ldots, K.$$

Therefore, we can set $E(p_{i,j})$ as the probability that $(\mathcal{U}_1, \mathcal{U}_2)$ belong to $Q_{i,j}$, that is

$$a_{i,j} = P_H\left(\mathcal{U}_1 \in \left(\frac{i-1}{K}, \frac{i}{K}\right], \mathcal{U}_2 \in \left(\frac{j-1}{K}, \frac{j}{K}\right]\right), \quad i, j = 1, \ldots, K. \tag{12}$$

Of course, marginally $a_i = a_{i+} = a_{+i} = P_H\left(\mathcal{U}_1 \in (\frac{i-1}{K}, \frac{i}{K}]\right) = \frac{1}{K}$. Then, $\alpha_{i,j} = \alpha P_H(Q_{i,j};\tau)$ and $\alpha_i = \alpha/K$ as desired.

Next, we introduce spatial dependence through a "copula argument". Let V be say, a mean 0 GP on D with covariance function $\gamma^2 \rho_\psi(s,s')$. For locations s and s' with $\mathbf{V} = (V(s), V(s')$ let $\mathcal{U}_1 = \Phi_1\left(\frac{V(s)}{\gamma}\right)$, $\mathcal{U}_2 = \Phi_2\left(\frac{V(s')}{\gamma}\right)$ and

$$\begin{aligned} H_V(u_1,u_2;\tau) &= P(\mathcal{U}_1 \le u_1, \mathcal{U}_2 \le u_2) \\ &= P(V(s) \le \gamma\,\Phi_1^{-1}(u_1), V(s') \le \gamma\,\Phi_2^{-1}(u_2)) \\ &= \Phi^{(2)}\left(\gamma\,\Phi^{-1}(u_1), \gamma\,\Phi^{-1}(u_2)\right). \end{aligned} \tag{13}$$

The distribution function $H_V(u_1,u_2;\tau)$ is the copula of the distribution of V. If $\hat{Q}_{i,j} = \left(\gamma\Phi^{-1}\left(\frac{i-1}{K}\right), \gamma\Phi^{-1}\left(\frac{i}{K}\right)\right] \times \left(\gamma\Phi^{-1}\left(\frac{j-1}{K}\right), \gamma\Phi^{-1}\left(\frac{j}{K}\right)\right]$, for any $i,j = 1,\ldots,K$ then, $a_{i,j} = P_{H_V}(\boldsymbol{\mathcal{U}} \in Q_{i,j}) = P_V\left((V(s),V(s')) \in \hat{Q}_{i,j}\right)$. Extension to the case of n locations is straightforward. It is enough to consider a distribution $H(\cdot)$ on the n-dimensional unit hypercube having uniform marginals coupled with the associated n-dimensional multivariate normal, arising from the GP.

A remaining question concerns the behavior of the $GSDP_K$ as $K \to \infty$. Clearly, the limiting behavior of the $GDP_K(\boldsymbol{\alpha}^{(K)} G_0(s,s'))$ for $K \to \infty$ depends on the specification of the Dirichlet parameters $\boldsymbol{\alpha}^{(K)} = (\alpha_{i,j}^{(K)}, i,j = 1,\ldots,K)$. The limiting behavior of the vector $\boldsymbol{p}^{(K)}$ of the K^n random probabilities $p_{i,j}^{(K)}$ can be obtained from results in Kingman (1975). That is, if $\max(\alpha_{i,j}^{(K)}, i,j = 1,\ldots,K)$ converges to zero and $\sum_{i=1}^K \sum_{j=1}^K \alpha_{i,j}^{(K)} \to \alpha$, $0 < \alpha < \infty$, then the *ordered* vector of the $p_{i,j}^{(K)}$'s converges in law to a Poisson–Dirichlet distribution. However, the limiting distribution depends more specifically on the behavior of partial sums such as $\sum_{j=1}^K \alpha_{i,j}^{(K)}$. Furthermore, since the support points of $G \sim GSPD_K(\boldsymbol{\alpha}^{(K)}, G_0)$ are dependent, the limit of the random process G does not follow directly from that of the vector $\boldsymbol{p}^{(K)}$. For instance, when the $GSDP_K$ is an SDP_K it converges weakly to a SDP with scale parameter α and base measure G_0. At the other extreme, with independent surface selection, the $GDP_K(\boldsymbol{\alpha}^{(K)} G_0(s,s'))$ converges weakly to a DP with scale parameter α and base measure $G_0(s)G_0(s')$. Finally, an intermediate choice might be $\alpha_{i,i} = a/k + b/k^2, \alpha_{i,j} = a/k^2, i \neq j$. In this case, we can show that the GDP_K converges weakly to a DP with base measure given by the mixture $aG_0(s,s') + bG_0(s)G_0(s')$ (general discussion is in Petrone *et al.*, 2006).

6. SIMULATED EXAMPLE COMPARING SDP, SDP_K AND $GSDP$, $GSDP_K$

We compare the SDP and our generalizations by means of a simulated data set (see Duan *et al.*, 2005). Data are generated from a finite mixture model of GPs, whose weights are assumed spatially varying. Let $\mathbf{Y}_t = (Y_t(s_1),\ldots,Y_t(s_n))^T, t = 1,2,\ldots,T$ be a set of independent observations at a set of locations $(s_1,\ldots,s_n)$. Then, each $Y_t(s)$ arises from a mixture of two GPs, $G_{0,s}^1$ and $G_{0,s}^2$, respectively, with mean ξ_i and covariance function $\sigma_i^2\,\rho_{\psi_i}(s,s')$, $i = 1,2$, $s,s' \in D$, such that $Y_t(s) \sim \alpha(s)G_{0,s}^1 + (1-\alpha(s))G_{0,s}^2$. The marginal weight is $\alpha(s) = P(Z(s) \ge 0)$, where $Z(s)$ is a mean zero stationary GP with covariance function $\rho_\eta(s-s')$. Therefore, we choose $Y_t(s)$ from $G_{0,s}^1$ if $Z_t(s) \ge 0$ or from $G_{0,s}^2$ if $Z_t(s) < 0$. Since $Z(s)$ is centered at zero, marginally we have $Y_t(s) \sim \frac{1}{2}N_n\left(\xi_1,\sigma_1^2\right) + \frac{1}{2}N_n\left(\xi_2,\sigma_2^2\right)$. However, the joint

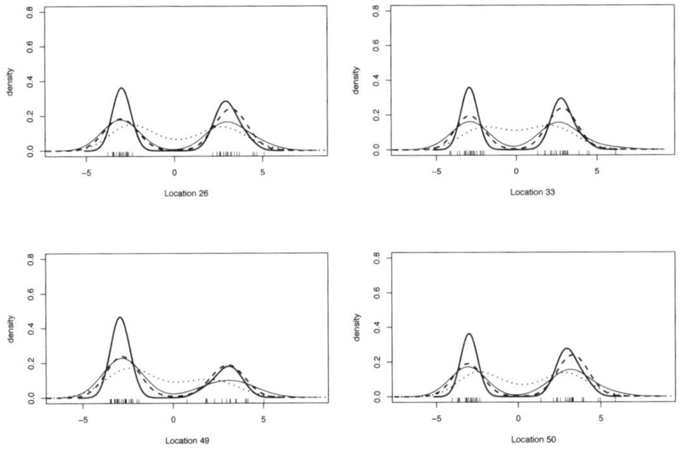

Figure 5: *True density (lighter line –) and predictive posterior density, respectively, under the SDP (light dotted line), the GSDP (thick dotted line) and the $GSDP_K$ models for the simulation example in Section 6.*

distribution for a pair s, s' in D is

$$(Y_t(s), Y_t(s')) \sim \begin{array}{l} \alpha_{1,1}(s,s')\, G^1_{0,s,s'} + \alpha_{2,2}(s,s')\, G^2_{0,s,s'} + \\ +\alpha_{1,2}(s,s')\, G^1_{0,s}\, G^2_{0,s'} + \alpha_{2,1}(s,s')\, G^2_{0,s}\, G^1_{0,s'}, \end{array} \tag{14}$$

where $\alpha_{i,j} = P\big((-1)^{i+1} Z(s) > 0,\, (-1)^{j+1} Z(s') > 0\big)$, $i, j = 1, 2$. Therefore, when s and s' are near to each other, it is very likely that $Y_t(s)$ and $Y_t(s')$ are from the same component. On the other hand, if s and s' are distant, the linkage between $Z(s)$ and $Z(s')$ is weak, so that $Y_t(s)$ and $Y_t(s')$ are chosen almost independently.

In our experiment, we have $n = 50$ random sites in a rectangle and $T = 40$. We set the parameters as follows: $\xi_1 = -\xi_2 = 3$, $\sigma_1 = 2\,\sigma_2 = 2$, $\phi_1 = \phi_2 = 0.3$, and $\eta = 0.3$. We fit the data using the SDP, the GSDP and the $GSDP_K$ with $K = 20$. To focus on the modeling of the spatial association, we assume $\mu(s) = 0$ in 3.

In Fig. 5, we plot the true density and the posterior predictive densities under the three models for four selected locations $(s_{26}, s_{33}, s_{49}, s_{50})$. The values of the 40 observations at each of these 4 locations are shown along the x-axis. The nonparametric models with spatially varying weights provide estimates closer to the true densities of the model and the data. This is confirmed by the bivariate plots in Fig. 6 (locations 26 and 50 are close, 49 and 50 are far apart) where we plot the probability contours of the true density along with the posterior distribution of the mean for the other models. It's interesting to compare these contours with 14. The posterior distributions of the $GSDP$ and $GSDP_K$ capture the expected behavior of the true density; explicit spatial modeling of the weights enables us to capture local details, as is also revealed by the heights of the local modes.

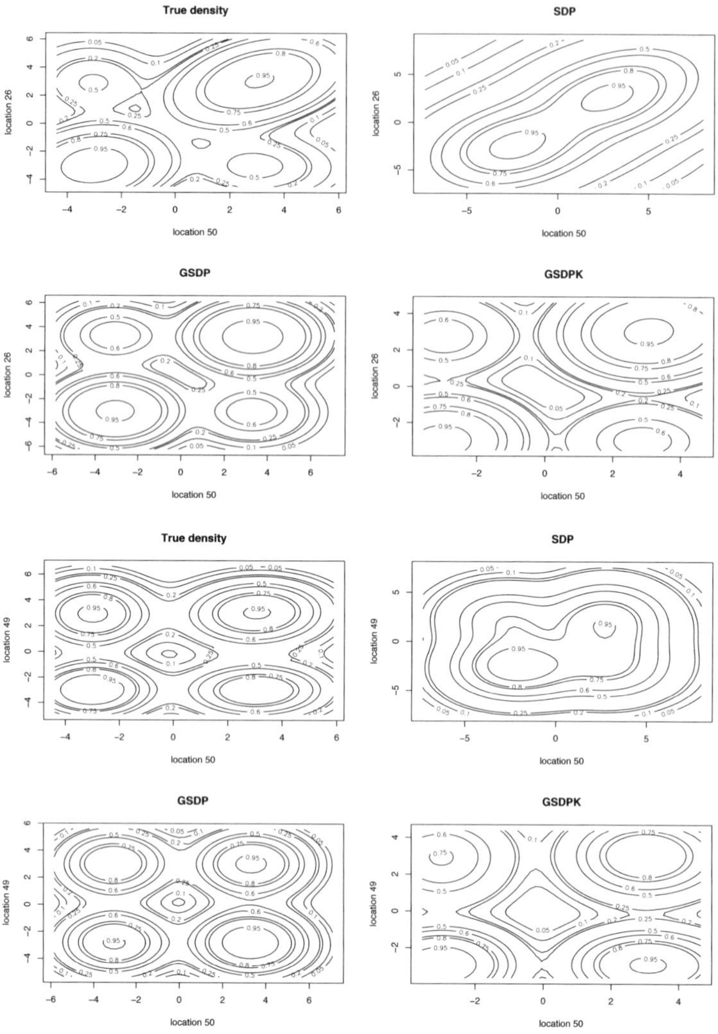

Figure 6: *Bivariate contour plots of the true densities and the posterior distributions of the mean of the SDP, GSDP and* $GSDP_K$ *models for locations* 26 *and* 50 *(a) and* 49 *and* 50 *(b). See Section 6.*

7. SUMMARY AND FURTHER ISSUES

We have developed spatial Dirichlet process specifications that can be used as an alternative to customary Gaussian process choices to model spatial random effects. Such models provide nonstationary, nonGaussian processes that can have locally varying DP marginals and can both marginally and jointly capture multimodalities. We have demonstrated this with both real and simulated datasets.

In the case where the replications might be time dependent, we can embed any of the foregoing specifications within a dynamic model. See Gelfang *et al.* (2005) for the SDP case and Duan *et al.* (2005) for the GSDP.

Multivariate spatial process modelling using GP's is reviewed in Gelfand et al. (2004) but the nonparametric setting has not been addressed. Gelfand et al. advocate coregionalization (random linear transformation of independent univariate process models). Here, coregionalization could be applied to the base measure or, possibly, by random linear transformation of independent spatial DP's.

Finally, as noted in the Introduction, an alternative role for this modelling is for functional data analysis (FDA). Here, we replace geographic space, $s \in D$, with covariate space, $z \in Z$, seeking to model a random function of z to explain responses Y. For a collection of individuals, modelling these functions as individual-level process realizations leads to DP's where the atoms are random functions. In further extension to spatial FDA, model development would use DP specifications for both the spatial and the functional aspects of the modelling.

REFERENCES

Agarwal, D. and Gelfand, A. (2005). Slice gibbs sampling for simulation based fitting of spatial data models. *Statist. Computing* **15**, 61–69.

Albert, J. and Chib, S. (1993). Bayesian analysis of binary and polychotomous response data. *J. Amer. Statist. Assoc.* **88**, 669–679.

Antoniak, C. E. (1974). Mixtures of Dirichlet processes with applications to Bayesian nonparametric problems. *Ann. Statist.* **2**, 1152–1174.

Banerjee, S., Carlin, B. and Gelfand, A. (2004). *Hierarchical Modeling and Analysis for Spatial Data.* London: Chapman and Hall

Banerjee, S., Gelfand, A. and Sirmans, C. (2003). Directional rates of change under spatial process models. *J. Amer. Statist. Assoc.* **98**, 946–954.

Barry, R. P. and Ver Hoef, J. M. (1996). Blackbox kriging: Spatial prediction without specifying variogram models. *J. Agricultural, Biological and Envir. Statistics*, **1**, 297–322.

Brown, P., Diggle, P. and Henderson, R. (2003). A non-Gaussian spatial process model for opacity of flocculated paper. *Scandinavian J. Statist.* **30**, 355–368.

Cifarelli, D. and Regazzini, E. (1978). Problemi statistici non parametrici in condizioni di scambiabilità parziale. Impiego di medie associative. *Quaderni dell'Istituto di Matematica Finanziaria dell'Universita' di Torino*, **Serie III, 12**, 1–13.

Clyde, M. and Wolpert, R. L. (2006). Bayesian modelling with overcomplete representations. *In this volume.*

Damian, D., Sampson, P. and Guttorp, P. (2001). Bayesian estimation of semi-parametric non-stationary spatial covariance structures. *Environmetrics*, **12**, 161–178.

De Iorio, M., Müller, P., Rosner, G. and MacEachern, S. (2004). An anova model for dependent random measures. *J. Amer. Statist. Assoc.* **99**, 205–215.

Diggle, P. J., Liang, K. and Zeger, S. L. (1994). *Analysis of Longitudinal Data.* Oxford: Clarendon Press.

Duan, J. A., Guindani, M. and Gelfand, A. E. (2005). Generalized Spatial Dirichlet Process Models. *Tech. Rep.*, Duke University, USA.

Ferguson, T. S. (1973). A Bayesian analysis of some nonparametric problems. *Ann. Statist.* **1**, 209–230.

Ferguson, T. S. (1974). Prior distributions on spaces of probability measures. *Ann. Statist.* **2**, 615–629.

Fuentes, M. and Smith, R. (2001). A new class of nonstationary spatial models. *Tech. Rep.*, North Carolina State University, USA.

Gelfand, A. E., Schmidt, A. M, Banerjee, S. and Sirmans, C .F. (2004) Nonstationary multivariate process modelling through spatially varying coregionalization. *Test* **13**, 1–50 (with discussion).

Gelfand, A., Banerjee, S. and Gamerman, D. (2005a). Spatial process modelling for univariate and multivariate dynamic spatial data. *Environmetrics*, **16**, 1–15.

Gelfand, A., Kottas, A. and MacEachern, S. (2005b). Bayesian nonparametric spatial modeling with Dirichlet processes mixing. *J. Amer. Statist. Assoc.* **100**, 1021–1035.

Griffin, J. and Steel, M. (2004). A class of dependent Dirichlet processes. *Tech. Rep.*, University of Kent at Canterbury, UK.

Higdon, D., Swall, J. and Kern, J. (1999). Non-stationary spatial modeling. *Bayesian Statistics 6* (J. M. Bernardo, J. O. Berger, A. P. Dawid and A. F. M. Smith, eds.) Oxford: University Press, 287–302.

Hjort, N. (2000). Bayesian analysis for a generalized Dirichlet process prior. *Tech. Rep.*, University of Oslo, Norway.

Ishwaran, H. and James, L. (2001). Gibbs sampling methods for stik breaking priors. *J. Amer. Statist. Assoc.* **96**, 161–173.

Ishwaran, H. and Zarepour, M. (2000). Markov chain Monte Carlo in approximate Dirichlet and beta two-parameter process hierarchical models. *Biometrika* **87**, 371–390.

Ishwaran, H. and Zarepour, M. (2000a). Dirichlet prior sieves in finite normal mixtures. *Statistica Sinica* **12**, 941–963.

Ishwaran, H. and Zarepour, M. (2000b). Exact and approximate sum-representations for the Dirichlet process. *Can. J. Statist.* **30**, 269–283.

Jain, S. and Neal, R. M. (2004). A split-merge Markov chain Monte Carlo procedure for the Dirichlet process mixture model. *J. Comp. Graphical Statist.* **13**, 158–182.

Kent, J. (1989). Continuity properties of random fields. *Ann. Prob.* **17**, 1432–1440.

Kingman, J. F. C. (1975). Random discrete distributions. *J. Roy. Statist. Soc. B* **37**, 1–22.

Kottas, A. , Duan, J. and Gelfand, A. E. (2006). Modeling Disease Incidence Data with Spatial and Spatio-temporal Dirichlet Process Mixtures *Tech. Rep.*, ISDS, Duke University, USA.

Le, N. and Zidek, J. (1992). Interpolation with uncertain spatial covariances: a Bayesian alternative to kriging. *J. Multivariate Analysis* **43**, 351–374.

MacEachern, S. (2000). Dependent Dirichlet processes. *Tech. Rep.*, The Ohio State University, USA.

Mardia, K. V. and Goodall, C. R. (1993). Spatial-temporal analysis of multivariate environmental monitoring data. *Multivariate Environmental Statistics*, (G. P. Patil and C. R. Rao, eds.) Amsterdam: North-Holland, 347–386.

Muliere, P. and Secchi, P. (1995). A note on a proper Bayesian bootstrap. *Tech. Rep.*, Università di Pavia, Italy.

Neal, R. M. (2000). Markov chain sampling methods for Dirichlet process mixture models. *J. Comp. Graphical Statist.* **9**, 249–265.

Ongaro, A. and Cattaneo, C. (2004). Discrete random probability measures: a general framework for nonparametric Bayesian inference. *Statist. Prob. Lett.* **67**, 33–45.

Palacios, M. and Steel, M. (2007). Non-gaussian Bayesian geostatistical modeling. *J. Amer. Statist. Assoc.* (to appear).

Petrone, S., Guindani, M. and Gelfand, A. (2006). Finite mixtures of spatial processes. *Tech. Rep.*, ISDS, Duke University, USA.

Petrone, S. and Raftery, A. (1997). A note on the Dirichlet process prior in Bayesian nonparametric inference with partial exchangeability. *Statist. Prob. Lett.* **36**, 69–83.
Pitman, J. and Yor, M. (1997). The two-parameter poisson-Dirichlet distribution derived from a stable subordinator. *Ann. Prob.* **25**, 855–900.
Richardson, S. and Green, P. (1997). On Bayesian analysis of mixtures with an unknown number of components. *J. Roy. Statist. Soc. B* **59**, 731–792.
Sampson, P. and Guttorp, P. (1992). Nonparametric estimation of nonstationary spatial covariance structure. *J. Amer. Statist. Assoc.* **87**, 108–119.
Scaccia, L. and Green, P. J. (2003). Bayesian growth curves using normal mixtures with nonparametric weights. *J. Comp. Graphical Statist.* **12**, 308–331.
Schmidt, A. and O'Hagan, A. (2003). Bayesian inference for non-stationary spatial covariance structure via spatial deformations. *J. Roy. Statist. Soc. B* **65**, 743–758.
Sethuraman, J. (1994). A constructive definition of Dirichlet priors. *Statistica Sinica* **4**, 639–650.
Sethuraman, J. and Tiwari, R. (1982). Convergence of Dirichlet measures and the interpretation of their parameter. *Statistical Decision Theory and Related Topics III* (S. S. Gupta and J. O. Berger, eds.) New York: Academic Press, 305–315.
Shapiro, A. and Botha, J. D. (1991) Variogram fitting with a general class of conditionally nonnegative definite functions. *Comput. Statist. Data Anal.* **11**, 87–96.
Stein, M. L. (1999). *Interpolation of Spatial Data – Some Theory of Kriging.* Berlin: Springer-Verlag.
Stein, M. L. (2005) Space-time covariance functions. *J. Amer. Statist. Assoc.* **100**, 310–321.

DISCUSSION

NILS LID HJORT (*University of Oslo, Norway*)

I have followed the work of Alan Gelfand and various co-authors on Bayesian spatial statistics over the past few years with interest. There are many aspects and subthemes, from detailed technicalities of modelling and simulation strategies to the challenges of actual applications, and I very much welcome the present survey paper.

It's about time. We have had spatial statistics since around 1950 (starting with Bertil Matérn and Danie Krige) and nonparametric Bayesian statistics since around 1970 (with the early work of Thomas Ferguson, Kjell Doksum, David Blackwell and Charles Antoniak). These brooks have grown into strong rivers, and are according to current estimates of spatial-temporal derivatives in the process of extending themselves into veritable floods. So it was a question of time until the twain met, and so they did: I view the spatial Dirichlet processes of Gelfand and co-authors as among the first serious attempts at intersecting 'nonparametric Bayes' with 'spatial statistics'.

Both of these areas are inherently large, and need to be so: Nonparametrics lives in infinite-dimensional spaces, and the ratio of the number of ways in which variables can be dependent divided by the number of ways they can be independent is also infinite. It would follow that the new, fledgling intersection area of 'spatial Bayesian nonparametrics' also will need to grow big. In this perspective I think we must realise that spatial Dirichlet processes, although a broadly versatile machinery, must not rule alone, or for too long. I expect Bayesian nonparametric spatial statistics to grow healthily in many directions over the coming ten years (and beyond), and view the spatial Dirichlet process direction as one of several.

It is perhaps unexpected that Dirichlet processes should turn out to be such a broadly flexible tool also in spatial contexts. Perhaps parallel developments may be expected in the model world of Gaußian and nearly Gaußian processes. Consider the basic application example of Gelfand, Guindani and Petrone, which takes the form

$$Y_t(s) = \mu_t(s) + \theta_t(s) + \epsilon_t(s) \tag{15}$$

for modelling say monitoring stations over time, at time points $t_1, \ldots, t_{40}$ and spatial positions $s_1, \ldots, s_{45}$, where $\mu_t(s)$ is trend, $\epsilon_t(s)$ is white noise, and $\theta_t(s)$ is the main spatial object. I wish to express a modicum of polite surprise that statisticians after staring at (15) for a while would say, 'Of course! Dirichlet processes!'. But, apparently, they do. I would expect there to be alternative fruitful models, e.g. built around Gaußian processes or relatives.

For a given spatial correlation function, giving rise to an $n \times n$ variance matrix for $\theta(s_1), \ldots, \theta(s_n)$ of the form $\sigma^2\Sigma_n(\phi)$, say, the pure Gaußian model amounts to

$$\begin{pmatrix} \theta(s_1) \\ \vdots \\ \theta(s_n) \end{pmatrix} = \sigma\Sigma(\phi)^{1/2} \begin{pmatrix} N_1 \\ \vdots \\ N_n \end{pmatrix} \tag{16}$$

in terms of independent standard normals. There are several ways in which nonparametric envelopes, or sausages, can be built around the normal, with a parameter dictating the degree of concentration around the given normal distribution. Thus different forms of spatial nonparametric models would emerge by taking

$$N_1, \ldots, N_n \sim \text{envelope}(\mathrm{N}(0,1); \delta),$$

in concert with (16). Here δ indicates some parameter that governs the tightness around the standard normal. It would be interesting to have such modelling attempts contrasted with those of Gelfand *et al.*, see also Gamerman *et al.* (2007).

Different versions of spatial Dirichlet processes. How can one make two Dirichlet processes G and G' dependent? Suppose both of them are required to have parameters aG_0, say. Again, there must be several different fruitful ways of achieving this. Gelfand et al. work with the Tiwari–Sethuraman representation, where

$$G = \sum_{j=1}^{\infty} p_j \delta(\theta_j) \quad \text{and} \quad G' = \sum_{j=1}^{\infty} p'_j \delta(\theta'_j),$$

say, where the $\{\theta_j\}$ sequence needs to be i.i.d. from G_0, as does the $\{\theta'_j\}$ sequence. One class of models emerges by keeping $p_j = p'_j$ but allowing dependence between θ_j and θ'_j; cf. MacEachern's (2000) early work that set off some of the later developments. One may also simultaneously work with dependence between the $\{p_j\}$ and $\{p'_j\}$ sequences. In the usual set-up,

$$p_j = \bar{q}_1 \cdots \bar{q}_{j-1} q_j \quad \text{and} \quad p'_j = \bar{q}'_1 \cdots \bar{q}'_{j-1} q'_j,$$

with $q_j \sim \text{Beta}(1, a)$ and $q'_j \sim \text{Beta}(1, a)$, writing $\bar{q}_j = 1 - q_j$ and $\bar{q}'_j = 1 - q_j$. A natural recipe is now to take

$$q_j = \text{Beta}^{-1}(\Phi(N_j), 1, a) \quad \text{and} \quad q'_j = \text{Beta}^{-1}(\Phi(N'_j), 1, a),$$

where the (N_j, N'_j)s are independent pairs, but with internal correlation ρ. This creates two dependent Dirichlet processes, with dependence both from θ_j to θ'_j and from p_j to p'_j, and the construction is easily generalised to full random fields, via normal processes. Here $\text{Beta}^{-1}(q, 1, a)$ is the inverse of the cumulative distribution function $\text{Beta}(x, 1, a)$ for the $\text{Beta}(1, a)$ distribution.

A different construction that also looks fruitful on paper is as follows. Let

$$G = \sum_{j=1}^{K} p_j \delta(\theta_j) \quad \text{and} \quad G' = \sum_{j=1}^{K} p'_j \delta(\theta'_j),$$

for a moderate or large K. As long as

$$p \sim \text{Dir}(a/K, \ldots, a/K) \quad \text{and} \quad p' \sim \text{Dir}(a/K, \ldots, a/K), \tag{17}$$

it is known that G and G' converge in distribution to the $\text{Dir}(aG_0)$ process; see discussion about this in Hjort (2003). The problem is therefore to make p and p' dependent without losing (17). Here one method is to define $p_j = A_j/A$ and $p'_j = A'_j/A'$, where A and A' are the sums of A_js and A'_js, respectively, and where

$$A_j = \text{Gam}^{-1}(\Phi(N_j), a/K, 1) \quad \text{and} \quad A'_j = \text{Gam}^{-1}(\Phi(N'_j), a/K, 1),$$

and again $\text{corr}(N_j, N'_j) = \rho$. Here $\text{Gam}^{-1}(q, a/K, 1)$ is the quantile inverse of the distribution function for a $\text{Gam}(a/K, 1)$ variable. Other copulae (also called 'the emperor's new clothes', see Mikosch, 2007) can also be employed here.

One might also contemplate versions of this where only the biggest Dirichlet process jumps are allowed to enter the approximations, say

$$G = \sum_{j=1}^{K} b_j \delta(\theta_j) \Big/ \sum_{j=1}^{K} b_j,$$

where $b_1 > \cdots > b_K$ are the K biggest jumps in a $\text{Dir}(aG_0)$ process. This is moderately awkward, but doable, via results in Hjort and Ongaro (2006).

Is the spatial component really required? This section title is not meant too provocatively, since I of course acknowledge that the spatial part of the model often needs to be there. But I wish to have in the toolbox various checks and tests for helping me decide whether I need the spatial component $\theta_t(s)$ in (15) or not. It is important to realise that in (15), both zero-mean terms $\theta_t(s)$ and $\epsilon_t(s)$ change values and interpretation with what we put into the trend part $\mu_t(s)$. Specifically, adding one more clever covariate with strong explanatory power will sometimes carry enough spatial information to make $\theta_t(s)$ become too small in size to really matter. I do suppose Gelfand et al.'s machinery can give me $\pi(\sigma \mid \text{data})$ relatively easily, about the standard deviation of the $\theta_t(s)$ part, or perhaps even more informatively the joint posterior of (σ, τ). But it would be nice to have a formalised procedure specifically for answering the question 'can I just as well set $\theta_t(s)$ to zero?' for each given application. This could perhaps be in the form of a post-processed posterior predictive p-value, as in Hjort, Dahl and Steinbakk (2006), or via a suitable Bayesian test, as in Rousseau (2007).

There might be Bernshteĭn–von Mises theorems to prove in these models, to the effect that with enough data things will go very well and different Bayesians will be in decent agreement a posteriori. Very often there would not be overwhelmingly much data, however, particularly in view of the somewhat elusive character of parameters relating to spatial correlation. This is an argument for being very careful with the prior construction. Gelfand et al. use reasonably generic prior descriptions, but they seem to consistently take σ (noise level for $\theta_t(s)$) and τ (noise level for the white part $\epsilon_t(s)$) as independent. This worries me, since it would clash both with Jeffreys' type priors and with other work about setting priors in variance component models?

Side effects. Bayesian and empirical Bayesian constructions are often successful in terms of achieving better overall accuracy of predictions, better average precision for estimating a flock of parameters, etc. This is achieved by intelligent amounts of smoothing and shrinking. There are sometimes less fortunate side effects, however; one might more easily miss 'alarm situations' and anomalies, which in many practical applications would be of great concern, e.g. for pollution monitoring. Or are there good fixes by setting up suitable loss functions, to be combined and assessed from the posterior calculations?

The last point I wish to make is that there is a potential problem with the Dirichlet process as the basic Lego building block in these contexts, related to the two roles of the 'a' parameter in the Dir(aG_0). As we know, the size of a seriously influences two different aspects of the distribution, (i) the degree of tightness around its prior mean G_0, and (ii) the level of clustering in repeated samples from G. These are not quite reconcilable. This might point to the need for extending the infinite-dimensional prior processes with one more parameter.

STEVEN N. MACEACHERN (*The Ohio State University, USA*)

The authors have written a stimulating paper that introduces a new element to models based on dependent nonparametric processes. A selection surface allows some portions of a realized surface to come from one component of a mixture and other portions of the realized surface to come from other components of the mixture. This innovation generates tremendous flexibility for modelling and will prove useful in many applications.

The description of a formal hierarchical model often lends additional insight to a modelling effort. The subsequent description (at the risk of modestly warping the authors' model) casts the model in a more traditional framework. In doing so, connections to existing processes are seen, and the importance of qualitative features of the selection surface become apparent.

The authors write the model (3) at the observed sites as

$$Y(\mathbf{s}) = \mu(\mathbf{s}) + \theta(\mathbf{s}) + \epsilon(\mathbf{s}).$$

This can be divided into three portions–one portion that is far-removed from the data, a second portion that provides the guts of the model and a third portion that connects the model to the data. The model includes a potential regression structure for $\mu(\cdot)$, including dependence on covariates, and it allows for distributions on hyperparameters that govern this regression structure. This is essential for modelling, although not the focus of this work.

The middle of the hierarchical model consists of a dependent nonparametric process. To generate the SDP or the GSDP, this mid-level process is a dependent Dirichlet process (DDP) as defined, for example, in MacEachern (2000). The DDP is defined by two countable sets of independent and identically distributed stochastic processes. The first set of processes is for the locations: Here, $\theta^*(\cdot)$, with $\theta_1^*(\cdot), \theta_2^*(\cdot), \ldots$ a random sample of these processes. These $\theta_l^*(\cdot)$ lead directly to $\theta_l^*(\mathbf{s})$ for vectors of sites in D through the usual finite-dimensional specifications (although there is often little loss in thinking of the entire surface, $\theta^*(\cdot)$ as having been realized).

The second set of processes determines the probabilities associated with the locations. Following MacEachern (2000), $V(\cdot)$ follows a stochastic process where $V(s)$ is distributed as Beta$(1, M)$ with $M > 0$ a mass parameter. From the surfaces $V_1(\cdot), V_2(\cdot), \ldots$, and following Sethuraman (1994), arithmetic takes one to the $p_1(\cdot), p_2(\cdot), \ldots$. Again, $V_l(\mathbf{s})$ is a finite dimensional realization from $V_l(\cdot)$. The choices for the distribution on $V(\cdot)$ are endless. A convenient choice for $V(\cdot)$ maps a Gaussian process with standard normal marginals into the Beta$(1, M)$ with a pair of transformations. Provided the covariance function of the Gaussian process allows $V(s) \neq V(s')$, this produces the 'multivariate stick-breaking' that is needed to ensure independence of the random distributions at s and s' (or independence in the limit) s and s' become distant.

At this point, the middle stage of the model is a DDP. The DDP includes multivariate as well as univariate $\theta^*(\cdot)$ (DeIorio et al., 2004). It can be specialized (e.g., $V_l(s) = V_l$ for all s and all l results in the single-p DDP which falls under Sethuraman's definition of the Dirichlet process), or it can be generalized to DDPs with spatially varying mass parameters (see MacEachern, 1999) or, more generally, to a wide variety of dependent nonparametric processes (for example, Hjort, 2000, or MacEachern, Kottas and Gelfand, 2001). In essence, the DDP and its variants describe distribution-valued stochastic processes indexed by covariate values, in this case the spatial location, s.

The final stage of the model presents the novelty. It consists of a likelihood to smooth out the discreteness inherent in mixture models, and it includes a 'selector surface'. The selector surface is driven by a latent process. Under one view, the surface, say $U(\cdot)$, has the property that $U(s)$ is uniformly distributed on the interval $(0, 1)$ for each s. The value of $U(s)$ determines which component of the mixture is active at site s through the rule that assigns the site to component k if both $\sum_{l=1}^{k} p_l \geq U(s)$ and $\sum_{l=1}^{k-1} p_l < U(s)$. Alternatively, one can pursue an approach whereby spatial regions are assigned to mixture components successively. A latent surface, $Z_1(\cdot)$, with appropriate marginal distribution at site s, is compared to the cutoff determined by $V_1(\cdot)$ at the same site, s. If $Z_1(s) \leq V_1(s)$, the site is assigned to the first component of the mixture; if $Z_1(s) > V_1(s)$, the site is not assigned to the first component of the mixture. A similar process unfolds for further components of the mixture, with comparisons of $Z_l(\cdot)$ to $V_l(\cdot)$. It is evident that the latter approach has several advantages over typical implementations of the former approach.

The selector surfaces have many interesting implications. If there is no variation in either $V(\cdot)$ or $Z(\cdot)$ across sites, the model reduces to the SDP model, a version of the single-p DDP model, and also a version of the Dirichlet process. If the realizations of $V(\cdot)$ and $Z(\cdot)$ show variation and are relatively smooth, one would expect to find edges, where there would be a nice curve separating one mixture component from another. If the realizations are very jagged, one would expect 'transition zones' without a clean jump from one component to another. In these zones, one would expect many switches back and forth between components. With the structure of the mixture model, where $\theta_l^*(s)$ and $\theta_{l'}^*(s)$ will typically be very different, one would expect discontinuities at the transition between mixture components. Properties such as clean breaks in the surface or transition zones are important in applications such as image analysis, where one typically searches for the edges (clean breaks) in the surface and expects discontinuities at these edges. Nicely, the authors' pursuit of the approach with the $Z(\cdot)$ rather than the $U(\cdot)$ approach with continuous spatial paths avoids the problem of always moving to adjacent components when switching from one component to another.

I would like to make a plea to the authors and to others working in the area to refrain from overloading notation, as is sometimes done for the Dirichlet process. For the Dirichlet process, the time-honored tradition (see Ferguson's early work, for example) uses the symbol α for the base measure. With respect for this tradition, and adapting notation to the spatial setting, α would represent a common (across sites) base measure while α_s would represent the base measure at the site s. Making use of an already taken symbol to represent a different quantity can only lead to confusion and make reading the literature more difficult for those entering the area.

Acknowledgement. This material is based upon work supported by the National Science Foundation under Award No. SES-0437251 and the National Security Agency under Award No. H98230-05-1-0065.

REPLY TO THE DISCUSSION

We appreciate the positive comments from Hjort and MacEachern and agree that, because there are so many ways to bridge spatial process modelling and Bayesian nonparametric modelling, we can expect to see growth in various directions in the near future.

A key point that we feel needs to be made in this rejoinder is the distinction in perspective and perhaps in specification, between MacEachern's DDP's and our SDP's. The issue centers on concern with joint or marginal models.

MacEachern's DDP in the spatial setting specifies the distribution of $\theta(s)$ as F_s, *i.e.*, a collection of random distributions indexed by spatial location. Focusing on the random distributions F_{s_1} and F_{s_2}, we have

$$v_l = \begin{pmatrix} v_l(s_1) \\ v_l(s_2) \end{pmatrix}, \quad l = 1, 2, \ldots, \Leftrightarrow p_l = \begin{pmatrix} p_l(s_1) \\ p_l(s_2) \end{pmatrix}, \quad l = 1, 2, \ldots,$$

and $\{\theta_l^*\}$ where $\theta_l^* = \begin{pmatrix} \theta_l^*(s_1) \\ \theta_l^*(s_2) \end{pmatrix}$ and $F_{s_j} \Leftrightarrow \{p_l(s_j)\}, \{\theta_l^*(s_j)\}, j = 1, 2$. Immediately, we can see that F_{s_1} and F_{s2} come from a DP. If the components of v_l and θ_l^* are dependent then F_{s_1} and F_{s2} are dependent. But what about the joint distribution of $\theta(s_1)$ and $\theta(s_2)$? If we assume conditional independence given F_{s_1} and F_{s_2}, as in, e.g., Cifarelli and Regazzini (1978), we have $P(\theta(s_1) = \theta_l^*(s_1), \theta(s_2) = \theta_{l'}^*(s_2)) = p_l(s_1)p_{l'}(s_2)$ with marginal dependence through mixing. Introducing dependence for the v's and θ^*'s across $s, s \in D$ is straightforward, e.g., suppose we have $v_l(s) = G^{-1}(\Phi(Z_l(s)))$ where G^{-1} is the inverse Beta(1,α) c.d.f., $Z_{l,D}$ are iid realizations from a mean 0 GP and, independently, the $\theta_{l,D}^*$'s are iid realizations from a GP. Indeed, this strategy is also mentioned by Hjort along with a different construction using transformation of a Gaussian process to gamma variables.

Amplifying MacEachern's subsequent suggestion, suppose we introduce a usual stickbreaking, $\{v_l\} \Leftrightarrow \{p_l\}$. Additionally, Z_D a mean 0, variance 1 GP realization yields $U(s) = \Phi(Z(s))$.

$$\begin{aligned} P(\theta(s_1) &= \theta_l^*(s_1), \theta(s_2) = \theta_{l'}^*(s_2)) \\ &= P(U(s_1) \in (\textstyle\sum_{j=1}^{l-1} p_j, \sum_{j=1}^{l} p_j), U(s_2) \in (\sum_{j=1}^{l'-1} p_j, \sum_{j=1}^{l'} p_j)) \\ &= P(Z(s_1) \in (\Phi^{-1}(\textstyle\sum_{j=1}^{l-1} p_j), \Phi^{-1}(\sum_{j=1}^{l} p_j)), Z(s_2) \\ &\qquad\qquad \in (\Phi^{-1}(\textstyle\sum_{j=1}^{l'-1} p_j), \Phi^{-1}(\sum_{j=1}^{l'} p_j)) \\ &\equiv p_{l,l'}(s_1, s_2). \end{aligned}$$

Evidently, we are constructing joint distributions as opposed to adopting the conditional independence assumption above. However, it is still the case that $F(\theta(s))$ comes from a DP.

Then, we come to the GSDP. Here, we introduce iid copies $\{Z_{l,D}\}$ of a GP with mean μ_D to create

$$p_{l,l'}(s_1, s_2) = P(Z_j(s_1) < 0, j < l, Z_l(s_1) > 0, Z_j(s_2) < 0, j < l', Z_{l'}(s_2) > 0).$$

We need μ_D to be random in order that the p's are and if $\mu(s)$ varies with s, $p_l(s)$ varies with s. Again, we are constructing joint distributions.

Finally, the GSDP-k introduces $Z_D \Leftrightarrow U_D = \Phi(Z_D)$. Now, $l, l' = 1, 2, ...k$ and

$$\{p_{l,l'}(s_1, s_2\} \sim Dir(\{a_{l,l'}(s_1, s_2)\})$$

with

$$\begin{aligned} a_{l,l'}(s_1, s_2) &= P(U(s_1) \in (\tfrac{l-1}{k}, \tfrac{l}{k}), U(s_2) \in (\tfrac{l'-1}{k}, \tfrac{l'}{k})) \\ &= P(Z(s_1) \in (\Phi^{-1}(\tfrac{l-1}{k}), \Phi^{-1}(\tfrac{l}{k})), Z(s_2) \in (\Phi^{-1}(\tfrac{l'-1}{k}), \Phi^{-1}(\tfrac{l'}{k}))). \end{aligned}$$

We see that automatically $\{p_l(s)\}$ varies with s by the random selection from the Dirichlet. Again, by definition, we are constructing joint probabilities for surface selection, hence joint distributions.

Next, we take up Hjort's discussion around his expression (16). Caution is needed here; (16) provides the joint distribution for a set of n locations under a GP but this is, in fact, a property of the GP. In general, for a given collection of locations, starting with i.i.d. random variables from some distribution, say f, using (16) produces a joint distribution. However, outside of the normal (or mixtures of normals), such transformation need not uniquely determine finite dimensional distributions. Hence, in general, we can not use (16) to define a process.

One aspect MacEachern helps to illuminate is the nature of the expected surface selection. However, we confess to being unclear as to his concern about 'moving to adjacent components' with the $U(\cdot)$ approach. The $Z(\cdot)$ approach will tend to yield similar adjacent selection with regard to the θ_l^*'s. Moreover, with the independence of the θ^*'s across l, why is this a worry?

Hjort raises some general issues regarding spatial process modelling (that are apart from the nonparametric aspect). First, he questions the need for the spatial component in the model. Here, he joins the ongoing debate regarding the mean-covariance trade-off – the size of the mean specification will, for a given dataset, determine the need for spatial random effects. Next, he expresses concern about prior assumptions. Here, we face the usual issues of parameter independence, choice of parametrization, identifiability of parameters, etc. Then, he raises the matter of smoothing away potentially interesting extremes in the spatial surface. Here, there are, in fact, two issues. One is the inherent smoothness associated with spatial process realizations under customary covariance functions and its impact with regard to spatial interpolation. The second is how, in particular, a Bayesian implementation of such interpolation imposes smoothness. Lastly, the dual role for the precision parameter in the DP is important to note. Here, we would see the issue as being distinguished by whether we mix or not, whether we introduce random effects or not. Without mixing, as in say customary Bayesian bioassay examples (see, e.g., Gelfand and Kuo, 1991), we would be looking at the DP through Ferguson's original perspective whence the precision parameter really is intended to reflect proximity of the random DP realization to some baseline CDF. Under mixing, interest turns to the random effects which, seen through the stickbreaking perspective, encourage clustering of individual observations into common populations. Furthermore, we can enrich the stickbreaking specification beyond a single precision parameter as we alluded to in generalizations above our expression (2) and also in Section 5.5.

Finally, we thank the discussants for their thoughtful input. Practical use of spatial and spatio-temporal models requires care in implementation and interpretation. Adding a nonparametric aspect certainly exacerbates this requirement.

ADDITIONAL REFERENCES IN THE DISCUSSION

Gamerman, D., Salazar, E. and Reis, E. (2007). Dynamic Gaussian process priors, with applications to the analysis of space-time data. *In this volume.*

Gelfand, A. E. and Kuo, L. (1991) Nonparametric Bayesian bioassay including ordered polytomous response. *Biometrika* **78**, 657–666.

Hjort, N. L. (2003). Topics in nonparametric Bayesian statistics . *Highly Structured Stochastic Systems* (P. Green, N.L. Hjort, S. Richardson, eds.) Oxford: University Press455–478, (with discussion).

Hjort, N. L. and Ongaro, A. (2006). On the distribution of random Dirichlet jumps. *Metron* **64**, 61–92.

Hjort, N. L, Dahl, F. A. and Steinbakk, G. H. (2006). Post-processing posterior predictive p-values. *J. Amer. Statist. Assoc.* **101**, 1157–1174.

MacEachern, S. N. (1999). Dependent nonparametric processes. *ASA Proceedings of the Section on Bayesian Statistical Science.* Alexandria, VA: American Statistical Association, 50–55.

MacEachern, S. N., Kottas, A, and Gelfand, A. E. (2001). Spatial nonparametric Bayesian models *ASA Proceedings of the Joint Statistical Meetings.* Alexandria, VA: American Statistical Association.

Mikosch, T. (2007). Copulas: tales and facts. *Extremes* (to appear, with discussion contributions and a rejoinder).

Rousseau, J. (2007). Approximating interval hypothesis: p-values and Bayes factors. *In this volume.*

BAYESIAN STATISTICS 8, pp. 201–226.
J. M. Bernardo, M. J. Bayarri, J. O. Berger, A. P. Dawid,
D. Heckerman, A. F. M. Smith and M. West (Eds.)

Bayesian Nonparametric Latent Feature Models

ZOUBIN GHAHRAMANI
University of Cambridge, UK
zoubin@eng.cam.ac.uk

THOMAS L. GRIFFITHS
Univ. of California at Berkeley, USA
tom_griffiths@berkeley.edu

PETER SOLLICH
King's College London, UK
peter.sollich@kcl.ac.uk

SUMMARY

We describe a flexible nonparametric approach to latent variable modelling in which the number of latent variables is unbounded. This approach is based on a probability distribution over equivalence classes of binary matrices with a finite number of rows, corresponding to the data points, and an unbounded number of columns, corresponding to the latent variables. Each data point can be associated with a subset of the possible latent variables, which we refer to as the latent *features* of that data point. The binary variables in the matrix indicate which latent feature is possessed by which data point, and there is a potentially infinite array of features. We derive the distribution over unbounded binary matrices by taking the limit of a distribution over $N \times K$ binary matrices as $K \to \infty$. We define a simple generative processes for this distribution which we call the Indian buffet process (IBP; Griffiths and Ghahramani, 2005, 2006) by analogy to the Chinese restaurant process (Aldous, 1985; Pitman, 2002). The IBP has a single hyperparameter which controls both the number of feature per object and the total number of features. We describe a two-parameter generalization of the IBP which has additional flexibility, independently controlling the number of features per object and the total number of features in the matrix. The use of this distribution as a prior in an infinite latent feature model is illustrated, and Markov chain Monte Carlo algorithms for inference are described.

Keywords and Phrases: NON-PARAMETRIC METHODS; MCMC; INDIAN BUFFET PROCESS; LATENT VARIABLE MODELS.

Zoubin Ghahramani is in the Department of Engineering, University of Cambridge, and the Machine Learning Department, Carnegie Mellon University; Thomas L. Griffiths is in the Psychology Department, University of California at Berkeley; Peter Sollich is in the Department of Mathematics, King's College London.

1. INTRODUCTION

Latent or hidden variables are an important component of many statistical models. The role of these latent variables may be to represent properties of the objects or data points being modelled that have not been directly observed, or to represent hidden causes that explain the observed data.

Most models with latent variables assume a finite number of latent variables per object. At the extreme, mixture models can be represented via *a single* discrete latent variable, and hidden Markov models (HMMs) via a single latent variable evolving over time. Factor analysis and independent components analysis (ICA) generally use more than one latent variable per object but this number is usually assumed to be small. The close relationship between latent variable models such as factor analysis, state-space models, finite mixture models, HMMs, and ICA is reviewed in (Roweis and Ghahramani, 1999).

Our goal is to describe a class of latent variable models in which each object is associated with a (potentially unbounded) vector of latent features. Latent feature representations can be found in several widely-used statistical models. In Latent Dirichlet Allocation (LDA; Blei, Ng and Jordan, 2003) each object is associated with a probability distribution over latent features. LDA has proven very successful for modelling the content of documents, where each feature indicates one of the topics that appears in the document. While using a probability distribution over features may be sensible to model the distribution of topics in a document, it introduces a conservation constraint–the more an object expresses one feature, the less it can express others–which may not be appropriate in other contexts. Other latent feature representations include binary vectors with entries indicating the presence or absence of each feature (e.g., Ueda and Saito, 2003), continuous vectors representing objects as points in a latent space (e.g., Jolliffe, 1986), and factorial models, in which each feature takes on one of a discrete set of values (e.g., Zemel and Hinton, 1994; Ghahramani, 1995).

While it may be computationally convenient to define models with a small finite number of latent variables or latent features per object, it may be statistically inappropriate to constrain the number of latent variables a priori. The problem of finding the number of latent variables in a statistical model has often been treated as a model selection problem, choosing the model with the dimensionality that results in the best performance. However, this treatment of the problem assumes that there is a single, finite-dimensional representation that correctly characterizes the properties of the observed objects. This assumption may be unreasonable. For example, when modelling symptoms in medical patients, the latent variables may include not only presence or absence of known diseases but also any number of environmental and genetic factors and potentially unknown diseases which relate to the pattern of symptoms the patient exhibited.

The assumption that the observed objects manifest a sparse subset of an unbounded number of latent classes is often used in nonparametric Bayesian statistics. In particular, this assumption is made in Dirichlet process mixture models, which are used for nonparametric density estimation (Antoniak, 1974; Escobar and West, 1995; Ferguson, 1983; Neal, 2000). Under one interpretation of a Dirichlet process mixture model, each object is assigned to a latent class, and each class is associated with a distribution over observable properties. The prior distribution over assignments of objects to classes is specified in such a way that the number of classes used by the model is bounded only by the number of objects, making Dirichlet process mixture models 'infinite' mixture models (Rasmussen, 2000). Recent work has extended these methods to models in which each object is represented by a distribution over features (Blei, Griffiths, Jordan and Tenenbaum, 2004; Teh, Jordan, Beal and Blei, 2004). However, there are no equivalent methods for dealing with other feature-based representations, be they binary vectors, factorial structures, or vectors of continuous feature values.

In this paper, we take the idea of defining priors over infinite combinatorial structures from nonparametric Bayesian statistics, and use it to develop methods for unsupervised learning in which each object is represented by a sparse subset of an unbounded number of features. These features can be binary, take on multiple discrete values, or have continuous weights. In all of these representations, the difficult problem is deciding which features an object should possess. The set of features possessed by a set of objects can be expressed in the form of a binary matrix, where each row is an object, each column is a feature, and an entry of 1 indicates that a particular objects possesses a particular feature. We thus focus on the problem of defining a distribution on infinite sparse binary matrices. Our derivation of this distribution is analogous to the limiting argument in (Neal 2000; Green and Richardson, 2001) used to derive the Dirichlet process mixture model (Antoniak, 1974; Ferguson, 1983), and the resulting process we obtain is analogous to the *Chinese restaurant process* (CRP; Aldous, 1985; Pitman, 2002). This distribution over infinite binary matrices can be used to specify probabilistic models that represent objects with infinitely many binary features, and can be combined with priors on feature values to produce factorial and continuous representations.

The plan of the paper is as follows. Section 2 discusses the role of a prior on infinite binary matrices in defining infinite latent feature models. Section 3 describes such a prior, corresponding to a stochastic process we call the *Indian buffet process* (IBP). Section 4 describes a two-parameter extension of this model which allows additional flexibility in the structure of the infinite binary matrices. Section 6 illustrates several applications of the IBP prior. Section 7 presents some conclusions.

2. LATENT FEATURE MODELS

Assume we have N objects, represented by an $N \times D$ matrix $\mathbf{X}$, where the ith row of this matrix, $\mathbf{x}_i$, consists of measurements of D observable properties of the ith object. In a latent feature model, each object is represented by a vector of latent feature values $\mathbf{f}_i$, and the properties $\mathbf{x}_i$ are generated from a distribution determined by those latent feature values. Latent feature values can be continuous, as in principal component analysis (PCA; Jolliffe, 1986), or discrete, as in cooperative vector quantization (CVQ; Zemel and Hinton, 1994; Ghahramani, 1995). In the remainder of this Section, we will assume that feature values are continuous. Using the matrix $\mathbf{F} = \left[\mathbf{f}_1^T\, \mathbf{f}_2^T\, \cdots\, \mathbf{f}_N^T\right]^T$ to indicate the latent feature values for all N objects, the model is specified by a prior over features, $p(\mathbf{F})$, and a distribution over observed property matrices conditioned on those features, $p(\mathbf{X}|\mathbf{F})$. As with latent class models, these distributions can be dealt with separately: $p(\mathbf{F})$ specifies the number of features, their probability, and the distribution over values associated with each feature, while $p(\mathbf{X}|\mathbf{F})$ determines how these features relate to the properties of objects. Our focus will be on $p(\mathbf{F})$, showing how such a prior can be defined without placing an upper bound on the number of features.

We can break the matrix $\mathbf{F}$ into two components: a binary matrix $\mathbf{Z}$ indicating which features are possessed by each object, with $z_{ik} = 1$ if object i has feature k

and 0 otherwise, and a second matrix $\mathbf{V}$ indicating the value of each feature for each object. $\mathbf{F}$ can be expressed as the elementwise (Hadamard) product of $\mathbf{Z}$ and $\mathbf{V}$, $\mathbf{F} = \mathbf{Z} \otimes \mathbf{V}$, as illustrated in Fig. 1. In many latent feature models, such as PCA and CVQ, objects have non-zero values on every feature, and every entry of $\mathbf{Z}$ is 1. In *sparse* latent feature models (e.g., sparse PCA; d'Aspremont, Ghaoui, Jordan and Lanckriet, 2004; Jolliffe and Uddin, 2003; Zou, Hastie and Tibshirani, 2004) only a subset of features take on non-zero values for each object, and $\mathbf{Z}$ picks out these subsets. Table 1 shows the set of possible values that latent variables can take in different latent variable models.

Table 1: *Some different latent variable models and the set of values their latent variables can take. DPM: Dirichlet process mixture; FA: factor analysis; HDP: Hierarchical Dirichlet process; IBP: Indian buffet process (described in this paper). Other acronyms are defined in the main text. 'Derivable from IBP' refers to different choices for the distribution of* $\mathbf{V}$.

Latent variable	Finite model ($K < \infty$)	Infinite model ($K = \infty$)
$f_i \in \{1...K\}$	finite mixture model	DPM
$\mathbf{f}_i \in [0,1]^K, \sum_k f_{ik} = 1$	LDA	HDP
$\mathbf{f}_i \in \{0,1\}^K$	factorial models, CVQ	IBP
$\mathbf{f}_i \in \Re^K$	FA, PCA, ICA	derivable from IBP

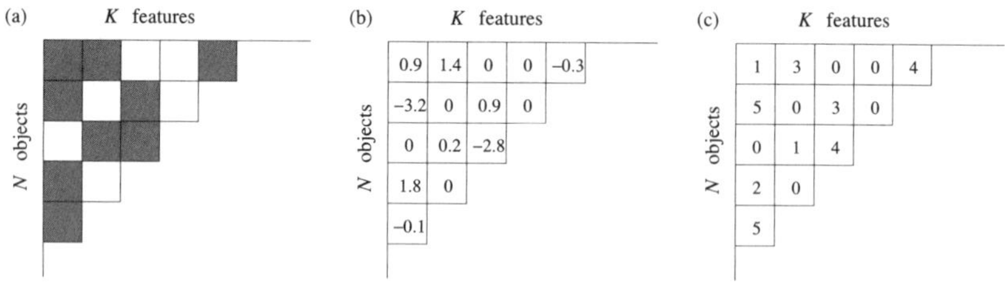

Figure 1: *Feature matrices. A binary matrix* $\mathbf{Z}$*, as shown in (a), can be used as the basis for sparse infinite latent feature models, indicating which features take non-zero values. Elementwise multiplication of* $\mathbf{Z}$ *by a matrix* $\mathbf{V}$ *of continuous values gives a representation like that shown in (b). If* $\mathbf{V}$ *contains discrete values, we obtain a representation like that shown in (c).*

A prior on $\mathbf{F}$ can be defined by specifying priors for $\mathbf{Z}$ and $\mathbf{V}$ separately, with $p(\mathbf{F}) = P(\mathbf{Z})p(\mathbf{V})$. We will focus on defining a prior on $\mathbf{Z}$, since the effective dimensionality of a latent feature model is determined by $\mathbf{Z}$. Assuming that $\mathbf{Z}$ is sparse, we can define a prior for infinite latent feature models by defining a distribution over infinite binary matrices. We have two desiderata for such a distribution: objects should be exchangeable, and inference should be tractable. The literature on nonparametric Bayesian models suggests a method by which these desiderata can be satisfied: start with a model that assumes a finite number of features, and consider the limit as the number of features approaches infinity (Neal, 2000; Green and Richardson, 2001).

3. A DISTRIBUTION ON INFINITE BINARY MATRICES

In this section, we derive a distribution on infinite binary matrices by starting with a simple model that assumes K features, and then taking the limit as $K \to \infty$. The resulting distribution corresponds to a simple generative process, which we term the Indian buffet process.

3.1. *A Finite Feature Model*

We have N objects and K features, and the possession of feature k by object i is indicated by a binary variable z_{ik}. Each object can possess multiple features. The z_{ik} thus form a binary $N \times K$ feature matrix, $\mathbf{Z}$. We will assume that each object possesses feature k with probability π_k, and that the features are generated independently. The probabilities π_k can each take on any value in $[0, 1]$. Under this model, the probability of a matrix $\mathbf{Z}$ given $\pi = \{\pi_1, \pi_2, \ldots, \pi_K\}$, is

$$P(\mathbf{Z}|\pi) = \prod_{k=1}^{K} \prod_{i=1}^{N} P(z_{ik}|\pi_k) = \prod_{k=1}^{K} \pi_k^{m_k}(1-\pi_k)^{N-m_k}, \tag{1}$$

where $m_k = \sum_{i=1}^{N} z_{ik}$ is the number of objects possessing feature k.

We can define a prior on π by assuming that each π_k follows a beta distribution. The beta distribution has parameters r and s, and is conjugate to the binomial. The probability of any π_k under the Beta(r, s) distribution is given by

$$p(\pi_k) = \frac{\pi_k^{r-1}(1-\pi_k)^{s-1}}{B(r,s)}, \tag{2}$$

where $B(r, s)$ is the beta function,

$$B(r,s) = \int_0^1 \pi_k^{r-1}(1-\pi_k)^{s-1}\, d\pi_k = \frac{\Gamma(r)\Gamma(s)}{\Gamma(r+s)}. \tag{3}$$

We take $r = \frac{\alpha}{K}$ and $s = 1$, so equation (3) becomes

$$B(\tfrac{\alpha}{K}, 1) = \frac{\Gamma(\frac{\alpha}{K})}{\Gamma(1+\frac{\alpha}{K})} = \frac{K}{\alpha}, \tag{4}$$

exploiting the recursive definition of the gamma function. The effect of varying s is explored in Section 4.

The probability model we have defined is

$$\begin{aligned} \pi_k \,|\, \alpha &\sim \text{Beta}(\tfrac{\alpha}{K}, 1) \\ z_{ik} \,|\, \pi_k &\sim \text{Bernoulli}(\pi_k). \end{aligned}$$

Each z_{ik} is independent of all other assignments, conditioned on π_k, and the π_k are generated independently. Having defined a prior on π, we can simplify this model by integrating over all values for π rather than representing them explicitly. The marginal probability of a binary matrix $\mathbf{Z}$ is

$$P(\mathbf{Z}) = \prod_{k=1}^{K} \int \left(\prod_{i=1}^{N} P(z_{ik}|\pi_k) \right) p(\pi_k)\, d\pi_k \tag{5}$$

$$= \prod_{k=1}^{K} \frac{B(m_k + \frac{\alpha}{K}, N - m_k + 1)}{B(\frac{\alpha}{K}, 1)} \tag{6}$$

$$= \prod_{k=1}^{K} \frac{\frac{\alpha}{K}\Gamma(m_k + \frac{\alpha}{K})\Gamma(N - m_k + 1)}{\Gamma(N + 1 + \frac{\alpha}{K})}. \tag{7}$$

This result follows from conjugacy between the binomial and beta distributions. This distribution is exchangeable, depending only on the counts m_k.

This model has the important property that the expectation of the number of non-zero entries in the matrix $\mathbf{Z}$,

$$E\left[\mathbf{1}^T \mathbf{Z} \mathbf{1}\right] = E\left[\sum_{ik} z_{ik}\right],$$

has an upper bound for any K. Since each column of $\mathbf{Z}$ is independent, the expectation is K times the expectation of the sum of a single column, $E\left[\mathbf{1}^T \mathbf{z}_k\right]$. This expectation is easily computed,

$$E\left[\mathbf{1}^T \mathbf{z}_k\right] = \sum_{i=1}^{N} E(z_{ik}) = \sum_{i=1}^{N} \int_0^1 \pi_k p(\pi_k)\, d\pi_k = N \frac{\frac{\alpha}{K}}{1 + \frac{\alpha}{K}}, \tag{8}$$

where the result follows from the fact that the expectation of a Beta(r, s) random variable is $r/(r + s)$. Consequently,

$$E\left[\mathbf{1}^T \mathbf{Z} \mathbf{1}\right] = KE\left[\mathbf{1}^T \mathbf{z}_k\right] = \frac{N\alpha}{1 + (\alpha/K)}.$$

For any K, the expectation of the number of entries in $\mathbf{Z}$ is bounded above by $N\alpha$.

3.2. *Equivalence Classes*

In order to find the limit of the distribution specified by equation (7) as $K \to \infty$, we need to define equivalence classes of binary matrices. Our equivalence classes will be defined with respect to a function on binary matrices, lof$(\cdot)$. This function maps binary matrices to *left-ordered* binary matrices. lof$(\mathbf{Z})$ is obtained by ordering the columns of the binary matrix $\mathbf{Z}$ from left to right by the magnitude of the binary number expressed by that column, taking the first row as the most significant bit. The left-ordering of a binary matrix is shown in Fig. 2. In the first row of the left-ordered matrix, the columns for which $z_{1k} = 1$ are grouped at the left. In the second row, the columns for which $z_{2k} = 1$ are grouped at the left of the sets for which $z_{1k} = 1$. This grouping structure persists throughout the matrix.

The *history* of feature k at object i is defined to be $(z_{1k}, \ldots, z_{(i-1)k})$. Where no object is specified, we will use *history* to refer to the full history of feature k, $(z_{1k}, \ldots, z_{Nk})$.

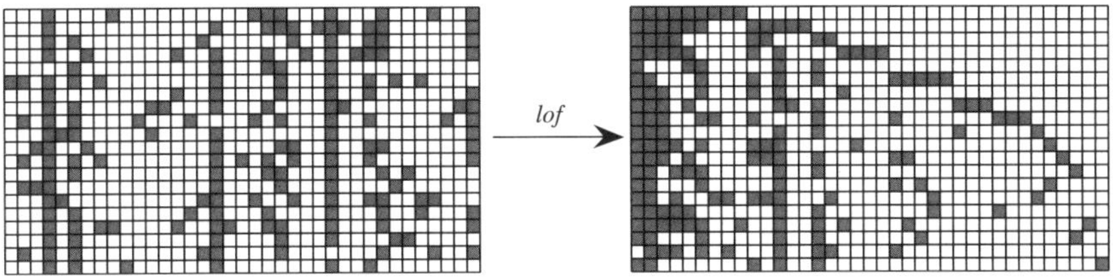

Figure 2: *Binary matrices and the left-ordered form. The binary matrix on the left is transformed into the left-ordered binary matrix on the right by the function lof(·). This left-ordered matrix was generated from the exchangeable Indian buffet process with $\alpha = 10$. Empty columns are omitted from both matrices.*

We will individuate the histories of features using the decimal equivalent of the binary numbers corresponding to the column entries. For example, at object 3, features can have one of four histories: 0, corresponding to a feature with no previous assignments, 1, being a feature for which $z_{2k} = 1$ but $z_{1k} = 0$, 2, being a feature for which $z_{1k} = 1$ but $z_{2k} = 0$, and 3, being a feature possessed by both previous objects. K_h will denote the number of features possessing the history h, with K_0 being the number of features for which $m_k = 0$ and

$$K_+ = \sum_{h=1}^{2^N-1} K_h$$

being the number of features for which $m_k > 0$, so $K = K_0 + K_+$. This method of denoting histories also facilitates the process of placing a binary matrix in left-ordered form, as it is used in the definition of lof(·).

The function lof(·) is many-to-one: many binary matrices reduce to the same left-ordered form, and there is a unique left-ordered form for every binary matrix. We can thus use lof(·) to define a set of equivalence classes. Any two binary matrices $\mathbf{Y}$ and $\mathbf{Z}$ are lof-equivalent if $\text{lof}(\mathbf{Y}) = \text{lof}(\mathbf{Z})$, that is, if $\mathbf{Y}$ and $\mathbf{Z}$ map to the same left-ordered form. The lof-equivalence class of a binary matrix $\mathbf{Z}$, denoted $[\mathbf{Z}]$, is the set of binary matrices that are lof-equivalent to $\mathbf{Z}$. lof-equivalence classes are preserved through permutation of either the rows or the columns of a matrix, provided the same permutations are applied to the other members of the equivalence class. Performing inference at the level of lof-equivalence classes is appropriate in models where feature order is not identifiable, with $p(\mathbf{X}|\mathbf{F})$ being unaffected by the order of the columns of $\mathbf{F}$. Any model in which the probability of $\mathbf{X}$ is specified in terms of a linear function of $\mathbf{F}$, such as PCA or CVQ, has this property.

We need to evaluate the cardinality of $[\mathbf{Z}]$, being the number of matrices that map to the same left-ordered form. The columns of a binary matrix are not guaranteed to be unique: since an object can possess multiple features, it is possible for two features to be possessed by exactly the same set of objects. The number of matrices in $[\mathbf{Z}]$ is reduced if $\mathbf{Z}$ contains identical columns, since some re-orderings of the columns of $\mathbf{Z}$ result in exactly the same matrix. Taking this into account, the cardinality of $[\mathbf{Z}]$ is

$$\binom{K}{K_0 \ldots K_{2^N-1}} = \frac{K!}{\prod_{h=0}^{2^N-1} K_h!},$$

where K_h is the count of the number of columns with full history h.

The binary matrix $\mathbf{Z}$ can be thought of as a generalization of class matrices used in defining mixture models; since each object can only belong to one class, class matrices have the constraint $\sum_k z_{ik} = 1$, whereas the binary matrices in latent feature models do not have this constraint (Griffiths and Ghahramani, 2005).

3.3. *Taking the Infinite Limit*

Under the distribution defined by equation (7), the probability of a particular lof-equivalence class of binary matrices, $[\mathbf{Z}]$, is

$$P([\mathbf{Z}]) \quad = \quad \sum_{\mathbf{Z}\in[\mathbf{Z}]} P(\mathbf{Z}) = \frac{K!}{\prod_{h=0}^{2^N-1} K_h!} \prod_{k=1}^{K} \frac{\frac{\alpha}{K}\Gamma(m_k + \frac{\alpha}{K})\Gamma(N - m_k + 1)}{\Gamma(N+1+\frac{\alpha}{K})}. \tag{9}$$

In order to take the limit of this expression as $K \to \infty$, we will divide the columns of $\mathbf{Z}$ into two subsets, corresponding to the features for which $m_k = 0$ and the features for which $m_k > 0$. Re-ordering the columns such that $m_k > 0$ if $k \leq K_+$, and $m_k = 0$ otherwise, we can break the product in Equation 9 into two parts, corresponding to these two subsets. The product thus becomes

$$\prod_{k=1}^{K} \frac{\frac{\alpha}{K}\Gamma(m_k + \frac{\alpha}{K})\Gamma(N - m_k + 1)}{\Gamma(N+1+\frac{\alpha}{K})}$$

$$= \left(\frac{\frac{\alpha}{K}\Gamma(\frac{\alpha}{K})\Gamma(N+1)}{\Gamma(N+1+\frac{\alpha}{K})}\right)^{K-K_+} \prod_{k=1}^{K_+} \frac{\frac{\alpha}{K}\Gamma(m_k + \frac{\alpha}{K})\Gamma(N - m_k + 1)}{\Gamma(N+1+\frac{\alpha}{K})} \tag{10}$$

$$= \left(\frac{\frac{\alpha}{K}\Gamma(\frac{\alpha}{K})\Gamma(N+1)}{\Gamma(N+1+\frac{\alpha}{K})}\right)^{K} \prod_{k=1}^{K_+} \frac{\Gamma(m_k + \frac{\alpha}{K})\Gamma(N - m_k + 1)}{\Gamma(\frac{\alpha}{K})\Gamma(N+1)} \tag{11}$$

$$= \left(\frac{N!}{\prod_{j=1}^{N}(j+\frac{\alpha}{K})}\right)^{K} \left(\frac{\alpha}{K}\right)^{K_+} \prod_{k=1}^{K_+} \frac{(N-m_k)!\prod_{j=1}^{m_k-1}(j+\frac{\alpha}{K})}{N!}. \tag{12}$$

Substituting equation (12) into equation (9) and rearranging terms, we can compute our limit

$$\lim_{K\to\infty} \frac{\alpha^{K_+}}{\prod_{h=1}^{2^N-1} K_h!} \frac{K!}{K_0!\,K^{K_+}} \left(\frac{N!}{\prod_{j=1}^{N}(j+\frac{\alpha}{K})}\right)^{K}$$

$$\times \prod_{k=1}^{K_+} \frac{(N-m_k)!\prod_{j=1}^{m_k-1}(j+\frac{\alpha}{K})}{N!}$$

$$= \frac{\alpha^{K_+}}{\prod_{h=1}^{2^N-1} K_h!} \exp\{-\alpha H_N\} \prod_{k=1}^{K_+} \frac{(N-m_k)!(m_k-1)!}{N!}, \tag{13}$$

where H_N is the Nth harmonic number, $H_N = \sum_{j=1}^{N} \frac{1}{j}$. The details of the steps taken in computing this limit are given in the appendix of (Griffiths and Ghahramani, 2005). Again, this distribution is exchangeable: neither the number of identical columns nor the column sums are affected by the ordering on objects.

3.4. *The Indian Buffet Process*

The probability distribution defined in equation (13) can be derived from a simple stochastic process. This stochastic process provides an easy way to remember salient properties of the probability distribution and can be used to derive sampling schemes for models based on this distribution. This process assumes an ordering on the objects, generating the matrix sequentially using this ordering. Inspired by the Chinese restaurant process (CRP; Aldous, 1985; Pitman, 2002), we will use a culinary metaphor in defining our stochastic process, appropriately adjusted for geography. Many Indian restaurants in London offer lunchtime buffets with an apparently infinite number of dishes. We can define a distribution over infinite binary matrices by specifying a procedure by which customers (objects) choose dishes (features).

In our Indian buffet process (IBP), N customers enter a restaurant one after another. Each customer encounters a buffet consisting of infinitely many dishes arranged in a line. The first customer starts at the left of the buffet and takes a serving from each dish, stopping after a Poisson(α) number of dishes as his plate becomes overburdened. The ith customer moves along the buffet, sampling dishes in proportion to their popularity, serving himself with probability m_k/i, where m_k is the number of previous customers who have sampled a dish. Having reached the end of all previous sampled dishes, the ith customer then tries a Poisson(α/i) number of new dishes.

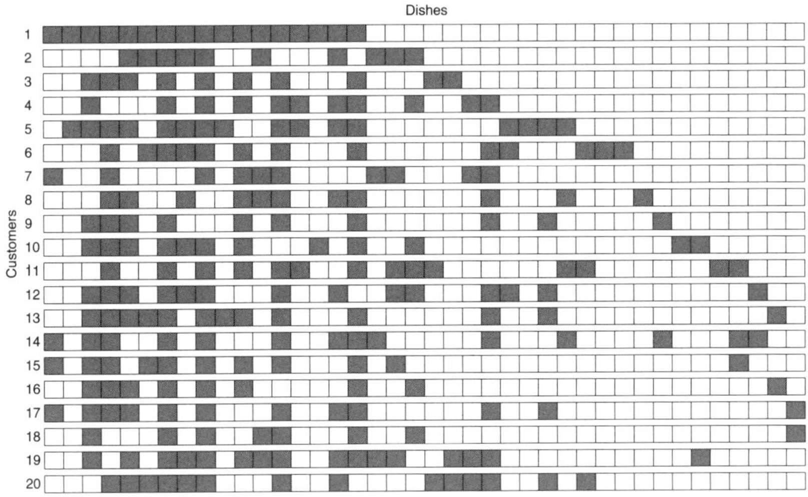

Figure 3: *A binary matrix generated by the Indian buffet process with* $\alpha = 10$.

We can indicate which customers chose which dishes using a binary matrix $\mathbf{Z}$ with N rows and infinitely many columns, where $z_{ik} = 1$ if the ith customer sampled the kth dish. Fig. 3 shows a matrix generated using the IBP with $\alpha = 10$. The first customer tried 17 dishes. The second customer tried seven of those dishes, and then tried three new dishes. The third customer tried three dishes tried by both previous customers, five dishes tried by only the first customer, and two new dishes. Vertically concatenating the choices of the customers produces the binary matrix shown in the figure.

Using $K_1^{(i)}$ to indicate the number of new dishes sampled by the ith customer, the probability of any particular matrix being produced by this process is

$$P(\mathbf{Z}) = \frac{\alpha^{K_+}}{\prod_{i=1}^{N} K_1^{(i)}!} \exp\{-\alpha H_N\} \prod_{k=1}^{K_+} \frac{(N-m_k)!(m_k-1)!}{N!}. \tag{14}$$

As can be seen from Fig. 3, the matrices produced by this process are generally not in left-ordered form. However, these matrices are also not ordered arbitrarily because the Poisson draws always result in choices of new dishes that are to the right of the previously sampled dishes. Customers are not exchangeable under this distribution, as the number of dishes counted as $K_1^{(i)}$ depends upon the order in which the customers make their choices. However, if we only pay attention to the lof-equivalence classes of the matrices generated by this process, we obtain the exchangeable distribution $P([\mathbf{Z}])$ given by equation (13):

$$\frac{\prod_{i=1}^{N} K_1^{(i)}!}{\prod_{h=1}^{2^N-1} K_h!}$$

matrices generated via this process map to the same left-ordered form, and $P([\mathbf{Z}])$ is obtained by multiplying $P(\mathbf{Z})$ from equation (14) by this quantity. It is also possible to define a similar sequential process that directly produces a distribution on left-ordered binary matrices in which customers are exchangeable, but this requires more effort on the part of the customers. We call this the *exchangeable IBP* (Griffiths and Ghahramani, 2005).

3.5. *Some Properties of This Distribution*

These different views of the distribution specified by Equation 13 make it straightforward to derive some of its properties. First, the effective dimension of the model, K_+, follows a Poisson(αH_N) distribution. This is easily shown using the generative process described in previous Section, since under this process K_+ is the sum of Poisson(α), Poisson($\alpha/2$), Poisson($\alpha/3$), etc. The sum of a set of Poisson distributions is a Poisson distribution with parameter equal to the sum of the parameters of its components. Using the definition of the Nth harmonic number, this is αH_N.

A second property of this distribution is that the number of features possessed by each object follows a Poisson(α) distribution. This also follows from the definition of the IBP. The first customer chooses a Poisson(α) number of dishes. By exchangeability, all other customers must also choose a Poisson(α) number of dishes, since we can always specify an ordering on customers which begins with a particular customer.

Finally, it is possible to show that $\mathbf{Z}$ remains sparse as $K \to \infty$. The simplest way to do this is to exploit the previous result: if the number of features possessed by each object follows a Poisson(α) distribution, then the expected number of entries in $\mathbf{Z}$ is $N\alpha$. This is consistent with the quantity obtained by taking the limit of this expectation in the finite model, which is given in Equation 8:

$$\lim_{K\to\infty} E\left[\mathbf{1}^T\mathbf{Z}\mathbf{1}\right] = \lim_{K\to\infty} \frac{N\alpha}{1+\frac{\alpha}{K}} = N\alpha.$$

More generally, we can use the property of sums of Poisson random variables described above to show that $\mathbf{1}^T\mathbf{Z}\mathbf{1}$ will follow a Poisson$(N\alpha)$ distribution. Consequently, the probability of values higher than the mean decreases exponentially.

3.6. *Inference by Gibbs Sampling*

We have defined a distribution over infinite binary matrices that satisfies one of our desiderata – objects (the rows of the matrix) are exchangeable under this distribution. It remains to be shown that inference in infinite latent feature models is tractable, as was the case for infinite mixture models. We will derive a Gibbs sampler for latent feature models in which the exchangeable IBP is used as a prior. The critical quantity needed to define the sampling algorithm is the full conditional distribution

$$P(z_{ik}=1|\mathbf{Z}_{-(ik)},\mathbf{X}) \propto p(\mathbf{X}|\mathbf{Z})P(z_{ik}=1|\mathbf{Z}_{-(ik)}), \tag{15}$$

where $\mathbf{Z}_{-(ik)}$ denotes the entries of $\mathbf{Z}$ other than z_{ik}, and we are leaving aside the issue of the feature values $\mathbf{V}$ for the moment. The likelihood term, $p(\mathbf{X}|\mathbf{Z})$, relates the latent features to the observed data, and will depend on the model chosen for the observed data. The prior on $\mathbf{Z}$ contributes to this probability by specifying $P(z_{ik}=1|\mathbf{Z}_{-(ik)})$.

In the finite model, where $P(\mathbf{Z})$ is given by equation (7), it is straightforward to compute the full conditional distribution for any z_{ik}. Integrating over π_k gives

$$P(z_{ik}=1|\mathbf{z}_{-i,k}) = \int_0^1 P(z_{ik}|\pi_k)p(\pi_k|\mathbf{z}_{-i,k})\,d\pi_k = \frac{m_{-i,k}+\frac{\alpha}{K}}{N+\frac{\alpha}{K}}, \tag{16}$$

where $\mathbf{z}_{-i,k}$ is the set of assignments of other objects, not including i, for feature k, and $m_{-i,k}$ is the number of objects possessing feature k, not including i. We need only condition on $\mathbf{z}_{-i,k}$ rather than $\mathbf{Z}_{-(ik)}$ because the columns of the matrix are generated independently under this prior.

In the infinite case, we can derive the conditional distribution from the exchangeable IBP. Choosing an ordering on objects such that the ith object corresponds to the last customer to visit the buffet, we obtain

$$P(z_{ik}=1|\mathbf{z}_{-i,k}) = \frac{m_{-i,k}}{N}, \tag{17}$$

for any k such that $m_{-i,k} > 0$. The same result can be obtained by taking the limit of equation (16) as $K\to\infty$. Similarly the number of new features associated with object i should be drawn from a Poisson(α/N) distribution.

4. A TWO-PARAMETER EXTENSION

As we saw in the previous section, the distribution on the number of features per object and on the total number of features are directly coupled, through α. This seems an undesirable constraint. We now present a two-parameter generalization of the IBP which lets us tune independently the average number of features each

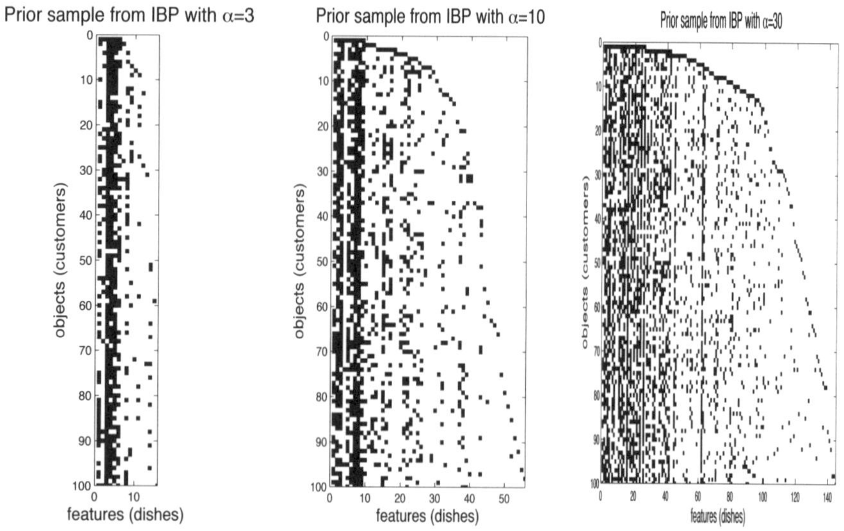

Figure 4: *Draws from the Indian buffet process prior with* $\alpha = 3$ *(left),* $\alpha = 10$ *(middle), and* $\alpha = 30$ *(right).*

object possesses and the overall number of features used in a set of N objects. To understand the need for such a generalization, it is useful to examine some samples drawn from the IBP. Figure 4 shows three draws from the IBP with $\alpha = 3$, $\alpha = 10$, and $\alpha = 30$ respectively. We can see that α controls both the number of latent features per object, and the amount of overlap between these latent features (i.e. the probability that two objects will possess the same feature). It would be desirable to remove this restriction, for example so that it is possible to have many latent features but little variability across objects in the feature vectors.

Keeping the average number of features per object at α as before, we will define a model in which the overall number of features can range from α (extreme stickiness/herding, where all features are shared between all objects) to $N\alpha$ (extreme repulsion/individuality), where no features are shared at all. Clearly neither of these extreme cases is very useful, but in general it will be helpful to have a prior where the overall number of features used can be specified.

The required generalization is simple: one takes $r = (\alpha\beta)/K$ and $s = \beta$ in equation (2). Setting $\beta = 1$ then recovers the one-parameter IBP, but the calculations go through in basically the same way also for other β.

Equation (7), the joint distribution of feature vectors for finite K, becomes

$$P(\mathbf{Z}) = \prod_{k=1}^{K} \frac{B(m_k + \frac{\alpha\beta}{K}, N - m_k + \beta)}{B(\frac{\alpha\beta}{K}, \beta)} \tag{18}$$

$$= \prod_{k=1}^{K} \frac{\Gamma(m_k + \frac{\alpha\beta}{K})\Gamma(N - m_k + \beta)}{\Gamma(N + \frac{\alpha\beta}{K} + \beta)} \frac{\Gamma(\frac{\alpha\beta}{K} + \beta)}{\Gamma(\frac{\alpha\beta}{K})\Gamma(\beta)}. \tag{19}$$

The corresponding probability distribution over equivalence classes in the limit $K \to \infty$ is (compare equation 13):

$$P([\mathbf{Z}]) = \frac{(\alpha\beta)^{K_+}}{\prod_{h\geq 1} K_h!} e^{-K_+} \prod_{k=1}^{K_+} B(m_k, N - m_k + \beta) \qquad (20)$$

with the constant K_+ defined below. As the one-parameter model, this two-parameter model also has a sequential generative process. Again, we will use the Indian buffet analogy. Like before, the first customer starts at the left of the buffet and samples Poisson(α) dishes. The ith customer serves himself from any dish previously sampled by $m_k > 0$ customers with probability $m_k/(\beta + i - 1)$, and in addition from Poisson($\alpha\beta/(\beta + i - 1)$) new dishes. The customer-dish matrix is a sample from this two-parameter IBP. Two other generative processes for this model are described in the Appendix.

The parameter β is introduced above in such a way as to preserve the average number of features per object, α; this result follows from exchangeability and the fact that the first customer samples Poisson(α) dishes. Thus, also the average number of nonzero entries in $\mathbf{Z}$ remains $N\alpha$.

More interesting is the expected value of the overall number of features, i.e. the number K_+ of k with $m_k > 0$. One gets directly from the buffet interpretation, or via any of the other routes, that the expected overall number of features is

$$K_+ = \alpha \sum_{i=1}^{N} \frac{\beta}{\beta + i - 1},$$

and that the distribution of K_+ is Poisson with this mean. We can see from this that the total number of features used increases as β increases, so we can interpret β as the feature *repulsion*, or $1/\beta$ as the feature *stickiness*. In the limit $\beta \to \infty$ (for fixed N), $K_+ \to N\alpha$ as expected from this interpretation. Conversely, for $\beta \to 0$, only the term with $i = 1$ contributes in the sum and so $K_+ \to \alpha$, again as expected: in this limit features are infinitely sticky and all customers sample the same dishes as the first one.

For finite β, it may be verified that the asymptotic behavior of K_+ for large N is $K_+ \sim \alpha\beta \ln N$, because in the relevant terms in the sum one can then approximate $\beta/(\beta+i-1) \approx \beta/i$. If $\beta \gg 1$, on the other hand, the logarithmic regime is preceded by linear growth at small $N < \beta$, during which $K_+ \approx N\alpha$.

We can confirm these intuitions by looking at a few sample matrices drawn from the two-parameter IBP prior. Figure 5 shows three matrices all drawn with $\alpha = 10$, but with $\beta = 0.2$, $\beta = 1$, and $\beta = 5$ respectively. Although all three matrices have roughly the same number of 1s, the number of features used varies considerably. We can see that at small values of β, features are very sticky, and the feature vector variance is low across objects. Conversely, at high values of β there is a high degree of feature repulsion, with the probability of two objects possessing the same feature being low.

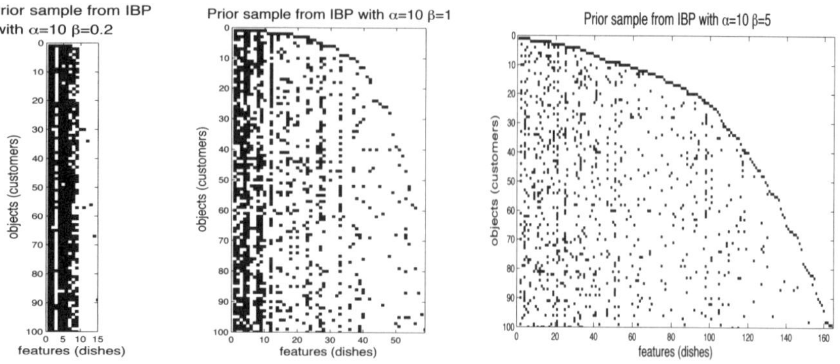

Figure 5: *Draws from the two-parameter Indian buffet process prior with* $\alpha = 10$ *and* $\beta = 0.2$ *(left),* $\beta = 1$ *(middle), and* $\beta = 5$ *(right).*

5. AN ILLUSTRATION

The Indian buffet process can be used as the basis of non-parametric Bayesian models in diverse ways. Different models can be obtained by combining the IBP prior over latent features with different generative distributions for the observed data, $p(\mathbf{X}|\mathbf{Z})$. We illustrate this using a simple model in which real valued data $\mathbf{X}$ is assumed to be linearly generated from the latent features, with Gaussian noise. This linear-Gaussian model can be thought of as a version of factor analysis with binary, instead of Gaussian, latent factors, or as a factorial model (Zemel and Hinton, 1994; Ghahramani 1995) with infinitely many factors.

5.1. *A Linear Gaussian Model*

We motivate the linear-Gaussian IBP model with a toy problem of modelling simple images (Griffiths and Ghahramani, 2005; 2006). In the model, greyscale images are generated by linearly superimposing different visual elements (objects) and adding Gaussian noise. Each image is composed of a vector of real-valued pixel intensities. The model assumes that there are some unknown number of visual elements and that each image is generated by choosing, for each visual element, whether the image possesses this element or not. The binary latent variable z_{ik} indicates whether image i possesses visual element k. The goal of the modelling task is to discover both the identities and the number of visual elements from a set of observed images.

We will start by describing a finite version of the simple linear-Gaussian model with binary latent features used here, and then consider the infinite limit. In the finite model, image i is represented by a D-dimensional vector of pixel intensities, $\mathbf{x}_i$ which is assumed to be generated from a Gaussian distribution with mean $\mathbf{z}_i\mathbf{A}$ and covariance matrix $\mathbf{\Sigma}_X = \sigma_X^2\mathbf{I}$, where $\mathbf{z}_i$ is a K-dimensional binary vector, and $\mathbf{A}$ is a $K \times D$ matrix of weights. In matrix notation, $E[\mathbf{X}] = \mathbf{ZA}$. If $\mathbf{Z}$ is a feature matrix, this is a form of binary factor analysis. The distribution of $\mathbf{X}$ given $\mathbf{Z}$, $\mathbf{A}$, and σ_X is matrix Gaussian:

$$p(\mathbf{X}|\mathbf{Z}, \mathbf{A}, \sigma_X) = \frac{1}{(2\pi\sigma_X^2)^{ND/2}} \exp\{-\frac{1}{2\sigma_X^2} \operatorname{tr}((\mathbf{X} - \mathbf{ZA})^T(\mathbf{X} - \mathbf{ZA}))\} \qquad (21)$$

where tr(·) is the trace of a matrix. We need to define a prior on $\mathbf{A}$, which we also take to be matrix Gaussian:

$$p(\mathbf{A}|\sigma_A) = \frac{1}{(2\pi\sigma_A^2)^{KD/2}} \exp\{-\frac{1}{2\sigma_A^2} \text{tr}(\mathbf{A}^T\mathbf{A})\}, \tag{22}$$

where σ_A is a parameter setting the diffuseness of the prior. This prior is conjugate to the likelihood which makes it possible to integrate out the model parameters $\mathbf{A}$.

Using the approach outlined in Section 3.6, it is possible to derive a Gibbs sampler for this finite model in which the parameters $\mathbf{A}$ remain marginalized out. To extend this to the infinite model with $K \to \infty$, we need to check that $p(\mathbf{X}|\mathbf{Z}, \sigma_X, \sigma_A)$ remains well-defined if $\mathbf{Z}$ has an unbounded number of columns. This is indeed the case (Griffiths and Ghahramani, 2005) and a Gibbs sampler can be defined for this model.

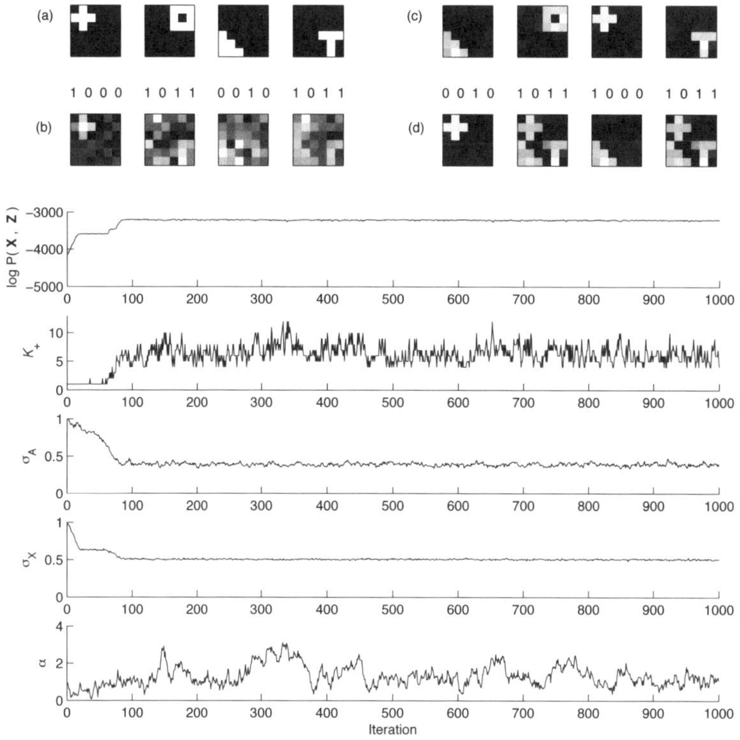

Figure 6: *Stimuli and results for the demonstration of the infinite binary linear-Gaussian model. (a) Image elements corresponding to the four latent features used to generate the data. (b) Sample images from the dataset. (c) Image elements corresponding to the four features possessed by the most objects in the 1000th iteration of MCMC. (d) Reconstructions of the images in (b) using the output of the algorithm. The lower portion of the figure shows trace plots for the MCMC simulation, which are described in more detail in the text.*

We applied the Gibbs sampler for the infinite binary linear-Gaussian model to a simulated dataset, $\mathbf{X}$, consisting of 100 6×6 images. Each image, $\mathbf{x}_i$, was represented as a 36-dimensional vector of pixel intensity values[1]. The images were generated from a representation with four latent features, corresponding to the image elements shown in Fig. 6(a). These image elements correspond to the rows of the matrix $\mathbf{A}$ in the model, specifying the pixel intensity values associated with each binary feature. The non-zero elements of $\mathbf{A}$ were set to 1.0, and are indicated with white pixels in the figure. A feature vector, $\mathbf{z}_i$, for each image was sampled from a distribution under which each feature was present with probability 0.5. Each image was then generated from a Gaussian distribution with mean $\mathbf{z}_i\mathbf{A}$ and covariance $\sigma_X \mathbf{I}$, where $\sigma_X = 0.5$. Some of these images are shown in Fig. 6(b), together with the feature vectors, $\mathbf{z}_i$, that were used to generate them.

The Gibbs sampler was initialized with $K_+ = 1$, choosing the feature assignments for the first column by setting $z_{i1} = 1$ with probability 0.5. σ_A, σ_X, and α were initially set to 1.0, and then sampled by adding Metropolis steps to the MCMC algorithm (see Gilks et al., 1996). Figure 6 shows trace plots for the first 1000 iterations of MCMC for the log joint probability of the data and the latent features, $\log p(\mathbf{X}, \mathbf{Z})$, the number of features used by at least one object, K_+, and the model parameters σ_A, σ_X, and α. The algorithm reached relatively stable values for all of these quantities after approximately 100 iterations, and our remaining analyses will use only samples taken from that point forward.

The latent feature representation discovered by the model was extremely consistent with that used to generate the data (Griffiths and Ghahramani, 2005). The posterior mean of the feature weights, $\mathbf{A}$, given $\mathbf{X}$ and $\mathbf{Z}$ is

$$E[\mathbf{A}|\mathbf{X}, \mathbf{Z}] = (\mathbf{Z}^T\mathbf{Z} + \frac{\sigma_X^2}{\sigma_A^2}\mathbf{I})^{-1}\mathbf{Z}^T\mathbf{X}.$$

Fig. 6(c) shows the posterior mean of $\mathbf{a}_k$ for the four most frequent features in the 1000th sample produced by the algorithm, ordered to match the features shown in Fig. 6(a). These features pick out the image elements used in generating the data. Fig. 6(d) shows the feature vectors $\mathbf{z}_i$ from this sample for the four images in Fig. 6(b), together with the posterior means of the reconstructions of these images for this sample, $E[\mathbf{z}_i\mathbf{A}|\mathbf{X}, \mathbf{Z}]$. Similar reconstructions are obtained by averaging over all values of $\mathbf{Z}$ produced by the Markov chain. The reconstructions provided by the model clearly pick out the relevant features, despite the high level of noise in the original images.

6. APPLICATIONS

We now outline five applications of the IBP, each of which uses the same prior over infinite binary matrices, $P(\mathbf{Z})$, but different choices for the likelihood relating latent features to observed data, $p(\mathbf{X}|\mathbf{Z})$. These applications will hopefully provide an indication for the potential uses of this distribution.

6.1. *A Model for Choice Behavior*

Choice behavior refers to our ability to decide between several options. Models of choice behavior are of interest to psychology, marketing, decision theory, and computer science. Our choices are often governed by features of the different options.

[1] This simple toy example was inspired by the 'shapes problem' in (Ghahramani, 1995); a larger scale example with real images is presented in (Griffiths and Ghahramani, 2006)

For example, when choosing which car to buy, one may be influenced by fuel efficiency, cost, size, make, etc. Görür et al. (2006) present a non-parametric Bayesian model based on the IBP which, given the choice data, infers latent features of the options and the corresponding weights of these features. The IBP is the prior over these latent features, which are assumed to be binary (either present or absent). Their paper also shows how MCMC inference can be extended from the conjugate IBP models to non-conjugate models.

6.2. *A Model for Protein Interaction Screens*

Proteomics aims to understand the functional interactions of proteins, and is a field of growing importance to modern biology and medicine. One of the key concepts in proteomics is a *protein complex*, a group of several interacting proteins. Protein complexes can be experimentally determined by doing high-throughput protein-protein interaction screens. Typically the results of such experiments are subjected to mixture-model based clustering methods. However, a protein can belong to multiple complexes at the same time, making the mixture model assumption invalid. Chu et al. (2006) propose a Bayesian approach based on the IBP for identifying protein complexes and their constituents from interaction screens. The latent binary feature z_{ik} indicates whether protein i belongs to complex k. The likelihood function captures the probability that two proteins will be observed to bind in the interaction screen, as a function of how many complexes they both belong to, $\sum_{k=1}^{\infty} z_{ik} z_{jk}$. The approach automatically infers the number of significant complexes from the data and the results are validated using affinity purification/mass spectrometry experimental data from yeast RNA-processing complexes.

6.3. *A Model for the Structure of Causal Graphs*

Wood et al. (2006) use the infinite latent feature model to learn the structure of directed acyclic probabilistic graphical models. The focus of this paper is on learning the graphical models in which an unknown number of hidden variables (e.g. diseases) are causes for some set of observed variables (e.g. symptoms). Rather than defining a prior over the number of hidden causes, Wood et al. use a non-parametric Bayesian approach based on the IBP to model the structure of graphs with countably infinitely many hidden causes. The binary variable z_{ik} indicates whether hidden variable k has a direct causal influence on observed variable i; in other words whether k is a parent of i in the graph. The performance of MCMC inference is evaluated both on simulated data and on a real medical dataset describing stroke localizations.

6.4. *A Model for Dyadic Data*

Many interesting data sets are *dyadic*: there are two sets of objects or entities and observations are made on pairs with one element from each set. For example, the two sets might consist of movies and viewers, and the observations are ratings given by viewers to movies. The two sets might be genes and biological tissues and the observations may be expression levels for particular genes in different tissues. Models of dyadic data make it possible to predict, for example, the ratings a viewer might give to a movie based on ratings from other viewers, a task known as *collaborative filtering*. A traditional approach to modelling dyadic data is *bi-clustering*: simultaneously cluster both the rows (e.g. viewers) and the columns (e.g. movies) of the observation matrix using coupled mixture models. However, as we have discussed, mixture models provide a very limited latent variable representation of data. Meeds et al. (in press) present a more expressive model of dyadic data based on the infinite latent feature model. In this model, both movies and viewers are represented by binary latent vectors with an unbounded number of elements, corresponding to the features they might possess (e.g. 'likes horror movies'). The two corresponding infinite binary matrices interact via a real-valued weight matrix which links features of movies to features of viewers. Novel MCMC proposals are defined for this model which combine Gibbs, Metropolis, and split-merge steps.

6.5. *Extracting Features from Similarity Judgments*

One of the goals of cognitive psychology is to determine the kinds of representations that underlie people's judgments. In particular, a method called 'additive clustering' has been used to infer people's beliefs about the features of objects from their judgments of the similarity between them (Shepard and Arabie, 1979). Given a square matrix of judgments of the similarity between N objects, where s_{ij} is the similarity between objects i and j, the additive clustering model seeks to recover a $N \times K$ binary feature matrix $\mathbf{F}$ and a vector of K weights associated with those features such that

$$s_{ij} \approx \sum_{k=1}^{K} w_k f_{ik} f_{jk}.$$

A standard problem for this approach is determining the value of K, for which a variety of heuristic methods have been used. Navarro and Griffiths (in press) present a nonparametric Bayesian solution to this problem, using the IBP to define a prior on $\mathbf{F}$ and assuming that s_{ij} has a Gaussian distribution with mean

$$\sum_{k=1}^{K_+} w_k f_{ik} f_{jk}.$$

Using this method provides a posterior distribution over the effective dimension of $\mathbf{F}$, K_+, and gives both a weight and a posterior probability for the presence of each feature. Samples from the posterior distribution over feature matrices reveal some surprisingly rich representations expressed in classic similarity datasets.

7. CONCLUSIONS

We have derived a distribution on infinite binary matrices that can be used as a prior for models in which objects are represented in terms of a set of latent features. While we derived this prior as the infinite limit of a simple distribution on finite binary matrices, we also showed that the same distribution can be specified in terms of a simple stochastic process–the Indian buffet process. This distribution satisfies our two desiderata for a prior for infinite latent feature models: objects are exchangeable, and inference via MCMC remains tractable. We described a two-parameter extension of the Indian buffet process which has the added flexibility of decoupling the number of features per object from the total number of features. This prior on infinite binary matrices has been useful in a diverse set of applications, ranging from causal discovery, to choice modelling, and proteomics.

APPENDIX

The generative process for the one-parameter IBP described in Section 3.4, and the process described in Section 4 for the two-parameter model, do not result in matrices which are in left-ordered form. However, as in the one-parameter IBP (Griffiths and Ghahramani, 2005) an exchangeable version can also be defined for the two-parameter model which produces left-ordered matrices. In the *exchangeable* two-parameter Indian buffet process, the first customer samples a Poisson(α) number of dishes, moving from left to right. The ith customer moves along the buffet, and makes a single decision for each set of dishes with the same history. If there are K_h dishes with history h, under which m_h previous customers have sampled each of those dishes, then the customer samples a Binomial$[m_h/(\beta + i - 1), K_h]$ number of those dishes, starting at the left. Having reached the end of all previously sampled dishes, the ith customer then tries a Poisson$[\alpha\beta/(\beta + i - 1)]$ number of new dishes. Attending to the history of the dishes and always sampling from the left guarantees that the resulting matrix is in left-ordered form, and the resulting distribution over matrices is exchangeable across customers.

As in the one-parameter IBP, the generative process for the two-parameter model also defines a probability distribution directly over the feature histories (cf. Section 4.5 of Griffiths and Ghahramani, 2005). Recall that the history of feature k is the vector $(z_{1k}, \ldots z_{Nk})$, and that for each of the 2^N possible histories h, K_h is the number of features possessing that history. In this two-parameter model, the distribution of K_h (for $h > 0$) is Poisson with mean

$$\alpha\beta\, B(m_h, N - m_h + \beta) = \alpha\beta\,\Gamma(m_h)\Gamma(N - m_h + \beta)/\Gamma(N + \beta).$$

REFERENCES

Aldous, D. (1985). Exchangeability and related topics. *XIII École d'été de probabilités de Saint-Flour.* Lecture Notes in Mathematics **1117**, (A. Dold and B. Eckmann, eds.) Berlin: Springer-Verlag, 1–198.

Antoniak, C. (1974). Mixtures of Dirichlet processes with applications to Bayesian nonparametric problems. *Ann. Statist.* **2**, 1152–1174.

Bernardo, J. M. and Smith, A. F. M. (1994). *Bayesian Theory.* New York: Wiley

Blei, D. M., Griffiths, T. L., Jordan, M. I, Tenenbaum, J. (2004). Hierarchical topic models and the nested Chinese restaurant process. *NIPS: Neural Information Processing Systems* **16**.

Blei, D. M., Ng, A. Y. and Jordan, M. I. (2003). Latent Dirichlet Allocation. *J. Machine Learning Research* **3**, 993–1022.

Chu, W., Ghahramani, Z., Krause, R., and Wild, D. L. (2006). Identifying protein complexes in high-throughput protein interaction screens using an infinite latent feature model *Proc. Biocomputing 2006: Proceedings of the Pacific Symposium.* (Altman *et al.*, eds.) **11**, 231–242.

d'Aspremont, A., El Ghaoui, L. E., Jordan, M. I., Lanckriet, G. R. G. (2005). A Direct formulation for sparse PCA using semidefinite programming. *Advances in Neural Information Processing Systems* **17**, Cambridge: University Press.

Escobar, M. D. and West, M. (1995). Bayesian density estimation and inference using mixtures. *J. Amer. Statist. Assoc.* **90**, 577–588.

Ferguson, T. S. (1983). Bayesian density estimation by mixtures of normal distributions. *Recent advances in statistics* (M. Rizvi, J. Rustagi and D. Siegmund, eds.) New York: Academic Press, 287–302.

Ghahramani, Z. (1995). Factorial learning and the EM algorithm. *Advances in Neural Information Processing Systems* **7** (G. Tesauro, D. S. Touretzky and T. K. Leen, eds.) San Francisco, CA: Morgan Kaufmann. 617–624

Gilks, W., Richardson, S. and Spiegelhalter, D. J. (1996). *Markov Chain Monte Carlo in practice.* London: Chapman and Hall.

Görür, D., Jäkel, F. and Rasmussen, C. (2006). A Choice Model with Infinitely Many Latent Features. *ICML 2006: Proceedings of the 23rd International Conference on Machine Learning*, (W. W. Cohen and A. Moore, eds.) 361–368.

Green P. J. and Richardson S. (2001). Modelling heterogeneity with and without the Dirichlet process. *Scandinavian J. Statist.* **28**, 355–375.

Griffiths, T. L. and Ghahramani, Z. (2005). Infinite Latent Feature Models and the Indian Buffet Process. *Tech. Rep.*, 2005-001, Univ. of Camridge, UK.

Griffiths, T. L. and Ghahramani, Z. (2006). Infinite latent feature models and the indian buffet process *Advances in Neural Information Processing Systems* **18** (Y. Weiss, B. Schölkopf and J. Platt, eds.) Cambridge: University Press, 475–482

Jolliffe, I. T. (1986). *Principal Component Analysis.* New York: Springer-Verlag.

Jolliffe, I. T. and Uddin, M. (2003). A modified principal component technique based on the lasso. *J. Comp. Graphical Statist.* **12**, 531–547.

Meeds, E., Ghahramani, Z., Neal, R. and Roweis, S. T. (2007). Modeling Dyadic Data with Binary Latent Factors. *Advances in Neural Information Processing Systems* **19**. (B. Schölkopf, J. Platt and T. Hoffman eds.) Cambridge: University Press(to appear).

Navarro, D. J. and Griffiths, T. L. (2007). A nonparametric Bayesian method for inferring features from similarity judgments. *Advances in Neural Information Processing Systems*, (B. Schölkopf, J. Platt and T. Hoffman, eds.) Cambridge, MA: MIT Press. (to appear).

Neal, R. M. (2000). Markov chain sampling methods for Dirichlet process mixture models. *J. Comp. Graphical Statist.* **9**, 249–265.

Pitman, J. (2002). Combinatorial stochastic processes. *Tech. Rep.*, 621. Dept. Statistics, U.C. Berkeley, USA.

Rasmussen, C. (2000). The infinite Gaussian mixture model. In *Advances in Neural Information Processing Systems* **12** (S. A. Solla, T. K. Leen and K.-R. Muller, eds.) Cambridge, MA: MIT Press, 554–560.

Roweis, S. T. and Ghahramani, Z. (1999). A unifying review of linear gaussian models. *Neural Computation* **11**, 305–345.

Shepard, R. N. and Arabie, P. (1979). Additive clustering representations of similarities as combinations of discrete overlapping properties. *Psychological Review* **86**, 87–123.

Teh, Y. W., Jordan, M. I., Beal, M. J. and Blei, D. M. (2007) Hierarchical Dirichlet Processes. *J. Amer. Statist. Assoc.* (to appear).

Ueda, N. and Saito, K. (2003). Parametric mixture models for multi-labeled text. *Advances in Neural Information Processing Systems* **15**. (S. Becker, S. Thrun and K. Obermayer, eds.) Cambridge: University Press.

Wood, F., Griffiths, T. L. and Ghahramani, Z. (2006). A non-parametric Bayesian method for inferring hidden causes. *UAI 2006: Proc. 22nd Conference on Uncertainty in Artificial Intelligence.* AUAI Press, 536–543.

Zemel, R. S. and Hinton, G. E. (1994). Developing population codes by minimizing description length. *Advances in Neural Information Processing Systems* **6**. (J. D. Cowan, G. Tesauro and J. Alspector, eds.) San Francisco, CA: Morgan Kaufmann, 3–10.

Zou, H., Hastie, T. and Tibshirani, R. (2007). Sparse principal component analysis. *J. Comp. Graphical Statist.* (to appear).

DISCUSSION

DAVID B. DUNSON (*National Institute of Environmental Health Sciences, USA*)

Ghahramani and colleagues have proposed an interesting class of infinite latent feature (ILF) models. The basic premise of ILF models is that there are infinitely many latent predictors represented in the population, with any particular subject having a finite selection. This is presented as an important advance over models that allow a finite number of latent variables. ILF models are most useful when all but a few of the features are very rare, so that one obtains a *sparse* representation. Otherwise, one cannot realistically hope to learn about the latent feature structure from the available data. The utility of sparse latent factor models has been compellingly illustrated in large p, small n problems by West (2003) and Carvalho et al. (2006). Given that performance is best when the number of latent features represented in the sample is much less than the sample size, it is not clear whether there are practical advantages to the ILF formulation over finite latent variable models that allow uncertainty in the dimension. For example, Lopes and West (2004) and Dunson (2006) allow the number of latent factors to be unknown using Bayesian methods.

That said, it is conceptually appealing to allow additional features to be represented in the data set as additional subjects are added, and it is also appealing to allow partial clustering of subjects. In particular, under an ILF model, subjects can have some features in common, leading to a degree of similarity based on the number of shared features and the values of these features.

Following the notation of Ghahramani et al., the $K \times 1$ latent feature vector for subject i is denoted $\mathbf{f}_i = (f_{i1}, \ldots, f_{iK})'$, with $f_{ik} = z_{ik}v_{ik}$, where $z_{ik} = 1$ if subject i has feature k and $z_{ik} = 0$ otherwise, and v_{ik} is the value of the feature. There are then two important aspects of the specification for an infinite latent feature model: (1) the prior on the $N \times K$ binary matrix $\mathbf{Z} = \{z_{ik}\}$, with $K \to \infty$; and (2) the prior on the $N \times K$ matrix $\mathbf{V} = \{v_{ik}\}$.

The focus of Ghahramani et al. is on the prior for $\mathbf{Z}$, proposing an Indian Buffet Process (IBP) specification. The IBP follows in a straightforward but elegant manner from the following assumptions: (i) the elements of $\mathbf{Z}$ are independent and Bernoulli distributed given π_k, the probability of occurrence of the kth feature; and (ii) $\pi_k \sim \text{beta}(\alpha/K, 1)$. Because the features are treated as exchangeable in this specification, it is necessary to introduce a left ordering function, so that it is possible to base inference on a finite approximation focusing only on the more common features.

In this discussion, I briefly consider the more general problem of nonparametric modeling of both $\mathbf{Z}$ and $\mathbf{V}$, proposing an exponentiated gamma Dirichlet process (EGDP) prior. The exponentiated gamma (EG) is used as an alternative to the IBP, with some advantages, while the Dirichlet process (DP) (Ferguson, 1973; 1974) is used for nonparametric modeling of the feature scores among subjects possessing a feature.

Exponentiated Gamma Dirichlet Process. To provide motivation, I focus on an epidemiologic application in which an ILF model is clearly warranted. In the Agricultural Health Study (Kamel et al., 2005), interest focused on studying factors contributing to neurological symptom (headaches, dizziness, etc) occurrence in farm workers. Individual i is asked through a questionnaire to record the frequency of symptom occurrence for p different symptom types, resulting in the response vector, $\mathbf{y}_i = (y_{i1}, \ldots, y_{ip})'$. It is natural to suppose that the symptom frequencies, $\mathbf{y}_i$, provide measurements of latent features, $\mathbf{f}_i = (f_{i1}, \ldots, f_{iK})'$. Here, $f_{ik} = z_{ik}v_{ik}$, with

$z_{ik} = 1$ if individual i has latent risk factor k and 0 otherwise, while v_{ik} denotes the severity of risk factor k for individual i. For example, feature k may represent the occurrence of an undiagnosed mild stroke, while v_{ik} represents how severe the stroke is, with more severe stroke resulting in more frequent neurological problems.

Such data would not be well characterized with a typical latent class model, which requires individuals to be grouped into a single set of classes. However, the approach of Ghahramani et al. is also not ideal in this case, as there are two important drawbacks. First, the assumption of feature exchangeability makes inferences on the latent features awkward. Thus, across posterior samples collected using an MCMC algorithm, the feature index changes meaning. This label ambiguity also occurs in DPM models. A solution in the setting of ILF models is to choose a prior that explicitly orders the features by their frequency of occurrence, with feature one being the most common. Second, one can potentially characterize the data using fewer features by modeling the feature scores $\{v_{ik}\}$ nonparametrically. This also provides a more realistic characterization of the data. By assuming a parametric model, one artificially inflates the number of features needed to fit the data, making the latent features less likely to characterize a true unobserved risk factor.

An exponentiated gamma Dirichlet process (EGDP) prior can address both of these issues. I first define the exponentiated gamma (EG) component of the prior, which provides a probability model for the random matrix, $\mathbf{Z}$. Without loss of generality, the features are ordered, so that the first trait tends to be more common in the population, and the features decrease stochastically in population frequency with increasing index h. This is accomplished by letting

$$\pi_h = 1 - \exp(-\gamma_h), \quad \gamma_h \stackrel{ind}{\sim} \mathcal{G}(1, \beta_h), \quad \text{for } h = 1, \ldots, \infty, \tag{23}$$

where $\boldsymbol{g} = \{\gamma_h, h = 1, \ldots, \infty\}$ is a stochastically decreasing infinite sequence of independent gamma random variables, with the stochastic decreasing constraint ensured by letting $\beta_1 < \beta_2 < \cdots < \beta_\infty$. Marginalizing over the prior for $\boldsymbol{g}$, we obtain

$$\begin{aligned} \Pr(Z_{ih} = 1 \,|\, \boldsymbol{b}) &= 1 - \int_0^\infty \exp(-\gamma_h)\, \beta_h \exp(-\gamma_h \beta_h)\, d\gamma_h \\ &= \frac{1}{1 + \beta_h}, \end{aligned} \tag{24}$$

which is decreasing in h for increasing $\boldsymbol{b} = \{\beta_h, h = 1, \ldots, \infty\}$.

Note that, unlike for the IBP, the exponentiated gamma (EG) process defined in (23) does not result in a Poisson distribution for $S_i = \sum_{h=1}^\infty Z_{ih}$, the number of traits per subject. Instead S_i is defined as the convolution of independent but not identically distributed Bernoulli random variables. A convenient special case corresponds to

$$\beta_h = \exp\{\psi_1 + \psi_2(h-1)\}, \quad h = 1, 2, \ldots, \infty, \tag{25}$$

which results in a logistic regression model for the frequency of trait occurrence upon marginalizing out $\boldsymbol{g}$. In this case, two hyperparameters, ψ_1 and ψ_2, characterize the EG process, with ψ_1 controlling the frequency of trait one and ψ_2 controlling how rapidly traits decrease in frequency with the index h. The restriction $\psi_2 > 0$ ensures that $\beta_1 < \beta_2 < \cdots < \beta_\infty$. Assuming (23) and (25), it is straightforward to show that the distribution of S_i can be accurately approximated by the distribution of $S_{iT} = \sum_{i=1}^{T} Z_{ih}$ for sufficiently large T. In most applications, a sparse representation with few dominant features (expressed by choosing $\psi \geq 1$) may be preferred. In such cases, an accurate truncation approximation can be produced by replacing $\mathbf{F} = \mathbf{Z} \bigotimes \mathbf{V}$ with $\mathbf{F}_T = \mathbf{Z}_T \bigotimes \mathbf{V}_T$, with $\bigotimes$ denoting the element-wise product, and $\mathbf{A}_T$ denoting the submatrix of $\mathbf{A}$ consisting of the first T columns. Here, T is a finite integer, e.g., $T = 20$ or $T = 50$.

Expressions (23) and (25) provide a prior for the random binary matrix, $\mathbf{Z}$, allocating features to subjects. In order to complete the EGDP specification, we let $v_{ih} = 0$ if $z_{ih} = 0$ and otherwise

$$(v_{ih} \,|\, z_{ih} = 1) \sim G_h, \quad G_h \sim DP(\alpha G_0). \tag{26}$$

Here, G_h represents a random probability measure characterizing the distribution of the hth latent feature score among those individuals with the feature. This probability measure is drawn from a Dirichlet process (DP) with base measure G_0 and precision α.

Nonparametric Latent Factor Models. To illustrate the EGDP, we focus on a nonparametric extension of factor analysis. For subjects $i = 1, \ldots, n$, let $\mathbf{y}_i = (y_{i1}, \ldots, y_{ip})'$ denote a multivariate response vector. Then, a typical factor analytic model can be expressed as:

$$\mathbf{y}_i = \boldsymbol{m} + \boldsymbol{L}\mathbf{f}_i + \boldsymbol{e}_i, \quad \boldsymbol{e}_i \sim N_p(\mathbf{0}, \boldsymbol{S}), \tag{27}$$

where $\boldsymbol{m} = (\mu_1, \ldots, \mu_p)'$ is a mean vector, $\boldsymbol{L}$ is a $p \times K$ factor loadings matrix, $\mathbf{f}_i = (f_{i1}, \ldots, f_{iK})'$ is a $K \times 1$ vector of latent factors, and $\boldsymbol{e}_i$ is a normal residual with diagonal covariance $\boldsymbol{S}$ (see, for example, Lopes and West, 2004). In a parametric specification, one typically assumes $f_{ih} \sim N(0, 1)$, while constraining the factor loadings matrix $\boldsymbol{L}$ to ensure identifiability.

Instead we let $\mathbf{f}_i \sim F$, with $F \sim EGDP(\boldsymbol{\psi}, \alpha, G_0)$, where F denotes the unknown distribution of $\mathbf{f}_i$ and $EGDP(\boldsymbol{\psi}, \alpha, G_0)$ is shorthand notation for the exponentiated gamma Dirichlet process prior with hyperparameters $\boldsymbol{\psi} = (\psi_1, \psi_2)'$, α and G_0. Due to the constraint that the higher numbered factors correspond to rarer features that are less frequent in the population, we avoid the need to constrain $\boldsymbol{L}$. To remove sign ambiguity, we instead restrict G_0 to have strictly positive support, ensuring that $f_{ih} \geq 0$ for all i, h.

Note that this characterization has several appealing properties. First, the distributions of the factor scores are modelled nonparametrically, with subjects automatically clustered into groups separately for each factor. One of these groups corresponds to the cluster of subjects not having the factor, while the others are formed through the discreteness property of the DP. Second, the formulation automatically allows an unknown number of factors represented among the subjects in the sample. Thus, uncertainty in the number of factors is accommodated in a very different manner from Lopes and West (2004). Third, for G_0 chosen to be truncated normal, posterior computation can proceed efficiently via a data augmentation MCMC algorithm. Using a truncation approximation (say with $T = 20$), the algorithm alternately updates: (i) $\boldsymbol{m}, \boldsymbol{L}, \boldsymbol{S}$ conditionally on $\mathbf{F}$ using Gibbs sampling steps; (ii) the elements of $\mathbf{Z}$ by sampling from the Bernoulli full conditional posterior distributions; (iii) $\{\gamma_h, h = 1, \ldots, T\}$ with a data augmentation step (relying on an approach similar to Dunson and Stanford, 2005 and Holmes and Held, 2006); (iv) $\mathbf{V}$ using standard algorithms for DPMs (MacEachern and Müller, 1998). Details are excluded given space considerations.

REPLY TO THE DISCUSSION

We thank Dr. Dunson for a stimulating discussion of our paper. In his discussion, Dunson makes several comments about our paper, and then proposes an alternative approach to sparse latent feature modelling. We first address his comments, and then turn to his suggested approach.

The first comment is that although the utility of sparse latent factor models has been illustrated by West and colleagues, it is not clear whether there are practical advantages to allowing the number of latent factors to be unbounded, as in our approach, as opposed to defining a model with a finite but unknown number of latent factors.

There are two advantages, we believe, one philosophical and one practical. The philosophical advantage is what motivates the use of nonparametric Bayesian methods in the first place: If we don't really believe that the data was actually generated from a finite number of latent factors, then we should not put much or any of our prior mass on such hypotheses. It is hard to think of many real-world generative processes for data in which one can be confident that there are some small number of latent factors. On the practical side, a finite model with an unknown number of latent factors may be preferable to an infinite model if there were significant computational advantages to assuming the finite model. However, inference in finite models of unknown dimension is in fact more computationally demanding, due to the variable dimensionality of the parameter space. Our experience comparing sampling from the infinite model and using Reversible Jump MCMC to sample from an analogous finite but variable-dimension model suggests that the sampler for the infinite model is both easier to implement and faster to mix (Wood *et al.*, 2006).

Dunson also states that for West and colleagues 'performance is best when the number of latent features represented in the sample is much less than the sample size'. However, West's (2003) model is substantially different from ours; it is essentially a linear Gaussian factor analysis model with a sparse prior on the factor loading matrix, while our infinite latent feature models can be used in many different contexts and allow the factors themselves to be sparse. We do not feel that the results that West reports on a particular application and choice of model specification can be generalized to Bayesian inference in all sparse models with latent features.

A second comment is that the assumption of feature exchangeability makes inference in the latent feature space awkward. This is a similar problem to the one suffered by Dirichlet process mixture (DPM) models where feature indices can change across samples in an MCMC run. We agree that questions such as 'what does latent feature k represent' are meaningless in models with exchangeable features. We would never really be interested in such questions. However, there are plenty of meaningful inferences that can be derived from such a model, such as asking how many latent features two data points share. Rather than looking at averages of $\mathbf{Z}$ across MCMC runs, which makes no sense in an model with exchangeable features, one can look at averages of the $N \times N$ matrix $\mathbf{Z}\mathbf{Z}^T$, whose elements measure the number of latent features two data points share. Dunson's proposed solution, a prior that explicitly orders features by their frequency of occurence, is interesting but probably not enough to ensure that meaningful inferences can be made about $\mathbf{Z}$. For example, if two latent features have approximately the same frequency across the data, then any reasonably well-mixing sampler will frequently permute their labels, again muddling inferences about $\mathbf{Z}$ and the parameters associated with the two latent features.

A third comment made by Dunson is that one can define a more flexible model by having a non-parametric model for the features scores v_{ik}, rather than a parametric model. We entirely agree with this last point, and we did not intend to imply that v needs to come from a parametric model. A non-parametric model for v_{ik}, for example based on the Dirichlet process, is potentially very desirable in certain contexts. One possible disadvantage of such a model is that it requires additional bookkeeping and computation in an MCMC implementation. For certain parametric models for v_{ik}, one can analytically integrate out the $\mathbf{V}$ matrix, making the MCMC sampler over other variables mix faster.

We now turn to the proposed exponentiated gamma Dirichlet process (EGDP). This is an interesting model, well worth further study and elaboration.

Our first comment on this model is that the γ_h random variables defined in equation (1) of the discussion are rather unnecessary. Pushing through the transformation of variables, we can compute the distribution on π_h implied by assuming that γ_h follows a particular distribution. In the case of the exponentiated gamma model, this gives $\pi_h \sim \text{Beta}(1, \beta_h)$. This leads us to the question of why this way around and not, e.g., $\text{Beta}(\alpha_h, 1)$? The latter would be a more natural way to generalize our $\pi_h \sim \text{Beta}(\alpha/K, 1)$ to have non-exchangeable latent features. In this proposal, the α_h would get smaller for $h \to \infty$, with the mean frequency for feature h being $\frac{\alpha_h}{\alpha_h+1}$. Writing both models in terms of their Beta distributions over feature frequencies highlights the similarities and differences between the two proposals. The choice $\text{Beta}(\alpha_h, 1)$ provides an alternative method for producing sparseness. Of course one could also look at $\text{Beta}(\alpha_h, \beta, \beta)$, to generalize our two-parameter model.

Making the features inequivalent is attractive in some respects, but on the other hand may reduce flexibility. With exponentially decreasing β's, the higher index features will be so strongly suppressed that they will be hard to 'activate' even with large amounts of data.

For the factor model in equation (5) of the discussion, we disagree that making the f's all positive is necessarily a good thing–one then models data that lie in a (suitably affinely transformed) octant of the space spanned by the columns of $\mathbf{\Lambda}$, rather than in the whole space. This is not merely a method for fixing a sign indeterminacy, but makes quite a different assumption about the data than in an ordinary factor analysis model. This model with positive factors is similar to a large body of work on non-negative matrix factorization models (e.g. Paatero and Tapper, 1994; Lee and Seung 1999).

To summarize, we thank Dr. Dunson for his interesting discussion and we hope that our work, his discussion, and this rejoinder will stimulate further work on sparse latent feature models.

ADDITIONAL REFERENCES IN THE DISCUSSION

Carvalho, C. M., Lucas, J., Wang, Q., Nevins, J. and West, M. (2005). High-dimensional sparse factor models and latent factor regression. *Tech. Rep.*, ISDS, Duke University, USA.

Dunson, D. B. (2006). Efficient Bayesian model averaging in factor analysis. *Tech. Rep.*, ISDS, Duke University, USA.

Dunson, D. B. and Stanford, J. B. (2005). Bayesian inferences on predictors of conception probabilities. *Biometrics* **61**, 126–133.

Ferguson, T. S. (1973). A Bayesian analysis of some nonparametric problems. *Ann. Statist.* **1**, 209–230.

Ferguson, T. S. (1974). Prior distributions on spaces of probability measures. *Ann. Statist.* **2**, 615–629.

Holmes, C. C. and Held, L. (2006). Bayesian auxiliary variable models for binary and multinomial regression. *Bayesian Analysis* **1**, 145–168.

Lee, D. D. and Seung, H. S. (1999). Learning the parts of objects by non-negative matrix factorization. *Nature* 401, 788–791.

Lopes, H. F. and West, M. (2004). Bayesian model assessment in factor analysis. *Statistica Sinica* **14**, 41–67.

MacEachern, S. N. and Müller, P. (1998). Estimating mixture of Dirichlet process models. *J. Comp. Graphical Statist.* **7**, 223–238.

Paatero, P. and U. Tapper (1994). Positive matrix factorization: A non-negative factor model with optimal utilization of error estimates of data values. *Environmetrics* **5**, 111–126.

West, M. (2003). Bayesian factor regression models in the 'large p, small n' paradigm. *Bayesian Statistics 7* (J. M. Bernardo, M. J. Bayarri, J. O. Berger, A. P. Dawid, D. Heckerman, A. F. M. Smith and M. West, eds.) Oxford: University Press, 723–732.

BAYESIAN STATISTICS 8, pp. 227–252.
J. M. Bernardo, M. J. Bayarri, J. O. Berger, A. P. Dawid,
D. Heckerman, A. F. M. Smith and M. West (Eds.)

Objective Bayesian Analysis of Multiple Changepoints for Linear Models

F. JAVIER GIRÓN
Universidad de Málaga, Spain
fj_giron@uma.es

ELÍAS MORENO
Universidad de Granada, Spain
emoreno@ugr.es

GEORGE CASELLA
University of Florida, USA
casella@stat.ufl.edu

SUMMARY

This paper deals with the detection of multiple changepoints for independent but non identically distributed observations, which are assumed to be modeled by a linear regression with normal errors. The problem has a natural formulation as a model selection problem and the main difficulty for computing model posterior probabilities is that neither the reference priors nor any form of empirical Bayes factors based on real training samples can be employed.

We propose an analysis based on the *intrinsic priors*, which do not require real training samples and provide a feasible and sensible solution. For the case of changes in the regression coefficients very simple formulas for the prospective and the retrospective detection of changepoints are found.

On the other hand, when the sample size grows the number of possible changepoints also does and consequently the number of models involved. A stochastic search for finding only those models having large posterior probability is provided. Illustrative examples based on simulated and real data are given.

Keywords and Phrases: BAYES FACTORS; CHANGEPOINTS; INTRINSIC PRIORS; MODEL SELECTION; POSTERIOR MODEL PROBABILITIES; STOCHASTIC SEARCH.

1. INTRODUCTION

There is an extensive literature on the changepoint problem from both the prospective and retrospective viewpoint, for single and multiple changepoints, with parametric and nonparametric sampling models, mainly from a frequentist point of view. For a review see the paper by Lai (1995).

This paper has been supported by MCyT grant SEJ2004–2447 (E. Moreno and F. J. Girón) and NSF grant DMS04–05543 (G. Casella).

The prospective, or on line, changepoint problem consists in the 'sequential' detection of a change in the distribution of a set of time-ordered data. In the retrospective changepoint problem the inference of the change in the distribution of the set of time-ordered data is based on the whole data set. These are two related but different problems. Each of them can be formulated as a model selection problem, and while the latter assumes that multiple changes might occur, the former looks for the first time a change is detected.

Under the Bayesian viewpoint, the retrospective analysis have been considered by many authors, for instance Chernoff and Zack (1964), Bacon and Watts (1971), Ferreira (1975), Smith (1975), Choy and Broemeling (1980), Smith and Cook (1980), Menzefrique (1981), Raftery and Ackman (1986), Carlin *et al.* (1992), Stephens (1994), Kiuchi *et al.* (1995), Moreno, Casella and García-Ferrer (2005), among others. Except for the last one, these papers have in common that the prior distribution of the position of a single changepoint is assumed to be uniformly distributed, and for the parameters of the models before and after the change, conjugate distributions are considered. The hyperparameters are determined either subjectively or using empirical Bayes estimators. Sometimes the values of the hyperparameters are chosen to obtain flat priors. An exception is Raftery and Ackman (1986) and Stephens (1994) where objective improper priors were considered. In the former paper the arbitrary constant involved in the Bayes factor for the improper priors was determined by assigning the value one to the Bayes factor at a given sample point, which is subjectively chosen, as in Spiegelhalter and Smith (1982). In the latter paper the value one was assigned to the constant.

An alternative to these conventional methods is to use intrinsic priors (Berger and Pericchi, 1996; Moreno *et al.*, 1998) which are automatically derived from the structure of the model involved, do not depend on any tuning parameters, and have been proved to behave extremely well in a wide variety of problems, in particular for multiple testing problems involving normal linear models (Casella and Moreno, 2006; Girón *et al.*, 2006b; Moreno and Girón, 2006). We will argue in Section 3 that among the existing objective Bayesian procedures the one based on intrinsic priors seems to be the only one that can be employed for the changepoint problem.

This paper generalizes the paper by Moreno, Casella and García-Ferrer (2005) in two directions. First, multiple changepoints are deemed possible. Second, the observations between two consecutive changepoints are independent but not identically distributed. Here, a normal linear model with deterministic explanatory variables is assumed for the sample observations.

The remainder of the paper is organized as follows. Section 2 formulates the prospective and retrospective changepoint problems. For the normal linear regression model, in Section 3 we discuss the difficulties in assessing prior distributions for all the model parameters involved, propose uniform priors for the number of changepoints, and a conditional uniform prior for the position of these changes, and intrinsic priors for the model parameters. Section 4 focuses on several issues related to the objective analysis of the homoscedastic normal linear model, and its relation with maximum likelihood approach to the problem. Section 5 is devoted to computational issues, Section 6 illustrates the findings on real and simulated data, and Section 7 contains some extensions and concluding remarks.

2. FORMULATION OF THE CHANGEPOINT PROBLEM

Let Y be an observable random variable with sampling distribution $f(y|\theta)$, $\theta \in \Theta$, and $\mathbf{y}^t = (y_1, \ldots, y_n)$ a vector of sequential observations. A single changepoint in the sample means that there is a position r in the sample, $1 \leq r \leq n-1$, such that the first r observations come from $f(x|\theta_1)$ and the rest of the observations from $f(x|\theta_2)$, where $\theta_1 \neq \theta_2$. The extension of this definition to more than a single changepoint, *i.e.*, multiple changepoints, is straightforward.

Later on this sampling distribution will be taken as a multivariate normal distribution whose mean may depend on some deterministic covariates, one of which might be a time variable.

2.1. *Retrospective Formulation*

We approach the retrospective multiple changepoints problem as one of model selection in which for a fixed sample size n we admit, in the first instance, that all possible changepoints configurations might occur and want to order the associated models with respect to a given criterion. In section 3 we will discuss the case of ruling out some models *a priori*, for instance, those containing two or more successive changepoints. In our approach, the Bayesian criterion will be dictated by $0-1$ loss functions meaning that the models be compared according to their posterior probabilities.

It is convenient to think of the models involved as a hierarchy: first conditionally on a given number of changepoints, and then to allow the number of changepoints to vary.

Let p, $1 \leq p \leq n-1$, denote the number of changepoints in the sample, $\mathbf{r}_p = (r_1, \ldots, r_p)$ the positions at which the changes occur, and $S_{\mathbf{r}_p} = (\mathbf{y}_1^t, \ldots \mathbf{y}_{p+1}^t)$ the partition of the vector of observations $\mathbf{y}$ such that the order of appearance of the observations in the vector is preserved. Thus, this generic partition is given by

$$S_{\mathbf{r}_p} = (\mathbf{y}_1^t, \ldots \mathbf{y}_{p+1}^t) = \{(y_1, \ldots, y_{r_1}), (y_{r_1+1}, \ldots, y_{r_2}), \ldots, (y_{r_p+1}, \ldots, y_n)\}.$$

The sampling distribution for the partition $S_{\mathbf{r}_p}$ is

$$f(\mathbf{y}|\theta_{p+1}, \mathbf{r}_p, p) = \prod_{i=1}^{r_1} f(y_i|\theta_1) \prod_{i=r_1+1}^{r_2} f(y_i|\theta_2) \times \cdots \times \prod_{i=r_p+1}^{n} f(y_i|\theta_{p+1}),$$

where the number of changepoints p, their position $\mathbf{r}_p = (r_1, \ldots, r_p)$, and the associated parameters $\boldsymbol{\theta}_{p+1} = (\theta_1, \ldots \theta_{p+1})$ are unknown quantities. The integer vector variable $\mathbf{r}_p$ belongs to the set $\mathfrak{N}^p = \{(r_1, \ldots, r_p) : 1 \leq r_1 < r_2 < \ldots < r_p \leq n-1\}$. and the parameter $\boldsymbol{\theta}_{p+1}$ belongs to the set $\boldsymbol{\theta}_{p+1} = \Theta_1 \times \ldots \times \Theta_{p+1}$. The singular case of $p = 0$ corresponds to the case of no change; in this case we set $\mathbf{r}_0 = n$, and the corresponding partition is $S_0 = \{\mathbf{y}\}$. The sampling distribution in this case is given by

$$f(\mathbf{y}|\theta, n, 0) = \prod_{i=1}^{n} f(y_i|\theta_0).$$

In the changepoint problem, we are primarily interested in making inferences on p and $\mathbf{r}_p$ but inferences on $\boldsymbol{\theta}_{p+1}$, or some functions of θ_{p+1}, may also be of interest.

In what follows we suppress the subscripts of the $\boldsymbol{\theta}$'s and $\mathbf{r}$'s when there is no possibility of confusion due to their dependence on p.

All the possible sampling models can be classified into boxes $\{\mathfrak{B}_0, \ldots, \mathfrak{B}_{n-1}\}$, where box $\mathfrak{B}_0$ contains the model with no change, $\mathfrak{B}_1$ contains the models with one changepoint, and so on.

Assuming a prior distribution for the parameters $(\boldsymbol{\theta}, \mathbf{r}, p)$ such that $(\boldsymbol{\theta}, \mathbf{r}, p) \in \boldsymbol{\Theta}^{p+1} \times \mathfrak{N}^p \times \{0, 1, \ldots, n-1\}$, say $\pi(\boldsymbol{\theta}, \mathbf{r}, p)$ generally given by the hierarchical decomposition $\pi(\boldsymbol{\theta}|\mathbf{r})\pi(\mathbf{r}|p)\pi(p)$, we have a general Bayesian model for the changepoint problem. Note that referring to the parameter p is equivalent to referring to the generic box $\mathfrak{B}_p$ which contains all sampling models with exactly p changepoints, so that the prior probability $\pi(p)$ assigned to the occurrence of p changepoints in the sample, is equivalently the prior probability assigned to the box $\mathfrak{B}_p$.

The main interest in our setting is making inferences on three quantities, first on the number of changepoints p, second on the configuration $\mathbf{r}$ conditionally on p and, in the third place, on the configuration $\mathbf{r}$ on the whole set of models comprising the set of all boxes $\mathfrak{B} = \mathfrak{B}_0 \cup \mathfrak{B}_1 \cup \ldots \cup \mathfrak{B}_{n-1}$. Thus, all we need is to compute $\pi(p|\mathbf{y})$, $\pi(\mathbf{r}|\mathbf{y}, p)$ and $\pi(\mathbf{r}|\mathbf{y})$.

To single out the model with changepoints at vector $\mathbf{r}$ it may be convenient to refer to it as model $M_{\mathbf{r}}$. Note that every changepoint model $M_{\mathbf{r}}$ is contained in one and only one box $\mathfrak{B}_p$; therefore, $\pi(\mathbf{r}|p) = 0$ if $\mathbf{r} \notin \mathfrak{N}^p$.

Let $m(\mathbf{y}|M_{\mathbf{r}})$ denote the marginal of the data $\mathbf{y}$ given model $M_{\mathbf{r}}$, that is

$$m(\mathbf{y}|M_{\mathbf{r}}) = m(\mathbf{y}|\mathbf{r}) = \int f(\mathbf{y}|\theta, \mathbf{r}, p)\pi(\theta|\mathbf{r})\, d\theta,$$

and $m(\mathbf{y}|M_0)$ the marginal under the no change model M_0, that is

$$m(\mathbf{y}|M_0) = \int f(\mathbf{y}|\theta_0)\pi(\theta_0, n, 0)\, d\theta_0.$$

If $B_{\mathbf{r}n} = m(\mathbf{y}|M_{\mathbf{r}})/m(\mathbf{y}|M_0)$ denotes the Bayes factor for comparing model $M_{\mathbf{r}}$ against M_0, straightforward probability calculations render the required posterior probabilities in terms of the Bayes factor –as will prove more convenient in the sequel, instead of the marginal $m(\mathbf{y}|M_{\mathbf{r}})$–, as follows

$$\pi(p|\mathbf{y}) = \frac{\pi(p)\sum_{\mathbf{s}\in\mathfrak{N}^p} \pi(\mathbf{s}|p)B_{\mathbf{s}n}(\mathbf{y})}{\sum_{q=0}^{n-1}\pi(q)\sum_{\mathbf{s}\in\mathfrak{N}^q}\pi(\mathbf{s}|q)B_{\mathbf{s}n}(\mathbf{y})}, \text{ for } p \in \{0, 1, \ldots, n-1\}, \tag{1}$$

$$P(M_{\mathbf{r}}|\mathbf{y}, p) = \frac{\pi(\mathbf{r}|p)B_{\mathbf{r}n}(\mathbf{y})}{\sum_{\mathbf{s}\in\mathfrak{N}^q}\pi(\mathbf{s}|q)B_{\mathbf{s}n}(\mathbf{y})}, \text{ for } M_{\mathbf{r}} \in \mathfrak{B}_p, \tag{2}$$

$$P(M_{\mathbf{r}}|\mathbf{y}) = \frac{\pi(p)\pi(\mathbf{r}|p)B_{\mathbf{r}n}(\mathbf{y})}{\sum_{q=0}^{n-1}\pi(q)\sum_{\mathbf{s}\in\mathfrak{N}^q}\pi(\mathbf{s}|q)B_{\mathbf{s}n}(\mathbf{y})}, \text{ for } M_{\mathbf{r}} \in \mathfrak{B}, \tag{3}$$

where, by convention, the Bayes factor for comparing the no chage model M_0 with itself is $B_{0n} = 1$.

As said above, when using 0–1 loss functions, the optimal decision on the discrete parameters p and $\mathbf{r}$ is to choose the model having the highest posterior probability of their corresponding distributions.

If inferences on the parameters θ or functions of them are required –in this case neccessarilly conditional on p–, they are made from the following posterior distribution

$$\pi(\theta_{p+1}|\mathbf{y}, p) = \sum_{\mathbf{s}\in\mathfrak{N}^p} \pi(\theta_{p+1}|\mathbf{y}, \mathbf{s}, p)\pi(\mathbf{s}|\mathbf{y}, p). \tag{4}$$

2.2. *Prospective Formulation*

The prospective formulation consists in detecting the first time n in which we choose the box $\mathfrak{B}_1$ against the box $\mathfrak{B}_0$, thus indicating that an unexpected change or anomaly might occur in a neighborhood of n. In quality control, it is understood that in that case we must stop the experiment and make a decision.

2.2.1. *Stopping Rules*

For a fixed sample size n, the Bayesian models in box $\mathfrak{B}_1$ are given by

$$M_{r_1} : \{f(\mathbf{y}|\theta_1, \theta_2, r_1, 1), \pi(\theta_1, \theta_2|r_1)\pi(r_1)\}, \text{ for } r_1 = 1, \ldots, n-1,$$

and the Bayesian model for the no change model in box $\mathfrak{B}_0$ is given by

$$M_0 : \{f(y|\theta_0, n, 0), \pi(\theta_0)\}.$$

From expression (1) for $p = 0$ and $p = 1$, conditional on $p \le 1$, and assuming that $\pi(p = 0) = \pi(p = 1) = 1/2$, we have

$$\pi(p = 0|\mathbf{y}) = \frac{1}{1 + \sum_{r_1=1}^{n-1} \pi(r_1) B_{r_1 n}(\mathbf{y})}, \quad \pi(p = 1|\mathbf{y}) = \frac{\sum_{r_1}^{n-1} \pi(r_1) B_{r_1 n}(\mathbf{y})}{1 + \sum_{r_1=1}^{n-1} \pi(r_1) B_{r_1 n}(\mathbf{y})},$$

so that box $\mathfrak{B}_1$ is to be chosen if $\pi(p = 1|\mathbf{y}) > \pi(p = 0|\mathbf{y})$ for the corresponding sample size n. Therefore, the Bayesian stopping rule is to stop at time N given by

$$N = \inf \Big\{ n : \sum_{r_1=1}^{n-1} \pi(r_1) B_{r_1 n}(\mathbf{y}) > 1 \Big\}. \tag{5}$$

Note that the general stopping rule (5) depends on a simple statistic which is a weighted sum of Bayes factors, where the weights are the prior probabilities of the models in box $\mathfrak{B}_1$. Hence, there remains the problem of eliciting these prior probabilities. If we choose a uniform distribution for r_1, that is $\pi(r_1) = 1/(n-1)$, the stopping rule becomes

$$N^U = \inf \Big\{ n : \frac{1}{n-1} \sum_{r_1=1}^{n-1} B_{r_1 n}(\mathbf{y}) > 1 \Big\}.$$

Other choices for $\pi(r_1)$ are deemed possible; for example, for one-step ahead on line detection, if we set $\pi(r_1) = 0$ for $r_1 = 1, \ldots, n-2$ and $\pi(n-1) = 1$, then the stopping rule is

$$N^{osa} = \inf \{ n : B_{(n-1)n}(\mathbf{y}) > 1 \}.$$

This rule might be of interest for anticipating the first instance whether either an outlying observation or a possible changepoint, though it can not discriminate between both possibilities; therefore the need to develop a more comprehensive strategy to on-line detection, which we now describe.

2.2.2. *Monitoring*

For on line detection of the first true changepoint –one which lasts more than one or a small number of observation, usually larger than the dimension, say k, of the parameters θ_j– a Bayesian monitoring procedure would be required which is not only more informative than a stopping rule but it is capable of discerning on line between transient changepoints, usually outlying observation, and a permanent changepoint.

From our perspective, monitoring can be accomplished by applying the retrospective procedure sequentially to the available data at every time instant n. This produces an array of sequences of either Bayes factors $B_{r_1 n}$ for all possible values of $r_1 = 0, \ldots, n$ or, what is equivalent but much more convenient due to the use of a common probability scale, the posterior probabilities $P(M_{r_1}|\mathbf{y})$ given by (2) conditional on $p \leq 1$. Therefore, the array we compute is

$$\begin{array}{llll}
P(M_0|y_1) & & & \\
P(M_0|y_1,y_2) & P(M_1|y_1,y_2) & & \\
\cdots & \cdots & \cdots & \cdots \\
P(M_0|y_1,\ldots,y_n) & P(M_1|y_1,\ldots,y_n) & \ldots & P(M_{n-1}|y_1,\ldots,y_n) \\
\cdots & \cdots & \cdots & \cdots
\end{array}$$

This monitoring procedure, when applied to real and simulated data, provides a very satisfactory solution to the problem of detecting the first changepoint because it is able to discriminate between isolated changepoints –single or patches of outlying observations– and the first true permanent changepoint. Note that for starting the monitoring process, the minimum sample size required n has to be larger than the dimension of the parameter space.

3. OBJECTIVE BAYESIAN METHODS

Once a sampling model is established, the main difficulty for both the prospective and retrospective detection of a changepoint is to assess the prior distributions for all the parameters in the models. Ideally, we would like that our subjective degree of belief on the parameters were sufficient to define the prior distributions for the problem. However, in real applications there are many parameters, as in the linear case, and to subjectively elicit the prior distributions is a very hard and demanding task. Thus, we have to rely on automatic or objective methods for deriving priors for the analysis.

3.1. *Objective Priors for the Discrete Parameters*

On the discrete parameters p and $\mathbf{r}$ uniform priors are the common choice. We have, in principle, two ways of assessing a uniform prior on these parameters: first, a uniform prior on the set of all possible model $\mathfrak{B}$, meaning that $\pi(\mathbf{r}, p) = 1/2^{n-1}$ for $\mathbf{r} \in \cup_p \mathfrak{N}^p$; and, second, using the hierarchical nature of the prior $\pi(\mathbf{r}, p) = \pi(\mathbf{r}|p)\pi(p)$, assigning first a uniform prior on the set $\{0, 1, \ldots, n-1\}$, and then a uniform prior on each box $\mathfrak{B}_p$, that is

$$\pi(\mathbf{r}, p) = \frac{1}{n} \frac{1}{\binom{n-1}{p}} = \frac{p!(n-p-1)!}{n!}, \text{ if } \mathbf{r} \in \mathfrak{N}^p. \tag{6}$$

This second choice automatically takes into account the fact that boxes $\mathfrak{B}_p$ contains different number of models –note that the number of models in box $\mathfrak{B}_p$ is $\binom{n-1}{p}$ which also depends on the sample size n.

On the other hand, if a uniform prior on the set of all models were used as in the first choice, the marginal of p is a Binomial distribution with parameters $n-1$ and $1/2$ instead of a discrete uniform. This means that, a priori, models with either a small or a large number of changepoints have a very small probability when compared with models with a number of changepoints of about $n/2$; and, this situation worsens when the sample size n increases. As expected, the use of this prior in simulated and real data produces paradoxical results, while the second prior produces very sensible results. Consequently, the first prior is ruled out from both theoretical and practical reasons.

Using the prior given by (6), formulas (1), (2) and (3), respectively, simplify to

$$\pi(p|\mathbf{y}) = \frac{p!(n-1-p)! \sum_{\mathbf{s}\in\mathfrak{N}^p} B_{\mathbf{s}n}(\mathbf{y})}{\sum_{q=0}^{n-1} q!(n-1-q)! \sum_{\mathbf{r}\in\mathfrak{N}^q} B_{\mathbf{r}n}(\mathbf{y})}, \text{ for } p = 0, 1, \ldots, n-1, \tag{7}$$

$$P(M_{\mathbf{r}}|\mathbf{y}, p) = \frac{B_{\mathbf{r}n}(\mathbf{y})}{\sum_{\mathbf{r}\in\mathfrak{N}^p} B_{\mathbf{r}n}(\mathbf{y})}, \text{ for } \mathbf{r} \in \mathfrak{N}^p, \tag{8}$$

$$P(M_{\mathbf{r}}|\mathbf{y}) = \frac{p!(n-1-p)! B_{\mathbf{r}n}(\mathbf{y})}{\sum_{q=0}^{n-1} q!(n-1-q)! \sum_{\mathbf{r}\in\mathfrak{N}^q} B_{\mathbf{r}n}(\mathbf{y})}, \text{ for } \mathbf{r} \in \cup_p \mathfrak{N}^p. \tag{9}$$

So far, all possible configurations $\mathbf{r}$ have been given *a priori* a positive probability. However, in practice, a consecutive pair of changepoints, which implies a partition of the data containing a single observation, should not be accepted as a *true* changepoint for this would correspond to an abrupt change caused by a single outlier between two adjacent observations.

Further, for estimating the k-dimensional parameter θ_j of the corresponding partition $\mathbf{y}_j$ we need at least a sample size larger than or equal to the dimension of θ_j. Hence, reducing the number of configurations by taking into account the preceding restrictions seems realistic in most problems. Therefore, the prior on $\mathbf{r}$ and p should now depend on a certain set of restrictions, say $\mathfrak{R}_k$, on the space of all models $\cup_p \mathfrak{N}^p$. Instead of working with the expression of the prior conditional on $\mathfrak{R}_k$ which, in some circumstances, may be difficult to specify, a much better strategy (see Box and Tiao 1992, pp. 67–69) is to restrict the posterior in the space of all models $P(M_{\mathbf{r}}|\mathbf{y})$ conditioning to the set $(\cup_p \mathfrak{N}^p) \cap \mathfrak{R}_k$, *i.e.*, to consider

$$P(M_{\mathbf{r}}|\mathbf{y}, \mathfrak{R}_k) \propto p!(n-1-p)! B_{\mathbf{r}n}(\mathbf{y}) \quad \text{for } \mathbf{r} \in (\cup_p \mathfrak{N}^p) \cap \mathfrak{R}_k.$$

Note that, for example, if we only consider those configurations $\mathbf{r}$ that satisfy the restriction $r_j - r_{j-1} > k$ for all j, then it is easy to show, using a simple combinatorial calculation, that for all $p > (n+k-1)/(k+1)$ the sets $\mathfrak{N}^p \cap \mathfrak{R}_k$ are empty, and the remaining ones have fewer models, except boxes $\mathfrak{B}_0$ and $\mathfrak{B}_1$.

3.2. *Objective Priors for the Continuous Parameters*

For the conditional distribution of $\theta|\mathbf{r}$ either conjugate priors or vague priors, usually a limit of conjugate priors with respect to some of the hyperparameters, are typically used. A difficulty is that conjugate prior distributions depend on many hyperparameters which need be assessed, and vague priors hide the fact that they are generally improper and consequently model posterior probabilities are not well-defined. Thus, it seems to us that, for the parameters θ of the conditional distribution $\theta|\mathbf{r}$, objective Bayesian priors might be appropriate here. Unfortunately, the objective reference priors for normal linear models are also improper.

We also note that to replace $B_{\mathbf{r}n}(\mathbf{y})$ with an empirical Bayes factor based on real training samples –for instance some sort of intrinsic Bayes factors–, is ruled out since a changepoint may occur before we have a training sample of minimal size.

To compute the Bayes factor $B_{\mathbf{r}n}(\mathbf{y})$ we propose to use as priors for θ_0 and $\boldsymbol{\theta}|\mathbf{r}$ the intrinsic priors $(\pi^N(\theta_0), \pi^I(\boldsymbol{\theta}|\mathbf{r}))$ derived from the improper reference priors $\pi^N(\theta_0)$ and $\pi^N(\boldsymbol{\theta}|\mathbf{r})$. Intrinsic priors do not use real training samples but theoretical ones and hence the difficulty due to the absence of real training samples disappear.

Further, they are completely automatic and hence there is no need to adjust any hyperparameter. Moreover, the no changepoint model M_0 is nested into any other $M_{\mathbf{r}}$, so that intrinsic priors for comparing a model with changepoints at position $\mathbf{r}$ with the no changepoint model do exist. The formal expression of the intrinsic prior for the parameters of the distribution $\boldsymbol{\theta}|\mathbf{r}$ conditional on an arbitrary but fixed point of the parameter of the no changepoint model θ_0, say $\pi^I(\boldsymbol{\theta}|\theta_0,\mathbf{r})$, is given by

$$\pi^I(\boldsymbol{\theta}|\theta_0,\mathbf{r}) = \pi^N(\theta|\mathbf{r})\, E_{\mathbf{Y}(\ell)|\theta,\mathbf{r}} \frac{f(\mathbf{Y}(\ell)|\theta_0,n)}{\int f(\mathbf{Y}(\ell)|\theta,\mathbf{r})\pi^N(\boldsymbol{\theta}|\mathbf{r})d\boldsymbol{\theta}},$$

where the expectation is taken with respect to $f(\mathbf{Y}(\ell)|\theta,\mathbf{r})$, ℓ being the minimal training sample size such $0 < \int f(\mathbf{Y}(\ell)|\theta,\mathbf{r})\pi^N(\boldsymbol{\theta}|\mathbf{r})d\,\theta < \infty$, (Berger and Pericchi 1996).

This conditional intrinsic prior is a probability density, and the unconditional intrinsic prior for $\boldsymbol{\theta}|\mathbf{r}$ is given by

$$\pi^I(\boldsymbol{\theta}|\mathbf{r}) = \int \pi^I(\boldsymbol{\theta}|\theta_0,\mathbf{r})\pi^N(\theta_0)\,d\theta_0,$$

which is an improper prior if the mixing distribution $\pi^N(\theta_0)$ is also improper. However, the Bayes factor for intrinsic priors is a well-defined Bayes factor (Moreno *et al.*, 1998).

4. THE HOMOSCEDASTIC NORMAL LINEAR MODEL

Suppose that $(y_1,\ldots,y_n)$ follows the normal linear model

$$\mathbf{y} = \mathbf{X}\boldsymbol{\beta} + \boldsymbol{\epsilon}, \qquad \boldsymbol{\epsilon} \sim N_n(\boldsymbol{\epsilon}|\mathbf{0},\tau^2\mathbf{I}_n),$$

where $\mathbf{X}$ is a $n\times k$ design matrix of full rank, $\boldsymbol{\beta}$ a $k\times 1$ vector of regression coefficients, and τ^2 is the common variance of the error terms. This model corresponds to the situation of no changepoint in the sample.

We assume that the variance error does not change across the sample so that the changes only affect to the regression coefficients.

4.1. *The Case of One Changepoint*

For clarity of exposition we consider first the case where there is only one changepoint at some unknown position r_1. Let $S_1 = (\mathbf{y}_1^t, \mathbf{y}_2^t)$ be a partition of $\mathbf{y}$ where the dimension of $\mathbf{y}_1$ is $n_1 = r_1$, the dimension of $\mathbf{y}_2$ is $n_2 = n - n_1$. We also split the design matrix $\mathbf{X}$ as

$$\mathbf{X} = \begin{pmatrix} \mathbf{X}_1 \\ \mathbf{X}_2 \end{pmatrix},$$

where $\mathbf{X}_1$ has dimensions $n_1 \times k$ and $\mathbf{X}_2$ has $n_2 \times k$. In the notation of the preceding sections we now have

$$f(y|\theta_0,n,0) = N_n(\mathbf{y}|\mathbf{X}\boldsymbol{\beta},\tau^2\mathbf{I}_n),$$

and

$$f(\mathbf{y}|\theta,r_1,1) = N_{n_1}(\mathbf{y}_1|\mathbf{X}_1\boldsymbol{\beta}_1,\sigma_1^2\mathbf{I}_{n_1})N_{n_2}(\mathbf{y}_2|\mathbf{X}_2\boldsymbol{\beta}_2,\sigma_1^2\mathbf{I}_{n_2}).$$

The objective intrinsic Bayesian model for the no changepoint is

$$M_0 : \left\{N_n(\mathbf{y}|\mathbf{X}\boldsymbol{\beta},\tau^2\mathbf{I}_n), \pi^N(\boldsymbol{\beta},\tau) = \frac{c}{\tau}\right\},$$

and, conditional on one changepoint at position r_1, the objective intrinsic Bayesian model is

$$M_{r_1} : \{N_{n_1}(\mathbf{y}_1|\mathbf{X}_1\boldsymbol{\beta}_1, \sigma_1^2\mathbf{I}_{n_1})N_{n_2}(\mathbf{y}_2|\mathbf{X}_2\boldsymbol{\beta}_2, \sigma_1^2\mathbf{I}_{n_2}), \pi^I(\boldsymbol{\beta}_1, \boldsymbol{\beta}_2, \sigma)\},$$

where π^N represents the improper reference prior for $\boldsymbol{\beta}, \tau$ (Berger and Bernardo 1992), and $\pi^I(\boldsymbol{\beta}_1, \boldsymbol{\beta}_2, \sigma_1)$ the intrinsic prior for the parameters $\boldsymbol{\beta}_1, \boldsymbol{\beta}_2, \sigma_1$. Next theorem, stated without proof, provides the form of the intrinsic priors.

Theorem 1 *Conditional on a fix but arbitrary point $\boldsymbol{\beta}, \tau$, the conditional intrinsic prior distribution $\pi^I(\boldsymbol{\beta}_1, \boldsymbol{\beta}_2, \sigma_1|\boldsymbol{\beta}, \tau)$ can be shown to be*

$$\begin{aligned} \pi^I(\boldsymbol{\beta}_1, \boldsymbol{\beta}_2, \sigma_1|\boldsymbol{\theta}, \tau) &= \frac{2}{\pi\tau(1+\sigma_1^2/\tau^2)} \\ &\times N_k(\boldsymbol{\beta}_1|\boldsymbol{\beta}, (\tau^2+\sigma_1^2)\mathbf{W}_1^{-1}) \times N_k(\boldsymbol{\beta}_2|\boldsymbol{\beta}, (\tau^2+\sigma_1^2)\mathbf{W}_2^{-1}), \end{aligned}$$

where

$$\mathbf{W}_i^{-1} = \frac{n}{2k+1}(\mathbf{X}_i^t\mathbf{X}_i)^{-1}, \ i = 1, 2.$$

Using this theorem, we get the following expression for the Bayes factor.

Theorem 2 *The Bayes factor for the model with a changepoint at position r_1 against the no changepoint model is*

$$B_{r_1 n}(\mathbf{y}) = \frac{2}{\pi}(2k+1)^{k/2}\int_0^{\pi/2} \frac{\sin^k\varphi\ (n+(2k+1)\sin^2\varphi)^{(n-2k)/2}}{(n\mathcal{B}_{r_1}+(2k+1)\sin^2\varphi)^{(n-k)/2}}\, d\varphi$$

where, if $RSS_1 = \mathbf{y}_1^t(\mathbf{I}-\mathbf{H}_1)\mathbf{y}_1$, $RSS_2 = \mathbf{y}_2^t(\mathbf{I}-\mathbf{H}_2)\mathbf{y}_2$ and $RSS_0 = \mathbf{y}^t(\mathbf{I}-\mathbf{H})\mathbf{y}$ the residual sum of squares of the linear submodels induced by the partition S_1 and the no change model, then the statistic $\mathcal{B}_{r_1}$ is

$$\mathcal{B}_{r_1} = \frac{RSS_1 + RSS_2}{RSS_0}.$$

Proof. Denoting the marginal of the data under the models M_0 and M_{r_1} by $m(\mathbf{y}|M_0)$ and $m(\mathbf{y}|M_{r_1})$, respectively, it can be shown that

$$m(\mathbf{y}|M_{r_1}) = \frac{\Gamma(\frac{n-k}{2})}{\pi^{(n-k+2)/2}}\int_0^{\pi/2}\frac{d\varphi}{|\mathbf{D}_0(\varphi)|^{1/2}D_1(\varphi)D_2(\varphi)[H_1(\varphi)-H_2(\varphi)]^{(n-k)/2}},$$

where

$$\mathbf{D}_0(\varphi) = \sum_{i=1}^2 \frac{2k+1}{n+(2k+1)\sin^2(\varphi)}\mathbf{X}_i^t\mathbf{X}_i$$

$$D_i(\varphi) = \sin^{n_i}(\varphi)\left(1+\frac{n}{(2k+1)\sin^2(\varphi)}\right)^{k/2}, \ i = 1, 2,$$

$$H_1(\varphi) = \sum_{i=1}^2 \frac{1}{\sin^2\varphi}\left(\mathbf{y}_i^t\mathbf{y}_i - \frac{n}{n+(2k+1)\sin^2\varphi}\mathbf{y}_i^t\mathbf{X}_i(\mathbf{X}_i^t\mathbf{X}_i)^{-1}\mathbf{X}_i^t\mathbf{y}_i\right),$$

$$H_2(\varphi) = (2k+1)\left(\sum_{i=1}^{2}\frac{\mathbf{y}_i^t\mathbf{X}_i}{n+(2k+1)\sin^2\varphi}\right)\left(\sum_{i=1}^{2}\frac{\mathbf{X}_i^t\mathbf{X}_i}{n+(2k+1)\sin^2\varphi}\right)^{-1}\left(\sum_{i=1}^{2}\frac{\mathbf{X}_i^t\mathbf{y}_i}{n+(2k+1)\sin^2\varphi}\right)$$

and

$$m(\mathbf{y}|M_0) = \frac{\Gamma(\frac{n-k}{2})}{\pi^{(n-k)/2}(\mathbf{y}^t(\mathbf{I_n}-\mathbf{H})\mathbf{y})^{(n-k)/2}|\mathbf{X}^t\mathbf{X}|^{1/2}},$$

with $\mathbf{H} = \mathbf{X}(\mathbf{X}^t\mathbf{X})^{-1}\mathbf{X}^t$ the hat matrix of the no change model. After some cumbersome algebraic manipulations we finally get the simplified expression of the Bayes factor. □

From this expression we note that the Bayes factor depends on the data through the sum of square of the residuals associated to the partition of the vector of observations at the position of the changepoint. Furthermore, the partition S_1 for which the sum $RSS_1 + RSS_2$ is minimum is the partition with the highest Bayes factor. Therefore, inside the box $\mathfrak{B}_1$ the ordering provided by ranking the models according to their values of $RSS_1 + RSS_2$, and according to their model posterior probabilities

$$P(M_{r_1}|\mathbf{y}, p=1) = \frac{B_{r_1 n}(\mathbf{y})}{\sum_{r_1=1}^{n-1} B_{r_1 n}(\mathbf{y})}, \quad r_1 = 1, \ldots, n-1,$$

is the same.

However, in the class $\mathfrak{B} = \mathfrak{B}_0 \cup \mathfrak{B}_1$ the ordering of the models is given by the values of their models posterior probabilities conditional on $p \leq 1$, that is

$$P(M_{r_1}|\mathbf{y}, p \leq 1) = \frac{B_{r_1 n}(\mathbf{y})}{n-1+\sum_{r_1=1}^{n-1} B_{r_1 n}(\mathbf{y})},$$

and

$$P(M_0|\mathbf{y}, p \leq 1) = \frac{n-1}{n-1+\sum_{r_1=1}^{n-1} B_{r_1 n}(\mathbf{y})}.$$

From these formulas, it is clear that the new ordering of the models in the box $\mathfrak{B}_1$ is the same as before.

4.2. *The Case of Multiple Changepoints*

For the analysis of the general case when there are p changepoints located at positions $\mathbf{r} = (r_1, \ldots, r_p)$, let the corresponding partition of the data be $S_{\mathbf{r}_p} = (\mathbf{y}_1^t, \ldots \mathbf{y}_{p+1}^t)$ and for $i = 1, \ldots, p+1$, where $r_0 = 0$ and $r_{p+1} = n$, let the dimension of each $\mathbf{y}_i$ be $n_i = r_i - r_{i-1}$. Extending the analysis of the previous subsection, it is easy to see that the Bayes factor for the corresponding intrinsic priors for comparing the model with p changepoints at $\mathbf{r}$ and the model with no changepoint, $B_{\mathbf{r}n}(\mathbf{y})$, turns out to be

$$B_{\mathbf{r}n}(\mathbf{y}) = \frac{2}{\pi}((p+1)k+1)^{pk/2}\int_0^{\pi/2}\frac{\sin^{pk}\varphi\,(n+((p+1)k+1)\sin^2\varphi)^{(n-(p+1)k)/2}}{(n\mathcal{B}_{\mathbf{r}}+((p+1)k+1)\sin^2\varphi)^{(n-k)/2}}\,d\varphi \tag{10}$$

where

$$\mathcal{B}_{\mathbf{r}} = \frac{T_{\mathbf{r}}}{RSS_0} = \frac{RSS_1 + \cdots + RSS_{p+1}}{RSS_0},$$

RSS_i is the residual sum of squares associated to the linear submodel corresponding to the data $\mathbf{y}_i$, i.e, $\mathbf{y}_i = \mathbf{X}_i \boldsymbol{\beta}_i + \boldsymbol{\epsilon}_i$, and $T_{\mathbf{r}}$ is the total of the residual sum of squares.

We observe again that the Bayes factor (10), for fixed values of p, is a decreasing function of the total of the residual sum of squares of the partition at the positions of the changepoints. This result, together with expression (8), imply that, conditional on the occurrence of p changepoints, *i.e.*, within box $\mathfrak{B}_p$, the model $M_{\mathbf{r}}$ with the highest posterior probability $P(M_{\mathbf{r}}|\mathbf{y}, p)$ is the one that minimizes the total

$$T_{\mathbf{r}} = RSS_1 + \cdots + RSS_{p+1},$$

and both criteria render the same ordering.

On the other hand, the ordering of the models in the whole class of models $\cup_{p=0}^{n-1} \mathfrak{B}_p$ is obtained from the values of

$$P(M_{\mathbf{r}}|\mathbf{y}) = \frac{p!(n-p-1)! B_{\mathbf{r}n}(\mathbf{y})}{\sum_{q=0}^{n-1} q!(n-q-1)! \sum_{\mathbf{s} \in \mathfrak{N}^q} B_{\mathbf{s}n}(\mathbf{y})}, \text{ if } \mathbf{r} \in \cup_p \mathfrak{N}^p, \tag{11}$$

or, equivalently, from the values of $p!(n-p-1)! B_{\mathbf{r}n}(\mathbf{y})$.

Note that in the class of all models $\mathfrak{B}$ the ordering given by equation (11) restricted to any of the boxes $\mathfrak{B}_p$ is the same as the one given by minimizing $T_{\mathbf{r}}$.

One nice property of the Bayes factor for the intrinsic priors is the following. For $p \geq n/k - 1$, the minimum of $T_{\mathbf{r}_p}$ is clearly 0. It can then be shown that the Bayes factor for the intrinsic priors is a decreasing function of p, for fixed n and k, such that its value at $p = n/k - 1$ is equal to 1. Thus, for any integer p such that $p \geq n/k - 1$ the Bayes factor of any model in box $\mathfrak{B}_p$ is smaller than 1, which is the default Bayes factor of the no change model.

This property automatically penalizes models with too many changepoints p; namely, no model in any box $\mathfrak{B}_p$ such that $p \geq n/k - 1$ will ever be preferred to the no change model, and the posterior probability of p changepoints will always be smaller than that of the no change model.

On the other hand, this restriction on p is often included in the formulation of the multiple changepoint problem to avoid the problem of estimating the regression coefficients when the number of data is smaller than the number of regressor variables (see Subsection 4.4).

4.3. *Relationship with the Likelihood Approach*

It is straightforward to see that, conditional on p, the maximum of the profile likelihood function $L(\mathbf{r})$ is attained at the configuration $\mathbf{r}$ which minimizes the value of the statistic $T_{\mathbf{r}} = RSS_1 + \ldots + RSS_{p+1}$. Consequently, inside the box $\mathfrak{B}_p$ the optimal Bayesian model for the intrinsic priors and the one corresponding to the profile MLE of $\mathbf{r}$ are the same and further, from expression (10), the ordering of the models produced by the values of the profile likelihood is the same as the one given by the posterior probabilities.

The profile likelihood function, whose maximum as a function of $\mathbf{r}$ increases as p increases, goes to infinity when $T_{\mathbf{r}_p} = 0$. This happens for all $p \geq n/k - 1$, thus making impossible the task of estimating the number of changepoints or even of comparing different configurations of multiple changepoints unless some penalty function is included in the profile likelihood function or some additional criterion is used for comparing among the different boxes.

When none of the contemplated models is the true model –a realistic assumption when analysing real data– and the sample size is large, the values of the profile likelihood function for models in a given box, say $\mathfrak{B}_p$ rank the models as follows: the largest likelihood corresponds to the model in $\mathfrak{B}_p$ closest to the true model, the second largest corresponds to the second closest, and so on, where *close* is understood here in terms of the Kullback-Leibler pseudo-distance. Intrinsic model posterior probabilities also share this nice property. But, as soon as we have to rank models from different boxes, the likelihood approach fails, as also does the AIC correction; however, model posterior probabilities derived from the intrinsic Bayes procedure –which also includes the objective prior distribution for $\mathbf{r}$ and p– automatically take care of the differences in size among the boxes as seen from expressions (10) and (11).

Once that we have established the relationship between the Bayesian and the likelihood approaches to the changepoint problem, the following natural question arises: Is the use of the intrinsic model posterior probabilities a really objective Bayesian procedure for changepoint problems?

The answer, we believe, is yes. Our argument runs as follows: Conditional on the number of changepoints p the total of the residual sum of squares $T_{\mathbf{r}_p} = \sum_{i=1}^{p+1} RSS_{r_i}$ is the minimal sufficient statistic for estimating the changepoint configuration $\mathbf{r}$. Therefore, the vector statistic $(T_{\mathbf{r}_0}, \ldots, T_{\mathbf{r}_{n-1}})$ is the minimal sufficient statistic for making inferences on the set of all sampling models $\mathfrak{B}$. But, as the Bayesian procedure based on intrinsic model posterior probabilities depends on the Bayes factor for the intrinsic priors given by expression (10), and this, in turn, depends on $(T_{\mathbf{r}_0}, \ldots, T_{\mathbf{r}_{n-1}})$, and, further, does not either depend on any hyperparameters nor statistic but the ancillary n and k, we conclude that it is an objective procedure.

4.4. *Estimating the Magnitude of the Changes*

Sometimes, conditional on the ocurrence of p changepoints, the interest might be on estimating the variations produced by the changepoints in the parameters $\boldsymbol{\beta}_1, \ldots \boldsymbol{\beta}_{p+1}$, which can inform us on the magnitude of these changes. The use of the intrinsic priors for estimating the parameters does not provide simple analytical expressions and even the numerical computation of the estimates are cumbersome. Instead, and recalling that the intrinsic priors are also improper priors based on the reference priors, the use of these priors for estimation produces a simple and sensible posterior for the whole parameter set $\boldsymbol{\beta}_1, \ldots \boldsymbol{\beta}_{p+1}$.

Conditional on the occurrence of p changepoints in the sample, the reference prior for $\boldsymbol{\beta}_1, \ldots, \boldsymbol{\beta}_{p+1}, \sigma_p$, where we want to remind that σ_p also depends on p, is

$$\pi^N(\boldsymbol{\beta}_1, \ldots \boldsymbol{\beta}_{p+1}, \sigma_p) = c/\sigma_p$$

Multiplying this prior by the likelihood function

$$f_n(\mathbf{y}|\boldsymbol{\beta}, \sigma_p, \mathbf{s}) = \prod_{i=1}^{p+1} N_{n_i}(\mathbf{y}_i|\mathbf{X}_i\boldsymbol{\beta}_i, \sigma_p^2 \mathbf{I}_{n_i})$$

conditional on p changepoints at position vector $\mathbf{r} = (r_1, r_2 \ldots, r_p)$ and normalizing, after some algebra, it renders the posterior of $\boldsymbol{\beta}_1, \ldots, \boldsymbol{\beta}_{p+1}, \sigma_p$ conditional on $\mathbf{y}$, $\mathbf{r}$ and p, which is a multivariate normal-sqrt-inverted-gamma distribution.

Integrating out σ_p in this posterior, the resulting posterior distribution of the $p+1$ regression parameters, conditional on the data $\mathbf{y}$, and the occurrence of p changepoints at position vector $\mathbf{r}$, is the following multivariate Student t distribution

$$\begin{pmatrix} \boldsymbol{\beta}_1 \\ \vdots \\ \boldsymbol{\beta}_{p+1} \end{pmatrix} \sim t_{(p+1)k}\left[\begin{pmatrix} \tilde{\boldsymbol{\beta}}_1 \\ \vdots \\ \tilde{\boldsymbol{\beta}}_{p+1} \end{pmatrix}, s^2 \begin{pmatrix} (\mathbf{X_1}^t\mathbf{X_1})^{-1} & \cdots & \mathbf{O} \\ \mathbf{O} & \ddots & \mathbf{O} \\ \mathbf{O} & \cdots & (\mathbf{X_{p+1}}^t\mathbf{X_{p+1}})^{-1} \end{pmatrix}; \nu\right] \tag{12}$$

where $\tilde{\boldsymbol{\beta}}_1, \ldots, \tilde{\boldsymbol{\beta}}_{p+1}$ denote the least squares estimates of the corresponding regression coefficients, conditional on the partition induced by $\mathbf{r}$, $s^2 = T_{\mathbf{r}_p}/\nu$ is the usual estimate of the variance σ_p^2, and $\nu = n - (p+1)k$ are the degrees of freedom. Incidentally, note that for this posterior distribution to be proper and neither singular nor degenerate, the conditions $\nu > 0$, $s^2 > 0$ and $|\mathbf{X_i}^t\mathbf{X_i}| > 0$ must hold, and for this it is neccessary that the number p of changepoints be strictly smaller than $n/k - 1$ and that the partitions $\mathbf{y}_i$ of $\mathbf{y}$ should have a minimum size $n_i \geq k$.

This is the small price to be paid in order to estimate the regression parameters using the reference prior instead of the intrinsic one. Note that these constraints imply a reduction in the number of boxes $\mathfrak{B}_p$ and in the models within each box, except when $k = 1$.

Finally, the conditional posterior of the regression parameters on p is the following mixture of multivariate Student t distributions

$$\pi(\boldsymbol{\beta}_1, \ldots, \boldsymbol{\beta}_{p+1}|\mathbf{y}, p) \sim \sum\nolimits_{\mathbf{s}} \pi(\boldsymbol{\beta}_1, \ldots, \boldsymbol{\beta}_{p+1}|\mathbf{y}, \mathbf{s}, p) P(M_{\mathbf{s}}|\mathbf{y}, p), \tag{13}$$

where $\pi(\boldsymbol{\beta}_1, \ldots, \boldsymbol{\beta}_{p+1}|\mathbf{y}, \mathbf{s}, p)$ are given by equation (12) for configuration $\mathbf{s}$, the weights of the mixture are $P(M_{\mathbf{r}}|\mathbf{y},p) = B_{\mathbf{r}n}(\mathbf{y})/\sum_{\mathbf{s}} B_{\mathbf{s}n}(\mathbf{y})$, and the number of mixture terms and the sum are restricted to those configurations $\mathbf{s}$ satisfying the constraints $n_i > 0$ for $i = 1, \ldots, p+1$.

The most useful parameters of interest in changepoint problems are the successive differences $\delta_i = \boldsymbol{\beta}_{i+1} - \boldsymbol{\beta}_i$ for $i = 1, \ldots, p$, the corresponding distributions of which can be easily obtained from equations (12) and (13), using well known properties of the multivariate Student t distribution. In fact,

$$\delta_i|\mathbf{y}, \mathbf{r}, p \sim t_k(\tilde{\boldsymbol{\beta}}_{i+1} - \tilde{\boldsymbol{\beta}}_i, s^2((\mathbf{X_i}^t\mathbf{X_i})^{-1} + (\mathbf{X_{i+1}}^t\mathbf{X_{i+1}})^{-1}); \nu)$$

and

$$\delta_i|\mathbf{y}, p \sim \sum\nolimits_{\mathbf{s}} t_k(\tilde{\boldsymbol{\beta}}_{i+1} - \tilde{\boldsymbol{\beta}}_i, s^2((\mathbf{X_i}^t\mathbf{X_i})^{-1} + (\mathbf{X_{i+1}}^t\mathbf{X_{i+1}})^{-1}); \nu) P(M_{\mathbf{s}}|\mathbf{y}, p).$$

Thus, as a conclusion, for parameter estimation using the reference priors, the simplicity of working with well known distributions, from which it is also easy to sample if Monte Carlo estimates are needed for more complex functions of the parameters, compensates the very small numerical differences between the intrinsic and the reference Bayesian approaches. However, when model comparison is involved, as in making inferences on $\mathbf{r}$ or p, reference priors on the papameters can not be used: they simply do not work. Intrisic priors for changepoint problems involving normal linear models, on the other hand, have nice theoretical properties, behave as true objective priors, and, as we will see, they work very well in both simulated and real data sets.

5. COMPUTATIONAL ISSUES

By far, the main difficulty in the analysis of changepoint problems is the large number of possible models, which is 2^{n-1}. Computing Bayes factors for the intrinsic priors for all models is thus unfeasible. Unless the number of changepoints p be known and small, the computation of Bayes factors for all models can be very time consuming. Therefore, we need to devise estrategies to detect the most probable models in each box $\mathfrak{B}_p$ and also estimating the total or the mean of the Bayes factors within each box.

5.1. *Forward Search*

A simple procedure to tackle the first problem, which can be named or described as sequential forward search, is to visit sequentialy all boxes starting from box $\mathfrak{B}_0$, which contains a single model –the no change model, whose Bayes factor is always equal to 1– then box $\mathfrak{B}_1$ which contains $n-1$ models and selecting the changepoint with highest Bayes factor and so on, retaining at each step the preceding model and adding the new changepoint with highest Bayes factor within the correspoding box. In this way, and in the case of examining all boxes, we only have to compute $1+(n-1)+(n-2)+\ldots+1 = n(n-1)/2+1$ Bayes factors at most.

The proposed forward changepoint search is a fast algorithm for finding a relatively good changepoint configuration within each box, as will be shown in the examples, without the need to resort to an exhaustive all models search, which turns out to be unfeasible even for small sample sizes. On the other hand, it provides no answer to the problem of estimating the number p of changepoints. In addition, this algorithm may have the same similar drawbacks as the classical stepwise algorithms (forward selection and backward elimination) used for variable selection in regression, in its Bayesian counterpart, see Girón, Moreno and Martínez (2006a), as it is just a conditional sequential search estrategy. Notwithstanding these weaknesses, this algorithm usually locates within each box models with high Bayes factors.

By adapting the Gibbs sampling procedure proposed by Stephens (1994) in subsection 3.1, pp. 166–167, sampling from the discrete distribution of $M_{\mathbf{r}}|\mathbf{y}, p$ seems very easy as the model posterior probabilities are proportional to the Bayes factor for intrinsic priors $B_{\mathbf{r}n}$. On the other hand, sampling from the posterior $M_{\mathbf{r}}|\mathbf{y}$ seems a much more demanding task.

All this prompted us to devise an efficient stochastic search algorithm for the multiple changepoint problem.

5.2. *Retrospective Search*

For the retrospective search, we find that a random walk Metropolis-Hastings algorithm works very well. We choose a symmetric random walk, and use the posterior probability (11) as the objective function. This insures that, at convergence, the resulting Markov chain is a sample from the posterior probability surface. Hence, states of high posterior probability will be visited more often.

To now choose the 'best" model, or to examine a range of good models, we would like to rank the models by their posterior probabilities, but, as mentioned above, this is not possible, as the number of models can be prohibitively large. Moreover, it is also the case that calculation of the denominator in (11) is prohibitive. The solution is to construct an MCMC algorithm with (11) as the stationary distribution. Such an algorithm, if properly constructed, would not only visit every model, but would visit the better models more often. Thus, a frequency count of visits to the models is directly proportional to the posterior probabilities.

For the regression model, we keep track of changepoints with a $n \times 1$ vector

$$\mathbf{c} = (0, 0, 1, 0, 1, 0, \ldots, 0, 1, 0, 0, 0)^t$$

where "1"indicates a changepoint. At each step of the stochastic search we select an observation at random (actually an index $1, 2, \ldots, n$). If it is a 1 we evaluate whether to change it to a 0, and if it is a 0 we evaluate whether to change it to a 1. This is done with a Metropolis step as follows:

Corresponding to a vector $\mathbf{c}$ there is a vector $\mathbf{r}$ of changepoints and a model $M_{\mathbf{r}}$.

(i) Generate a new $\mathbf{c}'$ and $\mathbf{r}'$, and $U \sim \text{Uniform}(0, 1)$.

(ii) Calculate

$$\rho = \min\left\{1, \frac{P(M_{\mathbf{r}'}|\mathbf{y})}{P(M_{\mathbf{r}}|\mathbf{y})}\right\}.$$

(iii) Move to $\mathbf{c}'$ if $U < \rho$, otherwise remain at $\mathbf{c}$.

(iv) Return to 1.

6. ILLUSTRATIVE EXAMPLES: REAL AND SIMULATED DATA

The exact procedure described above, and the Metropolis algorithm of Section 5.2 were tested and compared on a number of examples, both real and simulated.

6.1. *Simulated Data*

Example 1. We first tested the search algorithm on the data given in Fig. 1, which were simulated with the following model

$$y_i = \begin{cases} \frac{1}{4}x + \varepsilon & \text{if } x = 1, \ldots, 6; \\ 4 + \varepsilon & \text{if } x = 7, \ldots, 12; \\ 12 - \frac{1}{2}x + \varepsilon & \text{if } x = 13, \ldots, 18, \end{cases} \tag{14}$$

where $\varepsilon \sim N(0, 1)$. We ran the simulations for $\sigma = 0.25, 0.5, 0.75$, and typical data are shown in Fig. 1.

Note that it is very difficult to see the three different models for $\sigma = 0.75$. In fact, it is sometimes the case that, for large σ, the true model does not have the highest intrinsic posterior probability. In such cases, there are many competing models that are candidates for the "best", as we will now describe.

For each of $\sigma = 0.25, 0.5, 0.75$, the algorithm was run on 100 datasets, with 20,000 iterations of the Metropolis algorithm. The performance is summarized in Table 1, where we measured the number of times that the true model was in the top 5, top 10, or top 25, ranked on posterior probabilities.

Comparing the performance to the typical data sets, we see that the procedure always finds the true model when the error is reasonable, and does worse as the error term increases. But it is quite surprising that for $\sigma = 0.5$, where our eye cannot see the changepoint at 6, the true model is in the top 5 34% of the time, and in the top 10 almost 60% of the time.

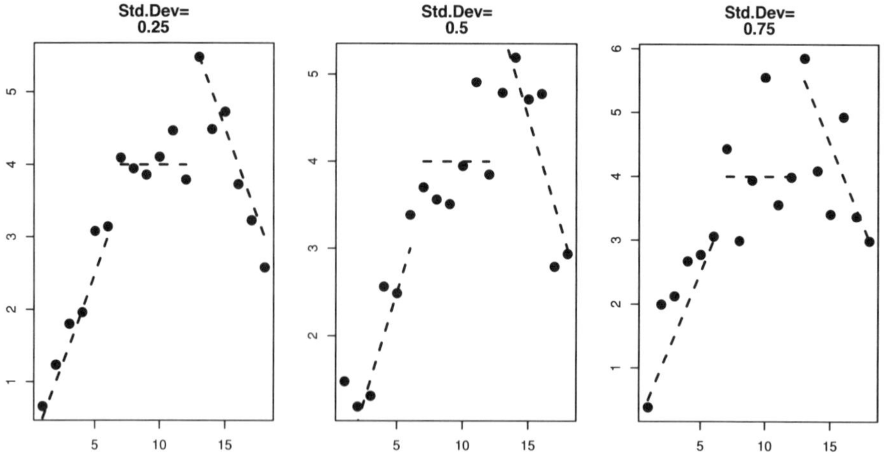

Figure 1: *Typical datasets and true model for data from Model (14). Note that the changepoint at $x = 6$ is barely discernible for $\sigma \geq 0.5$.*

Table 1: *Results of the simulations of model (14). Proportion of times that the true model is in the corresponding top category. The models are ranked by their intrinsic posterior probabilities.*

σ	Top 5	Top 10	Top 25
0.25	94	99	99
0.50	34	59	88
0.75	14	28	69

Example 2. **Quandt's Data**. A second simulated data set that we look at is from Quandt (1958), which consists of two simulated linear regressions with a slight change at time 12. That data are given in Fig. 2, and it is clear by looking at this Figure that the changepoint is barely discernible.

The simulated data come from the two regression models

$$y_i = \begin{cases} 2.5 + 0.7x_i + \varepsilon_i & i = 1, 2, \ldots, 12; \\ 5 + 0.5x_i + \varepsilon_i & i = 13 \ldots, 20. \end{cases}$$

The results from the exact analysis conditional on there being up to five changepoints, *i.e.*, , $p \leq 5$ can be summarized as follows: As seen from Table 2, the posterior mode of $\pi(p|\mathbf{y})$ is $\tilde{p} = 1$, pointing out to the existence of a single changepoint. This agrees with the true origin of the simulated data. Notice that this posterior probability does not have a very pronounced mode and that posterior probabilites to the right of the mode decrease very slowly. This behavior is typical of small data sets where, in addition, the models before and after the changepoint do not differ much in the range where the covariates lay as seen from Fig. 2.

On the other hand, in the space of all models, the six most probable models, are displayed in Table 3 in decreasing order.

The analysis of these data reveals that there is a –not clearcut– single changepoint at position 12 but, on the other hand, the no change model has higher posterior probability than that model. The rest of models have smaller posterior probability, but most of them are single changepoint models in the neiborhood of M_{12} as seen from Table 4. It is worthwhile remarking that, for these data, the forward selection procedure computes the best models within each box $\mathfrak{B}_p$ as the exact method for all $p = 0, 1, 2 \ldots, 5$. These results are also confirmed by the monitoring procedure, showing that the first permanent changepoint occurs at position 12.

Table 2: *Posterior probabilities of the number of changepoints p.*

p	0	1	2	3	4	5
$\pi(p\|\mathbf{y})$	0.162	0.229	0.180	0.154	0.142	0.132

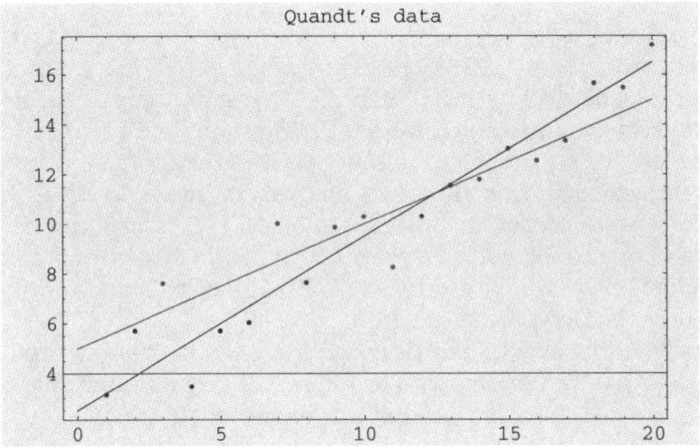

Figure 2: *Quandt's simulated data from two regression lines.*

Table 3: *Posterior probabilities of the most probable models $M_{\mathbf{r}}$.*

Models $\mathbf{r}$	0	12	10	11	9,12	9
$P(M_{\mathbf{r}}\|\mathbf{y})$	0.162	0.116	0.017	0.017	0.013	0.013

6.2. *Real Data*

The following example refers to a famous real data set, which is also a favorite in the changepoint literature.

Example 3. The Nile River Data. The data appearing in Fig. 4 are measurements of the annual volume of discharge from the Nile River at Aswan for the years 1871 to 1970.

This series was examined by Cobb (1978), Carlstein (1988), Dümbgen (1991), Balke (1993), and Moreno *et al.* (2005) among others, and the plot of the data reveals a marked and long-recognized steady decrease in annual volume after 1898.

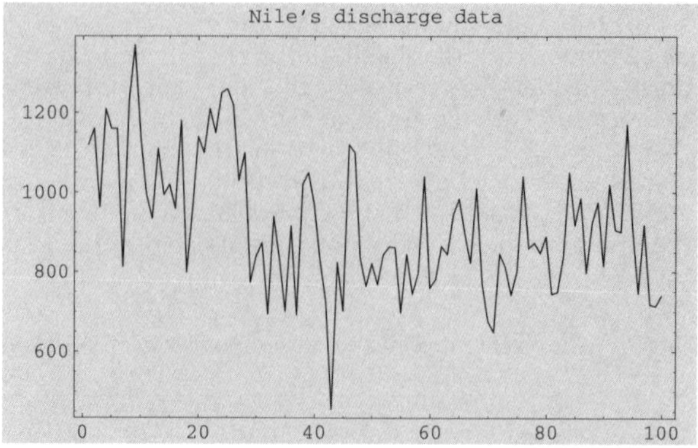

Figure 3: *Nile's discharge data for the years 1871–1970.*

Some authors have associated the drop to the presence of a dam that began operation in 1902, but Cobb (1978) cited independent evidence on tropical rainfall records to support the decline in volume. A Bayesian analysis using the jump diffusions, is found in Phillips and Smith (1996), though their results are not very accurate probably due to the small number of iterations of the algorithm. Denison *et al.* (2002) also consider the Bayesian analysis of these data. Both analysis use broadly uninformative conjugate priors. Another Bayesian analysis utilizing fractional Bayes factors and allowing for correlation in the error terms, is that of Garish and Groenewald (1999), which produces results closer to ours, and provides further references on other analysis of these data.

The results from the exact analysis conditional on there being up to three changepoints, *i.e.*, , $p \leq 3$ can be summarized as follows: As seen from Table 4, the posterior mode of $\pi(p|\mathbf{y})$ is $\tilde{p} = 1$, clearly pointing out to the existence of a single changepoint.

On the other hand, in the space of all models, the eight most probable models are displayed in Table 5, in decreasing order.

Table 4: *Posterior probabilities of the number of changepoints p.*

p	0	1	2	3
$\pi(p\|\mathbf{y})$	0.000	0.615	0.258	0.127

Table 5: *Posterior probabilities of the most probable models $M_{\mathbf{r}}$.*

Models $\mathbf{r}$	28	27	26	29	19, 28	21, 28	20, 28	28, 97
$P(M_{\mathbf{r}}\|\mathbf{y})$	0.466	0.076	0.036	0.029	0.006	0.006	0.006	0.005

The forward search produces the same results as the exact procedure for $p = 0, 1, 2$. For $p = 3$, the forward search finds model $M_{10,19,28}$ with posterior probability 0.00052 which is the fifth in position within box $\mathfrak{B}_3$, the first being model $M_{28,83,95}$ with posterior probability 0.00082, which is slightly more probable than the former. The stochastic search mostly confirms the above results as the most visited model occurs at position 28.

Summarizing our findings in analyzing these data: there is a single clearcut changepoint at position 28. The four most favored models are single changepoint models in the neighborhood of 28, and the probability of models with two or three changepoints are very small. The analysis of the remaining data, once the first 28 data points are deleted, clearly confirms that the point detected by the monitoring procedure at position 28 is the only one changepoint.

7. FURTHER DISCUSSION

In this paper we regard the detection of changepoints in the distribution of a sequential sample of observations as a model selection problem or, equivalently, as a multiple testing problem; consequently, we feel that a Bayesian formulation is the simplest and accuratest way to deal with it. Changepoint problems can be analysed from two different but otherwise complementary perspectives: sequential and retrospective. The former focuses on the first time a changepoint occurs and the second investigates the number and position of the changes in the sample based on the knowledge of the whole sample. However, by treating the first problem as a retrospective analysis at each step, we get a monitoring procedure which seems to be an excellent tool for constructing a sensible stopping rule.

We feel that monitoring should be preferred to the two stopping rules described in subsection 2.2.1 based on Bayes factors, which generalize the classical ones based on the likelihood estimators (Moreno *et al.*, 2005). Nevertheless, we want to stress the main difference between the sequential analysis —monitoring– in which the class of sampling models consists of the no change model and the set of models with exactly one changepoint, and the retrospective analysis where we consider the class of all possible models.

The main difficulties to carry out a retrospective analysis come from two sources: (i) the fact that for the continuous parameters involved in the problem –the regression coefficients and the variance errors–, the usual objective priors are unsuitable for computing model posterior probabilities as they are improper, and (ii) the huge number of models involved for large, or even moderate, sample size makes unfeasible the computation of all model posterior probabilities.

These difficulties are solved by using intrinsic priors for the continuous model parameters, and a Metropolis-like stochastic search algorithm for selecting those models with highest posterior probabilities. For the integer parameters, we have assumed a uniform prior for the number of changepoints and, conditional on them, a uniform prior for their models, justified by both theoretical and practical arguments. When some locations are subjectively excluded, for instance we do not allow consecutive changepoints, the posterior distribution on the models or configurations is truncated accordingly; this is the simplest computational strategy.

The above priors provide an automatic and quite simple Bayesian analysis in which there are no hyperparameters to be adjusted. Numerical examples indicate that this objective Bayesian formulation behaves extremely well for detecting changepoints and finding the distribution of their locations. We have also realized that when the number of possible models is very large the Metropolis algorithm is able to find those models with highest posterior probability in a reasonable computing time.

An additional property of the intrinsic priors is that –as they are proper distributions conditionally on the parameters of the no change model and, further, have a close form in terms of multivariate normal and half-Cauchy distributions– they are also amenable to the implementation of a simple Gibbs algorithm involving sampling from standard distributions plus a simple additional Metropolis step. Although this property has not been exploited in the paper, it will be of the utmost importance for the extension of our model to the heteroscedastic case.

REFERENCES

Bacon, D. W. and Watts, D. G. (1971). Estimating the transition between two intersecting straight lines. *Biometrika* **58**, 525–534.

Balke, N. S. (1993). Detecting level shifts in time series. *J. Business Econ. Statist.* **11**, 81–92.

Berger, J.O. and Bernardo, J.M. (1992). On the development of the reference prior method. *Bayesian Statistics 4* (J. M. Bernardo, J. O. Berger, A. P. Dawid and A. F. M. Smith, eds.) Oxford: University Press, 35–60.

Berger, J.O. and Pericchi, L.R. (1996). The intrinsic Bayes factor for model selection and prediction. *J. Amer. Statist. Assoc.* **91**, 109–122.

Box, G. and Tiao, G. (1992). *Bayesian Inference in Statistical Analysis. 2nd Edition.* New York: Wiley.

Carlstein, E. (1988). Nonparametric change-point estimation. *Ann. Statist.* **16**, 188–197.

Casella, G. and Moreno, E. (2006). Objective Bayesian variable selection, *J. Amer. Statist. Assoc.* **101**, 157–167.

Carlin, B. P., Gelfand, A. E. and Smith, A. F. M. (1992). Hierarchical Bayesian Analysis of changepoint Problems. *Applied Statist.* **41**, 389–405.

Chernoff, H. and Zacks, S. (1964). Estimating the current mean of a normal distribution which is subjected to changes in time. *Ann. Math. Statist.* **35**, 999–1018.

Choy, J. H. and Broemeling, L. D. (1980). Some Bayesian inferences for a changing linear model. *Technometrics* **22**, 71–78.

Cobb, G. W. (1978). The problem of the Nile: conditional solution to a change-point problem. *Biometrika* **65**, 243–251.

Denison, D. G. T., Holmes, C. C., Mallick, B. K. and Smith, A. F. M. (2002). *Bayesian Methods fro Nonlinear Classification and Regression.* Wiley: Chichester.

Dümbgen, L. (1991). The asymptotic behavior of some nonparametric change-point estimators. *Ann. Statist.* **19**, 1471–1495.

Ferreira, P.E. (1975). A Bayesian analysis of a switching regression model: known number of regimes. *J. Amer. Statist. Assoc.* **70**, 370–374.

Garisch, I. and Groenewald, P. C. (1999). The Nile revisited: Changepoint analysis with autocorrelation. *Bayesian Statistics 6* (J. M. Bernardo, J. O. Berger, A. P. Dawid and A. F. M. Smith, eds.) Oxford: University Press, 753–760.

Girón, F. J., Moreno, E. and Martínez, M. L. (2006a). An objective Bayesian procedure for variable selection in regression. *Advances on Distribution Theory, Order Statistics and Inference.* (N. Balakrishnan *et al.*, eds.) Boston: Birkhauser, 393–408.

Girón, F. J., Martínez, M. L., Moreno, E. and Torres, F. (2006b). Objective testing procedures in linear models: Calibration of the p-values. *Scandinavian J. Statist.* **33**, 765–784.

Kiuchi, A. S., Hartigan, J. A., Holford, T. R., Rubinstein, P. and Stevens, C. E. (1995). changepoints in the series of T4 counts prior to AIDS. *Biometrics* **51**, 236–248.

Lai, T. L. (1995). Sequential changepoint detection in quality control and dynamical systems. *J. Roy. Statist. Soc. B* **57**, 613–658.

Moreno, E., Bertolino, F. and Racugno, W. (1998). An intrinsic limiting procedure for model selection and hypothesis testing. *J. Amer. Statist. Assoc.* **93**, 1451–1460.

Moreno, E., Casella, G. and García-Ferrer, A. (2005). Objective Bayesian analysis of the changepoint problem. *Stochastic Environ. Res. Risk Assessment* **19**, 191–204.

Moreno, E. and Girón, F.J. (2006). On the frequentist and Bayesian approaches to hypothesis testing (with discussion). *Sort* **30**, 3–54.

Menzefrike, U. (1981). A Bayesian analysis of a change in the precision for a sequence of independent normal random variables at an unknown time point. *Applied Statist.* **30**, 141–146.

Phillips, D. V. and Smith, A. F. M. (1996). Bayesian model comparison via jump diffusions. *Markov Chain Monte Carlo in Practice* (W. R. Gilks, S. Richardson, and D. J. Spiegelhalter, eds.) London: Chapman and Hall, 215–239.

Quandt, R. E. (1958). The estimation of the parameters of a linear regression system obeys two separate regimes. *J. Amer. Statist. Assoc.* **53**, 873–870.

Raftery, A. E. and Akman, V. E. (1986). Bayesian analysis of a Poisson process with a change-point. *Biometrika* **73**, 85–89.

Smith, A. F. M. (1975). A Bayesian approach to inference about a change-point in a sequence of random variables. *Biometrika* **62**, 407–416.

Smith, A. F. M. and Cook, D. G. (1980). Straight lines with a changepoint: a Bayesian analysis of some renal ransplant data. *Applied Statist.* **29**, 180–189.

Spiegelhalter, D. J. and Smith A. F. M. (1982). Bayes factors for linear and log-linear models with vague prior information. *J. Roy. Statist. Soc. B* **44**, 377–387.

Stephens, D. A. (1994). Bayesian retrospective multiple-changepoint identification. *Applied Statist.* **43**, 159–168.

DISCUSSION

RAÚL RUEDA (*IIMAS, UNAM, Mexico*)

This paper is one more of a series of papers by the authors and their collaborators about model selection using intrinsic Bayes factors and related concepts. It is well written and easy to follow. It was a pleasure to read it.

The paper deals with the following situation:

Sampling from a normal linear model,

$$y_1, \ldots, y_n \sim N(y|\boldsymbol{\beta}, \tau^2\mathbf{I}_n),$$

the problem is to infer the number p of potential changepoints and their corresponding positions in the sample, $\mathbf{r}_p = \{r_1, \ldots, r_p\}$, as well as to make inferences on the regression parameters $\boldsymbol{\beta}$ or transformations thereof.

There are three main tasks in this model selection problem:

(i) Prior assignation. Intrinsic Bayes factors cannot be used here, since a changepoint can occur in the *training* sample; besides, the order is relevant.

(ii) Implementation. The number of models, and therefore, the number of comparisons in retrospective analysis increases with the sample size.

(iii) Monitoring. In prospective analysis, where the search for a changepoint is carried out sequentially, the proposed strategy for on-line detection has advantages over a stopping rule.

Concerning the *implementation* and *monitoring* issues, the authors propose a forward searching and a Metropolis-Hasting algorithm for the retrospective search which at the same time, provide a monitoring plan for the prospective case without any additional effort.

The need for automatic procedures for model selection, including hypotheses testing, has produced a variety of 'objective Bayes factors', which in some cases behave like real Bayes factors, at least asymptotically. Almost everybody agrees that the use of improper priors must be avoided when computing a Bayes factor.

This is because improper priors do not, in general, define unambiguously the Bayes factor. Unfortunately, 'automatic' priors are typically improper.

In almost all the proposals, the basic idea is 'to pay' with some sample information in order to obtain proper posterior distribution from the improper prior and use this posterior as a prior to define a real Bayes factor. Thus, we have *training samples, minimum training samples, partial samples* and so on. But the intrinsic priors, introduced by Berger and Pericchi (1996), use *imaginary samples*, so in the end we do not have to pay for anything! An advantage of the automatic Bayes factor resulting from using intrinsic priors is that 'it is close to an 'actual' Bayes factor as desired'.

However, it seems that the need for automatic or objective procedures has produced precisly what we criticise about frequentist statisitcs: a collection of *ad hoc* methods out of Bayesian principles. As an example of this, the authors use *intrinsic* priors for model selection and *reference* priors to estimate the magnitude of the changes. Morever, Bayes factors, including *actual* Bayes factors, can produce incoherent results, as Lindley's examples in the discussion of Aitkin's paper show (Aitkin, 1991).

A question for the authors: what about robustness? It is known that Bayes factors, as a device for model selection, arise from the $\mathcal{M}$-closed perspective, so what happens if, for example, the errors are Student-t instead of normal?. On a related note, I understand that the *intrinsic* prior depends on a 'standard' noninformative prior. How sensitive is it to this choice?

I congratulate the authors for a very interesting paper. I thank the authors because they have forced me to take an intensive course on Bayes factors and all their *automatic* extensions. When I first read his paper, a thought appeared almost immediately in my mind. . . I hope that The Beatles will forgive me:

When I was younger, so much younger than today,
I only needed random samples to use factors of Bayes
But now these days are gone and I feel so insecure
Now I am also confused with all this new stuff

Help me if you can, with the training samples
And before they become imaginary
Help me get my feet back on the ground
Won't you please, please help me.

NICOLAS CHOPIN and PAUL FEARNHEAD
(*University of Bristol, UK and Lancaster University, UK*)

Change-point modelling is a 'treat' for Bayesians, as frequentist methods are particularly unsatisfactory in such settings: asymptotics do no make sense if a given model is assumed to be true only for a finite interval of time. Moreover, practitioners are increasingly turning their attention to simple models, which are allowed to change over time, as an alternative to overly complicated time series models (e.g. long memory processes) that are difficult to interpret. Any contribution to this interesting field is therefore most welcome.

Our first comments relate to the choice of prior. The objective prior chosen for the number of changepoints appears slightly inconsistent in the following sense.

Imagine analysing a data set with $2n$ observations, but where you are first given the first n observations to analyse, then the second half later. The uniform prior on the number of changepoints within the first n observations, together with the same prior on the number of changepoints in second n observation implies a prior on the number of changepoints within the full data set that is not uniform. In more practical terms, the choice of a uniform prior appears unrealistic: we cannot think of any application where length of segments (distance between consecutive changepoints) should be as small as 2. A more coherent approach is to model directly the length of each segment (resulting in a product-partition prior structure, see Barry and Hartigan, 1993). This produces both consistent priors (in the sense described above), and is modelling directly an important feature of the data.

In that respect, trying to be non-informative may not always be relevant or useful for change point models. As for any parameter with a physical dimension, there is almost always prior information on these durations which can easily extracted: for daily data for instance, common sense is enough to rule out durations larger than a few years. This kind of consideration is essential in on line applications, where the sample size n of the complete data is not known in advance.

Secondly, an alternative approach to using intrinsic priors is to have a hierarchical model structure where you introduce hyperpriors on the hyperparameters of the priors for the parameters of each segment. This enables the data to inform the choice of prior for the segment parameters. It is possible to choose the hyperpriors to be improper in many situations (see Punskaya *et al.*, 2002 and Fearnhead, 2006). One advantage of this is that the priors on the parameters of each segment can be interpreted in terms of describing the variation in features of the model (e.g. means) across the different segments. It also seems mathematically and computationally simpler than using intrinsic priors.

Finally we would like to point out some other work on Bayesian analysis of multiple changepoint models. Perfect (iid) sampling from the posterior is possible for certain classes of changepoint models (Barry and Hartigan, 1993; Liu and Lawrence, 1998; Fearnhead , 2005, 2006). Otherwise, standard and trans-dimensional MCMC algorithms can be derived (Green, 1995; Gerlach *et al.*, 2000), as well as particle filters (Fearnhead and Clifford, 2003; Fearnhead and Liu, 2006; Chopin, 2006) for on-line inference. These particle filters can also be used for off-line inference, and at a smaller cost than MCMC, *i.e.*, $O(n)$ instead of $O(n^2)$. The MCMC algorithm proposed in the discussed paper is also implicitly $O(n^2)$: each iteration is $O(n)$, but proposing at random a new change means that $O(n)$ iterations are required to visit a given location. In fact, since the considered linear model allows for conjugacy, it would be interesting to see if one of the exact method mentioned above could be used as an importance sampling proposal.

REPLY TO THE DISCUSSION

Rueda. We thank Raúl Rueda for his comments on the paper and answer the questions he raises in his discussion.

We note that the conditional intrinsic prior is proper, and the unconditional intrinsic prior is improper. Therefore, this impropriety is inherited from the impropriety of the objective reference prior that we start with. However, the Bayes factor for intrinsic priors is a well defined limit of Bayes factors for proper priors.

The use of intrinsic priors for testing and reference priors for estimation is not at all *ad hoc.* We know that reference priors are perfectly reasonable objective priors for estimation, but they are not well-calibrated so cannot be used for testing. The intrinsic approach calibrates the reference prior for testing so, in fact, these two priors are connected in this way, with each being suited for its particular task.

We have taken another look at the Lindley/Aitkin discussion, and find that we neither agree with Lindley nor with Rueda. As Aitkin points out in his rejoinder, it is not Bayes factors, but Lindley's criterion, 'that is ridiculous". To illustrate, Lindley asserts that if we prefer model $\mathcal{M}_1$ to $\mathcal{M}_2$ and $\mathcal{M}_3$ to $\mathcal{M}_4$, then we should prefer $\mathcal{M}_1 \cup \mathcal{M}_3$ to $\mathcal{M}_2 \cup \mathcal{M}_4$. But this does not always follow. Suppose that in a four-variable regression problem, the best model with two regressors is $\{x_2, x_4\}$, and the univariate models are ordered, from best to worst, $\{x_1\}$, $\{x_2\}$, $\{x_3\}$, $\{x_4\}$. Then we prefer $\{x_1\}$ to $\{x_2\}$ and $\{x_3\}$ to $\{x_4\}$, but $\{x_2, x_4\}$ beats any other model with two regressors. So Lindley's preference ordering is not self-evident.

In our approach to changepoint problems we have assumed that the underlying regression models are normal. This assumption allows –in the monitoring procedure– for distinguishing between outliers and permanent changepoints. If, instead, we would consider Student errors, apart from the problem of deriving the corresponding intrinsic priors and Bayes factors, as there is no sufficient statistic of fixed dimension, it is not at all clear that outliers and changepoints could be distinguised.

The question about how sensitive the Bayes factor is to the choice of the non-informative prior when intrinsic priors are used is discussed at length in subsection 3.5 of Girón *et al.* (2006b) where, in the normal regression setting, we compare the influence of using priors of the form

$$\pi^N(\theta, \tau) \propto \frac{1}{\tau^q} \quad \text{for} \quad q = 1 \ldots, k,$$

in the posterior probability of the null.

Note that this class of priors includes the reference, when $q = 1$, and the Jeffreys' prior when $q = k$. The conclusion was that the choice of the usual reference prior results in a a much more stable procedure for comparing models. Box and Tiao (subsection 2.4.6, pp. 101–102, 1992) also discuss the sensitivity of the posterior with respect to the choice of the prior within this class in the standard normal setting.

Finally, we thank Raúl Rueda for his thoughtful comments, and add, in a spirit similar to his:

Hey Raúl, don't make it bad
We take a sad prior, and make it better
Remember to let it into your heart
Then you can start to make it better.

Chopin and Fearnhead. We thank Chopin and Fearnhead for their thoughtful comments. We will reply to each separately.

Choice of Prior. You are right in saying that the prior on the set of models should not be uniform. A uniform prior penalizes the model with small number of changes and this is not reasonable. However, we have not used a uniform prior on the set of models, but rather have used a uniform prior on the set of boxes, and then spread mass uniformly among the models within each box. Moreover, this prior only depend on the ancillary sample size statistic n, so that we do not believe it suffers from the slight inconsistency that you mention.

An objective analysis of a statistical problem is appropriate when you do not have subjective prior information. If you have such information, by all means use it!

For example, in analyzing the famous Coal-Mining Disaster Data, Fearnhead (2006) chooses a Gamma prior (with specific values for the hyperparameters) for the mean of the Poisson sampling distribution, and a Poisson distribution with mean 3 for the number of possible change points. We suspect that there are good reasons for choosing such specific priors.

Consecutive changepoints. Although there may be cases when there is reason to model changepoints through segmental length, recall that we are considering objective methods that do not take into account special features of the data.

However, if there is reason to impose the constraints on the length of the segments between two changepoints, this can be accomplished through restrictions on some configurations. For example, if $\mathfrak{R}_k$ is a set of restrictions requiring segments to have length greater than k, a simple combinatorial calculation shows that, once those configurations have been deleted, the objective prior $\pi(p|\mathfrak{R}_k)$ is

$$\pi(p|\mathfrak{R}_k) \propto \frac{\binom{n-1-k(p-1)}{p}}{\binom{n-1}{p}} \quad \text{for } p = 0, 1, \ldots, n-1.$$

It then follows that $\pi(p|\mathfrak{R}_k) = 0$ for $p \geq (n-1)/k$ and all such boxes $\mathfrak{B}_p$ are empty. Furthermore, $\pi(p|\mathfrak{R}_k)$ is a decreasing function of p with $\pi(0|\mathfrak{R}_k) = \pi(1|\mathfrak{R}_k)$ for all k. This last property implies that for monitoring, where we only consider boxes $\mathfrak{B}_0$ and $\mathfrak{B}_1$, each box has probability 1/2.

Lastly, we have also commented in Section 3.1 about the possibility of incorporating additional information which results in constraints on the *posterior* instead of on the prior, which may be technically easier in some cases.

Hierarchical models. Models based on intrinsic priors are inherently hierarchical, and we believe that, in general, hierarchical models are very useful. However, using improper priors in the last stage of a hierarchy results in improper marginals and can result in improper posterior distributions (Hobert and Casella, 1996). This means that the Bayes factors are not well defined. However, intrinsic priors always provide well defined Bayes Factors.

We are convinced that the natural formulation of the change point problem is as a model selection problem, that is, it is a *testing problem* rather than an estimation problem. As such, improper priors cannot be used in this context. However, intrinsic priors can be used for both testing and estimation. From the details in Section 4.4, and using the hierarchical structure of the intrinsic priors, it is easy to set up a Gibbs Sampler to estimate all of the posterior parameters.

Computational issues. You are right to point out that the algorithm used in the paper is $O(n^2)$, however, it should also be pointed out that the calculations are very fast in R, and large scale searches are feasible. Also, we have also developed another algorithm, b ased on an independent Metropolis-Hastings scheme that exploits the box structure, (see Casella and Moreno 2006) that is uniformly ergodic.

The exact methods that you discuss are indeed faster, but we have decided to use methods that are based on simpler algorithms in order to allow our methods to apply to more complex models. It seems that, for example, the faster recursions in Fearnhead (2006) do not apply to very general models without embedding them in a Gibbs sampler.

We also point out that our search algorithm is driven by a very specific objective function, the posterior probability of the models. In searching such spaces, we have found that a Metropolis-Hastings algorithm with a well-chosen candidate is hard to beat. Even with isolated modes, where we may need transition kernels based on a tempering structure, the Metropolis-Hastings Algorithm is typically an excellent choice (Jerrum and Sinclair, 1996).

ADDITIONAL REFERENCES IN THE DISCUSSION

Aitkin, M. (1991). Posterior Bayes factors. *J. Roy. Statist. Soc. B* **53**, 131–138 (with discussion).

Barry, D. and Hartigan, J. A. (1993). A Bayesian analysis for change point problems. *J. Amer. Statist. Assoc.* **88**, 309–319.

Chopin, N. (2006). Dynamic detection of change points in long time series. *Ann. Inst. Statist. Math.* (to appear).

Fearnhead, P. (2005). Exact Bayesian curve fitting and signal segmentation. *IEEE Trans. Signal Processing* **53** ,2160–2166.

Fearnhead, P. (2006). Exact and efficient inference for multiple changepoint problems. *Statist. Computing* **16**, 203–213.

Fearnhead, P. and Clifford, P. (2003). On-line inference for hidden Markov models. *J. Roy. Statist. Soc. B* **65**, 887–899.

Fearnhead, P. and Liu, Z. (2006) On-line inference for multiple changepoint problems. *Submitted*; available from `www.maths.lancs.ac.uk/~fearnhea/publications`.

Gerlach, R., Carter, C., and Kohn, R. (2000) Efficient Bayesian inference for dynamic mixture models. *J. Amer. Statist. Assoc.* **88**, 819–828.

Green, P.J. (1995) Reversible jump Markov chain Monte Carlo computation and Bayesian model determination. *Biometrika* **82**, 711–732.

Hobert, J. P. and Casella, G. (1996). The effect of improper priors on Gibbs sampling in hierarchical linear models. *J. Amer. Statist. Assoc.* **91**, 1461–1473.

Jerrum, M. and Sinclair, A. (1996). The Markov chain Monte Carlo method: An approach to approximate counting and integration. *Approximation Algorithms for NP-hard Problems* (D. S. Hochbaum, ed.) Boston: PWS Publishing, 482–520.

Liu, J. S. and Lawrence, C. E. (1998). Bayesian inference on biopolymer models.*Bioinformatics* **15**, 38–52.

Punskaya, E., Andrieu, C., Doucet , A. and Fitzgerald, W. J. (2002). Bayesian curve fitting using MCMC with applications to signal segmentation. *IEEE Trans. Signal Processing* **50**, 747–758.

BAYESIAN STATISTICS 8, pp. 253–282.
J. M. Bernardo, M. J. Bayarri, J. O. Berger, A. P. Dawid,
D. Heckerman, A. F. M. Smith and M. West (Eds.)

Bayesian Relaxation: Boosting, The Lasso, and Other L_α Norms

Chris C. Holmes
University of Oxford, UK
c.holmes@stats.ox.ac.uk

Alexandre Pintore
MRC Harwell, UK
pintore@stats.ox.ac.uk

Summary

Modern genomics has reinforced the need for statistical methods which can explore low-dimensional structure in very high dimensional data. Classical Relaxation methods, including Boosting and L_α Regularisation, have proved remarkably successful in addressing such problems and have arguably had an important influence on the field of modern regression analysis. Relaxation methods are best described as iterative model fitting procedures which start from a simple null model and then walk along a path of increasing model complexity by incrementing parameter values at each stage. Such methods are ideally suited to the analysis of genomics data because they usually provide 'automatic' selection and shrinkage of predictors. In this paper, we consider relaxation methods from a fully probabilistic standpoint, accommodating model uncertainty through prior distributions on the set of relaxation paths. We suggest to make inference by Importance Sampling over paths which retains the principle characteristics of Relaxation. This new probabilistic relaxation methods are built upon a connection between Regularisation and Generalised Ridge Regression. Results seem to suggest that these algorithms are of practical importance as alternatives to standard Bayesian methods and add weight to our belief that modelling the variance components of Bayesian GLMs provides a powerful and attractive approach to model determination in modern data analysis.

Keywords and Phrases: Boosting; Genomics; The Lasso; Machine Learning; Path Averaging; Relaxation; Variable Shrinkage & Selection.

1. INTRODUCTION

Modern genomics has reinforced the need for statistical methods which can explore low-dimensional structure in very high dimensional data. In such applications one is typically confronted with a great number of potential covariates so that the problem falls in the class of 'large p, small n' problems (West, 2002). Traditional methods such as least squares are not well suited to such problems since they lead to poor prediction accuracy, due to large variance, and do not facilitate qualitative interpretation in the structure governing the response mechanism. Thus, a number of methods have been developed in order to better tackle the 'small n, large p' problems in the sense just defined.

In this paper, we explore a general class of classical algorithms termed 'relaxation' approaches. These all have the characteristic of forward model building in which one starts from a null model and builds up complexity (relaxing a penalty) along a path through model space. Here, we will develop these methods within a fully probabilistic framework. In particular, our main findings and contributions are:

- To show that there is a formal correspondence between regularisation solutions for general L_α norms with $\alpha \in [0, 2)$ (such as the Lasso, L_1, and Ridge, L_2) and Bayesian Generalised Ridge Regression maximum a posteriori (MAP) estimates.

- To construct efficient approximation algorithms using Bayesian Generalised Ridge Regression, for fitting L_α regularisation models.

- To provide a general probabilistic framework for considering 'relaxation' methods.

- To highlight the use of Importance Sampling as an efficient inference procedure for implementing the relaxing property within a Bayesian framework.

- To show that Boosting has a formal representation as a sequential path of Bayesian MAP estimates.

- To provide a stochastic Boosting algorithm in order to incorporate a measure of uncertainty within standard Boosting.

The layout of the paper is as follows. Section 2 overviews the main statistical methods for tackling the 'large p, small n' problem. This is done by considering classical methods as well as their Bayesian counterparts. Section 3 highlights Generalised Ridge Regression (GRR) as a unifying framework for most of the approaches described in Section 2. In particular we show that there is a one-to-one correspondence between L_α regularisation and GRR. This suggests a fast stepwise model search algorithm, which we term Ridge Regression with Replacement (RwR) for approximating L_α solutions. In Section 4, we consider the embedding of RwR within a stochastic framework and derive two probabilistic models, the fully probabilistic Ridge with Replacement and the hybrid stochastic Ridge with Replacement. We describe their properties, as well as some of their limitations, especially with respect to inducing sparsity. Section 5 considers a development to RwR in another direction, which results in computationally efficient algorithms, that are very close to standard Boosting, but with measures of uncertainty. The paper closes with a discussion.

Throughout the paper we shall restrict ourselves to Generalised Linear Models (McCullagh and Nelder, 1989), that is, to models such that $E(\boldsymbol{Y}) = g(X\boldsymbol{\beta})$, where $g(\cdot)$ is a link function, X the $n \times p$ design matrix and $\boldsymbol{\beta}$ a $p \times 1$ vector of regression coefficients. Note that this model is quite general in that X can be built using non-linear basis functions such as regression splines for instance (Denison *et al.*, 2000, Smith and Kohn, 1996).

2. METHODS FOR TACKLING ILL-POSED REGRESSION PROBLEMS

We describe in this section the statistical methods that have been developed for handling 'small n, large p' problems and highlight some of the connections that characterise them. We start with traditional methods before describing their Bayesian counterparts.

2.1. *Traditional Non-Bayesian Methods for 'Small n, Large p' Problems*

The methods that have been developed in the classical literature can be grouped into four main categories:

- *Variable Selection.* This approach consists in selecting only a subset of the covariates, while eliminating the others from the model. The resulting model is then assumed to be 'true' and statistical inference, model checking and prediction is applied in the usual way (e.g. least squares for instance). A great number of methods exist for selecting a particular subset, the main ones being *best subset regression* and *forward stepwise selection*, which consists in finding a good path through the set of all possible subsets by sequentially adding into the model the predictor that most improves the fit according to some criterion, starting with only the intercept. Note that one can also use *backward elimination* by sequentially removing covariates from the full model though this is not generally applicable for $p \gg n$. With respect to all of these methods, more details can be found in Hastie *et al.*, (2001).

- *Regularisation.* Although variable selection methods are efficient in the sense that interpretation of the model is easy and prediction error usually small, they can still retain high variability (in that very different models can emerge for small changes to the data – to see this you can Bootstrap the data and examine the resulting model list), especially as the model becomes more and more complex. An alternative consists in imposing a penalty on the model regression coefficients. These methods are known as regularisation methods (as they were originally introduced as solutions to ill-posed inference problems) and are best cast as penalised likelihood approaches. The corresponding regression coefficients are obtained as

$$\hat{\beta} = \text{Arg Max}_{\beta}[l(\beta|\mathcal{D}) + P(\beta)] \tag{1}$$

  where $l(\beta|\mathcal{D})$ is a log-likelihood function which records fidelity to the data $\mathcal{D}$, monotone in the complexity of the model, and $P(\beta)$ denotes a penalty term on complexity. The Lasso and Ridge Regression penalise the vector of coefficients via an L_1 or L_2 norm respectively,

$$\begin{aligned} P_{(\text{Lasso})} &= -\eta \sum_j |\beta_j| \\ P_{(\text{Ridge})} &= -\eta \sum_j (\beta_j)^2, \end{aligned}$$

  where η is a smoothing parameter that governs the trade-off between the terms in (1).

  Ridge Regression dates back to the seminal paper of Hoerl and Kennard (1970) and was originally introduced as a solution for dealing with co-linearity in the column space of the covariate matrix X. More recently, the Lasso (Tibshirani, 1996) has become popular. The attractiveness of the Lasso stems from its well known property to produce sparse solutions via thresholding in that some/many of the coefficient values in $\hat{\beta}_{(Lasso)}$ may be zero, effectively removing the corresponding covariate (basis) from the model. In contrast, Ridge Regression provides solutions with all parameters non-zero, $\hat{\beta}_{(Ridge)_j} \neq 0, \forall j$.

More generally, a full array of regularisation penalty functions can be defined by considering L_α norms,

$$P_{(\alpha)} = -\eta \sum_j |\beta_j|^\alpha$$

with $\alpha > 0$, which was referred to as 'bridge regression' by Fu (1998). Throughout this paper, we will refer to these penalty functions as L_α regularisation. They offer a wide range of alternative solutions to the Lasso and Ridge Regression, while containing $\alpha = 1$ and $\alpha = 2$ as special cases.

- *Boosting.* The Boosting algorithm is widely recognised as one of the most important algorithms to have emerged from Machine Learning in the past 10 years. It was initially developed as a classification procedure (Y. Freund, 1995; Y. Freund and R. E. Schapire, 1996a, 1996b) aiming at reducing the bias of so-called 'weak learners' (i.e. simple models), in order to improve on the prediction performance of the models considered separately. It has gained increasing popularity due to the high levels of empirical predictive accuracy shown on disparate data sets as well as it's apparent, rather mysterious, robustness to overfitting. It has generated considerable interest and debate within the Statistics community, questioning traditionally held opinions on overfitting and even joint modelling (see for example comments by Buja in Friedman *et al.*, 2001). Important contributions to the statistical connections of Boosting can be found in Friedman *et al.* (2001), Buhlmann and Yu (2001) and Efron *et al.* (2001).

  In its standard form, Boosting, also referred to as *Forward Stagewise* by some authors (Efron *et al.*, 2004), can be described as an iterative model fitting procedure whereby at each step in the algorithm, the current set of residuals from the model fit are regressed on a (simple) additional component, usually a single explanatory variable or a step-function sometimes referred to as a '*stump*' or (very) shallow decision tree. Each step is characterised by a two stage procedure: first, the additional component is chosen by selecting the model fit with 'best' measure (usually likelihood). Secondly, once a direction has been chosen, rather than adding in the full parameter value for the chosen component, a 'shrunken' estimate is added by shaving the unconstrained estimate by some value fixed $\nu \in (0, 1)$ (Friedman, 2001). The final model is then written as a linear combination of the simple components fitted along the way. More details can be found in Buhlmann and Yu (2001). Thus, it appears to be a Forward Selection type of algorithm but where many small steps are taken as the model increases in complexity.

- *Dimension reduction via projection.* These are methods for which one seeks to produce a small number of linear combinations of the original covariates and then regress the response variable on the former rather than the latter. A number of methods exist for constructing such linear combinations, the best known ones being *Principal Components Regression* and *Partial Least Squares*, as described for instance in Hastie *et al.* (2001). We will not consider these further though we note that they are important methods in modern regression analysis and genomics.

We now review the close connections between the methods described above and emphasise that they all fall under a particular class of models, deemed 'relaxation' models.

2.2. *Connection between Subset Selection, Regularisation and Boosting*

The connections between Forward Selection and Boosting are clear since, as already mentioned above, the latter is a special case of the former, for which tiny steps rather than full steps in each covariate direction are taken. With respect to the connections between regularisation and Boosting, a major contribution to the statistical understanding of Boosting was provided by Friedman *et al.* (2001), where they illustrate that Boosting can be viewed as a gradient descent algorithm on a logistic loss function (for binary response data). This is highlighted further by Peng and Yu (2004) for normal linear regression. In turn, this allows for direct connections to be made between Boosting and other popular model fitting procedures using regularisation, such as the Lasso and Forward Selection. These connections as well as other results are discussed in Efron *et al.* (2001), where a new method, *least angle regression* (LARS), is developed which appears to be at midway between standard Boosting and standard Forward Selection. In particular, it can be shown that the Boosting and Lasso paths are equivalent under certain conditions (T. Hastie *et al.*, 2006)

We note that Forward Selection, Boosting and Regularisation can be referred to as *relaxation* methods, in that they have the form of a model fitting procedure which starts from a simple null model and then travels along a path of increasing model complexity by adding components to the model at each stage. It is useful for us to consider these relaxation methods as particular Markov decision processes, where conditional on the current state of the model a step is taken to increase the model complexity by taking one of p potential options. This property is essential in characterising these types of methods and we refer to it as the *relaxing* property. Finally, we note that Dimension-Reduction methods via Projection can also be viewed as satisfying the relaxing property and are therefore directly related to the other methodologies.

One of the main deficiencies with classical relaxation methods is their inability to account for uncertainty. For instance, standard Boosting consists in taking at each step the path with highest improvement in the particular score function under study. Often differences in score improvement between different directions can be so small as to render each direction near equally likely from a probabilistic point of view, yet only one direction, always the same, will be explored within the standard Boosting methodology. The non-Bayesian deals with this by assessing uncertainty to data sets which might have been observed but weren't, say through the use of Bootstrapping (Friedman and Popescu, 2003). In the Bayesian paradigm we condition on the observed Y and account for uncertainty via a prior on the model space.

2.3. *Bayesian Relaxation Methods*

We now describe the various approaches that have been used in order to incorporate Variable Selection, Regularisation and Boosting within a fully probabilistic framework.

- *Bayesian Variable Selection Methods.* Traditional variable selection are usually set into a Bayesian framework via a general hierarchical mixture formulation, which in the normal linear regression case, is of the form

$$P(Y|\beta, \sigma^2) \sim N(X\beta, \sigma^2 I) \tag{2}$$

where Y is a $n \times 1$ vector of responses, X is a $n \times p$ matrix of covariates, β the $p \times 1$ vector of unknown regression coefficients, and σ^2 the noise variance. Now, to each β_i, $i = 1, ..., p$ is attached an indicator $\gamma_i \in \{0, 1\}$, which indicates whether covariate i is in the model or not. Then, with $\gamma = (\gamma_1, ..., \gamma_p)'$, the uncertainty in model selection is modelled via a mixture model, that is $\pi(\beta, \sigma^2, \gamma) = \pi(\beta|\sigma, \gamma)\pi(\sigma^2|\gamma)\pi(\gamma)$ and, generally, one takes

$$\pi(\beta|\sigma^2, \gamma) = \pi(\beta|\gamma) \sim (1 - \gamma_i) I_{\beta_i = 0} + \gamma_i N(0, v(\gamma_i)) \tag{3}$$

and some prior, usually Inverse-Gamma and Binomial respectively, is taken for $\pi(\sigma^2|\gamma)$ and $\pi(\gamma)$, with all the γ_i's assumed independent. It is usual to specify $v(\gamma_i) = v$, $\forall i = 1, ..., p$. Inference is generally carried out using Reversible Jump Markov Chain Monte Carlo (Green, 2005) or Gibbs Sampling via stochastic search variable selection (George and McCulloch, 1996). More details on these approaches and efficient computational procedures for implementing them can be found in Smith and Kohn (1995), Geweke (1996), Clyde (1999) and more recently by Nott and Kohn (2005). Holmes and Denison (1999) considered a joint shrinkage and selection scheme for (3) with a different $v(\gamma_i)$ for each $i = 1, ..., p$. Note that this description generalises with little difficulty to the generalised linear regression case, $E(\boldsymbol{Y}) = g(X\boldsymbol{\beta})$.

- *Bayesian Regularisation.* The penalised likelihood approach sits naturally within a Bayesian framework since P in (1) is a subjective measure implying *a priori* preference for parsimony in the posterior model space of β. It is well known that Ridge Regression and the Lasso correspond to particular cases of (2) where the β_i's are independent and identically distributed with distribution $N(0, v_i)$ for some common parameter v_i. Ridge Regression then corresponds to $v_i = 1/\lambda$, $\forall i = 1, ..., p$, while the Lasso corresponds to

$$\pi(v_i) \sim Exp(\lambda), \; i = 1, ..., p \tag{4}$$

where $Exp(\cdot)$ denotes the exponential distribution. Note also that L_α norms correspond to power exponential families, for $\alpha \in [0, 2)$ (Box and Tiao, 1973). For further examples and recent extensions see Griffin and Brown (2005) and Park and Casella (2005), who both study the Lasso within a fully Bayesian framework.

Other developments of ridge type priors include Neal (1996), who developed the *automatic relevance determination* (ARD) framework, which was further particularised by Tipping (2001) under the name *relevance vector machine*

(RVM). These consist in assuming an Inverse Gamma $IG(a, b)$ distribution on the v_i's, rather than an Exponential distribution as in the Lasso case, and then, to adopt a *type II maximum likelihoods* criterion for the v_i's. Figueiredo (2001) considered the particular $IG(0, 0)$ distribution, that is, Jeffrey's prior. More generalisations, as well as a broad analysis can be found in Griffin and Brown (2005). Inference is generally undertaken using Markov Chain Monte Carlo.

- *Bayesian Boosting.* Unlike Variable Selection and Regularisation, Boosting seems to have received little attention within a Bayesian framework. Recently however, Chipman *et al.* (2005) have drawn connections between Boosting and Bayesian model fitting for a sum-of-decision trees model, where the sum of trees rather than each tree is treated as the model itself. The authors advocate a type of backfitting Markov Chain Monte Carlo algorithm for iteratively constructing and fitting successive residuals. This method is somewhat based on the characterisation by Friedman *et al.* (2001) of Boosting as a gradient descent algorithm on a logistic loss function.

- *Bayesian dimension reduction via projection* Similarly to Boosting, these methods seem to have received relatively less attention compared to Regularisation and Variable Selection. It is worth mentioning though the important contributions of Clyde *et al.* (1996) and West (2002).

It is interesting to note, with respect to comparing the different approaches of Variable Selection, Regularisation and Boosting, that the differences between all the latter are somewhat blurred within a Bayesian framework. Indeed, although different priors are used, all rely on hierarchical linear models, with a conditional normal distribution on the β_i's which are further assumed to be conditionally independent and identically distributed. This is perhaps not so true for Boosting though little work has been done in this area.

It is also instructive at this point to highlight the main difference between traditional methods and their Bayesian counterparts. Whereas the former are truly relaxation methods, in that one obtains *paths* of models of increasing complexity, the latter are not since one generally obtains a sequence of independent and thus possibly very different models. Indeed, the relaxing property is blurred within a Bayesian framework, where algorithms such as Markov Chain Monte Carlo are used.

3. A UNIFIED FRAMEWORK USING GENERALISED RIDGE REGRESSION AND VARIANCE COMPONENTS MODELS

In this section, we highlight Generalised Ridge Regression, and more generally variance components models, as a unifying framework for most of the traditional approaches described in Section 2. In particular, we show below that regularised estimators for L_α, $\alpha \in [0, 2)$ are equal to Bayesian maximum a posteriori (MAP) estimates under GRR. We also draw a link between GRR and Boosting in Section 5. This enables us to develop an alternative insight into these methods by which they can be described as selecting specific paths along their variance components.

GRR was mentioned in the original paper by Hoerl and Kennard (1970) and its Bayesian interpretation was discussed in Smith and Goldstein (1974) and Walker and Laud (1999). Consider the following constrained GRR Bayes MAP estimator,

$$\hat{\beta}_{(GRR)} = \text{Arg Max}_{\beta} \left[l(\beta) - \sum\nolimits_j v_j^{-1} \beta_j^2 \right], \quad \text{such that} \quad \sum\nolimits_j \{v_j\}^r = K \tag{5}$$

so that the GRR has a positive valued individual ridge parameter, v_j^{-1}, for each coefficient subject to the constraint that their rth power sum is a constant K for some $r > 0$. We choose to use the reciprocal notation, v_j^{-1}, as the parameters v_j then have an interpretation as the variance component of a mean zero independent normal prior on β_j.

Theorem 1 *Let $\alpha \in (0, 2)$ be any positive real number. For every L_α regularised solution $\hat{\beta}_{(\alpha)}$ determined by η, as given in Section 1, there exists a GRR solution $\hat{\beta}_{(\mathrm{GRR})}$ defined for some K, under the constraint*

$$\sum\nolimits_{i=1}^{p} v_i^r = K, \; r > 0, \tag{6}$$

with $r = \alpha/(2-\alpha)$ such that $\hat{\beta}_{(\alpha)} = \hat{\beta}_{(\mathrm{GRR})}$. The relationship is one-to-one and monotone so that $\eta' > \eta$ implies $K' > K$ so that the continuum of α solutions for varying η is played out by varying K on $[0, \infty)$.

The proof of this result is given in Appendix A. The implications of this theorem are that all the L_α regularised solutions have a Bayesian GRR representation under mean zero normal priors with constraints on the variance components of the priors.

Remark 1. Note that in the particular case of the Lasso (i.e. $\alpha = 1$), this equivalence was noted independently by Yves Grandvalet (1996), though the main motivation in that paper was to derive an EM algorithm for deriving the Lasso solution. Unfortunately the EM is not guaranteed to obtain the optimal solution and hence the approach is perhaps less attractive than the standard dynamic programming.

Remark 2. The uniqueness of the solution depends on that of the corresponding L_α regularisation problem.

Remark 3. It is clear from the proof of Theorem 1 given in Appendix A that the result remains valid if the residual sum of squares is replaced by any other loss function with similar properties, such as the logistic loss function for instance.

Remark 4. An equivalence between standard Boosting and GRR can also be stated. The corresponding result will be given in Section 5.

We can see that one recovers the ordinary ridge as $\alpha \to 2$, that is, $r \to \infty$. This result is derived in Appendix B. Another important limiting case corresponds to $r \to 0$ which leads to $\alpha \to 0$ and to the important result that the GRR tends to the ordinary forward stepwise algorithm.

3.1. *Properties*

Considering relaxation methods from a variance component (GRR) point of view offers a number of advantages over the standard methods based solely on penalised regression coefficients:

- First, one is able to define all L_α norms within a unified framework. This leads to the possibility of developing GRR algorithms able to span all L_α solutions. In the next section, we suggest a computationally efficient stepwise model search algorithm, termed ridge regression with replacement (RwR), which can be seen to approximate all L_α solutions.
- Second, we are able to approximate the degrees of freedom of any L_α solution. This consists in using a linear approximation to the degrees of freedom of the nonlinear estimate obtained. Note that this equivalence can also be used if an L_α solution is obtained using a different algorithm. Details are given for the Lasso case in Appendix C. These can be generalised to any $\alpha \in (0, 2)$.
- Thirdly, in the normal regression case we are able to calculate the marginal likelihood for different values of variance components thus obtaining a measure of the evidence in support of a model.
- Finally, we are able to retain conditional Gaussianity properties of our coefficients and predictions, when working with any L_α norm.

Although the GRR framework is very appealing since it enables one to use well-known machinery, it is important to note that a drawback of considering the variance components rather than the regression coefficients lies in the extra parameterisation that the GRR framework evolves. This is mainly a computational problem though.

3.2. *Stepwise Model Search using Ridge with Replacement*

The Lasso solution can be computationally hard to compute for large numbers of potential basis functions or for non-normal likelihoods. This has motivated people to consider stagewise variable selection as an approximation to the Lasso, and as a model search algorithm in its own right. This was already mentioned in Section 2. For example, the Blasso approach outlined in Zhao and Yu (2005) demonstrates an efficient forward-backward stepwise algorithm that matches the Lasso solution path to an arbitrary level of approximation.

Motivated by the paper of Zhao and Yu (2005) and more importantly by the theoretical results derived above, we propose a stepwise GRR model search algorithm, which works as follows. To begin, the vector of ridge parameters is set to 0, that is, each $v_j = 0$ for all j (and hence all β's are shrunk out of the model). Then at each iteration we seek to add a small constant Δ to one of the elements of v_j. The choice is determined by maximum penalised likelihood, as described below. The procedure is then repeated, at each stage adding an extra Δ to one variance component, so that we obtain a path through the space of regularisation parameters. We term this method Ridge Regression with Replacement as it is equivalent to variable selection with a fixed prior, $N(0, \Delta)$, where any covariate is allowed to enter into the model multiple times. We describe the algorithm further, initially for normal linear regression.

For given $V = \text{diag}\{v_1, \ldots, v_p\}$, Let $C(\cdot)$, define the GRR cost function to be maximised as

$$C(V) \quad = \quad l(\hat{\beta}|V) + P_{(\text{GRR})}(\hat{\beta}|V) \tag{7}$$

where $\hat{\beta}$ is defined as

$$\hat{\beta} \quad = \quad (X'X + V)^{-1}X'Y \tag{8}$$

so that $\hat{\beta}$ denotes the MAP estimate of β for given V. It is clear that (7) corresponds to maximum penalised likelihood for a set of $v_1, ..., v_p$ or to a Bayes MAP solution. Note: for other GLM likelihoods the estimates of $\hat{\beta}$ can be obtained efficiently using iterative reweighted least squares.

From above, we propose to build an algorithm for model search along maximum gradient path for (7) as follows,

`Initialise:` $v_j^{(0)} \leftarrow 0; \ \forall j$

`For i=1 to T`

$v_j^{(i)} \leftarrow v_j^{(i-1)}$; `for` $j = 1 \ldots, p$

$\hat{j} \leftarrow \arg\max_j C(v_1, \ldots, v_j + \Delta, \ldots, v_p)$

$v_{\hat{j}}^{(i)} \leftarrow v_{\hat{j}}^{(i)} + \Delta$

`End`

The above returns the Lasso solution (as $\Delta \to 0$). The more general algorithm is obtained if at each step, given the current values $(v_1, ..., v_p)$ one proposes to add

$$\Delta_j = \left(v_j^r + \Delta\right)^{1/r} - v_j, \ j = 1, ..., p$$

to each component, so that again, K, which is now defined as $\sum v_i^r$, increases in steps of size Δ, where $r = 2/(2 - \alpha)$, as described in the previous section. The resulting algorithm therefore takes the following form,

`Initialise:` $v_j^{(0)} \leftarrow 0; \ \forall j$

`For i=1 to T`

$v_j^{(i)} \leftarrow v_j^{(i-1)}$; `for` $j = 1 \ldots, p$

$\Delta_j^{(i)} = \left(\{v_j^{(i)}\}^r + \Delta\right)^{1/r} - v_j^{(i)}$; `for` $j = 1 \ldots, p$

$\hat{j} \leftarrow \arg\max_j C(v_1, \ldots, v_j + \Delta_j^{(i)}, \ldots, v_p)$

$v_{\hat{j}}^{(i)} \leftarrow v_{\hat{j}}^{(i)} + \Delta_{\hat{j}}^{(i)}$

`End`

An example for different L_α norms is provided in the next subsection.

The paths obtained provide good approximations to the exact solution at each stage as can be checked from the conditions for the optimal solution as given in Appendix A. See related comments in Efron *et al.* (2004).

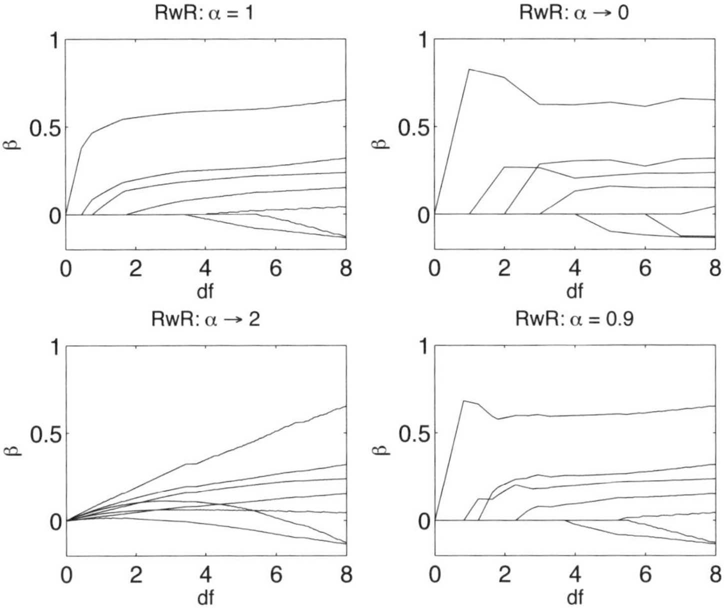

Figure 1: *RwR algorithm (Section 3.3) for $\alpha = 1$ (Lasso), $\alpha \to 2$ (RR) and $\alpha \to 0$ (FSS) and $\alpha = 0.9$, plotted against the degrees of freedom calculated as in Appendix C. The vertical axis denotes the value of the regression coefficients as a function of the approximate degrees of freedom. Note that all β values start at 0 and asymptote towards their MLEs at the right hand side.*

3.3. *Example*

In this section, we consider the application of the RwR Algorithm described above. To do so, we focus on the well known prostate data set, which has already been considered, among others, by Tibshirani (1996) and Hastie *et al.* (2001). We apply the algorithm for the lasso case (i.e. $\alpha = 1$), the forward stepwise search algorithm (i.e. $\alpha \to 0$), ridge regression ($\alpha \to 2$) and for $\alpha = 0.9$. Results are given in Fig. 1.

It is clear that our algorithm is able to approximate the ridge regression path, as well as the forward stagewise selection path as shown in Fig. 1. It can also be seen from Fig. 1 that $\alpha = 0.9$ offers a compromise between forward stagewise selection and the Lasso. Finally, we note that the paths evolve non-linearly as α moves away from 1.

4. PROBABILISTIC RELAXATION METHODS

One of the main drawbacks of the classical Ridge with Replacement algorithm using Bayes MAP lies in its lack of a measure of uncertainty. Indeed, similarly to the traditional non-probabilistic approaches developed in Section 2, one is unable to account for the uncertainty in the derived path. Thus, there is a need to incorporate RwR within a fully stochastic framework.

In the previous section, we have highlighted the link between traditional methods for handling 'large p, small n' problems and GRR. A similar link can be drawn within the Bayesian framework. Indeed, recalling the description of Bayesian methods given in Section 2, it is seen that most of them can be set under the GRR framework, in the sense that, within the normal linear regression case, one assumes the following model,

$$\begin{aligned} \boldsymbol{Y}|\boldsymbol{\beta} &\sim N(\boldsymbol{X}\boldsymbol{\beta}, \sigma^2 I) \\ \boldsymbol{\beta}|\boldsymbol{v} &\sim N(\boldsymbol{0}, V) \end{aligned}$$

where V is a $p \times p$ diagonal matrix with diagonal $\boldsymbol{v}$. Methods then usually differ with respect to the prior imposed on the $p \times 1$ vector v and on the inference procedure. Thus, they lead to stochastic walks on $\boldsymbol{v}$ and $\boldsymbol{\beta}$, where the probabilistic measures for the walks depend on the prior chosen. This suggests an easy way to incorporate ridge with replacement within a Bayesian framework using a Markov Chain Monte Carlo algorithm in order to infer posterior distributions on the regression coefficients $\boldsymbol{\beta}$ and the variance components $\boldsymbol{v}$ space.

This solution is perhaps a little unsatisfactory, especially with respect to the analysis of genomics data since, as already mentioned in the Section 2, the relaxing property is usually lost when Markov Chain Monte Carlo is used and with it the possibility of obtaining paths of models with increasing complexity. In particular, we note that sparse models, although exhibiting small probability, can be of great interest from an exploratory point of view.

4.1. *Fully Probabilistic RwR – Ridge Regression with Resampling – FPRwR*

The natural embedding of the previous models within a fully Bayesian framework takes the form of a model we term *ridge regression with resampling.* That is, we let

$$\begin{aligned} E(Y) &= \sum_{j}^{M} X_{d_j}\beta_j \\ \beta_j &\sim N(0, \Delta) \end{aligned}$$

where Δ is a small shrinkage constant and $d_j \sim M_p$ is a Multinomial (path) indicator variable which indicates which covariate the jth β acts on. Fully probabilistic approaches can be derived by specifying priors on $\pi(M)$ and $\pi(d_j)$.

Typical representations may be considered with $\pi(M) \sim Ge(\cdot)$ a Geometric distribution and $d_j \sim U[1, \ldots, p]$ a uniform on the p covariates. Note that unlike the traditional SSVS models (Clyde, 1999; George and McCulloch, 1993, 1999) each predictor variable can enter into the model *multiple times*, hence the name ridge with resampling. Note that this is formally equivalent to allowing for individual prior variance components in a Bayes linear model.

One can see that, in the limit as $\Delta \to 0$, $M \to \inf$ we can obtain limiting cases of well known representations such as the Lasso and ARD. One advantage of working in the discrete case is that we have extra flexibility to consider joint distributions on $\pi(d_1, \ldots, d_M)$ which favour sparsity in the number of zero coefficients.

One issue with the scheme above is that the Bayesian solutions do not map out paths, but rather gravitate to high-probability regions in the model space. For example, in Fig. 2 we see the results from running an MCMC stratergy to sample from the above model with flat priors on M and a uniform prior on d. What we observe is that the Bayesian solution finds a trade off between complexity and model fit as we would expect. We also see that all of the parameters are 'in the model'. This is also to be expected. However, as stressed above, regions of the model space with low probability may have high utility in terms of their qualitative information on the structural dependencies in the data. Clearly we could simply increment M above and run a separate simulation at each stage. However, setting up this procedure is time consuming and knowing *a priori* a suitable range for M and the increments is difficult to achieve.

In the remaining of the paper we explore pseudo-Bayesian methods which are motivated by full Bayesian Relaxation, but arguably posses certain computational and qualitative benefits. We would stress however that these methods need to be used with caution as they no longer posses the insurance of Bayesian coherency.

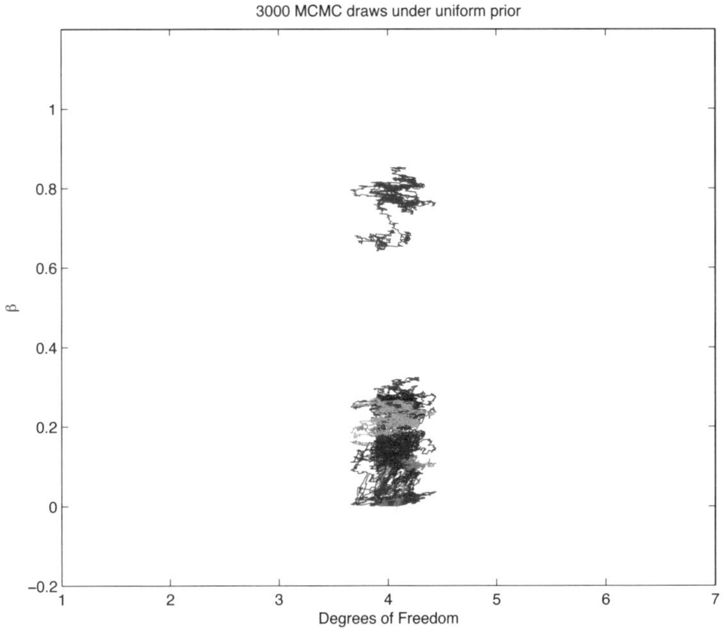

Figure 2: *MCMC sample paths run on Ridge with Resampling using flat priors on the number of components and* $\Delta = 0.01$ *for the Prostrate data of Fig. 1.*

4.2. *A Hybrid Approach – HRwR*

We now describe an alternative approach which is based on a Bayes MAP likelihood argument. Thus, we focus on the uncertainty in the $\boldsymbol{d}_i$'s and in M only; ignoring the uncertainty in β. That is, rather than considering paths on both the variance components and their regression coefficients, we consider paths solely on variance components, so that two paths leading to the same final estimate $v = (v_1, ..., v_p)$ lead to the same likelihood.

We consider a path, P, to be an ordered collection of directions chosen, $P = \{d_1, d_2, \ldots, d_M\}$. As before we take the Markov uniform distribution

$$Pr(\boldsymbol{d}_i|\boldsymbol{d}_{i-1}) = 1/p. \quad (9)$$

We note however that (9) can be modified in order to allow for sparser models by paying an extra penalty for a predictor variable which first enters into the model. We discuss this further in the Section 4.5.

With respect to the prior distribution on the number of iterations, M, it is important to note first that the former has no influence on the direction chosen at each step since all directions result in a step forward. Thus, its main influence lies in the prediction properties of our model. Since the model complexity increases with the number of iterations along a path, we must account for this increased complexity when comparing different model along a path, similarly to traditional model selection criteria. Thus, we take

$$\log Pr(M = m) \propto \sum\nolimits_{i=1}^{p} \{v_i\}^r, \quad m = 1, ..., I_{max} \quad (10)$$

where $v_i = \Delta \sum_j I(d_j = i)$, where $I(\cdot)$ is the identity function, so that

$$Pr(M = m) \propto exp\{K \sum\nolimits_{i=1}^{p} v_i^r\}, \quad m = 1, ..., I_{max} \quad (11)$$

for some constant K. It is clear from (10)–(11) that increases in model complexity are penalised. In the Lasso case, (11) results in

$$Pr(M = m) \propto exp\{K^{'}\Delta\}, \ m = 1, ..., I_{max}$$

for some constant $K^{'}$, so that in the limit, as $\Delta \to 0$, this can be viewed equivalently as taking an exponential distribution for $\sum v_i$. Note also that if one takes

$$\log Pr(M = m) \propto \frac{mp}{2I_{max}} \log n \quad (12)$$

then the resulting weight attached to each model along a path would correspond to a criterion very close to the BIC criterion, with the degrees of freedom replaced by $m \times p/I_{max}$.

With respect to the posterior distribution, we now define the pseudo-likelihood to be

$$Pr(Y|P, M) = Pr(Y|F) \sim N(F, \sigma^2) \quad (13)$$

for some variance parameter σ^2, that we assume known in what follows, without loss of generality, and where F is calculated by

$$\hat{F} = X(X'X + V)^{-1}X'Y \quad (14)$$

where, given a path $\boldsymbol{d}_i$, $i = 1, ..., M$, V is determined as explained in the previous section.

4.3. *Inference*

We now consider the task of inference. We wish to explore the use of Importance Sampling rather than MCMC since the former appears to possess a number of advantages over the latter, the more important and relevant ones being:

- Sampling paths rather than sequences of models as with the MCMC enables us to retain the main characteristics of standard relaxation algorithms, that is, the relaxing property. In other words, we wish to obtain collections of stochastic regularisation paths of models rather than sequences of models, as would be the case with MCMC, in order to mimic the behaviour of standard relaxation algorithms.
- MCMC does not naturally lend itself to forcing moves through the null space in order to obtain a path with strictly increasing complexity as is the case with general relaxation methods. This is an important property, especially from an exploratory point of view since it enables one to gain qualitative information from each run, something that MCMC runs don't provide naturally unless we fix $M = m$.
- It enables us to explore a greater number of simpler models than would usually be the case for a reasonable $\pi(M)$ and MCMC since the latter have very low numerical probability and thus are rarely explored by the chain, though they may have high qualitative merit. Note that one could clearly also re-run MCMC many times with more stringent penalties though this is time-consuming and hard to calibrate.
- Different priors can easily be examined without re-runs.

We now describe our Importance Sampler further. First, recall that we have

$$Pr[(P, M)|Y] = Pr(P|Y, M)Pr(M|Y). \tag{15}$$

It is easy to simulate from $Pr(P|M, Y)$, that is, to simulate a posterior regularisation path given the number of iterations. Indeed, we have

$$\begin{aligned} Pr(P|M, Y) = &= Pr(\boldsymbol{d}_i,\ i = 1, ..., M|Y) \\ &= \prod_{k=2}^{M} Pr(\boldsymbol{d}_k|\boldsymbol{d}_j,\ j = 1, ..., k-1,\ Y)Pr(\boldsymbol{d}_1|Y) \end{aligned} \tag{16}$$

and, for $k = 2, ..., M$,

$$\begin{aligned} Pr(\boldsymbol{d}_k|\boldsymbol{d}_j,\ j = 1, ..., k-1,\ Y) &\propto Pr(Y|\boldsymbol{d}_j,\ j = 1, ..., k)Pr(\boldsymbol{d}_k|\boldsymbol{d}_{k-1}) \\ &\propto Pr(Y|F_k) \end{aligned} \tag{17}$$

following (9) and where

$$F_k = X(X'X + V_k)^{-1}X'Y \tag{18}$$

with V_k the diagonal matrix with diagonal v^k, obtained from the values of $\boldsymbol{d}_j$, $j = 1, ..., k$. It is then clear that simulating from the conditional posterior $Pr(P|M, Y)$ is easily done via sequential sampling.

It thus only remains to be able to sample from the marginal posterior distribution $Pr(M|Y)$. The latter is given by

$$\begin{aligned} Pr(M=m|Y) &\propto \sum_{P_m} Pr(M=m, P_m|Y) \\ &\propto \sum_{P_m} Pr(Y|P_m)Pr(P_m|M=m)Pr(M=m) \end{aligned} \tag{19}$$

where P_m corresponds to the set of paths P such that $(P, M) \in \mathcal{R}$ and $M = m$, $Pr(M = m) = 1/I_{max}$, $Pr(Y|P_m)$ is given by (13) and $Pr(P_m|M = m)$ is described in the previous subsection.

4.4. *An Importance Sampler Probabilistic RwR*

Because one is generally interested in using Regularisation for prediction as well as qualitative interpretation purposes, we present a procedure that enables one to incorporate prediction and inference using the previously described procedure in a computationally efficient way. Thus, assume that one wishes to carry out prediction using a number N of stochastic RwR. One can then proceed using the following procedure:

(i) Simulate a stochastic path $P_{I_{max}}$ from $Pr(P|M, Y)$, with the number of iterations taken to be $M = I_{max}$.

(ii) Repeat the previous step N times, thus obtaining N independently and identically distributed stochastic paths $P^{(i)}_{I_{max}}$, $i = 1, ..., N$.

(iii) Using (19) we approximate $Pr(M = j|Y)$ for $j = 1, ..., I_{max}$ by

$$Pr(M=j|Y) = \frac{\sum_{i=1}^N Pr(Y|P_j^i)Pr(P_j^i|M=j)}{\sum_{i=1}^N Pr(P_j^i|M=j)}, \quad j=1,...,I_{max} \tag{20}$$

where $P_j^{(i)}$ corresponds to the truncation of $P^i_{I_{max}}$ after j steps, and (20) is normalised. Here, under the uniform distribution (9) we get

$$Pr(M=j|Y) = \frac{\sum_{i=1}^N Pr(Y|P_j^i)}{N} \tag{21}$$

Note that we use (20) because although each path has less probability as M increases, one must account for the fact that the number of paths over which to take the sum increases with M.

(iv) Now with respect to the task of

- inference: for each path $i = 1, ..., N$ in turn, select a number of iterations M_i using the approximation to $Pr(M = j|Y)$ determined above. This results in N sample paths from the true posterior distribution.

- prediction: predict at a new point $x*$ using

$$y^* = \frac{1}{N} \sum_{j=1}^{I_{max}} \sum_{i=1}^{N} y^*_{j,i} w^{j,i}_{IS} \tag{22}$$

where $y^*_{j,i}$ corresponds to the prediction using P^i_j and $w^{j,i}_{IS}$ is the (normalised) importance weight with $M = j$ and with $Pr(M = j|Y)$ approximated using the procedure described in the previous subsection.

Note that the efficiency of this algorithm stems from the fact that the samples we obtain are independent and identically distributed.

4.5. *Examples and Difficulties*

We return to the prostate data set considered in Section 3.3 and show some runs from the stochastic RwR. We consider $\alpha = 1$ and $\alpha \to 0$ as these correspond respectively to the Lasso and Forward selection algorithms. The other cases, although interesting, highlight the same characteristics as the former. The stochastic paths were obtained for $M = 10000$ iterations and steps of size $\Delta = 0.01$. Results are given in Fig. 3.

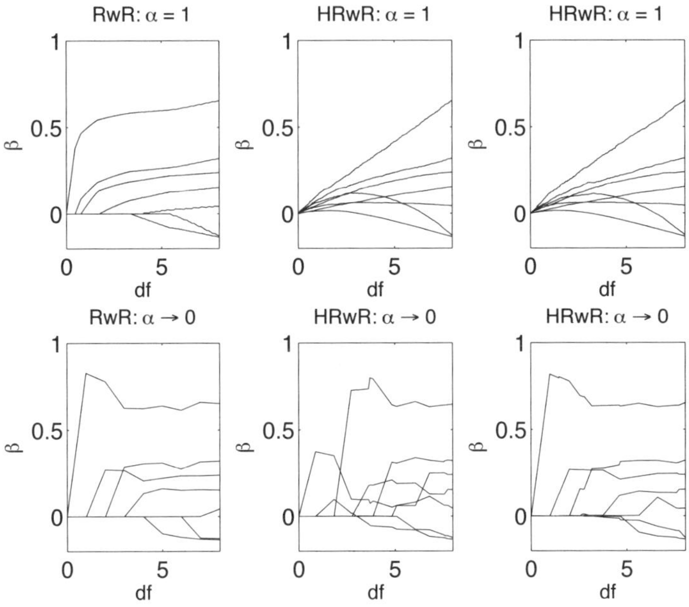

Figure 3: *RwR and HRwR algorithm. Above: $\alpha = 1$ (Lasso). Left plot: RwR; middle and right plots: simulations from HRwR. Below: $\alpha \to 0$ (FSS). Left plot: RwR; middle and right plots: simulations from HRwR. Section 4.6.*

It is clear from Fig. 3 that the Bayesian solutions for $\alpha = 1$ do not lead to sparse paths. Indeed, because of its stochastic nature, as the step size becomes smaller, the differences in likelihood for different directions become so small as to not discriminate between them. This therefore results in stochastic paths that resemble very much those of Ridge Regression. This feature is also shared by the Fully Probabilistic RwR of Section 4.1. We note that in simulations, using 10-fold cross-validation, the predictive ability of the full Bayesian model to the Hybrid approach was very similar.

The ridge like property gives us insight into why Bayesian versions of the Lasso, as described for instance in Park and Casella (2005) and Griffin and Brown (2003), do not result in sparse models with high probability, unlike the non-stochastic Lasso. The reason is clearly because the non-Bayesian models take a path of maximum probability at each stage. In order to get round this problem, one must therefore ensure sparsity within the model and especially within the prior. This is done in the next section.

4.6. *Encoding Sparsity for FPRwR and HRwR*

In the previous subsection, we highlighted the fact that HRwR does not lead to sparse paths (neither does FPRwR). Thus if one wishes to obtain sparser stochastic paths, that is, put more probability on sparser models, then two solutions could be used. First, clearly one could choose a smaller α. Indeed, it was shown in Section 2, that as α tends to zero, the paths tend towards those of a Forward Stepwise Selection (FSS). This tendency towards strong sparsity is quite stringent since simulations showed that for $\alpha = 0.75$, the paths resemble a FSS much more closely than the Lasso.

Secondly, we could change our prior on the directions $\boldsymbol{d}_i$, $i = 1, ..., M$ in order to emphasise the fact that we favour sparsity. This can be done by changing the uniform distribution (9) in Section 4.2, by incorporating an extra cost to be paid for entering the model, that is, for moving into a direction for which the current variance component is equal to zero. An example is given for the prostate data described previously in Fig. 4. Here the model includes a log penalty of p for the initial inclusion of a variance component larger than 0. This amounts to an extra spike of probability at $P(v_j = 0)$. It is clear from Fig. 4 that we are now able to obtain paths that exhibit sparsity, while retaining $\alpha = 1$.

5. A COMPUTATIONALLY EFFICIENT STOCHASTIC BOOSTING ALGORITHM

From a practical point of view, although the methods described in the previous section offer great flexibility, qualitative interpretation, as well as an alternative to standard Bayesian regularisation methods, they suffer from an important drawback. Indeed, they require one to invert a $K \times K$ matrix at each step in order to obtain the regression coefficients, where K is the dimension of the current model being examined. Now, inverting a $K \times K$ matrix is of order $O(K^3)$ and thus this can become prohibitive as K increases.

In this section, we therefore draw a link between RwR and standard Boosting and use this link in order to derive a computationally efficient algorithm that is close to Boosting in essence, while retaining a stochastic flavour.

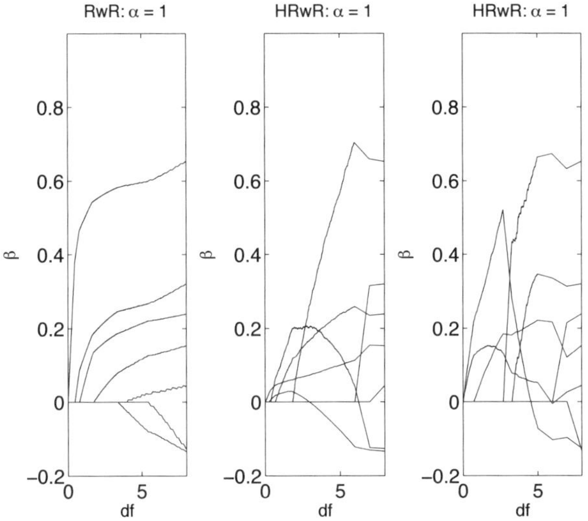

Figure 4: *RwR and two simulations from the Hybrid RwR algorithm for $\alpha = 1$ (Lasso), with sparse priors. Section 4.7. The Bayes solutions are now sparse, due to the inclusion of an entry penalty into the model.*

5.1. *A Link between RwR and Boosting*

Recalling the principles of Boosting as described in Section 2 and of RwR as described in Section 3, we note that Boosting is a particular type of RwR by which one chooses the best direction at each step based on some criterion but then, rather than updating the regression coefficients jointly, one simply updates the coefficient corresponding to the direction for the selected move, the others being held fixed. That is,

- RwR chooses a direction at each step according to $\hat{\beta}_{(MAP)}$, while Boosting chooses a direction according to $\hat{\beta}_{k(MLE)}$, $k = 1, ..., p$.

- Once a direction has been chosen, RwR updates all the coefficients to $\hat{\beta}_{(MAP)}$, while Boosting updates the coefficient corresponding to the selected direction, $\hat{j}$ say, to $\nu\hat{\beta}_{j(MLE)}$, for a shrinkage parameter $\nu \in [0, 1]$.

From the two properties just derived, it can be seen that Boosting is a stepwise GRR algorithm, where at each step i of the algorithm, one assumes

$$\begin{aligned} \boldsymbol{Y} - X\boldsymbol{\beta}^{(i-1)} &\sim N(X(:, \hat{j})\beta_{\hat{j}}^{(i)}, \sigma^2) \\ \beta_{\hat{j}}^{(i)} &\sim N(\boldsymbol{0}, g) \end{aligned}$$

for $i = 1, ..., M$, where $g > 0$ is a fixed parameter, and $\hat{j}$ the selected direction at the ith step. Then, it is clear that selecting a new variable to add via the MLE estimate or the MAP estimate is equivalent, since the priors are identical. Moreover the MAP estimate for the selected covariate is simply

$$\beta^{(i)}_{j,MAP} = \frac{g}{1+g}\beta^{(i)}_{j,MLE},\ j = 1, ..., p$$

so that the MAP estimate corresponds to a shaving of the MLE estimate with a shaving value given by $g/(1+g)$. Thus, this leads to the following result

Theorem 2 (Boosting Theorem). *For normal linear regression standard Boosting with shaving parameter ν is equivalent to a sequential Bayes (MAP) estimator with common variance $g = \nu/(1+\nu)$, with the MAP estimate selected at each step.*

Because Boosting only consists in univariate updating at each step, its computational efficiency is much greater than most other methods, especially with respect to problems involving a large number of covariates. Thus, with applications in genomics in mind, we propose a stochastic Boosting algorithm next, as an alternative to the methods of the previous section.

5.2. *A Stochastic Boosting Algorithm*

We consider the linear regression setting, as described in Buhlmann and Yu (2001), but extensions to generalised linear model can be implemented with no added difficult.

Consider the likelihood framework of the Hybrid RwR, given in Section 4.2. With respect to choosing a direction, we proceed along the same lines as for Hybrid RwR, that is, we consider a stochastic path $P = \{\boldsymbol{d_i}\}$, $i = 1, ..., M$ of directions, but rather than taking the likelihood to be given by (13)–(14), we take

$$P(Y|P) \sim N(F_B, \sigma^2) \tag{23}$$

with

$$F_B = \sum_{i=1}^{M-1}\left\{\sum_{j=1}^{p} x_j \beta^{(j)}_{S,i} d^{(j)}_i\right\} + \sum_{j=1}^{p} \beta^{(j)}_{mle,M} d^{(j)}_M \tag{24}$$

where each $\beta^{(j)}_{S,i}$, $i = 1, ..., M$ corresponds to the maximum likelihood estimator at step i if direction j is chosen, shaved by an amount ν, with the previous values obtained being held fixed. With respect to the last direction to be chosen, note that the unshaved maximum likelihood estimator, $\beta^{(j)}_{mle,M}$ is chosen for each possible direction j. Recall as well that at each step i, $\boldsymbol{d_i}$ follows a multinomial distribution with parameters $(1, \alpha_1, ..., \alpha_p)$ so that only one direction is chosen at each step.

The setting of this model is then similar to that of Section 4.2, with (13)–(14) simply replaced by (23)-(24). Updating proceeds similarly with the univariate MAP in the selected direction replacing the full MAP estimate in Section 4.2. The final estimate, given an entire path of length M, is thus given by

$$\hat{F} = \sum_{i=1}^{M}\left\{\sum_{j=1}^{p} x_j \beta^{(j)}_{S,i} d^{(j)}_i\right\} \tag{25}$$

Inference and Prediction follow similarly as in Sections 4.3 and 4.4, and an efficient Importance Sampler for stochastic Boosting can be constructed along the same lines as in Section 4.5.

5.3. *Examples*

In this section we consider some applications of the stochastic Boosting algorithm. In particular, we show that it is able to lead to better predictions than standard Boosting in some cases. We consider two applications: the Indian pima data, which was described in Ripley (1996) and the Hearts disease data is available at: http://www-stat.stanford.edu/ hastie/swData.htm. Both examples correspond to $0-1$ responses and so we use a Logistic stochastic Boosting algorithm.

In order to implement the Logistic stochastic Boosting algorithm, we use an iterative reweighted least squares algorithm, as given for instance in McCullagh and Nelder (1989), for calculating the maximum likelihood in each direction, at each step. This is straightforward to implement and highlights the advantages of working on the variance components. Moreover, because this results in only p univariate iterative reweighted least squares runs at each step, the resulting algorithm is computationally very efficient. With respect to the degrees of freedom, we approximate the latter via a linear approximation at each stage,

$$df_k = \sum_{j=1}^{M} \nu(1 - df_k^{(j-1)}) I_{d_j=k}, \ k = 1, ..., p$$

where M is the number of iterations, $df_k^{(j-1)}$ denotes the degrees of freedom for covariate k after $j-1$ steps, with $df_l^{0)} = 0$, $l = 1, ..., p$, and $I_{d_j=k}$ is an indicator function indicating which direction was chosen at step j.

Before considering the task of prediction, we plot the Boosting path, as well as two simulated stochastic paths in Fig. 5. The stochastic paths were obtained with $M = 1000$ iterations and a shaving parameter $g/(1+g)$ equal to 0.01.

Credible Intervals for the paths are given in Fig. 6. These were obtained by simulation, using $N = 200$ stochastic paths.

It is clear from Fig. 5 that our methodology is able to account for uncertainties in the data with respect to the relative importance of covariates one with another at each step of the algorithm.

Turning back to prediction, we consider two methods:

- First, we apply the Boosting and stochastic Boosting (SB) to the original data, that is, directions in which we move at each step relate directly to the original covariates.
- In the second case, we apply both algorithms using non-linear stumps derived using potential split-points at 8 quantiles of each covariate.

To carry out prediction, we approximate $\log Pr(M = m)$ by the approximated degrees of freedom (26) rather than by $M * \Delta$.

The Indian pima data in Ripley (1996) contains both a training and test data set, which we use to assess the predictive performances of both algorithms in both cases. For each method in turn, we use $N = 100$ stochastic paths with $I_{max} = 500$ and a shaving parameter $g/(1+g) = 0.1$. With respect to standard Boosting, we use two methods to assess prediction: the first, average Boosting (Ave Boost) for which we average the predictions for all the models along the path using a AIC criterion. The second, maximum Boosting (Max Boost), consists in selecting a model using the AIC criterion and predicting using this sole model. Results are given in Table 1.

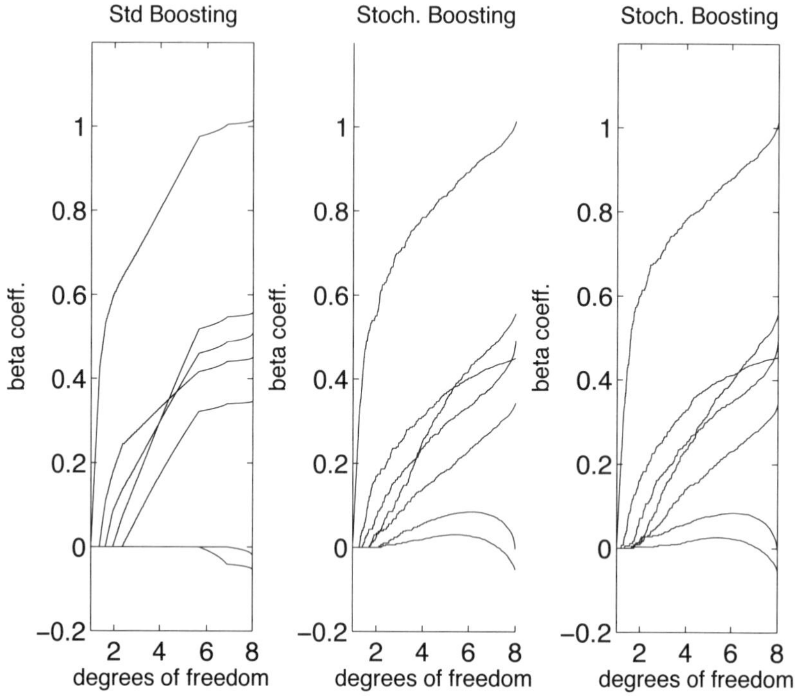

Figure 5: *Indian pima data. Standard boosting path and simulated stochastic paths for $M = 5000$ iterations, $g/(1+g) = 0.01$. Section 5.4.*

As can be seen from Table 1, we are able to compete with the standard Boosting algorithm but with the added advantage that we are able to account for uncertainty within our predictions.

Table 1: *Out of sample test error for Indian Pima data set. Section 5.4.*

Method	Ave. Boost	Max Boost	Bay. Boost.
1 (original)	65	65	66
2 (stumps)	69	70	65

With respect to the Hearts Disease data, we assess the prediction performances of Boosting and Bayesian Boosting by applying both methodologies to the original data, as with the first method described in the previous subsection. We use 10-fold cross-validation as a measure of the prediction properties of each method. Again, with respect to Bayesian Boosting we use $N = 100$ stochastic paths with $I_{max} = 500$ and a shaving parameter $g/(1+g) = 0.1$. We use average Boosting and maximum Boosting as prediction methods with the standard Boosting path. Results are given in Table 2.

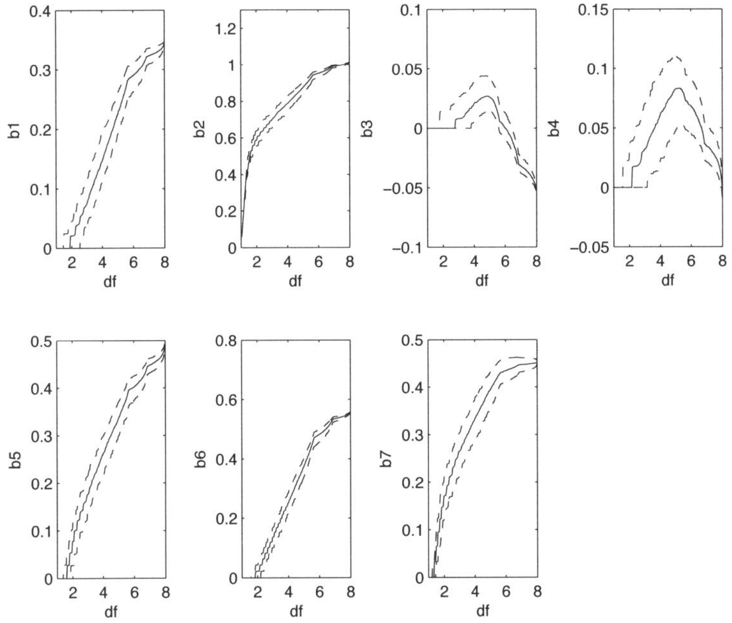

Figure 6: *Indian pima data. 95% Credible Intervals obtained using* $N = 200$ *stochastic paths for* $M = 5000$ *iterations,* $g/(1+g) = 0.01$. *Section 5.4.*

Table 2: *Out of sample test errors for* 10*-fold CV on Hearts disease data. Section 5.4.*

	Ave. Boost	Max Boost	Bay. Boost.
Tot	128	128	127

The results here seem to confirm those observed in Table 1, thus highlighting the efficiency of Bayesian Boosting as a stochastic classification method.

6. FINAL COMMENTS

In this paper we have sought to develop a framework for Bayesian Relaxation, mapping models along paths of increasing complexity. The corresponding stochastic relaxation methods offer an alternative to classical methods. Indeed, by focusing on paths rather than models, and using Importance Sampling rather than Markov Chain Monte Carlo, we have derived methodologies that are somewhat closer to exhibiting the relaxation property of their standard counterparts. This provides an advantage, especially with respect to exploring 'interesting' regions of the posterior model space, in the same spirit as relaxation methods. Our results suggest that modelling the variance components of Bayesian GLMs provides a powerful and attractive approach to model specification in complex regression analysis.

APPENDIX

A: Formal Equivalence Between the Generalised Ridge and L_α Regularisation

Let us consider the following generalised Ridge estimation problem: find $\beta = (\beta_1, ..., \beta_p)$ and $v = (v_1, ..., v_p)$ to minimise

$$L(\beta, v) = \sum_{i=1}^{n} \left(\sum_{j=1}^{p} x_{ij}\beta_j - y_i \right)^2 + \sum_{i=1}^{p} v_i^{-1}\beta_i^2 \tag{26}$$

subject to $\sum_{i=1}^{p} v_i^r = K$, for some constant $K \geq 0$ and any real $r > 0$. Using a Lagrange multipliers framework, this optimisation problem is equivalent to finding β, v and λ to minimise

$$C(\beta, v, \lambda) = \sum_{i=1}^{n} \left(\sum_{j=1}^{p} x_{ij}\beta_j - y_i \right) + \sum_{i=1}^{p} v_i^{-1}\beta_i^2 + \lambda \left(\sum_{i=1}^{p} v_i^r - K \right). \tag{27}$$

From (27), we obtain

$$\frac{\partial C(\beta, v, \lambda)}{\partial \beta_k} = \sum_{i=1}^{n} x_{ik} \left(\sum_{j=1}^{p} x_{ij}\beta_j - y_i \right) + v_k^{-1}\beta_k, \; k = 1, ..., p \tag{28}$$

$$\frac{\partial C(\beta, v, \lambda)}{\partial v_k} = -\frac{\beta_k^2}{v_k^2} + \lambda r v_k^{r-1}, \; k = 1, ..., p \tag{29}$$

$$\frac{\partial C(\beta, v, \lambda)}{\partial v_k} = \sum_{k=1}^{p} v_k^r - K. \tag{30}$$

Thus, solutions to (27) are obtained by setting (28), (29) and (30) to zero. From (29), letting $\mu = \lambda r$, we obtain

$$v_k^{r+1} = \frac{\beta_k^2}{\mu}, \; k = 1, ..., p, \quad v_k^r = \frac{|\beta_k|^{2r/(r+1)}}{\mu^{r/(r+1)}}, \; k = 1, ..., p. \tag{31}$$

From (31) and (30), we thus obtain, $K = \sum_k |\beta_k|^{2r/(r+1)}/\mu^{r/(r+1)}$, so that

$$\mu^{1/(r+1)} = \frac{\left(\sum_k |\beta_k|^{2r/(r+1)}\right)^{1/r}}{K^{1/(r+1)}} \tag{32}$$

and letting $K' = K^{1/(r+1)}$, (31) and (30) lead to

$$v_k = \frac{K' |\beta_k|^{2/(r+1)}}{\left(\sum_k |\beta_k|^{2r/(r+1)}\right)^{1/r}}. \tag{33}$$

Substituting (33) in (28), we therefore obtain that β minimises (27) if it is a solution of

$$\sum_{i=1}^{n} x_{ik} \left(\sum_{j=1}^{p} x_{ij}\beta_j - y_i \right) + \frac{1}{K'} \frac{\beta_k}{|\beta_k|^{2/(r+1)}} \left(\sum_{k} |\beta_k|^{2r/(r+1)} \right)^{1/r}. \tag{34}$$

Moreover,

$$\frac{1}{K'} \frac{\beta_k}{|\beta_k|^{2/(r+1)}} \left(\sum_k |\beta_k|^{2r/(r+1)} \right)^{1/r}$$
$$= \frac{1}{K'} sgn(\beta_k)|\beta_k|^{(r-1)/(r+1)} \left(\sum_k |\beta_k|^{2r/(r+1)} \right)^{1/r},$$

which simply corresponds to the derivative, with respect to β_k, $k = 1, ..., p$ of

$$\frac{1}{2K'} \left(\sum_k |\beta_k|^{2r/(r+1)} \right)^{(r+1)/r}.$$

Thus, β minimises (34) if it is a solution of the following optimisation problem

$$\sum_{i=1}^{n} \left(\sum_{j=1}^{p} x_{ij}\beta_j - y_i \right)^2 + \frac{1}{2K'} \left(\sum_{k=1}^{p} |\beta_k|^{2r/(r+1)} \right)^{(r+1)/r} \tag{35}$$

which is known to be equivalent to finding β to minimise

$$\sum_{i=1}^{n} \left(\sum_{j=1}^{p} x_{ij}\beta_j - y_i \right)^2 \tag{36}$$

subject to

$$\left(\sum_{k=1}^{p} |\beta_k|^{2r/(r+1)} \right)^{(r+1/r)} \leq C \tag{37}$$

for some constant C depending on $K^{'}$. And it is clear that the latter problem is equivalent to finding β to minimise (36) subject to $\sum_{k=1}^{p} |\beta_k|^{2r/(r+1)} \leq C^{\prime(r/(r+1)}$, which is exactly the general L_α regularisation problem, with $\alpha = 2r/(r+1)$. And, for increasing $r > 0$, α increases from 0 to 2 as $r \to \infty$.

Note that the proof is straightforward to generalise for α increasing from 2 to ∞, which corresponds to r increasing from $-\infty$ to -1^-.

B: Limit Case when $r \to \infty$

We now show that as $r \to \infty$, the generalised ridge tends to the ordinary ridge, that is, the constraint

$$\sum_{k=1}^{p} v_k^r = K \tag{38}$$

is equivalent in that case to

$$v_k = C, \ \forall k = 1, ..., p \tag{39}$$

for some constants K and C, with p an integer ≥ 1. This equivalence follows through with no difficulties once it is observed that, for any constant K and for any $r > 0$, one can write

$$C = \left(\frac{K}{p}\right)^{1/r} \tag{40}$$

so that

$$K = p\, C^r \tag{41}$$

where $C > 0$ if $K > 0$ and $C = 0$ otherwise. Thus, (38) is equivalent to

$$\sum_{k=1}^{p} \left(\frac{v_k}{C}\right)^r = p \tag{42}$$

and therefore, as $r \to \infty$, (42) leads to the condition that $v_k/C = 1$ for all $k = 1, ..., p$, that is, to (39). So that the generalised ridge is seen to correspond to the ordinary ridge regression problem in this case.

C: Approximating the degrees of freedom of the Lasso

A straightforward approach to approximating the degrees of freedom of the Lasso using the GRR equivalence is as follows.

Step 1. Given a value for η obtain the Lasso solution $\hat{\beta}_{(\text{Lasso})}$.

Step 2. Given $\hat{\beta}_{(\text{Lasso})}$ obtain the corresponding GRR parameters $\{\lambda_1, \ldots, \lambda_p\}$

Step 3. Given $\{\lambda_1, \ldots, \lambda_p\}$ calculate the degrees of freedom of the GRR linear smoother

To implement Step 1 we can use any of the methods discussed in Section 1, such as for example.

For Step 2 we make use of the following relationship

$$\lambda_j = K \times \frac{|\hat{\beta}_{(Lasso)_j}|}{\sum_i |\hat{\beta}_{(Lasso)i}|}.$$

The value of K corresponding to a fixed η is unknown and hence involves a simple line search.

For Step 3 to calculate the degrees of freedom of the GRR model we we note that

$$\hat{\beta}_{(\text{Lasso})} = \hat{\beta}_{(\text{GRR})} = (X'X + \Lambda)^{-1}X'Y,$$

where $\Lambda = \text{diag}(\lambda_1, \ldots, \lambda_p)$, a $p \times p$ diagonal matrix with elements $(\Lambda)_{jj} = \lambda_j$. Hence the predicted values $\hat{Y}$ are given by

$$\begin{aligned} \hat{Y} &= X\hat{\beta} \\ &= X(X'X + \Lambda)^{-1}X'Y \\ &= HY \end{aligned}$$

where H defines the $n \times n$ hat matrix the rows of which store the equivalent kernels, see .

The degrees of freedom is most easily calculated via the first equality in the formula

$$df(\eta) = p - \sum_{j=1}^{p} \frac{(\Lambda)^*_{jj}}{(\Lambda)_{jj}} = tr(H)$$

where Λ^* is the posterior variance-covariance matrix $\Lambda^* = (X'X + \Lambda)^{-1}$.

And thus, we obtain an approximation to the degrees of freedom.

ACKNOWLEDGEMENTS

The work of Alexandre Pintore was supported by grant number GR/S80615/01 from the EPSRC and the Faculty and Institute of Actuaries under their Quantitative Finance initiative. The work of Chris Holmes was supported by the MRC, Harwell, UK.

REFERENCES

Box, G. E. P. and Tiao, G. C. (1973). *Bayesian Inference in Statistical Analysis.* Reading, MA: Addison-Wesley.

Buhlmann, P. and Yu, B. (2001). Boosting with the L2 Loss: Regression and classification. *J. Amer. Statist. Assoc.* **98**, 324–340.

Chipman, H. A., George, E. I. and McCulloch, R. E. (2005). BART: Bayesian Additive Regression Trees. *Tech. Rep.*, Acadia University, Wolfville, Canada.

Clyde, M. (1999). Bayesian model averaging and model search strategies. *Bayesian Statistics 6* (J. M. Bernardo, J. O. Berger, A. P. Dawid and A. F. M. Smith, eds.) Oxford: University Press, 157–185 (with discussion).

Clyde, M., DeSimone, H. and Parmigiani, G. (1996). Prediction via orthogonalized model mixing. *J. Amer. Statist. Assoc.* **91**, 1197–1208.

Denison, D. G. T., Holmes, C. C., Mallik, B. K. and Smith, A. F. M. (2002). *Bayesian Nonlinear Methods for Classification and Regression.* Chichester: Wiley.

Efron, B., Hastie, T., Johnstone, I. and Tibshirani, R. (2004). Least angle regression. *Ann. Statist.* **2**, 407–499.

Freund, Y. (1995). Boosting a weak learning algorithm by majority. *Inform. and Comput.* **121**, 256–285.

Freund, Y. and Schapire, R. E. (1996a). Game theory, on-line prediction and boosting. *Proc. 9th Annual Conf. Computational Learning Theory*, 325–332.

Freund, Y. and Schapire, R. E. (1996b). Experiments with a new boosting algorithm. *Machine Learning: Proc. 13th Internat. Conf.*, Morgan Kaufman, San Francisco, 148–156.

Freund, Y. and Schapire, R. E. (1997). A decision-theoretic generalization of online learning and an application to boosting. *J. Comp. System. Sciences.* **55**.

Friedman, J., Hastie, T. and Tibshirani, R. (2000). Additive logistic regression: a statistical view of boosting. *Ann. Statist.* **28**, 307–337 (with discussion).

Friedman, J. and Popescu, B. (2003). Importance Sampled Learning Ensembles. *Tech. Rep.*, Stanford University, USA.

Frideman, J. and Popescu, B. (2004). Gradient directed regularization. *Tech. Rep.*, Stanford University, USA.

Fu, W. (1998). Penalized regression: the bridge versus the lasso. *J. Comp. Graphical Statist.* **7**, 397–416.

George, E. I. and McCulloch, R. E. (1993). Variable selection via Gibbs sampling. *J. Amer. Statist. Assoc.* **88**, 881–889.

George, E. I. and McCulloch, R. E. (1997). Approaches for Bayesian variable selection. *Statistica Sinica* **7**, 339–374.

Goldstein, M. and Smith, A. F. M. (1974). Ridge-type estimators for regression analysis. *J. Roy. Statist. Soc. B* **36**, 284–291.

Grandvalet, Y. (1998). Least absolute shrinkage is equivalent to quadratic penalization. *ICANN 98, Perspectives in Neural Computing.* Berlin: Springer-Verlag.

Griffin, J. E. and Brown, P. J. (2005). Alternative prior distributions for variable selection with very many more variables than observations. *Tech. Rep.*, University of Warwick, UK.

Hastie, T. and Tibshirani, R. (1990). *Generalized Additive Models.* London: Chapman and Hall.

Hastie, T., Tibshirani, R. and Friedman, J. (2001). *The Elements of Statistical Learning; Data Mining, Inference and Prediction.* New York: Springer-Verlag.

Hoerl, A. E. and Kennard, R. (1970). Ridge Regression: Biased estimation for nonorthogonal problems. *Technometrics* **12**, 55–67.

Holmes, C. C. and Denison, D. G. T. (1999). Bayesian wavelet analysis with a model complexity prior *Bayesian Statistics 6* (J. M. Bernardo, J. O. Berger, A. P. Dawid and A. F. M. Smith, eds.) Oxford: University Press, 769–776.

Ishwaran, H. and Rao, J. S. (2005). Spike and slab variable selection: Frequentist and Bayesian strategies. *Ann. Statist.* **33**, 730–773.

McCullagh, P. and Nelder, J. (1989). *Generalized Linear Models.* London: Chapman and Hall.

Nott D. J. and Kohn R. (2005). Adaptive sampling for Bayesian variable selection. *Biometrika* **92**, 747–763.

Park, T. and Casella, G. (2005). The Bayesian Lasso. *Tech. Rep.*, University of Florida, USA.

Ripley B. D. (1996). *Pattern Recognition and Neural Networks.* Cambridge: University Press.

Schapire, R. E. (1990). The strength of weak learnability. *Machine Learning* **5**, 197–227.

Schapire, R. E. (2003). The boosting approach to machine learning: An overview. *Nonlinear Estimation and Classification* (D. D. Denison, M. H. Hansen, C. Holmes, B. Mallick and B. Yu, eds.) Berlin: Springer-Verlag.

Schapire, R. E., Freund, Y., Bartlett, P. and Lee, W. (1998). Boosting the margin: a new explanation for the effectiveness of voting methods. *Ann. Statist.* **26**, 1651–1686.

Smith, M. and Kohn, R. (1996). Nonparametric regression using Bayesian variable selection. *J. Econometrics* **75**, 317–343.

Tibshirani, R. (1996). Regression shrinkage and selection via the lasso. *J. Roy. Statist. Soc. B* **58**, 267–288. *Encyclopedia of Statistical Sciences* **8** (S. Kotz, N. L. Johnson and C. B. Read, eds.) New York: Wiley, 129–136.

Tipping, M. E. (2001). Sparse Bayesian Learning and the Relevance Vector Machine. *Machine Learning* **1**, 211–244.

Walker, S. G. and Page, C. J. (2001). Generalized ridge regression and a generalization of the C_p statistic. *Appl. Statist.* **28**, 911–922.

West, M. (2002). Bayesian factor regression models. *Bayesian Statistics 7* (J. M. Bernardo, M. J. Bayarri, J. O. Berger, A. P. Dawid, D. Heckerman, A. F. M. Smith and M. West, eds.) Oxford: University Press, 733–742.

Zhao, P. and Yu, B. (2005). Boosted Lasso. *Tech. Rep.*, University of California, Berkeley, USA.

Zou, H., Hastie, T. J. and Tibshirani, R. (2004). On the degrees of freedom of the Lasso. *Tech. Rep.*, Stanford University.

DISCUSSION

ROBERT KOHN (*Fac. Business, Univ. New South Wales, Australia*)

This is an ambitious paper that provides the first steps in thinking about probabilistic relaxation methods from a Bayesian perspective. The primary motivation is to provide Bayesian analogues of methods that attempt to solve large p (number of variables) and small n (number of observations) problems.

Section 2 the paper provides an insightful overview of various approaches to ill-posed problems. Section 3 of the paper gives a theorem that connects the solution of each regularized problem to the solution of a corresponding generalized ridge regression problem. It is inherently valuable to demonstrate how different approaches lead to the same solution. While the authors give an elegant algebraic proof of their result, is it possible to prove Theorem 1 using a statistical argument that makes the result more intuitive?

I now comment on the probabilistic relaxation method. It seems that the authors measure of uncertainty is with respect to the probabilistic nature of the method itself. It would be interesting for the reader to understand the connection between the uncertainty discussed by the authors and the uncertainty obtained using Bayesian methods or the bootstrap. This difference in uncertainties could be discussed in terms of functionals of the regression surface.

Another issue that could be more fully explored is the performance of the probabilistic methods for the large p and small n problems that motivate the paper. Both the Pima Indian diabetes data and the Heart disease data do not seem to be of this form. Finally, and this seems an issue related to the previous one, the authors seem to use very simple parametric functional forms in their real applications. The authors may wish to extend their analyses more complex nonparametric forms as well as interactions in their analyses.

REPLY TO THE DISCUSSION

We would like to thank Professor Kohn for his insightful comments. We are aware that our method for model exploration presented in our article is somewhat raw, a little under-developed and even somewhat speculative. We hope the paper will spur interest in the area of Bayesian Boosting and drive further research in methods which dynamically explore parsimonious models in interesting regions of parameter space.

In relation to Professor Kohn's specific comments. Our Theorem in Section 3 shows how by constraining the variance components we are lead to various L_α norm penalties. The intuition behind this result is perhaps best explained via a diagram. In Fig. 7 we show the contours, in variance components space, of the penalty function for differing L_α norms. If the sum of the variance components is related to the degrees of freedom of the model, then we see that for $\alpha > 1$ the model (*a priori*) gains most in terms of degrees of freedom by setting the variance components equal, while for $\alpha < 1$ the opposite occurs. The strength of this balance/imbalance being proportional to $|\alpha - 1|$. For example, for $\alpha = 0.25$ we see that, for the same penalty, the model could 'choose' either, say, $\{v_1 = v_2 \approx 1.2\}$ or $\{v_1 = .01, v_2 \approx 0.5\}$. The latter, in some sense represents a sparser solution with (approximately) greater degrees of freedom and hence is better placed to obtain a higher likelihood; the opposite being true for $\alpha > 1$. For reference we have also shown a contour of Jeffrey's prior.

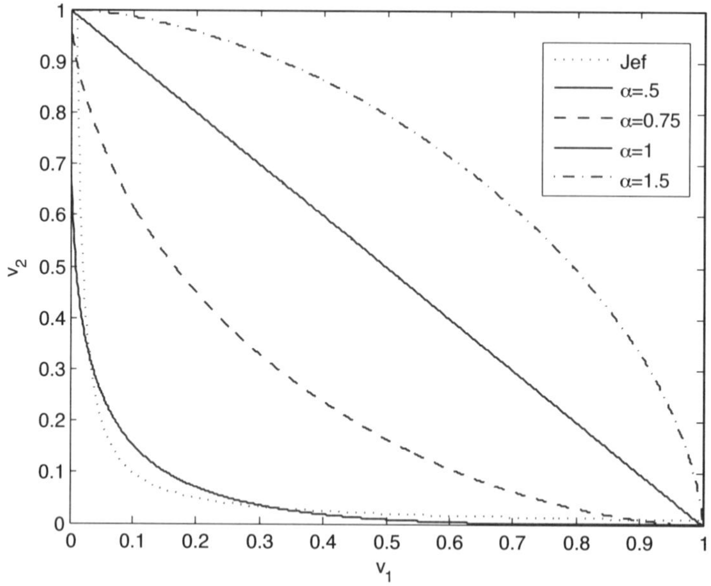

Figure 7: *Contours of equal penalty in variance components space for differing settings of α. Jeffrey's prior is also shown for reference.*

In regard to the comment on the method. The uncertainty measures we have developed are somewhat analogous to the Bootstrap techniques used by Friedman and Popescu (2003). In their paper they account for uncertainty in the Boosting paths by re-sampling the data. In a Bayesian setting we condition on the data and account for uncertainty by adopting a prior on the relaxation paths. The 'full' Bayesian solution could be obtained by placing a hyper-prior on the penalty parameter and then sampling over this. However, as we show, this produces non-sparse solutions; for in Boosting and the Lasso sparsity is obtained by taking paths of minimum cost at each step. It is then a non-trivial task to ensure the method explores interesting regions of the model space. Of course the proper, coherent, approach is to frame this all within a decision theoretic setting and define a loss function which encapsulates our utility in the solution space. The Bayesian solution then falls out, though enumerating this may be prohibitively troublesome.

Finally with regards large p small n examples. We have found our method useful in the exploratory analysis of gene expression microarray data, though this is not reported here. Recently we have also demonstrated the technique in the area of cluster analysis for genome-wide gene expression studies. A technical report on this will be available from the first author's website. The issue of investigations in nonparametric and interaction models is an interesting area for future research.

BAYESIAN STATISTICS 8, pp. 283–302.
J. M. Bernardo, M. J. Bayarri, J. O. Berger, A. P. Dawid,
D. Heckerman, A. F. M. Smith and M. West (Eds.)

The Bayesian Approach to the Analysis of Finite Population Surveys

RODERICK J. A. LITTLE
University of Michigan, USA
rlittle@umich.edu
HUI ZHENG
Massachusetts General Hospital and Harvard Medical School, USA
hzheng1@partners.org

SUMMARY

There are many Bayesian statisticians, and many sampling statisticians, but very few Bayesian sampling statisticians. This article considers why this is the case. We argue that the Bayesian approach is very attractive in the survey sampling setting, and offer a recipe for successful application of Bayesian ideas: Bayesian models for surveys need to take into account features of the survey design like clustering and stratification, limit strong parametric assumptions, and incorporate weakly informative prior distributions. Examples are provided to show that Bayesian inferences with these features can have good repeated sampling properties. We also describe a new Bayesian approach for the situation where the selection probabilities are known only for the sampled units, a case where some authors have argued that the Bayesian approach is problematic.

Keywords and Phrases: HORVITZ–THOMPSON ESTIMATOR, *P*-SPLINE MODEL

1. THE BAYESIAN APPROACH TO SAMPLE SURVEY INFERENCE

This paper considers the Bayesian approach to statistical inference for sample surveys of finite populations. In this introduction we provide a brief overview of what distinguishes sample survey data from others kinds of data that statisticians analyze, and how the Bayesian approach is applied in this setting. We then consider why the Bayesian approach is not common in sample survey practice, describe the key elements in the application of the Bayesian approach in this setting, and give some examples where gains are achieved over more classical frequentist approaches. We conclude with a proposed model for probability proportional to size samples that yields Bayes inferences for the case where sample weights are only available for the cases in the sample, a case where some have argued that the Bayesian approach is problematic.

1.1. *Why are Survey Data Different from Other Kinds of Data?*

The two most popular approaches to finite population survey inference are 1. *Design-based* inference, where probability statements are based on the distribution of sample selection, with the population quantities treated as fixed; and 2. *Frequentist superpopulation modeling*, where the population values are assumed to be generated from a superpopulation model that conditions on fixed parameters, and inferences are based on repeated sampling from this model. The Bayesian approach is flourishing in many other areas of applied statistics, and the basic theory of the Bayesian approach to surveys has been clear for many years (Ericson 1969, 1988; Basu 1971; Scott 1977; Binder 1982; Rubin 1983, 1987; Ghosh and Meeden 1997). Despite these facts, the Bayesian approach has been almost totally ignored by practitioners of finite population survey sampling.

A distinctive feature of survey inference is its primary (though not exclusive) focus on estimands that are functions of values in a population with a finite number of elements, often simple quantities like overall or subgroup means or totals. Bayes seems particularly suited to inference about such quantities, since they require predicting the values of variables for non-sampled items; the Bayesian approach is particularly natural for predictive inference.

However, other aspects of finite population sampling (discussed below) seem to survey practitioners to be less compatible with the Bayesian approach. Can the Bayesian approach be responsive to these features? We believe that it can, and indeed we think that when appropriately applied, the Bayesian paradigm provides the most satisfying inferential approach to survey inference (Little, 2004). Here are the features, with our view of the appropriate Bayesian response to them:

(i) *The use of a probability mechanism to determine which units in the population are selected.* In the design-based approach this sampling distribution forms the basis of the inference. Some Bayesians have argued that probability sampling is not useful, since the statistical inference is based on the model, not the selection mechanism. On the contrary, as discussed in the next section, the use of probability sampling ensures that the sampling mechanism is ignorable for Bayesian inference, thus simplifying the inference and making it less dependent on questionable assumptions.

(ii) *Survey samplers believe the analysis needs to take into account 'complex' sampling design features* such as stratification, differential probabilities of selection, clustering and multistage sampling. Consequently, they reject theoretical arguments suggesting such design features can be ignored if the model is correctly specified; models are always misspecified, they argue, and model answers are suspect even when model misspecification is not easily detected by model checks. Important examples of this line of reasoning include Kish and Frankel (1974), Holt, Smith and Winter (1980), Hansen, Madow and Tepping (1983) and Pfeffermann and Holmes (1985). The fact is that models that ignore these design features will not be considered seriously by survey samplers. We believe that aspects of the design like clustering and stratification can and should be explicitly incorporated in the Bayesian model to avoid sensitivity of inference to model misspecification. This point is illustrated in the examples below.

(iii) *An environment where many statistics are generated in a production mode that precludes detailed modeling.* Bayesian inference, with careful eliciting of models and associated prior distributions, is often perceived as 'too much work' in a production environment (e.g. Efron 1986). While some attention to model fit is needed to do good statistics under Bayesian or frequentist paradigms, 'off-the-shelf' Bayesian models can be developed that incorporate survey sample design features, and for a given problem the computation of the posterior distribution is prescriptive, via Bayes Theorem. This aspect would be aided by a Bayesian software package focused on survey applications. As for the choice of prior distribution, surveys often involve large amounts of data, and conventional reference priors that are dominated by the likelihood often work well, with appropriate care about their well-known quirks. See also the next point.

(iv) *Antipathy towards methods that involve strong subjective elements or assumptions.* This concern is rational, since surveys are often conducted by agencies that need to be viewed as objective and shielded from policy biases. It can be addressed by using models that make relatively weak assumptions, and noninformative priors that are dominated by the likelihood. The latter yields Bayesian inferences that are often similar to superpopulation modeling, with the usual differences of interpretation of probability statements.

(v) *Concern about repeated sampling properties of the inference.* There is a long tradition in the survey world of assessing properties of survey inferences in repeated sampling under the chosen design, and this aspect cannot be ignored by Bayesians. Our view is that the frequentist performance of Bayesian procedures can and should be evaluated, and indeed such assessments plays an important role in a calibrated Bayesian approach to inference, which we find attractive (Box, 1980; Rubin, 1984; Little, 2006).

1.2. *Basics of Bayes for Surveys*

Using the notation in Chambers and Skinner (2003), let $U = \{1, ..., N\}$ denote a finite population of N elements, let $y_U = (y_{tj}), j = 1, ..., J; t \in U$ denote the values of J variables in the population, and let $Q = Q(y_U)$ denote the finite population quantity of interest. Let $i_U = (i_1, ..., i_N)$ denote the vector of inclusion indicators, where $i_t = 1$ if unit t is selected in the sample and $i_t = 0$ otherwise. Write $y_U = (y_{inc}, y_{exc})$ where y_{inc} is the set of sampled values and y_{exc} is the set of non-sampled values. Let z_U be a set of design variables available for all units in the population. We assume initially that these design variables are recorded and available for analysis; the case where they are only recorded for the sample is discussed in Section 3. The full Bayesian approach (*e.g.*, Gelman *et al.*, 1995) specifies a model for both the survey data and the sample inclusion indicators. Parametric models can be formulated as:

$$p(y_U, i_U|z_U, \theta, \phi) = p(y_U|z_U, \theta)p(i_U|z_U, y_U, \phi),$$

where θ, ϕ are unknown parameters indexing the distributions of y_U and i_U given z_U, respectively. The likelihood of θ, ϕ based on the observed data is then:

$$L(\theta, \phi|z_U, y_{inc}, i_U) \propto p(y_{inc}, i_U|z_U, \theta, \phi) = \int p(y_U, i_U|z_U, \theta, \phi)dy_{exc}.$$

In the Bayesian approach the parameters (θ, ϕ) are assigned a prior distribution $p(\theta, \phi|z_U)$. Analytical inference about the parameters is based on the posterior distribution:

$$p(\theta, \phi | z_U, y_{inc}, i_U) \propto p(\theta, \phi | z_U) L(\theta, \phi | z_U, y_{inc}, i_U).$$

Descriptive inference about $Q = Q(y_U)$ is based on its posterior distribution given the data, which is conveniently derived by first conditioning on, and then integrating over, the parameters:

$$p(Q(y_U)|z_U, y_{inc}, i_U) = \int p(Q(y_U)|z_U, y_{inc}, i_U, \theta, \phi) p(\theta, \phi | z_U, y_{inc}, i_U) d\theta d\phi.$$

This general formulation requires a model for the inclusion process. When the units are selected by probability sampling, $p(i_U|z_U, y_U, \phi) = p(i_U|z_U)$ is a known distribution, and by Rubin's (1976) theory, the posterior distribution reduces to the simpler form

$$p(Q(y_U)|z_U, y_{inc}) = \int p(Q(y_U)|z_U, y_{inc}, \theta) p(\theta | z_U, y_{inc}) d\theta d\phi,$$

which does not involve or require a model for the inclusion indicators i_U. This simplification is an important motivation for a Bayesian to favor probability sampling as a method of selection. For it to happen, the model for y_U and prior distribution θ needs to condition on the design variables z_U. For simple random sampling this conditioning is not important, but features such as stratification and clustering need to be built into the model via the design variables z_U to obtain a satisfactory Bayesian inference. We consider first the simple example of stratified random sampling for a continuous outcome Y, previously discussed in Little (2003):

2. APPLICATIONS TO COMMON SURVEY DESIGNS

Example 1 (Stratified Random Sampling). For stratified random sampling, the population is divided into H strata and n_h units are selected by simple random sampling from the population of N_h units in stratum h. The sampling rates n_h/N_h may vary across the strata, and the sampling weight in stratum h is the inverse of the sampling rate, scaled to sum to the total sample size, namely $w_h = (n/N)(N_h/n_h)$. Let z_U represent stratum, with $z_t = h$ if unit t is in stratum h. This selection mechanism is ignorable providing the model for Y conditions on the stratum variable z_t. Robust models need to reflect the variation in the distribution of Y across the strata. A normal model that does this via stratum-specific means and variances is

$$\begin{aligned} [y_t | z_t &= h, \theta_h, \sigma_h^2] \sim_{ind} N(\theta_h, \sigma_h^2), \\ p(\theta_h, \log \sigma_h^2) &\propto const. \end{aligned} \tag{1}$$

where $N(a, b)$ denotes the normal distribution with mean a, variance b, and we have assumed the standard Jeffreys' reference prior for the parameters. Suppose we are interested in the population mean $\bar{Y} = \sum_{h-1}^{H} P_h \bar{Y}_h$, where $P_h = N_h/N$ is the population proportion in stratum h and $\bar{Y}_h$ is the population mean in stratum h. Routine Bayesian calculations lead to the following posterior distribution for $\bar{Y}$:

$$p(\bar{Y}|z_U, data, \{\sigma_h^2 : h = 1, ...H\}) \sim N(\bar{y}_{st}, \sigma_{st}^2); \quad p(\sigma_h^2 | data) \sim_{ind} Inv\chi^2(ss_h, n_h - 1)$$

where $\bar{y}_{st}$ is the stratified sample mean $\bar{y}_{st} = \sum_{h=1}^{H} p_h \bar{Y}_h, p_h = n_h/n$, the posterior variance is $\sigma_{st}^2 = \sum_{h=1}^{H} P_h^2(1 - n_h/N_h)\sigma_h^2/n_h, ss_h = (n_h - 1)s_h^2$, the sum of squares of the n_h sampled values of Y in stratum h, and the posterior distribution of σ_h^2 is such that ss_h/σ_h^2 is chi-squared with $n_h - 1$ degrees of freedom. Some notable features of this Bayesian solution are the following:

(i) The factors $(1 - n_h/N_h)$ in σ_{st}^2 are 'finite population corrections', and send the posterior variance of $\bar{Y}$ to zero as the sampling fractions n_h/N_h tend to one. These corrections emerge automatically from the Bayesian calculation.

(ii) $\bar{y}_{st}$ is the usual design-based stratified mean, and weights sampled cases in stratum h by the sampling weight w_h, as in the Horvitz–Thompson estimator (Horvitz and Thompson 1952). Thus, this is a simple situation where the Bayesian estimator does incorporate a feature of the sample design, namely the sampling weights.

(iii) σ_{st}^2 is the usual design-based variance of the stratified mean. Thus for known σ_h^2, the posterior distribution of $\bar{Y}$ gives credibility intervals equivalent to the usual design-based confidence intervals.

(iv) For unknown σ_h^2, the posterior distribution of $\bar{Y}$ is a mixture of t distributions, and provides a useful small-sample correction for estimating variances. This 't-correction' is not available in classical design-based results, which are essentially asymptotic. Draws from this posterior distribution are trivial to compute by direct simulation – no Markov Chain Monte Carlo required!

(v) Suppose that we adopted a simpler version of this model that assumed no stratum effects, that is $\theta_h = \theta, \sigma_h^2 = \sigma^2, p(\theta, \log \sigma^2) = const.$ The posterior mean of $\bar{Y}$ is then the unweighted sample mean, and is potentially very biased for $\bar{Y}$ if the sampling rates vary across the strata. The problem is that inferences from this model are non-robust to violations of the assumption of no stratum effects. Since stratum effects are often plausible, robustness considerations dictate against this simpler model. As another example of unwisely adopting a model that ignores a design feature, the stratified model assumes independence of units within the stratified model. This would be a poor assumption if the data were sampled in clusters within strata, since the posterior distribution will overstate precision if outcomes within clusters are correlated.

(vi) With many strata and small samples in each stratum, more efficient inferences are obtained by giving the strata means and/or variances proper prior distributions. The prior distribution

$$[\theta_h|\chi, \tau^2] \sim N(\mu, \tau^2), p(\mu, \tau^2) = const.,$$

which assumes exchangeability of the stratum means, leads to a posterior mean that shrinks the sampling weights w_h towards one. (Here and elsewhere a uniform prior is chosen for τ^2 rather than $\log \tau^2$ to avoid an improper posterior distribution for that parameter). The exchangeability assumption is important, but less crucial than the assumption of no stratum effects in the model of comment (v), since the shrinkage goes to zero with increasing n.

Elliott and Little (2000) discuss elaborations of this model that weaken the exchangeability by putting more structure on the stratum means.

Example 2 (Two-stage Random Sampling). A common feature of many complex survey designs is clustering of the units, and two-stage selection of clusters and units within clusters. This feature is incorporated into a Bayesian model by including random effects for the clusters. In particular, suppose the population is divided into C clusters, such as geographical areas. A basic form of two-stage sampling first selects a simple random sample of c clusters, and then selects a simple random sample of n_k of the units in each sampled cluster k. The sampling mechanism is ignorable conditional on cluster information, but the Bayesian model needs to account for within-cluster correlation in the population. A normal model that does this is:

$$\begin{aligned} &[y_{kt}|\theta_k, \sigma^2] \sim_{ind} N(\theta_k, \sigma^2) \\ &[\theta_k|\mu, \tau^2] \sim_{ind} N(\mu, \tau^2) \\ &p(\mu, \log\sigma^2, \tau^2) \sim const. \end{aligned} \tag{2}$$

Unlike the stratified sampling case, the cluster means θ_k cannot be assigned a flat prior distribution, $p(\theta_k) = const$, because this does not allow the information from sampled clusters to be used to predict the means of non-sampled clusters. The exchangeability of the random effects $\{\theta_k\}$, and the outcomes $\{y_{kt}\}$ within clusters, can be justified by the random sampling of clusters and individuals within clusters, using De Finetti's ideas. Maximum likelihood estimation for this model was first proposed for multistage sampling by Scott and Smith (1969), and can be justified from the Bayesian viewpoint as large sample Bayes. For small samples a fully Bayesian analysis (e.g. Gelfand et al., 1990) is preferred.

Example 3 (Probability Proportional to Size (PPS) Sampling). Often with a population of units of differing sizes, large units contribute more to population quantities of interest than smaller units, and it makes sense to include large units with higher probability. A popular design for doing this is sampling with probability to size (PPS). Specifically, suppose a size measure Z is known for all units in the population, and unit t is selected with probability π_t proportional to its size z_t. PPS samples can be selected in a number of ways (Hanif and Brewer 1980), but a practical and common design is the following: from a random starting point, units are selected systematically from a randomly-ordered list, at regular intervals on a scale of cumulated sizes (Kish 1965, Ch. 7); units that would be selected with probability one are removed and put in a 'certainty' stratum. We consider statistical inference for the finite population total T of a continuous outcome Y.

The standard design-based estimate for PPS samples is the Horvitz–Thompson (HT) (1952), which weights sampled units by the inverse of their probability of selection:

$$\widehat{T}_{HT} = \sum_{t=1}^{n} y_t/\pi_t,$$

where the summation is over n sampled units. This is also the projective estimator (Firth and Bennett 1998) for the following HT model

$$y_t|z_t \sim_{ind} N(\mu z_t, \sigma^2 z_t^2).$$

If units selected with certainty are excluded, the same model is obtained by replacing z_t here by the selection probability π_t, since they differ by a constant multiple. It is well known that the HT estimator is design unbiased, but can be inefficient when this HT model is not a good approximation to reality. An extreme parody is the famous circus elephant example in Basu (1971).

Zheng and Little (2003, 2005) propose to replace the HT model with the following penalized spline model (Ruppert, Wand and Carroll 2003),which assumes a much weaker spline relationship between the outcome and the design variable z_t:

$$(y_t|\beta,\sigma^2) \sim_{ind} N(f(z_t,\beta), z_t^{2k}\sigma^2) \quad (3)$$

$$f(z_t,\beta) = \beta_0 + \sum_{j=1}^{p}\beta_j z_t^j + \sum_{j=p+1}^{p+m}\beta_j(z_t-\kappa_{j-p})_+^m \quad (4)$$

$$p(\beta_{p+1},...,\beta_{p+m}|\tau^2) \sim_{ind} N(0,\tau^2) \quad (5)$$

$$p(\beta_0,...,\beta_p,\log\sigma^2,\tau^2) = const. \quad (6)$$

Here, k is a known constant reflecting assumptions about the error variance, $\kappa_1, ..., \kappa_m$ are known constants representing positions of knots for the spline, and $(u)_+^m = u^m$ if $u > 0$, and 0 otherwise. In the spirit of Ruppert (2002), and Ruppert, Wand, and Carroll (2003), we favor a modeling strategy that places a large number m of knots (for example, 15 or 30) at pre-specified locations, and then achieves smoothness by treating $\beta_{p+1}, ..., \beta_{p+m}$ as random effects centered at 0. With a large sample, prediction under this model is conveniently carried out by restricted maximum likelihood (REML), and can be implemented with readily available mixed model software; with small samples a fully Bayesian approach is preferable.

Zheng and Little (2003, 2005) assess the ability of inferences from this weak model to match or improve on design-based inferences (specifically, the HT estimator and more sophisticated generalized regression estimators, with a variety of associated variance estimates). Five artificial populations were simulated by adding independent errors with variance 0.2 to the following mean functions relating outcome and the inclusion probabilities π_t:

(NULL) $f(\pi_t) = 0.3$; no relationship;
(LINUP) $f(\pi_t) = 3\pi_t$, linearly increasing function with a zero intercept;
(LINDOWN) $f(\pi_t) = 0.58 - 3\pi_t$, linearly decreasing function with positive intercept;
(EXP) $f(\pi_t) = \exp(-4.64 + 26\pi_t)$, an exponentially increasing function;
(SINE) $f(\pi_t) = 1 + \sin(35.69\pi_t)$,a sinusoidal shaped function;
A sixth population was generated to yield an "S shaped function with heteroskedastic errors:
(ESS) $y_t = 0.6\,\text{logit}^{-1}(50\pi_t - 5 + \varepsilon_t)$, $\varepsilon_t \sim_{ind} N(0,1)$.

Table 1 presents root mean squared errors (RMSE) of point estimates from the following methods: HT, the Horvitz–Thompson estimator of the mean; GR, the generalized regression estimator with predictions from a simple linear regression, assuming a constant error variance; and Spline, a p-spline prediction estimator assuming a constant residual variance and 15 knots. Empirical Bayes was used to predict the non-sampled Y's here, and variances were computed by a jackknife to avoid sensitivity to the assumption of constant variance of the residuals. Thus the method is not strictly Bayesian, but similar in spirit to a full Bayesian implementation. For each of the six mean structures, the RMSE's are based on estimates for 500 PPS systematic samples of size 96. Table 1 suggests that Spline has smaller empirical RMSE than HT or GR for the populations with nonlinear mean structures (SINE, EXP and ESS), similar RMSE to GR when the mean function is linear (NULL, LINUP and LINDOWN), and similar RMSE as HT for the population LINUP, which favors the HT estimator.

Table 2 shows that Spline, with standard errors computed using the jackknife, yields narrower confidence intervals with coverage properties comparable to that of HT and GR. The only case where Spline has poor coverage is the SINE model, and this problem is resolved by increasing the number of knots for the spline. For more details and additional simulation results, see Zheng and Little (2003, 2005), and for extensions to two-stage samples, see Zheng and Little (2004).

Table 1: *Root mean squared errors of three estimates of the population yotal from probability proportional to size sampling.*
HT=Horvitz–Thompson Estimator; GR = generalized regression estimator; Spline = prediction estimator from penalized spline model.

	HT	GR	Spline
NULL	35	24	22
LINUP	27	34	26
LINDOWN	63	35	27
SINE	113	95	45
EXP	35	54	27
ESS	11	30	10

Table 2: *Average width (AW) and noncoverage rate (NC) of 95% confidence or credibility intervals over 1000 samples (target* 50 ± 20*).*
HT = Horvitz Thompson estimate with random groups variance estimate, GR = Generalized Regression estimate with Yates-Grundy variance estimate, Spline = Penalized Spline with Jackknife variance estimate. N = 1000, n = 100.

	HT		GR		Spline	
	AW	*NC*	AW	*NC*	AW	*NC*
NULL	131	*68*	88	*80*	89	*28*
LINUP	109	*42*	123	*64*	98	*48*
LINDOWN	230	*82*	124	*82*	94	*62*
SINE	446	*60*	340	*74*	145	*86*
EXP	135	*42*	193	*96*	105	*54*
ESS	48	*14*	109	*84*	37	*66*

3. BAYESIAN ANALYSIS WHEN DESIGN INFORMATION IS AVAILABLE ONLY FOR THE SAMPLED UNITS

So far we have assumed the design variables, such as the size variable in Example 3, were recorded for all units in the population. This information is needed to implement PPS sampling, and hence in principle is available for analysis. However, standard design-based estimators like the Horvitz–Thompson estimator only need the sizes (more specifically the sampling weights) of the sampled units, and as a consequence, design information on the non-sampled cases is often not included on public use data files. Some authors (Pfeffermann, Krieger and Rinott, 1998; Pfeffermann and Sverchkov, 2003) have argued that the fully Bayesian approach is too hard in this situation, and have advocated instead an analysis based on the likelihood that conditions on the cases t that are included in the sample, for which $i_t = 1$. Here the full Bayesian approach requires a supplemental model for the design variables, in order to predict their values for the non-selected cases. With some additional work, we believe the Bayesian approach can still be applied successfully. We conclude by showing how this can be done in the case of PPS sampling, via a Bayesian bootstrap model for the size variable, modified to account for PPS selection.

Example 4 (Bayes Inference for PPS Sampling where Sizes of Nonsampled Units are not Recorded). We consider PPS sampling as in Example 3, in the situation where the selection probabilities π_t (or the size measures z_t) are only available for the sampled cases, but the number of non-sampled cases $N - n$ is known. Bayesian inference then requires a Bayesian model for the size variable probability Z, with prior distribution $p_Z(\cdot)$. Draws from the posterior distribution of $\bar{Y}$ can then be simulated by the following procedure:

(a) For non-selected units t, draw predictions $z_t^{(d)}$ of the sizes from the posterior distribution $p(z|\text{data}, i_t = 0)$ of the size, given that these units are not selected.

(b) For non-selected units t, draw $y_t^{(d)}$ from the posterior distribution of Y given the drawn value of the size, $z_t^{(d)}$. We use the same penalized spline model as in Example 3.

(c) the resulting draw from the posterior distribution of $\bar{Y}$ is the average of the observed values y_t for units t in the sample and the predictions $y_t^{(d)}$ of units t not in the sample.

Steps (b) and (c) are identical to Bayesian simulation for Example 3, where the sizes are available for the nonsampled units, with drawn sizes $z_t^{(d)}$ replacing actual sizes z_t. Thus we confine discussion to Step (a). We note that when the sizes are unknown for the non-sampled units, the sample design becomes informative (Sugden and Smith 1984), and the draws of the sizes of non-sampled cases need to be corrected for the effects of selection. Given a model M_Z for the sizes with prior distribution $p_Z(\cdot)$, the required posterior distribution of the sizes given non-selection is related to the posterior distribution of the sizes given selection by

$$\begin{aligned} p(z|\mathsf{data}, i = 0) &= cp(z|\mathsf{data}, i = 1)p(i = 0|z, \mathsf{data})/p(i = 1|z, \mathsf{data}) \\ &= cp(z|\mathsf{data}, i = 1)(1 - \pi(z))/\pi(z), \end{aligned} \tag{7}$$

where c is a normalizing constant. We describe this predictive distribution for the (relatively nonparametric) Bayesian Bootstrap model. Let $\{x_1, .., x_K\}$ be the set of distinct sizes for the sampled units, and let n_k be the number of sampled cases with size x_k, $\sum_{k=1}^{K} n_k = n$. We assume that these counts are multinomial with probabilities $\{\phi_1, .., \phi_K\}$, which are assigned a $Dirichlet(0, \ldots, 0)$ prior distribution:

$$\begin{aligned} (n_1, ..., n_K|\phi_1, ..., \phi_K, i = 1) &\sim MNOM[n; (\phi_1, ..., \phi_K)], p(\phi_1, ..., \phi_K) \\ &\propto \prod_{k=1}^{K} \phi_k^{-1}. \end{aligned} \tag{8}$$

This model makes the assumption that only the selection probabilities that arise are those seen in the sampled cases; though unrealistic, this assumption does not seriously impact the quality of the resulting inferences. The posterior distribution of $(\phi_1, ..., \phi_K)$ is then $Dirichlet(n_1, \ldots, n_K)$:

$$p(\phi_1, ..., \phi_K) \propto \prod_{k=1}^{K} \phi_k^{n_k - 1}. \tag{9}$$

Let n_k^* be the number of nonsampled cases with size measure x_k, with $\sum_{k=1}^{K} n_k^* = N - n$. The posterior predictive distribution of these counts is then multinomial:

$$(n_1^*, ..., n_K^* | \phi_1, ..., \phi_K, i = 0) \sim MNOM[N - n; (\phi_1^*, ..., \phi_K^*)], \tag{10}$$

with $\phi_k^* = c\phi_k(1 - \pi_k)/\pi_k$, where $\pi_k = nx_k/N\bar{X}$ is the selection probability for units with size x_j and c is chosen so that $\sum_{k=1}^{K} \phi_k^* = 1$. The transformation from ϕ_k to ϕ_k^* is the required modification of the standard Bayesian bootstrap model to account for PPS selection. The modified BB consists of drawing values $(\phi_k^{(d)})$ from the distribution of Equation (9) and then drawing predicted counts $(n_k^{*(d)})$ from the Equation (10), with ϕ_k replaced by drawn value $\phi_k^{(d)}$.

We now present a preliminary simulation study to illustrate the repeated sampling properties of inferences from this model – future work will examine confidence coverage. Five populations of 3000 units were simulated as follows:

Thirty distinct values of size z_t are simulated from a Dirichlet distribution with parameters (1,1,...,1). Each distinct value of z_t is represented 100 times in the simulated populations.

Outcome values $y_t|z_t$ are drawn as normal with standard deviation 0.1 and mean given by one of the following functions:

NULL: $f(z_t) = 0.3$
LINUP: $f(z_t) = 3z_t$
LINDOWN: $f(z_t) = 0.6 - 3z_t$
EXP: $f(z_t) = 0.5 + 0.25\exp(8z_t)$
SINE: $f(z_t) = 0.5\sin(25z_t)$

Plots of these populations are shown in Fig. 1. Five hundred samples of size 300 were drawn from a population of 3000 units by systematic PPS sampling from a randomly ordered list. The following estimates of the population total T were computed for each sample:

HT: the design-weighted Horvitz–Thompson estimator;

Spline: average of sampled values of Y and empirical Bayes predictions of non-sampled values from the Penalized Spline Model, assuming sizes z_t are known for all units in the population (as in Example 3);

Spline/BB-A: as for spline, with sizes z_t drawn using the BB model described above; resulting estimates of $\bar{Y}$ were averaged over 100 draws of the BB predicted sizes.

Spline/BB-B: as for spline, with sizes z_t drawn using the BB model described above, and then scaled so that they average to their known mean $\bar{Z}^*$.; resulting estimates of $\bar{Y}$ were averaged over 100 draws of the BB predicted sizes.

Spline/HT: the Horvitz Thompson estimator applied to the empirical Bayes predictions from the Penalized Spline model for the respondents. This implements the method of Pfeffermann, Krieger and Rinott (1998) for the penalized spline prediction model.

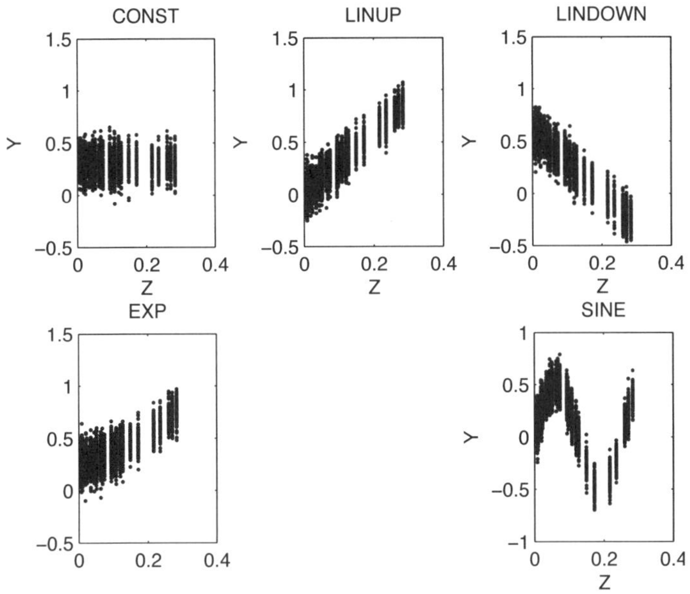

Figure 1: *Plots of five populations for Example 4.*

Table 3 shows the root mean squared error (RMSE) of these estimates, expressed as a percentage of the total, for each of the simulated populations. The Spline and Spline/BB methods display negligible bias (data not shown) and considerable reductions in RMSE over the HT and Spline/HT estimators. It is interesting that for the NULL, LINUP and LINDOWN populations, the RMSE's of Spline/BB-B are only slightly higher than that for Spline, suggesting that predicting the sizes of the non-sampled cases does not entail much information loss, assuming the number of non-sampled cases $N - n$ and mean size $\bar{Z}^*$ is known. This is plausible, because the prediction of the total depends on the sum of the sizes of non-sampled units, which is deducible from the fact that $\sum_{t=1}^{N} z(t)$ is known.

The RMSE of Spline/BB-B is lower than that of spline/BB-A in populations other than SINE, indicating a gain from constraining the drawn sizes to average to the known mean for non-sampled cases. Scaling seems less effective for the more complex SINE mean structure.

Spline/HT may be viewed as an approximate Bayes for the situation where $N - n$ is not known – full Bayes in this case requires prediction of N as well as the other quantities, which is not attempted here. Note that unlike Spline/BB-A or Spline/BB-B, Spline/HT does not achieved reduced RMSE's compared with HT, suggesting that a gain over HT requires use of knowledge of $N - n$ and $\bar{Z}^*$. The similar performance of Spline/HT and HT is also seen in the previous simulations in Zheng and Little (2003).

To assess the sensitivity of our method to a different choice of simulated distribution of z_t, we also simulated five populations with z_t following a log-normal distribution, with mean zero and standard deviation 0.6 after log-transformation. The estimation of sizes still assume the Dirichlet model for z_t. The mean functions

Table 3: *Root mean squared errors (as percentage of total) of five estimates of the population total from probability proportional to size sampling. HT = Horvitz–Thompson Estimator; Spline = P-Spline posterior mean; Spline/BB-A= P-Spline posterior mean with non-sampled sizes predicted using BB model, posterior values of Z not rescaled; Spline/BB-B= P-Spline posterior mean with non-sampled sizes predicted using BB model, posterior values of Z rescaled; Spline/HT = HT estimator with observed values of Y replaced by predictions from P-Spline model.*

	HT	Spline	Spline/BB-A	Spline/BB-B	Spline/HT
NULL	9.1	2.4	2.4	2.4	8.7
LINUP	3.2	2.7	7.9	2.6	2.7
LINDOWN	16.8	3.2	7.7	3.2	16.7
SINE	12.3	5.5	10.0	10.4	12.2
EXP	4.8	0.8	3.4	2.2	4.8

of Y are the same as those used in Table 3, except the SINE function is modified into $f(z_t) = 0.5\sin(30z_t)$ to accommodate the change in the distribution of z_t. Table 4 gives the root mean squared error of the same estimates, expressed as a percentage of the total, for each of the simulated populations using misspecified model for z_t. Spline/BB-B still has lower RMSE's than HT for all populations, although gains seem smaller. Comparative RMSE's of Spline/BB-B relative to Spline/BB-A are more mixed for these populations.

Table 4: *Root mean squared errors (as percentage of total) of five estimates of the population total from probability proportional to size sampling, for population sizes with a log-normal distribution.*
HT = Horvitz–Thompson Estimator; Spline = P-Spline posterior mean; Spline/BB-A= P-Spline posterior mean with non-sampled sizes predicted using BB model, posterior values of Z not rescaled; Spline/BB-B= P-Spline posterior mean with non-sampled sizes predicted using BB model, posterior values of Z rescaled; Spline/HT = HT estimator with observed values of Y replaced by predictions from P-Spline model.

	HT	Spline	Spline/BB-A	Spline/BB-B	Spline/HT
NULL	5.3	3.8	3.7	5.2	5.2
LINUP	4.3	3.9	4.8	3.9	3.9
LINDOWN	8.1	4.1	5.4	7.9	7.9
SINE	26.7	11.9	23.2	16.1	26.6
EXP	2.3	1.1	1.6	1.2	2.3

4. CONCLUSION

We have suggested some features of a successful application of Bayes methods to survey sampling problems involving stratification, clustering and PPS sampling – specifically, models with weak parametric assumptions and noninformative priors that take account of features of the survey design. In the case of PPS sampling, previous work based on a penalized spline model has been shown to yield inferences with superior repeated sampling properties to the standard design-based Horvitz–Thompson estimate. We have shown how the Bayesian approach can be extended to the case where sizes of units are only observed for the sampled cases, using a Bayesian bootstrap model for the sizes. Simulations suggest that Bayesian prediction of the sizes of non-sampled cases using a Bayesian Bootstrap model can yield gains in efficiency, by exploiting knowledge about the number of non-sampled cases and their average size. In future work we plan to study inference properties of this procedure in more detail, and extend it to other sample designs involving clustering of units.

REFERENCES

Basu, D. (1971). An essay on the logical foundations of survey sampling. *Foundations of Statistical Inference* (V. P. Godambe and D. A. Sprott, eds.) Toronto: Holt, Rinehart and Winston, 203–242.

Binder, D. A. (1982). Non-parametric Bayesian models for samples from finite populations. *J. Roy. Statist. Soc. A* **44**, 388–393.

Box, G. E. P. (1980). Sampling and Bayes inference in scientific modeling and robustness. *J. Roy. Statist. Soc. A* **143**, 383–430 (with discussion).

Chambers, R. L. and Skinner, C. J. (2003). *Analysis of Survey Data.* New York: Wiley

Efron, B. (1986). Why isn't everyone a Bayesian? *Amer. Statist.* **40**, 1–11 (with discussion).

Elliott, M. R. and Little, R. J. A. (2000). Model-based alternatives to trimming survey weights. *J. Official Statistics* **16**, 191–209.

Ericson, W.A. (1969). Subjective Bayesian models in sampling finite populations. *J. Roy. Statist. Soc. B* **31**, 195–233 (with discussion).

Ericson, W. A. (1988). Bayesian inference in finite populations. *Handbook of Statistics* **6**. Amsterdam: North-Holland, 213–246.

Firth, D. and Bennett, K. E. (1998). Robust models in probability sampling. *J. Roy. Statist. Soc. B* **60**, 3–21.

Gelman, A., Carlin, J. B, Stern, H. S., and Rubin, D. B. (1995). *Bayesian Data Analysis.* London: Chapman and Hall

Gelfand, A. E., Hills, S. E., Racine-Poon, A. and Smith, A. F. M. (1990). Illustration of Bayesian inference in normal data models using Gibbs sampling. *J. Amer. Statist. Assoc.* **85**, 972–985

Ghosh, M. and Meeden, G. (1997). *Bayesian Methods for Finite Population Sampling.* London: Chapman and Hall

Hanif, M. and Brewer, K. R. W. (1980). Sampling with unequal probabilities without replacement: A review. *Internat. Statist. Rev.* **48**, 317–335.

Holt, D., Smith, T. M. F., and Winter, P. D. (1980). Regression analysis of data from complex surveys. *J. Roy. Statist. Soc. A* **143**, 474-487.

Horvitz, D. G., and Thompson, D. J. (1952). A generalization of sampling without replacement from a finite universe. *J. Amer. Statist. Assoc.* **47**, 663–685.

Kish, L. (1965). *Survey Sampling.* New York: Wiley

Kish, L. and Frankel, M.R. (1974). Inferences from complex samples. *J. Roy. Statist. Soc. B* **36**, 1–37, (with discussion).

Little, R. J. A. (2003). The Bayesian approach to sample survey inference. *Analysis of Survey Data* (R.L. Chambers and C.J. Skinner, eds.) New York: Wiley, 49–57.

Little, R. J. A. (2004). To model or not to model? Competing modes of inference for finite population sampling. *J. Amer. Statist. Assoc.* **99**, 546–556.

Little, R. J. A. (2006). Calibrated Bayes: a Bayes/frequentist roadmap. *Amer. Statist.* **60**, 213–223.

Pfeffermann, D. and Holmes, D. J. (1985). Robustness considerations in the choice of method of inference for regression analysis of survey data. *J. Roy. Statist. Soc. A* **148**, 268–278.

Pfeffermann, D., Krieger, A. M. and Rinott, Y. (1998). Parametric distributions of complex survey data under informative probability sampling. *Statistica Sinica* **28**, 1087–1114.

Pfeffermann, D. and Sverchkov, M. (2003). Fitting generalized linear models under informative probability sampling. *Analysis of Survey Data.* New York: Wiley, 175–195.

Rubin, D.B. (1976). Inference and missing data. *Biometrika* **53**, 581-592.

Rubin, D. B. (1983). Comment on Hansen, Madow and Tepping. *J. Amer. Statist. Assoc.* **78**, 803–805.

Rubin, D. B. (1984). Bayesianly justifiable and relevant frequency calculations for the applied statistician. *Ann. Statist.* **12**, 1151–1172.

Rubin, D. B. (1987). *Multiple Imputation for Nonresponse in Surveys.* New York: Wiley

Ruppert, D. (2002). Selecting the number of knots for penalized splines. *J. Comp. Graphical Statist.* **11**, 735–757.

Ruppert, D., Wand, M. P., and Carroll, R. J. (2003). *Semiparametric Regression.* Cambridge: University Press.

Scott, A. J. (1977). Large-sample posterior distributions for finite populations. *Ann. Math. Statist.* **42**, 1113–1117.

Scott, A. J. and Smith, T. M. F. (1969). Estimation in multistage samples. *J. Amer. Statist. Assoc.* **64**, 830–840.

Sugden, R. A. and Smith, T. M. F. (1984). Ignorable and informative designs in survey sampling inference. *Biometrika* **71**, 495–506.

Zheng, H. and Little, R. J. (2003). Penalized spline model-based estimation of the finite population total from probability-proportional-to-size samples. *J. Official Statistics* **19**, 99–117.

Zheng, H. and Little, R. J. (2004). Penalized spline nonparametric mixed models for inference about a finite population mean from two-stage samples. *Survey Methodology* **30**, 209–218.

Zheng, H. and Little, R. J. (2005). Inference for the population total from probability-proportional-to-size samples based on predictions from a penalized spline nonparametric model. *J. Official Statistics* **21**, 1–20.

DISCUSSION

FABRIZIO RUGGERI (*CNR-IMATI, Italy*)

I would like to thank Little and Zheng for a very insightful paper in an important area where Bayesian methods are hardly used (as I experienced when looking for related works). I hope their paper can stimulate further research in the field and spread the use of Bayesian methods in survey sampling settings.

Robustness. The original title of the paper was mentioning a *robust* Bayesian approach. The idea of the authors is about robustness with respect to (w.r.t.) model misspecification, i.e. when inferences are not strongly tied to assumptions about the distribution of survey data y_U given the design variables z_U, as discussed in Chambers (2003). This aspect of robustness, thoroughly addressed in classical statistics, is definitely important but the choice of a Bayesian approach leads to a new class of sensitivity problems, induced by the choice of priors and loss functions, besides models.

As an example about finite population sampling, consider the problem of estimating the population mean when it is known a priori that the population median belongs to some interval, as studied in Nelson and Meeden (1998), who proposed a Bayesian non informative approach based on a Polya posterior. According to a robust Bayesian approach (see Rios Insua and Ruggeri, 2000, for a thorough review), the problem can be formalized with a *generalized moments constrained* class of priors

$$\Gamma = \{\pi : \int_0^m f(x \mid \theta)\pi(\theta)\, d\theta = F_X(m) = .5, a \leq m \leq b\},$$

or, equivalently,

$$\Gamma = \{\pi : F_X(a) \leq .5, F_X(b) \geq .5\}.$$

When interested in the behavior of $Eh(\theta)$ and its robustness w.r.t. misspecification of the prior, then the range, $\sup_\Gamma Eh(\theta) - \inf_\Gamma Eh(\theta)$, is analyzed. It is known that computation of sup and inf is obtained by considering discrete distributions over, at most, three points.

Noninformative priors. The robust Bayesian approach could be more effective than the choice of noninformative priors, as suggested by the authors, to address agencies' antipathy towards subjective judgements, as mentioned in the paper (point 4 of Section 1.1.). In general, the paper relies a lot on noninformative priors, e.g. in the examples of Section 2. Without entering in the discussion about the appropriateness of choice of noninformative priors, I would like to know what happens when proper prior information is available, both in terms of posterior distributions and parameters elicitation. I expect MCMC simulation would be needed and I wonder if simple, easy-to-use, codes could be provided using, *e.g.*, WinBUGS; this could definitely help the diffusion of Bayesian methods among people involved in survey sampling. Last, but not least, I think it is difficult to convince people to accept Bayesian methods proposing them to use noninformative priors: where is the gain w.r.t. classical approaches?

Dynamic (adaptive) survey sampling. I am currently working on e-democracy problems and, in particular, I am interested in processes about e-negotiation and, more in general, e-participation. In these processes, different stakeholders (e.g. Trade Unions and Industrialists Associations) are discussing about a common theme (e.g. workers' salary) and they are negotiating until an agreement is found. At different stages of the negotiation, the two associations (as in our example) might need to known their affiliates' opinions on the current proposals and, therefore, a survey is conducted.

This scenario differs from the ones presented by the authors, but I think a Bayesian approach could be taken here, specifying a model (e.g. a Dynamic Linear Model, as described in West and Harrison, 1997) able to describe the evolution of y_U (stakeholder's opinion) w.r.t. z_U (terms of the proposal). Since people can change opinions and aggregate according to different criteria over time, then evolution of strata should be considered as well. Instead of sampling anew any time, combination of experts' opinions and past data could lead to forecast of opinions without (or with smaller) survey sampling. In the same spirit, past data could be used to identify the *most representative* sample and use just it in the future.

Some theoretical questions. Some authors have been discussing about differences between sampling from infinite population and from finite one. In particular, someone claims that the theoretical foundations of finite sampling differ from infinite sampling ones, whereas others argue that both samplings are based on the same axioms of probability theory, so that they should have the same theoretical foundations. I would like to know a comment by the authors on the dispute and if (and how) it can affect what presented in the paper.

In the paper, sampling is, in general, without replacement. In some papers I read about finite population surveys, the issue of sampling with replacement was raised. I wonder how the proposed Bayesian approaches would be affected by the different sampling scheme.

Estimates are compared via RMSE (root mean square error) for simulated data. The approach is widely used to compare empirically different estimators in many fields, e.g. wavelets coefficient estimation. Looking at the wavelets literature, where tests are performed w.r.t. a standard, commonly used set of functions, here comparisons are made using six population proposed for the first time, I guess, in the current paper. I am not advocating that these functions are less valid than doppler, heavysine, bumps and blocks functions used in the wavelet literature; I am just pointing out that no widely recognized test exists to compare estimates and it is difficult to have a clearcut acceptance of the superiority of the proposed method. An important contribution to the acceptance of Bayesian methods could be provided by some theoretical results, e.g. about their optimality w.r.t. some criterion. I wonder if any had been proved.

Bayesian analysis is naturally linked to Decision analysis and it would be important to study the optimal (w.r.t. some loss function) allocation of sample sizes to each stratum (e.g. same proportion as stratum in the population) and optimal total sample size. Both quantities are relevant when surveys are subject to budget constraints.

Some practical questions. In Example 3, a penalized spline model with fixed knots is used. I wonder if a random number of knots (obtained, *e.g.*, using a Reversible Jump MCMC method) could provide a more adequate model, although I am aware that it is in contrast with the agencies' concerns about complex models. Mixture models, either w.r.t. a parametric distribution or, say, a Dirichlet process, are other complex, but very flexible, models worth being used.

Methods like partition models (see e.g. Quintana and Iglesias, 2003) and BART (see e.g. Chipman, George and McCulloch, 2007) could be used to partition the sample space and classify population into strata.

Related issues. Many other, related problems are (or could be) addressed following a Bayesian approach; I just want to mention few of them.

Imputation of missing data is a natural environment for the use of Bayesian methods, combining experts' opinions and similar data. A related issue is about nonresponse in sample surveys: I wonder if there is any update w.r.t. Little (1982).

Web surveys are very popular in the Internet era. They are, in general, characterized by unknown and evolving population size, with sample characteristics (sex, age, culture, etc.) different from total population. A recent unpublished proposal, due to Monica Pratesi, is an adjusted poststratified Horvitz–Thompson estimator, able to take care of distortions. I wonder if an equivalent Bayesian estimator is conceivable.

Finally, Bayesian methods are used for disclosure limitations, addressing problems like data aggregation to mask data or re-identification via combination of rare characteristics.

Conclusion. I hope my comments will be useful and will allow the authors to illustrate in their rejoinder further applications of Bayesian methods in finite population surveys. This is in the spirit of the paper which raises issues, illustrates results and, above all, sets directions for further investigations.

STEPHEN E. FIENBERG (*Carnegie Mellon University, USA*)

The presentation by Little and Zheng (LZ) was clear and convincing in its broad structure and in most of its arguments. Being a Bayesian sampling statistician should be as natural to a Bayesian as to a sampling statistician and I hope this paper serves to reintroduce the Valencia audience to a rich vein of statistical problems of much practical import. I say reintroduce, since the framework Little uses is essentially the same as the one used earlier by Sugden (1985) and by Smith (1988) at *Valencia* 2 and *Valencia* 3, respectively. See also Sugden and Smith (1984,1988).

In particular, the basic formulation of finite population sampling in terms of population stochastic structures and selection mechanisms makes clear the role of random selection of sample units and parellels the Rubin (1978) argument of why a Bayesian should randomize in experimental setting, cf. Fienberg and Tanur (1987). What went unsaid in the paper but becomes clear from the formulation is that is that not all forms of sampling have ignorable selection mechanisms, even in the absence of information missingness. For example if we have a logistic regression structure and sample by stratifying on the response variable, *e.g.*, see Hoem (1989), or if we only have information for the sampled units, the problem described by LZ in Section 3, and much earlier in Sugden and Smith (1984).

Where I differ with the authors is in whether it is appropriate to emulate the design-based frequentist approach of model-free statistical analysis. That is not an unreasonable approach when the goal of analysis is to produce estimates of simple population aggregates such as means or totals, especially in the presence of heterogeneity of unknown origin. But if we has as a goal the understanding of a stochastic phenomenon based on some highly-multivariate latent quantity, we cannot afford to adopt a model-free posture whether we are Bayesian or frequentist. Indeed we are likely to invoke hierarchical multi-level models involving large numbers of parameters.

The problems of what to do with weights and and design information become much more acute when one moves from simple one shot surveys to panels with data collected over time. Consider the example of the National Long Term Care Survey which assesses disability in U.S. elderly population. Erosheva (2003) reported on the analysis of a 2^{16} contingency table where the variables are activities of daily living(ADL) and instrumental activities of daily living (IADL)–individuals can be either disabled or healthy. She used a hierarchical Bayesian mixed membership model with $K = 5$ subpopulations or extreme profiles for the 16 ADL/IADL measures, pooled across four survey waves of NLTCS, 1982, 1984, 1989, and 1994. Thus the data form a 2^{16} contingency table. The table has 65,536 cells, only 3,152 of which are non-zero and there are a total of $N = 21,574$ observations. The weights on individuals across waves vary substantially. Here it is unclear what a design-based weighted analysis would even mean. In one of the examples Little described in the oral presentation, he mentioned using a 'non-informative' Dirichlet uniform prior. In the case of the data from the NLTCS this would clearly be highly informative with the prior adding more than three times the value of the sample from the likelihood.

I see another challenge for Bayesians in this paper. As LZ suggest, sampling statisticians are enthralled with the Horvitz–Thompson (HT) estimator as a mechanism for handling many different design and non-response problems, indeed some have used this as a primary device for integrating approaches across designs, *e.g.*, see Sarndal, et al. (1992). I've often been struck by this reliance on HT and I have concluded that the heuristic interpretation of the estimator is a very powerful metaphor. So I think we need to ask: Can thousands of sampling statisticians be wrong? LZ suggest that this might be the case. While the example they present is clearly suggestive that a systematic Bayesian approach can improve upon naive applications of HT, I keep wondering whether there might still be a simpler Bayesian variation on HT that will have at least some of the improvements that LZ find in their setting and at same time the heuristic interpretation that makes HT so appealing. This would clearly require a principled approach to incorporating the sampling process into the model in some kind of generative form.

REPLY TO THE DISCUSSION

We thank Fabrizio Ruggeri and Steve Fienberg for their wise and stimulating discussions, which provide useful additional references and broaden the scope of our presentation. We organize our rejoinder by topic.

Noninformative priors, nonparametric models. The discussants question whether relatively 'non-parametric' models with weak noninformative priors are appropriate in all survey situations, and we agree that the answer is clearly 'no'. We think it is good that the Bayesian approach encompasses classical frequentist answers when those answers work well, as in the case of large stratified samples in Example 1; but Bayes also offers better answers when classical frequentist methods work less well, as when there are many sparse strata with variable weights, and pooling strength across strata is achieved by imposing a hierarchical model on the stratum effects. If the structure of the population is of major interest, as in Steve's contingency table example, then that structure needs to be explicitly modeled to obtain useful inferences. This is more akin to what survey samplers would call 'analytical' rather than 'descriptive' inference. Steve's warning about the uniform prior adding spurious information in the contingency table example is important, and we were unclear in early versions of the paper that we chose the $Dirichlet(0, ..., 0)$ prior for the probabilities over the Jeffreys' $Dirichlet(1/2, ..., 1/2)$ prior in order to avoid that problem in our setting. The larger point in Steve's example is that a more structured model is needed to make sense of the very sparse contingency table, and we entirely agree.

Fabrizio notes that a noninformative prior is not appropriate when prior information is available, as in the example where the population median is known to lie in an interval. We agree that such information should be incorporated in the prior distribution, and it is much easier to do so using a Bayesian approach than under the frequentist paradigm. For an extremely simple illustration of this point, see Little (2006, Example 2). The known constraint on the median is relatively non-standard in the finite population setting, and we focused attention on more standard survey problems.

Fabrizio's dynamic sampling problem involving trade union negotiations is related to the analysis of panel or repeated cross-sectional surveys where similar information is collected repeatedly over time. It seems very reasonable in such settings to consider using information from prior surveys to inform the prior distribution.

Theoretical issues. Others are better equipped to answer Fabrizio's questions about theory, but under the superpopulation modeling approach we advocate, we do not see any major differences in Bayesian theory for survey sampling than for other branches of statistics. Inferences about superpopulation parameters are no different, and inferences about finite population quantities are not fundamentally any different than predictions in other applications of statistics, such as in time series modeling.

It is true that the design-based tradition has its own theory, with notions such as design consistency that are relatively 'model-free' and derive from the probability distribution that governs the sampling. The interplay between design-based and model-based asymptotics strikes us as quite subtle, and we commend it as a topic of interest to those with a theoretical bent.

More complex models. We have deliberately kept our modeling strategies simple, for example by adopting a parametric spline model with many fixed knots, and the simple Bayesian Bootstrap model for the sizes in Examples 3 and 4. One reason for doing this is that computation is feasible with existing standard software packages, and ease of computing seems an important practical issue. Fabrizio asked whether Win-Bugs can be used in the survey setting, and indeed the standard multilevel models we consider are readily amenable to analysis by that program.

Many elaborations of our models can be envisaged, for example, replacing the BB model by a Dirichlet process prior, or our penalized spline model with more sophisticated spline modeling with variable knot placement. It remains to be seen whether the added complexity is justified by substantial improvements in the answers – a pleasing feature of finite population sampling is that the objective of the inference is a real quantity rather than a parameter in an idealized model, so realistic simulations using real populations can be conducted to address such questions.

Fabrizio mentions survey nonresponse, which is readily amenable to a Bayesian approach. For more recent discussions that go beyond the Little (1982) paper, readers might consult Rubin (1987), Little and Rubin (2002), or Little (2003).

Finally, like Steve we are charmed by the heuristic interpretation of the Horvitz–Thompson estimator – each sampled case represents a number of cases in the population given by the HT weight. On the other hand the Bayesian approach has its own a powerful heuristic – build a model to predict values for the non-sampled cases – and this heuristic provides extensions of the HT estimator within a principled inferential framework, the framework for this conference.

ADDITIONAL REFERENCES IN THE DISCUSSION

Chambers, R. L. (2003). Which sample survey strategy? A review of three different approaches. *Tech. Rep.*, University of Southampton, UK.

Chipman, H. A., George, E. and McCulloch, R. E. (2007). BART: Finding low dimensional structure in high dimensional data. *In this volume.*

Erosheva, E. A. (2003). Bayesian estimation of the grade of membership model. *Bayesian Statistics 7* (J. M. Bernardo, M. J. Bayarri, J. O. Berger, A. P. Dawid, D. Heckerman, A. F. M. Smith and M. West, eds.) Oxford: University Press, 501–510.

Fienberg, S. E. and Tanur, J. M. (1987). Experimental and sampling structures: Parallels diverging and meeting. *Internat. Statist. Rev.* **55**, 75–96.

Hoem, J. M. (1989). The issue of weights in panel surveys of individual behavior. *Panel Surveys* (D. Kasprzyk, G. Duncan, G. Kalton, and M. P. Singh, eds.) New York: Wiley, 539–565.

Little, R. J. A. (1982). Models for nonresponse in sample surveys. *J. Amer. Statist. Assoc.* **77**, 237–250.

Little, R. J. A. (2003). Bayesian Methods for Unit and Item Nonresponse. *Analysis of Survey Data*, (R. L. Chambers and C .J. Skinner, eds.) New York: Wiley, 289–306.

Little, R. J. A. and Rubin, D. B. (2002). *Statistical Analysis with Missing Data*, 2nd edn. New York: Wiley

Nelson, D. and Meeden. G. (1998). Using prior information about population quantiles in finite population sampling. *Sankhyā A* **60**, 426–445.

Quintana, F. A. and Iglesias, P. L. (2003). Bayesian clustering and product partition models. *J. Roy. Statist. Soc. B* **65**, 557–574.

Rios-Insua, D. and Ruggeri, F. (eds.) (2000). *Robust Bayesian Analysis.* Berlin: Springer-Verlag.

Rubin, D. B. (1978). Bayesian inference for causal effects: The role of randomization. *Ann. Math. Statist.* **6**, 34–58.

Rubin, D. B. (1987). *Multiple Imputation for Nonresponse in Surveys.* New York: Wiley

Sarndal, C.-E., Swenssen, B. and Wretman, J. (1992). *Model Assisted Survey Sampling.* Berlin: Springer-Verlag

Sugden, R. A. and Smith, T. M. F. (1984). Ignorable and informative designs in survey sampling inference, *Biometrika* **71**, 495–506.

Sugden, R. A. (1985). A Bayesian view of ignorable designs in survey sampling inference, *Bayesian Statistics 2* (J. M. Bernardo, M. H. DeGroot, D. V. Lindley and A. F. M. Smith, eds.), Amsterdam: North-Holland, 751–754.

Sugden, R. A. and Smith, T. M. F. (1988). Sampling and assignment mechanisms in experiments, surveys and observational studies, *Internat. Statist. Rev.* **56**, 165–180.

Smith, T. M. F. (1988). To weight or not to weight, that is the question. *Bayesian Statistics 3* (J. M. Bernardo, M. H. DeGroot, D. V. Lindley and A. F. M. Smith, eds.) Oxford: University Press, 437–451.

West, M. and Harrison, J. (1997). *Bayesian Forecasting and Dynamic Models.* Berlin: Springer-Verlag.

BAYESIAN STATISTICS 8, pp. 303–324.
J. M. Bernardo, M. J. Bayarri, J. O. Berger, A. P. Dawid,
D. Heckerman, A. F. M. Smith and M. West (Eds.)

Detecting Selection in DNA Sequences: Bayesian Modelling and Inference

DANIEL MERL
Cornell University, USA
dmm229@cornell.edu

RAQUEL PRADO
Univ. of California Santa Cruz, USA
raquel@ams.ucsc.edu

SUMMARY

Recent developments in Bayesian modelling of DNA sequence data for detecting natural selection at the amino acid level are presented. This article summarizes and discusses empirical model-based approaches. Key features of the modelling framework include the incorporation of biologically meaningful information via structured priors, posterior detection of sites under selection, and model validation via posterior predictive checks and/or estimation of gene and species trees. In addition, model selection is handled using a minimum posterior predictive loss criterion. The models presented here can incorporate relevant covariates such as amino acid properties, extending in this way previous approaches. Applications include the analysis of two DNA sequence alignments with different characteristics in terms of evolutionary divergences among the sequences: an abalone sperm lysin alignment with a strong underlying phylogenetic structure and a low divergence sequence alignment encoding the Apical Membrane Antigen-1 (AMA-1) in the *P. falciparum* human malaria parasite.

Keywords and Phrases: BAYESIAN GENERALIZED LINEAR MODELS; DNA SEQUENCE DATA; NATURAL SELECTION; MODEL COMPARISON; STRUCTURED PRIORS; TREE ESTIMATION.

1. INTRODUCTION

Determining the effect of natural selection in DNA sequence data is a key subject in the areas of computational biology and population genetics. Several approaches have been developed in recent years to detect positive selection at the amino acid level. Relevant references include, among others, Goldman and Yang (1994); Muse and Gaut (1994); Nielsen and Yang (1998); Suzuki and Gojorobi (1999); Yang *et al.* (2000); Huelsenbeck and Dyer (2004); Suzuki (2004); Yang *et al.* (2005); Kosakovsky Pond and Frost (2005a, 2005b). All these approaches have focused on analyzing

Daniel Merl is Visiting Assistant Professor, Department of Mathematics, Cornell University. Raquel Prado is Assistant Professor, Department of Applied Mathematics and Statistics, Baskin School of Engineering, University of California Santa Cruz.

'phylogenetic data', *i.e.*, data in which each sequence in the alignment represents a unique species. When several sequences representing different individuals from one or more populations of the same species are considered, the approaches mentioned above may produce unreasonable results due to the lack of evolutionary divergence among the sequences. In this paper we summarize and discuss recent modelling frameworks specifically designed to analyze this latter type of data, referred to as 'polymorphic data'. This model-based methodology permits the incorporation of biologically meaningful prior information, while simultaneously allowing maximum flexibility in modelling substitution rates at the amino acid level. We therefore extend previous approaches presented in Prado *et al.* (2006) in order to include the following features: incorporation of relevant covariates, such as amino acid properties, and clustering functions on model parameters that can describe population, or geography specific effects. Additionally, we show how phylogenetic posterior estimation and posterior predictive checks can be used as model validation tools.

Section 2 summarizes the biological terminology that will be used throughout the paper and describes the models. Section 3 discusses different ways of identifying positively and negatively selected amino acid sites. The definitions are based on posterior distributions of the model parameters. In addition, a description of how to obtain estimates of phylogenies and gene trees is included. Although the main objective of the methodology presented here is detection of amino acid sites under selection, estimates of phylogenies – even if such estimates may only describe crude topological features underlying the sequence alignments – can be useful as model checking tools. In Section 4, model comparison and model validation procedures are discussed. Section 5 illustrates various aspects of the models in the analyses of two data sets with different evolutionary characteristics. First, analyses of an alignment coding a 122 residue region of the sperm lysin protein in 25 species of California abalone are presented. This data set has been extensively studied and it is considered to be a good example of how positive selection can act on individual amino acid sites. The alignment also displays a relatively strong phylogenetic signal. Then, analyses of sequences encoding AMA-1, a candidate antigen for malaria vaccine development, are presented. This data set is in some ways orthogonal to the lysin data set, as it consists of multiple sequences encoding AMA-1 within a single species, the human *P. falciparum* malaria parasite. Because of this the sequences display relatively low evolutionary divergence. Finally, Section 5 concludes with a summary of remarks, as well as current and future directions for research.

2. MODELS

We begin by summarizing some biological concepts that will be used throughout the paper.

Codon. This is the term given to a three nucleotide sequence codifying one of the 20 amino acids that serve as the building blocks of proteins. The universal genetic code has 64 possible codons of which 61 encode amino acids and the remaining three are stop codons, designating the end of the coding region.

Synonymous and non-synonymous substitutions. Given that there are 20 amino acids and 61 possible codons, multiple codons codify the same amino acid. Synonymous substitutions are those between codons that specify the same amino acid; e.g., a substitution of TTA for CTC is a synonymous one, since both codons encode the amino acid Leucine (L). Non-synonymous substitutions are those between codons specifying different amino acids; e.g., a substitution of TTC, encoding Phenylalanine (F), for TTA (L).

Neutral, negative and positive selection. Synonymous substitutions are expected to be neutral since they do not affect the amino acid composition of proteins. Non-synonymous substitutions may have negative effects on the protein function and so, they are expected to be eliminated by negative selection. In the event that such substitutions are selectively favorable, the frequency of the gene containing the new amino acid is increased until it becomes fixed in the population. This process is known as positive natural selection or adaptive evolution.

Transitions and transversions. Transitions are nucleotide substitutions between purines (Adenine (A) and Guanine (G)) or between pyrimidines (Cytosine (C) and Thymine (T)). Transversions include all the other nucleotide substitutions.

Phylogeny and genealogy. Phylogenies are evolutionary trees that describe the pattern of divergences by which a single common ancestral sequence evolved, over time, into the descendant sequences comprising a given alignment. They are used to represent evolutionary relationships among species using sequence data in which each sequence represents a species (phylogenetic data). If the sequences are from different individuals of the same species (polymorphic data), the information is genealogical and so, genealogies or gene trees can be used to show which sequences are most closely related.

In order to describe the general model formulation we follow the notation of Prado *et al.* (2006). Specifically, let $\mathbf{Y}$ denote the sequence alignment consisting of N sequences with $3 \times I$ nucleotides (*i.e.*, I codons). Let $\boldsymbol{Z}$ denote the substitution count data obtained from $\mathbf{Y}$ as follows. Define $\boldsymbol{y}_{i,j}$ to be the pair of homologous codons at site i, $i = 1, \ldots, I$, for the sequence pair indexed by j, with $j = 1, \ldots, J$ and $J = \binom{N}{2}$, the total number of possible pairs of sequences. Typically, only the polymorphic sites, *i.e.*, only those sites i that display at least one substitution in one pair of sequences, are included in the model. $\boldsymbol{Z}$ can be obtained in a number of ways, depending on how many types of substitutions will be represented in the model, and depending on whether or not phylogenetic or genealogical information will be used. For instance, Merl *et al.* (2005) used phylogenies to estimate ancestral nucleotide sequences, and then used the reconstructed sequences to count the total number of substitutions between two codons using a method similar to that proposed in Kosakovsky Pond and Frost (2005b). Prado *et al.* (2006) averaged the different numbers of substitutions per site over all possible one-substitution pathways that could have been followed between any two codons without allowing back-substitutions, self-canceling loops, and eliminating the pathways including stop codons. Regardless of which methodology is used to obtain $\boldsymbol{Z}$, $\boldsymbol{z}_{i,j}$ is a K-dimensional vector of counts, where each component, $z_{k,i,j}$, represents the number of substitutions of type k at site i between the two codons in the pairwise sequence comparison indexed by j.

For each $\boldsymbol{z}_{i,j}$ we define $\boldsymbol{z}_{i,j}$, a K-dimensional vector where each component, $\theta_{k,i,j}$, denotes the probability of substitution type k for site i and pair j. The model is then described by

$$\begin{aligned} \boldsymbol{Z}_{i,j} &\sim \text{Multinomial}(n_{i,j}, \boldsymbol{Z}_{i,j}), \quad n_{i,j} = \sum_k z_{k,i,j}, \\ \theta_{k,i,j} &= \frac{\exp(\eta_{k,i,j})}{\sum_l \exp(\eta_{l,i,j})}, \qquad (1) \\ \eta_{k,i,j} &= \alpha_k + \beta_{k,i} + \gamma_{k,h(j)} + \boldsymbol{\delta}'_{k,i}\mathbf{x}_{i,j}. \end{aligned}$$

The K-dimensional parameter vector $\boldsymbol{\alpha} = (\alpha_1, \ldots, \alpha_K)'$ models the baseline effects for each substitution type. The K-dimensional vectors $\boldsymbol{\beta}_i = (\beta_{1,i}, \ldots, \beta_{K,i})'$ capture site-specific departures from baseline substitution effects, *i.e.*, they account for different strengths of selective pressures expressed for each substitution type, for each amino acid site i. The K-dimensional vectors $\gamma_{h(j)} = (\gamma_{1,h(j)}, \ldots, \gamma_{K,h(j)})'$ describe pairwise, or more generally groupwise, departures from baseline and site-specific effects. In the particular case of $h(j) = j$, $\gamma_{h(j)}$ models pairwise effects due to evolutionary divergence between the two sequences indexed by j. These parameters would be associated with phylogenetic or gene tree effects. Other choices of $h(j)$ will be discussed later. Finally, the $\mathbf{x}_{i,j}$'s are C-dimensional vectors of covariates and the $\boldsymbol{\delta}_{k,i}$'s are C-dimensional parameter vectors. Covariates can include measures of amino acid properties such as polarity and hydrophobicity, or amino acid score matrices that measure distances between amino acids in terms of various properties (e.g., the Grantham matrix, see Grantham, 1974). The inclusion of these covariates may be useful in determining whether very radical substitutions are being encouraged by natural selection.

2.1. *Sub-Models*

Choosing K, the number of categories. Models with $K = 3$, where only synonymous, non-synonymous and no-change categories are considered, are useful in identifying amino acid sites under positive selection. Models with $K = 5$ in which synonymous transitions and transversions, non-synonymous transitions and transversions and no-substitutions are considered, are also used to detect sites under selection in sequences for which discriminating between transitions and transversions is key to determine if the observed rates of substitutions may be the result of codon bias. Codon bias is the tendency for a species to use a given set of codons more than others to encode a particular amino acid. Prado *et al.* (2006) used a 5-category model to analyze alignments of the AMA-1 antigen from the *P. falciparum* human malaria parasite. Accounting for transitions and transversions in these data is important to assess whether the increased rates of non-synonymous substitutions estimated at some amino acid sites are the result of the A+T richness in the genome (Escalante *et al.* 1998). Other model possibilities that would be useful for this purpose include, for example, $K = 4$, with a single category for synonymous substitutions and two categories of non-synonymous substitutions – e.g., those between G and C nucleotides, and all remaining substitutions – and a no-substitution category.

Pairwise and group effects. Models with $h(j) = j$ for all j specifically include all possible pairwise effects. These effects are often relevant in the analysis of phylogenetic sequences, *i.e.*, in alignments for which the sequences were obtained from distinct species. This approach is followed in the analyses of the sperm lysin sequences presented here. In cases of polymorphic data, more parsimonious choices can be considered. For instance, the AMA-1 alignment analyzed in Prado et al. (2006), and revisited here, consists of 23 sequences in total, with 12 isolates from Kenya, five isolates from India, and six isolates from Thailand. In this case, it may be reasonable to assume that we have samples from two distinct populations: an African population represented by the 12 sequences from Kenya, and an Asian population represented by the 11 sequences from India and Thailand. Then, we can write $h(j) = o$, for $o = 1, 2, 3$ defined in such a way that $h(j) = 1$ if j indexes a pair of sequences from Africa, $h(j) = 2$ if j indexes a pair of sequences from Asia, and $h(j) = 3$ if j indexes a pair of sequences from different populations, *i.e.*, one from

Africa and the other one from Asia. Another potentially interesting model is that in which $\gamma_{h(j)} = \mathbf{0}$ for all j. This model does not include any group effect (pairwise or other type) and it is sometimes useful in modelling sequences for which the genetic variability is mostly explained by the site-specific effects and so, phylogenetic or population effects are considered negligible.

Covariates. Sub-models include those with $\boldsymbol{\delta}_{k,i} = \boldsymbol{\delta}_k$ for all i, models with $\boldsymbol{\delta}_{k,i} = \boldsymbol{\delta}_i$ for all k, and those with $\boldsymbol{\delta}_{k,i} = \boldsymbol{\delta}$ for all k, i. Note that the inclusion of covariates that carry information on amino acid properties only affects non-synonymous substitutions and so, $\boldsymbol{\delta}_{k,i} = \mathbf{0}$ for all k indexing a synonymous or no-substitution category in any model that includes covariates.

2.2. *Prior Structure*

The prior structure used here is similar to that proposed in Prado *et al.* (2006). We now summarize the main features of such structure, directing the reader to the previous reference for details on the prior construction and a discussion about its biological implications.

The prior distributions on $\boldsymbol{\alpha}$, $\boldsymbol{\beta}$ and $\boldsymbol{\gamma}$ are all Gaussian. This is, $N(\boldsymbol{\alpha}|\mathbf{m}_\alpha, \sigma^2_\alpha \mathbf{I})$, $N(\boldsymbol{\beta}_k|\mathbf{m}_{\beta_k}, \sigma^2_\beta \mathbf{I})$ and $N(\boldsymbol{\gamma}_k|\mathbf{m}_{\gamma_k}, \sigma^2_\gamma \boldsymbol{W}_{\gamma_k})$, where $\mathbf{I}$ denotes the identity matrix with the appropriate dimension in each case. In addition, constraints on $\boldsymbol{\alpha}$, $\boldsymbol{\beta}$ and $\boldsymbol{\gamma}$ are set to guarantee identifiability (see Prado *et al.* 2006).

In absence of prior information about site-specific effects we set $\mathbf{m}_{\beta_k} = \mathbf{0}$ for all k. When phylogenetic/genealogical information is available, it may be incorporated in the prior structure by first translating the phylogeny/genealogy into a matrix of distances D, as well as a matrix of 'distances between distances' $\tilde{\mathrm{D}}$ (see Prado et. al., 2006). Then, $\mathbf{m}_\alpha$, $\mathbf{m}_{\gamma_k}$ and $\mathbf{W}_{\gamma_k}$, are expressed as functions of the elements of D, $\tilde{\mathrm{D}}$ and hyper-parameters $\boldsymbol{w}$, χ, χ^*, ζ and ζ^*. Here $\boldsymbol{w}$ is a vector of dimension $K - 1$ containing prior estimates for the relative frequencies of the first $K - 1$ substitution types, while χ, χ^*, ζ and ζ^* are used to control the strength of the phylogenetic/genealogy effects in the prior structure. These hyper-parameters can be given fixed values a priori or estimated a posteriori. In the latter case, the hyper-parameters are assumed independent a priori with χ and χ^* following uniform priors and ζ, ζ^*, as well as each element of $\boldsymbol{w}$ following exponential priors. In addition, the parameters σ^2_α, σ^2_β and σ^2_γ can be fixed a priori or estimated a posteriori under inverse-gamma priors. Alternatively, if the alignment includes sequences with relatively low evolutionary divergence, for which none or only very weak phylogenetic information is available, $\mathbf{m}_{\gamma_k} = \mathbf{0}$ and $\mathbf{W}_{\gamma_k} = \mathbf{I}$ are typically used.

Finally, Gaussian priors are also specified for the C-dimensional parameter vectors $\boldsymbol{\delta}_{k,i}$. Specifically, $N(\boldsymbol{\delta}_{k,i}|\mathbf{0}, \sigma^2_{\boldsymbol{\delta}} \mathbf{I})$ are used for all the categories k that model non-synonymous substitutions and all polymorphic sites indexed by i. As it was the case with other variance parameters, $\sigma^2_{\boldsymbol{\delta}}$ can be set to some fixed value or estimated a posteriori under an inverse-gamma prior.

3. POSTERIOR ESTIMATION

Posterior estimation is achieved via standard MCMC methods using the Poisson formulation of the multinomial model (see Baker, 1994).

3.1. *Identifying Sites Under Selection*

Once samples from the posterior distribution of the model parameters are obtained, sites under negative or positive selection can be identified by investigating the behavior of specific functions of such parameters.

3.1.1. *Definitions Based on* θ

Let $\mathcal{I}^*$ be a specific set of sites in the alignment. For instance, $\mathcal{I}^*$ can be the set of all the polymorphic sites in the alignment, or the set of all the sites that display at least one non-synonymous substitution. Let $\theta_S^{\mathcal{I}^*}$ and $\theta_{NS}^{\mathcal{I}^*}$ be the average synonymous and non-synonymous substitution probabilities, respectively, for the sites in $\mathcal{I}^*$. This is,

$$\theta_S^{\mathcal{I}^*} = \frac{1}{(|\mathcal{I}^*| \times J)} \sum_{i \in \mathcal{I}^*} \sum_{j=1:J} \sum_{l \in \mathcal{C}_S} \theta_{l,i,j}, \text{ and } \theta_{NS}^{\mathcal{I}^*} = \frac{1}{(|\mathcal{I}^*| \times J)} \sum_{i \in \mathcal{I}^*} \sum_{j=1:J} \sum_{l \in \mathcal{C}_{NS}} \theta_{l,i,j},$$

where $\mathcal{C}_S$ and $\mathcal{C}_{NS}$ are the sets of indexes of all the categories that involve synonymous and non-synonymous substitutions, respectively. Then, we say that there is evidence of positive selection in the gene if

$$P(\omega^* > \omega_0 | \boldsymbol{Z}) \equiv P\left(\frac{\theta_{NS}^{\mathcal{I}^*}}{\theta_S^{\mathcal{I}^*}} > \omega_0 \middle| \boldsymbol{Z} \right) \geq (1 - \alpha_1), \tag{2}$$

with $\alpha_1 \in [0, 1)$ and typically, $\alpha_1 \in [0, 0.05]$. The value of ω_0 is fixed and often set at $\omega_0 = 1$. A non-synonymous to synonymous substitution probabilities ratio equal to one is considered indicative of neutral selection, ratios smaller than one indicate negative selection, while ratios greater than one are linked to positive selection (e.g., Yang *et al.* 2005). When (2) holds, we can proceed to identify sites under positive selection. Specifically, we say that a site i is a positively selected site if

$$P(i^+ | \boldsymbol{Z}) \equiv P\left(\frac{\theta_{NS,i}}{\theta_S^{\mathcal{I}^*}} > \frac{\theta_{NS}^{\mathcal{I}^*}}{\theta_S^{\mathcal{I}^*}} \middle| \boldsymbol{Z} \right) = P(\theta_{NS,i} > \theta_{NS}^{\mathcal{I}^*} | \boldsymbol{Z}) \geq (1 - \alpha_2), \tag{3}$$

with $\alpha_2 \in [0, 1)$ and usually $\alpha_2 \leq 0.05$. If (2) does not hold then we say that the alignment is under neutral and/or negative selection.

Sites under negative selection can also be identified. Specifically, a site i is said to be under negative selection if

$$P\left(\frac{\theta_S^{\mathcal{I}^*}}{\theta_{NS}^{\mathcal{I}^*}} > \frac{1}{\omega_1} \middle| \boldsymbol{Z} \right) \geq (1 - \alpha_4), \tag{4}$$

and if

$$P(i^- | \boldsymbol{Z}) \equiv P\left(\frac{\theta_{S,i}}{\theta_{NS}^{\mathcal{I}^*}} > \frac{\theta_S^{\mathcal{I}^*}}{\theta_{NS}^{\mathcal{I}^*}} \middle| \boldsymbol{Z} \right) = P\left(\theta_{S,i} > \theta_S^{\mathcal{I}^*} | \boldsymbol{Z} \right) \geq (1 - \alpha_3), \tag{5}$$

with $\alpha_3, \alpha_4 \in [0, 1)$ and typically α_3, $\alpha_4 \in [0, 0.05]$. The value of ω_1 is set by the practitioner. For instance, if sites under very strong negative selection must be identified, $1/\omega_1$ is fixed at a value greater than 2.0.

3.1.2. *A Definition of Positive Selection Based on* $\boldsymbol{\beta}$ *and* $\boldsymbol{\delta}$

Prado *et al.* (2006) considers another definition for detecting sites under positive selection by writing $P(i^+|\boldsymbol{Z})$ in terms of the $\boldsymbol{\beta}$ parameters. In general, it has been found that such definition is more conservative that the definition in (3). Simulation studies suggest that a definition based on $\boldsymbol{\beta}$ produces less false positives but also has less power than the definition in (3) (see Prado *et al.* 2006).

Here, we extend the definition of Prado *et al.* (2006) to account for possible covariates added to the model. Once again, we first determine whether there is evidence of positive selection in the alignment by looking at $P(\omega^* > \omega_0|\boldsymbol{Z})$. In other words, if (2) holds for some fixed values ω_0 and α_1, we then proceed to identify which sites are under positive selection. Then, a site i is said to be under positive selection if

$$P(i^+|\boldsymbol{Z}) \equiv P\left(\beta^*_{NS,i} + f(\boldsymbol{\delta}'_{NS,i}\mathbf{x}_i) > \beta^*_{S,i}|\boldsymbol{Z}\right) \geq (1-\alpha_2), \tag{6}$$

with

$$\beta^*_{k,i} = \beta_{k,i} - \frac{1}{|\mathcal{I}^*|}\sum_{i\in\mathcal{I}^*}\beta_{k,i}, \quad \text{and} \quad f(\boldsymbol{\delta}'_{NS,i}\mathbf{x}_i) = \frac{1}{J}\sum_{j=1}^{J}\boldsymbol{\delta}'_{NS,i}\mathbf{x}_{i,j}.$$

3.2. *Phylogenetic and Gene Tree Estimation*

Although the models considered here were not developed for the purpose of phylogenetic/genealogical inference, it is possible to obtain posterior estimates of phylogenies or gene trees based on posterior distance matrices as follows.

Let $d_{h(j)}$ be an estimate of the distance between the pair of sequences indexed by j computed as follows

$$d_{h(j)} = \frac{1}{|\mathcal{I}^*|}\sum_{i\in\mathcal{I}^*}\sum_{k\in\mathcal{C}}\theta_{k,i,h(j)}, \tag{7}$$

with $\mathcal{C}$ including a particular set of substitution categories. For example, $\mathcal{C}$ can be the set containing all the substitution types (e.g., synonymous and non-synonymous), only the synonymous substitutions, or only the non-synonymous substitutions. Here, $\theta_{k,i,o}$, for a given o, is computed as

$$\theta_{k,i,o} = \frac{1}{|o|}\sum_{j;h(j)=o}\theta_{k,i,h(j)},$$

where $|o|$ is the number of indexes j such that $h(j) = o$. The distances (pseudo-distances) in (7) can be used to build a matrix D_h whose dimension depends on the structure defined by the function $h(j)$. For example, if $h(j) = j$ for all pairs indexed by j, the matrix D_h will have dimension $N \times N$, with off diagonal elements computed as $g(d_j)$, with g a particular function, such as the identity or the exponential. The diagonal elements of D_h can be computed using $g(0)$ (the distance between a sequence and itself is zero). In the example of the malaria sequences from two populations discussed above, $h(j) = o$ with $o = 1, 2, 3$, and so, D_h would be a 2×2 matrix. The off-diagonal elements would measure the average distances between populations (or a function of such distances), while the diagonal entries would contain average within population distances (or a function of such distances).

The matrix D_h can then be used as an input to one of the distance-based algorithms often used in practice to estimate phylogenies or gene trees such as the neighbor joining algorithm (see for example Felsenstein, 2004 for a detailed explanation of this and other related algorithms for phylogenetic estimation). Therefore, posterior estimates/samples of genealogies based on the distances defined in (7) can be obtained.

4. MODEL SELECTION AND MODEL VALIDATION

4.1. *Model Selection*

Model selection among different models, such as the various sub-models discussed in Section 2, is performed via the minimum posterior predictive loss approach of Gelfand and Ghosh (1998). This criterion can be computed easily using MCMC output and it has a decision theoretical justification given that it is obtained by minimizing a posterior predictive loss function within a particular family of models, and then, selecting the model that minimizes such criterion.

For each model $\mathcal{M}_m$ from a collection of M models, $\mathcal{M}_1, \ldots, \mathcal{M}_M$, the following quantity is computed,

$$D_\kappa(m) = \sum_{i,j} \min_{\mathbf{a}_{i,j}} \left\{ E_{\boldsymbol{z}^{rep}_{i,j}|\boldsymbol{Z}^{obs},m} \left[L(\boldsymbol{z}^{rep}_{i,j}, \mathbf{a}_{i,j}) + \kappa L(\boldsymbol{z}^{obs}_{i,j}, \mathbf{a}_{i,j}) \right] \right\}, \quad \kappa \geq 0, \tag{8}$$

where $\boldsymbol{Z}^{obs} \equiv \boldsymbol{Z}$ are the observed count data, $\boldsymbol{z}^{rep}_{i,j}$ is a K-dimensional vector of counts that replicates $\boldsymbol{z}^{obs}_{i,j}$, $L(\cdot,\cdot)$ is a loss function, $\mathbf{a}_{i,j}$ is a 'guess', representing a compromise between $\boldsymbol{z}^{rep}_{i,j}$ and $\boldsymbol{z}^{obs}_{i,j}$ and κ is a constant that weights the discrepancy between $\mathbf{a}_{i,j}$ and $\boldsymbol{z}^{obs}_{i,j}$. In other words, when $\kappa = 0$, $\mathbf{a}_{i,j}$ is chosen as a guess for $\boldsymbol{z}^{rep}_{i,j}$ and if $\kappa \neq 0$ the closeness of $\mathbf{a}_{i,j}$ to $\boldsymbol{z}^{obs}_{i,j}$ is also rewarded, and so, a compromised choice is required.

Various loss functions can be considered. Prado *et al.* (2006) computed (8) for the model formulation (1) with five categories using two loss functions: a squared error loss function and a loss function written in terms of the logarithm of a ratio of likelihoods, *i.e.*,

$$L(\boldsymbol{z}_{i,j}, \mathbf{a}_{i,j}) = 2 \log \frac{f(\boldsymbol{z}_{i,j}|\mathbf{q}(\boldsymbol{z}_{i,j}))}{f(\boldsymbol{z}_{i,j}|\mathbf{q}(\mathbf{a}_{i,j}))},$$

for some function $\mathbf{q}$. This loss function takes into account the GLM structure of the model. Details regarding the specific form of $\mathbf{q}$ and the calculation of $D_\kappa(m)$ appear in Prado *et al.* (2006).

4.2. *Model Checking*

We follow a posterior predictive approach to model checking. After obtaining R posterior samples of the K-dimensional probability vectors $\boldsymbol{\theta}_{i,j}$, for each i, j, *i.e.*, $\boldsymbol{\theta}^{(r)}_{i,j}$ for $r = 1, \ldots, R$, we can obtain R replicates $\boldsymbol{z}^{rep}_{i,j}$ for each i, j. This is

$$\boldsymbol{z}^{rep,r}_{i,j} \sim \text{Multinomial}(n_{i,j}, \boldsymbol{\theta}^{(r)}_{i,j}). \tag{9}$$

Then, we can summarize posterior distributions of relevant functions of $\boldsymbol{z}^{rep}_{i,j}$ and compare them with functions of the actual count values $\boldsymbol{z}_{i,j}$. For instance, we could derive the distribution of the number of transitions, transversions, synonymous and/or non-synonymous substitutions based on the replicates $\boldsymbol{Z}^{rep}$ for all, or a few, of the sites indexed by i, and determine whether the corresponding observed values are plausible values under such distributions.

In order to appropriately simulate $\boldsymbol{z}_{i,j}^{rep}$ in (9), we need to take into account the process used to obtain the count data $\boldsymbol{Z}$ from the alignment. For instance, when the procedure described in Prado *et al.* (2006) is followed to obtain $\boldsymbol{Z}$ in a 5-category model for which the categories are synonymous transitions and transversions, non-synonymous transitions and transversions and no substitutions, the resulting $z_{5,i,j}$ is a binary entry. In this case $\boldsymbol{z}_{i,j}^{rep,r}$ is simulated as follows. First $z_{5,i,j}$ is simulated from a Bernoulli distribution with probability $\boldsymbol{\theta}_{5,i,j}^{(r)}$. Then, $\boldsymbol{z}_{1:4,i,j}^{rep,r}$ is simulated from a multinomial distribution with parameters $(n_{i,j} - z_{5,i,j}^{rep,r})$ and $\boldsymbol{\theta}_{1:4,i,j}^{*,(r)}$, with $\theta_{k,i,j}^{*,(r)} = \theta_{k,i,j}^{(r)} / \sum_{k=1}^{4} \theta_{k,i,j}^{(r)}$. This will be illustrated in Section 5.

Another way of assessing model fit in phylogenetic data is by looking at the phylogenies obtained from the posterior estimates (or posterior samples) of the distance matrix D_h. This can only be done if no phylogenetic structure has been included in the prior specification. Close inspection of posterior tree estimates/samples can then be used as a tool for model validation. We can compare such estimates to substantive knowledge about the evolutionary process underlying the sequences whenever such information is available.

5. DATA ANALYSES

5.1. *Abalone Sperm Lysin Alignment*

This alignment codes for a 122 residue region of the sperm lysin protein for 25 species of California abalone. Abalone reproduction involves species specific sperm-egg recognition in which the sperm lysin binds and dissolves a complementary vitelline envelope (VE) surrounding the egg cell. This species-specific interaction is subjected to positive selection of some 23 residues in the lysin protein, as it compensates for ongoing genetic drift in the VE receptor (Galindo *et al.*, 2003; Lee *et al.*, 1995; Yang *et al.*, 2000). These data have been extensively studied and provide a good example of positive selection acting on individual amino acid sites. The data are included in the `PAML` software distribution (Yang, 1997). The sequences are sufficiently divergent, with a total tree length of 8.2 nucleotide substitutions per codon. Additionally, the crystalline structure of the molecule can be used to support or refute claims of positively selected amino acid sites (Yang *et al.* 2000).

Various models were fit to the count data obtained from the alignment using the procedure described in Prado *et al.* (2006). We focus on the results drawn from two 3-category models: a model with a prior specification that includes the phylogenetic structure shown in Fig. 3, and another model where such structure is not incorporated. The three categories correspond, respectively, to synonymous substitutions, non-synonymous substitutions and no substitutions. The phylogeny in Fig. 3 is that of Lee *et al.* (1995). For both models, the results presented here are based on 1000 MCMC samples obtained after convergence. Also, both models incorporate a single covariate $x_{i,j}$, where $x_{i,j}$ corresponds to the normalized Grantham matrix score (see Grantham, 1974) between the two amino acids indexed in the pairwise comparison j at site i. For each amino acid substitution, the Grantham score represents a physicochemical distance between the two amino acids involved in such substitution. The normalization of the matrix involved dividing each entry in the matrix by the maximum score value.

The phylogenetic prior was specified using the procedure described in Prado *et al.* (2006), taking into account that we are fitting a 3-category model instead of a 5-category one, and setting $\sigma_\alpha^2 = \sigma_\beta^2 = \sigma_\gamma^2 = 10$, $\sigma_\delta^2 = 1$, $\mathbf{m}_\beta = \mathbf{m}_\delta = \mathbf{0}$, and the hyperparameters needed to define $\mathbf{m}_\alpha$, $\mathbf{m}_{\gamma_k}$ and $\mathbf{W}_{\gamma_k}$ to $\boldsymbol{w} = (0.5, 0.5)$, $\chi = \chi^* = 1$, $\zeta = 1$ and $\zeta^* = 1$. The independent prior was specified by setting $\mathbf{m}_\alpha = \mathbf{m}_\beta = \mathbf{0}$, $\mathbf{m}_\gamma = \mathbf{m}_\delta = \mathbf{0}$, $\sigma_\alpha^2 = \sigma_\beta^2 = \sigma_\gamma^2 = 10$ and $\sigma_\delta^2 = 1$, and all the prior variance-covariance matrices equal to identities. Posterior results, in terms of which sites were detected as sites under selection, were not very sensitive to changes in the prior values of these parameters.

Table 1 shows various model selection criteria values for the two types of models fitted to the count data. Three criteria were considered, two of which correspond to the posterior predictive criteria of Gelfand and Ghosh (1998) under the log-likelihood ratio and the squared error loss functions, denoted by D_{dev}^κ and D_{se}^κ, respectively. The values displayed on the table correspond to $\kappa = 100$. Other values of κ were considered, leading to the same conclusions. The deviance information criteria (DIC), was also computed for both models (Spiegelhalter *et al.* 2002). Based on these values, the model with structured phylogenetic priors is preferred by all the criteria.

Table 1: *Model selection criteria. Lysin alignment.*

Model	D_{dev}^{100}	D_{se}^{100}	DIC
Independent	65413	37558	65142
Phylogenetic	65349	37495	64929

Even though the preferred model, according to the model selection criteria discussed above, is the one with phylogenetic priors, we now focus on the results obtained from the model with independent priors in order to illustrate some modelling features, particularly those related to model validation. Fig. 1 shows the posterior distributions of the non-synonymous to synonymous probabilities ratios for the sites identified as positively selected by both of the GLM-based definitions described in Section 3. The dark boxplots correspond to sites that were also identified as positively selected by at least one of the methods implemented in `PAML` (Yang, 1997) or in `HYPHY` (Kosakovsky Pond et al, 2005). Specifically, model M8b in `PAML` was fitted to the alignment. This is a model in which a discretized beta distribution is used to describe the non-synonymous to synonymous rates ratio –denoted by ω in the computational biology literature– between 0 and 1, and an additional category with $\omega > 1$. Three different methodologies were considered in `HYPHY`: the single ancestor counting method (SLAC), the fixed-effects likelihood method (FEL) and the random-effects likelihood method (REL). Details about how these models are used to detect sites under positive selection can be found in Kosakovsky Pond and Frost (2005a, 2005b). Fig. 1 shows that there is good agreement between the proposed GLM-based methodology and the existing methodologies based on stochastic models of molecular evolution and phylogenetic-based models implemented in `PAML` and `HYPHY`.

Fig. 2 displays the lysin crystal structure for one of the abalone species in the alignment (*H. Rufenses*). This is the same crystal structure shown in Yang *et al.* (2000). The sites positively identified by the GLM-based methodology do not exactly match those identified by model M8b in `PAML`, however, as shown in Fig. 2, there is

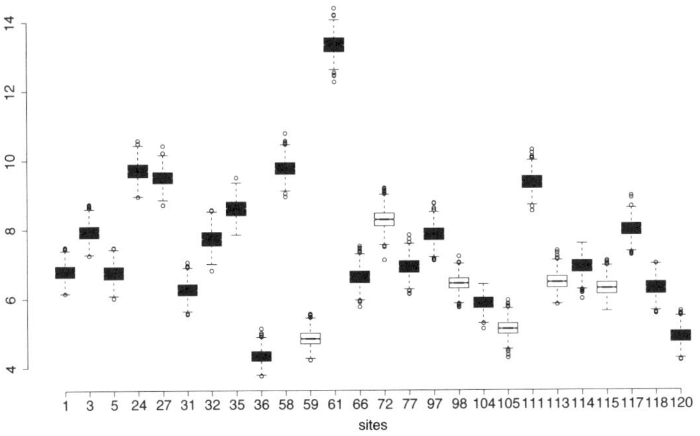

Figure 1: *Posterior distributions of the non-synonymous to synonymous probabilities ratios for the sites positively selected by GLM-based definitions. Dark boxplots indicate that those sites were also identified by at least one of the* `PAML` *or* `HYPHY` *methods.*

a great deal of agreement in terms of the locations of positively selected sites in the folded protein. Fig. 2 shows that the GLM positively selected sites cluster at the top and at the bottom of the molecule. These findings are in agreement with the results presented in Yang *et al.* (2000). Most of the conserved sites lie in the internal portions of the alpha helices of the protein. Such sites are involved in interhelical interactions and are functionally constrained in all lysins.

Fig. 4 shows an estimated phylogeny constructed using the posterior mean estimate of a distance matrix $\mathrm{D}_{h(j)} = \mathrm{D}_j$, whose entries are given by

$$d_j = \frac{1}{|\mathcal{I}^*|} \sum_{i \in \mathcal{I}^*} \theta_{1,i,j} + \theta_{2,i,j},$$

where $\theta_{1,i,j}$ and $\theta_{2,i,j}$ are the probabilities of synonymous and non-synonymous substitutions, respectively, for the pair of sequences indexed by j and the residue indexed by i. The estimated phylogeny displayed in the figure is based on posterior samples from the model with independent priors. A neighbor joining algorithm was used for the tree construction. Although the phylogenies in figures 3 and 4 are not identical, they show many topological similarities, indicating that the GLM-based approach provides a good model fit. In particular, most of the species that appear clustered in Fig. 3 are also clustered in the phylogeny displayed in Fig. 4.

5.2. *AMA-1 Alignment*

Prado *et al.* (2006) presents analyses of alignments comprising 23 sequences of the AMA-1 antigen in the human *P. falciparum* malaria parasite. We consider additional analyses of this alignment here, extending the approach of Prado *et al.* (2006) in order to add covariates related to amino acid properties. In addition, a collection of

Figure 2: *Lysin crystal structure from the red abalone* H. Rufenses. *Sites identified as positively selected are in black.*

models that make various assumptions about the underlying evolutionary characteristics of the sequences are fitted to the data, and compared via posterior predictive model selection criteria.

Malaria is a major public health problem, with approximately 300 to 500 million clinical cases and 1 to 3 million deaths estimated per year (Sachs and Malaney, 2002) and so vaccine development against the parasites that produce the disease is a priority. AMA-1 is one of the candidate antigens currently being considered for use in vaccine development. AMA-1 has been extensively studied, and there is convincing evidence that this antigen elicits a protective immune response against malaria (Polley *et al.* 2004). In addition, some residues have been associated with various clinical manifestations of the disease (Cortes *et al.* 2003). Genetic evolutionary studies have also shown that there is evidence that the gene is under positive selection (Escalante *et al.* 2001; Polley *et al.* 2003).

The alignment considered here consists of 23 sequences, each encoding a total number of 620 residues. From these 23 sequences, 12 sequences were taken from subjects in Kenya, five from subjects in India and six from subjects in Thailand. The sequences display 84 polymorphic sites and are available in the GenBank. Prado *et al.* (2006) analyzed this alignment using 5-category models with synonymous transitions and transversions, non-synonymous transitions and transversions and no-substitutions. Here, we consider 3-category models accounting for synonymous substitutions, non-synonymous substitutions and no-substitutions. In addition, three types of models were chosen by considering three different choices of the function $h(j)$ that defines the groupwise effects among the sequences. First, we consider models with $h(j) = j$ for all j, *i.e.*, models in which the γ parameters describe evolutionary distances among pairs of sequences. We refer to these models

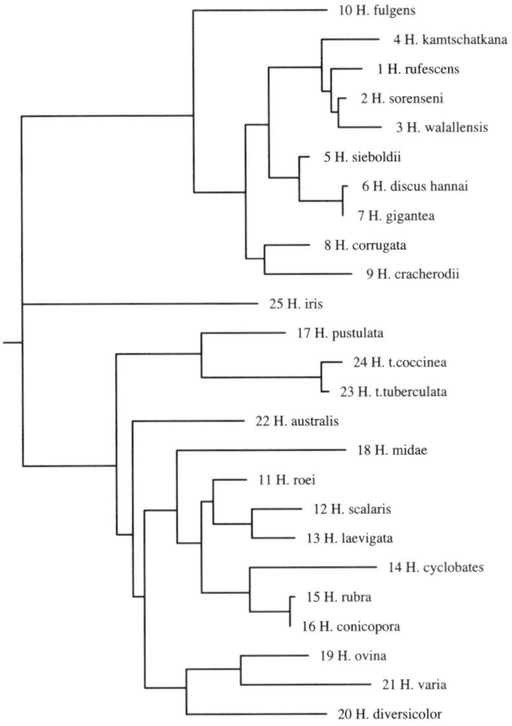

Figure 3: *Abalone sperm lysin phylogeny from Lee* et al. *1995.*

as 'full' models. Prado *et al.* (2006) only fitted these types of models. Second, we consider models in which $h(j) = 1$ for all the pairs j in which both sequences were from Africa, $h(j) = 2$ for all the pairs j in which both sequences were from Asia and $h(j) = 3$ for all the pairs j in which one sequence was from Africa and the other one was from Asia. We refer to these models as 'continent' models. Finally, we also fit models in which $h(j) = 1$ for all the pairs j in which both sequences were from Kenya, $h(j) = 2$ for all the pairs in which both sequences were from India, $h(j) = 3$ for all the pairs in which both sequences were from Thailand. Additionally, $h(j) = o$, with $o = 4, 5, 6$, for all the pairs j in which the sequences were from different countries (Kenya and India, Kenya and Thailand and India and Thailand, respectively). We refer to these models as the 'country' models. For each of the three types of models described above we fit all possible combinations of covariate-specific effects and site-specific effects, namely: models with and without a covariate that represents amino acid substitutions scores ($\boldsymbol{\delta}$) – again, the normalized Grantham matrix was used as a covariate – and models with and without the site-specific effects $\boldsymbol{\beta}$. Finally, models that do not include any group effects $\boldsymbol{\gamma}$ are also considered.

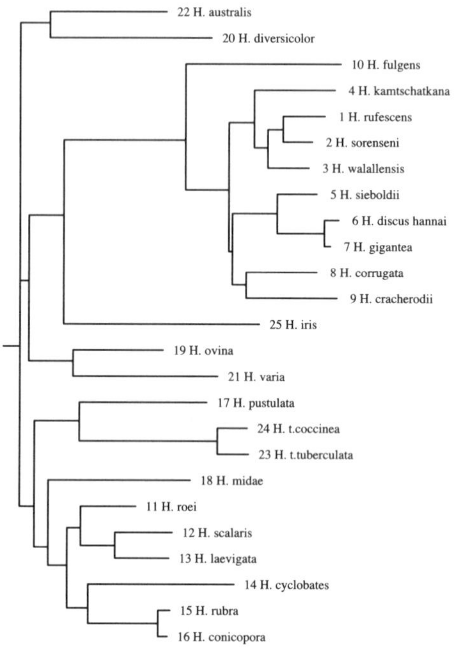

Figure 4: *Posterior estimation of an abalone sperm lysin phylogeny obtained from GLM approach.*

Table 2 shows various model selection criteria values for the all the model types described above, denoted by $(\boldsymbol{\alpha})$, $(\boldsymbol{\alpha}+\boldsymbol{\beta})$, $(\boldsymbol{\alpha}+\boldsymbol{\gamma})$, $(\boldsymbol{\alpha}+\boldsymbol{\delta})$, $(\boldsymbol{\alpha}+\boldsymbol{\beta}+\boldsymbol{\gamma})$, $(\boldsymbol{\alpha}+\boldsymbol{\beta}+\boldsymbol{\delta})$, $(\boldsymbol{\alpha}+\boldsymbol{\gamma}+\boldsymbol{\delta})$ and $(\boldsymbol{\alpha}+\boldsymbol{\beta}+\boldsymbol{\gamma}+\boldsymbol{\delta})$. Notice that four of these models, $(\boldsymbol{\alpha})$, $(\boldsymbol{\alpha}+\boldsymbol{\beta})$, $(\boldsymbol{\alpha}+\boldsymbol{\delta})$ and $(\boldsymbol{\alpha}+\boldsymbol{\beta}+\boldsymbol{\delta})$, are equivalent under any choice of $h(\cdot)$, since they do not involve a group-specific $\boldsymbol{\gamma}$ term. For purposes of model selection, we compare minimum posterior predictive deviance (D^{κ}_{dev}), minimum posterior predictive squared-error loss (D^{κ}_{se}), and the deviance information criterion (DIC). For D^{κ}_{dev} and D^{κ}_{se}, various values of κ were used, yielding virtually identical results. Values reported here were obtained using $\kappa = 100$. Superior model fit is indicated by smaller values of each model selection criterion. The value corresponding to the best-fitting model under each criterion is shown in bold. All model selection criteria indicate that the full model provides the best fit to the data. Adding the site-specific effects $\boldsymbol{\beta}$ to a given model produces the largest decrease in the model selection criteria values. For instance, compare the decrease in D^{100}_{dev}, D^{100}_{se} and DIC values for the models $(\boldsymbol{\alpha}+\boldsymbol{\beta})$ and $(\boldsymbol{\alpha}+\boldsymbol{\delta})$ with respect to $(\boldsymbol{\alpha})$, and the decrease in the criteria values of models $(\boldsymbol{\alpha}+\boldsymbol{\beta}+\boldsymbol{\gamma})$ and $(\boldsymbol{\alpha}+\boldsymbol{\gamma}+\boldsymbol{\delta})$ with respect to $(\boldsymbol{\alpha}+\boldsymbol{\gamma})$ in the country, continent and full models. Adding the group-specific effects $\boldsymbol{\gamma}$ to a given model produces the smallest decrease in the model selection criteria values. In fact, the model selection criteria values for the continent and country models are virtually equivalent. These findings are important since they imply that the variables that have the largest effects in the subsitution probabilities are the site-specific effects, while those related to various assumptions on the evolutionary distances among the sequences have the least impact. This indicates that, if clustering of the sequences in terms of their evolutionary distances is feasible, such clustering would not be related to geographical location.

Table 2: *Model selection criteria. AMA-1 alignment.*

Model	D^{100}_{dev}	D^{100}_{se}	DIC
$(\boldsymbol{\alpha})$	42961	19039	30063
$(\boldsymbol{\alpha}+\boldsymbol{\beta})$	38156	16713	25433
$(\boldsymbol{\alpha}+\boldsymbol{\delta})$	40601	17921	27675
$(\boldsymbol{\alpha}+\boldsymbol{\beta}+\boldsymbol{\delta})$	36161	15817	23675
continent $(\boldsymbol{\alpha}+\boldsymbol{\gamma})$	42966	19033	30065
continent $(\boldsymbol{\alpha}+\boldsymbol{\beta}+\boldsymbol{\gamma})$	38118	16700	25439
continent $(\boldsymbol{\alpha}+\boldsymbol{\gamma}+\boldsymbol{\delta})$	40574	17915	27682
continent $(\boldsymbol{\alpha}+\boldsymbol{\beta}+\boldsymbol{\gamma}+\boldsymbol{\delta})$	36124	15803	23681
country $(\boldsymbol{\alpha}+\boldsymbol{\gamma})$	42950	19030	30063
country $(\boldsymbol{\alpha}+\boldsymbol{\beta}+\boldsymbol{\gamma})$	38108	16703	25441
country $(\boldsymbol{\alpha}+\boldsymbol{\gamma}+\boldsymbol{\delta})$	40579	17915	27679
country $(\boldsymbol{\alpha}+\boldsymbol{\beta}+\boldsymbol{\gamma}+\boldsymbol{\delta})$	36123	15802	23681
full $(\boldsymbol{\alpha}+\boldsymbol{\gamma})$	41722	18528	29857
full $(\boldsymbol{\alpha}+\boldsymbol{\beta}+\boldsymbol{\gamma})$	37027	16258	25249
full $(\boldsymbol{\alpha}+\boldsymbol{\gamma}+\boldsymbol{\delta})$	39377	17436	27540
full $(\boldsymbol{\alpha}+\boldsymbol{\beta}+\boldsymbol{\gamma}+\boldsymbol{\delta})$	**35158**	**15439**	**23631**

We now discuss some posterior results obtained from the full model with the Grantham scores covariate. Fig. 5 displays the posterior predictive distributions of the number of synonymous substitutions $z_{34,1}$, the number non-synonymous substitutions $z_{34,2}$, and the number of no-substitutions $z_{34,3}$, adding over all the pairwise sequences for site 34. The dots in the histograms correspond to actual count values in $\boldsymbol{Z}$. These types of graphs are helpful for determining which features of the data are not well captured by the model. We looked at several of these graphs, focusing in particular on sites that displayed above average numbers of non-synonymous substitutions, as those are important for assessing the effect of positive selection in these antigen sequences. In all cases the graphs did not suggest major discrepancies between the posterior predictive distributions and the observed values.

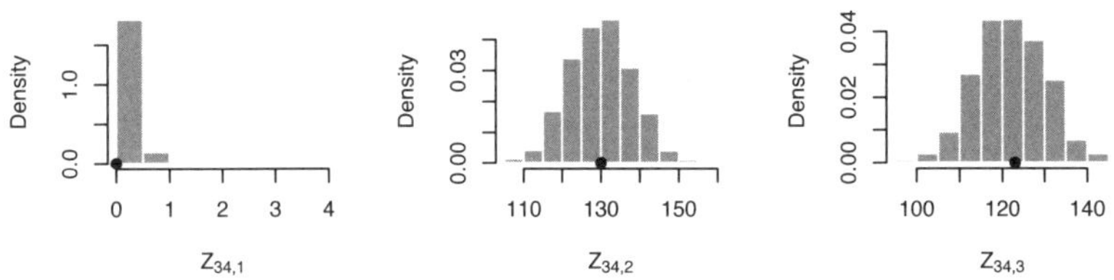

Figure 5: *Posterior predictive distributions for site 34. Dots indicate observed values.*

Table 3 lists the positively selected sites detected by the two methods described in Section 3. Method 1 uses a definition of positive selection based on the $\boldsymbol{\theta}$ parameters, while Method 2 uses a definition based on $\boldsymbol{\beta}$ and $\boldsymbol{\delta}$. Most of the sites listed here were previously identified as positively selected using a 5-category model without covariates (see results in Prado *et al.* 2006), except for those sites marked with (*). The full 3-category model that includes the amino acid scores detects the same sites previously detected by the 5-category model with no covariate when Method 2 is used. Four additional sites are identified at the level $\alpha_2 = 0.05$ when Method 1 is used. Further analyses need to be performed in order to assess whether these additional sites appear due to the reduction in the number of categories, or due to the incorporation of the amino acid scores.

Table 3: *Positively selected sites.*

Method	Sites identified
Method 1 ($p > 0.99$, $\alpha_2 = 0.01$)	**34 39 52** 162 167 172 187 190 197 200 201 204 225* 230 242 243 **267** 282 **283 285** 296 300 308 393 404 405 435 439 485 493 496 503 512 544 **581 584** 589
Method 1 ($p > 0.95$, $\alpha_2 = 0.05$)	+ 36* 175* 196* 245*
Method 2 ($p > 0.99$, $\alpha_2 = 0.01$)	187 200 243 405
Method 2 ($p > 0.95$, $\alpha_2 = 0.05$)	+ **34 39 52** 167 172 190 204 230 242 **267 282 285** 296 308 393 404 435 485 493 496 503 512 **581 584**

Sites in bold correspond to residues located in previously reported epitopes (Escalante *et al.* 2001). Epitopes are antigenic portions to which antibodies bind, and are therefore immunologically relevant. Further analyses that include several more sequences of AMA-1 in *P.falciparum* are needed to determine if the increased non-synonymous substitutions for the residues listed in Table 3 – particularly for those residues located in epitopes – are associated with specific clinical manifestations of the disease or with specific immune responses.

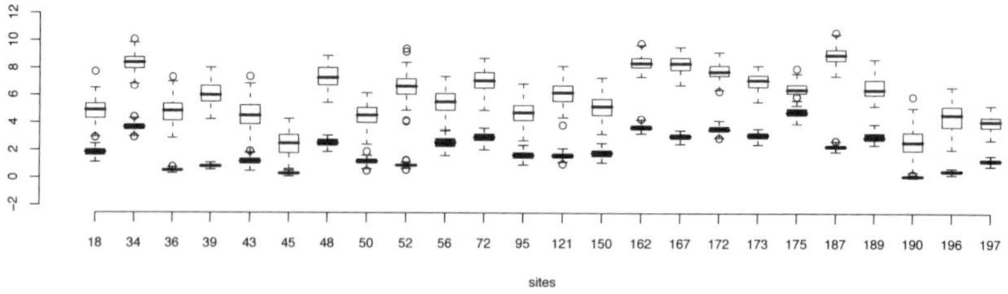

Figure 6: *Effects of the covariate in the AMA-1 antigen sequences. Sites 18–197.*

Figures 6–8 depict site-specific effects related to particular amino acid substitutions. Sites displaying large $\boldsymbol{\delta}_i$ effects that are also associated with relatively

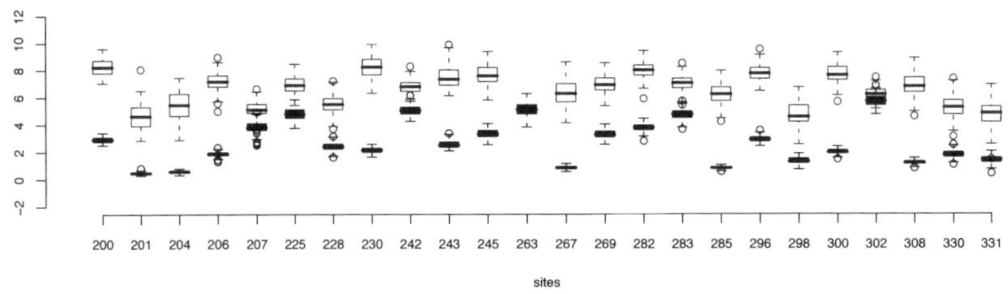

Figure 7: *Effects of the covariate in the AMA-1 antigen sequences. Sites 200–331.*

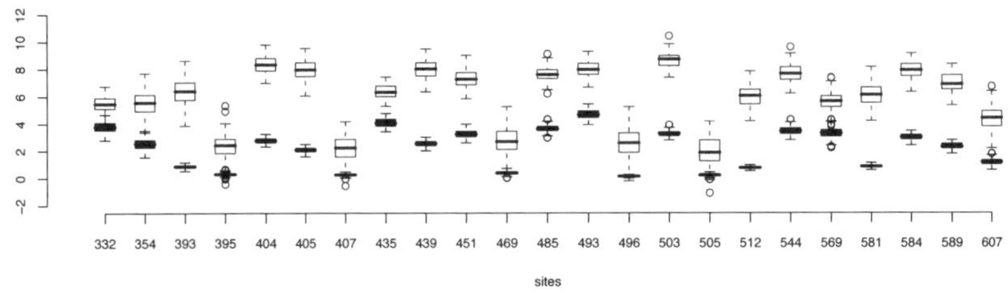

Figure 8: *Effects of the covariate in the AMA-1 antigen sequences. Sites 332–607.*

high normalized scores are of particular interest. Radical substitutions between two amino acids may be associated with the effects of positive selection. Each figure shows two sets of boxplots. The light boxplots summarize the posterior distributions of the $\boldsymbol{\delta}_i$ effects for the polymorphic sites in the AMA-1 alignment, while the dark boxplots summarize the posterior of $\boldsymbol{\delta}_i \, \bar{x}_i^*$, with $\bar{x}_i^*$ the average normalized Grantham score over all the non-synonymous substitutions for a particular site i. Some of the sites display relatively large $\boldsymbol{\delta}_i$ effects, however, they are associated with low Grantham scores, indicating that the observed substitutions in such sites are between largely similar amino acids, and are less likely to produce changes to the protein morphology that would be targeted by positive selection. This is the case of site 187, for example. Other sites, such as 175, 225, 242, 263 and 302, display moderately large $\boldsymbol{\delta}_i$ effects associated with large or moderately large Grantham scores, indicating that the observed substitutions in these sites were between relatively different amino acids in terms of a distance based on their physicochemical properties.

6. CURRENT RESEARCH AND FUTURE DIRECTIONS

The class of GLMs for substitutions count data derived from a given alignment constitutes a novel approach to modelling and describing genetic variability at the molecular level in DNA sequence data with relatively low evolutionary divergence. These models provide an empirical framework for detecting molecular adaptation at the amino acid level by expressing the observed genetic variability in a DNA sequence alignment in terms of species/population effects, residue-specific effects and possible covariates. The methods and models of Prado *et al.* (2006) are extended here to incorporate covariates and population and/or geographic-specific effects, as well as model validation tools that are based on comparing posterior results to substantial biological information. The GLM-based methodology also provides a way to include structured prior information on the underlying evolutionary processes that describe the substitution patterns in the sequences, whenever such information is available.

Future research will involve incorporating the $\boldsymbol{Z}$'s as latent variables in the models. The analyses of Prado *et al.* (2006), as well as some simulation studies included in Merl *et al.* (2005), suggest that the posterior results, at least in terms of which sites are identified as positively selected a posteriori, are not sensitive to the various methods used to obtain $\boldsymbol{Z}$ from the sequence alignments in the data analyzed here. However, a fully Bayesian approach that can quantify the uncertainty related to these underlying evolutionary processes should be considered. Simulation studies have also been performed to compare simpler versions of the GLM model in (1) to currently available methods for detecting positive selection such as those implemented in `PAML`, `HYPHY` and `MrBayes` (Huelsenbeck and Dyer, 2004). Such studies (see Merl *et al.* 2005) suggest that the GLM-based methodology has higher power than other methods to detect sites under selection in sequences with relatively low evolutionary divergences. Further simulation studies will be performed in the future to assess the effect of covariates in detecting positive selection.

The models presented here allow us to determine, from a statistical viewpoint, which sites are more likely to be under positive selection in malaria antigens. Many more sequences of AMA-1 for human *P. falciparum* and *P. vivax* are available, as well as alignments of two more candidate antigens for both species of the parasite. We expect to carry out extensive GLM-based analyses for these data. The preliminary results presented here, as well as future results from GLM-based analyses, will assist immunologists in the identification of specific residues that may be relevant to determining if, in fact, the candidate antigens are appropriate for vaccine development.

Other future research directions relate to developing a variable selection approach to detecting which amino acid properties are significant. Recently, `Tree SAAP` (Woolley *et al.*, 2003), a software that measures selective influences on 31 structural and biochemical amino acid properties during phylogenesis, was developed. Many additional amino acid properties can be included. We expect to extend the approach presented here to appropriately tackle the problem of selecting relevant covariates from a large pool of covariates that describe amino acid properties.

ACKNOWLEDGEMENTS

The work was supported by grant R01GM072003-02 from the National Institutes of Health/National Institute of General Medical Sciences. The authors acknowledge useful discussions with Ananías Escalante, School of Life Sciences, Arizona State University.

REFERENCES

Baker, S. (1994). The multinomial-Poisson transformation. *The Statistician* **43**, 495–504.

Escalante, A. A., Grebert, H. M., Chaiyaroj, S. C., Magris M., Biswas S., Nahlen, B. L. and Lal, A. A. (2001). Polymorphism in the gene encoding the apical membrane antigen 1 (AMA-1) of Plasmodium falciparum. X Asembo Bay Cohort Project. *Mol. Biochem Parasitol* **113**, 279–287.

Escalante, A. A., Lal, A. A. and Ayala, F .J. (1998). Genetic polymorphism and natural selection in the malaria parasite *Plasmodium falciparum*. *Genetics* **149**, 189–202.

Felsenstein J. (2004). *Inferring Phylogenies*. Sinauer Associates, Inc.

Galindo, B., Vacquier V. and Swanson, W. J. (2003). Positive selection in the egg receptor for abalone sperm lysin. *PNAS* **100**, 4639–4643.

Gelfand, A. E. and Ghosh, S. K. (1998). Model choice: A minimum posterior predictive loss approach. *Biometrika* **85**, 1–11.

Goldman, N. and Yang, Z. (1994). A codon-based model of nucleotide substitution for protein-coding DNA sequences. *Mol. Biol. and Evol.* **11**, 725–736.

Grantham, R. (1974) Amino acid different formula to help explain protein evolution. *Science* **185**, 862–864.

Huelsenbeck, J. P. and Dyer, K. A. (2004). Bayesian estimation of positively selected sites. *J. Mol. Evol.* **58**, 661–672.

Kosakovsky Pond, S. L. and Frost, S. D. W. (2005a) A simple hierarchical approach to modelling substitution rates. *Mol. Biol. and Evol.* 22, 223–234.

Kosakovsky Pond, S. L., Frost, S. D. W and Muse, S. V. (2005) HyPhy: Hypothesis testing using phylogenies. *Bioinformatics* **21**, 676–679.

Kosakovsky Pond, S. L. and Frost, S. D. W. (2005b) Not so different after all: A comparison of methods for detecting amino acid sites under selection. *Mol. Biol. and Evol.* 22, 1208–1222.

Lee, Y. H., Ota T. and Vacquier V. D. (1995). Positive selection is a general phenomenon in the evolution of abalone sperm lysin. *Mol. Biol. and Evol.* 12, 231–238.

Merl, D., Prado, R. and Escalante, A. A. (2005). Bayesian estimation of differential selection using generalized linear models. *Tech. Rep.*, University of California Santa Cruz, USA.

Muse, S. V. and Gaut, B. S. (1994). A likelihood approach for comparing synonymous and nonsynonymous substitution rates, with application to the Chloroplast genome. *Mol. Biol. and Evol.* 11, 715–724.

Nielsen, R. and Yang, Z. (1998). Likelihood models for detecting positively selected amino acid sites and applications to the HIV-1 envelope gene. *Genetics* **148**, 929–936.

Polley, S. D., Chokejindachai, W. and Conway, D. J. (2003). Allele frequency based analyses robustly map sequence sites under balancing selection in a malaria vaccine candidate antigen. *Genetics* **165**, 555–561.

Polley, S. D., Mwangi T., Kocken C. H., Thomas A. W., Dutta S., Lanar D. E., Remarque E., Ross A., Williams T. N., Mwanbingu G., Lowe B., Conway D. J. and Marsh K. (2004). Human antibodies to recombinant protein constructs of Plasmodium falciparum Apical Membrane Antigen 1 (AMA1) and their associations with protection from malaria. *Vaccine* 23, 718–728.

Prado, R., Merl, D. and Escalante, A. A. (2006). Detecting selection in DNA sequences: A model-based approach. *Tech. Rep.*, University of California Santa Cruz, USA.

Schwartz, R. M. and Dayhoff, M. O. (1978). *Atlas of Protein Sequence and Structure, 5 suppl.* (Vol. 3:353-358). Washington D.C.: Nat. Biomed. Res. Found.

Spiegelhalter, D. J., Best, N. G., Carlin B. P. and van der Linde, A. (2002). Bayesian measures of model complexity and fit. *J. Roy. Statist. Soc. B* **44**, 377–387, (with discussion).

Suzuki, Y. (2004). New methods for detecting positive selection at single amino acid sites. *J. Mol. Evol.* 59, 11–19.

Suzuki, Y. and Gojorobi, T. (1999). A method for detecting positive selection at single amino acid sites. *Mol. Biol. and Evol.* 16, 1315–1328.

Woolley S., Johnson J., Smith M. J., Keith A. Crandall K. A. and McClellan D. A. (2003). TreeSAAP: Selection on amino acid properties using phylogenetic trees. *Bioinformatics* **19**, 671–672.

Yang, Z. (1997). PAML: a program package for phylogenetic analysis by maximum likelihood. *Comput. Appl. Bisci.* **13**, 555–556.

Yang, Z., Nielsen, R., Goldman, N. and Pedersen, A. K. (2000). Codon substitution models for heterogeneous selection pressure at amino acid sites. *Genetics* **155**, 431-449.

Yang, Z., Swanson, W. J. and Vacquier V. D. (2000). Maximum-likelihood analysis of molecular adaptation in abalone sperm lysin reveals variable selective pressures among lineages and sites. *Mol. Biol. and Evol.* 17, 1446–1455.

Yang, Z., Wong, W. S. W. and Nielsen R. (2005). Bayes empirical Bayes inference of amino acid sites under positive selection. *Mol. Biol. and Evol.* 22, 1107–1118.

DISCUSSION

SINAE KIM and MARINA VANNUCCI
(*University of Michigan, USA* and *Texas A&M University, USA*)

It is a pleasure to discuss the work of Merl and Prado on detecting positive selection. Much of the work in evolutionary biology in the past years has focused on two main concepts. On one hand phylogenetics has looked at methods for inferring the history of life, based on the idea that all species are related to one another. Population genetics, on the other hand, has attempted to capture the phenomenon of natural selection by elaborating mathematical models that describe mutations in populations. Natural selection is the process in which individuals with beneficial mutations leave more offspring. The present paper focuses on detecting positive selection at the molecular level. In this discussion we will briefly review existing approaches, highlight the contribution of the current paper and comment on possible extensions.

Comparative methods based on alignment of sequences are the most widely used in evolutionary biology and have provided much of the evidence for natural selection. High relative rates of non-synonymous to synonymous changes are a clear indicator of positive selection. Among the existing methods in the literature the most successful are those providing codon specific estimates of substitution rates that incorporate phylogeny in the analysis. Bias can in fact result from failing to take into account the similarity in features across species that is caused by a common history. Some of the early approaches have assumed constant substitution rates across the sequence. Nielsen and Yang (1998) first developed a method that allows rates to vary across the sequence and adopted an MLE-based estimation method. Suzuki and Gojobori (1999) compared substitutions rates at each codon site using ancestral codons at interior nodes of the phylogenetic tree, as inferred by the maximum parsimony method. Huelsenbeck *et al.* (2001, 2004) first developed a fully Bayes approach that incorporates the uncertainty in the phylogeny. Stochastic search MCMC methods are used to sample the posterior space of possible trees.

The contribution of the work by Merl and Prado relies on an explicit GLM modeling of the evolution rates and a full Bayes inferential treatment. The approach is similar in spirit to the work of Huelsenbeck and collaborators. The specific interest of the authors, however, is in modeling the substitution rates, rather than in the estimation of the phylogeny tree. An additional feature of the proposed approach is that the authors do not assume a fixed number of possible values for the substitution rates. The authors present applications of their model to multiple alignments of sequences from same species, a particularly challenging scenario. The modeling framework proposed by the authors is quite flexible and allows a number of possible extensions. For example, uncertainty about k, the number of substitution types, and about the substitution counts, $\boldsymbol{Z}$, could be taken into account. The model also allows for the inclusion of a number of possible covariates. Bayesian variable selection techniques could be easily applied. Finally, intuitively it is possible that multiple sites can have the same substitution rate. This could motivate the use of clustering approaches via Dirichlet process priors, an idea recently pursued by Huelsenbeck *et al.* (2006).

We conclude this discussion by briefly mentioning our recent work in the detection of protein dissimilarities in multiple alignments using Bayesian variable selection. In Kim *et al.* (2006) we consider the dimer interface of chemokine families and identify positions in the amino acids sequence that show significant differences among the different families. Some of the positions we identify have turned out to be indicative of positive and negative selection. In particular, we have located a two-residue deletion which negative selects for protein interaction (so, inverting, it suggests that adding the two residues would positively select for the type of dimer we have considered). Results have been confirmed experimentally. The investigation of chemokines has been recently very active in relation to the design of treatments for HIV-related diseases.

REPLY TO THE DISCUSSION

We thank Kim and Vannucci for their kind comments and suggestions about future research topics.

We believe that using clustering approaches via Dirichlet process priors is an attractive and potentially powerful tool to identify residues that display the same substitution rate. As Kim and Vannucci have pointed out, this idea has already been developed by Huelsenbeck *et al.* (2006) in the context of stochastic models of sequence evolution. Specifically, Huelsenbeck *et al.* (2006) allowed the sites to be in one of a number of classes, with each class having a different non-synonymous to synonymous substitution rates ratio, and where the prior for the number of classes and the substitution rates ratio for each class is described by a Dirichlet process. In our GLM context a Dirichlet process prior can be used on the $\boldsymbol{\beta}$ parameters, allowing in this way an infinite number of substitution rate classes.

Another feature of the GLM models is that within this framework it is straightforward to compare a posteriori the ratio of non-synonymous (replacement) to synonymous fixed differences between species and the ratio of non-synonymous (replacement) to synonymous polymorphisms within species in an alignment that consist of several sequences from two closely related species. If the observed substitutions are neutral, the two ratios should be the same. Comparing the posterior distributions of these two ratios obtained from the GLM-based analysis would result in a Bayesian alternative to the McDonald and Kreitman test (see McDonald and Kreitman, 1991). McDonald and Kreitman (1991) proposed a G-test of independence for counts in the form of a 2×2 contingency table to test the null hypothesis that the proportion of replacement substitutions is independent of whether the subtitutions are fixed or polymorphic. We are currently working on analyses of DNA alignments of malaria antigens of *P. falciparum* and the chimpanzee malaria parasite *P. reichenowi*, which is the closest related species to the human parasite *P. falciparum*. The goal of these studies is to determine whether the non-synonymous to synonymous rates ratio within *P.falciparum* is significantly greater than that between the two species, implying that there is evidence of positive selection in the AMA-1 *P. falciparum* sequences.

ADDITIONAL REFERENCES IN THE DISCUSSION

Huelsenbeck, J. P., Ronquist, F., Nielsen, R. and Bollback, J. P. (2001). Bayesian inference of phylogeny and its impact on evolutionary biology. *Science* **294**, 2310–2314.

Huelsenbeck, J. P., Jain, S., Frost, S. W. D. and Pond, S. L. K. (2006). A Dirichlet process model for detecting positive selection in protein-coding DNA sequences. *PNAS*, **103**, 6263–6268.

Kim, S., Tsai, J., Kagiampakis, I., LiWang, P. and Vannucci, M. (2006). Detecting protein dissimilarities in multiple alignments using Bayesian variable selection. *Bioinformatics* (to appear).

McDonald J. H. and Kreitman M. (1991) Adaptive protein evolution at the *Adh* locus in *Drosophila*. *Nature* **351**, 652–654.

BAYESIAN STATISTICS 8, pp. 325–348.
J. M. Bernardo, M. J. Bayarri, J. O. Berger, A. P. Dawid,
D. Heckerman, A. F. M. Smith and M. West (Eds.)

Deriving Bayesian and Frequentist Estimators from Time-Invariance Estimating Equations: A Unifying Approach

ANTONIETTA MIRA
Univ. of Insubria, Varese, Italy
antonietta.mira@uninsubria.it

ADRIAN BADDELEY
Univ. of Western Australia, Australia
adrian@maths.uwa.edu.au

SUMMARY

Time-invariance estimating equations (Baddeley, 2000) are a recipe for constructing estimators of the parameter of any stochastic model, using properties of auxiliary Markov chains associated with the model. In this paper we extend the time-invariance framework to a Bayesian context. We recover some well known Bayesian estimators and construct new ones. We find some interesting relationships between Bayesian and frequentist estimators and set up a unifying approach.

Keywords and Phrases: ADMISSIBILITY; EXPONENTIAL FAMILIES; GIBBS SAMPLER; METROPOLIS–HASTINGS ALGORITHM; MAXIMUM PSEUDOLIKELIHOOD.

1. INTRODUCTION

There are many fascinating connections between parameter estimation in a stochastic model and auxiliary Markov processes associated with the model. This paper explores two of these connections.

Firstly, the admissibility of an estimator can be related to the recurrence of a Markov process. Brown (1971) showed that the (least squares) best invariant estimator of the mean of a multivariate normal distribution is admissible if and only if an associated diffusion is recurrent. Johnstone (1984, 1986) related the admissibility of estimators of Poisson means to the recurrence of birth-death Markov chains. General connections between admissibility of Bayesian estimators and recurrence of auxiliary Markov processes were found by Eaton (1992, 1997, 2001) and Hobert and Robert (1999).

Secondly, the generator of a Markov chain (or infinitesimal generator of a Markov process) is a ready-made supply of functions which have zero mean under the stationary distribution of the chain. This suggests a recipe (Baddeley, 2000) for deriving estimators of the parameter in a completely general stochastic model. We choose an auxiliary Markov chain whose stationary distribution is the model of interest. Then the generator of this chain provides unbiased estimating equations for the parameter (in a frequentist setting). When this recipe is applied to chains which give rise to popular Markov Chain Monte Carlo (MCMC) methods, it yields several familiar *ad hoc* estimators. For example, applying this method to the single-site Gibbs sampler yields Besag's maximum pseudolikelihood estimator. Kurtz and Li (2004) studied the performance of time-invariance estimators for spatial processes.

These two approaches have complementary strengths and weaknesses. The first one ('admissibility if and only if recurrence') starts with a given estimator, constructs an auxiliary Markov process, and relates the performance of the estimator to properties of the Markov process. The connection between the estimator and the Markov process is however regarded as 'obscure' by most writers (Eaton, 1992). Conversely, in the second approach ('equate the generator to zero'), we are free to choose the auxiliary Markov process, and the time-invariance estimator is derived from this process via the generator. The performance of the resulting estimator is usually unknown and it seems hard to determine an optimal estimator in this setting (Baddeley, 2000).

This paper attempts to reconcile these two points of view and to combine their advantages. We extend the results by Baddeley (2000) to a Bayesian setting and recover the formal Bayes estimator under squared error loss function (which we will assume unless otherwise specified) as an instance of the time-invariance recipe. We also examine issues of optimality for the resulting estimators.

In Baddeley (2000) a frequentist setting is considered and a Markov process is constructed on the sample space, $\mathcal{X}$, having the model $l(x|\theta)$ as its stationary distribution. In a Bayesian setting the parameter of interest, θ, becomes the random variable and the data is considered as given (fixed). This naturally leads to the idea that, in order to recover Bayesian estimates via time-invariance estimating equations (EE), we can construct a Markov chain on the parameter space, Θ, having the posterior, $\pi(\theta|x)$, as its stationary distribution. If the Markov chain considered is i.i.d. sampling from the posterior we indeed recover Bayes estimators. We show that a Metropolis–Hastings sampler having the prior as the proposal distribution also leads to the same estimating equation using a modified version of time-invariance EE which we will refer to as Type II EE (Section 4).

As an alternative, in the attempt of providing a unifying inferential framework, we can enlarge the state space and construct a Markov chain on $(\Theta, \mathcal{X})$. We will show (Section 5) that, in this setting, depending on the Markov process we choose, we recover both the Bayes estimate (BE) and the method of moments (MM) estimate for θ. In this latter case we are back in a frequentist framework even if we have implicitly defined a prior.

In the setting of Section 5 the marginal chain on the Θ subspace has the prior $\nu(\theta)$ as its stationary distribution. The marginal chain on the $\mathcal{X}$ subspace has the marginal likelihood, $m(x)$, as its stationary distribution. This suggests we mimic the reasoning followed in Section 4 and look at a Metropolis–Hastings sampler on the $\mathcal{X}$ portion of the state space having $l(x|\theta)$ as its stationary distribution and using $m(x)$ as the proposal. This brings us back to a frequentist setting and we again recover the MM estimate via Type II EE (Section 6).

In Section 8 we study other Bayes estimates that we can recover from EE by selecting specific underlying Markov. In particular we will consider the case when the parameter of interest is d-dimensional and a Gibbs sampler Markov chain is used.

2. BACKGROUND

This section establishes notation and gives relevant background from the literature.

2.1. *Time-Invariance Estimating Equations*

Given a probability distribution π on a sample space $\mathcal{X}$, it is of interest to find functions H on $\mathcal{X}$ satisfying

$$\mathbb{E}\, H(X) = 0 \qquad \text{when } X \sim \pi. \tag{1}$$

Applications include the construction of estimating equations, distributional approximations, and iterative estimation algorithms.

We recall that the generator, $\mathcal{A}$, of a discrete time Markov chain, (X_n), on $\mathcal{X}$, is the operator which maps any function $S : \mathcal{X} \to \mathbb{R}^d$, to another function $\mathcal{A}S$ defined by

$$(\mathcal{A}S)(x) = \mathbb{E}\left[S(X_{n+1}) - S(X_n) \mid X_n = x\right]. \tag{2}$$

The concept of generator can be generalized to a continuous time Markov chain. A standard fact about Markov chains is that all images of the generator have mean zero under the stationary distribution. This provides a ready-made source of solutions to (1). Given π (the likelihood, in a frequentist setting, or the posterior, in a Bayesian setting), choose a Markov chain whose stationary distribution is π. Let $\mathcal{A}$ be the generator of the chain. Then, for all functions, S, integrable with respect to π,

$$\mathbb{E}\,(\mathcal{A}\,S)(X) = 0 \qquad \text{when } X \sim \pi. \tag{3}$$

That is, the function $H = \mathcal{A}\,S$ satisfies (1). This principle appears frequently in the literature but is usually not highlighted.

This leads to a recipe for deriving estimators of the parameter for a completely general stochastic model (Baddeley, 2000). In a frequentist setting, suppose π and $\mathcal{A}$ in (3) depend on the parameter θ. In particular we identify π with the model of interest, $l(x|\theta)$, where $x \in \mathcal{X} \subset \Re^n$ is the data and $\theta \in \Theta \subset \Re^d$ is the parameter. Then

$$(\mathcal{A}_\theta\, S)(x) = 0 \tag{4}$$

is an unbiased estimating equation and often leads to a useful estimator $\widehat{\theta}$ provided the solution to (4) exists and is unique. Interestingly, when this recipe is applied to conditional independence models, it yields many of the standard *ad hoc* estimators in a unified way. For example (Baddeley, 2000), if π is an exponential family of Markov random fields, (X_n) the single-site Gibbs sampler, and S the canonical sufficient statistic, then $\widehat{\theta}$ is the maximum pseudolikelihood estimator. The method of moments estimator is derived by choosing (X_n) to be the i.i.d. sampler, in which each X_n is an independent realization from π so that (2) becomes

$$(\mathcal{A}S)(x) = \mathbb{E}\left[S(X_{n+1})\right] = S(x).$$

If the model for the data belongs to a regular exponential family, by taking S to be the canonical sufficient statistics, say V, and Y to be again i.i.d. sampling from the model, the time-invariance estimator is the MLE since (2) becomes

$$(\mathcal{A}S)(x) = \mathbb{E}\left[V(X_{n+1})\right] = V(x).$$

3. EXTENSION OF TIME-INVARIANCE APPROACH

3.1. *Estimating Equations of Type II*

The time-invariance estimating equations of Baddeley (2000) have connections with the Stein–Chen method of distributional approximation (Stein, 1986; Arratia et al. 1990; Barbour, 1997). In the latter, equations of the form (1) are used to characterize a distribution π. The closeness of another distribution P to π is measured by evaluating $\mathbb{E}_P|H(X)|$ or finding a simple upper bound on it.

In the Stein–Chen method, suitable functions H are also derived using a Markov chain (in discrete or continuous time) whose equilibrium distribution is π. Barbour (1997) distinguishes two different versions of Stein's method. Type I uses functions of the form $H = \mathcal{A}S$ as in (3). Type II uses functions of the form

$$H(x) = (\mathcal{F}T)(x) = \mathbb{E}[T(X_0, X_1)|X_0 = x] \tag{5}$$

where $T : \mathcal{X} \times \mathcal{X} \to \mathbb{R}$ is an *antisymmetric* function, $T(x, y) = -T(y, x)$; the chain is in discrete time and is *reversible* with respect to π. It is easy to prove that H, as defined in (5), satisfies (1), since

$$\begin{aligned}
\mathbb{E}_\pi(\mathcal{F}T)(X) &= \mathbb{E}_\pi[\mathbb{E}[T(X_0, X_1)|X_0 = X]] \\
&= \mathbb{E}[T(X_0, X_1)] \\
&= \mathbb{E}[T(X_1, X_0)] \\
&= \mathbb{E}[-T(X_0, X_1)]
\end{aligned}$$

where X_n denotes the equilibrium state of the chain. The third line follows from the reversibility of the Markov chain and the fourth from the antisymmetry of T. Comparing the third and fourth lines we conclude $\mathbb{E}_\pi(\mathcal{F}T)(X) = 0$.

Developing the analogy, we may extend the method of time-invariance estimating equations. In a frequentist setting, suppose π and $\mathcal{F}$ depend on the parameter θ. Then

$$(\mathcal{F}_\theta T)(x) = 0 \tag{6}$$

is an unbiased estimating equation for θ. We shall call it a **Type II estimating equation.**

Type II is a generalization of Type I, since if $T(x, y) = S(y) - S(x)$ we recover

$$\begin{aligned}
(\mathcal{F}T)(x) &= \mathbb{E}[S(X_1) - S(X_0)|X_0 = x] \\
&= \mathbb{E}[S(X_1)|X_0 = x] - S(x) \\
&= (\mathcal{A}S)(x),
\end{aligned}$$

at least for reversible discrete time chains.

3.2. *Type II for Metropolis–Hastings*

Consider a Metropolis-Hasting algorithm stationary with respect to some distribution $\pi_\theta(x)$ and denote with $\alpha_\theta(x, Z)$ the corresponding acceptance probability of a proposal Z given that the current position of the chain is x. We have, for a general antisymmetric $T(x, y)$,

$$(\mathcal{F}_\theta T)(x) = \mathbb{E}\left[T(x, Z)q(x, Z)\alpha_\theta(x, Z)\right].$$

Notice that there is no contribution from rejected proposals since $T(x,x) = 0$ by antisymmetry so that we get:

$$(\mathcal{F}_\theta T)(x) = \mathbb{E}_\theta\left[T(x,Z)q(x,Z)\left(\frac{q(x,Z)}{\pi_\theta(Z)} \wedge \frac{q(Z,x)}{\pi_\theta(x)}\right)\right].$$

Consider now special cases of T, in particular, an interesting choice is

$$T(x,y) = H_\theta(x,y)S(y) - H(y,x)S(x),$$

where S is arbitrary and

$$H_\theta(x,y) = \frac{\pi_\theta(x)}{q(y,x)\alpha_\theta(y,x)},$$

then T is clearly antisymmetric. But by detailed balance, H is symmetric, so

$$T(x,y) = H_\theta(x,y)\{S(y) - S(x)\}$$

and substituting

$$\begin{aligned}(\mathcal{F}_\theta T)(x) &= \int T(x,y)\alpha_\theta(x,y)q(x,y)\,\mathrm{d}y \\ &= \int [S(y) - S(x)]\,\pi_\theta(y)\,\mathrm{d}y \\ &= \mathbb{E}_\theta S(X) - S(x)\end{aligned}$$

where the first identity follows from the detailed balance condition:

$$\pi(x)q(x,y)\alpha(x,y) = \pi(y)q(y,x)\alpha(y,x).$$

We thus recover the method of moments/MLE from this choice of T.

Notice that choosing a function $H(x,y)$ not depending on θ and approximately equal to $H_\theta(x,y)$ would give us an approximation to the MLE.

3.3. *Backward Estimating Equations*

We can regard the original Type I time-invariance EE (4) as a forward equation since it requires computation of the expected value of the function S one step ahead, given the current value of the function: indeed (4) is equivalent to

$$\mathbb{E}[S(X_{n+1})|X_n = x] - S(x) = 0. \tag{7}$$

The backward counterpart of (7) is

$$\mathbb{E}[S(X_n)|X_{n+1} = x] - S(x) = 0, \tag{8}$$

which also provides an unbiased estimating equation. In order that this equation not depend on n, the sampler (X_n) should be in its equilibrium state.

Similarly, the backward counterpart of Type II EE is

$$(\mathcal{B}T)(x) = \mathbb{E}[T(X_0, X_1)|X_1 = x] \tag{9}$$

and by taking $T(x,y) = S(x) - S(y)$ we recover the backward counterpart of Type I EE.

Note that (7) and (8) coincide if the underlying Markov chain is reversible and likewise for (5) and (9).

Most Markov chains used for MCMC purposes are reversible. For example, we obtain non-reversible Markov chain by combining Gibbs steps in a fixed scan while a random scan Gibbs sampler is reversible and so is a Metropolis–Hastings algorithm. In Section 3.4 and 5 we use a two-step, fixed scan, Gibbs sampler where each one of the two sets of variables is updated in turn. Since estimating equations only consider a single update of the variables (from time t to time $t+1$, for example) we will see that the order in which the variables are updated makes a difference in the resulting estimate.

3.4. *Missing Observations*

Next we consider the extension of time-invariance estimating equations (Types I and II) to the context of missing data.

Suppose X is observable, Z is unobservable, and the joint distribution of $Y = (X, Z)$ under a parameter θ is π_θ. Time-invariance estimating equations for this problem would read

$$(\mathcal{A}_\theta S)(x,z) = 0 \tag{10}$$
$$(\mathcal{F}_\theta T)(x,z) = 0 \tag{11}$$

where x is the observed realization of X and z is the **unobservable** realization of Z.

Notice that these equations give rise to usable estimators $\widehat{\theta}$ only if *the left hand sides do not depend on* z. We can guarantee this by choosing a chain whose transition kernel $Q((x,z),\cdot)$ depends only on x. A natural choice for such chains is a Gibbs sampler, with a site updating scheme which begins by updating Z conditional on X.

Consider the simplest discrete-time Gibbs sampler in which Z and X are each updated *en bloc*.

The sampler updates $Z \sim \mathbb{P}_\theta(Z \mid X)$ at step 1 of each cycle, and then updates $X \sim \mathbb{P}_\theta(X \mid Z)$ at step 2 of the cycle. Then for Type I

$$\begin{aligned}(\mathcal{A}_\theta S)(x,z) &= \int_{\mathcal{Z}}\int_{\mathcal{X}} [S(x',z') - S(x,z)]\, \mathbb{P}_\theta(\,\mathrm{d}x' \mid z')\, \mathbb{P}_\theta(\,\mathrm{d}z' \mid x) \\ &= \mathbb{E}_\theta[U_\theta(Z) \mid x] - S(x,z)\end{aligned}$$

where

$$U_\theta(z) = \mathbb{E}_\theta[S(X,z) \mid z].$$

Here and in the sequel the 'non-prime' variable (z,x) is the current position of a Markov process while the 'prime' variable (z',x') is the position of the next step.

Then (10) reads

$$\mathbb{E}_\theta[U_\theta(Z) \mid x] = S(x,z). \tag{12}$$

This is an unbiased estimating equation in the sense that the expectations of both sides, with respect to the joint distribution of (X,Z) under θ, are equal.

Equation (12) is usable for the estimation of θ from x provided $S(x,z)$ depends only on x, say $S(x,z) = V(x)$. Then

$$U_\theta(z) = \mathbb{E}_\theta[V(X) \mid z]$$

and (12) becomes

$$\mathbb{E}_\theta[U_\theta(Z) \mid x] = V(x)$$

equating the observed value of $V(x)$ to the expectation of $V(X)$ where X is the updated value after one full cycle of the Gibbs sampler. This is an unbiased estimating equation with respect to the marginal distribution of X under θ.

However note that we may also use (12) for the *prediction* of Z from X, for example if Z is the unobserved part of a realization of a stochastic process. In that case we would take $S(x, z) = V(z)$ for some function $V : \mathcal{Z} \to \mathbb{R}$ which it is desired to predict. Then $U_\theta(z) = V(z)$, and (12) becomes

$$\mathbb{E}_\theta[V(Z) \mid x] = V(z);$$

this may be interpreted as indicating that the missing value $V(z)$ should be estimated by the conditional expectation $\mathbb{E}_\theta[V(Z) \mid x]$. This is an unbiased estimating equation with respect to the conditional distribution of Z given x under θ, hence also with respect to the joint distribution of (X, Z) under θ.

3.5. *Bayesian inference*

From the foregoing discussion of inference for missing values, it is a simple step to consider Bayesian inference. Suppose now that X and θ are both random, with $\theta \sim \nu(\theta)$ where $\nu(\theta)$ is the prior distribution for the parameter and $(X \mid \theta) \sim \ell(x \mid \theta)$. Thus $(\theta \mid X) \sim \pi(\theta \mid x) \propto \nu(\theta) l(x|\theta)$, the posterior.

Consider the Gibbs sampler which updates θ (by sampling from the posterior $\pi(\theta \mid x)$) in step 1 of the cycle, and updates x (by sampling from the model $\ell(x \mid \theta)$) in step 2. For Type I EE we consider a function $S(x, \theta)$ and obtain

$$(\mathcal{A}S)(x, \theta) = \int_\Theta \int_\mathcal{X} S(x', \theta')\, \ell(x' \mid \theta')\, \pi(\theta' \mid x)\, \mathrm{d}x'\, \mathrm{d}\theta' - S(x, \theta).$$

This leads to a family of unbiased estimating equations in the sense that

$$\mathbb{E}\left[(\mathcal{A}S)(X, \Theta)\right] = 0$$

where (X, Θ) is drawn from the joint distribution with density proportional to $\nu(\theta)\ell(x \mid \theta)$.

We can, at this stage, extend the concept of unbiased estimating equations as follows. We call $g(x, \theta) = 0$ an *unbiased frequentist estimating equation* if

$$\mathbb{E}_{X|\theta}[g(X, \theta)] = 0, \qquad \forall \theta.$$

Notice that the resulting estimator for θ is not necessarily unbiased but is usually consistent.

Similarly, define an *unbiased Bayesian estimating equation* if

$$\mathbb{E}_{\theta|x}[g(x, \theta)] = 0, \qquad \forall x.$$

Finally, define an *unbiased general estimating equation* if

$$\mathbb{E}_{\theta,x}[g(x, \theta)] = 0.$$

Clearly if g is an unbiased estimating equation in the frequentist or Bayesian sense, it will also be unbised in the extended sense but the converse is not generally true.

4. PARAMETER SPACE Θ

The formal Bayes estimate under squared error loss function is the posterior mean of the function of the parameter θ we want to estimate, say $g(\theta)$. Assume that g is bounded.

We recover this estimate via Type I EE by taking the special Markov chain consisting of *i.i.d. sampling from the posterior* and letting $S(\theta) = g(\theta)$. The resulting EE is

$$g(\theta) = \int g(\theta')\pi(\theta'|x)d\theta' \tag{13}$$

which we solve for $g(\theta)$ and conclude that an unbiased (relative to the stationary distribution of the chain i.e. the posterior) estimate of $g(\theta)$ is the posterior mean of $g(\theta)$ itself. Thus (13) is a Bayesian unbiased estimating equation.

Alternatively consider an *independence Metropolis–Hastings sampler* (i.e. the proposal does not depend on the current position of the chain) having the posterior as its stationary distribution and the prior as the proposal. The acceptance probability is $\alpha(\theta,\theta') = 1 \wedge \frac{l(x|\theta')}{l(x|\theta)}$ and the resulting Type II EE is

$$\int T(\theta,\theta')\nu(\theta')\alpha(\theta,\theta')d\theta' + T(\theta,\theta)\int \nu(\theta')[1-\alpha(\theta,\theta')]d\theta' = 0.$$

By the detailed balance condition

$$\pi(\theta|x)\nu(\theta')\alpha(\theta,\theta') = \pi(\theta'|x)\nu(\theta)\alpha(\theta',\theta), \qquad \forall\theta,\theta'$$

taking

$$T(\theta,\theta') = \frac{g(\theta')\pi(\theta|x)}{\nu(\theta)\alpha(\theta',\theta)} - \frac{g(\theta)\pi(\theta'|x)}{\nu(\theta')\alpha(\theta,\theta')} \tag{14}$$

and noting that $T(\theta,\theta) = 0$, $\forall\theta$, we again obtain (13).

Since, in the above reasoning, we made no use of the actual form of $\alpha(\theta,\theta')$ but only of the fact that it preserves reversibility, the previous argument would go through even if we used an acceptance probability that is 'less efficient' in the sense of Peskun (1973). This shows that it is not possible to deduce the relative efficiency of time-invariance estimates from an ordering defined on the transition kernels of the Markov chains (having a common stationary distribution), at least in terms of asymptotic variance (such as the Peskun ordering or the Covariance ordering of Geyer and Mira, 1999). The fact that Peskun ordering would not have been of much use here might also be related to the fact that $S(\theta,\theta) = 0$, $\forall\theta$; to improve a transition Kernel in the Peskun sense one has to reduce the probability of the corresponding chain to retain the same position over time.

The other concept we could use to order Markov chain transition kernels is their speed of convergence to stationarity.

So, let us follow this line related to speed of convergence to stationarity and consider the T function used for MH chain when the chain is instead i.i.d. sampling from the posterior. Likewise we could use the S function used for the i.i.d. case (namely the identity function $S(\theta) = \theta$) when the underlying Markov process is a MH sampler having the posterior as its stationary distribution and the prior as the proposal.

When doing i.i.d. sampling we get a chain that converges to stationarity in one step. Thus i.i.d. sampling beats everything else in terms of speed of convergence to stationarity.

The above reasoning can be performed in the classical setting as well to see what we get. But also in the classical setting the form of the resulting estimate seem to be the same and it is hard to interpret.

4.1. *Inhomogeneous Chains*

Time-inhomogeneous chains arise e.g. from the Gibbs sampler with deterministic update schedule. Take the case of a Gibbs sampler for a joint distribution of two variables (X, Z) which updates X en bloc followed by updating Z en bloc.

If we regard one time transition of the chain as consisting of two steps (an update of both Z and X) then it is important to realize that the generator is not the same if we start from an X update or a Z update. For the sequence

$$(x, z) \to (x, z') \to (x', z')$$

the generator is

$$(\mathcal{A}S)(x, z) = \mathbb{E}[U(Z)|X = x]$$

where $U(z) = \mathbb{E}[S(X, Z)|Z = z]$. For the sequence

$$(x, z) \to (x', z) \to (x', z')$$

the generator is

$$(\mathcal{A}S)(x, z) = \mathbb{E}[V(X)|Z = z]$$

where $V(x) = \mathbb{E}[S(X, Z)|X = x]$.

These are different since, in the first case,

$$\begin{aligned}(\mathcal{A}S)(x, z) &= \int\int S(x', z') f_{X|Z}(x'|z') f_{Z|X}(z'|x) \,\mathrm{d}x' \,\mathrm{d}z' \\ &= \int\int S(x', z') \frac{f_{X,Z}(x', z')}{f_Z(z')} \frac{f_{X,Z}(x, z')}{f_X(z)} \,\mathrm{d}x' \,\mathrm{d}z' \\ &= \frac{1}{f_X(x)} \int \left(\frac{f_{X,Z}(x, z')}{f_Z(z')} \int S(x', z') f_{X,Z}(x', z') \,\mathrm{d}x' \right) \mathrm{d}z';\end{aligned}$$

while in the second case

$$\begin{aligned}(\mathcal{A}S)(x, z) &= \int\int S(x', z') f_{Z|X}(z'|x') f_{X|Z}(x'|z) \,\mathrm{d}z' \,\mathrm{d}x' \\ &= \int\int S(x', z') \frac{f_{X,Z}(x', z')}{f_X(x')} \frac{f_{X,Z}(x', z)}{f_Z(z)} \,\mathrm{d}z' \,\mathrm{d}x' \\ &= \frac{1}{f_Z(z)} \int \left(\frac{f_{X,Z}(x', z)}{f_X(x')} \int S(x', z') f_{X,Z}(x', z') \,\mathrm{d}z' \right) \mathrm{d}x'.\end{aligned}$$

This fact becomes relevant when we consider different updating schemes over an enlarged state space (Section 5).

5. ENLARGED SPACE $(\Theta, \mathcal{X})$

A common trick used in MCMC simulation is to enlarge the original state space using auxiliary variables and to construct a Markov chain on the augmented space having the marginal distribution of interest as its stationary distribution when restricted back to the original state space. The slice sampler (Swendsen and Wang, 1987; Edwards and Sokal, 1988; Besag and Green, 1993), is a clever example of use of this strategy. The auxiliary variable may or may not have a physical interpretation (as in the Ising model) in terms of the original problem.

In this section we will use a similar trick and consider a Markov chain on the enlarged space $(\Theta, \mathcal{X})$ that updates the variables in the following order

$$(x, \theta) \rightarrow (x, \theta') \rightarrow (x', \theta')$$

and having the following transition kernel:

$$K(x', \theta'|x, \theta) = \pi(\theta'|x)l(x'|\theta').$$

This is a Gibbs sampler with stationary distribution proportional to $l(x|\theta)\nu(\theta)$. The marginal chain on the Θ portion of the state space is also Markovian and is stationary with respect to the prior, $\nu(\theta)$. The marginal Markov chain on the $\mathcal{X}$ subspace has $m(x) = \int \nu(\theta)l(x|\theta)d\theta$, the marginal likelihood (which coincides with the posterior normalizing constant), as its stationary distribution. In Eaton (1992, 1997) the marginal Θ chain is studied to determine whether the formal Bayes estimate is v-admissible. In Hobert and Robert (1999) the Authors look at the marginal $\mathcal{X}$ chain instead. If either of these chains is null recurrent the Bayes estimate is v-admissible. If ν is improper the chains cannot be positive recurrent. In Hobert and Robert (1999) the joint chain is called the *extended chain* while the two marginal sub-chains are called *conjugate chains*. We will adopt this terminology in the sequel.

Equating to zero the generator of the Gibbs sampler defined above applied to some function S defined on $(\Theta, \mathcal{X})$, gives the following Type I unbiased estimating equation:

$$\int S(x', \theta')\pi(\theta'|x)l(x'|\theta')d(x', \theta') = S(\theta, x). \tag{15}$$

If we are interested in estimating $g(\theta)$ let $S(x, \theta) = g(\theta)$. From (15) we obtain:

$$g(\theta) = \int g(\theta')\pi(\theta'|x)d(\theta')$$

which we solve for $g(\theta')$ and conclude that the estimator of $g(\theta)$ is the posterior expectation. Equation (15) shows that the formal Bayesian estimate can be derived from an unbiased EE where 'unbiasedness' is interpreted relative to integration with respect to both θ and x. In the unbiased EE (13) integration is instead performed only with respect to θ, that is, we are averaging relative to the posterior distribution.

If, instead of updating θ we update x first:

$$(x, \theta) \rightarrow (x', \theta) \rightarrow (x', \theta')$$

the following Type I EE is obtained:

$$\int S(x', \theta')l(x'|\theta)\pi(\theta'|x')d(x', \theta') = S(\theta, x). \tag{16}$$

Setting $S(x, \theta) = h(x)$ for some function h, we recover the frequentist MM estimate by solving for θ the following:

$$\int h(x')l(x'|\theta)dx' = h(x).$$

6. SAMPLE SPACE $\mathcal{X}$

In this section we restrict ourselves to the $\mathcal{X}$ portion of the state space and assume a frequentist framework. Construct a Markov chain having $l(x|\theta)$ as its stationary distribution via a Metropolis–Hastings independence sampler with $m(x)$ as the proposal. This choice is suggested by mimicking the parallelism existing between Section 4 and Section 5: see Table 1.

Table 1: *Overview of updating regimes and the resulting estimators.*

State space	Stationary distrib.	Markov process
$(\mathcal{X}, \Theta)$	$l(x\|\theta)\nu(\theta)$	$\pi(\theta'\|x)l(x'\|\theta')$
$(\mathcal{X}, \Theta)$	$\pi(\theta\|x)m(x)$	$l(x'\|\theta)\pi(\theta'\|x')$
Θ	$\pi(\theta\|x)$	i.i.d. from π
Θ	$\pi(\theta\|x)$	M-H with prop. $\nu(\theta)$
Θ	$\nu(\theta)$	$l(x'\|\theta)\pi(\theta'\|x')$
$\mathcal{X}$	$l(x\|\theta)$	i.i.d. from l
$\mathcal{X}$	$l(x\|\theta)$	M-H with prop. $m(x)$
$\mathcal{X}$	$m(x)$	$\pi(\theta'\|x)l(x'\|\theta')$
$(\mathcal{X}, \Theta)$	$\pi(\theta\|x)m(x)$	$\frac{1}{n}\sum_i \pi(\theta'\|x')l(x'_i\|x_{-i}, \theta)$
$(\mathcal{X}, \Theta)$	$l(x\|\theta)\nu(\theta)$	$\frac{1}{d}\sum_j l(x'\|\theta')\pi(\theta'_j\|\theta_{-j}, x)$
$(\mathcal{X}, \Theta)$	$\pi(\theta\|x)m(x)$	$\frac{1}{n}\sum_i \frac{1}{d}\sum_j l(x'_i\|x_{-i}, \theta)\pi(\theta'_j\|\theta_{-j}, x')$
$(\mathcal{X}, \Theta)$	$l(x\|\theta)\nu(\theta)$	$\frac{1}{n}\sum_i \frac{1}{d}\sum_j \pi(\theta'_j\|\theta_{-j}, x)l(x'_i\|x_{-i}, \theta')$
State space	Function	Estimate
$(\mathcal{X}, \Theta)$	$S(x, \theta) = g(\theta)$	BE
$(\mathcal{X}, \Theta)$	$S(x, \theta) = h(x)$	MM
Θ	$S(\theta) = g(\theta)$	BE
Θ	$T(\theta, \theta') = (14)$	BE
Θ	$T(\theta, x) = h(x) - h(\theta)$	MM
$\mathcal{X}$	$S(x) = h(x)$	MM
$\mathcal{X}$	$T(x, x') = (17)$	MM
$\mathcal{X}$	$T(\theta, x) = g(\theta) - g(x)$	BE
$(\mathcal{X}, \Theta)$	$S(x, \theta) = h(x)$	Pseudo-LHD
$(\mathcal{X}, \Theta)$	$S(x, \theta) = g(\theta)$	Pseudo-POST
$(\mathcal{X}, \Theta)$	$S(x, \theta) = h(x)$	Pseudo-LHD
$(\mathcal{X}, \Theta)$	$S(x, \theta) = g(\theta)$	Pseudo-POST

Taking

$$T(x, x') = \frac{h(x')l(x|\theta)}{m(x)\alpha(x', x)} - \frac{h(x)l(x'|\theta)}{m(x')\alpha(x, x')} \tag{17}$$

and using the Type II EE we recover the MM estimates.

As pointed out by Hobert and Robert (1999), the $\mathcal{X}$, the Θ and the $(\mathcal{X}, \Theta)$ chains are always either all (positive) recurrent or all transient. Also, m is proper if and only if ν is proper (Hobert and Robert, 1999). Thus, we can either simulate both the M-H samplers introduced or we cannot simulate any of them since the suggested proposal will be improper.

7. TWO MORE CHAINS TO CONSIDER

7.1. *Eaton's chain on* Θ

Consider the following updating scheme:

$$(\theta) \to (x') \to (\theta').$$

The corresponding transition kernel is:

$$\int l(x'|\theta)\pi(\theta'|x')dx'$$

which is reversible with respect to $\nu(\theta)$. The resulting Type I estimating equation for the function $S(\theta) = g(\theta)$ is:

$$\int g(\theta') \int l(x'|\theta)\pi(\theta'|x')dx'\ d\theta' = g(\theta)$$

$$\int l(x'|\theta) \int g(\theta')\pi(\theta'|x')d\theta'\ dx' = g(\theta).$$

If instead, as in the Type II framework, we consider $T(\theta, x) = h(x) - h(\theta) = -T(x, \theta)$ we recover the MM estimate. See the similar derivation at the end of the next section.

7.2. *Hobert's Chain on* $\mathcal{X}$

Consider now the following updating scheme:

$$(x) \to (\theta') \to (x')$$

The corresponding transition kernel is:

$$\int \pi(\theta'|x)l(x'|\theta')d\theta'$$

which is reversible with respect to $m(x)$. The resulting Type I estimating equation for $S(x) = h(x)$ is:

$$\int h(x') \int \pi(\theta'|x)l(x'|\theta')dx'\ d\theta' = h(x) \tag{18}$$

$$\int \pi(\theta'|x) \int h(x')l(x'|\theta')dx'\ d\theta' = h(x)$$

If instead, as in the Type II framework, we consider $T(\theta, x) = g(\theta) - g(x) = -T(x, \theta)$ we recover the BE.

$$\int\int [S(\theta', x') - S(\theta, x)]l(x'|\theta')\pi(\theta'|x')dx'd\theta'$$

$$
\begin{aligned}
&= \int\int [g(\theta') - g(x) - g(\theta) + g(x)] l(x'|\theta')\pi(\theta'|x') dx' d\theta' \\
&= \int\int [g(\theta') - g(\theta) l(x'|\theta')\pi(\theta'|x') dx' d\theta' \\
&\qquad + \int\int [g(x) - g(x')] l(x'|\theta')\pi(\theta'|x') dx' d\theta' \\
&= \int g(\theta')\pi(\theta'|x) d\theta' - g(\theta) + 0
\end{aligned}
$$

where we use (18) with $h = g$ to get that the last term is zero.

8. PSEUDO-LIKELIHOOD AND PSEUDO-POSTERIOR ESTIMATES

Suppose the parameter of interest is multidimensional: $\underline{\theta} = (\theta_1, \cdots, \theta_d)$ and likewise suppose $x = (x_1, \cdots, x_n)$. Let $x_{-i} = (x_1, \cdots x_{i-1}, x_{i+1}, \cdots x_n)$ and similarly define θ_{-i}.

8.1. *Maximum Pseudo-Likelihood*

Consider the following Markovian updating scheme: Choose i with equal probability and update x_i from $l(x_i|x_{-i}, \theta)$. Then update θ from $\pi(\theta|x')$, where $x' = (x_1 \cdots x_{i-1}, x_i', x_{i+1} \cdots x_n)$. The updating scheme just outlined is a Gibbs sampler on the enlarged $(\mathcal{X}, \Theta)$ space where θ and x are updated in a fixed scan. When updating x a single site random scan Gibbs sampler is used while the vector θ is updated in a single block. The Type I EE is

$$\frac{1}{n}\sum_i \int_{\mathcal{X}_i}\int_{\Theta} [S(x', \theta') - S(x, \theta)]\pi(\theta'|x') l(x_i'|x_{-i}, \theta) d\theta' \, dx_i'$$

and if we take $S(x, \theta) = h(x)$ the EE gives:

$$\frac{1}{n}\sum_i \int_{\mathcal{X}_i} [h(x') - h(x)] l(x_i'|x_{-i}, \theta) \, dx_i'. \tag{19}$$

Notice that we would get the same result by running a Gibbs sampler on the $\mathcal{X}$ space having the likelihood as its stationary distribution (cfr Baddeley, 2000).. Following Baddeley (2000) let the model be a discrete Markov random field i.e.

$$l(x|\theta) \propto \exp\{-\theta' v(x)\}$$

with $d = n$ and $v : \mathcal{X} \to \mathbb{R}^d$ being the potential function. Then, taking $h(x) = v(x)$ in (19) we recover the set of normal equations for the maximum pseudolikelihood estimator.

8.2. *Maximum Pseudo-Posterior*

Consider the following Markovian updating scheme: Choose j with equal probability and update θ_j from $\pi(\theta_j|\theta_{-j}, x)$. Then update x from $l(x|\theta')$ where $\theta' = (\theta_1 \cdots \theta_{j-1}, \theta_j', \theta_{j+1} \cdots \theta_d)$. Again, the updating scheme outlined is a Gibbs sampler where θ and x are updated in a fixed scan. The difference with the schemes analyzed in Section 8.1 is that here we use a single site random scan Gibbs sampler for $\underline{\theta}$ while x is updated in a single block. The Type I EE is

$$\frac{1}{d}\sum_j \int_{\mathcal{X}} \int_{\Theta_j} [S(x',\theta') - S(x,\theta)] l(x'|\theta') \pi(\theta'_j|\theta_{-j}, x) dx'\, d\theta'_j$$

and if we take $S(x,\theta) = g(\theta)$ the EE gives:

$$\frac{1}{d}\sum_j \int_{\Theta_j} [g(\theta') - g(\theta)] \pi(\theta'_j|\theta_{-j}, x)\, d\theta'_j.$$

Notice that we would get the same result by running a Gibbs sampler on the Θ space having the posterior as its stationary distribution. Mimicking the reasoning followed in the frequentist setting (Section 8.1), assume that the posterior is a discrete time Markov random field i.e.

$$\pi(\theta|x) \propto \exp\{-\beta' k(\theta)\}$$

and take $g(\theta) = k(\theta)$ the potential function for the posterior. Assume the loss function is of the form

$$L(\theta, \hat{\theta}) = |\theta - \hat{\theta}|$$

then the Bayes estimator relative to this loss function is the mode of the posterior. If it is difficult to maximize the posterior we could approximate its mode by maximizing the pseudo-posterior $\prod_j \pi(\theta_j|\theta_{-j}, x)$.

Finally consider the following two updating schemes:

Scheme 1: Choose i with equal probability and update x_i from $l(x_i|x_{-i}, \theta)$. Then choose j with equal probability and update θ_j from $\pi(\theta_j|\theta_{-j}, x')$.

Scheme 2: Choose j with equal probability and update θ_j from $\pi(\theta_j|\theta_{-j}, x)$. Then choose i with equal probability and update x_i from $l(x_i|x_{-i}, \theta')$.

The updating schemes just outlined are both Gibbs samplers on the enlarged $(\mathcal{X}, \Theta)$ space where θ and x are updated in a fixed scan. In particular in Scheme 1 we first update x and then θ while in Scheme 2 we reverse the updating order. When updating either x or θ a single site random scan Gibbs sampler is used. The resulting Type I EE are:

Scheme 1:

$$\frac{1}{n}\frac{1}{d}\sum_i \sum_j \int_{\mathcal{X}_i} \int_{\Theta_j} [S(x',\theta') - S(x,\theta)] l(x'_i|x_{-i}, \theta) \pi(\theta'_j|\theta_{-j}, x')\, d\theta'_j\, dx'_i;$$

Scheme 2:

$$\frac{1}{n}\frac{1}{d}\sum_i \sum_j \int_{\mathcal{X}_i} \int_{\Theta_j} [S(x',\theta') - S(x,\theta)] \pi(\theta'_j|\theta_{-j}, x) l(x'_i|x_{-i}, \theta')\, d\theta'_j\, dx'_i.$$

Consider now, for Scheme 1, $S(x,\theta) = h(x)$ and for Scheme 2, $S(x,\theta) = g(\theta)$. For a discrete time Markov random field, the resulting EE yields the maximum pseudolikelihood estimator for Scheme 1 and the maximum pseudo-posterior estimator for Scheme 2.

9. CONCLUSIONS

The time-invariance EE framework is particularly relevant when the model of interest is highly structured or when the likelihood or the posterior are not known analytically or are intractable. In these situations it is often natural to express the likelihood or the posterior as the equilibrium distributions of an associated Markov process in discrete or continuous time. The time-invariance EE idea takes advantage of this convenient representation. This is the same setting where MCMC algorithms become helpful. In both settings (EE and MCMC) the time index of the auxiliary Markov process is a artificial dimension added to the original problem with, typically, no intrinsic meaning related to the original problem. The difference is that in MCMC one needs to simulate the Markov chain while this is not required for EE.

Issues related to the statistical performance of time-invariance EE are discussed briefly in Baddeley (2000). In particular the author focuses on invariance, consistency, asymptotic normality and optimality. Unfortunately little can be said with high level of generality on these issues. In Mira and Baddeley (2001) an attempt is made to relate efficiency of the auxiliary Markov process (measured by Peskun order) with efficiency of the resulting time-invariance estimators (measured by Godambe-Heyde variance optimality criterion).

For spatial processes, Kurtz and Li (2004) obtained a law of large numbers and a central limit theorem for time-invariance estimators in the large domain limit (i.e. as the spatial domain increases to R^k).

As Baddeley (2000) points out 'the time-invariance approach can perhaps best be regarded as a useful way of generating a variety of estimators for further study' with the advantage that the computational complexity of the resulting estimator can be controlled by the choice of the selected underlying Markov process . We can now add that time-invariance EE offer a unifying framework where Bayesian and frequentist estimators can be embedded and possibly compared in terms of their performance. Further study is needed in this direction.

REFERENCES

Arratia, R., Goldstein, L. and Gordon, L. (1990). Poisson approximation and the Chen–Stein method. *Statist. Science* **5**, 403–434.

Baddeley, A. J. (2000). Time-invariance estimating equations. *Bernoulli* **6**, 783–808.

Barbour, A. D. (1997). Stein's method. *Encyclopedia of Statistical Sciences, Update* **1**, 513–521. New York: Wiley.

Besag, J. and Green, P. J. (1993). Spatial statistics and Bayesian computation. *J. Roy. Statist. Soc. B* **55**, 25–37.

Brown, L. D. (1971). Admissible estimators, recurrent diffusions, and insoluble boundary value problems. *Ann. Math. Statist.* **42**, 855–904.

Eaton, M. L. (1992). A statistical diptych: admissible inferences—recurrence of symmetric Markov chains. *Ann. Statist.* **20**, 1147–1179.

Eaton, M. L. (1997). Admissibility in quadratically regular problems and recurrence of symmetric Markov chains: Why the connection? *J. Statist. Planning and Inference* **64**, 231–247.

Eaton, M. L. (2001). Markov chain conditions for admissibility in estimation problems with quadratic loss. *State of the Art in Probability and Statistics: Festschrift for Willem R. van Zwet*, (M. Gunst, C. Klaassen, and A. Vaart, eds.) Hayward, CA: IMS, 223–243.

Edwards, R. G. and Sokal, A. D. (1988). Generalization of the Fortium–Kasteleyn–Swendsen–Wang representation and Monte Carlo algorithm. *Phys. Rev. D* **38**, 2009–2012.

Hobert, J. P. and Robert,C. P. (1999). Eaton's Markov chain, its conjugate partner and P-admissibility. *Ann. Statist.* **27**, 361–373.

Johnstone, I. (1984). Admissibility, difference equations, and recurrence in estimating a Poisson mean. *Ann. Statist.* **12**, 1173–1198.

Johnstone, I. (1986). Admissible estimation, Dirichlet principles, and recurrence of birth-death chains on Z_t^p. *Probab. Theory and Rel. Fields* **71**, 231–269.

Kurtz T.G. and Li S. (2004). Time-invariance modeling and estimation for spatial point processes: General theory. *Tech. Rep.*, Univ. of Wisconsin, Madison, USA.

Mira, A. and Geyer, C. J. (1998). Ordering Monte Carlo Markov chains. *Tech. Rep.*, School of Statistics, Univ. of Minnesota.

Mira, A. (2001). Ordering and improving the performance of Monte Carlo Markov chains. *Statist. Science* **16**, 340–350.

Mira A. and Baddeley A. (2001). Performance of time-invariance estimators. *Tech. Rep.*, University of Western Australia, 2001/15.

Stein, C. (1986). *Approximate computation of expectations.* Hayward, CA: IMS.

Swendsen, R. H. and Wang, J. S. (1987). Non-universal critical dynamics in Monte Carlo simulations. *Phys. Rev. Lett.* **58**, 86–88.

DISCUSSION

RICHARD L. SMITH (*University of North Carolina, U.S.A.*)

This very impressive paper brings together an enormous number of ideas from seemingly disconnected fields. There are frequentist ideas associated with estimating equations for general stochastic processes, and approaches to optimality associated with the names of Godambe and Heyde; connections with the method of maximum likelihood and the method of moments, also pseudolikelihood, general Bayes estimators and a new concept called the pseudo-posterior; connections with admissibility theory; connections with ideas of ordering Markov chains in terms of their rates of convergence in different metrics; and there is even a passing reference to the Stein (or Stein–Chen) method of probability approximation, which is usually regarded as a means of obtaining explicit upper bounds in rate-of-convergence problems.

With so many ideas interacting, it is difficult to identify any one specific theme to concentrate on for a discussion. In my discussion, I would like to give my own perspective on what is going on. I hope that by so doing, I can persuade other to read the paper and form their own perspective of what are the most important themes.

Connection with Baddeley (2000). It is important to understand Baddeley's paper to set the present work in context. Baddeley's formulation was a classical one: we assume a random variable X from a density $f(x \mid \theta)$ where x or X lies in some sample space $\mathcal{X}$, and θ lies in a parameter space Θ. However we are thinking in terms of general stochastic models for which an explicit representation of f may be hard to compute.

Baddeley assumes that for each θ it is possible to construct a Markov chain $\{X_n,\ n = 0, 1, 2, ...\}$ for which $f(\cdot \mid \theta)$ is the stationary distribution. Assume it has a generator $\mathcal{A}_\theta S(x) = \mathrm{E}\{S(X_{n+1}) - S(X_n) \mid X_n = x\}$. Then for $X \sim f(\cdot \mid \theta)$,

$$\mathrm{E}_\theta\{\mathcal{A}_\theta S(X)\} \quad = \quad 0$$

for any sensible function $S(X)$.

It follows that setting

$$\mathcal{A}_\theta S(X) \quad = \quad 0$$

defines an *unbiased estimating equation for* θ in considerable generality. This approach is likely to be particularly appealing in cases where the exact formula for $f(\cdot \mid \theta)$ is intractable but it is relatively easy to define a Markov chain for which f is the stationary distribution — precisely the situation from which most MCMC methods begin. Thus although the viewpoint in Baddeley's paper is purely frequentist, it already contains a number of elements that are familiar to Bayesians who use MCMC.

First example. Suppose $X = \{X_i,\ i \in \mathcal{I}\}$ is a Markov random field with density of the form $f(x \mid \theta) \propto \exp\{\theta V(x)\}$ where the normalizing constant is intractable.

Besag (1975) proposed to estimate θ by maximizing

$$\prod_{i \in \mathcal{I}} f(x_i \mid x_{-i}, \theta) \tag{20}$$

where x_{-i} denotes the set of all elements in x excluding x_i. Equation (20) is called the *pseudolikelihood* and the resulting estimator is the MPLE.

Baddeley showed that this is the time-invariance estimating equation for this model when the embedding Markov chain is Gibbs sampling.

Thus in this case a well-known estimator (the MPLE) has been derived as a time-invariance estimating equation. This suggests that the time-invariance estimating principle may be used to derive alternative estimators for a Markov random field or for more complicated stochastic models where suitable estimators either do not exist at all, or have been previously derived only through *ad hoc* arguments.

Second example. Consider a stationary point process on a compact subset of $\mathbb{R}^d$ that has a density $f(x \mid \theta)$ defined for all possible realizations x with respect to a unit Poisson process. For this process, Besag *et al.* (1982), Jensen and Møller (1991) defined the pseudolikelihood function by extending the definition for Markov random fields.

However for this process there is another estimator called the Takacs–Fiksel estimator which depends on an arbitrary function h defined on the sample space $\mathcal{X}$. So this raises the question of what relationship exists between the two estimators.

Baddeley answered the question as follows: It is possible to define a spatial birth-death process of which X is the stationary distribution. If f is an exponential family and $S(x) = V(x)$ (the canonical statistic) then the time-invariance estimator is MPLE. On the other hand if $S = h$ (apparently we don't need an exponential family assumption in this case) then the time-invariance estimator is the Takacs–Fiksel estimator.

I believe this discussion highlights some of the major motivations for the approach. In some problems there are estimators (such as Takacs–Fiksel) that don't appear to be motivated by any likelihood-based approach. Time-invariance estimating equations provide a unifying perspective that allows such estimators to be related to maximum likelihood and pseudolikelihood. There are also instances where the time-invariance approach leads to new estimators that have not been studied previously. Baddeley provides several other examples and makes a start on discussing properties such as consistency and asymptotic optimality.

New approaches to estimation. Given that background, what's new in the present paper?

I believe one can highlight two essentially new ideas:

- Type II estimating equations: Choose an antisymmetric function $T(x, y) = -T(y, x)$ and define $(\mathcal{F}_\theta T)(x) = \mathrm{E}_\theta\{T(X_n, X_{n+1}) \mid X_n = x\}$.

  Then for a reversible Markov chain, $\mathrm{E}_\theta\{(\mathcal{F}_\theta T)(X)\} = 0$. Hence setting $(\mathcal{F}_\theta T)(X) = 0$ defines an unbiased estimating equation for θ.

  If $T(x, y) = S(x) - S(y)$, this reduces to the Type I estimating equation.

  This extension is especially useful in cases where the embedding Markov chain is Hastings-Metropolis, because the acceptance probability for Hastings-Metropolis is a complicated function of both the existing and proposed new states, that cannot be reduced to a simple difference of two functions. However, there are a number of instances when an appropriate estimating equation can be expressed as a Type II EE.

- The second extension is to a Bayesian EE approach, in which we embed the *joint* distribution of X and θ in a Markov chain on $(\mathcal{X}, \Theta)$ — many examples of this lead to standard Bayes estimators.

As several examples in the paper show, these ideas significantly extend Baddeley's original approach. In particular, the Bayesian extension allows many Bayes estimators to be expressed as time-invariance estimators.

Connection with admissibility theory. Brown (1971) showed that it is possible to characterize admissibility or inadmissibility of estimators of a multivariate normal mean in terms of recurrence or transience of an associated diffusion process.

Johnstone (1984, 1986) developed a similar characterization for the estimation of Poisson means, where in his case the associated Markov chain was a birth and death process.

Although these are celebrated papers, they are also highly technical and difficult to understand in even an intuitive way. On the other hand, the approach of Eaton (1982, 1992, 1997) is a less powerful but far more straightforward theory for determining when the generalized Bayes estimator derived from an improper prior ν on Θ is almost-ν-admissible, a slightly weaker concept than traditional admissibility. A simplified version of Eaton's recipe is as follows:

(i) For $\theta,\ \eta \in \Theta$, define

$$r(\theta \mid \eta) \quad = \quad \int_{\mathcal{X}} q(\theta \mid x) p(x \mid \eta) dx$$

where $p(x \mid \eta)$ is the likelihood and $q(\theta \mid x)$ is the posterior density given $X = x$ when the prior is ν.

(ii) Think of $r(\theta \mid \eta)$ as the transition density of a Markov chain on Θ.

(iii) Under suitable regularity conditions, the recurrence of this Markov chain implies almost-ν-admissibility of the generalized Bayes estimator.

(Eaton also comments on the converse property, i.e. when is it true that transience of the Markov chain leads to inadmissibility of the estimator? Apparently there are no general theorems in this direction, but it nevertheless seems to be assumed that the result is generally true.)

Hobert and Robert (1999) defined an alternative Markov chain through the transition kernel

$$\tilde{r}(y \mid x) = \int_{\Theta} p(y \mid \theta) q(\theta \mid x) d\theta.$$

Note that this defines a Markov chain on $\mathcal{X}$. If we couple the $\mathcal{X}$ and Θ updatings together, we also get a joint Markov chain defined on $\mathcal{X} \times \Theta$.

Hobert and Robert showed that all three Markov chains (including Eaton's) are positive recurrent, null recurrent or transient together. In particular, in some cases it is possible to prove recurrence of the Markov chain on $\mathcal{X}$-space when the corresponding result on Θ-space would not follow from any known stochastic process results — a major motivation and justification for their approach.

The present paper shows how these Markov chains can be used to re-derive a number of known (Bayesian and frequentist) estimators. But it's unclear to what extent it leads to really new estimators. The maximum pseudo-posterior estimator (Section 8.2) is one example that clearly exploits this idea of updating both the $\mathcal{X}$ and Θ spaces in succession, but my impression is that further study will be needed to decide whether this really is a good idea.

Asymptotic properties of estimators. The approach of this paper offers potentially a large number of estimators for a given stochastic model. Whether the estimator is derived from the original estimating equation approach of Baddeley (2000), or from one of the more Bayes-oriented schemes of the present paper, it is still natural to use frequentist properties such as asymptotic variance as a means of discriminating among different point estimators. Sections 4 and 9 of the paper refer to attempts to relate asymptotic properties of the estimators to order properties of the generating Markov chains; I would like to make some comments about that and to propose a small extension to one of the results of Mira and Baddeley (2001).

If we consider the estimator $\tilde{\theta}$ defined by solving $(\mathcal{A}_\theta S)(X) = 0$, then the Godambe–Heyde formula leads to the approximation

$$\mathrm{Var}(\tilde{\theta}) \approx \mathrm{E}\left\{\nabla_\theta(\mathcal{A}_\theta S)(X)\right\}^{-T} \mathrm{Cov}\{(\mathcal{A}_\theta S)(X)\} \mathrm{E}\left\{\nabla_\theta(\mathcal{A}_\theta S)(X)\right\}^{-1}.$$

This has numerous alternative names, including the 'information sandwich formula" (Freedman, 2006).

If Θ is one-dimensional, the formula reduces to

$$\mathrm{Var}(\tilde{\theta}) \approx \frac{\mathrm{Var}\{(\mathcal{A}_\theta S)(X)\}}{\left[\mathrm{E}\left\{\frac{\partial(\mathcal{A}_\theta S)(X)}{\partial\theta}\right\}\right]^2}. \tag{21}$$

Suppose now that $\{Y_n\}$ and $\{Z_n\}$ are two reversible Markov chains with the same stationary distribution π_θ. With obvious notation, we also let $\mathcal{A}_{Y,\theta}$ and $\mathcal{A}_{Z,\theta}$ denote the generators indexed by θ, and $\tilde{\theta}_Y$, $\tilde{\theta}_Z$ the resulting time-invariance estimators. If Y_n dominates Z_n in *covariance ordering* (Mira 2001), one of the consequences is

$$\mathrm{E}_\theta\{S(X)(\mathcal{A}_{Y,\theta} S)(X)\} \leq \mathrm{E}_\theta\{S(X)(\mathcal{A}_{Z,\theta} S)(X)\} \leq 0 \tag{22}$$

for any $S \in L_0^2(\pi)$, the class of square integrable functions having zero mean with respect to π (Mira and Baddeley, 2001).

Consider the case $S(x) = \frac{\partial}{\partial\theta} \log f(x;\theta)$. Then

$$\mathrm{E}_\theta\{S(X)(\mathcal{A}_{Y,\theta}S)(X)\} \quad = \quad -\mathrm{E}_\theta\left\{\frac{\partial}{\partial\theta}(\mathcal{A}_{Y,\theta}S)(X)\right\} \cdot \tag{23}$$

To see (23), we merely need to note that for a statistic $T(x,\theta)$ which is uniformly differentiable in θ,

$$\frac{\partial}{\partial\theta}E_\theta\left\{T(X,\theta)\right\} \quad = \quad E_\theta\left\{T(X,\theta)\frac{\partial \log f(X;\theta)}{\partial\theta}\right\} + E_\theta\left\{\frac{\partial T(X,\theta)}{\partial\theta}\right\}. \tag{24}$$

However for $T(x,\theta) = \mathcal{A}_\theta(S)(x)$, the left hand side of (24) is 0, and we then deduce (23).

By combining (21), (22) and (23), we deduce the following:

Proposition. If

(i) $\{Y_n\}$ dominates $\{Z_n\}$ in covariance ordering, for each θ, and

(ii) $\mathrm{Var}\{(\mathcal{A}_{Y,\theta}S)(X)\} \leq \mathrm{Var}\{(\mathcal{A}_{Z,\theta}S)(X)\}$,

then (under uniform differentiability conditions) $\tilde{\theta}_Y$ is more efficient than $\tilde{\theta}_Z$ in the Godambe-Heyde sense.

This result differs from that in Mira and Baddeley (2001) only in that they assumed an exponential family, and that assumption seems to me unnecessary. This distinction could be important for extending the results to other kinds of spatial processes (such as those that arise in geostatistics) where an exponential family assumption would be unduly restrictive.

Nevertheless, this is still a limited result, since as shown in several examples by Mira and Baddeley (2001), assumption (ii) cannot be dispensed with. It seems that quite a bit more work is needed to understand exactly how properties of the embedded Markov chain translate to those of the time-invariance estimator.

Conclusions. This is a very stimulating paper that brings together numerous Bayesian and frequentist concepts and provides a very general perspective on estimation of stochastic processes. I congratulate the authors, and look forward to seeing further developments of their work.

FREEDOM GUMEDZE (*University of Cape Town, South Africa*)

The authors present a time-invariance estimating equations-approach to generating a plethora of estimators. Their estimators are equivalent to known estimators such as Bayes, maximum likelihood and method of moment estimators and therefore do not seem to be anything new. I would like to know in which situations are their estimators applicable. I look forward to the results of the comparative studies on the performance of the new estimators mentioned in the discussion of the paper.

JAMES P. HOBERT (*University of Florida, USA*) and
CHRISTIAN P. ROBERT (*Université Paris Dauphine and CREST-INSEE, France*)

We are obviously pleased to see a renewed interest in the use of Markov chains for deriving estimators and their properties. Before reading the paper, we were actually expecting more optimality results than a general construction principle and we find the output quite interesting from a conceptual point of view.

Evaluating estimators or prior distributions?. The authors describe the results in Eaton (1992) as a method of evaluating estimators. For example, the authors write:

> 'The first one ('admissibility if and only if recurrence') starts with a given estimator, constructs an auxiliary Markov process, and relates the performance of the estimator to properties of the Markov process.'

While Eaton's (1992) results can certainly be used to establish that a particular estimator is admissible, they are perhaps more accurately viewed as a way of evaluating prior distributions. Indeed, the precursor to Eaton (1992) is Eaton (1982), the title of which is 'A Method for Evaluating Improper Prior Distributions.' We now explain this alternative perspective.

Suppose we are to observe X from the statistical model $P(dx|\theta)$ where $\theta \in \Theta$ is an unknown parameter. Suppose further that we desire an estimator of $g(\theta)$ where $g : \Theta \to \mathbb{R}$ is bounded. For an extension that can handle unbounded functions of θ, see Eaton (2001). Let $\nu(d\theta)$ be an improper prior measure on θ; i.e., $\nu(\Theta) = \infty$, and assume that the marginal measure, defined as

$$M(dx) = \int_\Theta P(dx|\theta)\nu(d\theta) ,$$

is a σ-finite measure on $\mathcal{X}$. In this case, there exists a formal posterior distribution $Q(d\theta|x)$ which satisfies

$$Q(d\theta|x)M(dx) = P(dx|\theta)\nu(d\theta) .$$

This equality means that the two measures on $\mathcal{X} \times \Theta$ are equivalent. The formal Bayes estimator of $g(\theta)$ under squared error loss is, of course,

$$\hat{g}(x) = \int_\Theta g(\theta)Q(d\theta|x) .$$

Eaton's (1992) results imply that $\hat{g}$ is (almost-ν-) admissible if the Markov chain on Θ with Markov transition function

$$R(d\theta|\eta) = \int_\mathcal{X} Q(d\theta|x)P(dx|\eta)$$

is (locally-ν-) recurrent. This is the Markov transition function (Mtf) discussed in Section 7.1 of Mira and Baddeley. (Mira and Baddeley suggest in the quotation above that recurrence is actually a necessary and sufficient condition for admissibility, but this has not been established.) Note that R depends only on the model $P(dx|\theta)$ and the prior $\nu(d\theta)$ and does not involve the function g. Consequently, if the Markov chain is recurrent, then the formal Bayes estimator of *every* bounded function of θ is admissible. Thus, recurrence of the Markov chain implies that the formal Bayes estimators (of bounded functions) derived using the prior ν have good frequentist properties. One might therefore endorse ν as a good 'all purpose' prior to use in conjunction with $P(dx|\theta)$. The argument for using probability matching priors as default priors is based on similar reasoning.

Hobert and Robert (1999) and Eaton *et al.* (2007) have shown that (under mild regularity conditions) the Markov chain driven by R is (locally-ν-) recurrent if and only if the Markov chain on $\mathcal{X}$ with Mtf

$$\tilde{R}(dx|y) = \int_{\Theta} P(dx|\theta)Q(d\theta|y)$$

is (locally-M-) recurrent. This is the Mtf discussed in Section 7.2 of Mira and Baddeley. The correspondence between R and $\tilde{R}$ provides a theoretical connection between the results of Eaton (1992) and those of Brown (1971) and Johnstone (1984, 1986), who established relationships between admissibility of estimators and recurrence of Markov processes on the *sample space*, $\mathcal{X}$. Moreover, the relationship between R and $\tilde{R}$ has turned out to be quite useful from a practical standpoint as it is often much easier to analyze $\tilde{R}$ than R; see, *e.g.*, Hobert and Robert (1999), Hobert and Schweinsberg (2002), Hobert *et al.* (2004).

MCMC connections and computations. In addition to using Markov chains to establish good properties of priors or of estimators or even to derive new estimators which is what is done in the paper, we could think of using estimating equations in convergence assessment for Markov chain Monte Carlo algorithms. If there exists one or several functions H such that eqn. (1) holds, we can expect that the empirical average of the $H(X_n)$'s is close to 0 when convergence is achieved (i.e. when the chain is in its stationary regime). Therefore, any non-trivial H that satisfies eqn. (1) can be used as a score in the spirit of Brooks and Gelman (1998a,b), following the use of *control variates* in Monte Carlo simulation (Rubinstein, 1981). This perspective is obviously weakly related to the theme of the paper but could open new avenues. However, it may be an empty proposal if we cannot find other functions H than those equal to $\mathcal{A}S$ as in eqn. (3).

Solving [in practice] an EE is another issue not mentioned in the paper. Since Bayes estimators and MLEs are particular examples of EE's, we assume there is no particular computational advantage in using this representation. (Or are we wrong?)

Busy like a BEE? The Bayesian perspective is of course present of the paper but one may wonder if the Bayesian Estimating Equation (BEE) perspective is really useful from an operational point of view. In other words, saying that the posterior expectation is solution of eqn. (13) is not helpful for the derivation of this estimator, even though it can bring new perspectives on its optimality properties.

As pointed out by the authors at the end of the talk, it would be helpful to relate the BEE principle with more traditional decision-theoretic approaches like loss functions. In particular, the approach of Eaton (1992) is clearly linked with a specific loss function.

REPLY TO THE DISCUSSION

Discussion by R. L. Smith. We are grateful to Prof. Smith for giving his illuminating prospective on the topic, pointing out connections with the recent literature and highlighting the new contributions of the paper. Furthermore in his discussion Prof. Smith partially extends one of the results presented during the conference, by removing one of the assumptions. The original result (that appears in a technical report by Mira and Baddeley, 2001) was restricted to the exponential family setting, this assumption has been relaxed and this will allow for wider applicability.

Discussion by F. Gumedze. It is true that some of the estimators derived from the time-invariance framework are equivalent to well known estimators, both in the classical and in the Bayesian paradigm. This is indeed one of the strength of the approach in that a unifying framework is proposed that will allow a comparison in terms of performance of the resulting estimators. On the other hand, it is also true that new estimators are derived from the suggested approach, by properly selecting the underlying Markov process and the function S. Just to give a few examples we recall the maximum pseudo-posterior estimator that mimics, in the Bayesian setting, the idea of the maximum pseudolikelihood estimator. In the frequentist setting, several new estimators for parameters were derived by time-invariance in Baddeley (2000). The ability to derive new estimators from the time-invariance framework is recognized by the other two discussions where we read 'In addition to using Markov chains to establish good properties of priors or of estimators or even to derive new estimators, which is what is done in the paper, ...' (Hobert and Robert) and 'There are also instances where the time-invariance approach leads to new estimators that have not been studied previously' (Smith).

Discussion by J. P. Hobert and C. P. Robert. We thank the Authors for the interesting comments, in particular the search of functions H, other than $\mathcal{A}S$, satisfying (1) for convergence diagnostic purposes in MCMC is valuable and solutions to the Poisson equation could be a good source for such functions.

In the frequentist setting, we believe that the idea of deriving estimating equations from the generator of an MCMC sampler originates with Baddeley (2000). We value the parallel interpretation, highlighted by the discussants, of Eaton's results (1992) in terms of evaluation of prior distributions, even if, in the joint Bayesian-frequentist framework we propose, this alternative interpretation has minor added value. It would be much more interesting the possibility to relate both the Bayesian and the frequentist estimating equation framework to the loss function and, more generally, to the decision theory approach. More research along this direction could potentially have high impact.

ADDITIONAL REFERENCES IN THE DISCUSSION

Besag, J.E. (1975), Statistical analysis of non-lattice data. *The Statistician* **24**, 179–195.

Besag, J., Milne, R. and Zachary, S. (1982), Point process limits of lattice processes. *Journal of Applied Probability* **19**, 210–216.

Brooks, S. P. and Gelman, A. (1998a). General methods for monitoring convergence of iterative simulations. *J. Comp. Graphical Statist.* **7**, 434–455,

Brooks, S. P. and Gelman, A. (1998b). Some issues in monitoring convergence of iterative simulations. *Tech. Rep.*, Univiversity of Bristol.

Brown, L. D. (1971). Admissible estimators, recurrent diffusions, and insoluble boundary value problems. *Ann. Math. Statist.* **42**, 855–904.

Eaton, M. L. (1982). A method for evaluating improper prior distributions. *Statistical Decision Theory and Related Topics III* **1** (S. S. Gupta and J. O. Berger, eds.) New York: Academic Press, New York: Academic Press

Eaton, M. L., Hobert, J. P. and Jones, G. L. (2007). On perturbations of strongly admissible prior distributions. *Annales de l'Institut Henri Poincaré, Probabilités et Statistiques*, (to appear).
Freedman, D. A. (2006). On the so-called 'Huber sandwich estimator" and 'Robust standard errors". *Amer. Statist.* **60**, 299–302.
Hobert, J. P. and Schweinsberg, J. (2002). Conditions for recurrence and transience of a Markov chain on Z^+ and estimation of a geometric success probability. *Ann. Statist.* **30**, 1214–1223.
Hobert, J. P., Marchev, D. and Schweinsberg, J. (2004). Stability of the tail Markov chain and the evaluation of improper priors for an exponential rate parameter. *Bernoulli* **10**, 549–564.
Jensen, J. L. and Møller, J. (1991). Pseudolikelihood for exponential family models of spatial point processes. *Annals of Applied Probability* **1**, 445–461.
Rubinstein, R. Y. (1981). *Simulation and the Monte Carlo Method.* New York: Wiley.

BAYESIAN STATISTICS 8, pp. 349–370.
J. M. Bernardo, M. J. Bayarri, J. O. Berger, A. P. Dawid,
D. Heckerman, A. F. M. Smith and M. West (Eds.)

FDR and Bayesian Multiple Comparisons Rules

PETER MÜLLER
M.D. Anderson Cancer Center, USA
pmueller@mdanderson.org

GIOVANNI PARMIGIANI
Johns Hopkins University, USA
gp@jhu.edu

KENNETH RICE
Univ. of Washington, Seattle, USA
kenrice@u.washington.edu

SUMMARY

We discuss Bayesian approaches to multiple comparison problems, using a decision theoretic perspective to critically compare competing approaches. We set up decision problems that lead to the use of FDR-based rules and generalizations. Alternative definitions of the probability model and the utility function lead to different rules and problem-specific adjustments. Using a loss function that controls realized FDR we derive an optimal Bayes rule that is a variation of the Benjamini and Hochberg (1995) procedure. The cutoff is based on increments in ordered posterior probabilities instead of ordered p-values. Throughout the discussion we take a Bayesian perspective. In particular, we focus on conditional expected FDR, conditional on the data. Variations of the probability model include explicit modeling for dependence. Variations of the utility function include weighting by the extent of a true negative and accounting for the impact in the final decision.

Keywords and Phrases: DECISION PROBLEMS; MULTIPLICITIES; FALSE DISCOVERY RATE.

1. INTRODUCTION

We discuss Bayesian approaches to multiple comparison problems, using a Bayesian decision theoretic perspective to critically compare competing approaches. Multiple comparison problems arise in a wide variety of research areas. Many recent discussions are specific to massive multiple comparisons arising in the analysis of high throughput gene expression data. See, for example, Storey *et al.* (2004) and references therein. The basic setup is a large set of comparisons. Let r_i denote the unknown truth in the i-th comparison, $r_i = 0$ (H_0) versus $r_i = 1$ (H_1), $i = 1, \ldots, n$. In the context of gene expression data a typical setup defines r_i as an indicator for gene i being differentially expressed under two biologic conditions of interest. For each gene a suitably defined difference score z_i is observed, with $z_i \sim f_0(z_i)$ if $r_i = 0$, and $z_i \sim f_1(z_i)$ if $r_i = 1$. This is the basic setup of the discussions in Benjamini and Hochberg (1995); Efron *et al.* (2001); Storey (2002); Efron and Tibshirani (2002); Genovese and Wasserman (2002, 2003); Storey *et al.* (2004); Newton *et al.* (2004); Cohen and Sackrowitz (2005) and many others. A traditional approach to address the multiple comparison problem in these applications is based on controlling false discovery rates (FDR), the proportion of false rejections relative to the total number of rejections. We discuss details below. A similar setup arises in the analysis of high throughput protein expression data, for example, mass/charge spectra from MALDI-TOF experiments, as described in Baggerly *et al.* (2003).

Many other applications lead to similar massive multiple comparison problems. Clinical trials usually record data for an extensive list of adverse events (AE). Comparing treatments on the basis of AEs takes the form of a massive multiple comparison problem. Berry and Berry (2004) argue that the hierarchical nature of AEs, with AEs grouped into biologically different body systems, is critical for an appropriate analysis of the problem. Another interesting application of multiple comparison and FDR is in classifying regions in image data. Genovese *et al.* (2002) propose an FDR-based method for threshold selection in neuroimaging. Shen *et al.* (2002) propose an enhanced procedure that takes into account spatial dependence, specifically in a wavelet based spatial model. Another traditional application of multiple comparisons arises in record linkage problems. Consider two data sets, A and B, for example billing data and clinical data in a clinical trial. The record matching problem refers to the question of matching data records in A and B corresponding to the same person. Consider a partition of all possible pairs of data records in A and B into matches versus non-matches. A traditional summary of a given partition is the Correct Match Rate (CMR), defined as the fraction of correctly guessed matches relative to the number of true matches. See, for example, Fortini *et al.* (2001, 2002). Another interesting related class of problems are ranking and selection problems. Lin *et al.* (2004) describe the problem of constructing league tables, i.e., reporting inference on ranking a set of units (hospitals, schools, etc.). Lin et al. explicitly acknowledge the nature of the multiple comparison as a decision problem and discuss solutions under several alternative loss functions.

To simplify the remaining discussion we will assume that the multiple comparison problem arises in a microarray group comparison experiment, keeping in mind that the discussion remains valid for many other massive multiple comparison. A microarray group comparison experiment records gene expression for a large number of genes, $i = 1, \ldots, n$, under two biologic conditions of interest, for example tumor tissue and normal tissue. For each gene we are interested in the comparison of the two competing hypotheses that gene i is differentially expressed versus not differentially expressed. We will refer to a decision to report a gene as differentially expressed as a discovery (or positive, or rejection), and the opposite as a negative (fail to reject).

2. FALSE DISCOVERY RATES

Many recently proposed approaches to address massively multiple comparisons are based on controlling false discovery rates (FDR), introduced by Benjamini and Hochberg (1995). Let δ_i denote an indicator for rejecting the i-th comparison, for example flagging gene i as differentially expressed and let $D = \sum \delta_i$ denote the number of rejections. Let $r_i \in \{0, 1\}$ denote the unknown truth, for example an indicator for true differential expression of gene i. We define $\text{FDR} = (\sum(1 - r_i)\delta_i)/D$ as the fraction of false rejections, relative to the total number of rejections. The ratio defines a summary of the parameters (r_i), the decisions (δ_i) and the data (indirectly, through the decisions). As such it is neither Bayesian nor frequentist. How we proceed to estimate and/or control it depends on the chosen paradigm. Traditionally one considers the (frequentist) expectation $E(\text{FDR})$, taking an expectation over repeated experiments. This is the definition used in Benjamini and Hochberg (1995). Applications of FDR to microarray analysis are discussed, among many others, in Efron and Tibshirani (2002). Storey (2002, 2003) introduces the positive FDR (pFDR) and the q-value and improved estimators for the FDR. In the pFDR the expectation is defined conditional on $D > 0$. Efron and Tibshirani (2002) show the connection between FDR and the empirical Bayes procedure proposed in Efron *et al.* (2001) and the FDR control as introduced in Benjamini and Hochberg (1995). Genovese and Wasserman (2003) discuss more cases of Bayes-frequentist agreement in controlling various aspects of FDR. Let p_i denote a p-value for testing $r_i = 1$ versus $r_i = 0$. They consider rules of the type $\delta_i = I(p_i < t)$. Controlling the posterior probability $P(r_i = 0 \mid Y, p_i = t)$ is stronger than controlling the expected FDR for a threshold t. Specifically, let FDR(t) denote FDR under the rule $\delta_i = (p_i < t)$, let $\hat{q}(t) \approx P(r_i = 0 \mid Y, p_i = t)$ denote an approximate evaluation of the posterior probability for the $i-$th comparison, and let $Q(t) \approx E(\text{FDR}(t))$ denote an asymptotic approximation of expected FDR. Then $q(t) \leq Q(t)$. The argument assumes concavity of the c.d.f. for p-values under the alternative $r_i = 1$. Genovese and Wasserman (2002) also show an agreement of confidence intervals and credible sets for FDR. They define the realized FDR process FDR(T) as a function of the threshold T and call T a (c, α) confidence threshold if $P(\text{FDR}(T) < c) \geq 1 - \alpha$. The probability is with respect to a sampling model that includes an (unknown) mixture of true and false hypotheses. The Bayesian equivalent is a posterior credible set, i.e., controlling $P(\text{FDR}(T) \leq c \mid Y) \leq 1 - \alpha$. Genovese and Wasserman (2003) show that controlling the posterior credible interval for FDR(T) is asymptotically equivalent to controlling the confidence threshold.

Let z_i denote some univariate summary statistic for the i-th comparison, for example a p-value. Many discussions are in the context of an assumed i.i.d. sampling model for z_i, from a mixture model $f(\cdot)$ with terms f_0 and f_1 corresponding to subpopulations of differentially and not-differentially expressed genes, respectively:

$$z_i \sim p_0\, f_0(z_i) + (1 - p_0)\, f_1(z_i) \equiv f(z_i).$$

Using latent indicators $r_i \in \{0, 1\}$ introduced earlier, the mixture is equivalent to the hierarchical model:

$$p(z_i \mid r_i = j) = f_j(z_i) \text{ and } Pr(r_i = 0) = p_0. \tag{1}$$

Let F_0 and F denote the c.d.f. for f_0 and f. Efron and Tibshirani (2002) define FDR for rejection regions of the type $\{z_i \leq z\}$,

$$\text{Fdr}(z) \equiv p_0 F_0(z)/F(z)$$

and denote it 'Bayesian FDR'. The Bayesian label is justified by the use of Bayes theorem to find the probability of false discovery given $\{z_i \leq z\}$, which they show is equivalent to the defined Fdr statistic. The probability statement is in the context of the assumed mixture model, for assumed known f_0, f_1 and p_0. In particular, there is no learning about p_0. However, using reasonable data-driven point estimates for the unknown quantities f_0, f_1 and p_0, the Fdr statistic provides an good approximation for what $P(r_{n+1} = 1 \mid z_{n+1} \leq z, Y)$ would be in a full Bayesian model with flexible priors on f_1 and f_0. Here and throughout this paper we use Y to generically indicate the observed data. Efron *et al.* (2001) introduce local FDR, as

$$\text{fdr}(z) \equiv p_0 f_0(z)/f(z).$$

Under the mixture model, and conditioning on f_0, f_1 and p_0, the fdr statistic is the probability of differential expression, so $\text{fdr}(z) = Pr(r_i = 1 \mid z_i = z, Y, f_0, f_1, p_0)$. As before, one can argue that under a sufficiently flexible prior probability model on f_0, f_1, p_0, reasonable point estimates can be substituted for the unknown quantities, allowing us to interpret values of fdr as posterior probabilities, without reference to a specific prior model (subject to identifiability constraints). In the following sections we argue that posterior inference for the multiple comparison should consider more structured models. Inference should not stop with the marginal posterior probability of differential expression. Rules of the type $\delta_i = I(z_i \leq z)$ are intuitive, but not necessarily optimal. In the context of a full probability model, and assuming a reasonable utility function, it is straightforward to derive the optimal rule.

Considering frequentist expectations of FDR, i.e., expectations over repeated sampling, we need to consider expectations over a ratio of random variables. Short of uninteresting trivial decision rules, the decision $\delta_i = \delta_i(Y)$ is a function of the data and appears in both, numerator and denominator of the ratio. The discussion significantly simplifies under a Bayesian perspective. The only unknown quantity in $\text{FDR} = \sum \delta_i(1 - r_i)/D$ is the unknown r_i in the numerator. Let $v_i = P(r_i = 1 \mid Y)$ denote the marginal posterior probability of gene i being differentially expressed and define

$$\overline{\text{FDR}} = E(\text{FDR} \mid Y) = \sum(1 - v_i)\delta_i/D. \tag{2}$$

Newton *et al.* (2004) consider decision rules that classify a gene as differentially expressed if $v_i > \gamma^*$, fixing γ^* to achieve a certain pre-set false discovery rate, $\text{FDR} \leq \alpha$. Newton *et al.* (2004) comment on the dual role of v_i in decision rules like $\delta_i = I(v_i > \gamma^*)$. It determines the decision, and at the same time already reports the probability of a false discovery as $1 - v_i$ for $\delta_i = 1$ and the probability of a false negative as v_i for $\delta_i = 0$.

3. POSTERIOR PROBABILITIES ADJUST FOR MULTIPLICITIES

'Posterior inference adjusts for multiplicities, and no further adjustment is required.' The statement is only true with several caveats. First, the probability model needs to include a positive prior probability of non-differential expression for each gene i. Second, the model needs to include a hyperparameter that defines the prior probability mass for non-differential expression. For example, consider the mixture model (1), with independence across i, conditional on p_0. The statement requires that p_0 be a parameter with a hyperprior $p(p_0)$, rather than fixed. Scott and Berger (2003) discuss the nature of this adjustment and show some examples. In the context of microarray data analysis, Do *et al.* (2005) carry out the same simulation experiment in the context of a mixture model as in (1). Results are shown in Table 1. The table shows marginal posterior probabilities v_i for differential expression. The nature of the model is such that v_i depends on the gene only through an observed difference score z_i, making it meaningful to list v_i by observed difference score z_i. The marginal posterior probability of differential expression adjusts for the multiplicities. If there are many truly negative comparisons, as in the third row of the table, then the model reduces the marginal probabilities of differential expression. If on the other hand there are many truly positive comparisons, as in the first row, then the model appropriately increases the marginal probabilities.

Table 1: *Posterior probabilities of differential expression, as a function of the observed difference score z_i, under three different simulation truths, using $p_0 = 0.4$ (first row), 0.8 (second row) and 0.9 (third row) for the proportion of false comparisons. Probabilities $v_i > 0.4$ are marked in bold face.*

	Observed z scores								
p_0	−5.0	−4.0	−3.0	−2.0	−1.0	0.0	1.0	2.0	3.0
0.4	**1.0**	**1.0**	**1.0**	**0.9**	**0.5**	0.2	**0.4**	**0.9**	**1.0**
0.8	**0.9**	**0.9**	**0.8**	**0.4**	0.1	0.1	0.1	**0.4**	**0.8**
0.9	**0.5**	0.4	0.3	0.1	0.1	0.0	0.0	0.1	0.3

The probability model need not be i.i.d. sampling. Any probability model that includes a positive prior probability for $r_i = 0$ and $r_i = 1$, i.e., any model that allows inference on how comparisons between units are true or false leads to a similar adjustment. Berry and Hochberg (1999) discuss this perspective. An interesting probability model is proposed in Gopalan and Berry (1998). They consider the problem of comparing group means in an ANOVA setup. They introduce a prior probability model on all possible partitions of matching group means using the probability model on random partitions that is implied by sampling from a random probability measure with a Dirichlet process prior.

Berry and Berry (2004) discuss inference for adverse events (AE) in clinical trials, proposing a hierarchical model to address the multiplicity issue. The data are occurrences of a large set of adverse events reported in a two arm clinical trial comparing standard versus experimental therapy. The authors argue that the conclusion about a set of AEs with elevated occurrence under the experimental therapy should be different, depending on whether these AEs cluster in the same body system, or are scattered across different body systems. In the latter case it should be considered more likely that the reported AEs are due to random occurrence, whereas in the earlier case it seems more likely that the increased AEs are caused by the drug. Berry and Berry (2004) develop a three-level hierarchical model with levels corresponding to AEs, body systems, and the collection of all body systems. The proposed hierarchical model leads to the desired inference. Due to borrowing of strength in the hierarchical model AEs that cluster in the same body system lead to higher posterior probability of an underlying true difference than if the same AE counts were observed across different body systems.

4. DECISION THEORETIC APPROACHES

In a review of a Bayesian perspective on multiple comparisons Berry and Hochberg (1999) comment that 'finding posterior distributions of parameters is only part of the Bayesian solution. The remainder involves decision analysis.' Computing posterior probabilities of differential expression only estimates parameters in the probability model. It does not yet recommend a specific decision about flagging genes as differentially expressed or not. Reasonable solutions are likely to follow some notion of monotonicity. All else being equal, genes with higher marginal probability of differential expression should be more likely to be reported as differentially expressed. However, differing levels of differential expression, focused interest in some subsets of genes, and inference about dependence might lead to violations of monotonicity. More importantly, this argument, without refinement, does not provide the threshold beyond which comparisons should be rejected.

It can be shown (Müller *et al.*, 2004) that under several loss functions that combine false negative and false discovery counts and/or rates the optimal decision rule is of the following form. Recall that δ_i is an indicator for the decision to report gene i as differentially expressed and $v_i = Pr(r_i = 1 \mid Y)$ denotes the marginal posterior probability of differential expression for gene i. The optimal decision is to declare all genes with marginal probability beyond a threshold as differentially expressed:

$$\delta_i^* = I(v_i > t). \tag{3}$$

The choice of loss function determines the specific threshold. In Müller *et al.* (2004) we consider four alternative loss functions. Similar to FDR we define $\text{FD} = \sum(1 - r_i)\delta_i$ and $\text{FN} = \sum r_i(1-\delta_i)$ as the false positive and false negative counts, and $\text{FNR} = \text{FN}/(n-D)$ as the false negative ratio. We use $\overline{\text{FD}}$, $\overline{\text{FN}}$ and $\overline{\text{FNR}}$ for the posterior expectations. All are easily evaluated. For example, $\overline{\text{FN}} = \sum v_i(1-\delta_i)$. Considering various combinations of these statistics we define alternative loss functions. Since the posterior expectation is straightforward, we specify the loss functions already as posterior expected loss. The first two loss functions are linear combinations of the false negative and positive counts and ratios. We define

$$L_N(\delta, z) = c\,\overline{\text{FD}} + \overline{\text{FN}}, \tag{4}$$

and $L_R(\delta, z) = c\,\overline{\text{FDR}} + \overline{\text{FNR}}$. The loss function L_N is a natural extension of $(0,1,c)$ loss functions for traditional hypothesis testing problems (Lindley, 1971). From this perspective the combination of error rates in L_R seems less attractive. The loss for a false discovery and false negative depends on the total number of discoveries or negatives, respectively. Genovese and Wasserman (2002) interpret c as the Lagrange multiplier in the problem of minimizing FNR subject to a bound on FDR. They compare the Benjamini and Hochberg (1995) rule (BH) and the optimal rule under L_R and show that BH almost achieves the optimal risk, in particular for a large fraction of true nulls.

Alternatively, we consider bivariate loss functions that explicitly acknowledge the two competing goals:

$$L_{2R}(\delta, z) = (\overline{\text{FDR}}, \overline{\text{FNR}}), \quad L_{2N}(\delta, z) = (\overline{\text{FD}}, \overline{\text{FN}}).$$

We need to define the minimization of the bivariate functions. A traditional approach to select an action in multicriteria decision problems is to minimize one dimension of the loss function while enforcing a constraint on the other dimensions (Keeney *et al.*, 1976). We thus define the optimal decisions under L_{2N} as the minimization of $\overline{\text{FN}}$ subject to $\overline{\text{FD}} \le \alpha_N$. Similarly, under L_{2R} we minimize $\overline{\text{FNR}}$ subject to $\overline{\text{FDR}} \le \alpha_R$.

Under all four loss functions the optimal rule is of the form (3). See Müller *et al.* (2004) for a statement of the optimal cutoffs t. The result is true for any probability model with non-zero prior probability for differential and non-differential expression. In particular, the probability model could include dependence across genes.

One of the assumptions underlying these loss functions is that all false negatives are equally undesirable, and all false positives are equally undesirable. This is inappropriate in most applications. A false negative for a gene that is truly differentially expressed, but with a small difference across the two biologic conditions, is surely less of a concern than a false negative for a gene that is differentially expressed with a large difference. The large difference might make it more likely that follow up experiments will lead to significant results. Assume now that the probability model includes for each gene i a parameter m_i that can be interpreted as the level of differential expression, with $m_i = 0$ if $r_i = 0$, and $m_i > 0$ if $r_i = 1$. For example, in the hierarchical gamma/gamma model proposed in Newton *et al.* (2001) this could be the absolute value of the log ratio of the gamma scale parameters that index the sampling distributions under the two biologic conditions. A log ratio of $m_i = 0$ implies equal sampling distributions, i.e., no differential expression. In the mixture of Dirichlet process model of Do *et al.* (2005), m_i would be the absolute value of the latent variable generated from the random probability measure with Dirichlet process prior. A natural extension of the earlier loss functions is to

$$L_m(m, \delta, z) = -\sum \delta_i \, m_i + k \sum (1 - \delta_i) m_i + c\, D.$$

A similar weighting with the relative magnitude of errors is underlying Duncan's (1965) multiple comparison procedure. Since $r_i = 0$ implies $m = 0$, the summations go only over all true positives, $r_i = 1$. The loss function includes a reward proportional to m_i for a correct discovery, and a penalty proportional to m_i for each false negative. The last term encourages parsimony, without which the optimal decision would be to trivially flag all genes. Straightforward algebra shows that the optimal decision is similar to (3). Let $\overline{m}_i = E(m_i \mid Y)$ denote the posterior expected level of differential expression for gene i. The optimal rule is

$$\delta_i^* = I\{m_i \geq c/(1 + k)\}.$$

Flag all genes with m_i greater than a fixed cutoff. The optimal rule remains essentially the same if we replace m_i in the loss function by some function of m, allowing in particular for the loss to be a non-linear function of the true level of differential expression.

$$L_f(m, \delta, z) = -\sum \delta_i f_D(m_i) + \sum (1 - \delta_i) f_N(m_i) + c\, D. \tag{5}$$

The functions $f_D(m)$ and $f_N(m)$ would naturally be S-shaped, monotone functions with a minimum at $m = 0$, and perhaps level off for large levels of m. Let $f_{N_i} = E(f_N(m) \mid Y)$ denote the posterior expectation for $f_N(m_i)$, and similarly for f_D. The optimal decision is

$$\delta_f^* = I\{f_{D_i} + k f_{N_i} > c\}.$$

Flag all genes with sufficiently large expected reward for discovery and/or penalty for a false negative. The rule δ_f^* follows from the fact that the choice of m_i in L_m was arbitrary.

The introduced loss functions are all generic in the sense of being reasonable loss functions without reference to a specific decision related to the multiple comparisons. If the goal of the inference is a very specific decision with a clearly recognizable implication, a problem-specific loss function should be used as the relevant criterion for the multiple comparison. For example, Lin *et al.* (2004) and Shen and Louis (1998) consider the problem of ranking units like health care providers. Ranking is a specific form of a multiple comparison problem. It could be described as all pairwise comparisons, subject to transitivity. They introduce loss functions that formalize the implications of a specific ranking, relative to the true ranks, and show the optimal rules for several such loss functions.

Example 1. Epithelial Ovarian Cancer (EOC)
Wang et al. (2004) report a study of epithelial ovarian cancer (EOC). The goal of the study is to characterize the role of the tumor microenvironment in EOC. To this end the investigators collected tissue samples from patients with benign and malignant ovarian pathology. Specimens were collected, among other sites, from peritoneum adjacent to the primary tumor. RNA was co-hybridized with reference RNA to a custom-made cDNA microarray including a combination of the Research Genetics RG_HsKG 031901 8k clone set and 9,000 clones selected from RG Hs seq ver 070700. A complete list of genes is available at

`http://nciarray.nci.nih.gov/gal files/index.shtml`

(The array is listed as custom printing Hs-CCDTM-17.5k-1px).

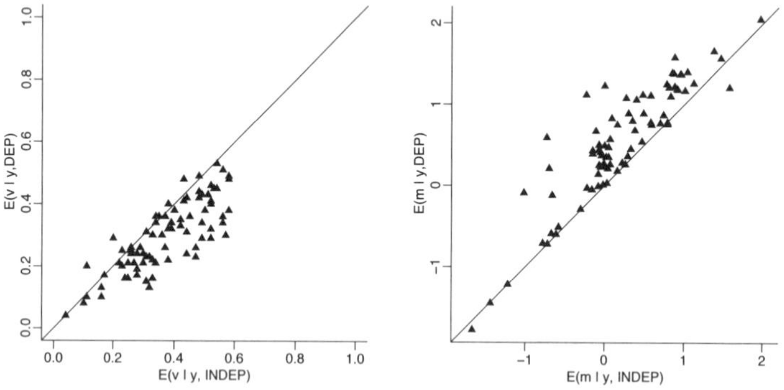

(a) v_i (b) m_i

Figure 1: *Inference with dependent prior (y-axis) vs. indep prior (x-axis). The changes are large enough to change the decisions.*

We focus on the comparison of 10 peritoneal samples from patients with benign ovarian pathology versus 14 samples from patients with malignant ovarian pathology. The raw data was pre-processed using BRB ArrayTool (`http://linus.nci.nih.gov/BRB-ArrayTools.html`). In particular, spots with minimum intensity less than 300 in both fluorescence channels were excluded from further analysis. See Wang et al. (2004) for a detailed description.

We computed probabilities of differential expression using the POE model proposed in Parmigiani *et al.* (2002). Inference is summarized by marginal probabilities of differential expression v_i. One parameter in the model is interpretable as the level of differential expression. Briefly, the basic POE model includes a trinary indicator e_{it} for gene i and sample t, with $e_{it} \in \{-1, 0, 1\}$ for under-expression, normal and over-expression. In a variation of the original POE model we use a probit prior for

e_{it}. The probit prior includes a regression on an indicator for malignant ovarian pathology. We denote the corresponding coefficient in the model by m_i, and interpret it as the level of differential expression for gene i. The original model does not include a gene-specific parameter r_i that can be interpreted as differential expression for gene i. We define $r_i = I(|m_i| > \epsilon)$, using $\epsilon = 0.5$.

Fig. 1 shows the selected lists of reported genes under the loss functions L_N and L_m (marked L_N and L_m). To facilitate the comparison we calibrated the tradeoff parameter c in both loss functions to fix $D = 20$. The difference in the two solutions are related to the difference between statistical significance and biologic significance. Because of varying precisions, it is possible that a gene with a very small level of differential expression reports a high posterior probability of differential expression, and vice versa.

5. APPROXIMATING BENJAMINI AND HOCHBERG'S RULE

The earlier introduced loss functions and the corresponding Bayes rules control various aspects of false discovery and false negative counts and rates. While similar in spirit, the rules are different from methods that have been proposed to control frequentist expected FDR, for example the rule defined in Benjamini and Hochberg (1995), henceforth BH.

It is not possible to justify BH (applied to the sorted p-values) as an optimal Bayes rule under a loss function that includes a combination of FD(R) and FN(R). This is shown in Cohen and Sackrowitz (2005) and extended in Cohen and Sackrowitz (2006). BH can be described as a step-up procedure that compares ordered observed difference scores $z_{(i)}$ with pre-determined critical cutoffs C_i. Let j denote the smallest integer i with $z_{(i)} > C_i$. All comparisons with difference scores beyond $z_{(j)}$ are rejected. Cohen and Sackrowitz (2005) show that such rules are inadmissible. The discussion includes a simple example that makes the inadmissibility easily interpretable. Consider a set of (ordered) p-values $p_{(i)}$ with $p_{(i)} = n\frac{\alpha}{n} - \epsilon$, i.e., all equal values. In particular, the largest p-value, $p_{(n)}$, falls below the BH boundary $(j\,\alpha)/n$. The BH rule would lead us to reject all comparisons. Now consider $p_{(i)} = i\frac{\alpha}{n} + \epsilon$. The p-values $p_{(i)}$, $i = 1, \ldots, n-1$ are substantially smaller, and $p_{(n)}$ is only slightly larger. Yet, we would be lead not to reject any comparison.

But interestingly, it is possible to still mimic the mechanics of the popular BH method as the Bayes rule under a specific loss function. The rule replaces the p-values with increments in posterior probabilities. The correspondence is not exact, and can not be in the light of Cohen and Sackrowitz' inadmissibility results.

Recall that $\delta_i(z) \in \{0,1\}$ denotes the decision rule for the i-th comparison, $r_i \in \{0,1\}$ is the (unknown) truth, $v_i = Pr(r_i = 1 \mid Y)$ are the marginal posterior probabilities, $\text{FD} = \sum \delta_i(1-r_i)$ are the false discovery count, $\overline{\text{FD}} = \sum \delta_i(1-v_i) = E(\text{FD} \mid Y)$, and $D = \sum \delta_i$ is the number of rejections. Let $w_i = 1 - v_i$ denote the marginal probability of the i-th null model.

Consider the loss function $\ell_B(\delta, z, r) = I(\text{FD} > \alpha D) - g_D$, with a monotone reward g_D for the number of discoveries. Marginalizing w.r.t. r, conditional on the data, we find the expected loss $L_B(\delta, z) = P(\text{FD} > \alpha D \mid Y) - g_D$. By Chebycheff's inequality, $P(\text{FD} > \alpha D) \leq \overline{\text{FD}}/(\alpha D)$. Using this upper bound, we define $L_B \approx$

$$L_U(\delta, z) \equiv \frac{\overline{\text{FD}}}{\alpha D} - g_D = \overline{\text{FDR}}/\alpha - g_D.$$

Without loss of generality assume $w_1 \leq w_2 \leq w_3 \ldots$ are ordered. We show that under L_U with $g_D = D/n$, the optimal decision is to use a threshold equal to the largest j with (appropriately defined) increment in posterior probability w_j less than $(j\alpha)/n$. See below for the appropriate definition of increment in w_j.

For fixed D, the optimal rule selects the D largest probabilities v_i. Let $\delta_i^D = I(i \leq D)$ denote this rule. To determine the optimal rule we still need to find the optimal $D = j$. Consider the condition $L_U(\delta^j, z) \leq L_U(\delta^{j-1}, z)$ for preferring δ^j over δ^{j-1}:

$$\frac{1}{\alpha j}\sum_{i=1}^{j} w_i - g_j \leq \frac{1}{\alpha(j-1)}\sum_{i=1}^{j-1} w_i - g_{j-1}.$$

After some simplification, and letting $w_j = \frac{1}{j}\sum_{i=1}^{j} w_i$ denote the average across comparisons 1 through j, and $\Delta w_j = w_j - w_{j-1}$, the condition becomes $L_U(\delta^j, z) < L_U(\delta^{j-1}, z)$ if $\Delta w_j < (g_j - g_{j-1})\alpha j$. A similar condition is true for lag k comparisons. Let $w_{ij} = \frac{1}{j-i+1}\sum_{h=i}^{j} w_i$ and $\Delta w_{ij} = w_{i+1,j} - w_i$.

$$L_U(\delta^j, z) \leq L_U(\delta^{j-k}, z) \ \text{ if } \ \Delta w_{j-k,j} \leq (g_j - g_{j-k})\alpha j/k \tag{6}$$

The earlier condition was the special case for $k = 1$. For $g_j = j/n$ the condition becomes $\Delta w_{j-k,j} \leq \frac{\alpha j}{n}$. Condition (6) characterizes the optimal rule δ^*. Let $B(2) \equiv 1$ and $B(j) = \max_{i<j}\{i : \Delta w_{B(i),i} < \frac{\alpha i}{n}\}$. In words, $B(j)$ is the best rule δ^i, $i < j$. The optimal rule is δ^j for

$$j = \max_i \left\{ i : \ \Delta w_{B(i),i} \leq \frac{\alpha i}{n} \right\}.$$

This characterizes the optimal rule by an algorithm like BH, applied to the increments in posterior probabilities $\Delta w_{B(i),i}$.

An alternative justification of a BH type procedure is the following approximation. Recall that under the loss function L_B, the optimal rule must be of the type $\delta_i = I(v_i \geq v_j)$ for some optimal j. If we knew the number n_0 of true null hypotheses, then we would find FD $\leq (1-v_j)n_0$, and thus FDR $\leq (1-v_j)n_0/j$. Assume that the probabilities v_i are ordered. To minimize v_j, while controlling FDR we would determine the cutoff by the maximum j with $(1 - v_j) \leq j\alpha/n_0$. Finally, replacing n_0 by the conservative bound n we get a BH type rule on the posterior probabilities $(1 - v_j)$.

A fundamental difference of BH and the rule under L_U is the use of posterior probabilities instead of p-values. Of course, the two are not the same. The relationship of p-values and posterior probabilities is extensively discussed, for example, in Casella and Berger (1987), Sellke *et al.* (2001) and Bayarri and Berger (2000).

The loss function L_B serves to make the use of BH type rules plausible under an approximate expected loss minimization argument. We would not, however, recommend it for practical use. Assume we had fantastically good data, with $v_i \in \{0, 1\}$, i.e., we essentially know the truth for all comparisons. The rule $\delta_i = I(v_i = 1)$ that reports the known truth is not optimal. Under L_B it can be improved by knowingly reporting false positives with $v_i = 0$. This is possible since L_B rewards for large D, and only introduces a penalty for false positives if the set threshold αD is exceeded. A similar statement applies for L_U.

6. FDR AND DEPENDENCE

In previous sections we argued for the use of posterior probabilities to account for multiplicities, and for a decision theoretic approach to multiple comparisons. The two arguments are not competing, but naturally complement each other. A structured probability model helps us to identify genes that might be of more interest than others.

In particular, the dependence structure of expression across genes might be of interest. If the goal is to develop a panel of biomarkers to classify future samples, then it is desireable to have low correlation of the expression levels for the reported set of differentially expressed genes. For other applications one might want to argue the opposite. Recall the example about inference on adverse events mentioned in the introduction.

In Müller and Parmigiani (2006) we introduce a probability model for gene expression that includes dependence for subsets of genes. The dependent subsets are typically identified as genes with a common functionality or genes corresponding to the nodes on a pathway of interest. The probability model allows us to use known pathways to formulate informative prior probability models for dependence across genes that feature in that pathway. Alternatively, for a small to moderately large set of genes the model allows us to learn about dependence starting from relatively vague prior assumptions. We briefly outline the features of the model that are relevant for the decision about reporting differentially expressed genes. Dependence is introduced not on the observed gene expressions, but on imputed trinary indicators $e_{it} \in \{-1, 0, 1\}$ for under- and over-expression of gene i in sample t. We build on the POE model introduced in Parmigiani *et al.* (2002), and already briefly mentioned earlier. In a variation of the basic POE model, in Müller and Parmigiani (2006) we represent the probabilities for the trinary outcome by a latent normal random variable z_{it}, with

$$e_{it} = \begin{cases} -1 & \text{if } z_{it} < -1 \\ 0 & \text{if } -1 < z_{it} < 1 \\ 1 & \text{if } z_{it} > 1. \end{cases} \tag{7}$$

The latent variables z_{it} are continuous random variables that allow us to introduce the desired dependence on related genes, as well as a regression on biologic condition. Let x_t denote a sample-specific covariate vector including an indicator x_{t1} for the biologic condition of the sample t and other sample-specific covariates. For example, in the case of a two group comparison between tumor and normal tissues, x_{t1} could be a binary indicator of tumor. Also, let $\{e_{jt};\ j \in N_i\}$ denote the trinary indicators for other genes that we wish to include as possible parent nodes in the dependent prior model for z_{it}. We assume a regression

$$z_{it} = g(x_t, e_{jt},\ j \in N_i) + \epsilon_i, \tag{8}$$

with mean function $g(\cdot)$ and standard normal residuals ϵ_i. The regression on e_{jt} introduces the desired dependence, and the regression on x_t includes the regression on the biologic condition x_{t1}, as before. Let m_i denote the regression coefficient for x_{t1}, the biologic condition. Also, define Σ_1 as the correlation matrix of $\{z_{it};\ \delta_i = 1\}$, the latent scores corresponding to the reported genes. The model allows us to include a term in the loss function that penalizes the reporting of highly correlated genes. We modify L_f to

$$L_D(m, \delta, z) = -k_1 \log(|\Sigma_1|) - k_2 \sum \delta_i f_D(m_i) + k_3 \sum (1 - \delta_i) f_N(m_i) + k_4\, c\, D. \tag{9}$$

The loss function encourages the inclusion of few highly differentially expressed genes with low correlation. Correlation is formalized as the tetrachoric correlation of the trinary outcomes e_{it}. See, for example, Ronning and Kukuk (1996) for a discussion of polychoric correlations for ordinal outcomes.

Example 1 (continued). Epithelial Ovarian Cancer (EOC)
Earlier we reported inference using the POE model and the loss functions L_N and L_m. We reanalyze the data with a variation of the POE model that includes dependent gene expression. In the implementation we specified (9) with $k_1 = 1$, $k_2 = 0.01$, $k_3 = k_4 = 0$, restricting to $D = 20$ (for comparability with the results under L_N and L_m), and setting $f_D(m_i) = m_i^2$. The inference summaries v_i and $\overline{m}_i$ change slightly when adjusting for dependence. The change in the estimates is shown in Fig. 1.

Although the changes are minimal for most genes, the impact in the final decision is visible. The first four rows of Figure 2 show the reported set of genes under L_N using the independent POE model (row 1) versus the dependent model (row 2), under L_m using the independent (row 3) and dependent (row 4) model. The last row shows inference under the loss function L_D.

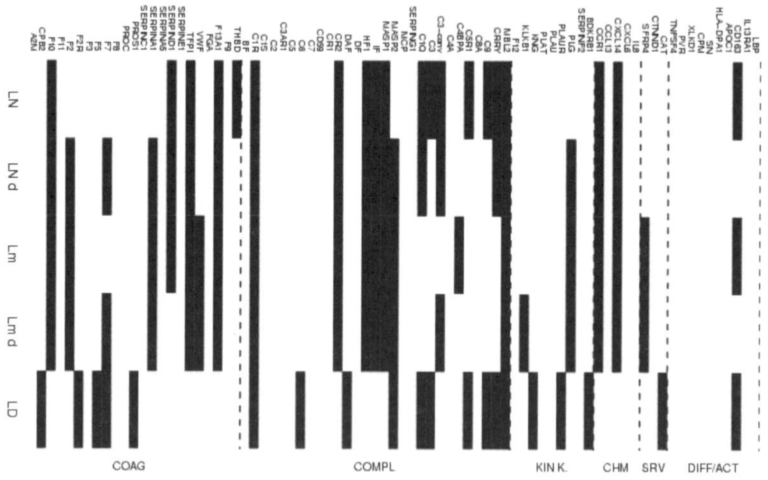

Figure 2: *Genes with $\delta_i = 1$, using L_N (top), L_N and dep model, L_m, L_m and dep model, and L_D (bottom).*

7. A PREDICTIVE LOSS FUNCTION

Microarray experiments are often carried out as hypothesis generating experiments, to screen for promising genes to be followed up in later experiments. Abruzzo et al. (2005) describe a setup, using RT-PCR to validate a list of differentially expressed genes found in a microarray group comparison experiment. In particular, they use TaqMan Low-Density Arrays for real-time RT-PCR (Applied Biosystems, Foster City, USA). They consider inference for nine samples from patients with chronic lymphocytic leukemia (CLL), using the microarray experiment for a first step screening experiment, and the real-time RT-PCR to validate the identified genes. With a setup like this experiment in mind we define a utility function that is based on the success of a future followup study. To define the desired loss function we need to introduce some more detail of a joint probability model for the microarray and real time RT-PCR experiments. Eventually we will use a stylized description of inference in this model to define a loss function.

Let z_{it} be a suitably normalized measurement for gene i in sample t, with approximately unit marginal variance, $var(z_{it}) \approx 1$. For example, z_{it} could be the latent probit score in (7). Let y_{it} denote the recorded outcome of the RT-PCR for gene i in sample t. The data are copy numbers, interpreted as base two logarithm of the relative abundance of RNA in the sample. Abruzzo et al. (2005) use a normal linear mixed effects model to calibrate the raw data against a calibrator sample and an endogenous control, chosen to reduce the variance of corrected responses across samples. Let y_{it} denote the calibrated pre-processed data. An important conclusion of the discussion in Abruzzo et al. (2005) is inference about the correlation of the microarray and RT-PCR measurements. They find a bimodal distribution of correlations across genes. About half the genes show a cross-platform correlation $\rho_i \approx 0.8$, and half show essentially zero correlation, $\rho_i = 0$.

We introduce a simple hierarchical model to represent the critical features of the cross-platform dependence, and a realistic distribution of the RT-PCR outcomes. We build on the POE model described earlier, in (7) and (7), without necessarily including the dependent model extension. Let z_{it} denote the latent score in (7). For y_{it} we assume:

$$p(y_{it} \mid z_{it}, \rho_i) = \begin{cases} z_{it} & \text{with prob. } \rho_i \\ N(0,1) & \text{with prob. } (1-\rho_i) \end{cases} \tag{10}$$

with $Pr(\rho_i = \rho^\star) = p_\rho$ and $Pr(\rho_i = 0) = 1 - p_\rho$. We use $\rho^\star = 0.8$ and $p_\rho = 0.5$ to approximately match the reported inference in Abruzzo et al. (2005). Also, after standardization Abruzzo et al. (2005) found standard deviations in the RT-PCR outcomes for each gene across samples in the range of approximately 0.5 through 1.5 (with some outliers below and above). We chose the unit variance in the normal term above to match the order of magnitude of these reported standard deviations.

Consider now the problem of reporting a list of differentially expressed genes in a microarray group comparison experiment. We assume that the selected genes will be validated in a future followup real-time RT-PCR experiment, using, for example, the described TaqMan Low-Density Arrays. We build a loss function designed to help us to construct a rule that identifies genes that are most likely to achieve a significant result in the followup experiment. For a stylized description we assume that the followup experiment is successful for gene i if we can report a statistically significant difference of expression across the two biologic conditions.

In words, the construction of the proposed loss function proceeds as follows. For each identified gene i we first select an alternative hypothesis. We then carry out a sample size argument based on achieving a desired power for this alternative versus the null hypothesis of no differential expression, using a traditional notion of power. Next we find the posterior predictive probability of a statistically significant outcome (R_i) for the future experiment. Finally, we define a loss function with terms related to the posterior predictive probability for R_i and the sampling cost for the future experiment. The stylized description is not a perfect reflection of the actual experimental process. It is not even a reasonable model for actual data analysis. But we believe it captures the critical features related to the desired decision of identifying differentially expressed genes. In particular, it includes a natural correction of statistical significance for the size of the effect.

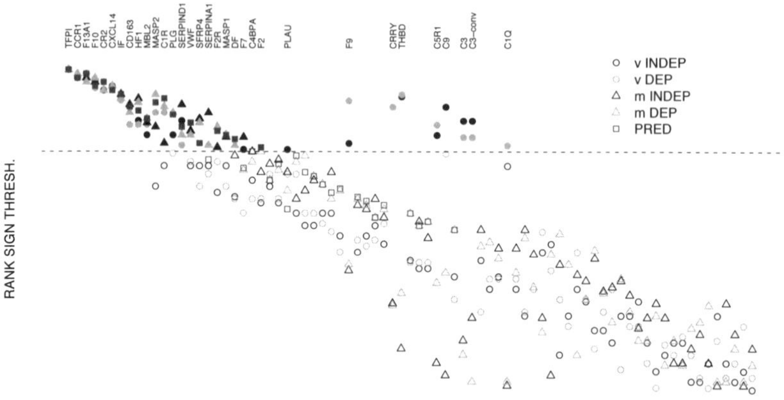

Figure 3: *Comparison of optimal rules for L_N (circles) under the independent model (black) and dependent model (light gray), L_m (triangle) under the independent (black) and dependent model (gray), and L_F (square). For each gene (horizontal axis) symbols are plotted against the rank of the corresponding significance threshold statistic (vertical axis). The reported genes under each rule are the top 20 ranked genes (above the dashed horizontal line). The symbols corresponding to selected genes are filled. The names of selected genes are shown on the vertical axis. Genes are sorted by average rank under the five criteria.*

Let $x_t \in \{-0.5, 0.5\}$ denote a (centered) indicator for the biologic condition of sample t. Recall from (7) that the level of differential expression for gene i, m_i, was defined as the probit regression coefficient for an indicator of the biologic condition. Let $(\overline{m}_i, s_i)$ denote the posterior mean and standard deviation of m_i. Let $\rho = p_\rho \rho^\star$ denote the assumed average cross-platform correlation, averaged across genes. Let $\mu_{i1} = E(y_{it} \mid x_t > 0)$ and $\mu_{i0} = E(y_{it} \mid x_t < 0)$ denote the mean expression under the two conditions in the followup experiment. For the upcoming sample size argument we assume a test of the null hypothesis $H_0 : \mu_{i1} - \mu_{i0} = 0$ versus the alternative hypotheses $H_1 : \mu_{i1} - \mu_{i0} = m^*_{yi}$, with $m^*_{yi} = \rho(m_i - s_i)$, the mean difference under an assumed alternative $m_i = m_i - s_i$. Let q_α denote the $(1-\alpha)$ quantile of a standard normal distribution. Assuming that upon conclusion of the followup experiment the investigators carry out a normal z-test, we find a required sample size

$$n_i(z) \geq 2\left[(q_\alpha + q_\beta)/m^*_{yi}\right]^2$$

for a given significance level α and power $(1-\beta)$. The sample size is a function of the data z, implicitly through the choice of the alternative m^*_{yi}. Here sample size refers to the number of samples under each biologic condition, i.e., the total number of samples is $2n$. Let y_{i0} and y_{i1} denote the sample average in the followup experiment, for gene i and the two conditions. Let $R_i = \{(y_{i1} - y_{i0})\sqrt{n/2} \geq q_\alpha\}$ denote the event of a statistically significant difference for gene i in the followup experiment. Let $\pi_i = Pr(R_i \mid Y)$ denote the posterior predictive probability of R_i. Let $\Phi(\cdot)$ denote the standard normal c.d.f.

$$\pi_i(z) = (1 - p_\rho)\alpha + p_\rho \Phi \left[\frac{\rho^\star \overline{m}_{i1} \sqrt{n_i/2} - q_\alpha}{\sqrt{1 + \frac{n}{2}\rho^{*2} s_i^2}} \right].$$

Combining n_i and π_i to trade off the competing goals of small sampling cost and high success probability we define a loss function

$$L_F(\delta, z) = \sum_{\delta_i = 1} [-c_1 \pi_i(z) + n_i(z)] + c_2 D. \tag{11}$$

Under the loss function L_F, the optimal rule is easily seen as

$$\delta_i^* = I(n_i + c_2 \leq c_1 \pi_i).$$

If we replace classical power by Bayesian power (Spiegelhalter *et al.*, 2004, Chapter 6.5.3) then π_i remains constant by definition, leaving only the bound on the sample size n_i. Also, c_2 could be zero if the size of the reported short-list is not an issue. Fig. 3 shows the optimal decision under L_F for the EOC example (squares). We used $c_1 = 3000$ and calibrated c_2 such that the optimal decision reports $D = 20$ genes, as before. The value for c_1 was chosen to have the reward c_1 match approximately 10 times the average sampling cost of a followup trial.

8. OVERVIEW

Table 2 summarizes the proposed loss functions and rules. For all except L_D the optimal rule can be described as a threshold for an appropriate gene-specific summary statistic of the data. Storey (2005) describes such rules as significance thresholding. The summaries are the marginal posterior probability of differential expression v_i, posterior mean and standard deviation of the level of differential expression (m_i, s_i), the sample size for a followup experiment n_i, the posterior predictive probability of a significant outcome π_i, and the increment in posterior probability of non-differential expression $\Delta w_{B(i),i}$. The last three are functions of (v_i, m_i, s_i) only.

Table 2: *Alternative loss functions and optimal rules*

Loss function	Rule
$L_N = c\text{FD} + \text{FN}$	$\delta_i = I(v_i > t)$
$L_m = -\sum \delta_i m_i + k \sum(1 - \delta_i) m_i + cD$	$\delta_i = I(m_i > t)$
$L_D = -k_1 \log(\lvert\Sigma_1\rvert) - k_2 \sum \delta_i f_D(m_i) + {}$ $+k_3 \sum(1 - \delta_i) f_N(m_i) + k_4 D$	no closed form
$L_F = \sum_{\delta_i=1}(-c_1\pi_i + n_i) + c_2 D$	$\delta_i = I(n_i + c_2 \leq c_1 \pi_i)$
$L_U = \text{FDR}/\alpha - g(D)$	$\delta_i = I(v_i \geq v_j)$ with $j = \max_i \{i :\ \Delta w_{i,B(i)} \leq \alpha i/n\}$

In summary, all optimal rules are computed on the basis of only a few underlying summaries. This makes it possible to easily consider multiple rules in a data analysis. Critical comparison of the resulting rules leads to a finally reported set of comparisons. In some cases the application leads to a different loss function. Good examples are the loss functions considered in Lin *et al.* (2004). If a specific loss function arises from a specific case study, it should be used.

The loss function L_D requires the additional summary $\Sigma_1(\delta)$. Let S denote a covariance matrix of the relevant latent variables for *all* genes that are considered for reporting in L_D. The desired $\Sigma_1(\delta)$ for any subset of selected genes is then computed as the marginal correlation matrix for that subset. Let S_δ denote the submatrix defined by choosing rows and columns selected by δ. Let $\lambda_i = 1/\sqrt{S_{ii}}$, and λ_δ denote the vector of λ_i corresponding to the reported genes. We use $\Sigma_1(\delta) = \lambda_\delta[S_\delta]\lambda'_\delta$. This reduces δ_D to a function of m and S only.

Fig. 3 compares the reported gene lists for the loss functions L_N, L_m and L_D, under the independent model and the dependent model. For many genes the decision remains unchanged across all loss functions. For some genes with high probability of differential expression, but small level of differential expression, and vice versa, the decision depends on whether or not the terms in the loss function are weighted by m_i.

9. FURTHER DISCUSSION

We have reviewed alternative approaches to addressing problems related to massive multiple comparisons. Starting from traditional rules that control expected FDR, with the expectation over repeated sampling, we have discussed the limitation of the interpretation of these decisions as Bayesian rules. We have argued for a solution of the massive multiple comparisons as a decision problem, and we have shown how this is implemented in structured probability models including dependence across genes. Most, but not all, loss functions lead to rules that can be defined in terms of a significance thresholding function, $S(data)$, as proposed in Storey (2005).

The proposed approaches are all based on casting the multiple comparison problem as a decision problem and thus inherit the limitations of any decision theoretic solution. In particular, we recognize that not all research is carried out to make a decision. The decision theoretic perspective might be inappropriate when an experiment is carried out for more heuristic learning, without being driven by specific decisions. Also, all arguments were strictly model-based. Even results that apply for any probability model still need a specific probability model to implement related approaches. Like any model-based inference, the implementation involves the often difficult tasks of prior elicitation, and the choice of appropriate parametric models. Additionally, our arguments require the choice of a utility function. A common feature of early stage hypothesis generating experiments is that they serve multiple needs. We might want to carry out a microarray experiment to suggest interesting genes and proteins for further investigation, to propose suitable candidates for a correlation study with clinical outcomes, and also to simply understand molecular mechanisms.

We caution against over-interpretation of results based on highly structured probability models and often arbitrary choices of utility functions. Data analysis for high-throughput gene expression data is particularly prone to problems arising from data pre-processing. Often it is more important to understand the pre-processing of the raw data, and correct it if necessary, than to spend effort on sophisticated modeling. Specifically related to the dependent probability model, it is important to acknowledge limitation of pathway information that is used to select the set of possible parent nodes N_i in (8) when constructing the dependent probability model.

Pathway information does not necessarily describe relations among transcript levels, although it carries some information about it.

We have focused on the inference problem of reporting lists of differentially expressed genes, and inference on massive multiple comparisons in general. A similar framework, using the same probability models and loss functions, can be used for other decision problems related to the same experiments. For example, one could consider choosing the sample size for a future experiment (sample size selection), ranking genes or selecting a fixed set of genes (for a panel of biomarkers).

ACKNOWLEDGMENTS

Research was supported by NIH/NCI grant 1 R01 CA075981 and by NSF DMS034211. Kenneth Rice was supported by Career Development Funding from the Department of Biostatistics, University of Washington

REFERENCES

Baggerly, K. A., Morris, J. S., Wang, J., Gold, D., Xiao, L. C. and Coombes, K. R. (2003). A comprehensive approach to analysis of MALDI-TOF proteomics spectra from serum samples. *Proteomics* **9**, 1667–1672.

Bayarri, M. J. and Berger, J. O. (2000). *P* values for composite null models. *J. Amer. Statist. Assoc.* **95**, 1127–1142.

Benjamini, Y. and Hochberg, Y. (1995). Controlling the false discovery rate: A practical and powerful approach to multiple testing. *J. Roy. Statist. Soc. B* **57**, 289–300.

Berry, D. A. and Hochberg, Y. (1999). Bayesian perspectives on multiple comparisons. *J. Statist. Planning and Inference* **82**, 215–227.

Berry, S. and Berry, D. (2004). Accounting for multiplicities in assessing drug safety: A three-level hierarchical mixture model. *Biometrics* **60**, 418–426.

Casella, G. and Berger, R. L. (1987). Reconciling Bayesian and frequentist evidence in the one-sided testing problem. *J. Amer. Statist. Assoc.* **82**, 106–111.

Cohen, A. and Sackrowitz, H. B. (2005). Decision theory results for one-sided multiple comparison procedures. *Ann. Statist.* **33**, 126–144.

Cohen, A. and Sackrowitz, H. B. (2006). More on the inadmissibility of step-up. *Tech. Rep.*, Rutgers University.

Do, K., Müller, P. and Tang, F. (2005). A bayesian mixture model for differential gene expression. *Appl. Statist.* **54**, 627–644.

Duncan, D.B. (1965). A Bayesian Approach to Multiple Comparisons, *Technometrics* **7**, 171-222.

Efron, B. and Tibshirani, R. (2002). Empirical Bayes methods and false discovery rates for microarrays. *Genetic Epidemiology* **23**, 70–86.

Efron, B., Tibshirani, R., Storey, J. D. and Tusher, V. (2001). Empirical Bayes analysis of a microarray experiment. *J. Amer. Statist. Assoc.* **96**, 1151–1160.

Fortini, M., Liseo, B., Nuccitelli, A. and Scanu, M. (2001). On Bayesian record linkage. *Research in Official Statistics*, **4**, 185–198.

Fortini, M., Liseo, B., Nuccitelli, A. and Scanu, M. (2002). Modelling issues in record linkage: a Bayesian perspective. *Proceedings of the ASA meeting.* ASA.

Genovese, C., Lazar, N. and Nichols, T. (2002). Thresholding of statistical maps in neuroimaging using the false discovery rate. *NeuroImage* **15**, 870–878.

Genovese, C. and Wasserman, L. (2002). Operating characteristics and extensions of the false discovery rate procedure. *J. Roy. Statist. Soc. B* **64**, 499–518.

Genovese, C. and Wasserman, L. (2003). Bayesian and Frequentist Multiple Testing. *Bayesian Statistics 7* (J. M. Bernardo, M. J. Bayarri, J. O. Berger, A. P. Dawid, D. Heckerman, A. F. M. Smith and M. West, eds.) Oxford: University Press, 145–162.

Gopalan, R. and Berry, D. A. (1998). Bayesian multiple comparisons using Dirichlet process priors. *J. Amer. Statist. Assoc.* **93**, 1130–1139.

Keeney, R. L., Raiffa, H. A. and Meyer, R. F. C. (1976). *Decisions With Multiple Objectives: Preferences and Value Tradeoffs.* New York: Wiley

Lin, R., Louis, T. A., Paddock, S. M. and Ridgeway, G. (2004). Loss function based ranking in two-stage hierarchical models. *Tech. Rep.*, Johns Hopkins University, USA.

Lindley, D. V. (1971). *Making Decisions.* New York: Wiley

Müller, P. and Parmigiani, G. (2006). Modeling dependent gene expression. *Tech. Rep.*, M.D. Anderson Cancer Center, USA.

Müller, P., Parmigiani, G., Robert, C. and Rouseau, J. (2004). Optimal sample size for multiple testing: the case of gene expression microarrays. *J. Amer. Statist. Assoc.* **99**, 990–1001.

Newton, M., Noueriry, A., Sarkar, D. and Ahlquist, P. (2004). Detecting differential gene expression with a semiparametric heirarchical mixture model. *Biostatistics* **5**, 155–176.

Newton, M. A., Kendziorsky, C. M., Richmond, C. S., R., B. F. and Tsui, K. W. (2001). On differential variability of expression ratios: improving statistical inference about gene expression changes from microarray data. *J. Computational Biology* **8**, 37–52.

Parmigiani, G., Garrett, E. S., Anbazhagan, R. and Gabrielson, E. (2002). A statistical framework for expression-based molecular classification in cancer. *J. Roy. Statist. Soc. B* **64**, 717–736.

Ronning, G. and Kukuk, M. (1996). Efficient estimation of ordered probit models. *J. Amer. Statist. Assoc.* **91**, 1120–1129.

Scott, J. and Berger, J. (2003). An exploration of aspects of Bayesian multiple testing. *Tech. Rep.*, ISDS, Duke University, USA.

Sellke, T., Bayarri, M. J. and Berger, J. O. (2001). Calibration of p-values for testing precise null hypotheses. *Amer. Statist.* **55**, 62–71.

Shen, W. and Louis, T. A. (1998). Triple-goal estimates in two-stage hierarchical models. *J. Roy. Statist. Soc. B* **60**, 455–471.

Shen, X., Huang, H.-C. and Cressie, N. (2002). Nonparametric hypothesis testing for a spatial signal. *J. Amer. Statist. Assoc.* **97**, 1122–1140.

Storey, J. D. (2002). A direct approach to false discovery rates. *J. Roy. Statist. Soc. B* **64**, 479–498.

Storey, J. D. (2003). The positive false discovery rate: A Bayesian interpretation and the q-value. *Ann. Statist.* **31**, 2013–2035.

Storey, J. D. (2005). The optimal discovery procedure I: a new approach to simultaneous signficance testing. *Tech. Rep.*, University of Washington, USA.

Storey, J. D., Taylor, J. E. and Siegmund, D. (2004). Strong control, conservative point estimation and simultaneous conservative consistency of false discovery rates: a unified approach. *J. Roy. Statist. Soc. B* **66**, 187–205.

DISCUSSION

TOM FEARN (*University College London, UK*)

The problem of multiple testing is getting a lot of attention at present because of the emergence in biomedical research of high-throughput devices such as microarrays. The analysis of data from such a device may involve many thousands of significance tests, focussing attention sharply on the problem of whether and how to allow for multiple testing. The approach of adjusting significance levels to control false discovery rate (FDR), the proportion of errors amongst results declared significant, has become popular. The cynic might suggest that this popularity may have more to do with the existence of the elegant result by Benjamini and Hochberg (1995) that permits such control than with any widespread conviction that FDR is more relevant than competing criteria.

The current paper is a wide-ranging look at this area from a Bayesian perspective, reviewing existing work, presenting interesting new results, and including much insightful discussion. Before adding my own brief discussion, I would like to illustrate some of the issues in multiple testing by describing a problem with lifts (elevators).

A problem with lifts. Standing in the hotel lobby, I am faced with five lifts. Each has a separate call button, and I may decide to press any or all of them. Some of the lifts may not be working. For each lift separately a test of the null hypothesis that the lift is not working has been carried out, and the resulting p-values are shown in Table 3.

Table 3: *Five hypothesis tests*

Lift	1	2	3	4	5
p_i	0.02	0.02	0.04	0.05	0.06

I am a frequentist who hates making mistakes, and do not wish to push the button for a lift that is not working, though I will of course allow myself one error in 20. Thus I may call the first four lifts, where $p \leq 0.05$, but not lift five. However, perhaps I should be worried not about my lift-wise error rate but about my experiment-wise error rate. With five tests, I can control this at 1 in 20 by invoking Bonferroni and reducing my critical p-value to $0.05/5 = 0.01$. Now, alas, I cannot press any buttons. At this point Benjamini and Hochberg (BH), who happen to be passing, come to the rescue by suggesting that I compare the conveniently ordered p-values with the sequence $0.05i/5$ or 0.01, 0.02, 0.03, 0.04, 0.05, and declare significant any observed p-value that is less than or equal to its corresponding critical value, together with all those to its left. This would allow me to call lifts one and two whilst still protecting my FDR. Less helpfully, another passer by points out that there are two more lifts behind me and, although they probably do not work, they ought strictly to be considered to be part of my experiment. The new set of seven p-values is shown in Table 4.

Table 4: *Seven hypothesis tests*

Lift	1	2	3	4	5	6	7
p_i	0.02	0.02	0.04	0.05	0.06	0.99	0.99

The new BH critical values are $0.05i/7$, giving 0.007, 0.014, 0.021, 0.029, 0.036, 0.043, 0.050, and now none of the observed p-values is small enough. I am just about to make my rather confused way up 42 flights of stairs when by good fortune a passing Bayesian presses all the buttons and allows me to share her lift.

I leave the reader to draw, according to taste, various possible morals from this parable.

Further Discussion. The key question is whether a Bayesian viewpoint can help the practitioner with more serious problems than the one just described. The answer must be yes. The longer answer comes in two parts.

Firstly, we should replace the p-values with posterior probabilities. These probabilities need to come from a realistic model, the more realistic the better. In Section 1 of the paper, the simple mixture model

$$f(z_i) = p_0 f_0(z_i) + (1 - p_0) f_1(z_i)$$

for the distribution of independent summary statistics z_i is discussed. The typical fequentist analysis assumes that the null-hypothesis density $f_0(z_i)$ is known and makes worst-case assumptions about the proportion p_0 of true nulls and the density $f_1(z_i)$ under the alternative. More realistic assumptions (and models) will lead to more realistic conclusions and, as part of this, will automatically 'adjust' the posterior probabilities for multiplicity.

As the authors remark, this adjustment is not the whole story. Experiments with microarrays, for example, generally lead to decisions and often to further experimentation. For me the most exciting parts of the paper describe attempts to analyse this situation using decision theory. Of course it is of interest to study, as the authors do, the extent to which the popular frequentist procedures can or cannot be justified in a decision theory framework. However my personal view is that the Bayesian community devotes rather too much effort to trying to reproduce frequency results, and it is the serious, context-dependent modelling efforts in this paper that earn my loudest applause. Once we start thinking properly about a real problem, I suspect we will typically want to throw out not just the p-values, but also the posterior probabilities. For practical purposes it is the size, or rather what we believe about the size, of an effect that matters, not its significance.

ALEX LEWIN and SYLVIA RICHARDSON (*Imperial College, London, UK*)

We congratulate the authors on an interesting paper, containing several new ideas. We would like to comment on one aspect. At the end of Section 4, the authors say that the difference between the lists found using loss functions L_N and L_m are due to the difference in biological significance and statistical significance. The loss function L_m is supposed to take account of biological significance where the more usual L_N does not.

It is not very clear what the term 'biological significance' means here however. The loss function L_m leads to a rule based only on the posterior mean $E(m_i|\text{ data}) \equiv \bar{m}_i$, *i.e.* $\delta_i^* = I[|\bar{m}_i| \geq c/(1+k)]$ where c and k are parameters in the loss function, chosen according to the importance given to numbers of true and false discoveries. This rule is therefore similar to the simple cut-offs used on log fold changes, which take no account of the variability in the data (though some smoothing will have occurred if the m_i are modelled as a mixture). There does not seem to be any specific quantity representing biological significance.

In Lewin et al. 2006, we proposed a straightforward way to combine biological and statistical significance in a clearly defined manner. Here the indicator for differential expression is defined as $r_{i1} \equiv I[|m_i| > m_{cut}]$, where m_{cut} is chosen to be a biologically interesting level of differential expression. The loss function L_N is used, so genes are selected as differentially expressed based on posterior probabilities, thus requiring statistical significance as well. The decision rule is

$$\text{Rule 1:} \quad \delta_{i1}^* \quad = \quad I[P(|m_i| > m_{cut}|\text{ data}) > v_{cut}],$$

where v_{cut} is determined by the parameters of the loss function.

We demonstrate this rule on simulated data from a mixture model, where the gene differential expression parameter m_i has a mixture prior (we simulate 2500 genes from a mixture of Normal and Gamma distributions, with 500 genes differentially expressed). We have used $m_{cut} = 0.3$. To compare with our proposed rule, we define an alternative indicator r_{i2} for differential expression which only uses statistical significance. This is the mixture allocation parameter ($r_{i2} = 1$ for genes not allocated to the null component). The decision rule based on this and the loss function L_N is

$$\text{Rule 2:} \quad \delta_{i2}^* = I[P(r_{i2} = 1 | \text{ data}) > v_{cut}].$$

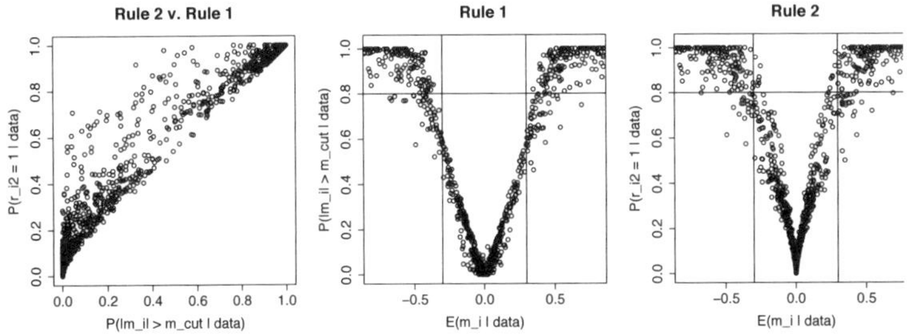

Figure 4: *Posterior probabilities of differential expression and posterior means of the differential expression parameter used in the three decision rules. Vertical lines indicate the biological cut-off parameter* m_{cut}*, horizontal lines show* $v_{cut} = 0.8$.

The first plot in Fig. 4 compares the posterior probabilities used in the two rules. As $P(|m_i| > m_{cut}|$ data) is less than or equal to $P(r_{i2} = 1|$ data), Rule 1 will select fewer genes than Rule 2, whatever the value of v_{cut} (thus including biological significance restricts the selection, as expected). The second and third plots show the posterior probabilities for the two rules versus the posterior means $\bar{m}_i$. Vertical lines indicate the cut-off m_{cut}. Even for values of v_{cut} as large as 0.8 (shown as horizontal lines in the plots), Rule 2 selects genes with low $|\bar{m}_i|$, while Rule 1 does not. Finally, note that any rule that uses cut-offs solely on $E(m_i|$ data) (such as rules based on L_m) will select genes with widely varying degrees of statistical significance.

REPLY TO THE DISCUSSION

We thank the discussants for their stimulating contributions. The discussions raise several important issues and pose interesting questions.

The elevator example in Professor Fearn's discussion is a delightful and concise illustration of the issues underlying a decision-theoretic multiplicity adjustment. We fully agree about the importance of realistic modeling. We are less critical about assuming a known model f_0 and substituting point estimates for the proportion p_0 of true nulls. Several recent references propose excellent point estimates for p_0.

Lewin and Richardson point out an interesting example of rules similar to those proposed in our paper. Lewin *et al.* (2006) consider for each gene the posterior probability that the level of differential expression is beyond a pre-determined level m_{cut}. Genes are identified as differentially expressed if this posterior probability is beyond a certain cutoff (Rule 1 in the discussion). The cutoff m_{cut} can be chosen to reflect biologic significance. Lewin and Richardson comment that the rule $\delta^{\star}$ under L_m proposed in our paper does not include a similar specification of biologic significance. We agree. To include a notion of biologic significance one would need to use the more general loss L_f. Using, for example, $f_D(m) = f_N(m) = I(|m| > m_{cut})$ would lead to Rule 1. While loss functions that include weights for the relative magnitude of errors have been proposed as early as Duncan (1965), we agree that such problem-specific formulations are important to consider. The model proposed in Lewin *et al.* (2006) illustrates another interesting extension. As a generalization of Rule 1 the paper proposes a rule that simultaneously includes cutoffs for the level of differential expression *and* the level of overall expression. Such rules can be derived as arising from L_f if we generalize m_i to a multivariate gene-specific summary.

ADDITIONAL REFERENCE IN THE DISCUSSION

Lewin, A., Richardson, S., Marshall, C., Glazier, A. and Aitman, T. (2006). Bayesian modelling of differential gene expression. *Biometrics* **62**, 1–9.

BAYESIAN STATISTICS 8, pp. 371–416.
J. M. Bernardo, M. J. Bayarri, J. O. Berger, A. P. Dawid,
D. Heckerman, A. F. M. Smith and M. West (Eds.)

Estimating the Integrated Likelihood via Posterior Simulation Using the Harmonic Mean Identity

ADRIAN E. RAFTERY
University of Washington, USA
raftery@u.washington.edu

MICHAEL A. NEWTON
University of Wisconsin, USA
newton@stat.wisc.edu

JAYA M. SATAGOPAN
Sloan-Kettering Cancer Center, USA
satagopj@mskcc.org

PAVEL N. KRIVITSKY
University of Washington, USA
pavel@stat.washington.edu

SUMMARY

The integrated likelihood (also called the marginal likelihood or the normalizing constant) is a central quantity in Bayesian model selection and model averaging. It is defined as the integral over the parameter space of the likelihood times the prior density. The Bayes factor for model comparison and Bayesian testing is a ratio of integrated likelihoods, and the model weights in Bayesian model averaging are proportional to the integrated likelihoods. We consider the estimation of the integrated likelihood from posterior simulation output, aiming at a generic method that uses only the likelihoods from the posterior simulation iterations. The key is the harmonic mean identity, which says that the reciprocal of the integrated likelihood is equal to the posterior harmonic mean of the likelihood. The simplest estimator based on the identity is thus the harmonic mean of the likelihoods. While this is an unbiased and simulation-consistent estimator, its reciprocal can have infinite variance and so it is unstable in general.

We describe two methods for stabilizing the harmonic mean estimator. In the first one, the parameter space is reduced in such a way that the modified estimator involves a harmonic mean of heavier-tailed densities, thus resulting in a finite variance estimator. The resulting estimator is stable. It is also

Adrian E. Raftery is Blumstein–Jordan Professor of Statistics and Sociology at the University of Washington in Seattle, Michael A. Newton is a Professor of Statistics and of Biostatistics and Medical Informatics at the University of Wisconsin in Madison, Jaya M. Satagopan is Associate Attending Biostatistician at the Memorial Sloan–Kettering Cancer Center in New York City, and Pavel N. Krivitsky is Graduate Research Assistant at the University of Washington Department of Statistics in Seattle. The research of Raftery and Krivitsky was supported by NIH grant 8R01EB 002137-02.

self-monitoring, since it obeys the central limit theorem, and so confidence intervals are available. We discuss general conditions under which this reduction is applicable.

The second method is based on the fact that the posterior distribution of the log-likelihood is approximately a gamma distribution. This leads to an estimator of the maximum achievable likelihood, and also an estimator of the effective number of parameters that is extremely simple to compute from the loglikelihoods, independent of the model parametrization, and always positive. This yields estimates of the log integrated likelihood, and posterior simulation-based analogues of the BIC and AIC model selection criteria, called BICM and AICM. We provide standard errors for these criteria. We illustrate the proposed methods through several examples.

Keywords and Phrases: AIC; AICM; BASKETBALL; BAYES FACTOR; BAYESIAN MODEL AVERAGING; BETA-BINOMIAL; BIC; BICM; DIC; EFFECTIVE NUMBER OF PARAMETERS; FAN-SHAPED DISTRIBUTION; GAMMA DISTRIBUTION; GENETICS; HIERARCHICAL MODEL; LATENT SPACE; MARGINAL LIKELIHOOD; MARKOV CHAIN MONTE CARLO; MAXIMUM LIKELIHOOD; MODEL COMPARISON MODEL SELECTION; NORMALIZING CONSTANT; POISSON–GAMMA; RANDOM EFFECTS MODEL; ROBUST LINEAR MODEL; SOCIAL NETWORKS; UNIT INFORMATION PRIOR

1. INTRODUCTION

The integrated likelihood, also called the marginal likelihood or the normalizing constant, is an important quantity in Bayesian model comparison and testing: it is the key component of the Bayes factor (Kass and Raftery 1995; Chipman, George, and McCulloch 2001). The Bayes factor is the ratio of the integrated likelihoods for the two models being compared. When taking account of model uncertainty using Bayesian model averaging, the posterior model probability of a model is proportional to its prior probability times the integrated likelihood (Hoeting, Madigan, Raftery, and Volinsky 1999).

Consider data y, a likelihood function $\pi(y|\theta)$ from a model for y indexed by a parameter θ, in which both y and θ may be vector-valued, and a prior distribution $\pi(\theta)$. The integrated likelihood of y is then defined as

$$\pi(y) = \int \pi(y|\theta)\pi(\theta)\, d\theta.$$

The integrated likelihood is the normalizing constant for the product of the likelihood and the prior in forming the posterior density $\pi(\theta|y)$. Furthermore, as a function of y prior to data collection, $\pi(y)$ is the prior predictive density.

Evaluating the integrated likelihood can present a difficult computational problem. Newton and Raftery (1994) showed that $\pi(y)$ can be expressed as an expectation with respect to the posterior distribution of the parameter, thus motivating an estimate based on a Monte Carlo sample from the posterior. By Bayes's theorem,

$$\frac{1}{\pi(y)} = \int \frac{\pi(\theta|y)}{\pi(y|\theta)}\, d\theta = E\left\{ \left. \frac{1}{\pi(y|\theta)} \right| y \right\}. \tag{1}$$

Equation (1) says that the integrated likelihood is the posterior harmonic mean of the likelihood, and so we call it the *harmonic mean identity*. This suggests that the integrated likelihood $\pi(y)$ can be approximated by the sample harmonic mean of the likelihoods,

$$\hat{\pi}_{\mathrm{HM}}(y) = \left[\frac{1}{B}\sum_{t=1}^{B}\frac{1}{\pi(y|\theta^t)}\right]^{-1}, \tag{2}$$

based on B draws $\theta^1, \theta^2, \ldots, \theta^B$ from the posterior distribution $\pi(\theta|y)$. This sample might come out of a standard Markov chain Monte Carlo implementation, for example. Though $\hat{\pi}_{\mathrm{HM}}(y)$ is consistent as the simulation size B increases, its precision is not guaranteed.

The simplicity of the harmonic mean estimator (2) is its main advantage over other more specialized techniques (Chib 1995; Green 1995; Meng and Wong 1996; Raftery 1996; Lewis and Raftery 1997; DiCiccio, Kass, Raftery, and Wasserman 1997; Chib and Jeliazkov 2001). It uses only within-model posterior samples and likelihood evaluations which are often available anyway as part of posterior sampling. A major drawback of the harmonic mean estimator is its computational instability. The estimator is consistent but may have infinite variance (measured by $\mathrm{Var}\{[\pi(y|\theta)]^{-1}|y\}$) across simulations, even in simple models. When this is the case, one consequence is that when the cumulative estimate of the harmonic mean estimate (2) based on the first B draws from the posterior is plotted against B, the plot has occasional very large jumps, and looks unstable.

In this article we describe two approaches to stabilizing the harmonic mean estimator. In the first method, the parameter space is reduced such that the modified estimator involves a harmonic mean of heavier-tailed densities, thus resulting in a finite variance estimator. We develop general conditions under which this method works. The resulting estimator obeys the central limit theorem, yielding confidence intervals for the integrated likelihood. In this way it is self-monitoring.

The second approach is based on the fact that the posterior distribution of the loglikelihood is approximately a shifted gamma distribution. This leads to an estimator of the maximum achievable likelihood, and also an estimator of the effective number of parameters that is very simple to compute, uses only the likelihoods from the posterior simulation, is independent of the model parametrization, and is always positive. This yields estimates of the log integrated likelihood, and posterior simulation-based analogues of the BIC and AIC model selection criteria, called BICM and AICM. Standard errors of these criteria are also provided. We illustrate the proposed methods through several examples.

In Section 2 we describe the parameter reduction method and in Section 3 we give several examples. In Section 4 we describe the shifted gamma approach and we report a small simulation study and an example. In Section 5 we discuss limitations and possible improvements of the methods described here, and we mention some of the other methods proposed in the literature.

2. STABILIZING THE HARMONIC MEAN ESTIMATOR BY PARAMETER REDUCTION

An overly simple but helpful example to illustrate our first method is the model in which $\theta = (\mu, \psi)$ records the mean and precision of a single normally distributed data point y. A conjugate prior is given by

$$\psi \sim \mathrm{Gamma}(\alpha/2, \alpha/2)$$

$$(\mu|\psi) \quad \sim \quad \text{Normal}(\mu_0, n_0\psi),$$

where α, n_0, and μ_0 are hyperparameters (e.g., Bernardo and Smith, 1994, page 268 or Appendix I). The integrated likelihood, $\pi(y)$, is readily determined to be the ordinate of a t density, $\text{St}(y|\mu_0, n_0/(n_0+1), \alpha)$ in the notation of Bernardo and Smith (1994, page 122 or Appendix I). Were we to approximate $\pi(y)$ using equation (2), instead of taking the analytically determined value, we could measure the stability of the estimator with the variance $\text{Var}\{[\pi(y|\theta)]^{-1}|y\}$. This variance, in turn, is determined by the second noncentral moment $\text{E}\{[\pi(y|\theta)]^{-2}|y\}$, which is proportional to

$$\int\int \psi^{\alpha/2} \exp\left\{\frac{\psi}{2}[(y-\mu)^2 - n_0(\mu-\mu_0)^2 - \alpha]\right\} d\psi d\mu,$$

and which is infinite in this example owing to the divergence of the integral in μ for each ψ. The reciprocal of the light-tailed normal density forms too large an integrand to yield a finite posterior variance, and hence the harmonic mean estimator is unstable.

An alternative estimator, supported equally by the basic equation (1), is

$$\hat{\pi}_{\text{SHM}}(y) = \left[\frac{1}{B}\sum_{t=1}^{B}\frac{1}{\pi(y|\mu^t)}\right]^{-1}, \tag{3}$$

which we call a stabilized harmonic mean. In (3), μ^t is the mean component of $\theta^t = (\mu^t, \psi^t)$, and thus is a draw from the marginal posterior distribution $\pi(\mu|y)$. The stabilized harmonic mean is formed not from standard likelihood values, but rather from marginal likelihoods obtained by integrating out the precision parameter ψ. It is straightforward to show that this integrated likelihood has the form of a t ordinate,

$$\pi(y|\mu) = \text{St}\left\{y|\mu, (\alpha+1)/[\alpha + n_0(\mu-\mu_0)^2], \alpha+1\right\}.$$

The intuition motivating (3) is that since $\pi(y|\mu)$ has a heavier tail than $\pi(y|\theta)$, averages of reciprocal ordinates become averages of less variable quantities than in (2). Measuring stability as above, we observe that

$$\text{E}\left\{[\pi(y|\mu)]^{-2}\Big|\, y\right\} \propto \int \frac{\{1 + [(y-\mu)^2 + n_0(\mu-\mu_0)^2]/\alpha\}^{\alpha/2+1}}{\{1 + n_0(\mu-\mu_0)^2/\alpha\}^{\alpha+1}}\, d\mu \tag{4}$$

is finite when $\alpha > 1$ and $n_0 > 0$. This result is proved in Appendix II.

Fig. 1 compares the harmonic mean $\hat{\pi}_{\text{HM}}(y)$ to the stabilized harmonic mean $\hat{\pi}_{\text{SHM}}(y)$ for various parameter settings of this simple normal example. For each case, both estimates use a common sample of $B = 5000$ independent and identically distributed posterior draws for the mean μ and precision ψ. Shown for each sample is the value of both estimators using ever larger amounts of the sample. Fig. 1 shows clearly how the infinite variance of the harmonic mean estimator manifests itself in practice. Every so often a parameter value with a very small likelihood is generated from the posterior, and this yields a very large value of the reciprocal of the likelihood, which in turn greatly reduces $\hat{\pi}_{\text{HM}}(y)$. Subsequently, $\hat{\pi}_{\text{HM}}(y)$

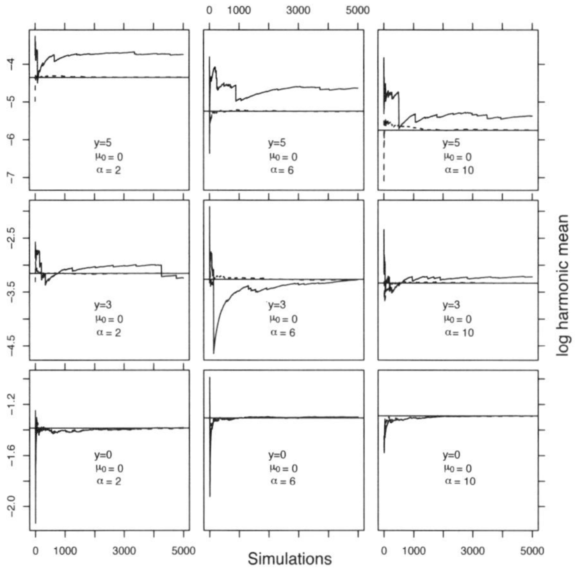

Figure 1: *Normal (bold line) and stabilized t-based (dotted line) harmonic mean estimates of the log integrated likelihood compared with the true value (dashed line), when the data y follow a univariate normal distribution as described in Section 2. The estimate based on the first B values simulated from the posterior distribution is plotted against B for one set of 5,000 values simulated from the posterior in each situation. The top row of the figure displays the harmonic mean estimates when* $y = 5$ *and* $\mu_0 = 0$*. The second row corresponds to* $y = 3$ *and* $\mu_0 = 0$*. The bottom row gives the figures for* $y = 0$ *and* $\mu_0 = 0$*. The three columns correspond to* α *values of 2, 6 and 10. The value of* n_0 *is 1. The plot shows that the normal estimate is unstable but the stabilized estimate is much more stable and converges rapidly to the correct value.*

increases gradually, until another very small likelihood is encountered. Improved performance of the stabilized harmonic mean is evident in Fig. 1.

The t-based estimator $\hat{\pi}_{\text{SHM}}(y)$ converges much more rapidly than the standard estimator, and does not exhibit the same pattern of occasional massive changes. To further validate this observation, we recomputed both final estimators on 1000 independent posterior samples of size $B = 1000$ (Fig. 2). Relative stability of the $\hat{\pi}_{\text{SHM}}(y)$ is clearly indicated.

The reciprocal estimator $\{\hat{\pi}_{\text{SHM}}(y)\}^{-1}$ is a sum of quantities that have finite variance, and so it has a limiting normal distribution by the central limit theorem. This fact can be used to obtain a confidence interval for the integrated likelihood. Table 1 gives the coverage probabilities and the average length of the confidence intervals for the parameter values in Figs 1 and 2, using 1000 independent Monte Carlo samples each of size $B = 1000$. The empirical coverage probabilities are close to their nominal levels. This makes the method a self-monitoring one, in that even if the estimate it provides is imprecise, this will be made clear to the user.

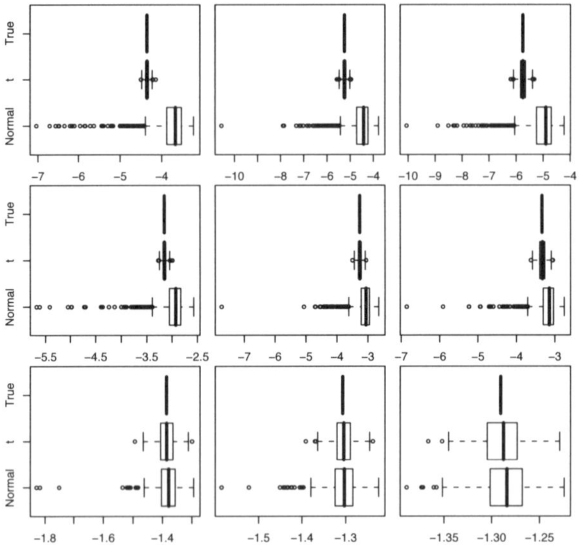

Figure 2: *Boxplots to assess the variability of the estimated integrated likelihood. Shown are the true integrated likelihood, and the normal and stabilized t-based harmonic mean estimators, both on the logarithmic scale. The estimates are obtained from 1000 Monte Carlo samples of size 1000. These estimates are shown for the same configurations of parameters as in Figure 1.*

The multivariate normal model is a direct extension of the univariate normal example discussed above. The standard estimator, obtained using equation (2), is a harmonic mean of multivariate normal densities. This can be easily shown to be an unstable estimator of the integrated likelihood. Integrating the precision parameter leads to a heavier tailed multivariate t density, which can be used to obtain a stable estimator analogous to equation (3).

The stabilized harmonic mean estimator was first reported in a statistical genetics application in which numerical stability of a t-based harmonic mean was observed (Satagopan, Yandell, Newton, and Osborn 1996). Section 3.1 presents a detailed study of this case. Although the genetical model used by these authors was rather specialized, the method to obtain a more stable estimate is quite general: approximate $\pi(y)$ by a harmonic mean of values $\pi[y|h(\theta^t)]$, where $\theta^1, \theta^2, \ldots, \theta^B$ form a sample from the posterior distribution $\pi(\theta|y)$. The function $h(\theta)$ must reduce the parameter space as much as possible, while not making the calculation of the marginal likelihood $\pi[y|h(\theta)]$ too difficult. In the examples we work out, $h(\theta)$ is of lower dimension than θ, typically obtained by integrating out one or several of the components. Taking $h(\theta)$ to be constant is an extreme case; $\pi[y|h(\theta)]$ then becomes the integrated likelihood $\pi(y)$. Of course, if this were computable there would be no need to calculate an approximation, and in any case, the harmonic mean estimator would have zero variance. To form harmonic means from reduced distributions is a general variance reduction technique.

Table 1: *Coverage probabilities for 50%, 80%, 90%, and 95% Confidence Intervals for the stablilized harmonic mean estimator, for the situations shown in Fig s 1 and 2, for 1000 Monte Carlo samples each of size 1000. The average lengths of the confidence intervals for the reciprocal of the likelhood are shown in parentheses. Column 1 shows the parameters used in the simulation, column 2 shows the true value of $\{\pi(y)\}^{-1}$, and columns 3, 4, 5, and 6 give the coverage probabilities.*

(y, μ_0, α)	True $\{\pi(y)\}^{-1}$	50%	80%	90%	95%
(5, 0, 2)	78.09	0.49	0.79	0.90	0.94
		(5.46)	(10.38)	(13.32)	(15.88)
(5, 0, 6)	190.19	0.50	0.81	0.90	0.95
		(23.87)	(45.36)	(58.22)	(69.37)
(5, 0, 10)	314.38	0.53	0.78	0.88	0.93
		(62.44)	(118.64)	(152.27)	(181.44)
(3, 0, 2)	23.44	0.49	0.82	0.90	0.95
		(1.29)	(2.44)	(3.14)	(3.74)
(3, 0, 6)	26.20	0.49	0.78	0.89	0.93
		(2.41)	(4.57)	(5.87)	(6.99)
(3, 0, 10)	28.05	0.48	0.79	0.88	0.93
		(3.57)	(6.78)	(8.71)	(10.37)
(0, 0, 2)	4.00	0.47	0.79	0.90	0.93
		(0.17)	(0.32)	(0.41)	(0.49)
(0, 0, 6)	3.70	0.48	0.77	0.87	0.93
		(0.12)	(0.22)	(0.28)	(0.34)
(0, 0, 10)	3.63	0.47	0.81	0.86	0.93
		(0.12)	(0.22)	(0.28)	(0.34)

Theorem 1 *If h is a measurable function of θ then*

$$\mathrm{Var}\left\{ \frac{1}{\pi[y|h(\theta)]} \middle| y \right\} \leq \mathrm{Var}\left\{ \frac{1}{\pi[y|\theta]} \middle| y \right\}.$$

Either variance may be infinite. If the left hand side is infinite, then the right hand side is infinite also.

To avoid measure-theoretic considerations, we prove Theorem 1 only under the additional condition that $h(\theta)$ is a dimension-reducing transformation: *i.e.*, $\theta = (\alpha, \beta)$, $h(\theta) = \alpha$, and both α and β range freely so that the prior density $\pi(\theta) = \pi(\alpha)\pi(\beta|\alpha)$ is well-defined. See Appendix III for a proof. In certain hierarchical models, where analytical integration is possible on one or two levels, it may be possible to identify useful reductions $h(\theta)$ to facilitate stable harmonic mean calculations.

Gelfand and Dey (1994) noted an extension of the basic identity (1) which justifies estimating the integrated likelihood by the harmonic mean of $\pi(y|\theta^t)\pi(\theta^t)/f(\theta^t)$

where, as before, the θ^t's are sampled from the posterior, but now $\pi(\theta)$ is the prior density and $f(\theta)$ is any (normalized) density on the parameter space. The idea is to choose f carefully so as to minimize Monte Carlo error. We show in Section 3.3 that our proposed stabilization can be combined with this technique for improved performance. Indeed there is some synergy in this combination because the proposed stabilization reduces the dimension of θ, thus making it simpler to identify a useful f function.

3. STABILIZED HARMONIC MEAN ESTIMATOR: EXAMPLES

3.1. *Statistical Genetics Example*

Linear models are used frequently in quantitative genetics to relate variation in a measured trait (phenotype) to variation in underlying genes affecting the trait (genotype). Doerge, Zeng, and Weir (1997) provide a useful review from a statistical perspective. We reconsider the particular model

$$y_i = \mu + \sum_{j=1}^{s} \alpha_j g_{i,j} + \epsilon_i, \qquad i = 1, \cdots, n, \tag{5}$$

used by Satagopan *et al.* (1996) to infer the genetic causes of variation in the time-to-flowering phenotype in the plant species *Brassica napus.* In (5), the i indicates different plants in a sample of size $n = 105$, the phenotypes $y = (y_i)$ are the logarithms of the times to flowering, and the decomposition on the right-hand-side characterizes the expected phenotype conditional on the genotype $g_i = (g_{i,j})$ at a set of s different genetic loci. Here ϵ_i is modeled as a mean zero normally distributed disturbance with variance σ^2 independent of genetic factors, μ is the marginal expected phenotype and α_j is the genetic effect of the jth quantitative trait locus (QTL). From the particular experimental design, each genotype $g_{i,j}$ takes one of two possible values, coded as $\{-1, 1\}$, with equal marginal probability.

The model (5) would be rather standard except that the genotypes $g = (g_i)$ are unobserved; in fact, for each i they represent the values of a random process defined over the whole genome and evaluated at s distinct positions $\lambda = (\lambda_1, \ldots, \lambda_s)$, the s putative QTLs. The number of QTLs, s, is unknown, as are their positions λ and their effects $\alpha = (\alpha_1, \ldots, \alpha_s)$. Indirect information about the QTL genotypes comes through genotype data $m = (m_i)$, obtained in this example from a panel of 10 molecular markers in the chromosomal region of interest. The statistical problem is to infer the unknown parameters $\theta = (\mu, \alpha, \lambda, \sigma^2)$ from marker and phenotype data (m, y), and considering missing genotypes g.

Satagopan *et al.* (1996) presented a Bayesian solution in which Markov chain Monte Carlo (MCMC) was used to sample the posterior distribution of all the unknowns conditional on s, the number of QTLs, separately for a range of values of s. To infer s, the integrated likelihood $\pi(y|m, s)$ was approximated for each s via a harmonic mean, and this enabled calculation of Bayes factors

$$\mathrm{BF}(s_1, s_2) = \pi(y|m, s_1)/\pi(y|m, s_2). \tag{6}$$

We reconsider this calculation in further detail. We can condition on marker information m because its marginal distribution $\pi(m)$ does not depend on any of the unknown parameters.

The prior for θ factorizes into a uniform prior over ordered loci $\lambda = (\lambda_1, \ldots, \lambda_s)$ within the chromosomal region under consideration and a conjugate prior for μ, $\alpha = (\alpha_j)$, and σ^2:

$$\begin{aligned} \pi(\mu|\sigma^2) &= \text{Normal}(\mu_0, \sigma^2/n_0), \\ \pi(\alpha_j|\sigma^2) &= \text{Normal}(\alpha_{0,j}, \sigma^2/n_{0,j}), \qquad j = 1, \cdots, s \\ \pi(\sigma^2) &= \text{Inverse Gamma}(\zeta/2, \zeta/2), \end{aligned}$$

where $\mu_0 = 5$, $n_0 = 1$, $\alpha_{0,j} = 5$, $n_{0,j} = 1$, for each j and $\zeta = 8$. Fixing the number of loci s, one complete scan of the MCMC sampler updates each element of θ and all the missing genotypes in g. See Satagopan *et al.* (1996) for further details on the component updates. A total of 3 chains, corresponding to $s = 1, 2$, and 3, were obtained. For a fixed s (= 1, 2, or 3), we report results below based on a chain of length 400,000 complete scans, subsampled every 100 scans, with the first 100 saved states removed as burn-in; diagnostics indicated that the resulting subsampled scans were close to being independent. Thus this corresponds to an effective independent sample size of about 3,900 for estimating the genetic effect parameters.

Unknowns (θ^t, g^t) are sampled from their posterior distribution conditional on observed phenotypes y, marker genotypes m and the model dimension parameter s. Invoking the standard harmonic mean argument, as in (2), we approximate $\pi(y|m, s)$ by

$$\hat{\pi}_{\text{HM}}(y|m, s) = \left[\frac{1}{B} \sum_{t=1}^{B} \frac{1}{\pi(y|m, \theta^t, g^t, s)} \right]^{-1}. \tag{7}$$

As in the simple normal example of Section 2, a problem arises with (7) because we are averaging reciprocals of normal ordinates. To stabilize the estimator, we integrate out the variance parameter σ^2 and obtain

$$\hat{\pi}_{\text{SHM}}(y|m, s) = \left[\frac{1}{B} \sum_{t=1}^{B} \frac{1}{\pi(y|m, h(\theta^t), g^t, s)} \right]^{-1}, \tag{8}$$

where $h(\cdot)$ returns all components of θ except the variance parameter. In (8), $\pi(y|m, h(\theta^t), g^t, s)$ is a scaled t density, $\text{St}_n(y|\mu + \alpha' g, I, \zeta)$.

Fig. 3 shows the cumulative Bayes factor estimates obtained from three chains, ($s = 1, 2$, and 3), based on integrated likelihood estimates in either (7) and (8). Evidently the stabilization has worked in this more complicated example: there are fewer massive changes in the estimate. Numerically, we obtain $BF(1, 2) = 0.368$ using (7), and $BF(1, 2) = 0.395$ using the stabilized estimator (8). The estimates of $BF(2, 3)$ are rather more disparate: 13.15 and and 4.39, respectively. In any case we would conclude that the two-locus model is most likely *a posteriori*.

Fig. 4 indicates the Monte Carlo sampling variability of the two estimators. The above computations were replicated 75 times. To reduce the computational burden of the simulation, we used a value of B equal to half of the earlier value. The side-by-side boxplots further confirm the success of the stabilization in the present example.

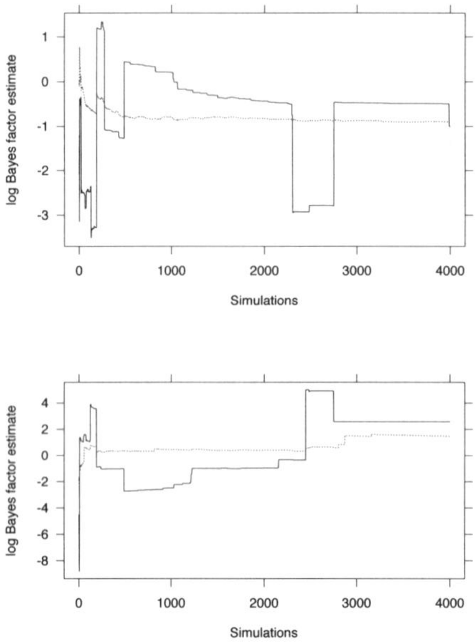

Figure 3: *Log Bayes Factor estimates for the flowering time data, based on MCMC. The log Bayes factor based on the first B saved scans of the MCMC run is plotted against B. The comparison between the one-locus and two-loci models is shown on the top. The bottom figure corresponds to the comparison between the two-loci and three-loci models. The bold line is the standard harmonic mean estimate of the log Bayes factor, and the dotted line is the stabilized t-based estimate. The plot shows that the stabilized estimate is much more stable than the standard one.*

We note that other dimension-reducing transformations $h(\cdot)$ could be used in this example. For example, we could sum out the genotype values g and thus average reciprocals of finite mixtures of normals (or t's). It may also be possible to integrate out the genetic effects α. Neither of these has been attempted here.

3.2. *Beta–Binomial Example*

A naturally occurring hierarchical model has observable counts $y = (y_i)$, $i = 1, \ldots, m$, arising as conditionally independent binomial random variables with numbers of trials (n_i) and success probabilities $p = (p_i)$. In turn, the p_i's are modeled as conditionally independent beta variables with canonical hyperparameters, a and b say, upon which some further prior distribution $\pi(a, b)$ is placed. To obtain the probability of y in this model, we must integrate out both the p_i's and the hyperparameters a and b. It is routine to sample the full parameter set $\theta = (p, a, b)$ from its posterior distribution (Gelman, Carlin, Stern and Rubin, 2003). For example, an MCMC simulation might update each p_i from its Beta full-conditional distribution, and then resort, perhaps, to a random-walk proposal to update a and b.

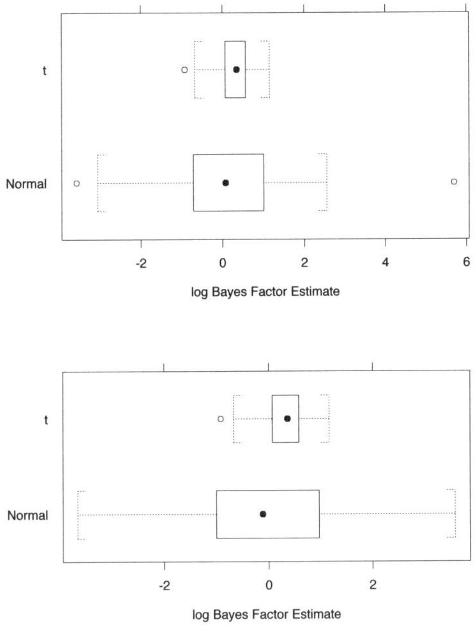

Figure 4: *Assessing the variability of the log Bayes Factor estimates for the flowering time data, using 75 replications of the MCMC run. The top panel shows the comparison between the one-locus and two-loci models, and the bottom panel shows the comparison between the two-loci and three-loci models. In each panel, the variability among the stabilized t-based estimates is shown on top, and that among the standard normal estimates is shown below.*

The basic harmonic mean combines reciprocals of binomial likelihoods from the posterior sample, and, it turns out, can be quite unstable. As before, stability is determined by the second noncentral moment

$$E\left\{[\pi(y|\theta)]^{-2}|y\right\} \propto \int\int\prod_i\left\{\int p^{a-1-y_i}(1-p)^{b-1-n_i+y_i}\,dp\right\}\pi(a,b)\,da\,db.$$

Unless we take an extreme prior $\pi(a,b)$ which ensures that $a > \max(y_i)$ and $b > \max(n_i - y_i)$, this integral can diverge. Typically, a prior extreme enough to avoid this divergence would be unrealistically peaked. This is unsatisfactory, ruling out the standard (unstabilized) harmonic mean estimator as a practical tool for the beta–binomial model.

It is straightforward to stabilize the harmonic mean by reducing the dimension of θ as in previous examples. One possibility is to take $h(\theta) = (a,b)$; *i.e.*, to integrate out all the binomial success probabilities. In this conjugate structure, we have a closed form beta–binomial expression for $\pi\{y|h(\theta)\}$, namely

$$\pi\{y|h(\theta)\} = \prod_i \frac{\Gamma(n_i+1)}{\Gamma(n_i-y_i+1)\Gamma(y_i+1)} \frac{\Gamma(a+b)}{\Gamma(a+b+n_i)} \frac{\Gamma(a+y_i)}{\Gamma(a)} \frac{\Gamma(b+n_i-y_i)}{\Gamma(b)}. \quad (9)$$

The harmonic mean of these beta–binomial probabilities, calculated from the (a, b)'s sampled from their posterior, is consistent for the integrated likelihood. We may expect this to be more stable since the beta–binomial distribution is more diffuse than the binomial, and so the reciprocals of the probabilities may not be as extreme. The stability of this estimator is determined by the second noncentral moment, which satisfies

$$E\left\{[\pi(y|a,b)]^{-2}|y\right\} \quad \leq \quad \int (a+b+n_{\max}-1)^m \pi(a,b)\, da\, db,$$

where $n_{\max} = \max n_i$. Stability is ensured when prior moments of a and b exist.

Data on free-throw percentages from the National Basketball Association (NBA) provide an interesting demonstration of the harmonic mean calculations. On March 9, 1999, there were 414 active NBA players of whom 374 had attempted at least one free throw by that point in the season. Among these 374 players, the numbers of attempts (n_i) ranged from 1 to 205, with a mean of about 35. We model y_i, the number of made free throws by player i, as Binomial with n_i trials and unknown success probability p_i. The average free throw percentage y_i/n_i is about 70% in the data reported at `www.yahoo.com` (and available from the authors).

We consider the problem of evaluating the integrated likelihood $\pi(y)$ under the hierarchical beta–binomial model given above. This would be useful when comparing this model with other hypothesized models for these data. We place independent standard exponential priors on $a - \epsilon$ and $b - \epsilon$ where $\epsilon = 1$ is a lower truncation point of the prior. MCMC was used to simulate the posterior. The following numerical results are based on a single chain of length 2.5 million complete scans, subsampled every 50 scans, and with the first 100 saved states removed as burn-in. Significant trends were not detected in the output and standard MCMC diagnostics indicated that little dependence remained in the saved states. Computations were done separately on a second run and we saw no appreciable differences.

We calculated natural logarithms of the product binomial likelihood and the product beta–binomial likelihood (9). From these values we obtained the standard harmonic mean estimate and the stabilized one. The log estimates were –817.0 and –942.9 respectively; these are quite different. The standard estimate is known to be unstable. Indeed the variance of the sampled loglikelihood values was 146.3 while that of the sampled log beta–binomial values was only 4.1. Variance on the log scale does not tell the whole story because we are averaging on the anti-log scale; it is outliers (having very low likelihood) that are particularly influential, but still variance gives some indication.

Suspecting that some additional improvements could be made, we combined the stabilization technique with the method of Gelfand and Dey (1994) discussed at the end of Section 2, using a Gaussian approximation to the posterior $\pi(a, b|y)$ as the density f. The estimate becomes a harmonic mean of the values $\pi(y|a, b)\pi(a, b)/f(a, b)$, with (a, b)'s sampled from their posterior. The main advantage of this adjustment is that now the influence of individual sample points is greatly diminished.

The estimated log integrated likelihood is -951.4, which matches a brute force grid-based numerical integration of $\pi(y|a,b)\pi(a,b)$ almost exactly. Thus we see that the initial stabilization method worked fairly well and was easily improved.

3.3. *Other Reductions: A Simple Poisson–Gamma Model*

Sometimes useful reductions are hard to find, and the natural approach we have considered of integrating out a parameter does not work. A simple example is when y has a Poisson distribution with mean $\gamma\lambda$, and γ is exponentially distributed with mean 1 and independent of λ *a priori*. The standard harmonic mean estimator of $\pi(y)$ uses samples $\theta^i = (\lambda^i, \gamma^i)$ from $\pi(\theta|y)$, and averages the reciprocals of Poisson probabilities. Stability depends on the second noncentral moment

$$\mathrm{E}\left\{ [\pi(y|\theta)]^{-2} \middle| y \right\} \quad \propto \quad \int\!\!\int \frac{1}{(\gamma\lambda)^y} \exp\{-\gamma(1-\lambda)\}\pi(\lambda)\, d\gamma d\lambda.$$

Note that the inner integral diverges for any $\lambda > 1$, so that the standard harmonic mean is unstable. The natural reduction would be to take $h(\theta) = \lambda$. Thus the marginal likelihood $\pi[y|h(\theta)] = \pi(y|\lambda)$ is a geometric distribution $\lambda^y/(1+\lambda)^{(y+1)}$. Stability here hinges upon

$$\mathrm{E}\left\{ [\pi(y|\lambda)]^{-2} \middle| y \right\} \quad \propto \quad \int \left(\frac{1+\lambda}{\lambda}\right)^y (1+\lambda)\,\pi(\lambda)\, d\lambda.$$

For small λ, the dominant term of the integrand is $\pi(\lambda)/\lambda^y$, and so stability of the modified harmonic mean depends on the prior, though for a standard Gamma prior, for example, this integral can diverge. In other words, both variances in Theorem 1 equal infinity. Thus integrating out γ does not produce a stabilized harmonic mean estimator in this case.

Another, further reduction does work, however. Consider the case where λ, like γ, has a prior exponential distribution with mean 1. Suppose that $h(\theta) = 0$ if $\lambda \le \epsilon$, and $h(\theta) = \lambda$ if $\lambda > \epsilon$, where ϵ is a small predetermined constant. Then $\pi[y|h(\theta) = 0] \approx \epsilon^{y+1}/(y+1)$ (better approximations are readily available if necessary), and it is easily shown that $E\{\pi[y|h(\theta)]^{-2}|y\} < \infty$. Thus, with this refinement, the modified harmonic mean estimator is stable.

4. SHIFTED GAMMA ESTIMATOR OF THE INTEGRATED LIKELIHOOD

4.1. *Shifted Gamma Estimator*

We now consider a different approach to stabilizing the harmonic mean estimate. If MCMC is used to simulate from the posterior, we suppose that the the output has been thinned in such as way that we have an approximately independent sequence of loglikelihoods $\{\ell_t : t = 1, \ldots, B\}$.

We use the fact that asymptotically (as the amount of data underlying the likelihoods increases to infinity, not the number of samples from the posterior), the posterior distribution of the loglikelihoods is given by

$$\ell_{\max} - \ell_t \sim \mathrm{Gamma}(\alpha, 1), \tag{10}$$

where $\ell_{\max}$ is the maximum achievable loglikelihood, and $\alpha = d/2$ where d is the dimension of the parameter θ, *i.e.*, the number of parameters in the underlying model

(Bickel and Ghosh 1990; Dawid 1991). In (10), a Gamma(α, λ^{-1}) distribution with shape parameter α and scale parameter λ has the density

$$f_X(x) = \frac{x^{\alpha-1}\exp(-x/\lambda)}{\Gamma(\alpha)\lambda^{\alpha}}. \tag{11}$$

With this definition, $E(X) = \alpha\lambda$, and $\text{Var}(X) = \alpha\lambda^2$. This can also be viewed as a scaled χ^2 distribution with $d = 2\alpha$ degrees of freedom. Fan, Hung, and Wong (2000) showed that (10) holds under more general conditions than the usual Wald-type conditions required for the likelihood ratio test statistic to be asymptotically χ^2.

In principle, we could use the asymptotic approximation (10) directly to approximate the posterior harmonic mean and hence the integrated likelihood. There are three main difficulties with this, however. First, in general we will not know $\ell_{\max}$ from a posterior sample, because the maximum likelihood will typically not be reached. In practice, the difference between $\ell_{\max}$ and the maximum observed loglikelihood in the MCMC sample can be quite large when the number of parameters is big. Second, in general, we will not know the effective number of parameters, d, especially in hierarchical and other random effects models of the kind often estimated using MCMC. Third, with the posterior distribution (10) of the loglikelihoods, the posterior harmonic mean, and hence the integrated likelihood, is infinite.

The first two difficulties can be resolved by noting that simple moment estimators of $\ell_{\max}$ and α are available. Under the assumption (10), $E[\ell_{\max} - \ell_t] = \alpha$ and $\text{Var}(\ell_t) = \alpha$. Replacing the expectation and variance of ℓ_t by their sample equivalents and solving, we thus get the moment estimators $\hat{\alpha} = s_\ell^2$ and $\hat{\ell}_{\max} = \bar{\ell} + s_\ell^2$, where $\bar{\ell}$ and s_ℓ^2 are the sample mean and variance of the ℓ_t's.

It is clear that $\ell_{\max}$ is at least as big as the largest observed loglikelihood, $\max_t \ell_t$. Thus we could refine the moment estimator of $\ell_{\max}$ to take account of this, as $\hat{\ell}^*_{\max} = \max\{\hat{\ell}_{\max}, \max_t \ell_t\}$, or $\hat{\ell}^{**}_{\max} = \max\{\hat{\ell}_{\max}, \max_t \ell_t + \delta\}$, where δ is some small positive number that is small on the typical scale of loglikelihoods, such as 0.01. We have found, however, that it rarely happens that $\hat{\ell}_{\max}$ is smaller than $\max_t \ell_t$, and that even when it does, the difference is very small. Thus we have not found this refinement of much use in practice.

The third difficulty implies that the approximation (10) is not accurate enough for any actual data that would be encountered. One possibility is to modify it by allowing a scale parameter that is not exactly equal to 1, so that the approximate posterior distribution becomes

$$\ell_{\max} - \ell_t \sim \text{Gamma}(\alpha, \lambda^{-1}), \tag{12}$$

where $\lambda < 1$. In practice, λ will be less than 1, but not much less than 1.

Given the approximation (12), we can find the integrated likelihood using the fact that if $X \sim \text{Gamma}(\alpha, \lambda^{-1})$, then the moment generating function of X is

$$m_X(t) = E[e^{tX}] = (1 - \lambda t)^{-\alpha}. \tag{13}$$

Combining the harmonic mean identity (1) with equations (12) and (13), we see that the integrated likelihood is given by

$$\log \pi(y) = \log E[e^{-\ell_t}|y] = \ell_{\max} + \alpha \log(1 - \lambda). \tag{14}$$

This has an interesting similarity to the BIC approximation to the log integrated likelihood,

$$\log \hat{\pi}_{\mathrm{BIC}}(y) = \ell(\hat{\theta}) - \frac{d}{2}\log(n), \tag{15}$$

where $\hat{\theta}$ is the maximum likelihood estimator, so that $\ell(\hat{\theta}) = \ell_{\max}$, the maximum achievable loglikelihood. In general, under regularity conditions,

$$\log \pi(y) = \log \hat{\pi}_{\mathrm{BIC}}(y) + O_P(1), \tag{16}$$

(Schwarz 1978), so that the relative error in $\log \hat{\pi}_{\mathrm{BIC}}(y)$ tends to zero asymptotically. If the prior $\pi(\theta)$ is a normal unit information prior, then the approximation is more accurate and the $O_P(1)$ term in (16) is replaced by $O_P(n^{-1/2})$ (Kass and Wasserman 1995, Raftery 1995). We have that $\alpha = d/2$, and so $-\log(1-\lambda)$ in (14) corresponds to $\log(n)$ in (15).

We already have estimates of $\ell_{\max}$ and α in (14), and so to obtain an estimate of the integrated likelihood it remains only to estimate λ. Unfortunately this is difficult, because λ is typically close to 1, and so the value of $\pi(y)$ is sensitive to its precise value. On the other hand, the loglikelihoods $\{\ell_t\}$ typically do not allow us to distinguish well between values of λ close to 1. We have experimented with Bayesian and other estimators of λ, but so far the estimates we have tried have not been very accurate. This is a topic of ongoing research.

In the meantime we suggest a posterior simulation-based version of BIC. BIC is defined by

$$\mathrm{BIC} = 2\ell(\hat{\theta}) - d\log(n), \tag{17}$$

and by analogy we define

$$\mathrm{BICM} = 2\hat{\ell}_{\max} - \hat{d}\log(n), \tag{18}$$

where BICM stands for BIC–Monte (Carlo). This yields the following approximation to the log integrated likelihood:

$$\begin{aligned} \log \hat{\pi}_{\mathrm{BICM}}(y) &= \hat{\ell}_{\max} - \frac{\hat{d}}{2}\log(n) && (19)\\ &= \bar{\ell} - s_\ell^2\left(\log(n) - 1\right). && (20) \end{aligned}$$

One difficulty with this criterion is that the sample size n is not always well-defined, particularly in the kind of models commonly estimated by MCMC. Volinsky and Raftery (2000) showed that in another context, when different choices are possible, they each give valid approximations to the integrated likelihood, corresponding to different unit information priors, that differed in the definition of a 'unit'. Thus a reasonable choice may follow by considering what a reasonable definition of a 'unit' is. Volinsky and Raftery (2000) gave an example of one way of determining this.

The definition (17) of BIC, on which BICM as defined by (18) is based, is adequate for fixed effects models, but not for hierarchical or random effects models, as shown for example by Pauler (1998) and Berger, Ghosh, and Mukhopadhyay (2003). Pauler (1998) in her equation (11) proposed a modified definition of BIC for hierarchical models, called S_M, and showed its validity in her Theorem 2. In this approach each parameter potentially has a different 'n' associated with it, corresponding to the effective sample size, or order of information involved in estimating it, and the definition of BIC becomes

$$\mathrm{BIC} = 2\ell(\hat{\theta}) - \sum_{k=1}^{K} \log(n_k), \tag{21}$$

where n_k is the effective sample size involved in the estimation of the k-th parameter, θ_k. We consider a slight modification of this, namely

$$\mathrm{BIC} = 2\ell(\hat{\theta}) - \sum_{k=1}^{K} \log(n_k + 1). \tag{22}$$

This is asymptotically equivalent to (21), but unlike (21) it assigns a nonzero, although small penalty even when $n_k = 1$. It also remains defined even when $n_k = 0$, i.e. when there are no data relevant to that parameter, assigning no penalty in that case, which seems appropriate. In general, determining n_k involves assessing the Fisher or observed information for θ_k (Pauler 1998), but we will take as a rough approximation the number of data points that participate in the estimation of θ_k.

This leads to a modified definition of BICM. Parameters are divided into classes according to the number of data points that participate in the estimation of each one, and are ordered according to the value of n_k. The random effects will be last, and will be assigned an effective number of parameters equal to $\hat{d} - K'$, where K' is the number of parameters already accounted for. Thus we have

$$\mathrm{BICM} = 2\hat{\ell}_{\max} - \sum_{k=1}^{K'} \log(n_k + 1) - (\hat{d} - K') \log(n_{K'+1}). \tag{23}$$

An example of the use of equation (23) is given in Section 4.3.

In a similar way, we can write down a posterior simulation-based version of AIC (Akaike 1973). AIC can be defined as

$$\mathrm{AIC} = 2\ell_{\max} - 2d, \tag{24}$$

which we can estimate by

$$\begin{aligned} \mathrm{AICM} &= 2\hat{\ell}_{\max} - 2\hat{d} & (25)\\ &= 2\hat{\ell}_{\max} - 4s_\ell^2 & (26)\\ &= 2(\bar{\ell} - s_\ell^2). & (27) \end{aligned}$$

Thus AICM is seen to be a very simply computed penalized version of the posterior mean of the loglikelihoods, using only the loglikelihoods from the posterior simulation. There is a substantial literature on the relative merits of AIC and BIC, and many of the same arguments could probably be made about AICM and BICM. Our derivation of BICM is as an approximation to the log integrated likelihood, but AICM does not have such an interpretation.

We can obtain standard errors of BICM and AICM using the facts that $\mathrm{Var}(\bar{\ell}) \approx d/(2B)$ and

$$\mathrm{Var}(s_\ell^2) \approx d(11d/4 + 12)/B,$$

together with the approximate posterior independence of $\bar{\ell}$ and s_ℓ^2. The criteria BICM (in both the forms we have given) and AICM are both of the form $a\bar{\ell} - bs_\ell^2$. The standard error of a criterion of this kind is thus

$$\sqrt{a^2\hat{d}/(2B) + b^2\hat{d}(11\hat{d}/4 + 12)/B}.$$

Note that this standard error takes account only of the Monte Carlo variation, *i.e.*, it is an estimate of the standard deviation of the criterion over repeated posterior simulation runs of the same length. It does not take account of error in the approximation to the log integrated likelihood or of sampling variation in the data themselves. It is a standard error of BICM, when BICM is viewed as an estimator of the BICM value that would be obtained asymptotically if the number of draws from the posterior grew without bound; similarly for AICM. Note also that it depends crucially on the assumption that the posterior simulation draws are approximately independent, so that if MCMC is used, the posterior sample would have to be thinned enough for this to be the case.

As we have noted, the moment estimator of α implies that $\hat{d} = 2s_\ell^2$ can be viewed as an estimator of the effective number of parameters. Spiegelhalter, Best, Carlin, and van der Linde (2002) proposed a different estimator of the effective number of parameters from posterior simulation, $p_D = 2(\log \pi(\bar{\theta}|y) - \bar{\ell})$, where $\bar{\theta}$ is the mean of the values of θ simulated from the posterior. In our limited experience, we have found that p_D and $\hat{d}$ are similar and that both work well in situations where the number of parameters is known.

However, Spiegelhalter *et al.* (2002) have pointed out that p_D is not invariant to the model's parameterisation because it involves the posterior mean of the parameters, $\bar{\theta}$, and that this noninvariance can be consequential. They also pointed out that p_D can be negative. In addition, p_D may not be well defined in situations where the meaning of $\bar{\theta}$ is not clear, such as multinomial parameters, or finite mixture models where the unobserved group memberships are included in the MCMC scheme (Diebolt and Robert 1994). A similar problem arises when there is near posterior nonidentifiability such as label-switching in mixture models or random effects without identifying constraints (Celeux, Hurn, and Robert, 2000; Stephens, 2000). One way around this is to use a posterior mode of θ instead of $\bar{\theta}$, but Richardson (2002) gave several examples of mixture models where p_D with this definition inadequately penalizes model complexity. The estimator $\hat{d}$ is defined simply and unambiguously in all those cases.

The estimator $\hat{d}$ was also derived by Gelman *et al.* (2003, Section 6.7), and used by them instead of p_D in their alternative definition of DIC, the measure of fit originally defined by Spiegelhalter *et al.* (2002). AICM is equivalent to Gelman *et al.* (2003)'s definition of DIC. Gelman *et al.* (2003) used this alternative definition because the deviance function was not available to their MCMC program and so their program could not compute p_D routinely, whereas it could compute $\hat{d}$; see also Sturtz, Ligges and Gelman (2005).

An interesting observation follows from the results of Fan *et al.* (2000). They consider the situation where, roughly speaking, the level-w contour of the likelihood function has the form $\hat{\theta} + a_n w^r S$, where $\hat{\theta}$ is the maximum likelihood estimator, $r > 0$ is a constant, $a_n \to 0$ is a sequence, and S is a surface in R^d. The standard situation where the likelihood contours are elliptical has $r = 1/2$, $a_n = O(n^{-1/2})$, and $S = \{\theta : \theta^T \Sigma \theta\}$ where Σ is the Fisher information matrix, so that S is an ellipse.

When the contours are not elliptical, they say that the distribution is 'fan-shaped.' They show that in general under these conditions

$$\ell_{\max} - \ell_t \sim \text{Gamma}(rd, 1). \tag{28}$$

In the standard, elliptical situation with $r = \frac{1}{2}$, this reduces to (10) as before.

They give several simple examples within their class where the likelihood contours are not elliptical. One is inference about the minimum of a shifted exponential distribution whose scale parameter is known. In that case they show that $r = 1$. Thus the 'effective number of parameters' in that case is 2, even though there is only one actual parameter. This illustrates the fact that the term 'effective number of parameters' is really just a figure of speech. It suggests that what is important for estimating the integrated likelihood is the shape parameter of the approximating gamma distribution, not a literal count of the parameters in the model. The arguments above suggest that the former may continue to be well approximated by $2s_\ell^2$ even when this does not coincide with a simple count of the number of parameters.

Finally, we note that when the number of parameters (not necessarily data points) becomes large, the shifted gamma approximation to the posterior distribution of the loglikelioods (12) becomes approximately normal. The posterior distribution of the reciprocal of the likelihood is then approximately lognormal, leading to the estimator

$$\log \hat{\pi}_{\text{LN}}(y) = \bar{\ell} - \tfrac{1}{2} s_\ell^2. \tag{29}$$

This was proposed by Pritchard, Stephens, and Donnelly (2000), who also noted that a better approximation might be available by using a gamma distribution for the loglikelihoods, thus prefiguring the present work, although they did not develop their observation further. Pritchard *et al.* (2000) proposed and used $\log \hat{\pi}_{\text{LN}}(y)$ as a model choice criterion rather than an estimator of the log integrated likelihood. It is interesting to note that

$$\log \hat{\pi}_{\text{LN}}(y) = \hat{\ell}_{\max} - \tfrac{3}{4}\hat{d},$$

so that $\log \hat{\pi}_{\text{LN}}(y)$ is a penalized version of the estimated maximum loglikelihood, with a penalty similar to but smaller than that of AICM, equal to $\frac{3}{4}\hat{d}$ rather than $\hat{d}$ as for AICM.

4.2. *Multivariate Normal Simulation Experiment*

In order to assess the estimators $\hat{d}$, $\hat{\ell}_{\max}$ and $\pi_{\text{BICM}}(y)$, we first carried out a small simulation study using a canonical multivariate normal situation. The data $y_1, \ldots, y_n$ are independent and identically distributed $\text{MVN}_d(\mu, I)$ random vectors, and the prior for μ is $\mu \sim \text{MVN}_d(0, I)$. The sufficient statistic is then just the d-dimensional $\bar{y} \sim \text{MVN}_d(\mu, I/n)$. We simulated values of μ from its posterior distribution $\mu|y \sim \text{MVN}_d(n\bar{y}/(n+1), I/(n+1))$. The loglikelihoods are then given by

$$\ell_t = \log p(\bar{y}|\mu^t) = \frac{d}{2}\log(n/2\pi) - \frac{n}{2}\sum_{j=1}^{d}(\bar{y}_j - \mu_j)^2.$$

The true maximum likelihood is $\frac{d}{2}\log(n/2\pi)$ and the true log integrated likelihood is

$$\pi(y) = \frac{d}{2}\log\left(\frac{n}{(n+1)2\pi}\right) - \tfrac{1}{2}\frac{n}{n+1}\sum_{j=1}^{d}\bar{y}_j^2.$$

Our goal was to see how the method worked under a wide range of values of d and n, so we fixed μ at $(0.15, \ldots, 0.15)$. We simulated values of the number of parameters d from a discrete uniform distribution on the integers from 1 to 100, and we simulated values of the sample size n from a discretized log-uniform distribution with $\log(n) \sim U[3, 9]$, so that approximately, n ranged from 20 to 8000, with a median of 400, subject to the constraint that $d < n$. Thus the simulation encompassed standard situations with a small number of parameters and a large sample size, and also situations where there were almost as many parameters as data points, ranging up to moderately large numbers of parameters (100). For each pair of values of d and n sampled, a dataset consisting of $\bar{y}$ was drawn, and then the posterior distribution was simulated. Altogether, 1000 datasets were simulated, and for each dataset a sample of size 100,000 was drawn from the posterior. This is a standard fixed effects model and so for BICM we used the definition (18).

The results are shown in Fig. 5. The upper left panel shows the histogram of loglikelihoods for one dataset with $d = 10$ and $n = 100$, together with the fitted gamma distribution superimposed. The fit is extremely good, and this was the case for all the datasets that we examined. The upper right panel shows the estimated maximum achievable loglikelihood plotted against the true maximum likelihood for the 1000 simulated datasets. The estimation was good, even in cases with larger number of parameters, where the largest loglikelihood among those sampled, $\max_t \ell_t$, was much smaller than the true maximum loglikelihood. The lower left panel shows the estimated number of parameters plotted against the true number; again the estimation was very good. Finally, the lower right panel shows the approximated and true log integrated likelihoods; again the estimation was good.

In the simulated situation, the prior used was a unit information prior, so it is of interest to see what happens if a different prior is used. We experimented with situations where the prior was $\mu \sim \text{MVN}_d(0, \sigma^2)$ where $\sigma^2 \neq 1$. Note that the unit information prior corresponds to $\sigma^2 = 1$. The good results for $\hat{d}$ and $\hat{\ell}_{\max}$ remained unchanged. As long as σ^2 was larger than about 0.2, *i.e.*, as long as the prior was not highly informative, the value of $\log \hat{\pi}_{\text{BICM}}(y)$ remained very highly correlated with the true value of $\log \pi(y)$. The slope of the line in the lower right panel of Fig. 5 was no longer unity, but the fact that the correlation remained very high means that model comparisons based on the estimated log integrated likelihoods would remain accurate. A more accurate approximation to the absolute value of $\pi(y)$ could be obtained by replacing $\log(n)$ by $\log(\sigma^2 n)$ in the expression (19) for $\hat{\pi}_{\text{BICM}}(y)$. However, this would be a model-specific adjustment and would take us beyond the generic estimates that we are aiming for here.

4.3. *Example: Latent Space Models for Social Networks*

Social network data consist of observations on relations between actors, for example whether one individual says she likes another. Often such data are binary, in which a directed or undirected relation between actor i and actor j either exists or does not. In this case, the data consist of values of y_{ij} for $i, j = 1, \ldots, n$, where i and j index the n actors, and $y_{ij} = 1$ if the relation from i to j exists and $y_{ij} = 0$ if it does not.

Hoff, Raftery, and Handcock (2002) introduced the latent position model for data such as these. In this model, each actor i is assumed to be associated with an observed or latent position in an unobserved q-dimensional Euclidean 'social space', denoted by z_i. Then the model says that the y_{ij} are conditionally independent given

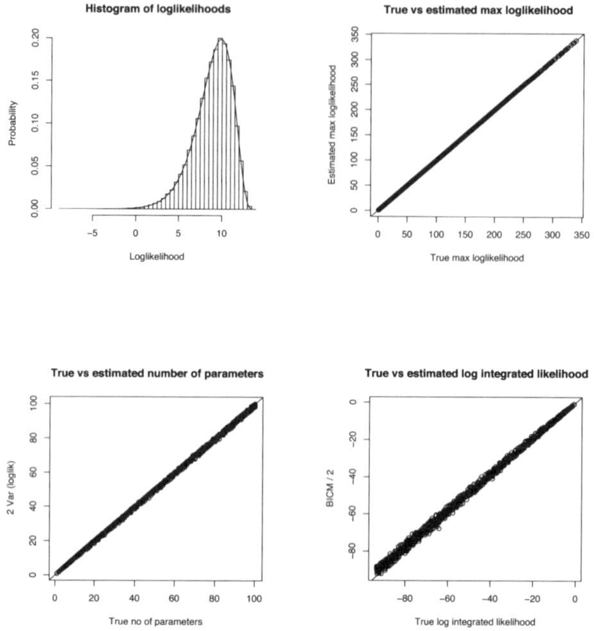

Figure 5: *Multivariate Normal simulation study of the shifted Gamma estimator. Upper left: Histogram of the loglikelihoods for one dataset with* $d = 10$ *parameters and* $n = 100$ *data points, with the fitted gamma density superimposed. Upper right: The estimated maximum achievable loglikelihood,* $\hat{\ell}_{\max}$*, plotted against the true maximum loglikelihood for the 1000 simulated datasets. Lower left: The estimated number of parameters,* $\hat{d}$*, plotted against the true number of parameters for the 1000 datasets. Lower right: The estimated log integrated likelihood,* $\log \hat{\pi}_{\text{BICM}}(y)$*, plotted against the true log integrated likelihood for the 1000 simulated datasets. In the last three plots, the solid line is the* $y = x$ *or identity line.*

the latent positions, with

$$\log\left(\frac{\Pr(y_{ij}=1)}{\Pr(y_{ij}=0)}\right) = \beta - |z_i - z_j|, \tag{30}$$

$$z_i \stackrel{\text{iid}}{\sim} \text{MVN}_q(0, \sigma^2 I). \tag{31}$$

There are just two parameters for which priors are needed, β and σ^2, and we use the priors $\beta \sim N(0, 10^2)$ and $\sigma^2 \sim \sqrt{10}\,\text{Inverse}\,\chi^2_3$. These priors are proper but reasonably spread out. Estimation is carried out by MCMC on β, σ^2 and the z_i's.

Here we consider a well-known dataset on the relations among 18 monks collected by Sampson (1968). Each monk was asked with which other monks he had positive relations. Based on extensive analyses of these and much other data, the 18 monks have traditionally been classified into three groups: the Loyal Opposition, the Young Turks, and the Outcasts. Hoff *et al.* (2002) analyzed a subset of these data, and the fuller dataset we analyze here was previously analyzed by Handcock, Raftery, and Tantrum (2005).

Interest focuses here on the choice of dimension, and MCMC estimation is carried out for each dimension $q = 1, 2, 3, 4$. In computing BICM, the issue of how to define the penalty term arises. This is a random effects model, and so we use the form (23). There are three 'groups' of parameters: β (one parameter) with associated effective sample size n_1, σ^2 (one parameter) with associated effective sample size n_2, and the latent positions z_i, which are random effects (all remaining parameters), with associated effective sample size n_3.

The number of actors in the data is 18 so that the number of potential links is $\binom{18}{2} = 306$, and the number of actual links is 88. The parameter β is estimated from a logistic regression with 306 cases and 88 'successes', and we take the associated effective sample size to be $n_1 = 88$, following arguments analogous to those of Volinsky and Raftery (2000). The parameter σ^2 is estimated from data on the 18 actors, and so we take the associated effective sample size to be $n_2 = 18$. Finally, most of the information about an actor's latent position z_i comes from the links to and from that actor. There are an average of $88/18 = 4.9$ links per actor, and so we take the effective sample size associated with the random effects to be $n_3 = 4.9$. BICM is then defined as

$$\text{BICM} = \hat{\ell}_{\max} - \log(n_1 + 1) - \log(n_2 + 1) - (\hat{d} - 2)\log(n_3 + 1). \tag{32}$$

Note that the values of n_1 and n_2 chosen do not affect model selection, because the corresponding terms cancel in computing differences between BICM values for different models, which are what matter for model comparisons.

The results are shown in Table 2. In addition to our estimates of the maximized likelihood, estimates of the maximized loglikelihood by numerical optimization are shown. These agree reasonably closely with our estimates. Also, the number of parameters involved in the MCMC simulation is shown, and this corresponds fairly well with $\hat{d}$, the estimated number of parameters. There is no reason to expect the effective number of parameters to be the same as the number of parameters over which the MCMC algorithm iterates in this kind of hierarchical latent variable model, but in this case they do line up rather well.

Table 2: *Comparing dimensions in the latent space social network model. q is the dimension of the latent space, $\ell_{\max}$ is the maximized loglikelihood from a numerical optimisation routine, and # par is the total number of parameters estimated, including the latent position coordinates. The best values of BICM and AICM are shown in bold.*

q	$\hat{\ell}_{\max}$	$\ell_{\max}$	$\hat{d}$	# par	$\log\hat{\pi}_{\text{BICM}}(y)$	SE	$\frac{1}{2}$AICM	SE
1	−128.6	−129.1	20.4	20	−148.6	0.3	−149.0	0.4
2	−109.6	−110.3	38.0	38	**−145.3**	0.9	**−147.6**	0.7
3	−87.8	−89.9	66.1	56	−148.4	1.0	−154.0	1.1
4	−79.3	−73.3	78.6	74	−151.0	1.2	−157.9	1.3

According to the $\hat{\pi}_{\text{BICM}}(y)$ estimate of the integrated likelihood, the preferred latent space model for these data is a two-dimensional one. These data have usually been visualized in two dimensions, so this agrees with previous practice, although we are not aware of any previous efforts to choose the dimension of the latent space in a formal way. AICM makes the same choice, although by a small margin over the one-dimensional model.

Fig. 6 shows the estimated latent positions for these data. The left panel shows the estimated two-dimensional positions. The three well-known groups are clearly delineated. It is clear that the density of links is highest within each group. However, the Young Turks have some links to both of the other groups, while the Loyal Opposition and the Outcasts are joined by very few links. This suggests that a one-dimensional arrangement with the Young Turks in the middle might represent the main features of the data adequately.

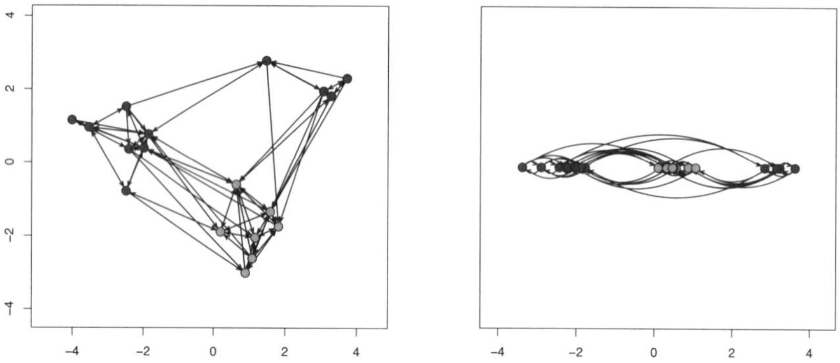

Figure 6: *Estimated latent positions of monks in social network example. Left panel: Two-dimensional latent positions with links also shown. Right panel: One-dimensional latent positions. In both plots, the known groupings of the monks are shown: Left = Loyal opposition; Midddle = Young Turks; Right = Outcasts. Both the one-dimensional and the two-dimensional latent position models give results that are consistent with the known groupings.*

The right panel of Fig. 6 shows the one-dimensional estimated latent positions. The three main groups are as well identified by the one-dimensional model as by the two-dimensional model. Again it seems reasonable that the Young Turks have a more central position, suggesting that a one-dimensional latent space captures most of the main features of the data, as suggested by the relatively small differences in BICM and AICM between the one- and two-dimensional models.

Our method provides standard errors for BICM and AICM, and these are also shown in Table 2. These increase rapidly, and roughly proportionally with the number of parameters. They can be used to calculate standard errors of the difference between the BICM values for two different models using the fact that the values for different models are independent. We use the standard formula

$$\text{SE}\ (\text{BICM}_1 - \text{BICM}_2) = \sqrt{\text{SE}(\text{BICM}_1)^2 + \text{SE}(\text{BICM}_2)^2}.$$

A similar formula holds for differences between AICM values. On the key model comparison, between the one- and two-dimensional latent space models, the standard error of the difference between the values of $\log \hat{\pi}_{\text{BICM}}(y)$ for the two models is about 0.95, suggesting that the observed difference of 3.3 would be unlikely to change sign if more MCMC runs were done. The standard error of the difference between the two values of $\frac{1}{2}$AICM is about 0.8, casting some doubt on whether the observed difference of 1.3 would persist in a longer MCMC run. If one wanted to select one model on the basis of AICM, this suggests that a longer MCMC run should be used.

5. DISCUSSION

Our final goal is a generic method that estimates the integrated likelihood using only the likelihoods given a set of draws from the posterior. We have investigated approaches to this based on the harmonic mean identity, which says that the integrated likelihood is the posterior harmonic mean of the likelihood. The most obvious esimator from this, the sample posterior harmonic mean of the likelihoods, is unbiased and simulation-consistent, but does not have finite variance in general and so is often unstable (Newton and Raftery 1994).

We have investigated two approaches to more stable estimation of the integrated likelihood using the harmonic mean identity. The first is to reduce the parameter space and then use the sample posterior harmonic mean; by judiciously choosing the likelihood to be used this can yield stable and finite variance estimators. The second approach involves modeling the posterior distribution of the loglikelihood by a shifted gamma distribution. This leads to estimates of the effective number of parameters and the true maximum likelihood that seem to work well, and hence to posterior-simulation-based analogues of the well-known BIC and AIC criteria, called BICM and AICM.

Our first approach takes advantage of dimension-reducing transformations on the parameter space. The proposed variance stabilizing method extends a very simple tool into a range of widely used hierarchical statistical models. As illustrated in Section 3, dimension reduction is straightforward in certain hierarchical models. Sometimes the natural approach of integrating out a nuisance parameter does not yield a stabilized estimator, however, and one must search farther. We have given one example in Section 3.3, a simple Poisson–Gamma model, where the natural approach does not work directly, but a slight refinement of the $h(\cdot)$ function does yield a stabilized estimator. The trick used there to find this refined h function was based on the fact that the estimator is stable if and only if $E\{\pi[y|h(\theta)]^{-2}|y\} < \infty$. We wrote this expectation as an integral, identified the part of the range of integration responsible for the integral being infinite, and effectively carried out the integration over that small part of the space via analytic approximation, thus defining a new h function. Dimension reduction for variance stabilization may not be an effective method to compute normalizing constants in certain very hard problems. In the cases we have studied, we have shown that it is possible to stabilize the harmonic mean estimator and obtain estimates that are much more accurate, but still easy to calculate.

Another application of our first stabilization approach includes robust linear models (Andrews and Mallows 1974; Carlin and Louis 1996). The robust linear model has an error term distributed as $Z/\sqrt{U}$, where Z and U are independent,

Z has a centered normal distribution, and U has a χ^2 distribution. The standard harmonic mean estimator can have infinite variance. A stabilized harmonic mean estimator can then be obtained by integrating out the denominator U.

Hierarchical models that involve standard distributions may be good candidates for our first approach. For one thing, MCMC is well understood for within-model posterior simulation. Furthermore, the integrations required for dimension reduction may be solved analytically. The simplicity of the resulting stabilized harmonic mean is its main advantage.

Our second approach involves modeling the posterior distribution of the log-likelihoods by a shifted gamma distribution. This fits the observed distribution of loglikelihoods well in some applications, and leads to very simple estimates of the effective number of parameters and the true maximum likelihood that seem of good quality. This in turn yields posterior-simulation-based analogues of the BIC and AIC criteria, BICM and AICM. It also provides simple standard errors for these criteria, which can be useful both for assessing the results and for deciding whether enough samples have been drawn from the posterior for model comparison purposes.

The BICM criterion we have defined requires the specification of sample size, and this may be problemmatical in some applications. The analogies with the results of Volinsky and Raftery (2000) suggest that in fixed effects models acceptable choices may be possible by considering what a reasonable choice of a unit of information for a unit information prior would be. Analogies with the results of Pauler (1998) suggest a corresponding approach for random effects or hierarchical models. In our examples, these approaches have worked fairly well.

It would be desirable, however, to have a fully automated solution where this parameter could be estimated from the posterior simulation output. We have investigated various possible solutions to this, mostly Bayesian estimates of the gamma distribution parameters that exploit the prior information that the scale parameter is less than 1, but not much less than 1. The results so far have not satisfied us fully, however, and so we did not present them here.

The general idea explored here, of estimating the posterior harmonic mean of the likelihood by modeling the loglikelihoods, may yield progress by using models other than the shifted gamma distribution. For example, it may be possible to make progress by recognizing that in regular models the posterior distribution of the loglikelihood can be approximated asymptotically by a shifted and scaled noncentral chi-squared distribution with a small noncentrality parameter, perhaps better than by the (central) shifted and scaled gamma distribution we have been using so far. The estimation of the scale and noncentrality parameters is delicate, however.

Another approach might take advantage of the work that has been done on approximating the posterior distribution of the loglikelihood using Edgeworth expansions. Bickel and Ghosh (1990) proposed such an expansion where the leading term is of the form (10). This expansion would not in itself be useful for the present purpose because the leading term still yields an infinite log integrated likelihood, but the basic idea may be fruitful in a modified form. Other expansions that have been proposed might also be useful; many of these are reviewed by Reid (2003).

A range of other methods for computing integrated likelihoods from posterior simulation have been proposed. Most of these methods are not generic algorithms that use only the output of the posterior simulation; in most cases they require additional simulations or model-specific calculations. Other methods have been proposed for estimating Bayes factors or posterior model probabilities, but not the underlying integrated likelihoods themselves. Subsets of the different methods have been reviewed and compared by DiCiccio, Kass, Raftery, and Wasserman (1997), Han and Carlin (2001), Bos (2002), Clyde and George (2004), Sinharay and Stern (2005), and Rossi, Allenby and McCulloch (2005, Chapter 6).

Newton and Raftery (1994) proposed modifications of the harmonic mean estimator using real or imaginary draws from the prior, and these have been applied, for example by Zijlstra, van Duijn, and Snijders (2005), with some success, but they are still somewhat unstable. As we discussed in Section 2, Gelfand and Dey (1994) proposed a method that can be viewed as a generalization of the harmonic mean estimator. It requires the careful choice of a function of the entire parameter vector, tailored for each application, and so is not as generic as the methods we have been discussing, although with a good choice of function it can perform well. As we have shown in Section 3.2, it can be combined with our first approach to achieve further improvements.

The method of Chib (1995) was developed for the specific case where posterior simulation is done by Gibbs sampling. It is based on the conditional probability formula for the normalizing constant, and requires running specially designed auxiliary conditional MCMC samplers. Chib and Jeliazkov (2001) extended this to the case of the Metropolis-Hastings algorithm, in which case it requires a different auxiliary simulation algorithm additional to the main MCMC algorithm. These methods have been successfully applied to specific models, for example by Albert and Chib (2001), Chib, Nardard, and Shephard (2002), and Basu and Chib (2003). However, Neal (1999) showed that Chib (1995)'s application of the idea to mixture models was incorrect, and Rossi et al (2005, Section 6.9) showed the instability of the method due to large outliers in the posterior simulation.

The method of Chib (1995) was developed for the specific case where posterior simulation is done by Gibbs sampling. It is based on the conditional probability formula for the normalizing constant, and requires running specially designed auxiliary conditional MCMC samplers. Chib and Jeliazkov (2001) extended this to the case of the Metropolis-Hastings algorithm, in which case it requires a different auxiliary simulation algorithm additional to the main MCMC algorithm.

Oh (1999) proposed a method based on an identity that requires knowledge of full conditional posterior densities. Lockwood and Schervish (2005) proposed two methods, one a brute force method, and the other a sequential approach that is related to the method of Oh (1999). Chen (2005), building on Chen (1994), proposed a method that uses another identity. It involves the use of latent variables and the proposed optimal version of the method requires knowledge of the full conditional posterior distribution of the parameters given the latent variables, including all normalizing constants.

A version of the Laplace method in which the required posterior modes and Hessian matrices are estimated from posterior simulation output, called the Laplace–Metropolis method, was proposed by Raftery (1996) and Lewis and Raftery (1997). This is a generic method but can depend on the model's parameterization, and may not work well for very high-dimensional models. Importance sampling based methods have also been proposed (Nandram and Kim 2002; Steele, Raftery, and Emond 2006), but these can also require model-specific computations.

Several methods have been proposed for estimating Bayes factors, or ratios of integrated likelihoods, but not the integrated likelihoods themselves. These include the Savage–Dickey ratio and a generalization of it (Verdinelli and Wasserman 1995), and bridge sampling (Meng and Wong 1996; Mira and Nicholls 2004). Johnson (1999) has proposed a method for estimating the integrated likelihood that involves simulating from a second density as well as the posterior; it seems that for its performance to be good the second density needs to be carefully chosen taking account of the situation at hand.

A general approach to estimating posterior model probabilities is to use transdimensional MCMC, pioneered by Green (1995) with his introduction of reversible jump MCMC; a review of this area is given by Sisson (2005). These methods can be used to estimate Bayes factors, but not the underlying integrated likelihoods. Bayes factors can be read off the output of transdimensional MCMC directly, and more efficient approaches to estimating Bayes factors from transdimensional MCMC have been discussed by Bartolucci, Scaccia, and Mira (2006). Godsill (2001) has pointed out that integrating out parameters analytically can improve the efficiency of transdimensional MCMC; this is analogous to our proposal here to stablize the harmonic mean estimator by parameter reduction.

ACKNOWLEDGEMENTS

The authors are grateful to Mark Handcock and Marijtje van Duijn for extensive and helpful discussions. They also thank Jim Berger, David Dunson, Peter Hoff, Donna Pauler, Matthew Stephens and Aki Vehtari for useful comments.

APPENDIX I: STUDENT'S t

Student t

Copying from Bernardo and Smith (1994, page 122),

$$\mathrm{St}(x|\mu,\lambda,\alpha) = c\left[1+\frac{\lambda}{\alpha}(x-\mu)^2\right]^{-(\alpha+1)/2}, \qquad c = \frac{\Gamma((\alpha+1)/2)}{\Gamma(\alpha/2)\,\Gamma(1/2)}\left(\frac{\lambda}{\alpha}\right)^{1/2}.$$

Multivariate Student t

Using the notation of Bernardo and Smith (1994, page 139),

$$\mathrm{St}_n(x|\mu,\lambda,\alpha) = c\left[1+\frac{1}{\alpha}(x-\mu)^T\lambda(x-\mu)\right]^{-(\alpha+n)/2},$$

where

$$c = \frac{\Gamma((\alpha+n)/2)}{\Gamma(\alpha/2)\;(\alpha\pi)^{n/2}}\det(\lambda)^{1/2}.$$

x and μ are of dimension n. λ is a symmetric, positive-definite $n \times n$ matrix, and $\alpha > 0$.

APPENDIX II: PROOF OF EQUATION (4)

Define

$$f(\mu) = \tfrac{n_0}{\alpha}(\mu-\mu_0)^2 \qquad \text{and} \qquad g(\mu) = \tfrac{1}{\alpha}(y-\mu)^2\,.$$

Set

$$a(\mu) = 1 + \frac{g(\mu)}{1 + f(\mu)} \; .$$

It can be easily shown that the maximum of the continuous function $a(\mu)$ occurs at $\mu^* = \mu_0 - \alpha/[n_0(y - \mu_0)]$, and the maximum value of the function is

$$a(\mu^*) = 1 + \frac{1}{n_0} + g(\mu_0) \; .$$

Further $a(\mu) \to 1+1/n_0$, as $\mu \to \pm\infty$. The expected value of interest can be written as

$$E\left\{\frac{1}{[\pi(y|\mu)]^2}|y\right\} \quad \propto \quad \int [a(\mu)]^{\alpha/2+1}[1 + f(\mu)]^{-\alpha/2} d\mu \; ,$$

where $[1 + f(\mu)]^{-\alpha/2}$ is proportional to a t-density of the form

$$\mathrm{St}(\mu|\mu_0, n_0(\alpha - 1)/\alpha, \alpha - 1) \; .$$

Since $1 \leq a(\mu) \leq a(\mu^*)$, the integral on the right hand side is finite by dominated convergence theorem when $\alpha > 1$ and $n_0 > 0$.

APPENDIX III: PROOF OF THEOREM 1

Define $\alpha = h(\theta)$, write $\theta = (\alpha, \beta)$, and set

$$a = \mathrm{E}\left\{\frac{1}{[\pi(y|\alpha)]^2}\bigg| y\right\} \qquad \text{and} \qquad b = \mathrm{E}\left\{\frac{1}{[\pi(y|\theta)]^2}\bigg| y\right\}.$$

Since both $1/\pi(y|\alpha)$ and $1/\pi(y|\theta)$ have common expectation $1/\pi(y)$, it suffices to show that $a \leq b$. Expanding b, we have

$$\begin{aligned} b \;&=\; \int\!\!\int \frac{1}{[\pi(y|\alpha,\beta)]^2}\, \pi(\alpha,\beta|y)\, d\beta\, d\alpha \\ &=\; \int\!\!\int \frac{1}{[\pi(y|\alpha,\beta)]^2}\, \pi(\beta|\alpha,y)\, p(\alpha|y)\, d\beta\, d\alpha \\ &=\; \int b(\alpha)\, \pi(\alpha|y)\, d\alpha \end{aligned}$$

where

$$b(\alpha) = \int \frac{1}{[\pi(y|\alpha,\beta)]^2}\, \pi(\beta|\alpha,y)\, d\beta.$$

By contrast,

$$a = \int a(\alpha)\pi(\alpha|y)\, d\alpha$$

where

$$a(\alpha) = \frac{1}{[\pi(y|\alpha)]^2}.$$

Therefore, it is sufficient to prove that $a(\alpha) \leq b(\alpha)$ for all α. Simplifying $b(\alpha)$, we have

$$\begin{aligned} b(\alpha) &= \int \frac{1}{[\pi(y|\alpha,\beta)]^2} \pi(\beta|\alpha, y)\, d\beta \\ &= \int \frac{1}{[\pi(y|\alpha,\beta)]^2} \frac{\pi(y|\alpha,\beta)\,\pi(\beta|\alpha)\,\pi(\alpha)}{\pi(y|\alpha)\,\pi(\alpha)}\, d\beta \\ &= \frac{1}{\pi(y|\alpha)} \int \frac{\pi(\beta|\alpha)}{\pi(y|\alpha,\beta)}\, d\beta. \end{aligned}$$

Cancelling one factor $1/\pi(y|\alpha)$, we have $a(\alpha) \leq b(\alpha)$ if

$$\frac{1}{\pi(y|\alpha)} \leq \int \frac{\pi(\beta|\alpha)}{\pi(y|\alpha,\beta)}\, d\beta.$$

This follows by Jensen's inequality using the distribution $\pi(\beta|\alpha)$. In the event that one or another of the integrals diverges, $a(\alpha) \leq b(\alpha)$ must continue to hold.

REFERENCES

Akaike, H. (1973). Information theory and an extension of the maximum likelihood principle. *Proc. 2nd Internat. Symp. Information Theory* (B. N. Petrov and F. Csáski, eds.), 267–281. Budapest: Akadémiai Kiadó.

Albert, J. H. and Chib, S. (2001). Sequential ordinal modeling with applications to survival data. *Biometrics* **57**, 829–836.

Andrews, D. F. and Mallows, C. L. (1974). Scale mixtures of normality. *J. Roy. Statist. Soc. B* **36**, 99–102.

Bartolucci, F., Scaccia, L. and Mira, A. (2006). Efficient Bayes factor estimation from the reversible jump output. *Biometrika* **3**, 41–52.

Basu, S. and Chib, S. (2003). Marginal likelihood and Bayes factors for Dirichlet process mixture models. *J. Amer. Statist. Assoc.* **98**, 224–235.

Berger, J. O., Ghosh, J. K. and Mukhopadhyay, H. (2003). Approximations and consistency of Bayes factors as model dimension grows. *J. Statist. Planning and Inference* **112**, 241–258.

Bickel, P. J. and Ghosh, J. K. (1990). Decomposition of the likelihood ratio statistic and the Bartlett correction – A Bayesian argument. *Ann. Statist.* **18**, 1070–1090.

Bos, C. S. (2002). A comparison of marginal likelihood computation methods. *Tech. Rep.*, Vrije Unversiteit Amsterdam, The Netherlands..

Carlin, B. P. and Louis, T. A. (1996). *Bayes and Empirical Bayes Methods for Data Analysis*, Chapter 6, pp. 209–211. London: Chapman and Hall

Celeux, G., Hurn, M. and Robert ,C. (2000). Computational and inferential difficulties with mixture posterior distribution. *J. Amer. Statist. Assoc.* **95**, 957–970.

Chen, M.-H. (1994). Importance-weighted marginal Bayesian posterior density estimation. *J. Amer. Statist. Assoc.* **89**, 818–824.

Chen, M.-H. (2005). Computing marginal likelihoods from a single MCMC output. *Statistica Neerlandica* **59**, 16–29.

Chib, S. (1995). Marginal likelihood from the Gibbs output. *J. Amer. Statist. Assoc.* **90**, 1313–1321.

Chib, S. and Jeliazkov, I. (2001). Marginal likelihood from the Metropolis-Hastings output. *J. Amer. Statist. Assoc.* **96**, 270–281.

Chib, S., Nardard, F. and Shephard, N. (2002). Markov chain Monte Carlo methods for stochastic volatility models. *J. Econometrics* **108**, 281–316.

Chipman, H., George, E. I. and McCulloch, R. E. (2001). The practical implementation of Bayesian model selection. *Model Selection* (IMS Lecture Notes) **38**) Hayward, CA: IMS, 65–134.

Clyde, M. and George, E. I. (2004). Model uncertainty. *Statist. Science* **19**, 81–94.

Dawid, A. P. (1991). Fisherian inference in likelihood and prequential frames of reference. *J. Roy. Statist. Soc. B* **53**, 79–109, (with discussion).

DiCiccio, T. J., Kass, R. E., Raftery, A. E. and Wasserman, L. (1997). Computing Bayes factors by combining simulation and asymptotic approximations. *J. Amer. Statist. Assoc.* **92**, 903–915.

Diebolt, J. and Robert, C. P. (1994). Bayesian estimation of finite mixture distributions. *J. Roy. Statist. Soc. B* **56**, 363–375.

Doerge, R. W., Zeng, Z.-B. and Weir, B. S. (1997). Statistical issues in the search for genes affecting quantitative traits in experimental populations. *Statist. Science* **12**, 195–219.

Fan, J., Hung, H.-N. and Wong, W.-H. (2000). Geometric understanding of likelihood ratio statistics. *J. Amer. Statist. Assoc.* **95**, 836–841.

Gelfand, A. E. and Dey, D. K. (1994). Bayesian model choice: asynptotics and exact calculations. *J. Roy. Statist. Soc. B* **56**, 501–514.

Gelman, A., Carlin, J. B., Stern, H. S. and Rubin, D. B. (2004). *Bayesian Data Analysis*, 2nd edn. Boca Raton, FL: Chapman and Hall/CRC Press.

Godsill, S. J. (2001). On the relationship between Markov chain Monte Carlo methods for model uncertainty. *J. Comp. Graphical Statist.* **10**, 230–248.

Green, P. J. (1995). Reversible Markov chain Monte Carlo computation and Bayesian model determination. *Biometrika* **82**, 711–732.

Han, C. and Carlin, B. P. (2001). Markov chain Monte Carlo methods for computing Bayes factors: A comparative review. *J. Amer. Statist. Assoc.* **96**, 1122–1132.

Handcock, M. S., Raftery, A. E. and Tantrum, J. (2005). Model-based clustering for social networks. *Tech. Rep.*, University of Washington, USA.

Hoeting, J. A., Madigan,D., Raftery, A. E. and Volinsky, C. T. (1999). Bayesian model averaging: A tutorial. *Statist. Science* **15**, 193–195 (with discussion).

Hoff, P., Raftery, A. E. and Handcock, M. S. (2002). Latent space approaches to social network analysis. *J. Amer. Statist. Assoc.* **97**, 1090–1098.

Johnson, V. E. (1999). Posterior distributions on normalizing constants. *Tech. Rep.*, ISDS, Duke University, USA.

Kass, R. E. and Raftery, A. E. (1995). Bayes factors. *J. Amer. Statist. Assoc.* **90**, 773–795.

Kass, R. E. and Wasserman, L. (1995). A reference Bayesian test for nested hypotheses and its relationship to the Schwarz criterion. *J. Amer. Statist. Assoc.* **90**, 928–934.

Lewis, S. M. and Raftery, A. E. (1997). Estimating Bayes factors via posterior simulation with the Laplace-Metropolis estimator. *J. Amer. Statist. Assoc.* **92**, 648–655.

Lockwood, J. R. and Schervish, M. J. (2005). MCMC strategies for computing Bayesian predictive densities for censored multivariate data. *J. Comp. Graphical Statist.* **14**, 395–414.

Meng, X.-L. and Wong, W.-H. (1996). Simulating ratios of normalizing constants: a theoretical exploration. *Statistica Sinica* **6**, 831–860.

Mira, A. and Nicholls, G. (2004). Bridge estimation of the probability density at a point. *Statistica Sinica* **4**, 603–612.

Nandram, B. and Kim, H. (2002). Marginal likelihood for a class of Bayesian generalized linear models. *J. Statist. Computation and Simulation* **72**, 319–340.

Neal, R. M. (1999). Erroneous results in 'Marginal likelihood from Gibbs output'. http://www.cs.utoronto.ca/~radford.

Newton, M. A. and Raftery, A. E. (1994). Approximate Bayesian inference by the weighted likelihood bootstrap. *J. Roy. Statist. Soc. B* **56**, 3–48 (with discussion).

Oh, M.-S. (1999). Estimation of posterior density functions from a posterior sample. *Comput. Statist. Data Anal.* **29**, 411–427.

Pauler, D. K. (1998). The Schwarz criterion and related methods for normal linear models. *Biometrika* **85**, 13–27.

Pritchard, J. K., Stephens, M. and Donnelly, P. (2000). Inference of population structure using multilocus genotype data. *Genetics* **155**, 945–959.

Raftery, A. E. (1995). Bayesian model selection in social research (with discussion). *Sociological Methodology* **25**, 111–196.

Raftery, A. E. (1996). Hypothesis testing and model selection. In W. R. Gilks, D. J. Spiegelhalter, and S. Richardson (Eds.), *Markov Chain Monte Carlo in Practice*, pp. 163–188. London: Chapman and Hall.

Reid, N. (2003). Asymptotics and the theory of inference. *Ann. Statist.* **31**, 1695–1731.

Richardson, S. (2002). Discussion of Spiegelhalter *et al. J. Roy. Statist. Soc. B* **64**, 626–627.

Rossi, P. E., Allenby, G. M. and McCulloch, R. (2005). *Bayesian Statistics and Marketing.* Chichester: Wiley

Sampson, S. F. (1968). *A Novitiate in a Period of Change: An Experimental and Case Study of Relationships.* Ph.D. Thesis, Cornell University, USA.

Satagopan, J. M., Yandell, B. S., Newton, M. A. and Osborn, T. C. (1996). A Bayesian approach to detect quantitative trait loci using Markov chain Monte Carlo. *Genetics* **144**, 805–816.

Schwarz, G. (1978). Estimating the dimension of a model. *Ann. Statist.* **6**, 497–511.

Sinharay, S. and Stern, H. S. (2005). An empirical comparison of methods for computing Bayes factors in generalized linear mixed models. *J. Comp. Graph. Statist.* **14**, 415–435.

Sisson, S. A. (2005). Transdimensional Markov chains: A decade of progress and future perspectives. *J. Amer. Statist. Assoc.* **100**, 1077–1089.

Spiegelhalter, D. J., Best, N. G., Carlin, B. P. and van der Linde, A. (2002). Bayesian measures of model complexity and fit. *J. Roy. Statist. Soc. B* **64**, 583–639 (with discussion).

Steele, R., Raftery, A. E. and Emond, M. (2006). Computing normalizing constants for finite mixture models via incremental mixture importance sampling (IMIS). *J. Comp. Graphical Statist.* **15**, 712–734.

Stephens, M. (2000). Dealing with label-switching in mixture models. *J. Roy. Statist. Soc. B* **62**, 795–809.

Sturtz, S., Ligges, U. and Gelman, A. (2005). R2WinBUGS: A package for running WinBUGS from R. *J. Statist. Software* **12**, 1–17.

Verdinelli, I. and Wasserman, L. (1995). Computing Bayes factors using a generalization of the Savage-Dickey density ratio. *J. Amer. Statist. Assoc.* **90**, 614–618.

Volinsky, C. T. and Raftery, A. E. (2000). Bayesian information criterion for censored survival models. *Biometrics* **56**, 256–262.

Zijlstra, B. J. H., van Duijn, M. A. J. and Snijders, T. A. B. (2005). Model selection in random effects models for directed graphs using approximated Bayes factors. *Statistica Neerlandica* **59**, 107–118.

DISCUSSION

NICHOLAS G. POLSON (*University of Chicago, USA*)

The authors are to be congratulated on their extension of the popular harmonic mean (HM) estimator for marginal likelihoods. They propose a stabilised harmonic mean (SHM) estimator for stabilising the possible infinite variance of the original harmonic mean estimator. A number of examples illustrating their approach are given. They also discuss an application to model comparison and provide a posterior simulation-based alternative to BIC which they term BICM. This measure requires an estimate of the maximum achievable log integrated likelihood $l_{\max}$ and they show how to use the MCMC draws to achieve this. A common theme throughout the paper is that there is extra information, particularly in the tails, from the MCMC output for harmonic means. This information can be thoughtfully used to provide better estimates than the usual ergodic averaging approach. In this discussion I will focus on three issues: (1) The Monte Carlo convergence rate for harmonic mean estimators based on a method described in Wolpert (2002), (2) Alternative approaches to marginal likelihoods based on extensions of the Savage density ratio, see Verdinelli and Wasserman (1995) and Jacquier and Polson (2002) and (3) an MCMC approach for computing the maximum achievable log-integrated likelihood, see Jacquier, Johannes and Polson (2006).

First, the basic problem that the harmonic mean estimator tackles is the estimation of the marginal likelihood defined by $\pi(y) = \int \pi(y|\theta)p(\theta)d\theta$. This is a central problem in Bayesian inference and forms the basis for model selection and comparison. The harmonic mean estimator has become popular due to its simplicity. It simply takes the MCMC draws $\{\theta^{(t)}\}_{t=1}^{B}$ and computes

$$\hat{\pi}_{HM}(y) = \frac{1}{\frac{1}{B}\sum_{t=1}^{B}\frac{1}{\pi(y|\theta^{(t)})}}$$

as an estimator. A caveat noted by the authors is that this estimator can have infinite variance. The proposal described here is a stabilised harmonic mean estimator of the form

$$\hat{\pi}_{SHM}(y) = \frac{1}{\frac{1}{B}\sum_{t=1}^{B}\frac{1}{\pi(y|\mu^{(t)})}}$$

where $\mu = h(\theta)$ is a dimensionality reduction. The intuition is that this marginalisation will lead to a heavier-tailed distribution $\pi(y|\mu)$ which in turn will lead to an estimator with finite variance.

Another caveat with these types of estimators is that they can have slow convergence properties in B. From a practical perspective this implies that a few large outliers can dominate the estimator which leads to large Monte Carlo errors. Wolpert (2002) discusses this issue and proposes a method to accelerate the convergence. Specifically, the issue is as follows: standard ergodic averaging yields

$$\frac{S_B}{B} \to \frac{1}{\hat{\pi}(y)}$$

where $S_B = \frac{1}{B}\sum_{t=1}^{B} \frac{1}{\pi(y|\theta^{(t)})}$. However, this occurs at a *very slow* rate given by

$$\frac{S_B}{B} \approx \frac{1}{\hat{\pi}(y)} + ZB^{\frac{1}{\alpha}-1}$$

where $\alpha \approx 1$. The asymptotic distribution Z can be characterised as a fully-skewed stable distribution $S_\alpha(\delta_B, \gamma_B)$. Unfortunately, the convergence rate can be poor. In a simple normal location problem for example, the convergence rate in B is

$$\frac{1}{\alpha} - 1 = \left(1 + \frac{n\tau^2}{\sigma^2}\right)^{-1} \approx 0.$$

Hence, the Monte Carlo error can be large. One solution is to use the information in the whole distribution of S_B/B rather than simple ergodic averaging. Specifically, we can use quantile information to estimate $(\alpha, \delta_B, \gamma_B)$ and hence estimate $1/\pi(y) = E(S_B/B)$ using the moment identity

$$E\left(\frac{S_B}{B}\right) = \hat{\delta} - \hat{\gamma}\tan\frac{\pi\hat{\alpha}}{2}\,.$$

Again the intuition is that there is extra information in the whole distribution of the MCMC draws.

It should also be noted that slow convergence can occur in other approaches for estimating marginal likelihoods. For example, the marginal likelihood approach in Chib (1995) where the estimator

$$\hat{\pi}_C(y) = \frac{\pi(y|\theta^\star)\pi(\theta^\star)}{\frac{1}{B}\sum_{t=1}^{B}\pi(\theta^\star|\phi^{(t)}, y)}$$

is used for any $\theta^\star$. From a purely Monte Carlo perspective it is always wise to avoid estimators than are ratios of averages

One class of problems where we can avoid the use of ratios of MC averages is the class of models where extensions of the Savage-density ratio approach to calculating Bayes factors ($\mathcal{BF}$) applies. There is a long history discussing the relationship between marginal likelihoods, Bayes factors and their use in model selection. A common approach for nested models is to estimate

$$\mathcal{BF} = \frac{1}{B}\sum_{t=1}^{B}\frac{p(\theta_1^0|\theta_2^{(t)}, y)}{p(\theta_1^0)}\,.$$

With the use of data augmentation many of the restrictive conditions for the original approach can be relaxed. For example, Verdinelli and Wasserman (1995) generalises this estimator to

$$\mathcal{BF} = \hat{C}\frac{p(\theta_1|y)}{p(\theta_1)} \quad \text{where } \hat{C} = \frac{1}{B}\sum_{t=1}^{B}\frac{p_1(\theta_2^{(t)})}{p(\theta_1, \theta_2^{(t)})}\,.$$

Jacquier and Polson (2002) provides an extension that relaxes the assumption the $p(\theta_2|\theta_1^0) = p(\theta_2|\mathcal{M}_1)$. Here we obtain as estimator of the form

$$\mathcal{BF} = \frac{1}{B}\sum_{t=1}^{B}\frac{p(y|\theta_1^1, \theta_2^{(t)})}{p(y|\theta_1^{(t)}, \theta_2^{(t)})}\frac{p_1(\theta_2^{(t)})}{p(\theta_2^{(t)}|\theta_1^{(t)})}\,.$$

When these approaches apply, the key is that they avoid Monte Carlo estimates that are reciprocals. Standard central limit theorem type convergence results in B also hold for these procedures.

Finally, there is a useful MCMC alternative for estimating the maximum achievable log integrated likelihood for model comparison, see Jacquier, Johannes and Polson (2006). Specifically, suppose that you are interested in an integrated likelihood of the form $\mathcal{L}(\theta_1) = \int_{\theta_2} f(y|\theta_1, \theta_2)p(\theta_2|\theta_1)d\theta_2$ for a given model specification.

Now one can consider a MCMC posterior analysis where we copy the variable θ_2 to be integrated over, J independent times. This leads to a joint posterior of the form

$$p_J\left(\theta_2^J, \theta_1|y\right) \propto \prod_{j=1}^{J} f(y|\theta_1, \theta_2^j)p(\theta_2^j, \theta_1).$$

This joint density has the property that on marginalising out the θ_2 parameter leads to a marginal $\pi_J(\theta_1|y) \propto \exp\left(J \ln \mathcal{L}(\theta_1)\right)$. This marginal collapses on $\hat{\theta}_1$ as $J \to \infty$ and so $l_{\max} = \ln \mathcal{L}(\hat{\theta}_1)$ is easily estimated.

Using MCMC to simulate from $p_J\left(\theta_2^J, \theta_1|y\right)$ is straightforward as we can iteratively simulate the conditionals

$$\begin{aligned} \theta_2^j|\theta_1, y &\sim \quad p(\theta_2^j|\theta_1, y) \\ \theta_1|\theta_2^J, y &\sim \quad \textstyle\prod_{j=1}^{J} p(\theta_1|\theta_2^j, y). \end{aligned}$$

In many situation this approach gives reasonable answers for small values of J.

In summary, this paper provides main insights and suggestions into tackling the hard problem of computing marginal likelihoods. One area where the literature is currently silent is in describing why Markov Chain MC sampling has been so successful for computing marginalisation constants as compared to more standard importance sampling techniques which have similar Monte carlo averaging properties. Maybe it is due to the authors' intuition that there's a lot of extra information in the whole distribution, particularly the tails, of harmonic mean estimators?

BRADLEY P. CARLIN (*University of Minnesota, USA*) and
DAVID J. SPIEGELHALTER (*MRC Biostatistics Unit Cambridge, UK*)

Congratulations to the authors for a fine extension and updating of their earlier ideas in Bayesian model choice using the harmonic mean estimator, AIC, and BIC. The first author is a long-time supporter and developer of BIC-related tools for Bayesian data analysis, and this paper makes a significant and welcome addition to this literature.

We wish to discuss the authors' estimate of the effective number of model parameters, $\hat{d} = 2s_\ell^2$, or twice the sample variance of the log-likelihood samples ℓ_t. The authors obtain this as the result of a simple moment estimate of α in their shifted gamma model (Section 4.1). In fact this same estimate appeared in an early version of the original DIC paper by Spiegelhalter *et al.* (2002); see for example equation (18) in the August 2001 version of the paper, available online at `ftp://muskie.biostat.umn.edu/pub/2001/rr2001-013.pdf.gz`. In that paper, the reasoning underlying the estimate was based on the approximation

$$D(\theta) \approx D(\hat{\theta}) + \chi_p^2\ , \tag{33}$$

where $D(\theta) = -2\ell(\theta)$, the *deviance*, which when prior information is weak is essentially identical to the shifted gamma model in (10). Approximation (33) holds provided the posterior distribution $p(\theta \,|\, y)$ can be reasonably well approximated by a multivariate normal distribution. Taking posterior expectations of both sides produces

$$E[D(\theta) \,|\, y] \approx D(\hat{\theta}) + p \,,$$

motivating the basic formula for our measure of effective model size, $p_D = \overline{D} - D(\hat{\theta})$, where $\overline{D}$ is the sample mean of the MCMC deviance samples, $D_t \equiv -2\ell_t$. If we however take the posterior *variance* of both sides of (33), we obtain

$$Var[D(\theta) \,|\, y] \approx 2p \,.$$

This suggests the alternative empirical estimate of model size $p_V = s_D^2/2$, where s_D^2 is the sample variance of the deviance samples. Clearly the relationship between D and ℓ means that $p_V = \hat{d}$. Incidentally, the use of half the variance of the deviance as the effective number of parameters is also recommended in Gelman *et al.* (2004, pp.181–182), and is the method used to compute DIC in the `R2WinBUGS` package for R; see `http://cran.r-project.org/src/contrib/Descriptions/R2WinBUGS.html`.

However, the main point here is not to claim credit for this variance-based estimate of effective model size, but instead to clarify why we deleted it from the final version of the DIC paper! We tried for a long time to convince ourselves that this attractive and parameterisation-invariant quantity was an appropriate measure, but failed. The crucial test for us was whether, in the normal hierarchical model, the effective number of parameters was given as $\text{tr}(H)$, where H is the 'hat' matrix that projects the observations onto their fitted values. As described in Spiegelhalter *et al.* (2002), this measure has been derived from numerous perspectives, in particular in a recent extension of AIC to normal mixed models (Vaida and Blanchard, 2005). (To be honest, our proposal for p_D was motivated primarily by working backwards from this desired result.)

Consider the simulation exercise carried out by the authors in Section 4.2 for the general case with prior variance σ^2. Denote $n_0 = 1/\sigma^2$, the effective sample size in the prior, and let $\rho = n/(n + n_0)$ so that $\mu_i|\overline{y}_i \sim \text{N}(\rho\overline{y}_i, \rho n_0)$. Hence ρ can be interpreted as the 'shrinkage', or the proportion of the posterior precision arising from the likelihood: in this situation $p_D = \text{tr}(H) = d\rho$.

The posterior distribution of the deviance can be derived in closed form in this situation, and is a sum of shifted non-central chi-square distributions (Spiegelhalter et al., 2002) with exact variance

$$V = 2\rho^2[d + 2(1-\rho)n_0 \sum_{j=1}^{d} \overline{y}_j^2] \,.$$

Since $\overline{y}_j \sim \text{N}(0, (\rho n_0)^{-1})$, V has a sampling expectation

$$E[V] = 2\rho^2[d + 2(1-\rho)d/\rho] = 2d\rho(2-\rho) \,.$$

Hence $p_V = V/2 \approx d\rho(2-\rho)$, compared to $p_D = d\rho$. When ρ is near 1 the two methods will closely agree: we note that ρ is 0.99 in Fig. 5 and generally very close to 1 in other situations explored by the authors. However in any interesting application of hierarchical models there will be non-negligible shrinkage and values of ρ closer to 0.5 may be more typical. In such cases we might expect p_V to substantially *overestimate* the effective number of parameters.

This is born out in the following analysis of the well-known Scottish lip cancer data, introduced by Clayton and Kaldor (1987) and most famously analyzed by Besag *et al.* (1991). Table 3 presents a side-by-side comparison of p_D and $\hat{d}$ for a variety of models, all using a non-Gaussian (Poisson) likelihood. The models we consider are easily fit in `WinBUGS 1.4` (`http://www.mrc-bsu.cam.ac.uk/bugs/welcome.shtml`) or `OpenBUGS` (`http://mathstat.helsinki.fi/openbugs/`) since this dataset is an example in the `GeoBUGS` User Manual: click on `Map` and pull down to `Manual`, then click on `Examples`. The full model for the log-relative risk of lip cancer in county i is given by

$$log(RR)_i = \beta_0 + \beta_1 x_i + \phi_i + \theta_i \ , \tag{34}$$

where β_0 is an intercept, β_1 is the effect of a single covariate (AFF, the percentage of county i's population engaged in agriculture, fishing, or forestry; essentially a surrogate for sunlight exposure), the ϕ_i are spatial clustering random effects assigned a conditionally autoregressive (CAR) prior, and the θ_i are pure heterogeneity random effects assigned an i.i.d. normal prior. Several authors have investigated whether these data support inclusion of either or both of the two sets of random effects, with the general consensus being that the clustering random effects are helpful, but the heterogeneity terms add little of additional value. The saturated model simply fits independent effects for each area and hence has no covariate structure.

Table 3: *Comparison of p_D and $\hat{d}$ for the Scottish lip cancer data.*

model	p_D	ΔDIC	$\hat{d}$	ΔAICM
full	33.1	—	42.9	—
clustering only	29.2	–0.5	45.0	5.5
heterogeneity only	40.0	11.6	59.4	21.3
fixed effects only	1.97	151.9	1.92	142.2
saturated model	52.8	18.1	57.8	13.4

Table 3 compares the effective model sizes p_D and $\hat{d}$, as well as the corresponding overall model choice statistics, DIC and AICM. Both of these latter two quantities are expressed in terms of change relative to the full model (34), with smaller values indicating preferred models. All of our computations were done in `WinBUGS`, using a single chain of 10,000 samples retained following a 1000-iteration burn-in period. We see that the performance of DIC and AICM are comparable, both obtaining rough equivalence between the full and clustering only models, which are slightly preferred over the heterogeneity only model and strongly preferred over the fixed effects only model. The two effective model size statistics also behave similarly in the simple fixed effects case, both obtaining an answer very close to the correct value of 2.0. However, the $\hat{d}$ values do emerge as significantly larger than p_D for all of the random effects models. The $\hat{d}$ value for the heterogeneity only model seems especially doubtful, since the upper bound here (obtained by simply counting parameters) is only 59: there are 56 county-level random effects, two fixed regression effects, and a single variance parameter (one could also argue that this upper bound should be 56, the number of data points). Also, $\hat{d}$ for the heterogeneity only model exceeds the value of $\hat{d}$ for the saturated model; it seems anomalous that a heterogeneity-only model can somehow be more complex than the saturated model. Much larger MCMC sample sizes did not eliminate these problems with $\hat{d}$. More broadly, while the model choice issue here is fairly clear-cut, in other hierarchical modeling examples AICM may tend to select somewhat simpler models than DIC, due to its significantly higher effective model size penalty.

Regarding the use of $\hat{d}$ in BIC: in our reply to the discussion of Spiegelhalter *et al.* (2002) we said we could find no strong reason to use p_D in BIC, attractive though it would be. The authors define BICM 'by analogy'; we wonder if there is any stronger theoretical justification for this assumption?

Naturally, far more extensive investigations of DIC, AICM, BICM, and related methods are warranted. Perhaps p_D's most embarrassing feature is that it can sometimes be negative, and clearly $\hat{d}$ does avoid this problem. Other recent work in this area includes that of Lu *et al.* (2007), who extended the approach of Hodges and Sargent (2001) to generalized linear mixed model settings, obtaining effective model size estimates that respect both the lower (0) and upper (raw parameter count) boundaries. Also noteworthy is the forthcoming *Bayesian Analysis* discussion paper by Celeux *et al.* (2006), who offered a variety of 'repaired' versions of p_D for missing data (especially mixture model) settings.

DAVID DRAPER (*University of California at Santa Cruz, USA*)

I would like to add to the discussion of this interesting and stimulating paper by posing two questions.

As noted, for example, by Draper (1995), two Laplace approximations for computing log Bayes factors—when comparing parametric models M_j indexed by parameter vectors θ_j of dimension k_j, on the basis of a data set y consisting of n conditionally exchangeable observations—are

$$\ln p(y|M_j) = \frac{1}{2}k_j \ln(2\pi) - \frac{1}{2}\ln|\hat{I}_j| + \ln p(y|\hat{\theta}_j, M_j) + \ln p(\hat{\theta}_j|M_j) + O(n^{-1}), \quad (35)$$

where $\hat{\theta}_j$ is either the mode of the posterior distribution $p(\theta_j|y, M_j)$ or the MLE and $\hat{I}_j$ is the observed information matrix evaluated at $\hat{\theta}_j$, and

$$\ln p(y|M_j) = -\frac{1}{2}k_j \ln n + \ln p(y|\hat{\theta}_j, M_j) + O(1), \quad (36)$$

the latter a large-n approximation to the former that is recognizable as the basis of BIC (cf. Equation (15) in the paper under discussion here). Raftery (1995) noted that 'it is possible to improve on [Equation (35) in this discussion contribution] in its MLE form by taking a single Newton step toward the posterior mode, starting at the MLE.' Equations (35) and (36) above (and the refinement suggested by Raftery 1995) have the advantage of relative computational simplicity and perhaps speed (if reliable maximization software is handy) when compared with some of the ideas explored in the paper under discussion here. Do the authors have any experiences they can share that would shed light on the speed-versus-accuracy tradeoffs inherent in a comparison of their methods with Laplace-based approaches, when both are appropriate to consider?

As is well known, Bayes factors involve comparing quantities of the form

$$\begin{aligned} p(y|M_j) &= \int \left[\prod_{i=1}^{n} p(y_i|\theta_j, M_j) \right] p(\theta_j|M_j)\, d\theta_j \\ &= E_{(\theta_j|M_j)} L(\theta_j|y, M_j), \end{aligned} \tag{37}$$

i.e., Bayes factors are based on comparisons of expectations of likelihoods with respect to the *priors* in the models under comparison, and this is why they behave so unstably as model selection criteria with diffuse priors, as a function of how the diffuseness is specified. Many ad hoc methods for attempting to cope with this instability have by now been suggested, including {partial, intrinsic, fractional} Bayes factors, well calibrated priors, conventional priors, intrinsic priors, expected posterior priors, and so on (e.g., Pericchi 2005); the list seems as endless as its ad-hockery is disheartening. It is arguable (e.g., Draper and Krnjajić 2006) that this is a good reason for shifting attention in Bayesian model specification away from Bayes factors and toward model selection criteria, such as the predictive log score, that do not suffer from such instabilities when diffuse prior information is all that is available; but (a) the dependence of the mixing weights in Bayesian model averaging (BMA) on ratios of Bayes factors and (b) the central role that BMA plays in appropriately propagating model uncertainty combine to leave the impression that evaluating Bayes factors with diffuse prior information cannot, regrettably, be entirely avoided. Do the authors see any possibility that any of the ideas they have explored for stabilizing MCMC-based *estimates* of Bayes factors might be used to help stabilize *the Bayes factors themselves* in the presence of diffuse prior information?

CHRIS SHERLOCK and PAUL FEARNHEAD (*Lancaster University, UK*)

We would like to congratulate the author on a stimulating paper. The goal of a simple and automatic procedure for estimating the integrated likelihood is an important one, and we are pleased to see some further work extending the harmonic mean estimator.

We decided to test out the BICM and AICM procedures described within the paper for the problem of model selection for Markov-modulated Poisson Processes. These are models for the occurence of events through time, and assume an underlying continuous-time Markov process $X(t)$ which has state space $\{1, \ldots, K\}$. The data consists of the times of events from a Poisson process of time-varying intensity $\lambda(t)$ which depends on the underyling $X(t)$ process. So conditional on $X(t) = k$ we have $\lambda(t) = \lambda_k$. The parameters of the model are the different intensities $\{\lambda_1, \ldots, \lambda_K\}$, and the entries in the rate matrix of $X(t)$.

We applied this model to analyse the occurence of Chi-sites along the lagging strand of the genome of Ecoli (see Fearnhead and Sherlock 2006 for details of this application). For $K > 1$, this is a non-trivial example, for example the likelihood is symmetric across re-labelling of the states of the $X(t)$ process whereas we have used a non-symetric informative prior so that we have a complex multi-modal posterior. However, Fearnhead and Sherlock 2006 describe how to implement a Gibbs sampler for this model, and it is thus possible to use the idea of Chib (1995) to get accurate estimates of the integrated likelihood for $K = 2$ and 3 (the $K = 1$ case can be calculated analytically). Thus we have a 'correct' answer to compare the results of BICM and AICM to.

As the paper suggests, calculating BICM or AICM values from MCMC output is simple and quick. The results we obtained, together with twice the integrated likelihood as calculated from the method of Chib (1995) are shown in Table 4. (We give twice the integrated likelihood values as it is this that BICM and AICM estimate.) As suggested by the theory, the plots of the log-likelihood values output from the MCMC run show that these do closely follow a shifted gamma distribution (results not shown).

For the $K = 1$ case BICM appears to give a better estimate of twice the integrated likelihood than AICM, but it gives substantial underestimates for the $K = 2$ and $K = 3$ case. As a result BICM incorrectly shows strong evidence for $K = 1$. By comparison AICM performs well: while it over-estimates twice the integrated likelihood for all three models, the relative estimates are very close to the truth. Based on AICM you would correctly choose $K = 2$, and have appropriate estimates of the strength of evidence for this model over $K = 1$ and $K = 3$ respectively.

For comparison we also tried the harmonic mean estimator, and this estimator also performed very well. Twice the harmonic mean estimates are −935.5, −927.2 and −929.9 for $K = 1$, 2 and 3 respectively. While these again are over-estimates of the true iterated likelihoods, the relative estimates are close to being correct.

Table 4: *BICM, AICM and (twice) integrated likelihood values (2× IL) for the Chi-site data.*

	$K = 1$	$K = 2$	$K = 3$
2× IL	−937.9	−927.7	−931.0
BICM	−937.6	−940.0	−946.3
AICM	−934.9	−925.3	−928.8

Finally we wonder if you could say something about the conditions required for the posterior distribution of the likelihood values (10) to hold – do these require the standard regularity conditions for the likelihood ratio statistic to asymptotically have a χ^2_d distribution? Also is it possible to get a higher-order result for this limiting distribution, which could then be integrated to give a (finite) estimate of the integrated likelihood? Such a result may give theoretical justification for choosing one of AICM or BICM over the other in different scenarios.

I. CLAIRE GORMLEY and T. BRENDAN MURPHY
(*Trinity College Dublin, Ireland*)

We would like to congratulate the authors on their excellent paper on computing integrated likelihoods which is a topic of great importance in Bayesian model comparison.

We would like to discuss our experiences in applying AICM, BICM and alternatives when selecting the dimensionality for a latent space model for rank data.

Background. Irish elections employ a voting system called proportional representation by means of a single transferable vote (PR-STV). In the PR-STV system, voters rank some or all of the candidates in order of preference. Votes are counted and subsequently transferred between candidates, using the voter preferences, in a series of counts to determine who gets elected. More details on the electoral system and the counting of votes is given in Gormley and Murphy (2005).

Hence, Irish elections generate rank data recording the preferences of the voters for candidates. We have recently developed mixture models for modeling rank data in the context of analyzing Irish election data (Gormley and Murphy, 2005) and Irish college application data (Gormley and Murphy, 2006a).

Gormley and Murphy (2006b) develops a latent space model for rank data where voters and candidates are located in a latent space $\mathbf{Z} \subseteq \mathbb{R}^D$. Each voter is located at position z_i $(i = 1, 2, \ldots, M)$ and each candidate at location ζ_j $(j = 1, 2, \ldots, N)$. Let $d(z_i, \zeta_j)$ be the squared Euclidean distance from voter i to candidate j. We would expect that voter i will give high preferences to the candidates that are closest and low preferences to those that are distant.

Let

$$p_{ij} = \frac{\exp\{-d(z_i, \zeta_j)\}}{\sum_{j'=1}^{N} \exp\{-d(z_i, \zeta'_j)\}}$$

be the probability of voter i selecting candidate j in first position. These probabilities can be used to give the probability of a vote (*i.e.*, candidate ordering) using the Placket–Luce model (Plackett, 1975)

$$\mathbf{P}\{\mathbf{x}_i|\mathbf{p}\} = \prod_{t=1}^{n_i} \frac{p_{ic(i,t)}}{\sum_{s=t}^{N} p_{ic(i,s)}},$$

where $c(i,1), c(i,2), \ldots, c(i,n_i)$ is the ordered list of the candidates selected by voter i and $c(i,n_i+1), c(i,n_i+2), \ldots, c(i,N)$ is an arbitrary ordering of the candidates not selected by voter i.

These models are fitted in a Bayesian framework and samples from the posterior distribution are generated using a random walk Metropolis-Hastings algorithm.

Choice of dimensionality. An important issue when fitting these models is the choice of D, the dimensionality of the latent space.

In this discussion, we will concentrate on data from an exit poll taken at the 1997 Irish Presidential Election. We fitted latent space models for D=1, 2 and 3 and computed the AICM, BICM (using equation (23) from the paper), DIC (Spiegelhalter et al, 2002) and the Pritchard et al (2000) criterion for each model; the values obtained are given in Table 5.

Table 5: *The values of AICM, BICM, DIC and Pritchard et al's criterion computed for the 1997 Presidential Election exit poll data for dimensionality D equal to 1, 2 and 3. The model selected is highlighted in bold.*

	AICM	BICM	DIC	Pritchard
$D = 1$	**−20171**	**−19244**	18270	**17855**
$D = 2$	−23217	−21221	17966	18246
$D = 3$	−26621	−23671	**17923**	19270

There is good consistency across the results for the AICM, BICM and the Pritchard et al criterion with each selecting $D = 1$ whereas DIC selects $D = 3$. A simple principal components analysis of the estimated candidate locations suggests that most of the variation in candidate locations is explained by a single dimension which captures the race between McAleese (winner of the election) and the other candidates (Scallon, Banotti, Roche and Nally).

Hence, AICM and BICM select a dimensionality that is consistent with the estimated candidate configurations. This contrasts with DIC which doesn't appear to have a strong enough penalty on dimensionality.

CHRISTIAN P. ROBERT (*Université Paris Dauphine, France*) and
NICOLAS CHOPIN (*University of Bristol, UK*)

Comparison. The issue of approximating marginal densities obviously remains an up-to-date concern for Bayesian Statistics, since two invited talks at this conference are centred around it, namely the present paper and the alternative proposal of Skilling *(this volume)*. While we are not completely convinced of the advantages of nested sampling (see our discussion in this volume), we would welcome the authors' opinion of the respective worths of both approaches. In particular, Skilling's *(this volume)* perspective is completely in line with the second approach of the present paper, that is, based on a (prior or posterior) distribution of the likelihood function, since Skilling's marginal is expressed as $\pi(y) = \mathbb{E}_\pi[L]$, while Raftery et al.'s marginal is $\pi(y) = \mathbb{E}[L^{-1}|y]$.

Potential dangers. Let us first state that we find the representation of Newton and Raftery (1994) quite interesting in that it allows for an approximation of the marginal density based on the output of the MCMC simulation of the posterior $\pi(\theta|y)$ (rather than from the prior as in nested sampling). Its major drawback is however the disastrous feature of a potential infinite variance, against which the Rao-Blackwellised solution proposed in this version does not always work. We can take for instance the case of the normal variance, $y|\sigma \sim \mathcal{N}(0, \sigma^2)$ and $\pi(\sigma^2|y) \sim \mathcal{IG}(3/2, 1+y^2/2)$ where nested sampling provides an approximation of $\pi(y)$ in agreement with a decrease of the error in $\sqrt{n}$ (see Fig. 7 (left), while the harmonic mean approximation has no variance and obviously varies much more for the same computational effort. A further difficulty is that, in complex settings, the infinite variance of the harmonic estimator may remain undetected. For instance, a run up to $5,000,000$ iterations produces an apparent decrease of the error in $\sqrt{n}$ in Fig. 7 (right).

Bayes factors. A feature also common to both Skilling's and Raftery *et al.* 's approaches is that they do not easily adapt to multiple models environments, as those encountered in model choice and the computation of Bayes factors, for which generic approximations methods like path sampling (Gelman and Meng, 1998) are readily available, being based on simulations from the (alternative) posterior distributions.

Asymptotic approximations. To go back to the Gamma approximation, we are a bit concerned with the $\log(1-\alpha)$ in the expression of $\log \pi(y)$ in Eqn. (14). Would the Gamma approximation in Eqn. (10) be *exact* (by chance, or because the model has some specific structure, *e.g.*, Gaussian), then $\log \pi(y)$ would be infinite. This does not seem intuitive, and it also casts some doubt on the practicality of this approach. Beyond this specific point, one must always be wary of approximations methods that do not provide a way of evaluating, even roughly, the approximation error (like BIC). In that respect, a sequence of increasingly accurate (and possibly increasingly expensive to compute) approximations, would be preferable, as the authors suggest briefly in the conclusion.

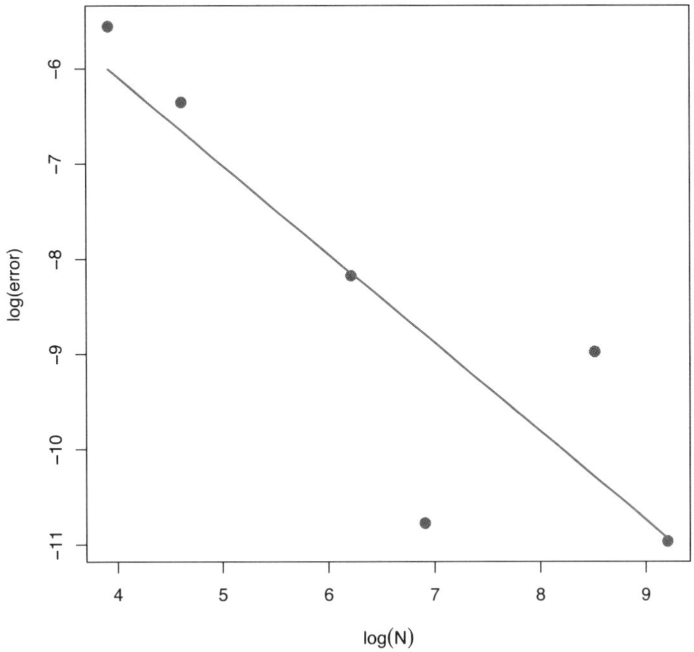

Figure 7: *Comparison of the error evolution for nested sampling and harmonic approximation:* (left) *Evolution of the error in Skilling's nested sampling approximation of the marginal density when* $y = 5$ *in a* $y|\sigma \sim \mathcal{N}(0, \sigma^2)$ *and* $\pi(\sigma^2|y) \sim \mathcal{IG}(3/2, 1 + y^2/2)$ *model, using* N *initial simulations from the exponential prior and* $j = N/2$ *replications.* (right) *Evolution of the error of the harmonic approximation for the same problem, using* 1000 *times more simulations from the posterior than in nested sampling.*

Remark. A final bibliographical remark is about the study of DIC in missing data models: Celeux *et al.* (2006) give a detailed analysis of the multiple possible interpolations of p_D and of DIC in such setups, agreeing with the authors about its instability.

REPLY TO THE DISCUSSION

We are very grateful to all the discussants for their stimulating discussions.

Basic idea. The basic idea of our paper is that the integrated likelihood can be estimated from the posterior distribution of the loglikelihood, using the harmonic mean identity. This reduces the problem of estimating the integrated likelihood to a one-dimensional one. The simplest way to do is via the harmonic mean estimator of Newton and Raftery (1994), but while this is simulation-consistent, it often has infinite variance.

We suggested two ways of getting integrated likelihood estimators with better properties. The first is to reduce the parameter space. The second, on which most of the discussants focused, is to model the posterior distribution of the log-likelihood parametrically, estimate the parameters of the resulting model from MCMC or other posterior simulation output, and then apply the harmonic mean identity to the resulting estimated model. We proposed a shifted scaled gamma distribution as a possible approximating model, and showed that it gave reasonable results in some examples.

However, this is not the only approximating model that could be used. One could instead use a shifted scaled noncentral χ^2 distribution; this is exact for Gaussian fixed effects models. One could also use a sum of shifted scaled noncentral χ^2 distributions; this is exact for an ANOVA-type Gaussian random effects model, as shown by Spiegelhalter *et al.* (2002). Our experience so far, including with random effects models, is that the shifted gamma distribution does provide a good approximation in a wide range of situations; this conforms to the experience reported by Sherlock and Fearnhead, for example.

Sherlock and Fearnhead asked whether one could obtain a higher-order expansion for the asymptotic gamma approximation, and then integrate this. Bickel and Ghosh (1990) proposed an expansion of this type. Their expansion couldn't be used directly for this purpose, because the terms in it lead to infinite estimates of the integrated likelihood. However, it does seem possible that an expansion of this kind could be obtained that would be useful in the present context.

A related possibility, not mentioned in our paper or by any of the discussants, would be to approximate the posterior distribution of the loglikelihood by a mixture of normals. Roeder and Wasserman (1997) showed that this can approximate a wide range of distributions, including gamma-like distributions, and proposed ways of estimating this model. This might have the advantage of leading to more stable estimates, as the tail declines quickly. The estimator of the integrated likelihood proposed by Pritchard *et al.* (2000) could be viewed as a special case of such a normal mixture model, with just one term.

Polson discussed some very interesting unpublished work of Wolpert (2002), who indicated that the posterior distribution of the harmonic mean estimator is asymptotically a stable law, with a stable index that is typically just large enough to ensure that the mean exists. This suggests the possibility of using an approximating stable distribution of the reciprocal likelihood (rather than the loglikelihood) as the basis for a stabilized harmonic mean estimator. Wolpert's paper doesn't report numerical experience, but it is worth investigating further. Nevertheless, it seems likely that it would be as hard to estimate the stable index of the stable distribution as it is to estimate the scale parameter of the gamma distribution that we suggest; this may only shift the difficulty to another arena.

The effective number of parameters. Carlin and Spiegelhalter suggest that $\hat{d}$ may be an overestimate of the effective number of parameters, referring to the normal random effects example. However, we introduce $\hat{d}$ primarily as an estimate of (twice) the scale parameter in the approximating gamma distribution rather than of the actual number of parameters; in regular fixed effects models these coincide asymptotically. However, even in fixed effects models they don't always coincide: for example in the exponential distribution example discussed by Fan *et al.* (2000), the number of degrees of freedom in the χ^2 distribution that approximates the posterior distribution of the loglikelihoods is twice the number of parameters.

In order to investigate this further, we simulated some data and posterior distributions from the normal random effects model discussed by Carlin and Spiegelhalter, as described by Spiegelhalter *et al.* (2002). For this model, p_D does not depend on the data, whereas $\hat{d}$ does. Carlin and Spiegelhalter suggested that $\hat{d}$ will tend to overestimate the effective number of parameters relative to p_D, but we found that $\hat{d}$ was often actually slightly smaller than p_D.

In this model, the true $\ell_{\max}$ is known, and this makes it easier to estimate both α and λ in the approximating gamma distribution, using a moment estimator, for example. We found that in this situation, moment estimators of the gamma distribution parameters were very similar to maximum likelihood or other estimators. This provides an alternative to both p_D and $\hat{d}$.

We found that (i) the shifted gamma distribution continued to fit well for a wide variety of situations including ones with large shrinkage; (ii) our moment estimates of α tended to be *larger* than either $p_D/2$ or $\hat{d}/2$; and (iii) the gamma distribution tended to fit the observed distribution better with the moment estimates than with either $\hat{d}$ or p_D. When n_i, the sample size in the i-th group, was the same for all groups, we found that $-\log(1-\hat{\lambda})$ was well approximated by $\log(n_i+1)$, lending some support to the random effects BICM approximation (22).

Theory. Carlin and Spiegelhalter asked what the theoretical rationale for BICM is, and, in a related question, Sherlock and Fearnhead asked what regularity conditions are needed for the asymptotic posterior distribution of the loglikelihood (10) to hold. The regularity conditions are given by Fan *et al.* (2000); these are weaker than the conditions required for the asymptotic χ^2 distribution of the likelihood ratio test statistic to hold. In particular, the contours of the loglikelihood do not need to be asymptotically elliptical.

The theoretical rationale for BICM is outlined in Section 4.1 of our paper. In brief, Raftery (1995, Section 4.1), drawing heavily on Kass and Wasserman (1995), showed that with a unit information prior, the integrated likelihood for a fixed effects model can be approximated using BIC, with error of order $O(n^{-1/2})$. BIC involves the maximized loglikelihood, $\ell_{\max}$, and the number of parameters, d. We assume that neither of these is available explicitly, and that they must be estimated from posterior simulation. Under the weak regularity conditions we have mentioned, the maximized loglikelihood (suitably normalized) can be estimated consistently from posterior simulation output by $\hat{\ell}_{\max}$, and the number of parameters can be consistently estimated by $\hat{d}$. Thus BICM provides a consistent estimator of BIC, and hence of twice the integrated likelihood.

This theoretical rationale applies to fixed effects models. However, Pauler (1998) has provided a theoretical derivation of an extension of BIC to random effects models, and it seems that this could be the basis for a theoretical rationale for BICM also for random effects models. More work on this is required, however.

Chopin and Robert asked what happens if the gamma approximation is exact. However, this will be the case only with an improper prior, and in that case the integrated likelihood is undefined in any event. Chopin and Robert also asked about standard errors for BICM and AICM. These are now in the paper (they were not in the version of the paper presented at the Valencia meeting).

The harmonic mean estimator. Our work was motivated by the observation that the harmonic mean estimator of the (reciprocal) integrated likelihood has infinite variance, a fact pointed out in the paper that introduced it (Newton and Raftery 1994). In spite of this apparently undesirable behavior, the original harmonic mean estimator has been widely used, and several researchers have reported satisfactory results with it, including Sherlock and Fearnhead in their discussion here.

The results of Wolpert (2002), reported by Polson, shed some light on this. Wolpert showed that the harmonic mean estimator does have a distribution. In spite of having infinite variance, most of the mass of this distribution may be fairly concentrated. If one is comparing two models, and the distributions of the harmonic mean estimators of their integrated likelihoods have little overlap, then the model comparison is clear. This is often the case. A simple way of assessing these distributions is by estimating the harmonic mean estimator from replicated posterior simulation runs, as done for example by Zijlstra *et al.* (2005).

The Wolpert results suggest that the distribution of the harmonic mean estimator from a subset of a posterior simulation run will be similar to that from the run as whole. Thus a reasonable approach could be to divide a posterior simulation run (suitably thinned) into, say, 9 or 19 batches, and use the resulting distribution as a rough sampling distribution of the harmonic mean estimator itself. This should be good enough at least for assessing whether the result of a comparison between two models is decisive. Thus this provides some theoretical support for continued use of the harmonic mean estimator in practice, at least in some situations.

Experience with BICM and AICM. Gormley and Murphy reported positive experience with BICM and AICM, and we were glad to see their results.

Sherlock and Fearnhead gave results for a modulated Poisson process in which accurate values of the integrated likelihood were available. AICM and the harmonic mean estimator worked well, but BICM did not. In particular, BICM supported a one-state model over a two-state model, unlike the true integrated likelihood. In the version of the paper on which they commented and that was presented at Valencia, only the fixed effects version (18) of BICM was given explicitly. If instead we use the version (22), with $n_k = n/K$, which seems more appropriate, then the second line of their table (for BICM) becomes -937.6, -936.6, -939.6. BICM is now more accurate than before and in particular supports the two-state model. It is still not fully satisfactory for this example, however.

Comparisons with other methods. Several discussants asked about comparisons with other methods. In our paper we gave a fairly extensive literature review of relevant methods. However, our goal here is specific: a generic method for estimating the integrated likelihood of a model from posterior simulation output that uses only the loglikelihoods of the simulated parameter values, and in particular does not involve model-specific algebra or computation. Several of the methods mentioned do not fall into this category, and so we have not viewed them as directly comparable for the present purposes.

Draper asked about a comparison with the Laplace method. The Laplace method can be efficient and highly accurate, but it is not generic, and indeed applying it to a specific model can require a great deal of work. Polson mentioned his very interesting work on estimation of the maximized likelihood and extensions of the Savage density ratio method. These methods are generally for computing ratios of integrated likelihoods rather than individual ones, and Polson's work is for specific models, and so has a somewhat different goal from ours.

Chopin and Robert asked about a comparison between our approach and that of Skilling in the present volume. Skilling's approach is also generic, and so comparable to ours. We agree with Chopin and Robert's comments. Skilling's method involves integrating over the prior, and in the low-dimensional examples he gives, this works well. However, in the much higher-dimensional models commonly tackled using MCMC, this may well not work so well. We would be interested to see further experience with this proposal.

Finally, Chopin and Robert suggested that our approach was not well suited to situations with multiple models. However, it yields an estimate of the integrated likelihood for each model estimated, and so provides a ready way to compare all the models estimated in a round of data analysis. Thus it does seem well adapted to situations where multiple models are being fitted.

ADDITIONAL REFERENCES IN THE DISCUSSION

Besag, J., York, J. C. and Mollié, A. (1991). Bayesian image restoration, with two applications in spatial statistics. *Ann. Inst. Statist. Math.* **43**, 1–59 (with discussion).

Celeux, G., Forbes, F., Robert, C. P. and Titterington, D. M. (2006). Deviance information criteria for missing data models. *Bayesian Analysis* **1**, 651–706 (with discussion).

Clayton, D. G. and Kaldor, J. M. (1987). Empirical Bayes estimates of age-standardized relative risks for use in disease mapping. *Biometrics* **43**, 671–681.

Draper, D. (1995). Assessment and propagation of model uncertainty. *J. Roy. Statist. Soc. B* **57**, 45–97 (with discussion).

Draper, D. and Krnjajić, M. (2006). Bayesian model specification. *Tech. Rep.*, University of California, Santa Cruz, USA.

Fearnhead, P. and Sherlock, C. (2006). An exact Gibbs sampler for the Markov modulated Poisson process. *J. Roy. Statist. Soc. B* **68**, 767–784.

Gormley, I. C. and Murphy, T. B. (2005). Exploring Heterogeneity in Irish Voting Data: A Mixture Modeling Approach. *Tech. Rep.*, 05/09, Department of Statistics, Trinity College Dublin, Ireland.

Gormley, I. C. and Murphy, T. B. (2006a). Analysis of Irish Third-Level College Applications Data. *J. Roy. Statist. Soc. A* **169**, 361–380.

Gormley, I. C. and Murphy, T. B. (2006b). A Latent Space Model for Rank Data. *Tech. Rep.*, 06/02, Department of Statistics, Trinity College Dublin, Ireland.

Hodges, J. S. and Sargent, D. J. (2001). Counting degrees of freedom in hierarchical and other richly-parameterised models. *Biometrika* **88**, 367–379.

Jacquier, E. and Polson, N. G. (2002) Odds Ratios for Stochastic Volatility Models. *Tech. Rep.*, University of Chicago, USA.

Jacquier, E., Johannes, M. and Polson, N. G. (2006) MCMC Maximum Likelihood for Latent State Models. *J. Econometrics* **137**, 615–640.

Lu, H., Hodges, J. and Carlin, B. P. (2007). Measuring the complexity of generalized linear hierarchical models. *Can. J. Statist.* **1**, 69–87.

Pericchi, L. (2005). Unifying concepts for methods of Bayesian model choice. *Tech. Rep.*, Universidad de Puerto Rico en Rio Piedras, Puerto Rico.

Plackett, R. L. (1975). The analysis of permutations. *Appl. Statist.* **24**, 193–202.

Raftery, A. E. (1995). Discussion of 'Assessment and propagation of model uncertainty,' by Draper D, *J. Roy. Statist. Soc. B* **57**, 78–79.

Vaida, F. and Blanchard, S. (2005). Conditional Akaike information for mixed-effects models. *Biometrika* **92**, 351–370.

Wolpert, R. (2002). Stable Limit laws for Marginal Probabilities from MCMC streams: Acceleration of Convergence. *Tech. Rep.*, ISDS, Duke University, USA.

BAYESIAN STATISTICS 8, pp. 417–452.
J. M. Bernardo, M. J. Bayarri, J. O. Berger, A. P. Dawid,
D. Heckerman, A. F. M. Smith and M. West (Eds.)

Approximating Interval Hypothesis: p-values and Bayes Factors

JUDITH ROUSSEAU
Université Paris Dauphine and *CREST, France*
rousseau@ceremade.dauphine.fr

SUMMARY

In this paper we study Bayes factors and special p-values in their capacity of approximating interval null hypotheses (or more generally neighbourhoods of point null) hypotheses by point null or eventually parametric families in the case of the goodness of fit test of a parametric family. We prove that when the number of observations is large Bayes factors for point null hypotheses can approximate Bayes factors for interval null hypotheses for extremely small intervals. We also interpret the significance of calibration goodness of fit tests using a p-value in terms of width of neighbourhoods of the point null hypothesis. Finally we study the consistency of Bayes factors for goodness of fit tests of parametric families, which enables us to shed light on the behaviour of the Bayes factors.

Keywords and Phrases: ASYMPTOTIC; BAYES FACTORS; POINT NULL; p-VALUES; GOODNESS OF FIT TESTS.

1. INTRODUCTION

Point null hypothesis are often justified as approximations of more realistic hypotheses which would typically be defined as neighbourhoods of some specific distribution or family of distributions. It is not so much the exactness of the point null which is of interest, but rather wether the data can be relatively well represented by the null model. The difficulty however is to decide what *relatively well* means. In some applied problems, this notion can be made precise enough for the analysis but in most problems it cannot. Statisticians and practitioners have therefore often replaced this *relatively well* by exactly, counting on the uncertainty in the data to account for it, or as Hodges and Lehman (1954) put it, on the lack of power of the test employed. However when the number of observations is large the tests often become very powerful, so that the point null hypotheses are bound to be rejected

---

Judith Rousseau is Professor of Statistics at the University Paris Dauphine.

under large samples. In this context understanding how a test reacts with the number of observations is crucial and has received huge attention for a very long time, both in the frequentist and in the Bayesian literature; see for instance Nemirowski (2000), Gouriéroux and Monfort (1996) for general theory, Robins and Wasserman (2000) for a discussion on these issue or Dass and Lee (2004) for the consistency of Bayes factors.

The testing problems we are going to investigate can be expressed in the following way: let the observation $X = (X_1, ..., X_n) \in \mathcal{X}^n$ with $\mathcal{X} \subset \mathbb{R}^d$, be a n-sample from a distribution P_θ, $\theta \in \Theta$.

- Interval null hypothesis:

$$H_0^\varepsilon : d(\theta, \theta_0) \leq \varepsilon, \quad \text{versus} \quad H_1^\varepsilon : d(\theta, \theta_0) > \varepsilon \quad \text{or} \tag{1}$$

$$H_0^\varepsilon : d(\theta, \Theta_0) \leq \varepsilon, \quad \text{versus} \quad H_1^\varepsilon : d(\theta, \Theta_0) > \varepsilon, \tag{2}$$

  where Θ_0 is a subspace of the parameter space Θ having null Lebesgue measure (relative to Θ), and $d(.,.)$ is a distance or pseudo-distance. This second case will be more specifically studied in the goodness of fit framework, where Θ_0 is the parametric family to be tested.

- Point null hypothesis:

$$H_0 : \theta = \theta_0, \quad \text{versus} \quad H_1 : \theta \neq \theta_0 \quad \text{or} \tag{3}$$

$$H_0 : \theta \in \Theta_0, \quad \text{versus} \quad H_1 : \theta \notin \Theta_0. \tag{4}$$

This covers a wide variety of problems: point null hypotheses in parametric models, variable selections in regression models, goodness of fit testing. In most cases the real question is defined in terms of H_0^ε versus H_1^ε, however as the choice of ε is often difficult to make a priori, people have been using the point null hypothesis as an approximation to the interval null hypotheses.

One of the most common Bayesian testing procedure is based on the Bayes factor, or the posterior probability of the null hypothesis. They are the Bayesian answers to the 0–1 types of losses. One of the main reasons for their popularity is due to their simplicity of interpretation. Posterior probabilities have an obvious meaning and Bayes factors are related to them. They can also be interpreted as automatic penalized likelihoods or ratios of penalized likelihoods where the penalization comes from the integration over the parameter spaces, so that Bayes factors naturally favor simpler models, see Berger and Jefferys (2001). This phenomenon called *Ockham's razor* is directly linked to the automatic penalization induced by the integration over the parameter space, allowing to take into account the uncertainty on the parameters in each model. This automatic penalization inherent to Bayes factors makes them very valuable, since the choice of penalization is often critical. On the other hand, their being the Bayesian answers to 0–1 types of losses indicates that they are not well adapted to testing *interval* null hypotheses approximated by point null hypotheses. In other words the 0–1 loss seems to be too abrupt to answer to a problem where the real aim is wether or not the null model is a reasonable approximation of the *true* one. However in a seminal paper, Berger and Delampady (1987) showed that Bayes factors for point null hypotheses, in the special case of the normal model, were reasonable approximations of Bayes factor for small interval null hypotheses. Verdinelli and Wasserman (1996) have extended their result by proving the the Bayes factor associated with the interval null test in the form $H_0^\varepsilon : |\theta - \theta_0| \leq \varepsilon$ versus $H_1^\varepsilon : |\theta - \theta_0| > \varepsilon$ could be approximated by the Bayes factor associated with the point null test $H_0 : \theta = \theta_0$ versus $H_1 : \theta \neq \theta_0$, if ε were small enough. However the question of how small ε need to be is not answered to, although in their paper Berger and Delampady give a bound in terms of ε. This question becomes even more important when the dimension of the parameter space is large or infinite as it can become extremely difficult to estimate a priori the level ε which can be tolerated as a distance from the point null model.

In this paper we investigate two types of Bayesian testing procedures with regard to this approximation of interval null hypothesis testing by point null hypothesis testing. The first type is the Bayes factor approach, in other words the answer to the 0–1 types of losses. The second type of testing procedure is the Bayesian answer to a distance based loss, calibrated by a p-value. This procedure has been used by Robert and Rousseau (2003) and McVinish *et al.* (2005) for goodness of fit tests. Following Berger and Delampady (1987) and Verdinelli and Wasserman (1996) we first try to answer to the question asked previously, i.e. how small does ε need to be to make sure that the Bayes factor for the point null is behaving similarly to the Bayes factor for the testing problem H_0^ε versus H_1^ε, in particular when the number of observations becomes large. Surprisingly the answer is very small, actually much too small, as will be shown in Section 2.

The distance approach is the Bayesian testing procedure associated with the loss function defined by:

$$L_\epsilon(\theta, \delta) = \begin{cases} (d(\theta, \Theta_0) - \epsilon)1\!\!1_{d(\theta,\Theta_0)>\epsilon} & \text{when} \quad \delta = 0 \\ (\epsilon - d(\theta, \Theta_0))1\!\!1_{d(\theta,\Theta_0)\leq\epsilon} & \text{when} \quad \delta = 1. \end{cases} \tag{5}$$

The Bayesian estimate associated with a prior π on the parameter space Θ, the above loss and data X is then defined by

$$\delta^\pi(X) = \begin{cases} 0 & \text{when} \quad E^\pi\left[d(\theta, \Theta_0) \,|\, X\right] \leq \epsilon \\ 1 & \text{when} \quad E^\pi\left[d(\theta, \Theta_0) \,|\, X\right] > \epsilon. \end{cases} \tag{6}$$

where $E^\pi[l(\theta) \,|\, X]$ is the posterior expectation of $l(\theta)$.

In cases where we have no prior knowledge of the tolerance threshold ϵ we can calibrate it by computing a p-value associated with the test statistic $H(X) = E^\pi\left[d(\theta, \Theta_0) \,|\, X\right]$. When the null set is a single point, $\Theta_0 = \{\theta_0\}$, the p-value is defined by $p_0(X) = P_{\theta_0}\left[H(Y) > H(X) \,|\, X\right]$, else we consider a Bayesian p-value proposed by Berger and Bayarri (2000), namely the conditional predictive p-value. These p-values are defined by:

$$p^U_{\text{cpred}}(X) = \int_{\Theta_0} P_\eta\left[H(Y) > H(X) \,|\, U, X\right] d\pi_0(\eta|U)$$

where U is an appropriate statistic, $\eta \in \Theta_0$ and π_0 is a prior probability on Θ_0. Berger and Bayarri (2000) advocate the use of the conditional maximum likelihood estimator given the test statistic $H(X)$, in the restricted model Θ_0 for U, but here, following Robert and Rousseau (2003), McVinish *et al.* (2005) and Fraser and Rousseau (2005) we simply consider $\hat{\eta}$, the maximum likelihood estimator in the restricted model Θ_0 for U. The p-value we consider is thus defined by

$$p_{\text{cpred}} = \int_{\Theta_0} P_\eta \left[H(Y) > H(X) \,|\, \hat{\eta}_x, X \right] d\pi_0(\eta|\hat{\eta}_x). \tag{7}$$

This p-value has been studied by Robert and Rousseau (2003) in terms of its first order properties and by Fraser and Rousseau (2005) for its higher order properties, showing that it is asymptotically uniform under the null and giving its asymptotic equivalent in general cases.

The p-value calibration therefore plays the same role as the point null Bayes factor discussed previously. The same question is thus asked: What is the range of values of ϵ corresponding to the use of the above p-value? In Section 2, we give answers to this question in three different setups. First we consider the case where $\Theta_0 = \{\theta_0\}$, then we consider composite null hypothesis in goodness of fit tests with and without encompassing the null hypothesis in the alternative. In a goodness of fit test of a parametric family $\mathcal{F} = \{f_\eta, \eta \in \Theta_0\}$, where the f_η's are densities with respect to some common measure μ, say the Lebesgue measure, the interval null formulation of the problem takes the form:

$$H_0^\varepsilon : d(f, \mathcal{F}) \leq \varepsilon, \qquad \text{versus} \qquad H_1^\varepsilon : d(f, \mathcal{F}) > \varepsilon$$

and the point null formulation of the problem:

$$H_0 : f \in \mathcal{F}, \qquad \text{versus} \qquad H_1 : f \notin \mathcal{F},$$

where f is the *true* density of the observations with respect to the same measure μ.

In the Bayesian framework the nonparametric alternative to $\mathcal{F}$ needs to be *represented* by a prior on the set of densities on $\mathcal{X}$. There are many ways to construct such priors. In the context of goodness of fit tests some people have embedded the parametric family in the alternative. Carota and Parmigiani (1996) embed the parametric model in a mixture of Dirichlet processes (over a parameter η), whose means are the cumulative distribution functions under the null hypothesis. These approaches are however not very appropriate for testing an absolutely continuous model. Berger and Guglielmi (2001) embed their parametric model using mixtures of Polya trees centered on the parametric model. As a third way of embedding the parametric model, Verdinelli and Wasserman (1998) and Robert and Rousseau (2003) reduce the problem to finding a prior distribution on $[0, 1]$, rather than on $\mathbb{R}$ or $\mathbb{R}^p$, through the use of the probability transform, that is, considering $F_\eta(X)$. Let g_ψ, $\psi \in \mathcal{S}$ be the support of the prior distribution on the set of densities on $[0, 1]$ (typically $\mathcal{S}$ is infinite dimensional). The nonparametric alternative is then represented as

$$\mathcal{H} = \{h_\theta(x) = f_\eta(x) g_\psi(F_\eta(x)), \eta \in \Theta_0, \psi \in \mathcal{S}\}.$$

In this paper, we focus on this way of embedding.

When there is no embedding, it goes back to constructing a probability distribution Π on the set of densities on $\mathcal{X}$ and to compare a simpler model namely $\mathcal{F}$ with a more complicated one represented by Π. The latter model needs not give some special meaning to the parametric model. This approach is to some extent considered by Florens, Richard and Rolin (1996), although they work with the Dirichlet process as a prior which is not adapted to estimating densities with respect to the Lebesgue measure. They also give a special meaning to the parametric model, since they construct their nonparametric prior such that the marginal densities for one observation under the parametric and the nonparametric models are the same. McVinish *et al.* (2005) consider no embedding at all. To some extent the approach with no-embedding is more adapted to a 0–1 type of losses than the approach with embedding, since the idea is to test two different models, where one is not quite included in the other one.

An interesting difference appears in the behaviour of the p-value wether we embed the null model in the alternative or not, as seen in Section 2.2.

To understand better the impact of embedding in a Bayesian goodness of fit we also study the asymptotic properties of the Bayes factor in both cases. This enables us also to shed some light on the behaviour of Bayes factors in large dimension problems. In particular, general conditions are given for the Bayes factors to be inconsistent. Using the Theorem of McVinish *et al.* (2005) on the consistency of Bayes factors in goodness of fit problems of a parametric family, we finally obtain conditions which are close to being sufficient and necessary in some contexts.

In Section 2 we compare the Bayes factors associated with precise null hypotheses as defined by (3) and with imprecise null hypotheses as defined by (1) respectively. We prove that the width of the interval must be very small for both Bayes factors to behave similarly. In Section 2.2 we study the calibrated distance approach in goodness of fit tests. In particular we determine the calibration induced by the use of p-values. In Section 3.1 we consider goodness of fit tests of a fixed distribution and in Section 3.2 We study goodness of fit tests of a parametric family. Section 4 is dedicated to consistency properties of the Bayes factors in goodness of fit tests of parametric families.

2. ARE BAYES FACTORS GOOD AT APPROXIMATING INTERVAL NULL HYPOTHESIS BY POINT NULL HYPOTHESIS?

In this section we study the difference between the Bayes factor associated with a point null hypothesis in the form (3) and the Bayes factor associated with the interval null hypothesis in the form (1). First we define some notations.

2.1. *Notations and Context*

Let $X = (X_1, ..., X_n)$ be a n-sample, each observation being distributed from P_θ, $\theta \in \Theta$. Assume that the model is dominated by a common measure μ on $\mathcal{X} \subset \mathbb{R}^d$, $d \geq 1$. We denote by $\mathcal{L}$ the set of probability densities on $\mathcal{X}$. Let d be a distance or pseudo-distance between two densities on $\mathcal{X}$ and denote by $S_{\epsilon_n}(\theta_0, d)$ the neighbourhood of f_{θ_0} defined by

$$S_{\epsilon_n}(\theta_0, d) = \{\theta \in \Theta; d(f_{\theta_0}, f_\theta) \leq \epsilon_n\}.$$

Let π be a prior on Θ and denote by $\lambda_n(\theta_0, d)$ the prior mass of $S_{\epsilon_n}(\theta_0, d)$: $\lambda_n(\theta_0, d) = \pi(S_{\epsilon_n}(\theta_0, d))$. In the sequel, when there is no ambiguity we will drop (θ_0, d) in our notations and consider only S_n and λ_n. Consider the interval null testing problem defined by:

$$H_0^n : \theta \in S_{\epsilon_n}(\theta_0, d), \quad \text{versus} \quad H_1^n : \theta \notin S_{\epsilon_n}(\theta_0, d)$$

and the sharp null hypothesis problem expressed as

$$H_0^\star : f_\theta = f_{\theta_0}, \quad \text{versus} \quad H_1^\star : f_\theta \neq f_{\theta_0}.$$

The Bayes factor associated with the first problem is then given by

$$B_n = \frac{\int_{S_{\epsilon_n}(\theta_0,d)} f_\theta(X) d\pi(\theta)(1-\lambda_n)}{\lambda_n \int_{S^c_{\epsilon_n}(\theta_0,d)} f_\theta(X) d\pi(\theta)}, \tag{8}$$

where A^c denotes the complementary of the set A. The Bayes factor associated with the second problem is given by

$$B_n^\star = \frac{f_{\theta_0}(X)}{\int_\Theta f_\theta(X) d\pi(\theta)}. \tag{9}$$

We also denote by $K(f, f')$ the Kullback–Leibler divergence between f and f', that is

$$K(f, f') = \int_{\mathcal{X}} f(x) \log \left(f(x)/f'(x)\right) d\mu(x).$$

Similarly $V(f, f')$ denotes the second moment of the log of the ratio of the densities f and f' under f, that is

$$V(f, f') = \int_{\mathcal{X}} f(x) \left(\log f(x)/f'(x)\right)^2 d\mu(x).$$

In the sequel $l_n(\theta)$ denotes the log-likelihood function associated with n observations and if $\theta_1, \theta \in \Theta$, $\Delta(\theta_1, \theta) = K(f_{\theta_1}, f_{\theta_0}) - K(f_{\theta_1}, f_\theta)$ and $V_{\theta_1}(\theta, \theta_0) = E_{\theta_1}\left[(l_1(\theta_0) - l_1(\theta))^2\right]$. We also consider the following pseudo-distance:

$$d_K(f, f') = K(f, f') + V(f, f').$$

This pseudo-distance will be used to obtain sufficient conditions on ϵ_n for both Bayes factors, B_n and $B_n^\star$ to behave similarly. In the following sections we will also consider the Hellinger distance on densities with respect to a common dominating measure μ:

$$h^2(f, f') = \int \left(\sqrt{f(x)} - \sqrt{f'(x)}\right)^2 d\mu(x),$$

and the L_1 distance:

$$\|f - f'\|_1 = \int \left|f(x) - f'(x)\right| d\mu(x).$$

The formulation considered above is oriented towards tests on densities rather than tests on parameters. It allows us to consider in the same framework parametric and nonparametric models. Note however that in regular parametric models both formulations are equivalent, see Remark 2.

2.2. *Comparison of Both Bayes Factors*

Theorem 1 *In the framework defined previously,*

- *(i) Let* $\epsilon_n = \epsilon_0/n$ *and* $d(.,.) = d_K(.,.)$. *Suppose that for all* $\theta_1 \in \Theta$, *there exists* $c, M > 0$ *such that*

$$\liminf_n \frac{\pi(S_{n,1})}{\pi(S_{\epsilon_n}(\theta_0, d_K))} > 0, \quad \textit{with } S_{n,1} = \left\{\theta; \Delta(\theta_1, \theta) > -\frac{c}{n}, V_{\theta_1}(\theta_0, \theta) \leq \frac{M}{n}\right\},$$

then for all $\theta_1 \in \Theta$

$$\lim_{C \to +\infty} \limsup_n P_{\theta_1}\left[\frac{B_n}{B_n^\star} \leq 1/C\right] \quad = \quad 0.$$

- *(ii) Let* $\epsilon_n = \epsilon_0 v_n/n^2$ *with* $v_n \to \infty$ *and* $d(.,.) = K(.,.)$. *If there exists* $\theta_1 \in \Theta$ *and* $a, V_0 > 0$ *and a sequence* $\rho_n > 0$ *decreasing to zero such that*

$$\liminf_n \frac{\pi[S_{n,2}]}{\pi[S_{\epsilon_n}(\theta_0, K)]} > 0,$$

with $S_{n,2} = \{f_\theta \in S_n; V_{\theta_1}(f_{\theta_0}, f_\theta) \leq V_0 v_n \rho_n/n; \Delta(\theta_1, \theta) > 2a\sqrt{v_n}/n\}$ *then*

$$\frac{B_n}{B_n^\star} \to +\infty, \quad P_{\theta_1}.$$

The proof is given in the Appendix.

Remark 1. The first part of the Theorem 1 implies that B_n is at least of the same order as $B_n^\star$ when ϵ, the tolerance threshold is small enough, but not too small since it corresponds to the parametric rate of convergence in smooth models. The second part of the theorem is more precise since it says that B_n can be actually much larger than $B_n^\star$ as soon as ϵ is greater than n^{-2}. This is far too small to be of any use in practice, for large sample sizes at least, since the typical parametric rate of convergence in smooth parametric families is much greater than that. It therefore appears that Bayes factors for interval null hypotheses cannot be well approximated by Bayes factors for point null hypotheses, at least when the number of observations is large.

Remark 2. Although the condition on $S_{n,2}$ seems a bit restrictive, note that it is not so. Indeed if $\theta_1 \neq \theta_0$, when n is large enough $\Delta(\theta_1, \theta) = O(\sqrt{K(f_{\theta_0}, f_\theta)})$ and is usually not of smaller order. Hence typically as soon as $K(f_{\theta_0}, f_\theta) \geq a_1 v_n/n^2$, for a large enough we will have $\Delta(\theta_1, \theta) \geq a\sqrt{v_n}/n$. To illustrate this consider the simpler case of a smooth parametric family. Then if θ' is close to θ,

$$K(f_{\theta'}, f_\theta) = i(\theta')\frac{(\theta' - \theta)^2}{2} + o((\theta - \theta')^2),$$

where $i(\theta)$ is the Fisher information matrix associated with one observation. Also, using a Taylor expansion around θ_0, we obtain

$$\Delta(\theta_1, \theta) \quad = \quad (\theta - \theta_0) \int f_{\theta_1}(x) \frac{\partial \log f_\theta(x)}{\partial \theta}\bigg|_{\theta\theta_0} dx + 0(\theta - \theta_0)^2.$$

Hence, if θ_1 is fixed and such that

$$E_{\theta_1}\left(\frac{\partial f_{\theta_0}(X)}{\partial\theta}\right) \neq 0$$

(say > 0), we have as long as $\theta - \theta_0 > \epsilon\sqrt{v_n}/n$, and if $\theta > \theta_0$, $\Delta(\theta_1, \theta) > 2a\sqrt{v_n}/n$, for ϵ large enough. Also, the set $S_{\epsilon_n}(f_{\theta_0}, K)$ corresponds to the subset of Θ defined by $\{|\theta - \theta_0| < \delta_0\sqrt{v_n}/n\}$. Moreover the condition on $V_{\theta_1}(f_{\theta_0}, f_\theta)$ is satisfied for instance as long as $\log f_\theta(x)$ is bounded uniformly for θ in a neighbourhood of θ_0, with $\rho_n = O(1/n)$. This leads to

$$\pi[S_{n,2}]/\pi[S_{\epsilon_n}(f_{\theta_0}, K)] > \pi[\delta_0\sqrt{v_n}/n > \theta - \theta_0 > \epsilon/n]/\pi[|\theta - \theta_0| < \delta_0\sqrt{v_n}/n] > 1/4,$$

when n is large enough.

Remark 3. This result is quite general in the sense that it can apply to parametric and nonparametric frameworks.

Example 1 (Smooth parametric families).

Assume that $\{f_\theta, \theta \in \Theta\}$, $\Theta \subset \mathbb{R}^d$ with d some positive integer is a smooth parametric family. Then calculations in Remark 2 imply that

$$S_{\epsilon_n}(\theta_0, d_K) \subset \{|\theta - \theta_0| < c\sqrt{\epsilon_n}\},$$

for some constant $c > 0$. If $\theta_1 \neq \theta_0$ there exists a constant $c(\theta_1, \theta_0) \neq 0$, say positive, such that

$$\Delta(\theta_1, \theta) = c(\theta_1, \theta_0)(\theta - \theta_0) + O((\theta - \theta_0)^2).$$

Let $\epsilon_n = \epsilon_0/n$, then as soon as $\theta > \theta_0$ $\Delta(\theta_1, \theta) = O(\epsilon_n^2) > -c/n$ for all $\theta > \theta_0$ and satisfying $\theta \in S_{\epsilon_n}(\theta_0, d_K)$. Using Remark 2 together with the above result we thus prove that the assumption in (i, Theorem 1) is satisfied. In remark 2, we obtained that the assumption in (ii, Theorem 1) is also satisfied, therefore Theorem 1 applies to smooth parametric families.

Example 2 (Nonparametric models: sieve models).

We consider the special case of exponential family models:

$$\log f_{\theta,K}(x) = \sum_{j=1}^{K} \theta_j B_j(x) - c(\theta),$$

where K is unknown and where the B_j's form a basis of some functional space like the set of squared integrable functions, or the set of polynomials, or the set of piecewise polynomial functions as in the log-spline models. The parameter has therefore the form (K, θ), where $K \in \mathrm{N}$ and, conditional on K, $\theta \in \mathbb{R}^{K-1}$, since for identifiability reasons we need a constraint in the form $\sum_{j=1}^{K} \theta_j = 0$ or $\theta_0 = 1$. Consider a prior on the probability space and let $(K_0, \boldsymbol{\theta}_0)$ be a specific value of the parameter. Assume also that the prior on the $\theta_j's$ has support included in $[-M, M]$ for each $j \geq 1$, with M large enough so that $\boldsymbol{\theta}_0$ belongs to the support of the prior on the subset $\{(\theta_1, ..., \theta_{K_0}), \theta_j \in \mathbb{R}\}$. Then if $\theta = (\theta_1, ..., \theta_{K_0})$ and $\boldsymbol{\theta}_0 = (\theta_{01}, ..., \theta_{0K_0})$, the model is smooth and the result obtained in Example 2 applies. If $\underline{\theta} = (\theta_1, ..., \theta_K)$ with $K > K_0$, then f_θ belongs to S_{ϵ_n} if there are extra constraints in the form $|\theta_j| < c\sqrt{\epsilon_n}$, for $j = K_0 + 1, ..., K$ to add to the constraints on the first K_0 parameters. This implies that λ_n can be written in the form $C\epsilon_n^{K_0/2}(1 + o(1))$ and the results obtained in Example 1 can be extended to this case.

3. UNDERSTANDING THE CALIBRATED DISTANCE APPROACH IN TERMS OF INTERVAL NULL HYPOTHESIS TESTING

In this section, we focus on goodness of fit tests. We could extend the following results to more parametric tests, but the interest of these tests is stronger in the framework of goodness of fit testing.

Let $X = (X_1, ..., X_n)$ be a n-sample from a probability distribution having density f. Again, we compare the interval null hypothesis tests: let $d(.,.)$ be a distance or a pseudo-distance between densities and $\epsilon_n > 0$

$$H_0^n : d(f, f_0) \leq \epsilon_n \quad \text{versus} \quad H_1^n : d(f, f_0) > \epsilon_n \tag{10}$$

with the sharp null hypothesis problem expressed as

$$H_0^\star : f = f_0, \quad \text{versus} \quad H_1^\star : f \neq f_0. \tag{11}$$

We will also consider goodness of fit tests against a parametric family, with both formulations:

$$H_0^n : d(f, \mathcal{F}) \leq \epsilon_n \quad \text{versus} \quad H_1^n : d(f, \mathcal{F}) > \epsilon_n, \tag{12}$$

where $\mathcal{F} = \{f_\eta, \eta \in \Theta_0\}$ and $d(f, \mathcal{F}) = \inf\{d(f, f_\eta), \eta \in \Theta_0\}$ and the sharp null hypothesis problem expressed as

$$H_0^\star : f \in \mathcal{F}, \quad \text{versus} \quad H_1^\star : f \notin \mathcal{F}. \tag{13}$$

In the Bayesian context we need to define both the null hypothesis and the alternative precisely, in other words to put a prior on the alternative and on the null model. Let π_0 be a prior on Θ_0 (under the null) and π be a prior under the alternative.

We consider the test statistic based on the loss function $L(f, \delta)$ defined by (5) with $\theta = f$, $\Theta_0 = \{f_0\}$ or $\mathcal{F}$ and $\epsilon = \epsilon_n$. When ϵ_n is known the Bayesian test is given by (??). When ϵ_n is not known or fixed a priori, we propose to calibrate it with a p-value. Hence we formally construct a test for the problem (11) (resp. (13)) instead of (10) (resp. (12).) However by using this p-value we define some implicit threshold values ϵ_n, or at least a range of implicit values. The aim of this section is to describe these implicit values. Note here that the p-value is not used in the usual frequentist sense but is used as a calibration tool.

First we consider the simpler case of goodness of fit of a specific distribution against any distribution. Then we extend the results to the goodness of fit of a parametric family. Two cases will be considered, the embedded approach and the non-embedded approach.

3.1. *Goodness of Fit Test of a Fixed Null Distribution*

We first consider the problem of a point null hypothesis $\{f_0\}$ or a neighbourhood of f_0. The p-value is the defined by

$$p_0(X) = P_0\left[H(Y) > H(X)|X\right],$$

where $P_0[A]$ defines the probability of A under f_0 and $H(Y) = E^{\pi}\left[d(f, f_0)|Y\right]$ is the posterior expectation of the distance between f and the null in the alternative model. The distance is typically going to be either the Hellinger or the L_1 distance.

Let $\alpha \in (0,1)$, then $p_0(X) = P_0\left[H(Y) > H(X)|X\right] = \alpha$ if $H(X) = G_{0n}^{-1}(1-\alpha)$, where G_{0n} is the cumulative distribution function of $H(Y)$ under f_0. Although we are not considering a Neyman–Pearson framework, we can use the quantities $G_{0n}^{-1}(1-\alpha)$ to understand the meaning of the calibration induced by the use of the p-value $p_0(X)$. In other words considering that the observed p-value is small when it is equal to α corresponds to considering that $G_{0n}^{-1}(1-\alpha)$ is large as a value of $H(X)$. The order of the implicit ϵ_n's can therefore be understood as the order the values $G_{0n}^{-1}(1-\alpha)$'s for fixed α's. This corresponds to the rate of convergence of the posterior distribution in the alternative model, under f_0. Let ε_n be such that

$$P^{\pi}\left[S_{\varepsilon_n}(f_0, d)^c|Y\right] = O_P(\varepsilon_n), \quad \text{and} \quad P^{\pi}\left[S_{\rho_n\varepsilon_n}(f_0, d)|Y\right] = o_P(1),$$

for all sequence $(\rho_n)_n$ decreasing to 0, then for any $\alpha \in (0,1)$,

$$\begin{aligned} G_{0n}^{-1}(1-\alpha) > C\varepsilon_n \quad &\Leftrightarrow \quad 1-\alpha > P_0\left[H(Y) < C\varepsilon_n\right] \\ &\Rightarrow \quad 1-\alpha > P_0\left[P^{\pi}\left[S_{\varepsilon_n}(f_0, d)^c|Y\right] < (C-1)\varepsilon_n\right] \end{aligned}$$

the latter term of the right hand side can be made as close to 1 as need be by choosing C and n large enough, implying that for C and n large enough,

$$G_{0n}^{-1}(1-\alpha) > C\varepsilon_n.$$

Moreover, let ρ_n be any decreasing sequence to 0, then

$$\begin{aligned} G_{0n}^{-1}(1-\alpha) \le \rho_n\varepsilon_n \quad &\Leftrightarrow \quad 1-\alpha \le P_0\left[H(Y) \le \rho_n\varepsilon_n\right] \\ &\Rightarrow \quad 1-\alpha \le P_0\left[2\rho_n\varepsilon_n P^{\pi}\left[S_{\varepsilon_n}(f_0, d) \cap S_{2\rho_n\varepsilon_n}(f_0, d)^c|Y\right] \le \rho_n\varepsilon_n\right] \\ &\Rightarrow \quad 1-\alpha \le P_0\left[P^{\pi}\left[S_{\varepsilon_n}(f_0, d) \cap S_{2\rho_n\varepsilon_n}(f_0, d)^c|Y\right] \le 1/2\right]. \end{aligned}$$

The latter term of the right hand side can be made as close to 0 as need be, so that when n is large enough

$$G_{0n}^{-1}(1-\alpha) > \rho_n\varepsilon_n.$$

Finally it appears that the calibration with a p-value of the distance based approach implies an ϵ_n of the same order as the rate of convergence of a Bayesian estimator (say the posterior mean) of f_0 under the nonparametric model. It often is the case that the nonparametric alternative is constructed around f_0, or more precisely as perturbations around the null model. In such cases, f_0 typically belongs to the support of π, the nonparametric prior, and the nonparametric rate of convergence for estimating f_0 is then often of order $n^{-1/2}$ up to a $\log n$ term.

Example 3 (Example 2 – Part 2: Exponential family).

Consider as a special case the exponential family model defined in Example 2, where the B_j's form the Legendre basis on $[0, 1]$ as defined by Verdinelli and Wasserman (1998) for instance and f_0 is the uniform on $[0, 1]$. Then by considering the identifying constraint on the θ_j's in the form $\theta_0 = 1$, f_0 can be written as a special case of this exponential family with $\theta_j = 0$, for all $j \geq 1$ or by choosing $K = 1$ and $\theta_1 = 0$. Hence the prior mass of $\log n/n$-Kullback–Leibler neighbourhoods of f_0 is bounded from below by $c\sqrt{\log n/n}$ leading to an upper bound for the rate of convergence of order $\sqrt{\log n/n}$ in terms of the L_1 or the Hellinger distance. A lower bound on the rate is of order $1/\sqrt{n}$, so that the implicit ϵ_n devised by the calibration method is of order $v_n/\sqrt{n}$ with $v_n \in (1, \sqrt{\log n})$ when d is either the Hellinger distance or the L_1 distance.

Remark 4. The above result is quite general. For instance if instead of the above collection of exponential families we consider a mixture model with unknown number of components and f_0 can be represented as a finite mixture, then the rate of convergence is often of order $O(\sqrt{\log n/n})$, see for instance Ghosal and Van der Vaart (2001) in the case of Gaussian mixtures.

Remark 5. As is often the case, it is a lot easier to find an upper bound on ϵ_n than a lower bound. In the following of the paper we will consider essentially upper bounds, although some remarks will be made on lower bounds as well.

The case of a goodness of fit tests of a fixed density is much easier than the case of a goodness of fit tests of a parametric family. In the following section we consider the goodness of fit tests of a parametric family. Two cases will be considered: The embedding approach of Verdinelli and Wasserman (1998) and Robert and Rousseau (2003) and the non-embedding approach as is in McVinish *et al.* (2005). The results obtained in the following section shade light on the meaning of embedding or not embedding in goodness of fit tests.

3.2. *Goodness of Fit Test of a Parametric Family*

In this section we consider the testing problems defined by (12) and (13). We use the same loss function as in Section 3.1, so that the Bayes estimate of the test, in the case of a known threshold ϵ_n is given by $\delta(X) = 1$ if $H(X) > c_n$, where $H(X) = E^\pi[d(f, \mathcal{F})|X]$. Similarly to before, when ϵ_n is not fixed a priori we calibrate it using a p-value. However, in this case there is no obvious p-value since under the null hypothesis the probability depends on the unknown value η. A certain number of p-values have been proposed in the literature, the most well known are the plug-in p-value and the posterior p-value. These have many defects due to their strong double use of the data, see Bayarri and Berger (2000) and Robbins, Van der Vaart and Ventura (2000) for a discussion of their properties. To improve on these Bayarri and Berger (2000) have proposed other p-values: the conditional predictive p-value and the partial posterior predictive p-value. Their asymptotic properties have been studied by Robbins *et al.* (2000). Higher order asymptotic properties of the conditional predictive p-value associated with the maximum likelihood estimator $\hat{\eta}$ have been studied by Fraser and Rousseau (2005). In this paper we consider the same p-value as in Robert and Rousseau (2003) and Fraser and Rousseau (2005), namely the conditional predictive p-value based on the maximum likelihood estimator defined by (7) in Section 1. Recall that this p-value can be written as:

$$p_{\text{cpred}} = \int_{\Theta_0} P_\eta \left[H(Y) > H(X) | \hat{\eta}, X \right] d\pi_0(\eta|\hat{\eta}).$$

The test statistic is based on the nonparametric prior π. This prior π can be constructed by either embedding the parametric model in the nonparametric one, or quite independently of the parametric model. We first study the embedding approach and more precisely Verdinelli and Wasserman's (1998) embedding.

3.2.1. *The Embedding Case*

Denote by F_η the cumulative distribution functions associated with f_η, $\eta \in \Theta_0$, consider a nonparametric family of densities on $[0,1]$ $\{g_\psi, \psi \in \mathcal{S}\}$ and assume that there exits $\psi_0 \in \mathcal{S}$ such that $g_{\psi_0}(u) = 1$, $\forall u \in [0,1]$. The alternative nonparametric family is represented by

$$\mathcal{H} = \{h_{\eta,\psi}(x) = f_\eta(x) g_\psi(F_\eta(x)), \eta \in \Theta_0, \psi \in \mathcal{S}\}. \tag{14}$$

Let π be a prior oh $\mathcal{H}$ which can be expressed in the form $d\pi(\eta,\psi) = d\pi_1(\psi) d\pi(\eta|\psi)$.

Using the same approach as in Section 3.1 we want to estimate the implicit values ϵ_n obtained by the calibration using the representation

$$\epsilon_n = O(G_n^{-1}(1-\alpha)), \quad \alpha \in (0,1),$$

where G_n is the cumulative distribution function of $H(Y)$ when Y is distributed from $m^\pi(y|\hat{\eta}_x) = \int_{\Theta_0} f(y|\eta, \hat{\eta}_x) d\pi(\eta|\hat{\eta}_x)$. We now characterize this implicit order.

Assume that the parametric model is regular and in particular that the regularity assumptions of Theorem 3.1 of Robert and Rousseau are satisfied. Then Theorem 3.1 of Robert and Rousseau implies that

$$p_{\text{cpred}} = P_{\eta^\perp} \left[H(Y) > H(X) | \hat{\eta}_x, X \right] + o_P(1), \tag{15}$$

where $\eta^\perp = \text{argmin}_\theta K(f, f_\theta)$ and f is the *true* density of the observations. Therefore $p_{\text{cpred}}(X) \le \alpha$ if and only if $H(X) \ge (G_{\eta^\perp}^{\hat{\eta}})^{-1}(1 - \alpha + o_P(1))$, where $G_{\eta^\perp}^{\hat{\eta}}$ is the conditional cumulative distribution function of $H(Y)$ given the $\hat{\eta}^y = \hat{\eta}$, relative to $\eta^\perp$. Then as in the case of the point null p-value, we can interpret this procedure as deciding which value of ϵ_n is small, by understanding ϵ_n as $(G_{\eta^\perp}^{\hat{\eta}_x})^{-1}(1-\alpha+o_P(1))$, where α is small in the uniform scale. We now determine the order of $(G_{\eta^\perp}^{\hat{\eta}_x})^{-1}(1-\alpha)$, for fixed α's. To do so, we first characterise the rate of convergence associated with the full model $\mathcal{H}$ when the true density is in the null model in terms of the rate of convergence associated with the nonparametric family on $[0,1]$, i.e. $\mathcal{S}$ when the true distribution if the uniform on $[0,1]$.

Theorem 2 *Let $(\rho_n)_n$ be a sequence decreasing to zero such that $\rho_n \ge \rho_0 \log n/n$, for some positive ρ_0. Let $\mathcal{S}_n \subset \mathcal{S}$, for all $n \ge 1$.*

- *(i) Assume that there exists $c_1 > 0$ such that for all*

$$K_n = \sum_{j \ge 0} \sqrt{\pi(A_{n,j})} \le e^{c_1 n \rho_n},$$

where $(A_{n,j}, j \ge 0)$ is a ρ_n−covering of $\mathcal{S}_n \cap \{\psi, h^2(g_\psi, g_{\psi_0}) > C\rho_n\}$, with $C \ge 1$.

- *(ii) Assume that there exists* $c_2 > 0$ *satisfying*

$$\pi(\mathcal{S}_n^c) \leq e^{-(c_2+1)n\rho_n}$$

such that $e^{-nc_2\rho_n/2} \leq C_1\rho_n$ *for some positive constant* C_1.

- *(iii) Suppose that for some* $c_2/2 > b > 0$

$$\pi\left[S_{\rho_n}(g_{\psi_0}, d_K)\right] \geq e^{-bn\rho_n}.$$

Assume also that Θ_0 *is compact, that for each* $\psi \in \mathcal{S}_n$, *there exists* $c, C_\psi > 0$ *such that*

$$\left|\log\left(\frac{g_\psi(u)}{g_\psi(v)}\right)\right| \leq C_\psi\left[\frac{|v-u|}{u \wedge v} + |v-u|^c\right], \tag{16}$$

and suppose that there exists $c_3 > 0$ *such that* $\sup_{\psi\in\mathcal{S}_n} C_\psi \leq e^{nc_3\rho_n}$. *Then if for all* $\eta \in \Theta$ *there exists* $\tau > 0$ *such that*

$$\int \sup_{|\eta-\eta'|<\tau}\left|\frac{\partial f_{\eta'}(x)}{\partial\eta}\right| dx < \infty \tag{17}$$

with

$$\int_{-\infty}^{F_\eta^{-1}(u)} \sup_{|\eta-\eta'|<\tau}\left|\frac{\partial f_{\eta'}(x)}{\partial\eta}\right| dx = O(u), \quad \int_{F_\eta^{-1}(u)}^{\infty} \sup_{|\eta-\eta'|<\tau}\left|\frac{\partial f_{\eta'}(x)}{\partial\eta}\right| dx = O(1-u), \tag{18}$$

there exists $C_0, C, c > 0$ *such that*

$$P^\pi\left[f_{\eta,\psi}; h^2(f_{\eta,\psi}, f_{\eta_0,\psi_0}) > C\rho_n|X^n\right] \leq e^{-nc\rho_n} \wedge C_0\rho_n, \quad P_0^\infty \ a.s.$$

The proof of Theorem 2 is given in the Appendix.

The above theorem essentially states that if ρ_n is the convergence rate of the posterior distribution (in hellinger distance) for the estimation of g in $[0,1]$, using π_1 when $g_0 = g_{\psi_0}$ then when $f_0 = f_{\theta_0}g_{\psi_0}(F_{\theta_0})$, the rate of convergence of the posterior in the full model under π is also equal to ρ_n. The first set of assumptions (i)–(iii) are here to ensure that ρ_n is the rate of convergence for the posterior distribution in the model g_ψ. We could have considered other sets of assumptions leading to the same results, such as those of Ghosal *et al.* (2001), the result would still be the same. These assumptions are adapted from Walker *et al.* (2007) to sieve sets, in order to accept unbounded densities as will be seen in the case of mixtures of betas. Note that the posterior rate of convergence for estimating f_η does not depend on η under the above assumptions. This is one of the interests of embedding, since it reduces the impact of the nuisance parameter η. We will see in Section 3.2.2 that η can have a lot more influence when there is no embedding.

Corollary 3 *Under the conditions of Theorem 2, we have*

$$P_{\psi_0}\left[E^\pi\left[h^2(1, g_\psi)|U^n\right] > 2\rho_n\right] \to 0$$

and

$$P_0\left[E^\pi\left[h^2(f_{\theta,\psi}, \mathcal{F})|Y^n\right] > C\rho_n\right] \to 0.$$

Proof of Corollary 3. The result comes from the following inequality: P_0^∞ almost surely, we have that

$$E^\pi\left[h^2(1,g_\psi)|U^n\right] \leq \rho_n + P^\pi\left[h^2(1,g_\psi) > \rho_n|U^n\right] \leq 2\rho_n$$

since

$$P^\pi\left[h^2(1,g_\psi) > \rho_n|U^n\right] \leq \rho_n$$

and similarly for the second inequality. □

This implies in particular that $(G_{\theta\perp}^{\hat\theta})^{-1}(1-\alpha) = 0_P(\rho_n)$, for all $\alpha \in (0,1)$. Indeed, since

$$P_{\theta\perp}\left[H(Y^n) > C\rho_n\right] = o(1),$$

$$P_{\theta\perp}\left[P_{\theta\perp}\left[H(Y^n) > C\rho_n|\hat\theta\right] > \epsilon\right] \leq \epsilon^{-1}P_{\theta\perp}\left[H(Y^n) > C\rho_n\right]$$

for any positive ϵ and it goes to 0 as n goes to infinity.

Remark 6. The above results give only upper bounds on the order of ϵ_n. Obtaining an exact order in the nonparametric case is often very difficult. We shall see in some examples however that this upper bound is close to the lower bound when ρ_n is also close to the lower bound on the posterior rate of convergence on $\mathcal{S}$.

Remark 7. The above results are valid for all (almost) true densities and not only under the null hypothesis. This comes from the fact that p_{cpred} can be approximated by a conditional p-value as described by (15), even underthe alternative.

Example 4 (Mixtures of Betas).

We consider the following example: $g_\psi = p_0 + (1-p_0)\sum_{j=1}^K p_j g_{a_j,b_j}$, where $g_{a,b}$ is a Beta density with parameters (a,b). In other words $\psi = (K, p_0, \underline{p}, a, b)$ where u designates the vector in $\mathbb{R}^K$: $(u_1, ..., u_K)$. We prove that if $\mathcal{S}$ is defined by mixtures of Betas distributions as described above, the rate of convergence for estimating g_{ψ_0} under the prior π_1 on $\mathcal{S}$ is of the form $\log n/n$ and that we obtain the same rate for estimating f_θ, $\theta \in \Theta$ under the prior π on the full model $\mathcal{H}$. More precisely, if we constrain the parameters of the betas to be bounded, we obtain $\log n/n$ otherwise if the prior on these parameters has light tails (Gaussian type) then the rate of convergence is $\log n^2/n$.

Consider ψ and ψ' corresponding to the same K and the same $\tilde A_j$: $|p_0 - p_0'| < \delta_1$, $|a_j - a_j'| < \delta_2$ $|p_j - p_j'| < \delta_3$. Let $\mathcal{F} = \{f_\theta, \theta \in \Theta\}$ be a regular model.

$$\begin{aligned} A &= \left|\log\left(\frac{g_{\psi'}(u)}{g_{\psi'}(u+\gamma(u))}\right)\right| \\ &= \left|\log\left(p_0' + (1-p_0')\sum_{j=1}^K p_j' g_{a_j',b_j'}(u)\right) - \log\left(p_0' + (1-p_0')\sum_{j=1}^K p_j' g_{a_j',b_j'}(u+r\gamma(u))\right)\right| \\ &\leq \frac{(1-p_0')\sum_{j=1}^K p_j'\left|\partial g_{a_j',b_j'}(u+r\gamma(u))/\partial u\right|}{p_0' + (1-p_0')\sum_{j=1}^K p_j' g_{a_j',b_j'}(u+r\gamma(u))}, \end{aligned}$$

with $r \in (0,1)$.

$$A \leq |\gamma(u)| \max_{j\leq K} C * (a_j' + b_j' + 1)(u + r\gamma(u))^{-1}(1-(u+r\gamma(u)))^{-1}.$$

Moreover, under the above hypothesis on the parametric model

$$|F_{\theta'}(F_\theta^{-1}(u)) - u| \le |\theta' - \theta| \int_{-\infty}^{F_\theta^{-1}(u)} \sup_{|\eta-\theta|<\tau} \left| \frac{\partial f_\eta(x)}{\partial \theta} \right| dx.$$

So that there exists $C_2 > 0$, independent of θ, ψ satisfying $|\gamma(u)| \le C_2 u(1-u)$. This implies that for $|\theta' - \theta| C_2 < 1/2$,

$$\int_0^1 g_\psi(u) A(u) du \quad \le \quad 2CC_2 |\theta' - \theta| \max_{j \le K} (a'_j + b'_j + 1).$$

Hence, by choosing $|\theta' - \theta|$ small enough, independently of ψ, ψ' if we assume that $a_j + b_j \le M$, we have that $f_{\theta',\psi'} \in A_j$ when $g_{\psi'}$ in $\tilde{A}_j$ and $|\theta' - \theta| \le C_0 \delta$. When Θ is compact, the number of such balls $(N(\delta))$ is bounded, for each δ. Hence

$$\sum_j \sqrt{\pi(A_j)} \le N(\delta) \sum_j \sqrt{\pi_1(\tilde{A}_j)} < \infty.$$

If the set of (a'_j, b'_j) is not bounded. Consider $\mathcal{S}_n = \{(K, a, b), 0 < a_j, b_j \le M_n\}$. Then if M_n is such that

$$\pi(|a_j| + |b_j| > M_n) \le e^{-n(c_2+1)\rho_n} = c_n,$$

$$\begin{aligned} \pi(\mathcal{S}_n^c) &= \sum_k p(k) \{1 - \pi_k(\forall j; a_j, b_j \le M_n)\} \\ &\le \sum_k p(k) \left\{1 - (1 - c_n)^k\right\} \\ &= 1 - \sum_k p(k)(1 - c_n)^k \\ &\le c_n \sum_k kp(k), \end{aligned}$$

where the last inequality comes from the Laplace transform of K, since $\sum_k p(k)(1 - c_n)^k = E[e^{-\alpha K}]$ with $\alpha = -\log(1 - c_n)$.

Then we obtain a ρ_n-covering of $\mathcal{H}_n = \Theta \times \mathcal{S}_n$ by considering coverings of H_n in the form

$$\tilde{A}_{n,j,l} = \{|\theta - \theta_l| \le \delta \rho_n / M_n\} \times A_{n,j}.$$

We thus have

$$\begin{aligned} K_n &\le \sum_{j,l} \sqrt{\pi(\tilde{A}_{n,j,l})} \\ &\le CN(\delta\rho_n/M_n) (\rho_n/M_n)^{k/2} \sum_j \sqrt{\pi(A_{n,j})} \\ &\le e^{nc_1\rho_n} e^{-d/2 \log(\rho_n/M_n)} \\ &\le e^{nc'_1\rho_n}, \end{aligned}$$

as soon as $-\log(\rho_n/M_n) \leq 2(c_1' - c_1)n\rho_n/d$, i.e. as soon as $\rho_n \geq \rho_0 M_n \log n/n$ with ρ_0 large enough. This implies that in the unbounded case the rate of convergence ρ_n cannot be better than say $\log n^2/n$. To obtain this rate of convergence it is necessary that

$$\pi(a_j + b_j > M_0 \log n) \leq e^{-(c_2+1)\log n^2},$$

which means that the tails of the distribution on the parameters of the beta densities are as light as the Gaussian tails.

In this example, it is to be believed that $1/n$ (in terms of of $h^2(.,.)$) is a lower bound on the rate of convergence, that is

$$P^{\pi}\left[S_{\rho_n/n}(f_0, h^2)\right] = o_{P_0}(1),$$

for any sequence ρ_n converging to 0. Hence, up to a $\log n$ term $\epsilon_n = 0(1/n)$, the parametric rate of convergence.

Remark 8. A particular case of interest here is when $\rho_n = \log n/n$ up to a multiplicative constant. This happens in many cases such as mixtures with unknown number of components, infinite dimensional exponential family, see Example 2. Then in most cases, up to the $\log n$ term this is also a lower bound on ϵ_n.

Remark 9. Although we have no rigorous proof of it, we believe that the posterior distribution shrinks more quickly around g_{ψ_0} under π_1 the nonparametric prior on $\mathcal{S}$ than does the posterior distribution around f_{θ_0} under π the nonparametric prior on $\mathcal{H}$.

3.2.2. *The Non-Embedding Case*

In this section we consider a testing procedure where we do not embed the parametric model into the nonparametric one. We still want to test

$$H_0: \quad d(f, \mathcal{F}) \leq \epsilon_n, \quad H_1: \quad d(f, \mathcal{F}) > \epsilon_n$$

using a nonparametric estimation procedure with no embedding. This has been done for instance in McVinish *et al.* (2005), where the authors consider mixtures of triangular distributions as a special case of nonparametric prior. It is also the approach considered in Goutis and Robert (1997), although they are considering a fully parametric framework, with estimation as the ultimate aim. One of the advantages of the approach with no embedding is its greater simplicity (compared to the embedded approach considered above).

We shall see in this section that strange phenomena might happen in such cases. To describe these phenomena, consider the framework of McVinish *et al.* (2005). We recall the nonparametric model: g_ψ is a mixture of triangular distributions with an unknown number of components which can be written in the following form: For a given $k \geq 1$, let the sequence $0 = t_{-1} = t_0 < t_1 < \ldots < t_{k-1} < t_k = t_{k+1} = 1$ be a partition of $[0,1]$. The function $h_i(x)$ is the triangular density function with support on the interval $[t_{i-1}, t_{i+1}]$ and mode at x_i. The mixture of triangular distributions has the density function

$$h(x) = \sum_{i=0}^{k} w_i h_i(x),$$

where $w_i \geq 0$, $\sum_{i=0}^{k} w_i = 1$. Note that $h(x)$ is a piecewise linear function on $[0,1]$ interpolating the points $(x_i, w_i h_i(x_i))$. The two cases considered by Perron and Mengersen (2001) and McVinish *et al.* (2005) are:

I For each k, the partition of $[0,1]$ is assumed fixed and the weights w_i are to be estimated.

II For each k, the weights w_i are fixed at $w_0 = w_k = \frac{1}{2k}$, $w_i = \frac{1}{k}$, $i = 1, \ldots, k-1$ and the partition is to be estimated.

In other words the parameter ψ is either (k,t) or (k,w). In both cases, the uniform distribution on $[0,1]$ plays a special role, as it can be represented as a mixture of triangulars for any number of components. The uniform distribution corresponds to $w = (1/2k, 1/k, ..., 1/k, 1/2k)$ and $t = (0, 1/k, ..., (k-1)/k, 1)$. As such if the true distribution of the observations with density f_0 is the uniform density on $[0,1]$ or any finite mixture of triangular distributions, then the posterior distributions associated with g_ψ described above shrinks towards f_0 at a rate $\sqrt{\log n}/\sqrt{n}$. Now, if f_0 is regular but cannot be represented as a finite mixture of triangulars then the posterior distribution shrinks to f_0 at a much slower rate. McVinish *et al.* (2005) prove for instance that if f_0 is Hölder with regularity $\beta \leq 2$ then the rate of convergence is – up to a $\log n$ term – the minimax rate : $n^{-\beta/(2\beta+1)}$. Now let $\mathcal{F}$ be a smooth parametric model containing the uniform distribution for instance (or any finite mixture of triangular distributions) but containing also densities which cannot be represented as finite mixtures of triangulars. Denote f_0 the true density and $f_{\eta^\perp}$ its projection onto $\mathcal{F}$ (in terms of Kullback–Leibler distance). If $f_{\eta^\perp}$ is the uniform then $\epsilon_n = O_P(\log n/n)$ and if $f_{\eta^\perp}$ is not the uniform nor any finite mixtures of triangular distributions, then ϵ_n can be much bigger, possibly of the order $O_P(n^{-2\beta/(2\beta+1)})$, where β is the regularity parameter of the Hölder class containing f_0.

Obviously, this phenomenon is not restricted to the choice of the nonparametric prior chosen in McVinish *et al.* (2005). Consider for instance a prior based on some kind of expansion such that any distribution which can be represented by a finite expansion of the same kind can be estimated by such a prior at a specific rate, say $\log n/n$. If the parametric model contains at least one such distribution and some distributions that are quite different, then the same phenomenon occurs.

This fact is slightly disturbing, since the value $\eta^\perp$ should typically be of no importance in the procedure, whereas it appears to be of crucial importance when there is no embedding. It affects the order of the calibration of ϵ_n in the goodness of fit testing problem. We will see in the following section that differences also appear when considering the Bayes factor approach instead of the calibrated distance approach, when there is no embedding.

4. CONSISTENCY PROPERTIES OF THE BAYES FACTOR: EMBEDDING VERSUS NON-EMBEDDING

In this section we study the behaviour of the Bayes factor in the goodness of fit test of a parametric family. In particular we consider conditions ensuring its consistency and conditions under which the Bayes factor is not consistent. Dass and Lee (2004) prove in a very elegant way that Bayes factors are essentially always consistent in a goodness of fit test of a fixed known distribution f_0. The testing problem of a null parametric family, defined by (4), is slightly more complicated. It is well known that Bayes factor can be interpreted as penalized likelihood ratio statistics, as illustrated by its approximation by the BIC criterion in regular cases (see for instance Kass and Raftery, 1995). Outside regular cases, the penalization is not always well understood. Berger, Ghosh and Mukhopadhyay (2003) determine the penalization induced by the Bayes factor in the case of general ANOVA problems in terms of efficient number of observations, Chambaz and Rousseau (2005) study the behaviour of the Bayes factor in general nested family problems. They show that the prior mass of Kullback–Leibler neighbourhoods of the true distribution in each model is a key factor in the penalization. McVinish *et al.* (2005) give conditions for consistency of the Bayes factor in goodness of fit testing problems of a parametric family.

In this section, we will give conditions for the Bayes factor to be inconsistent in goodness of fit problems. The results obtained prove the importance of the prior mass of Kullback neighbourhoods of the true distribution and its role in the penalization. We then apply this result to the embedding approach and to the approach without embedding. These results shade light both on the impact of embedding and on the way the Bayes factor induces the penalization. We will see in particular that inconsistencies appear more often in the case of non-embedding than in the case of embedding.

Note however that the reasons for *bizarre* phenomena associated with the no-embedding approach are not the same wether we use the calibrated p-value approach or wether we consider the Bayes factor. In both cases however the problems are related to the fact that the null hypothesis is *heterogeneous* from the point of view of the alternative. This disappears when embedding is considered, specially the kind of embedding studied here, since it all comes back to estimating the same density g_{ψ_0} in the nonparametric model. We have proved it, in the case of the p-value approach in Section 3.2. We now prove it for the Bayes factor approach.

4.1. *Inconsistency of the Bayes Factor*

In goodness of fit problems, or in any problem where the null can be regarded as a sub-model of the alternative, consistency of the Bayes factor under the alternative is usually much easier to obtain than under the null since the marginal distribution of the null model is typically exponentially small. In this section we therefore study consistency under the null.

Following Theorem 4.1 in McVinish *et al.* (2005), the Bayes factor associated with the goodness of fit test of a parametric family against a nonparametric alternative is consistent under the null if the posterior distribution in the nonparametric model is consistent under any distribution of the parametric family, if the parametric family is regular and most importantly if the nonparametric prior probability of ϵ_n-neighbourhoods of the true density is of order $o(n^{-d/2})$, where ϵ_n is the rate of convergence of the posterior towards $f_0 = f_\eta$, where $\eta \in \Theta_0$ and d is the dimension of the parameter under the null.

This latter condition is actually not very far, in regular cases, from being a necessary condition. The following theorem gives conditions leading to non-consistency of the Bayes factor under the null. Let π be the prior in the alternative model and π_0 the prior in the null model.

Theorem 4 *Assume that there exists $\eta_0 \in \Theta_0$ and $\epsilon_0 > 0$ such that*

$$\pi\left[S_{\epsilon_0^2/n}(f_{\eta_0}, d_K)\right] \geq Cn^{-d/2},$$

for some $C > 0$. Assume also that the parametric model $\mathcal{F} = \{f_\eta, \eta \in \Theta_0\}$ is regular. Then the Bayes factor for testing $H_0 : f \in \mathcal{F}$ against $H_1 : f \notin \mathcal{F}$, with the alternative being modeled by π does not converge to infinity with n under the null hypothesis.

Proof of Theorem 4. Recall that the Bayes factor can be written as

$$B_n = \frac{\int_{\Theta_0} \prod_i f_\eta(X_i) d\pi_0(\eta)}{\int \prod_i f(X_i) d\pi(f)}.$$

For notational ease, we denote $S_{\epsilon_0^2/n}(f_{\eta_0}, d_K)$ by S_n and f_{η_0} by f_0 throughout the proof. We also denote $l_n(f)$ the log-likelihood calculated at the density f. Under the usual regularity conditions on the parametric model,

$$\begin{aligned} B_n^{-1} &\geq \frac{\int_{S_n} e^{l_n(f) - l_n(f_0)} d\pi(f)}{c_n n^{-d/2}} \\ &\geq b_n e^{-c} n^{d/2} \pi[S_n \cap V_n], \end{aligned}$$

where $V_n = \{(f, X); l_n(f) - l_n(f_0) > -c\}$ and c_n, b_n^{-1} are $O_P(1)$. Therefore

$$\begin{aligned} P_{\eta_0}^n\left[B_n^{-1} > \delta\right] &\geq P_{\eta_0}^n\left[\pi[S_n \cap V_n] > \delta e^c b_n^{-1} n^{-d/2}\right] \\ &\geq P_{\eta_0}^n\left[\pi[S_n \cap V_n^c] < \pi[S_n] - \delta e^c b_n^{-1} n^{-d/2}\right]. \end{aligned}$$

Under the assumptions of Theorem 4, $\pi[S_n] > Cn^{-d/2}$ for some $C > 0$, hence by choosing δ small enough, with probability tending to 1,

$$\pi[S_n] - \delta e^c b_n^{-1} n^{-d/2} > \alpha \pi[S_n],$$

with α as close to 1 as need be. This probability is bounded away from 0 if $P_{\eta_0}\left[\pi[S_n \cap V_n^c] \geq \alpha\pi[S_n]\right]$ is bounded away from 1. We have:

$$\begin{aligned} P_{\eta_0}^n\left[\pi[S_n \cap V_n^c] \geq \alpha\pi[S_n]\right] &\leq \frac{\int_{S_n} P_f\left[V_n^c\right] d\pi(f)}{\alpha\pi(S_n)} \\ &\leq \frac{\epsilon_0^2}{(c - \epsilon_0^2)^2 \alpha}. \end{aligned}$$

By choosing $c > \epsilon_0^2/2$ and α close enough to 1, the above term is made strictly less than 1, which achieves the proof. $\square$

Remark 10. The main difference with the sufficient condition given in McVinish *al.* (2005) is on the prior mass of neighbourhoods of the true density in the nonparametric model described by π. To obtain consistency of the Bayes factor, McVinish *et al.* require the prior mass of L_1 neighbourhoods of f_0 of length ϵ_n to be of smaller order than $n^{-d/2}$, where ϵ_n is the rate of convergence (in terms of L_1) of the posterior distribution around f_0. Here we prove that the Bayes factor is inconsistent if the prior mass of ϵ_0^2/n Kullback–Leibler neighbourhoods of f_0 is greater than $n^{-d/2}$. The change of metric is often not a real problem, so the real significant difference is in the change between ϵ_n^2 (in terms of the square of the L_1 metric which is *equivalent* to the Kullback–Leibler divergence to some extent) to ϵ_0^2/n. Again a particular case of interest is when ϵ_n^2 is of order $1/n$ up to a $\log n$ term (see Examples 2 and 4), since this prior mass condition comes close to a sufficient and necessary condition. Note also that obtaining a lower bound on the prior mass is usually easier than an upper bound. This is particularly true for mixture models, where identifiability may not be verified.

Remark 11. Theorem 4 shades light on the penalization induced by the Bayes factor (when interpreted as a penalized likelihood). It finally reflects some kind of relative ease of estimation under each model of the true density. This ease of estimation is measured in terms of the way the priors (models) are able to approach the true density. More precisely the smaller the number of parameters in a model is needed to approach the true density the *easier* it is to estimate the density. This number of parameters needed to approach the density can be quite different from the real number of parameters involved by the prior, so that this notion of ease of estimation can be quite different from a real ease of estimation, such as computational ease or interpretational ease. The special cases of embedded and non-embedded models below illustrate this phenomenon.

We now apply this result together with Theorem 4.1 of McVinish *et al.* (2005) to comment on the effect of embedding on Bayes factors.

4.1.1. *Embedding*

We now study the consistency of the Bayes factor in the case where the parametric model is embedded in the nonparametric one, using the cumulative distribution function transform. Similarly to before, let $\{g_\psi, \psi \in \mathcal{S}\}$ be the nonparametric model on $[0,1]$ and $\mathcal{H} = \{f_{\eta,\psi} = f_\eta(x) g_\psi(F_\eta(x)), \eta \in \Theta_0, \psi \in \mathcal{S}\}$ be the full nonparametric model describing the alternative. Let π_1 be a prior on $\mathcal{S}$, π on $\Theta_0 \times \mathcal{S}$ and π_0 on Θ_0 under the null. Theorem 4.1 of McVinish *et al.* (2005) imply that the important condition to check is that for all $\theta_0 \in \Theta$,

$$\pi\left[f_{\eta,\psi}; \|f_{\eta,\psi} - f_{\eta 0}\|_1 \leq \epsilon_n\right] = o(n^{-d/2}).$$

Note that

$$\|f_{\eta,\psi} - f_{\eta 0}\|_1 = \|g_\psi - h_\eta\|_1, \tag{19}$$

with $h_\eta(u) = (f_{\eta 0}/f_\eta)(F_\eta^{-1}(u))$. Hence denote

$$\tilde{\mathcal{F}} = \{h_\eta; \eta \in \Theta_0\}.$$

In this section we do not give general conditions for consistency or inconsistency of the Bayes factor since this has been done previously. Instead we give specific conditions leading to either of these behaviours that we believe are common enough.

We assume that Θ_0 is compact and that the model $\tilde{\mathcal{F}} = \{h_\eta, \eta \in \Theta_0\}$ is regular. Assume similarly to before that $\mathcal{S} = \cup_k \mathcal{S}_k$, where $\mathcal{S}_k$ is a finite dimensional parameter space, with dimension d_k. We assume that for each k the submodel $\psi \in \mathcal{S}_k$ is regular.

We first give a sufficient condition for inconsistency of the Bayes factor.

Theorem 5 *Consider the conditions stated previously, and assume also that the conditional prior on η given ψ is uniformly bounded from below. If $\pi_1\left[\psi : g_\psi \in \tilde{\mathcal{F}}\right] > 0$, the Bayes factor does not converge to infinity when the true density is equal to f_{η_0}.*

Similarly, if $\pi\left[\eta; h_\eta \in \mathcal{S}\right] > 0$, then necessarily there exists $k \in \mathrm{N}$ such that $\pi\left[\eta; h_\eta \in \mathcal{S}_k\right] > 0$. If the dimension d_k of the smallest subset $\mathcal{S}_k$ satisfying the above inequality is smaller or equal to d then the Bayes factor is inconsistent under f_{η_0}.

Proof of Theorem 5. The proof follows directly from Theorem 4. Indeed, since Θ_0 is compact, $d_K(g_\psi, \tilde{\mathcal{F}}) = 0$ if and only if $g_\psi \in \tilde{\mathcal{F}}$. Moreover $d_K(f_{\eta_0}, f_{\eta,\psi}) = d_K(h_\eta, g_\psi)$ therefore,

$$\begin{aligned}\pi\left[(\eta,\psi); d_K(f_{\eta_0}, f_{\eta,\psi}) \leq \epsilon_0^2/n\right] &\geq \int_{\tilde{\mathcal{F}}} \pi\left[\eta; d_K(h_\eta, g_\psi) \leq \epsilon_0^2/n|\psi\right] d\pi_1(\psi)\\ &\geq cn^{-d/2}\pi_1[\tilde{\mathcal{F}}]\\ &\geq c'n^{-d/2},\end{aligned}$$

where the last two inequalities come from the regularity condition of $\tilde{\mathcal{F}}$ and from the condition on $\pi_1[\tilde{\mathcal{F}}]$.

The second part of the proof of Theorem 5 is proved in the same way. □

This condition is not a necessary condition, however we do believe that it is nearly so. To illustrate this, consider the same models in η and ψ with the same regularity conditions as above. We assume that the prior densities π_0 and π_k, $k \in \mathrm{N}$ (the conditional priors) are bounded. The rate of convergence for estimating f_{η_0} in the full model $\mathcal{H}$ is then bounded by $\epsilon_n = \log n/\sqrt{n}$. Assume that the conditions of Theorem 5 are not satisfied, in other words that $\pi_1\left[\psi : g_\psi \in \tilde{\mathcal{F}}\right] = 0$ and $\pi\left[\eta; h_\eta \in \mathcal{S}\right] = 0$. Since $h_{\eta_0} = g_{\psi_0} = 1$ the above set has null probability but is non empty, consider k_0 the smallest integer for which there exists η such that $h_\eta \in S_{k_0}$ and denote by $\mathcal{F} \cap \mathcal{S}_{k_0}$ the set of densities that can be written both as h_η and as g_ψ, with $\psi \in S_{k_0}$. For notational ease we indifferently write $\psi \in S_{k_0}$ or $g_\psi \in S_{k_0}$. Let $\eta \in \Theta_0, \psi \in S_{k_0}$ be such that $\|h_\eta - g_\psi\|_1 \leq \epsilon_n$ then for all ψ' such that $\|h_\eta - g_{\psi'}\|_1 \leq \epsilon_n$ we obtain that $\|g_{\psi'} - g_\psi\|_1 \leq 2\epsilon_n$. Hence

$$\begin{aligned}\pi_n &= \pi\left[(\eta,\psi) \in \Theta_0 \times \mathcal{S}; \|h_\eta - g_\psi\|_1 \leq \epsilon_n\right]\\ &\leq C\epsilon_n^d \pi_1(S_n^\psi),\end{aligned}$$

where $S_n^\psi = \{\psi; \exists\eta \in \Theta_0; \|g_\psi - g_\eta\|_1 \leq \epsilon_n\}$. If d_{k_0} the dimension of the set $\mathcal{S}_{k_0}$ is greater than d then using the same calculations but inverting ψ and η leads to $\pi_n = O(\epsilon_n^{d_{k_0}}) = o(n^{-d/2})$ and we can apply Theorem 4.1 of McVinish *et al.* (2005) which implies that the Bayes factor is consistent. To be more precise, care should be taken on dependence on $k \geq k_0$, however it usually is not so much of a problem so we shall not enter into these problems, and only pretend that the order $O(\epsilon_n^{d_{k_0}})$ is also valid for the subsets $\mathcal{S}_k$, $k \geq k_0$. If $d > d_{k_0}$, we need to prove that $\pi_k(S_n^\psi) = o(\log n^- d)$, for all $k \geq k_0$ satisfying $d_k \leq d$. This is a very natural result. We now prove it in a more restrictive case, where we assume that the subset $\mathcal{F} \cap \mathcal{S}_k$ can be characterized in terms of η by considering that some components say η_1 of η are null and in terms of ψ by considering that some components say ψ_1 are null. In other words there exist parametrisations $\eta = (\eta_1, \eta_2)$ and $\psi = (\psi_1, \psi_2)$ such that $h_\eta \in \mathcal{F} \cap \mathcal{S}_k$ if and only if $\eta_1 = 0$ and $g_\psi \in \mathcal{F} \cap \mathcal{S}_k$ if and only if $\psi_1 = 0$. We then prove that when n is large S_n^ψ is at most at a distance $C\epsilon_n$ from $\mathcal{F} \cap \mathcal{S}_k$, for some fixed constant $C > 0$.

First note that S_n^ψ is a sequence decreasing to $\mathcal{F} \cap \mathcal{S}_k$ so that $\pi(S_n^\psi)$ goes to zero as n goes to infinity. Hence if when n is large enough, if $\psi \in S_n^\psi$ and h_η is such that $\|h_\eta - g_\psi\|_1 \leq \epsilon_n$ then ψ and η are as close as need be to $\mathcal{F} \cap \mathcal{S}_k$, in other words η_1 and ψ_1 are close to 0. We can then consider without loss of generality that $h_{(0,\eta_2)} = g_{(0,\psi_2)}$. Since the models are regular we can work with the Hellinger distance instead of the L_1 distance, changing only ϵ_n by a fixed multiplicative constant. A Taylor expansion of h_η and g_ψ around $h_{(0,\eta_2)} = g_{(0,\psi_2)} = g_0$ leads to:

$$
\begin{aligned}
h^2(h_\eta, g_\psi) &= \int \left(\sqrt{h_\eta} - \sqrt{g_\psi}\right)^2 (x)dx \\
&= \frac{1}{4}\left[\psi_1^t \left(\int \frac{d_1^\psi g_0 (d_1^\psi g_0)^t}{g_0} dx\right)\psi_1 + \eta_1^t \left(\int \frac{d_1^\eta g_0 (d_1^\eta g_0)^t}{g_0} dx\right)\eta_1 \right. \\
&\quad \left. -2\psi_1^t \left(\int \frac{d_1^\psi g_0 (d_1^\eta g_0)^t}{g_0} dx\right)\eta_1\right] + o(|\psi_1|^2 + |\eta_1|^2),
\end{aligned}
$$

where $d_1^\psi g_0$ (resp. $d_1^\eta g_0$) is the derivative of g_ψ (resp. h_η) with respect to ψ_1 (resp. η_1). The latter term can be bounded from below by $c|\psi_1|^2|\eta_1|^2$ as long as the quadratic form is definite positive, that is as long as $d_1^\psi g_0$ and $d_1^\eta g_0$ are not linearly dependent as functions of x. In this case, $\psi \in S_n^\psi$ only if $\psi_1 = O(\epsilon_n)$. Hence the prior mass of S_n^ψ is bounded by a constant times $\epsilon_n^{d_{k1}}$. We finally obtain that the prior mass of $S_{\epsilon_n}(f_0, L_1)$ is bounded by a term in the form $\epsilon_n^{d+d_{k1}} = o(n^{-d/2})$ and the Bayes factor is consistent, according to McVinish *et al.* (2005).

Remark 12. The condition to get inconsistency of the Bayes factor in Theorem 5 indicate that the prior under the alternative is not well considered since it means that it puts positive mass to the null model, to some extent.

Example 5 (Mixtures of Betas (2)). Let π_1 be a prior defined on the set of finite mixtures of betas on $[0, 1]$, assume for ease of calculations that the parametric model is the exponential model : $f_\eta(x) = e^{-\eta x}\eta, \eta > 0$. Then $h_\eta(u) = \eta_0/\eta(1-u)^{\eta_0/\eta - 1}$, which is indeed a regular model, as a Beta (a, b) distribution constrained by $a = 1$. In this case $\mathcal{F} \subset \mathcal{S}_k$, for all $k \geq 1$, where $\mathcal{S}_k$ is the mixture of beta densities defined previously for the example of mixtures of betas. The constraints on the parameters for each $k \geq 1$ defining the subset $\mathcal{F}$ of $\mathcal{S}_k$ are thus of the form : $p_0 = 0, p_j = 0, j \geq 2, p_1 = 1, a_1 = 1$. It follows from the above discussion that the Bayes factor is consistent in this case. Note that if instead of the mixtures of betas defined previously we had considered mixtures of betas in the form:

$$g_{\psi_k}(u) = \sum_{j=1}^{k} p_j g_{a_j,b_j}(u), \tag{20}$$

so that the uniform does not have a special representation, then the Bayes factor would still be consistent. We will see that it is not so when we do not embed the parametric family in the nonparametric one.

4.1.2. *Non-Embedding*

If we do not embed the parametric model in the nonparametric model, Bayes factors might be inconsistent even with reasonable priors. We give no general theorem specific to the non-embedded case since Theorem 5 covers this case.

To make a comparison with the case where the parametric family is embedded in the nonparametric one, consider the mixture of Beta distributions example with the parametric family being a Beta distribution. If we constrain the Beta (a, b) of the null model to have $a = 1$ then this also covers the case where the null model is the exponential family model, transformed on $[0, 1]$ using the transform: $u = 1 - e^{-x}$. Assume first that the alternative is represented by (20). Then the prior mass of ϵ_0^2/n-Kullback–Leibler neighbourhoods of the true distribution, when the true is any beta$(1, b)$ distribution, is of order $O(n^{-1}) = o(n^{-1/2})$. Moreover $\epsilon_n = O(\sqrt{\log n}/\sqrt{n})$ and the prior mass of L_1 distance of the null is of order $O(n^{-1} \log n) = o(n^{-1/2})$ so that the Bayes factor is consistent at f_η, for any $\eta \in \mathcal{F}$. On the other hand if the mixture of Betas distributions is represented as in Section 3.2.1 (Example 4), then the prior mass of $(1/n)$-Kullback–Leibler neighbourhoods of the uniform distribution is exactly of order $1/\sqrt{n}$ so that the Bayes factor is inconsistent in this case. However the prior mass in the alternative model of L_1-neighbourhoods of the true distribution is still of order $o(n^{-1/2})$ if the true is not the uniform distribution and the Bayes factors are consistent at these distributions. In such situation we have this very strange phenomenon where the Bayes factor does not behave at all the same way depending on the value of the parameter η in the null model.

4.1.3. *Discussion*

In this paper we have studied some aspects of Bayesian testing of precise hypotheses, emphasizing on their capacity of approximating their counterparts in imprecise null hypotheses testing. The two approaches we have studied are the Bayes factors (or posterior probabilities) and the distance approach calibrated by p-values. The former is the most common Bayesian testing procedure, the latter is to our opinion better suited for goodness of fit tests. Bayes factors are the Bayesian answers to 0–1 types of losses and are therefore better suited for well separated hypotheses testing problems. In problems where the null belongs to the closure of the alternative, it seems too abrupt a loss function. We have proved that when the number of observations becomes large the Bayes factor associated with point null hypotheses is a poor approximation of Bayes factors of interval null hypotheses, unless the intervals are extremely small. They need to be a lot smaller than the typical estimation rate of convergence. This means that in large samples contexts Bayes factors for point null hypotheses cannot be understood as Bayesian answers to imprecise null hypotheses testing problems.

The above results on the Bayes factors for goodness of fit tests of parametric families versus nonparametric families show that Bayes factors can be interpreted as penalized likelihood ratio statistics, where the penalization indicates a certain complexity of the models. This notion of complexity is not the usual one. It is related to the capacity of each prior to approximate the *true* distribution. This capacity can be measured by the prior mass of neighbourhoods of the true distribution. If the mass is small then the prior cannot approximate well the true distribution and the associated model is penalized. This penalization has therefore very little to do with the complexity of the model in terms of number of parameters or ease of interpretation of the parameters. In special cases, such as regular finite dimensional models, it comes to the same thing, however in goodness of fit tests it usually does not.

The second Bayesian testing procedure we have studied in this paper is the distance approach, calibrated by a p-value. We show that such a calibration induces an imprecise null hypothesis with width of order the rate of convergence of posterior distribution in the alternative model, under f_0, the point null hypothesis. When the null hypothesis is a parametric family then this interpretation remains. However depending on the representation of the alternative model, very different phenomena occur. In particular, the results obtained in this paper shed light on the impact on embedding the parametric family into the nonparametric model.

The aim of this paper is not so much to determine which of the two procedures are best, although our preference goes to the distance approach at least in goodness of fit problems. Our aim is rather to understand better how these procedures react to large samples, when the real question is not whether the precise null hypothesis is true but whether it is a reasonable hypothesis. Obviously both methods are bound to reject the null under large samples, as are all methods based on precise null hypotheses. To answer to this question of practical significance versus statistical significance, when the number of observations becomes large it is necessary to assess more precisely what practical significance means, in each problem. In particular, the larger the number of observations, the more accurate this notion of practical significance must be assessed. One way to do this may be through the loss function, by defining carefully in the loss function the notion of practical significance.

ACKNOWLDGEMENT

The author expresses her deep appreciation to Edward George for initiating the present work and for many fruitful discussions of the material. She also thanks Christian Robert and Ghislaine Gayraud for helpful discussions on the problem.

APPENDIX

Proof of Theorem 1

For both parts of the theorem, we need to find a lower bound on $B_n/B_n^\star$. First note that, as long as $\lambda_n \leq 1/2$, *i.e.*, as n is large enough,

$$\frac{B_n}{B_n^\star} \geq \frac{m_n}{2\lambda_n}, \quad \text{where} \quad m_n = \int_{S_n} \frac{f_\theta(X)}{f_{\theta_0}(X)} d\pi(\theta).$$

and S_n corresponds to either $S_{\epsilon_n}(\theta_0, d_K)$ or $S_{\epsilon_n}(\theta_0, K)$ wether we consider part (i) or part (ii) of Theorem 1. The idea of the proof is that near θ_0, we can find a subset of S_n such that the log-likelihood ratio $l_n(\theta) - l_n(\theta_0)$ is large enough in probability with respect to P_{θ_1}.

We now prove part (i): Let $\theta_1 \in \Theta$

$$\begin{aligned} P_{\theta_1}\left[\frac{B_n}{B_n^\star} \le 1/C\right] &\le P_{\theta_1}\left[\frac{m_n}{\lambda_n} \le 2/C\right] \\ &\le P_{\theta_1}\left[\pi[S_{n,1} \cap A_n] \le 2\lambda_n e^a/C\right] \end{aligned}$$

where $A_n = \{(\theta, X); l_n(\theta) - l_n(\theta_0) > -a\}$ and $a > 0$. Choosing C large enough, we obtain that

$$\begin{aligned} P_{\theta_1}\left[\pi[S_{n,1} \cap A_n] \le 2\lambda_n e^a/C\right] &\le P_{\theta_1}\left[\Pi(S_{n,1} \cap A_n^c) > \pi(S_{n,1}) - 2\lambda_n e^a/C\right] \\ &\le \frac{\int_{S_{n,1}} P_{\theta_1}\left[A_n^c\right] d\pi(\theta)}{\pi(S_{n,1}) - 2\lambda_n e^a/C}. \end{aligned}$$

The latter term is well defined when n is large enough since there exists $\delta > 0$ such that $\pi(S_{n,1})/\lambda_n > \delta$. Hence it is enough to choose C such that $2e^a/C < \delta/2$. Moreover, if $\theta \in S_{n,1}$,

$$\begin{aligned} P_{\theta_1}\left[A_n^c\right] &= P_{\theta_1}\left[l_n(\theta_0) - l_n(\theta) > a\right] \\ &\le \frac{nV_{\theta_1}(f_{\theta_0}, f_\theta)}{(a + n\Delta(\theta_1, \theta))^2} \\ &\le \frac{4M}{a^2}, \end{aligned}$$

as long as $a > 2c$. By considering $a = \log C/2$ this goes to zero as C goes to infinity.

We now prove part (ii): Let $\theta_1 \neq \theta_0$ satisfying the hypotheses of part (ii) of Theorem 1. Let $C, b > 0$ and $A_n = \{(\theta, X); l_n(\theta) - l_n(\theta_0) > b\}$. Then similarly to before,

$$\begin{aligned} P_{\theta_1}\left[\frac{B_n}{B_n^\star} \le C\right] &\le P_{\theta_1}\left[\frac{m_n}{\lambda_n} \le 2C\right] \\ &\le \frac{\int_{S_{n,2}} P_{\theta_1}\left[A_n^c\right] d\pi(\theta)}{\pi(S_{n,2}) - 2\lambda_n e^{-b}C}. \end{aligned}$$

The assumption of part (ii) of Theorem 1 implies the existence of $\delta > 0$ such that $\pi(S_{n,2})/\lambda_n > \delta$, then if b satisfies $2e^{-b}C < \delta/2$ and if $\theta \in S_{n,2}$,

$$\begin{aligned} P_{\theta_1}\left[A_n^c\right] &= P_{\theta_1}\left[l_n(\theta_0) - l_n(\theta) > -b\right] \\ &\le \frac{nV_{\theta_1}(f_\theta, f_{\theta_0})}{(n\Delta(\theta_1, \theta) - b)^2} \\ &\le \frac{4\rho_n V_0}{a^2}, \end{aligned}$$

which goes to zero when n goes to infinity. This achieves the proof. □

Proof of Theorem 2

The main difficulty is to control $g_\psi(F_{\eta'}(F_\eta^{-1}(u)))$ when η' is close to η uniformly on $\mathcal{S}_n$. More precisely, we need to control the distance between $f_{\eta'}g_{\psi'}(F_{\eta'})$ and $f_\eta g_\psi(F_\eta)$ in terms of the distance between $g_{\psi'}$ and g_ψ and $|\eta' - \eta|$.

$$\begin{aligned}
\Delta &= h^2(f_\eta g_\psi(F_\eta), f_{\eta'}g_{\psi'}(F_{\eta'})) \\
&= h^2(g_\psi, g_{\psi'}) + 2\int \sqrt{f_\eta g_\psi(F_\eta)}\left(\sqrt{f_\eta g_{\psi'}(F_\eta)} - \sqrt{f_{\eta'}g_{\psi'}(F_{\eta'}))}\right) \\
&\leq h^2(g_\psi, g_{\psi'}) + 2h^2(f_\eta g_{\psi'}F_\eta), f_{\eta'}g_{\psi'}(F_{\eta'}))^{1/2} \\
&\leq h^2(g_\psi, g_{\psi'}) + 2K(f_\eta g_{\psi'}F_\eta), f_{\eta'}g_{\psi'}(F_{\eta'}))^{1/2}.
\end{aligned}$$

We therefore need only consider the second term of the right hand side, which we denote Δ_2.

$$\Delta_2 = \int_0^1 g_{\psi'}(u)\log\left(\frac{g_{\psi'}(u)}{g_{\psi'}(F_{\eta'}(F_\eta^{-1}(u)))}\right)du + (\eta - \eta')\int f_\eta g_{\psi'}(F_\eta)\frac{\partial \log f_{\bar\eta}}{\partial \eta}dx.$$

There are many possible types of conditions that might imply the result given above, we present here one set of conditions as it is convenient when dealing with mixtures of betas. Essentially what is needed is that there exists C, c such that for all ψ, ψ', with $d_H(g_\psi, g'_\psi) < \delta/2$

$$\left|\int_0^1 g_{\psi'}(u)\log\left(\frac{g_{\psi'}(u)}{g_{\psi'}(F_{\eta'}(F_\eta^{-1}(u)))}\right)du\right| \leq C|\eta - \eta'|^c. \tag{21}$$

Then condition (16) together with (17) and (18) imply that

$$|\Delta_2| \leq C_\psi|\eta - \eta'|^c \leq e^{c_3 n\rho_n}|\eta - \eta'|^c.$$

Finally $\Delta \leq 2\rho_n$ if $h^2(g_\psi, g_{\psi_0}) \leq \rho_n$ and $|\eta - \eta'| \leq e^{-c_3 n\rho_n/c}\rho_n^{1/c}$. This implies that we can construct a $2\rho_n$-covering of the complementary of the $C\rho_n$-Hellinger neighbourhood of f_0 by considering a ρ_n-covering of the complementary of the $C/2\rho_n$ Hellinger neighbourhood of g_{ψ_0} and a $e^{-c_3 n\rho_n/c}\rho_n^{1/c}$-covering of Θ_0. This leads to

$$\begin{aligned}
K_n &= \sum_{j,l}\sqrt{\pi(\tilde{A}_{n,j,l})} \\
&\leq \sum_j \sqrt{\pi(\tilde{A}_{n,j})}\sum_l \sqrt{\pi(|\eta - \eta_j| \leq e^{-c_3 n\rho_n/c}\rho_n^{1/c})} \\
&\leq C\sum_j \sqrt{\pi(\tilde{A}_{n,j})}\rho_n^{-cd/2}e^{dc_3 n\rho_n/2} \\
&\leq Ce^{\tilde{c}n\rho_n}.
\end{aligned}$$

Following the lines of Walker *et al.* (2007) we have that

$$\sum_n e^{-n\epsilon^2/8}K_n < \infty$$

as long as $\epsilon_n^2 = a_0\rho_n$ with a_0 large enough. We now consider the condition on the Kullback–Leibler neighbourhoods of the true density: We have

$$\pi_1\left[g_\psi; d_K(g_{\psi_0}, g_\psi) \leq \rho_n\right] \geq e^{-nC\rho_n}.$$

Then, using the same calculations as previously around g_{ψ_0}, since $f_0 = f_{\eta_0} g_{\psi_0}(F_{\eta_0})$, for all $|\eta - \eta_0| < a\rho_n^{1/c} e^{-c_3 n\rho_n/c}$, with a small enough, if

$$K(g_{\psi_0}, g_\psi) \leq \rho_n/2, V(g_{\psi_0}, g_\psi) \leq \rho_n/2$$

then

$$d_K(f_0, f_{\eta,\psi}) \leq 2\rho_n$$

so that

$$\pi[S_{\rho_n}(f_0, d_K)] \geq c' e^{d\log\rho_n - (C+c_3/c)n\rho_n} \leq e^{-C'n\rho_n},$$

as long as $\log\rho_n < -cn\rho_n$, i.e. $\rho_n \geq \log n/n$ which is the case. □

REFERENCES

Bayarri, M. J. and Berger, J. O. (2000). *p*-values for composite null models. *J. Amer. Statist. Assoc.* **95**, 1127–1142.

Berger, J. and Delampady (1987). Testing precise hypotheses. *Statist. Science* **2**, 317–335.

Berger, J., Ghosh, J. K., and Mukhopadhyay, N. (2003). Approximations to the Bayes factor in model selection problems and consistency issues. *J. Statist. Planning and Inference* **112**, 241–258.

Berger, J. and Guglielmi A. (2001). Bayesian testing of a parametric model versus nonparametric alternatives. *J. Amer. Statist. Assoc.* **96**, 174–184.

Carota, C. and Parmigiani, G. (1996). On Bayes factors for nonparametric alternatives. *Bayesian Statistics 5* (J. M. Bernardo, J. O. Berger, A. P. Dawid and A. F. M. Smith, eds.) Oxford: University Press, 507–511.

Chambaz, A. and Rousseau, J. (2005). Nonasymptotic bounds for Bayesian order identification with application to mixtures. *Tech. Rep.*, CEREMADE. Université Paris Dauphine, France.

Dass, S. C. and Lee, J. (2004). On the consistency of Bayes factors for testing point null versus nonparametric alternatives. *J. Statist. Planning and Inference* **119**, 143–152.

Florens, J. P., Richard, J. F. and Rolin, J. M. (1996). Bayesian encompassing specifications tests of a parametric model against a non parametric alternative, *Tech. Rep.*, Institut de Statistique, Université Catholique de Louvain, Belgique.

Fraser, D. and Rousseau, J. (2005). Developing *p*-values: a Bayesian-frequentist convergence. *Tech. Rep.*, CEREMADE. Université Paris Dauphine, France.

Ghosal, S., Ghosh, J.K. and van der Vaart, A. (2000). Convergence rates of posterior distributions, *Ann. Statist.* **28**, 500–531.

Ghosal, S. and Van der Vaart, A. (2001). Entropies and rates of convergence for maximum likelihood and Bayes estimation for mixtures of normal densities. *Ann. Statist.* **29** 1233–1263.

Gouriéroux C. and Monfort A. (1996). *Statistique et Modèles Econométriques* (2 tomes, 2nd ed.) Paris: Economica.

Goutis C. and Robert, C. P. (1997). Selection between hypotheses using estimation criteria. *Ann. d'Éco. Statist.*, **46**, 29–37.

Jefferys, W. and Berger, J. (1992) Ockham's razor and Bayesian analysis. *Amer. Scientist.* **80**, 64–72.

Kass, R. E. and Raftery, A. (1995). Bayes Factors. *J. Amer. Statist. Assoc.* **90**, 773–795.

McVinish, R., Rousseau, J. and Mengersen, K. (2005). Bayesian Mixtures of Triangular distributions with application to Goodness-of-Fit Testing. *Tech. Rep.*, CEREMADE. Université Paris Dauphine, France.

Perron, F. and Mengersen, K. (2001). Bayesian nonparametric modelling using mixtures of triangular distributions. *Biometrics* **57** 518–529.

Nemirovski, A. (2000). Topics in non-parametric statistics. *Lectures on Probability Theory and Statistics, Saint Flour, 1998.* Berlin: Springer-Verlag

Robert, C. P. and Rousseau, J. (2003). A mixture approach to Bayesian goodness of fit. *Tech. Rep.*, CEREMADE. Université Paris Dauphine, France.

Robins, J. M., van der Vaart, A. and Ventura, V. (2000). The asymptotic distribution of p-values in composite null models. *J. Amer. Statist. Assoc.* **95**, 1143–1156.

Robins, J. and Wasserman, L. (2000). Conditioning, likelihood, and coherence: A review of some foundational concepts. *J. Amer. Statist. Assoc.* **95**. 1340–1346.

Verdinelli, I. and Wasserman, L. (1996). Bayes factors, nuisance parameters and imprecise tests. *Bayesian Statistics 5* (J. M. Bernardo, J. O. Berger, A. P. Dawid and A. F. M. Smith, eds.) Oxford: University Press, 765–771.

Verdinelli, I. and Wasserman, L. (1998) Bayesian goodness-of-fit testing using infinite-dimensional exponential families. *Ann. Statist.* **26**, 1215–1241.

Walker, S. G., Lijoi, A. and Prünster, I. (2007). On rates of convergence for posterior distributions in infinite-dimensional models. *Ann. Statist.* (to appear).

DISCUSSION

SONIA PETRONE (*Bocconi University, Italy*)

In this interesting paper, several results are provided, sheding light on the behavior of Bayes factors for large n and high dimensional problems, and studying alternative, distance-based tests for goodness of fit; and professor Rousseau has a certain dose of courage in touching fundamental problems (testing point null hypothesis, goodness of fit, consistency) which have been debated at length in the Bayesian literature and on which, I think, the debate remains open. I found the paper quite demanding to read: for the many clever technical aspects, for the many delicate questions and the richness of suggestions and, possibly on a less positive side, for somehow mixing, at points, Bayesian and 'less-Bayesian' or frequentist notions, even if re-interpreted in a Bayesian framework. Thus, reading the paper stimulates, once more, a reflection on basic issues: can a Bayesian 'test' subjective probability assignments (including models)? Are models true, or useful, or honest? What are sensible loss functions? Should the loss function be subjectively chosen, or calibrated to the data? How does the choice of a model depend on the sample size (would we use the same model for small and large n)?

Reflections on Bayesian goodness of fit. The basic problem in the paper is goodness of fit of a model f_0 or of a parametric class $\mathcal{F}_0 = \{F_\theta, \theta \in \Theta_0\}$. In the Bayesian literature there are informal methods for model assessment and formal, decision-based methods for model comparison (see *e.g.*, Gelfand *et al.*, 1992, and references therein). However, there seems to be less work on Bayesian goodness of fit, that is in absence of specific alternative models. In 1999, I was studying a problem posed by Coram and Diaconis (see Coram and Diaconis, 2003), regarding a conjecture on the distribution of data related to the zero's of the zeta Riemann function. Coram and Diaconis had treated this problem using frequentist goodness of fit tests; but their question was: can we do anything more Bayesian? I wasn't able to find a convincing answer (see a brief discussion in Petrone, 2003) so I return again on that problem: what would be a Bayesian goodness of fit test? Using a nonparametric prior can be a coherent answer; yet, until recently there were very few studies about this in the Bayesian nonparametric literature. And at an even more fundamental level: what could the Bayesian approach offer 'more convincing' than the frequentist solutions, with specific regard to goodness of fit? Perhaps, the emphasis on the *utility* of a models, rather than on the model being true?

I think that both questions are still open, and the paper, giving a contribution to the debate, is certainly welcome. What the paper studies are Bayesian goodness of fit *tests*; i.e., decision rules for $H_0 : f = f_0$ versus $H_1 = f \neq f_0$, or (more difficult) $H_0 : f \in \mathcal{F}_0$ versus $H_1 = f \notin \mathcal{F}_0$. This rises the question about the meaning of statements such as $f = f_0$. In a decision based interpretation (Neyman and Pearson, or Bayesian-decisional), there is an underlying decision problem, that is, one has to choose between two terminal actions a_0 and a_1 (for example, in financial applications, selecting a portfolio) and the loss function formalizes the consequences of these actions (loss or gain with the selected portfolio). Yet, especially with zero-one loss functions, there remains somehow the idea of being investigating wether f_0 is the true density. A first objection is that in the Bayesian approach, probabilities are subjective judgments, and one cannot test if subjective judgments are true or not, but only if they are coherent. In practice, people feel the need of model checking, and in some sense this intuitive and practically relevant question is relevant for subjective probabilities too, if the observed frequencies play the role usually played by the *true* F_0.

Consider for simplicity the case of an exchangeable sequence $(X_n, n \geq 1)$, which is the framework studied in the paper. Suppose that we have expressed our information through the probability law P of the random process $(X_n, n \geq 1)$. By de Finetti representation theorem we can represent P as $P(X_1 \leq x_1, \ldots, X_n \leq x_n) = \int_{\mathcal{F}} \prod_{i=1}^n F(x_i) d\pi(F)$, $n \geq 1$, where F is the weak limit of the sequence of empirical d.f.'s $\hat{F}_n$. Then we might find easier to assign P by choosing a measure π on $\mathcal{F}$, that we interpret as our opinion on the limit of the empirical d.f.'s, $\hat{F}_\infty$ say. Of course, this limit is not observable so that π is just a tool for assigning the exchangeable law P; but if we take it seriously, then a natural requirement is that for $n \to \infty$, the posterior $\pi(F|x^{(n)})$ concentrates around the observed frequencies, or that the predictive distribution $F_n^{pred}(x) = E(F(x)|x^{(n)})$ is close to the empirical d.f., that is $\sup_x \mid \hat{F}_n(x) - F_n^{pred}(x) \mid \to 0$, for all sequences $(x_1, x_2, \ldots)$. This would express the idea of no surprise from the data.

But these requirements are not achievable in general with a σ-additive probability P, due to the constraints implied by σ-additivity to the conditional probabilities; yet, they hold almost-surely w.r.t. P (the first is Doob's consistency theorem; the second, Glivenko–Cantelli theorem for exchangeable sequences, is proved in Berti and Rigo (1997)). The exceptional set of P probability zero can be quite large if the prior π is parametric, that is its support is $\mathcal{F}_0$. If instead one is uncertain on the model, then $\pi(\mathcal{F}_0) = 1$ is clearly not a honest assignment: one should use a nonparametric prior, whose support is the whole class $\mathcal{F}$. This is the approach in the paper and I agree with it. There remains however an exceptional set of sequences

$(x_1, x_2, \ldots)$ having P-probability zero on which the above requirements do not hold. In this sense, and for these sequences, one could be surprised by the data. This is due to the fact that the σ-additive P is not honest but used only for convenience. If we do not want to abandon σ-additivity, then it seems natural to investigate 'how large' is the exceptional set, with 'as if' reasoning: if $(x_1, x_2, \ldots)$ is such that the observed $\hat{F}_\infty = F_0$, do the desired requirements hold? The notion of consistency of Bayesian posteriors, discussed by Diaconis and Freedman (1986), tries to control the exceptional sets (somehow motivated by the fact that, if $X_1, X_2, \ldots$ are i.i.d. with d.f. F_0 then $F_\infty = F_0$, a.s.-F_0^∞). However, this approach to consistency ultimately leads to substituting the observed frequency with the true d.f. F_0. A different condition is studied by Diaconis and Freedman (1990); for multinomial experiments, they show that, if the prior satisfies a certain requirement of positivity, then for all sequences $(x_1, x_2, \ldots)$ the posterior odds for a parameter interval of fixed length centered on the observed frequencies converge to infinity (more precisely, $\pi([\bar{x}_n - h, \bar{x}_n - h] | x_1, \ldots, x_n)/\pi([\bar{x}_n - h, \bar{x}_n - h]^c | x_1, \ldots, x_n) \geq ab^n$ for computable constants a and b, such that the bound grows exponentially fast as $n \to \infty$). This clearly uses the observed frequencies and no notions of *true* F_0 and eliminates exceptional sets. It does not seem easy to generalize from multinomial probabilities, however one could in principle study when analogous result hold for nonparametric priors: can we find lower bounds for the posterior odds of neighborhoods of the empirical d.f., with no exceptional sets?

In that case, we would be sure of having no 'surprise' from the data. If we do not use a nonparametric prior, instead, but for simplicity a parametric one, then the problem is posed outside the rules of coherent probability and, I think, cannot be treated formally. We look for auxiliary tools for guiding ourselves in striving to use all the information we have in assigning P. These could be proper scoring rules, computed *a posteriori*, after the data have been observed. They would be used informally to assess wether we tried hard enough and elicited a honest P, or if we should interrogate ourselves further to incorporate all the information we have in assigning P. See de Finetti (1980), page 1151. It is interesting however that formal tests, for example the Bayes factor, may correspond to specific choices of the scores (log predictive scores for the Bayes factor). I refer to Chakrabarti and Ghosh (2006) for further discussion.

What loss function for goodness of fit? Even when a nonparametric prior has been assigned, one might nevertheless consider 'using a parametric model', for its greater simplicity or since it results more *useful* than the more complex nonparametric one. We have a decision problem and a basic question addressed in the paper is what are appropriate loss functions. The action space considered in the paper is $\mathcal{A} = \{a_0, a_1\}$, with a_0 corresponding to $H_0 : f = f_0$ (or, in the case of testing a parametric model, $H_0 : f \in \mathcal{F}_0$). A first choice which is examined is the zero-one loss function, which leads to posterior odds and Bayes factors. The choice of a 0–1 loss has been criticized, since it does not consider the distance between f and f_0 (see Bernardo and Smith (1994) section 6.1.5). In the introduction of the paper, we read that, in goodness of fit, '*the question of interest is not the exactness of the point null hypothesis but wether the data can be relatively well represented by the null model*'. When using a Bayes factor, this question is addressed by studying the equivalence between the Bayes factor for a point null hypothesis ($H_0 : f = f_0$; analogous setting for the problem of testing a parametric model $\mathcal{F}_0$) and that for an interval null hypothesis ($H_0 : d(f, f_0) < \epsilon_n$). It results that ϵ_n should be very small for the two Bayes factors behaving equivalently, too small for this equivalence being of practical relevance. An alternative procedure is then proposed. The idea of using the model 'if it describes relatively well the data' is expressed through a distance-based loss function defined as $L(f, a_0) = d(f, f_0) - \epsilon_n$ if $d(f, f_0) > \epsilon_n$ and zero otherwise; and $L(f, a_1) = \epsilon_n - d(f, f_0)$ if $d(f, f_0) \leq \epsilon_n$ and zero otherwise. The resulting Bayesian decision rule chooses a_1 if $H(x^{(n)}) = E(d(f, f_0) \mid x^{(n)}) \geq \epsilon_n$. The choice of ϵ_n should be subjective, since it is part of the loss function. However, it is true that choosing ϵ_n is quite delicate, especially for large sample size. The suggested solution is to calibrate ϵ_n by a p-value, which in the simplest case is $P_{f_0}(H(X^{(n)}) \geq H(x^{(n)}))$. This suggestion, as shown in the paper, interestingly relates the implied choice of ϵ_n with the convergence rate of the nonparametric prior to f_0. The more complex case of a parametric null hypothesis $H_0 : f \in \mathcal{F}_0$ is cleverly reduced to this case, by nesting the parametric model in the class of distributions on which the nonparametric family is supported, developing an idea of Verdinelli and Wasserman (1998). Interesting results on the implied ϵ_n and on convergence rates are provided for this case, too.

Even if the p-value calibrated distance procedure is of interest, some perplexity remains. Treating the posterior expected loss $H(x^{(n)}) = E(d(f, f_0) \mid x^{(n)})$ as a test statistics and computing a p-value, as suggested, appears somehow weird from a Bayesian point of view. I do not mean to criticize the approach but I'll try to to discuss possible developments by considering other choices of the loss function. First note that other decision problems can be of interest, when we assign a nonparametric prior but investigate the possible advantages of a parametric model. One remark is that, in fact, the decision problem involves both the parametric model *and* the prior. Recently, Bassetti, Bodini and Regazzini (2002) and Walker and Gutiérrez-Peña (2007) have considered the problem when, despite having assigned a nonparametric prior, one wants to report the result by means of a parametric model, which could be a standard reference-model for the problem or just a simpler, more interpretable model. In this case the distribution on the parameters of the model does not express prior beliefs (we know that the model is not true but only used for convenience in reporting), and in fact it is obtained as the solution of a decision problem, which, given the parametric model f_θ, has action space $\mathcal{A} = \{\mu(\theta)\}$ i.e. the space of the possible 'prior' distributions on θ. Here the model is fixed and the decision is about the prior. If we want to evaluate the pair 'model and prior' instead, a possible suggestion is to look at the predictive density $p_M(y \mid x^{(n)}) = \int f_\theta(y) d\mu(\theta | x^{(n)})$, which depends both on the model and on the distribution of θ. Gutiérrez-Peña and Walker (2001) propose a Bayesian predictive approach of this nature for model comparison.

The idea is that one might want to compare several models, $M_1, \ldots, M_k$ say, being aware that none of them might be 'true' and therefore honestly assigning a nonparametric prior. In this context model selection is formalized as a decision problem with a logarithmic score as a utility function (Bernardo, 1979). More precisely, the utility function for model j is $U(M_j, F) = \int \log p_{M_j}(y|x^{(n)}) dF(y)$, where $p_{M_j}(y \mid x^{(n)})$ is the predictive density for model j. The goal is to maximize the posterior expected utility $\tilde{U}(M_j) = E(U(M_j, F)|x^{(n)}) = \int \log p_{M_j}(y|x^{(n)}) dF_n^{pred}(y)$, where $F_n^{pred}(y) = E(F(y)|x^{(n)})$ is the predictive distribution obtained from the nonparametric prior. Note that $\tilde{U}(M_i) > \tilde{U}(M_j)$ corresponds to $KL(F_n^{pred}, p_{M_i}) < KL(F_n^{pred}, p_{M_j})$, where $KL(\cdot, \cdot)$ denotes the Kullback–Leibler divergence. Here no evident need of calibration appears, since the problem is model comparison. But in goodness of fit, no explicit alternative is specified, or in other words the alternative is the nonparametric model; and clearly, we cannot define a decision rule which chooses the parametric model if $KL(F_n^{pred}, p_M) < KL(F_n^{pred}, F_n^{pred}) = 0$. Thus, we might slightly change the loss function so that we choose the model M if $\tilde{U}(M)$ is sufficiently large or if $KL(F_n^{pred}, M) < \epsilon$. This is closely related to the distance-based approach suggested in the paper, and the issue of calibrating ϵ arises. However, in practice we would be interested in using the parametric predictive not only if it is a good approximation of the nonparametric predictive distribution, but also because the parametric model is simpler, or more interpretable. Therefore one might think of introducing a penalization for complexity or a term related to precision in the utility function. In a parametric setting, these issues are discussed for example by Poskitt (1987); one might investigates similar issues for nonparametric priors. A utility function which is not only distance-based might lead to a Bayesian decision rule which selects the parametric model if its posterior expected utility is bigger than that of the nonparametric model, avoiding the need of calibration.

Finally, there are other issues that I discuss more briefly.

What nonparametric prior? Using a nonparametric prior is the honest choice when there is uncertainty on the model. But another question arises: what nonparametric prior? Some desirable properties might be asked for a nonparametric prior, such as consistency or the behavior of the posterior odds as in Diaconis and Freedman (1990). But can we give suggestions for choosing among different nonparametric priors, all possibly satisfying the desired properties? There are often some features of the nonparametric prior which make its use natural or interesting in the specific problem at hand. For example, the Dirichlet process, or more generally species sampling priors, are well suited as priors on the latent distribution in mixture or latent variables models, since they can naturally model ties and offer a probability law on random partitions (this is perhaps the main reason of the impressive growth of applications of the Dirichlet process in a wide variety of problems, as this Valencia meeting has shown). In other problems, and specifically in goodness of fit, the discrete nature of the Dirichlet process can be a drawback, and one might prefer a prior which selects absolutely continuous d.f.'s. One might argue that any nonparametric prior can be used successfully for goodness of fit. But the paper compares different choices and shows that they are not equivalent in terms of the properties of the resulting decision rules. In particular, a natural issue, which is discussed at length in the paper, is wether one should embed (or nest) the parametric model in the nonparametric prior or not. The non-nested choice is often simpler but, if f_0 is not in the model, one might have an undesirable asymptotic behavior of the tests. These phenomena are somehow related to convergence rates of the nonparametric prior to f_0. Note that a nonparametric prior has full support in the sense that it selects d.f.'s in a class $\mathcal{B} = \{f_\psi, \psi \in S\}$ which is dense (in the appropriate sense) in the class $\mathcal{F}$ of all the d.f.'s on the sample space. If $f_0 \in \mathcal{B}$, the rate of convergence for a nonparametric prior supported on $\mathcal{B}$ can be fast, basically the parametric rate. However, if f_0 does not belong to $\mathcal{B}$, the convergence rate is generally slower. This might be disturbing, even if, on the other hand, it seems natural that the convergence rate depends on the class of densities to be estimated. I do not know if the phenomena discussed in the paper are drawbacks of the nonparametric prior or of the procedures under study. I feel that a nonparametric prior should be considered for its features, including interpretability or simplicity. However, this is, I think, an open problem on which further reflection is worth. Note that Bayes factors for comparing nonparametric priors have been studied (Subhashis Ghosal, personal communication). Anyway, it is a positive fact that years ago the problem was how to define nonparametric priors and now we can wonder 'how to choose among nonparametric priors'!

Consistency of Bayes factors. The last part of the paper studies consistency of Bayes factors with nonparametric priors. The case of a simple null hypothesis $H_0 : f = f_0$ has been studied by Dass and Lee (2004): consistency of the Bayes factor is a fairly direct consequence of the consistency of the posterior. The Bayes factor behaves asymptotically like the posterior odds (up to the prior odds); in particular, if f_0 is true (that is, a.s. $-F_0^\infty$), the posterior concentrates around f_0, by Doob consistency; the numerator of the Bayes factor converges to one and the denominator to zero, so that the Bayes factor converges to infinity, as desired. The case of a parametric null hypothesis $H_0 : f \in \mathcal{F}_0$ is more complex since clearly neighborhoods of f_0 may contain distributions in $\mathcal{F}_0$ or in $\mathcal{F}_o^c$. In Section 3 of the paper it is shown that the Bayes factor goes to infinity a.s. F_0^∞ if the nonparametric prior π is consistent at f_0 (at rate ϵ_n), if the parametric model is regular and if π gives positive but sufficiently small probability to ϵ_n-neighborhoods of f_0. Otherwise, the Bayes factor can be inconsistent, in the sense that it does not converge to infinity when f_0 is the true density; the examples in the paper show that inconsistency can appear more often in the non-nested case. However, I wonder if in these examples the Bayes factor stays bounded away from zero, and if this means that, if the prior does not penalize enough for complexity, asymptotically one can do as well by using the *true* parametric model for prediction than the nonparametric one (I am thinking of the log-Bayes factor in terms of ratio of log predictive scores). Interestingly, the prior mass on neighborhoods of f_0 is regarded as a measure of the penalization for complexity implied in the Bayes factor. As recently shown by Walker *et al.* (2005), the prior mass on appropriate neighborhoods of f_0 also determines the convergence rates: I would ask to professor Rousseau if she has more intuition about the connection among convergence rates, complexity and the asymptotic behavior of goodness of fit tests.

Extensions to non-exchangeable sequences. In the paper, only the case of exchangeable sequences is studied (this is clearly the basic dependence structure to consider, and it is complicated enough). But goodness of fit is more interesting in more complex cases, for example for assessing the dependence structure itself of the observables (as in time series analysis), or for variable selection. In case of uncertainty on the dependence structure, similar ideas could be developed, and in the spirit of the paper one would assign a nonparametric prior on the random probability law P of the sequence $(X_n, n \geq 1)$. This appears crazy but in fact it is possible, at least in principle, as the recent literature on nonparametric priors for random curves shows; see, *e.g.*, Gelfand *et al.* (2006). Also, an appropriate version of Glivenko-Cantelli theorem hold (Berti and Rigo, 2002).

ROSS MCVINISH (*Queensland University of Technology, Australia*)

A side effect of embedding. My comments are restricted to the comparison of Bayes factors from embedding and non-embedding priors. The author has demonstrated why one may prefer a prior which is embedded within the parametric family when one is assessing goodness of fit. In this case, inconsistency of the Bayes factor is typically not a problem whereas for the non-embedded prior seemingly well constructed priors may produce inconsistency. However, it is worth considering other possible consequences of using an embedded prior. Consider a goodness of fit test for the family

$$f_\theta(x) = \exp\left(\theta' b(x) + \alpha(\theta)\right) c(x), \quad x \in [0,1],\ \theta \in \Theta \subset \mathbb{R}^2$$

using the mixtures of triangular distributions prior. Inconsistency of the Bayes factor occurs in the non-embedded prior due to the amount of probability this prior places in a neighbourhood of the uniform distribution, $\theta = (0,0)$. The embedded prior avoids the inconsistency by distributing the amount of probability placed on a neighbourhood of the uniform over a neighbourhood of the entire parametric family. The effect of this is that for cases other than $\theta = (0,0)$ the amount of probability placed on neighbourhoods of f_θ is increased substantially. Let A_ϵ be the set $\{f : \|f_\theta - f\| < \epsilon\}$ for $\theta \neq (0,0)$. One can show that the embedded prior places at least $O(\epsilon^3)$ probability on A_ϵ while the non-embedded prior places at most $O(\exp(-\epsilon^{-\kappa}))$, for some $0 < \kappa < \infty$, probability on A_ϵ. As shown in McVinish *et al.* (2005), these prior probabilities essentially control the rate at which the B_n^{-1} will converge to zero under the null hypothesis and so, with the exception of $\theta = (0,0)$, this rate of convergence will be slower with the embedded prior than with the non-embedded prior. Might it be preferable to simply restrict the number of components in the non-embedded mixture of triangulars to be $k \geq 3$ which would avoid the inconsistency? Are there any other benefits to using an embedded prior with the Bayes factor?

REPLY TO THE DISCUSSION

First I want to thank professor Petrone for her interesting and deep discussion on my paper. Her discussion touches many aspects of Bayesian nonparametric approaches and Bayesian testing procedures. Her discussion on the different notions of asymptotic is very enlightening, however I am very doubtful about the possibility of generalizing Diaconis and Freedman's result to general nonparametric priors on density. Indeed, when dealing with nonparametric rates of concentration of the posterior distribution around the *true* density, say f_0, we essentially try to bound posterior odds of neighbourhoods of f_0, and up to now every results I have seen had to control the size of sets on which the obtained inequalities were not satisfied. In other words the inequalities could not be proved to be satisfied for all sequences. However bounds are obtained and to some extent they can be considered as non asymptotic, although it is true that the constants entering these bounds might not be very useful. Now replacing f_0 by the empirical density function seems to me to make things worse and this is why I have doubt about the possible results of such an approach. Also, I believe that looking at frequentist properties of Bayesian procedures and proving that some Bayesian procedures share the same properties as frequentist procedures help in making the Bayesian procedures attractive.

The question of loss function in goodness of fit seems to me to be crucial and all the more when the number of observations gets large. The different propositions of Sonia Petrone are interesting and certainly using predictive distributions seems reasonable. I think that one has to keep in mind the real objectives of the statistical analysis when constructing the loss function for the goodness of fit test. Penalizing for complexity of the nonparametric model or introducing some notion in the utility as suggested by Sonia Petrone seems to me very similar to choosing the tolerance ϵ.

I believe that when the number of observations gets large it becomes necessary to assess this notion of precision a priori, which might be done in many ways depending on the context. Calibration as proposed in the paper only comes when there is no other options, to some extent, and this can happen in particular when the statistical analysis has more than one objective or is to lead to further investigations of various sorts. However the objective of the paper was not really to sell the calibration-distance approach or the Bayes factor approach but to try to understand better what they meant and how they reacted to certain problems.

I also think it is very encouraging that now people are wondering which nonparametric prior to choose, as a sign that they are actually willing to choose one. If I didn't comment on the positiveness of the properties of embedding or non embedding when constructing a nonparametric prior in goodness of fit problems, it is because I was not quite sure about what to think on these results. I believe now that if one makes the choice of non-embedding, that is to some extent the choice of simplicity, then one has to make sure that the parametric model behaves correctly with respect to the nonparametric model, and that no strange asymptotic behaviour as described in the paper would happen.

About the consistency of Bayes factors and the notion of penalization induced by Bayes factors: the penalization (or complexity) comes from the integration over the parameter space, which turns out to be sort of equivalent to integration over neighbourhoods of the *true* distribution, which explains why the penalization is of this nature and why it is the same as in the study of rates of convergence of posterior distributions. In both cases we are essentially looking at the same thing. Does this mean that the nonparametric prior is not penalizing enough for real complexity? I think it probably depends on what real complexity means? computational complexity? interpretability? obviously, looking at the posterior distribution does not take into account these notions of complexity and if they are the ones of interest then they need to be introduced in the loss function.

Finally, I agree with Sonia Petrone that a lot more remains to be done on Bayesian goodness of fit and in particular outside the exchangeable case.

Now I also want to thank R. McVinish for commenting on the choice between embedding and non-embedding. I personally prefer the embedded approach now because it forces all the possible densities of the null model to behave similarly, which might not be the case of the non-embedded approach, even though one might get consistency of the Bayes factor for any of them. However, it is true that the non-embedded approach is much simpler and the above drawback might be avoided by carefully choosing the nonparametric model with regards to the parametric one.

ADDITIONAL REFERENCES IN THE DISCUSSION

Bassetti, F., Bodini, A. and Regazzini, E. (2002). Outline of a new predictive and self-controlling methodology. *Technical report* 2002–33, Department of Statistics, Stanford University.

Bernardo, J. (1979). Expected information as expected utility. *Ann. Statist.* **7**, 686–690.

Bernardo, J. M. and Smith, A. F. M. (1994). *Bayesian Theory.* Chichester: Wiley

Berti, P. and Rigo, P. (1997). A Glivenko-Cantelli theorem for exchangeable random variables. *Statistics and Probability Lett.* **32**, 385–391.

Berti and Rigo (2002). A uniform limit theorem for predictive distributions. *Statistics and Probability Lett.* **56**, 113–120.

Chakrabarti, A. and Ghosh, J. (2006). Some aspects of Bayesian model selection for prediction. *In this volume.*

Coram M. and Diaconis P., (2003). New tests for the correspondence between unitary eigenvalues and the zeros of Riemann's zeta function. *J. of Physics A: Math. Gen.* **36**, 2883–2906.

de Finetti, B. (1980). Probabilità. *Enciclopedia Einaudi*, Torino: Einaudi, 1146–1187

Diaconis, P. and Freedman, D. (1990). On the uniform consistency of Bayes estimates for multinomial probabilities. *Ann. Statist.* **18**, 1317–1327.

Gelfand, A. E., Dey, D. K. and Chang, H. (1992). Model determination using predictive distributions with implementation via sampling-based methods. *Bayesian Statistics 4* (J. M. Bernardo, J. O. Berger, A. P. Dawid and A. F. M. Smith, eds.) Oxford: University Press,147–167 (with discussion).

Gelfand, A. E., Guindani, M. and Petrone, S. (2006). Bayesian nonparametric modelling for spatial data using Dirichlet processes. *In this volume.*

Gutiérrez-Peña, E. and Walker, S. G. (2001). A Bayesian approach to model selection. *J. Statist. Planning and Inference* **93**, 259–276.

Petrone, S. (2003). A predictive point of view on Bayesian nonparametrics. In *Highly Structured Stochastic Systems* (P. Green, N. Hjort and S. Richardson, eds.), Oxford: University Press

Poskitt, D. S. (1987). Precision, complexity and Bayesian model determination. *J. Roy. Statist. Soc. B* **49**, 199–208.

Walker, S. G. and Gutiérrez-Peña, E. (2007). Parametric inference in a nonparametric framework. *Test* (to appear).

BAYESIAN STATISTICS 8, pp. 453–470.
J. M. Bernardo, M. J. Bayarri, J. O. Berger, A. P. Dawid,
D. Heckerman, A. F. M. Smith and M. West (Eds.)

Bayesian Probability in Quantum Mechanics

RÜDIGER SCHACK
Royal Holloway, University of London, UK
r.schack@rhul.ac.uk

SUMMARY

This paper reviews arguments for the claim that all probabilities arising in quantum mechanics are Bayesian degrees of belief. The starting point is the observation that there is a one-to-one map from quantum states to points on the probability simplex. Bayes's rule remains valid in quantum theory, provided it is kept in mind that the outcomes of quantum measurements have no preassigned values. The probabilities of quantum measurement outcomes depend explicitly on a quantum prior whose general form will be discussed. Similar to the classical case, the quantum de Finetti theorem simplifies the analysis of repeated quantum trials if the prior is exchangeable. The Bayesian approach to quantum theory described here solves the problem of the 'collapse of the wave function'. It also provides almost trivial explanations of a number of phenomena in quantum information theory, including the no-cloning theorem and quantum teleportation.

Keywords and Phrases: BAYES'S RULE; NO CLONING; POSITIVE OPERATOR VALUED MEASURE; QUANTUM DE FINETTI THEOREM; QUANTUM STATES; QUANTUM TELEPORTATION.

1. INTRODUCTION

The Bayesian approach to classical statistical mechanics is well established (see, *e.g.*, Jaynes, 1957 and Balian, 1991). Each particle in a classical gas has a precise position and velocity. At each moment in time, the microstate of the gas is fully determined by the positions and velocities of all its constituent particles or, equivalently, by a point in phase space. Statistical mechanics is concerned with probability distributions on phase space. From a Bayesian perspective, these probabilities represent some agent's degrees of belief. For a clear understanding of statistical mechanics, it is important not to confuse probabilities with observable quantities such as, *e.g.*, the actual fraction of particles moving faster than 1 meter per second. The fraction of particles moving faster than 1 meter per second is a random variable, entirely determined by the microstate of the gas. The *probability* that a given particle moves faster than 1 meter per second, however, is not determined by the microstate. It is a function both of the agent's observations and his prior.

In quantum theory, this clear distinction between probabilities on the one hand and observable quantities on the other hand seems less clear-cut. Quantum theory is nondeterministic, and quantum probabilities are governed by physical law. There seems to be neither room nor a need for an agent's prior in fundamental quantum mechanics. Are quantum probabilities outside the realm of the Bayesian approach? Is the applicability of Bayesian theory in quantum mechanics limited to questions such as the analysis of experimental data, whereas more fundamental probabilities are objective physical parameters?

I am going to show in this paper that the answer to these questions is No. All probabilities arising in quantum mechanics can be interpreted as Bayesian degrees of belief. This is the core claim of the Bayesian approach to quantum mechanics that has been developed over the last few years by Caves, Fuchs and Schack (2002a, 2002b); see also Barnum *et al.* (2000), Fuchs and Peres (2000), Schack, Brun and Caves (2001), Brun, Caves and Schack (2001), Fuchs (2002), Schack (2003), Fuchs, Schack and Scudo (2004) and Caves and Schack (2005). It turns out that the Bayesian approach is not only fully consistent with the mathematical formalism of quantum theory and its applications to laboratory experiments, but leads to the solution of a number of major conceptual issues.

The starting point of the argument is Section 2, which reviews the basic rules of quantum mechanics in terms of density operators and positive operator valued measures (POVMs). This formulation is then used to show that quantum states can be represented by points on the standard probability simplex, *i.e.*, quantum states can be viewed as probability distributions.

The one-to-one correspondence between quantum states and probability distributions is not an isolated fact. The whole of Bayesian probability theory applies to quantum states without modification, as long as it is remembered that 'unperformed experiments have no results' (Peres, 1978). This has an important consequence for the use of conditional probabilities in quantum theory. In the expression $P(A \mid B)$, B must be a proposition. The conditional probability $P(A \mid B)$ is therefore not defined when B refers to the potential outcomes of an unperformed measurement. Section 3 expands on these points by discussing the relation between Bayes's rule and the quantum-mechanical updating rule.

Procedures for the preparation of quantum states play a major role in most formulations of quantum theory. It is often stated or implied that one can give, in unambiguous terms, a description of an experimental device that prepares a given quantum state. According to this view, a quantum state is determined by the preparation procedure. Therefore there would be no room for a prior; quantum states, and the probabilities derived from them, would not represent Bayesian degrees of belief, but objective facts about the preparation device. In Section 4 it is argued that this view neglects to take into account that quantum mechanics applies to preparation devices as well. If quantum state preparation is phrased in terms of Bayesian updating as explained in Section 3, it becomes clear that any quantum state assignment depends explicitly on a prior.

A particularly important class of priors are infinitely exchangeable priors for repeated trials. In complete analogy to classical Bayesian theory, if a quantum physicist's prior for an experiment is infinitely exchangeable, his analysis of the experimental data becomes relatively straightforward. This is a consequence of the quantum de Finetti theorem discussed in Section 5.

Up to this point, the discussion in this paper has concentrated on showing that, despite a common prejudice, it is natural to regard all probabilities arising in quantum mechanics as Bayesian degrees of belief. The remainder of the paper looks at some of the benefits of this view. One of the most important motivations for an agent-centred approach to quantum probability comes from the problem of the collapse of the wavefunction. The collapse of the wavefunction is a major problem for any interpretation of quantum mechanics that regards quantum states as physically real. Section 6 shows that this problem does not exist in the Bayesian approach.

The Bayesian view does not only eliminate major conceptual difficulties, it also demystifies phenomena. Two examples, discussed in Section 7, are the quantum no cloning theorem and quantum teleportation. For both of these, the Bayesian approach provides almost trivial explanations.

Since this paper is intended for a readership of statisticians, I do not assume any familiarity with quantum mechanics. On the other hand, without explaining or defending this view any further, I will treat all probabilities as Bayesian degrees of belief, guiding an agent's decisions as described in Bernardo and Smith (1994).

2. A QUANTUM STATE IS A PROBABILITY DISTRIBUTION

This section explains how quantum states can be understood as probabilities. The one-to-one correspondence between quantum states and probability distributions derived below is well known (see Hardy, 2001 and Fuchs, 2002 for two recent papers on quantum foundations making use of this approach). All the results from quantum theory quoted below are phrased in terms of linear algebra in a finite-dimensional vector space.

In standard introductions to quantum theory, quantum states are vectors ('state vectors') in a complex Hilbert space, and measurements are represented by projection operators ('projective measurements'). State vectors are also called *pure states*, and projective measurements are called *von Neumann measurements*. In this formulation, more general states appear as mixtures of pure states, and more general measurements are described as some interaction with an auxiliary system followed by a von Neumann measurement on the auxiliary system.

The alternative approach adopted in the present paper takes general states and generalized measurements as fundamental and treats state vectors and projective measurements as special cases. This approach is mathematically equivalent to the standard approach (see, *e.g.*, Nielsen and Chuang, 2000), but from a Bayesian perspective it is more natural.

Because most textbooks take classical mechanics as the starting point, position and momentum variables appear early on, leading to the introduction of infinite-dimensional Hilbert spaces with all their mathematical complications. For the question of probability in quantum mechanics, these complications are a distraction. We are going to restrict our discussion to the finite-dimensional case. Another potential source of difficulty is the fact that quantum theory uses complex numbers. With one notable exception (the quantum de Finetti theorem), everything in this paper remains valid if the term 'Hilbert space' is replaced by 'real vector space with an inner product.'

We thus associate with a physical system a d-dimensional Hilbert space, H. The state of a system is a positive semi-definite operator with trace 1, *i.e.*, a $d \times d$ matrix with nonnegative eigenvalues that sum to 1. Such an operator is called a *density operator* and usually denoted by ρ. If one of the eigenvalues of a density operator ρ

is equal to 1, *i.e.*, if ρ is a one-dimensional projection operator, then ρ is a pure state.

A measurement is described by a POVM or positive operator valued measure (see, *e.g.*, Peres, 1993), *i.e.*, a collection $\{E_k\}$ $(k = 1, \ldots, m)$ of positive semidefinite $d \times d$ matrices (called *effects*) acting on H, where the index k labels the possible outcomes of the measurement. The number, m, of possible outcomes is arbitrary. In particular, m is allowed to be greater than d, the dimension of Hilbert space. To form a POVM, the effects E_k must satisfy the completeness relation

$$\sum_{k=1}^{m} E_k = I\,, \tag{1}$$

i.e., they must form a resolution of the identity. The probability of obtaining the measurement outcome k is now given by

$$p_k = \mathrm{tr}(E_k \rho)\,. \tag{2}$$

Given the rule (2), the conditions defining a POVM are the most general conditions compatible with the requirement that the p_k are probabilities. The requirement $p_k \geq 0$ for all density operators ρ implies that E_k is positive semidefinite, and the requirement that the p_k sum to 1 for all density operators ρ implies the completeness relation (1). Any POVM describes a valid quantum measurement.

The special case of a von Neumann measurement is recovered if the effects, E_k, are mutually orthogonal projection operators. In this case, the number of possible measurement outcomes is clearly bounded by the dimension of Hilbert space, d. The approach advocated here does not give any special status to von Neumann measurements.

The set of all $d \times d$ matrices forms a d^2-dimensional vector space. If $E_1, \ldots, E_{d^2}$ are linearly independent $d \times d$ matrices, the system of equations

$$\begin{aligned} \mathrm{tr}(E_1\rho) &= p_1\,, \\ \mathrm{tr}(E_2\rho) &= p_2\,, \\ \ldots \\ \mathrm{tr}(E_{d^2}\rho) &= p_{d^2} \end{aligned} \tag{3}$$

has a unique solution ρ. If, in addition, the matrices E_k form a POVM, *i.e.*, are positive semidefinite and sum to the identity, the set $\{E_k\}$ is called a *minimal informationally complete POVM.* In this case, the right-hand side of the equations (3) lists the probabilities for the d^2 possible measurement outcomes if the POVM is measured for the state ρ. The vector $\mathbf{p} = (p_1, \ldots, p_{d^2})$, which is a point on the probability simplex in d^2 dimensions, specifies ρ uniquely.

Informationally complete POVMs are not difficult to construct. For an explicit construction, see Caves, Fuchs and Schack (2002a). For any given minimal informationally complete POVM, the set of all quantum states is represented by a convex subset of the probability simplex in d^2 dimensions. There is a one-to-one correspondence between quantum states and points on this convex subset of the simplex. A quantum state is a probability distribution for the outcomes of a suitably chosen measurement (see Fuchs, 2002). Such a measurement serves as a kind of reference frame for quantum states. Throughout this paper, we will use the term *fiducial POVM* for a minimal informationally complete POVM used as a reference measurement in this way.

The two representations of quantum states described in this section are completely equivalent. The density-operator formalism is usually more convenient for calculations. It tends to hide, however, the close relation between the quantum formalism and probability theory.

When one uses the representation of quantum states on the simplex, one has to pay attention to two important features of quantum theory. Firstly, not all points on the simplex correspond to valid quantum states (Fuchs, 2002). For example, for any informationally complete POVM and any quantum state, no outcome has probability 1.

Secondly, quantum measurement outcomes have no preassigned values. This is a consequence of the Kochen-Specker theorem (see, *e.g.*, Peres, 1993) and the Bell inequalities (see, *e.g.*, Peres, 1978; Mermin, 1985), two of the most fundamental results in quantum theory. For the Bayesian approach, this means the following. Let the random variable O be the outcome of some measurement. Then the statement $O = k$, for some integer k, has a truth value only if the measurement is actually performed. If the measurement is not performed, the statement $O = k$ is neither true nor false and therefore is not a proposition that can be used for conditioning. The conditional probability $P(\cdot \,|\, O = k)$ is undefined if the measurement is not performed.

The most important application of the above is in quantum interference. Assume that three measurements, described by three possibly different POVMs, are performed one after the other on the same system. Denote the respective outcomes by the random variables A, B and C. If measurement A is performed, we can write down conditional probabilities, $P(B = b \,|\, A = a)$ for the outcomes of measurement B. Similarly, if measurement B is performed, we can write down conditional probabilities, $P(C = c \,|\, B = b)$ for the outcomes of measurement C. So, if both measurements A and B are performed, the law of total probability gives

$$P(C = c \,|\, A = a) = \sum_b P(C = c \,|\, B = b)\, P(B = b \,|\, A = a)\,. \tag{4}$$

This is exactly what quantum mechanics predicts. The situation changes fundamentally, however, if measurement B is not performed. In this case, $B = b$ has no truth value, and therefore writing down the conditional probabilities in eq. (4) is illegal. The phenomenon of quantum interference is a manifestation of the fact that the law of total probability does not hold if the measurement B is not performed. In this case, the conditional probabilities $P(C = c \,|\, A = a)$ have to be computed directly. The general form of conditional probabilities and the role of Bayes's rule in quantum theory is the topic of the next section.

3. UPDATING QUANTUM STATES; BAYES'S RULE

The updating of a probability assignment in the light of new data is at the centre of Bayesian statistics. Usually, updating proceeds by conditioning. Let

$$P(A = a \,|\, B = b)$$

denote the conditional probability for a random variable A, given data in the form of a second random variable, B. If $B = b$ is observed, the updated probability for A is simply

$$P(A = a) = P(A = a \,|\, B = b)\,. \tag{5}$$

In quantum mechanics, updating proceeds in exactly the same way. Let the random variables A and B label the possible outcomes of two POVMs. If the POVM B is performed and the outcome b is observed, the updated probability for the outcomes of measurement A is

$$P(A = a) = P(A = a \,|\, B = b) \,. \tag{6}$$

Classically, the conditional probability $P(A = a \,|\, B = b)$ is linked to the prior, $P_{\text{prior}}(A = a)$, via Bayes's rule,

$$P(A = a \,|\, B = b) = \frac{P(B = b \,|\, A = a) P_{\text{prior}}(A = a)}{P(B = b)} \,. \tag{7}$$

This rule is valid in quantum mechanics provided both random variables A and B refer to measurements that are actually carried out.

For conceptual reasons, it can be convenient to describe measurements that are not carried out. The fiducial POVMs introduced in the previous section provide an important example of a class of such measurements (Fuchs, 2002). In terms of fiducial POVMs, quantum updating can be phrased as follows.

As before, we consider an arbitrary measurement, with m possible outcomes described by a random variable A. Our fiducial measurement is denoted F, with outcomes $F = 1, \ldots, d^2$, where d is the dimension of Hilbert space. The prior state is a vector $\mathbf{p}$, whose components are given by $P(F = f)$ $(f = 1, \ldots, d^2)$. We now carry out the measurement A and obtain the outcome $a \in \{1, \ldots, m\}$. Our posterior state is then a vector $\mathbf{p}_a$, with components given by $P(F = f \,|\, A = a)$ $(f = 1, \ldots, d^2)$. These probabilities are our degrees of belief about the outcomes of F, given that in a previous measurement of A we actually obtained the outcome a. We cannot use Bayes's rule to evaluate $\mathbf{p}_a$ because we do not carry out a measurement of F before measuring A, and therefore the conditional probabilities $P(A = a \,|\, F = f)$ needed in Bayes's rule are undefined.

In this situation, quantum mechanics provides a very general rule to link the prior to the conditional posterior states. The quantum rule associates with each outcome a of the measurement A a linear operator, $\mathcal{M}_a$ (called a *quantum operation*), which acts on $\mathbb{R}^{d^2}$, *i.e.*, which maps d^2-dimensional real vectors to d^2-dimensional real vectors. For each vector $\mathbf{p}$ in the convex subset of the simplex corresponding to valid quantum states, *i.e.*, for each $\mathbf{p}$ representing a quantum state, let

$$q_a = |\mathcal{M}_a(\mathbf{p})|_1 \tag{8}$$

denote the 1-norm of the vector $\mathcal{M}_a(\mathbf{p})$, *i.e.*, the sum of the absolute values of the components. For the linear operators $\mathcal{M}_a$ $(a = 1, \ldots, m)$ to describe updating following a measurement of A, they must satisfy, for each $\mathbf{p}$ representing a quantum state, the following conditions:

1. if q_a is nonzero, $q_a^{-1}\mathcal{M}_a(\mathbf{p})$ is a quantum state, *i.e.*, is in the subset of the simplex corresponding to valid quantum states;

2. $\sum_a q_a = 1$;

3. if the measurement is applied to part of a larger system, the updated probabilities for the outcomes of a fiducial POVM for the total system correspond to a valid state for the total system.

(The first and third condition mean that $\mathcal{M}_a$, when viewed as acting on the space of density operators, is a *completely positive map.*)

If the three conditions above are satisfied, the set of operators $\mathcal{M}_a$ describes a measurement that can be realized in principle. If the prior state is $\mathbf{p}$, the probability of outcome a is given by

$$P(A = a) = q_a \tag{9}$$

and the conditional state $\mathbf{p}_a$ is

$$\mathbf{p}_a = \frac{1}{q_a}\mathcal{M}_a(\mathbf{p}) \,. \tag{10}$$

This is the most general form of the quantum updating rule. Fuchs (2002) has given a reformulation of this rule which makes its relation to classical Bayesian updating very transparent. The need for a specifically quantum updating rule does not, however, arise because Bayesian theory is not fully valid in quantum mechanics. Bayesian inferences are valid in a quantum setting whenever all probabilities refer to well-defined events.

4. QUANTUM PRIORS

If Bayes's rule is used for updating, the posterior generally depends on the prior. The only exception is the case where, in eq. (7), the observed data, $B = b$, rule out all but one hypothesis, $A = a_0$. The posterior is then

$$P(A = a) = \begin{cases} 1 & \text{if } a = a_0\,, \\ 0 & \text{otherwise.} \end{cases} \tag{11}$$

In quantum mechanics, the posterior state $\mathbf{p}_a$ generally depends both on the prior $\mathbf{p}$ and the map $\mathcal{M}_a$. There is a special class of measurements for which the posterior state is essentially independent of the prior state $\mathbf{p}$. To be precise, for these measurements there is a state, $\mathbf{a}$, such that

$$\frac{1}{q_a}\mathcal{M}_a(\mathbf{p}) = \mathbf{a} \tag{12}$$

for all states $\mathbf{p}$ for which $q_a = |\mathcal{M}_a(\mathbf{p})|_1$ is nonzero, *i.e.*, for all states for which the outcome a has nonzero probability. In this case, one says that the measurement *prepares the state* $\mathbf{a}$. An example of state preparation is provided by a polarizing beamsplitter, which prepares a photon in one of two polarisation states, say vertical or horizontal, conditional on the direction in which the photon is found travelling after leaving the beamsplitter.

As mentioned before, all components of the vector $\mathbf{a}$ are less than 1. This means that, unlike the classical situation of Eq. (11), the prior-independent probabilities resulting from quantum state preparation are nontrivial, *i.e.*, are less than 1 and generally greater than 0.

State preparation plays a major role in most expositions of quantum mechanics. Because the prepared state does not depend on the prior state, it may be tempting to regard the former as objective, strictly determined by the preparation device and the observed outcome of the measurement. This view, however, is not mandated by the quantum formalism. Although the prepared state is independent of the prior state, it depends on prior beliefs in the form of the linear operator $\mathcal{M}_a$. It follows that all quantum states, including those prepared by a preparation device, can be regarded as representing Bayesian degrees of belief.

5. REPEATED QUANTUM TRIALS: DE FINETTI THEOREMS

An important part of experimental physics is concerned with drawing inferences from results of repeated trials. The treatment of repeated trials in Bayesian theory is enormously facilitated by de Finetti's representation theorem (see, *e.g.*, Bernardo and Smith, 1994) which shows that, given a prior judgment of exchangeability, frequency data are the only data relevant for updating. The de Finetti theorem states that an infinitely exchangeable probability distribution for N trials can be written as a weighted mixture of i.i.d. distributions, *i.e.*, a weighted mixture of products of identical single-trial distributions.

Similarly, repeated quantum trials can be analyzed in terms of the quantum de Finetti theorem (Hudson and Moody, 1976; Caves, Fuchs and Schack, 2002a; Koenig and Renner, 2004). If quantum states are represented as vectors of probabilities of outcomes for a fiducial POVM, *i.e.*, as points on the simplex, the de Finetti theorem for quantum states is very similar to the classical de Finetti theorem. When quantum states are represented in this way, the definition of exchangeability is the same for quantum states and for probability distributions. The quantum de Finetti theorem states that an infinitely exchangeable quantum state can be written as a weighted mixture of products of identical states. This differs from the original, classical, de Finetti theorem only in so far as an exchangeable multi-trial quantum state can be written as a weighted mixture of identical single-trial distributions chosen from the subset of the simplex representing valid quantum states, rather than the set of all single-trial distributions as in the classical case.

The quantum de Finetti theorem looks particularly elegant if it is written in terms of density operators. The statement is that any exchangeable state, $\rho^{(N)}$, of N systems, can be written in the form

$$\rho^{(N)} = \int \rho \otimes \ldots \otimes \rho \, P(\rho) d\rho \,, \tag{13}$$

where $P(\rho)d\rho$ is a measure on the space of single-system density operators and $\rho \otimes \ldots \otimes \rho$ is a product of N identical single-system density operators. This theorem solves the problem of what prior multi-trial quantum states allow one to make efficient use of frequency data. It is used implicitly in the analysis of many quantum experiments.

The classical de Finetti theorem does, of course, apply to conditional probability distributions. Given the similar role played by conditional probabilities in Bayes's rule on the one hand, and by quantum operations in the quantum updating rule on the other hand, one might hope that there exists a de Finetti-type representation theorem for quantum operations. This is indeed the case (Fuchs, Schack and Scudo, 2004; see also Fuchs and Schack, 2004). The de Finetti theorem for quantum operations is the basis for *quantum process tomography*. Quantum process tomography uses experimental data to update the quantum operations assigned to a measurement device. The de Finetti theorem for quantum operations solves the problem of what prior multi-trial quantum operations allow one to make efficient use of frequency data in quantum process tomography.

6. COLLAPSE OF THE WAVEFUNCTION

If a quantum particle is in a pure state, its state can be expressed in the form of a *wavefunction*, $\psi(\vec{x})$, whose square $|\psi(\vec{x})|^2$ is the probability density of finding the particle, in a position measurement, at $\vec{x} \in \mathbb{R}^3$. A wavefunction can be highly delocalized, especially if the particle in question is a photon. There is nothing in principle preventing a wavefunction from extending over distances of lightyears. If the particle is detected at a particular point $\vec{x}$, the updated wavefunction is localized near $\vec{x}$. Updating from a delocalized prior wavefunction to a localized posterior wavefunction is often called *collapse of the wavefunction*. Because detecting a particle at one place can instantly change the value of the wavefunction at a distant location, collapse of the wavefunction is a major problem for a view that regards wavefunctions as objectively real.

If wavefunctions are interpreted as representing Bayesian degrees of belief, however, wavefunction collapse is a perfectly natural consequence of correlations. Whenever there is a correlation between two random variables, A and B, describing the outcomes of observations at two different locations, x_A and x_B, an observation of the value of A at x_A can instantly change the probabilities for the possible outcomes of an observation of B at x_B.

In the simplest case, A and B are binary random variables. Assume, *e.g.*, the joint probabilities

$$\begin{aligned} P(A=0, B=0) = P(A=1, B=1) = 1/3\,, \\ P(A=0, B=1) = P(A=1, B=0) = 1/6\,. \end{aligned} \tag{14}$$

The prior (marginal) probability of observing $B = 0$ at x_B is $P_{\text{prior}}(B = 0) = 1/2$. If $A = 0$ is observed at x_A, the probability of observing $B = 0$ at x_B changes instantaneously to $P(B = 0) = P(B = 0|A = 0) = 2/3$. What changes here is an agent's degree of belief; nothing changes at x_B.

All of the above remains valid when A and B are the outcomes of quantum measurements at x_A and x_B. Any change in the (marginal) quantum state of the system at x_B resulting from the observation of the outcome $A = a$ at x_A is a change of a physicist's degrees of belief. Nothing changes at x_B.

This said, correlations between quantum measurement outcomes at x_A and x_B can have surprising and interesting properties, which are systematically studied in the theory of entanglement (see, e.g., Nielsen and Chuang, 2000). Most notably, quantum correlations can violate the *Bell inequalities* (see, e.g., Peres, 1978; Mermin, 1985). For the present paper, the most important consequence of the Bell inequalities is that outcomes of quantum measurements have no preassigned values (Peres, 1978). As long as this is taken into account, in the way explained in Sections 2 and 3, no paradoxes will arise in a Bayesian analysis of quantum correlations.

7. NO CLONING AND TELEPORTATION

The no cloning theorem (Wootters and Zurek, 1982) and quantum teleportation (Bennett *et al.* 1993) are famous results in quantum information theory. Both are counterintuitive if quantum states are regarded as having an objective reality, and both have natural explanations in the Bayesian approach. (For a systematic exploration of Bayesian explanations of quantum phenomena, see Spekkens, 2004.)

The no cloning theorem is an important result, with far-reaching consequences for quantum cryptography. From a Bayesian perspective, it is not surprising, however. A very simple classical version of the no cloning theorem goes as follows. Let A

and B be two independent binary random variables distributed as

$$P(A=0)=p\,,\quad P(B=0)=1, \tag{15}$$

where $0<p<1$. Then there do not exist functions f and g such that the random variables $F=f(A,B)$ and $G=g(A,B)$ are independent and distributed as

$$P(F=0)=p\,,\quad P(G=0)=p, \tag{16}$$

i.e., such that the joint distribution of F and G is a product of two identical copies of the original distribution of A (implying in particular that $P(F=0,G=0)=p^2$). It is thus impossible to clone a probability. This is easy to prove, but it is also clear from a Bayesian perspective why cloning would be absurd. Repeating the cloning process indefinitely would result in a sequence of cloned random variables. Making any single-trial probability assignment would then amount to assigning an i.i.d. to a sequence of clones, which would indeed be incomprehensible.

The quantum no cloning theorem is a straight generalization of the above. Given one system in an unknown quantum state, say represented by an unknown probability vector $\mathbf{p}$ in the subset of the simplex corresponding to valid quantum states, there is no physical process that puts two systems into the joint product state $\mathbf{p}\otimes\mathbf{p}$. Quantum states are represented by probability distributions, and probability distributions cannot be cloned. If one accepts the main thesis of this paper, namely that quantum states represent Bayesian degrees of belief, cloning of quantum states would be absurd.

Quantum teleportation provides another compelling example for the conceptual clarity of the Bayesian approach to quantum mechanics. Teleportation is a protocol involving three parties, often named Victor, Alice and Bob. It is usually described as follows. When Alice and Bob last met, they prepared a pair of particles in a particular highly correlated state (a *maximally entangled state*). Alice kept one of the particles, and Bob kept the other one. At present Bob is at a distant location, and Alice and Bob can communicate only via a classical channel, *i.e.*, they can exchange classical bits.

At this stage Victor prepares a further particle in a state $\mathbf{p}$ and hands it to Alice, without giving her any information about $\mathbf{p}$. Alice has now in her possession two particles, the one handed to her by Victor and the one correlated with Bob's particle. She performs a certain joint measurement (called a *Bell measurement*) on these two particles. The measurement has four equiprobable outcomes, labeled 00, 01, 10 and 11 for simplicity. If the measurement outcome is 00, the resulting state of Bob's particle is $\mathbf{p}$. For each of the three other outcomes, the resulting state of Bob's particle is related to $\mathbf{p}$ by a simple standard transformation. Alice now sends to Bob the two classical bits needed to inform him about the measurement outcome, and Bob performs the corresponding transformation on his particle. As a result of this procedure the state of Bob's particle is $\mathbf{p}$, irrespective of the outcome of Alice's measurement. It is then said that the state $\mathbf{p}$ has been *teleported* from Alice's location to Bob's location.

The term teleportation suggests that something has been transported from Alice's location to Bob's location. A moment's thought reveals, however, that the net effect of the teleportation process is that Victor's probability distribution $\mathbf{p}$, which initially encoded his degrees of belief about measurement outcomes for his original particle, now encodes his degrees of belief about measurement outcomes for Bob's particle. Viewed like this, the term teleportation is clearly misleading.

Here is a simple classical version of teleportation, which is adapted from a thought experiment invented by van Enk and Fuchs (private communication) and independently by Cohen (2006); see also Spekkens (2004) for a more elaborate toy version of teleportation. Consider three binary random variables, X, A and B. Initially, A and B are perfectly correlated,

$$P(A=0, B=0) = P(A=1, B=1) = 1/2\,. \tag{17}$$

Victor's beliefs about X are expressed by the probability

$$P(X=1) = p\,, \tag{18}$$

(the value of p is unknown to Alice and Bob) and his probability for B is

$$P(B=1) = 1/2\,. \tag{19}$$

Now Alice measures the value of the random variable

$$G = X \oplus A\,, \tag{20}$$

i.e., $G = 0$ if $X = A$ and $G = 1$ otherwise. Alice sends the result of her measurement, a single bit of information, to Bob. If Bob receives the result $G = 1$, he inverts B, *i.e.*, he applies the transformation $B \to 1 - B$ to his random variable; otherwise he does nothing.

Finally, Alice tells Victor that the procedure has been completed. This causes Victor to update his probability for B to $P(B = 1) = p$. The value p has been 'teleported' from Alice's location to Bob's location.

This toy version of teleportation is trivial, and closely analogous to quantum teleportation. Why does quantum teleportation generate so much interest? The reason should be this: For the toy version of teleportation given above, it is possible to construct an explanation in terms of preassigned values. If $G = 0$, the values preassigned to X and B are the same; if $G = 1$, inverting B will make them the same. Therefore Victor's posterior probability for B trivially equals his prior for X. This sort of explanation is not available in quantum mechanics. Quantum teleportation cannot be explained in terms of an underlying (local) realistic mechanism. In quantum mechanics, measurement outcomes have no preassigned values. As soon as one accepts this point, however, the Bayesian approach allows one to understand quantum teleportation as a straight generalization of the classical toy version given above.

Quantum mechanics is Bayesian probability theory without the safety net of preassigned values.

REFERENCES

Balian, R. (1991). *From Microphysics to Macrophysics.* Berlin: Springer-Verlag.

Barnum, H., Caves, C. M., Finkelstein, J., Fuchs, C. A. and Schack, R. (2000). Quantum probability from decision theory? *Proc. R. Soc. Lond. A* **456**, 1175–1182.

Bennett, C. H., Brassard, G., Crépeau, C., Jozsa, R., Peres, A. and Wootters, W. K. (1993). Teleporting an unknown quantum state via dual classical and Einstein–Podolsky–Rosen channels. *Phys. Rev. Lett.* **70**, 1895–1899.

Bernardo, J. M. and Smith, A. F. M. (1994). *Bayesian Theory.* Chichester: Wiley.

Brun, T. A., Caves, C. M. and Schack, R. (2001). Entanglement purification of unknown quantum states. *Phys. Rev. A* **63**, 042309.1–042309.10.
Caves, C. M., Fuchs, C. A. and Schack, R. (2002a). Unknown quantum states: The quantum de Finetti representation. *J. Math. Phys.* **43**, 4537–4559.
Caves, C. M., Fuchs, C. A. and Schack, R. (2002b). Conditions for compatibility of quantum state assignments. *Phys. Rev. A* **66**, 062111.1–06211.11.
Caves, C. M. and Schack, R. (2005). Properties of the frequency operator do not imply the quantum probability postulate. *Ann. Phys.* **315**, 123–146.
Cohen, O. (2006). Classical teleportation of classical states. *Fluctuation and Noise Letters* **6**, C1-C8.
Fuchs, C. A. (2002). Quantum mechanics as quantum information (and only a little more). *Arxiv.org e-print* quant-ph/0205039.
Fuchs, C. A. and Peres, A. (2000). Quantum theory needs no interpretation. *Physics Today* **53** (3), 70–71.
Fuchs, C. A. and Schack, R. (2004). Unknown quantum states and operations, a Bayesian view. *Quantum State Estimation* (M. G. A. Paris and J. Řeháček, eds.) Berlin: Springer-Verlag, 147–187.
Fuchs, C. A., Schack, R. and Scudo, P. F. (2004). A de Finetti representation theorem for quantum process tomography. *Phys. Rev. A* **69**, 062305.1–062305.6.
Hardy, L. (2001). Quantum theory from five reasonable axioms. *Arxiv.org e-print* quant-ph/0101012.
Hudson, R. L. and Moody, G. R. (1976). Locally normal symmetric states and an analogue of de Finetti's theorem. *Z. Wahrscheinlichkeitstheorie verw. Geb.* **33**, 343–351.
Jaynes, E. T. (1957). Information theory and statistical mechanics. *Phys. Rev.* **106**, 620–630.
Koenig, R. and Renner, R. (2005). A de Finetti representation for finite symmetric quantum states. *J. Math. Phys.* **46**, 122108.1–122108.23.
Mermin, N. D. (1985). Is the moon there when nobody looks? Reality and the quantum theory. *Physics Today* **38** (4), 38–47.
Nielsen, M. A. and Chuang, I. L. (2000). *Quantum Computation and Quantum Information.* Cambridge: University Press.
Peres, A. (1993). *Quantum Theory: Concepts and Methods.* Dordrecht: Kluwer.
Peres, A. (1978). Unperformed experiments have no results. *Am. J. Phys.* **46**, 745–747.
Schack, R. (2003). Quantum theory from four of Hardy's axioms. *Found. Phys.* **33**, 1461–1468.
Schack, R., Brun, T. A. and Caves, C. M. (2001). Quantum Bayes rule. *Phys. Rev. A* **64**, 014305.1–014305.4.
Spekkens, R. W. (2004). In defense of the epistemic view of quantum states: a toy theory. *Arxiv.org e-print* quant-ph/0401052.
Wootters, W. K. and Zurek, W. H. (1982). A single quantum cannot be cloned. *Nature* **299**, 802–803.

DISCUSSION

INGE S. HELLAND (*University of Oslo, Norway*)

We argue that there are similarities between the quantum mechanical formalism and concepts of statistics. Thus the fact that we have similar ideas about Bayesianism can be extended in many ways; some of this seems to be of practical value.

Introduction. Perhaps the clearest way to see that a quantummechanical state must be associated by some sort of subjective probability, is to give a brief description of Albert Einstein's EPR phenomenon, as modified by David Bohm. Let two electrons be bound together in some state where they have total spin 0, and let them thereafter depart to two distant places. By performing a measurement of spin component of particle A in some chosen direction a, the state of the other particle B can immediately be deduced. Note that different choices of direction a give different state predictions for particle B *as perceived by by the observer at* A.

To Einstein, this phenomenon was so astonishing that he declared quantum mechanics to be incomplete, and was searching for the rest of his life for an alternative theory. Today, most physicists believe in quantum theory on the basis of the long range of predictions by the theory that have been verified by experiments. However, to people outside the physics community, the abstract quantum formulation serves to alienate them from the whole area.

Thus Rüdiger Schack is to be welcomed for trying to formulate several phenomena of quantum theory in terms of simple probabilities, but I am afraid that most statistical readers of his paper will be a little confused by the abstract notions that he introduces: What is a density operator? How can one interprete a POVM? Can one understand a quantum operation in some way?

Let me first stress what we agree about: The probabilties of quantum mechanics can and must be interpreted in a Bayesian way. Fuchs and Peres (2000) expressed this in a clear manner: *Attributing reality to quantum states leads to a host of' 'quantum paradoxes'.* Rüdiger himself says in his conclusion: *Quantum mechanics is Bayesian probability theory without a safety net of preassigned values.* In fact, these ideas go back to Niels Bohr (cited from Petersen, 1985): *It is wrong to think that the task of physics is to find out how nature is. Physics concerns what we can say about nature.*

A statistical basis? Fuchs (2002) argues in various ways that all parts of the quantum formulation should be derivable from quantum information. This is certainly an interesting programme, especially in the light of the interest that has been aroused around the various parts of quantum information in the latter years (Nielsen and Chuang, 2000).

However, most parts of quantum information is derivable from the the ordinary quantum axioms, so in a way this is to derive the theory from some of its consequences. Personally, I have found it more interesting to strive for some unity of science through trying to derive the axioms of quantum mechanics from something related to ordinary statistical theory. The results of this endeavour are given in Helland (2005, 2006). I will sketch the main theory here; then I will use it in the next section to give a statistical characterization of some of Rüdiger's concepts.

First of all, in ordinary statistics it is essential to distinguish between observations and parameters. The parameters are the theoretical state quantities, usually written by Greek letters. The purpose of most empirical investigations is to arrive at statements concerning one or more parameter. The observations are ordinarily modelled by conditional probability distributions, given a set of parameters. The Bayesian element is a probability distribution of the parameters, either prior or posterior, the latter found by Bayes' formula.

In physics this can be implemented by the fact that every measurement needs a measurement apparatus, and the measurements have a probability distribution, given some parameters. If the measurement apparatus is perfect, the parameter is numerically equal to the observation. Nevertheless, it is useful to keep the distinction between the two: The parameter is the mental concept that we need in asking questions about nature. The fact that these questions concern theoretical parameters, not the data themselves, may be related to the Bayesian view as formulated in the introduction.

An important concept that we need in this way of arriving at quantum theory is that of complementary parameters. This is related to the concept of counterfactuals, and several largescale examples can also be given. A simple example is the expected survival time for a single patient at a single time under two different treatments. A quantum theoretical example may be the spin component of a single particle in two different directions. Mathematically, this can be formulated in terms of an inaccessible total parameter ϕ and observable parameters $\lambda^a(\phi)$. Rüdiger's statement about quantum mechanic's lack of a safety net of preassigned values may be related to inaccessible components of the total parameter.

To formulate quantum mechanical statements from this point of departure, one must include complementarity, use maximal focusing in defining parameters, and make use of group symmetry.

In fact, for this approach symmetry seems to be very important. We first assume a total, inaccessible parameter ϕ and a basic group G defined upon the space of this parameter. Then the maximal measurable parameters λ^a are assumed connected through group elements:

$$\lambda^b(\phi) = \lambda^a(\phi g_{ab}). \tag{21}$$

Let now $\mathbf{L}$ be $L^2(\Phi, \mu)$, where Φ is the total parameter space, and μ is the right Haar measure of G, and let $\mathbf{H^a}$ be the subspace of $\mathbf{L}$ consisting of functions of the form $\tilde{f}(\lambda^a(\phi))$. Then these spaces are connected together by

$$\mathbf{H}^b = U(g_{ab})\mathbf{H}^a, \tag{22}$$

where U is the regular representation of the group G.

This means that we can take one of these spaces as our Hilbert space. The unit vectors of this space will in the usual quantummechanical way be eigenvectors of some natural self-adjoint operators. It is assumed that the parameter is maximal in the sense that it is not a function of another observable parameter.

This implies the focus aspect. Specifically, in Helland (2005, 2006) it is proved under general assumptions that these unit vectors can be interpreted as (a) A question: What is the value of λ^a ? (b) together with an answer: $\lambda^a = \lambda^a_k$.

More precisely, (a) and (b) together are in one-to-one correspondence with a ray in the Hilbert space represented by a unit vector v^a_k. Note that every unit vector can be given such an interpretation; in the ordinary formulation one may assume that the relevant self-adjoint operator has some natural physical interpretation as an observable.

Finally, the quantum probabilities are introduced through Born's formula:

$$P(\lambda^b = \lambda^b_i | \lambda^a = \lambda^a_k) = |v^{a\dagger}_k v^b_i|^2. \tag{23}$$

This formula is proved by using symmetry arguments together with a recent version of Gleason's theorem due to Busch (2003) in Helland (2006). A related derivation replacing the symmetry assumptions by a Dutch-book argument can be found in Caves, Fuchs and Schack (2002c).

On the formalism. I have already claimed that unit vectors v_i^b can be interpreted as questions plus answers. These are called pure state vectors in quantum mechanics. Alternatively, pure states may be characterized by one-dimensional projection operators

$$\rho_i^b = v_i^b v_i^{b\dagger}. \tag{24}$$

In a true Bayesian spirit, states may be uncertain, that is, characterized by probabilities p_i^b. This gives a density operator

$$\rho^b = \sum_i p_i^b \rho_i^b. \tag{25}$$

A special case is when we have an earlier perfect experiment on a parameter λ^a, and the probabilities p_i^a are given as transition probabilities from Born's formula. Other cases are when these probabilities are prior probabilities, or posteriors, given some data.

A formally related, but in principle very different case is when we observe data and just have a statistical model $p(x|\lambda^b)$. Since we are working with discrete parameters here, it is natural to also let the data x be discrete. We can then define new operators, called effects, by

$$E_x = \sum_i p(x|\lambda_i^b)\rho_i^b. \tag{26}$$

Since the ρ_i^b are orthogonal operators, all effects will have eigenvalues in the interval $[0,1]$. Furthermore, since $\sum_x p(x|\lambda_i^b) = 1$, we have $\sum_x E_x = I$. Thus the effects constitute what in the quantum mechanical literature is called a positive operator valued measure (POVM). From a statistical point of view, E_x has a very definite interpretation: It is the uncertain answer to some focused question given some statistical model and some data.

Now we are ready to combine the concepts above. If the prior probability distribution of the parameter λ^b is given by p_i^b, thus characterized by the density matrix ρ, then the marginal distribution of the data x is given by

$$p(x) = \sum p(x|\lambda_i^b)p_i^b = \mathrm{tr}(E(x)\rho). \tag{27}$$

The proposal given in the paper by Rüdiger Schack is to characterize the state of a quantum system by these numbers. This is mathematically possible, but rather indirect from our point of view. The point is that for a quantum system described by a Hilbert space of dimension d, it is possible to find a measurement given by d^2 data points x so that the marginal probabilities of these data points uniquely determines the quantum state. The name fiducial measurement for this situation has interesting historical statistical connotations.

Assume now that the prior state is given by the density $\rho^b = \sum_j p_j^b \rho_j^b$. Then by Bayes' rule the posterior state will be

$$\sum_j p_{j|x}^b \rho_j^b = \sum \frac{p_j^b p(x|\lambda_j^b)}{p(x)}\rho_j^b = \frac{A\rho^b A^\dagger}{\mathrm{tr}(E^b(x)\rho^b)}, \tag{28}$$

where A is diagonal in the basis system $\{v_j^b\}$ with diagonal elements $\sqrt{p(x|\lambda_j^b)}$. In more complicated situations where measurement takes place via socalled ancilla systems, Fuchs (2002) and others have shown that the posterior state density takes the form

$$\frac{1}{\mathrm{tr}(E^b(x)\rho^b)} \sum_k A_k^b \rho^b A_k^{b\dagger}, \tag{29}$$

where

$$\sum A_k^{b\dagger} A_k^b = E^b(x). \tag{30}$$

This can be taken as a starting point for the theory of quantum operations, which in Nielsen and Chuang (2000) is developed in three equivalent forms. One of these forms leads to the linear operator $\mathcal{M}_a(\mathbf{p})$ introduced in the paper by Rüdiger.

A macroscopic example. Consider four different medicaments with expected responses μ_A, μ_B, μ_C and μ_D. Imagine also a mixture of the four with expected response $\frac{1}{4}(\mu_A + \mu_B + \mu_C + \mu_D)$. Look first at a randomized pairwise experiment A with parameter

$$\lambda_A = \mathrm{sign}[\mu_A - \frac{1}{4}(\mu_A + \mu_B + \mu_C + \mu_D)]. \tag{31}$$

For simplicity we will assume an ideal experiment without experimental error. Assume now that the above experiment has been performed with the result $\lambda^A = +1$. Consider next the counterfactual situation: Imagine that we with the same experimental units instead had performed an experiment B with parameter

$$\lambda_B = \mathrm{sign}[\mu_B - \frac{1}{4}(\mu_A + \mu_B + \mu_C + \mu_D)]. \tag{32}$$

Then what is the probability $P(\lambda_B = +1|\lambda_A = +1)$? The point is that this situation posesses enough symmetry to fit into the Born formula situation. Introducing vectors using the symmetry and performing the Born formula calculation, gives the answer that the probability is 1/3 (Helland, 2005).

Now we can leave the counterfactual situation by the following argument: Using experimental units from the same population should give the same experimental result when experimental error is disregarded. Thus we have two real (ideal) experiment connected by the probability 1/3, calculated by a quantum mechanical formula.

Note that the result is different from what is found from an ordinary calculation using natural Bayes type priors. For instance, if $\mu_A, ..., \mu_D$ are independent $N(\mu, \sigma)$, then the probability above is independent of μ and σ, and a numerical calculation gives 0.43. By a simple limiting argument, this result also holds for flat priors.

A natural conclusion may seem to be that ordinary parametric priors are not applicable for calculating probabilities between experiments involving the same parameters. It would would be interesting to explore this further in the context of experimental design.

Science and language. Quantum mechanics has been characterized as the greatest scientific success of our time, then measured in terms of its empirically verified predictions in several areas. In fact, this is what justifies the whole theory: It is made for empirical predictions.

Mathematical statistics is very different, but even so the two disciplines have one thing in common: Both have great success empirically, statistics in a large number of applications in very different fields.

Yet people speak completely different language in the two disciplines.

Scientific language and culture has developed though the effort of great human beings, but they were human beings. The founding fathers of quantum theory struggled through decades to find the best formulations. The founders of statistical science fought vigourously against each other on fundamental questions.

The language and the culture which exists today is a result of an historical development. So it is a valid question to ask: If the history had been different, could we then have developed a common culture for empirical prediction and inference? I don't mean that we necessarily should have used the same language; it seems very reasonable to use the Hilbert space language for the microworld and the parameter-and-observation language in large scale applications in most practical case. Yet we ought to have the possibility to translate in order to understand each other's language and learn from it. Much of what has been said above has been to argue for the statement that such translation is possible, interesting and fruitful.

It may be of some interest to contemplate the analogy with Gödel's incompleteness theorem: Every rich enough formalism contains questions that cannot be answered within the formalism. I do not know exactly where the boundaries for quantum theory go here, but I know it very well for mathematical statistics: Every practical application of a reasonable size contains questions for which one cannot get an answer by appealing to mathematical theory; one has to use intuition. A merging of cultures with exchange of ideas may in the long run lead to richer formalisms with broader boundaries.

In fact, I am not at all foreign for the idea that one can learn something from neighbouring cultures which is of value also in practical situations. For instance, look at the foundation of ordinary Bayesian theory: On one of the first pages of Bernardo and Smith (1994) you find the definition of a decision problem, where the available actions is a well-defined set. Many practical decisions are not taken in this way: Decisions are taken in a context, and part of the decision process may be to change the context, for instance by changing focus. The focusing part is not unrelated to a part of the suggested foundation of quantum mechanics, as sketched above.

REPLY TO THE DISCUSSION

I would like to thank Inge Helland for his thoughtful comments. I agree that it is an important goal to make the quantum formalism accessible to the statistics community. In his discussion paper, Inge Helland provides a more detailed introduction to some aspects of the quantum formalism, which I welcome. I certainly agree with him that representing quantum states as real vectors on the probability simplex is rarely convenient for actual calculations. The same is true for the abstract description of quantum updating in terms of completely positive maps.

Important as explaining the quantum formalism to statisticians is, however, it was not the purpose of my paper. The paper uses only as much formalism as is needed to make its main points. For instance, in Section 2, quantum states and informationally complete POVMs are introduced in order to show that any quantum state can be represented by a probability distribution. Similarly, Section 3 introduces the quantum updating rule in order to identify the prior in quantum state preparation.

What I have tried to demonstrate is that all probabilities in quantum mechanics can be viewed as Bayesian degrees of belief, and that the apparatus of Bayesian statistics is fully applicable to quantum mechanics as long as it is understood that it is illegal to condition on outcomes of unperformed measurements. To support this claim, my paper establishes that (i) quantum states are in one-to-one correspondence with points on a subset of the probability simplex; (ii) Bayes's rule is valid for the outcomes of measurements that are actually carried out; (iii) quantum probabilities always depend on prior beliefs; (iv) repeated quantum trials can be analyzed using the quantum de Finetti theorems. A great advantage of the Bayesian approach to quantum mechanics is its conceptual simplicity and clarity. This is exemplified by the almost trivial Bayesian account of wavefunction collapse, no-cloning and quantum teleportation.

ADDITIONAL REFERENCES IN THE DISCUSSION

Busch, P. (2003). Quantum states and generalized observables: A simple proof of Gleason's theorem. *Phys. Rev Lett.* **91** 120403.1–120403.4.

Caves, C. M., Fuchs, C. A. and Schack, R. (2002c). Quantum probabilities as Bayesian probabilities. *Physical Review* A **65** 022305.1–022305.6.

Helland, I.S. (2005). Quantum theory as a statistical theory under symmetry. *Foundations of Probability and Physics* **3** (A. Khrennikov, ed.), 127–149.

Helland, I. S. (2006). Extended statistical modeling under symmetry; the link towards quantum mechanics. *Ann. Statist.* **34**, 42–77.

Petersen, A. (1985). The philosophy of Niels Bohr. *Niels Bohr: A Centenary Volume* (A. P. French and P. J. Kennedy, eds.). 299–310.

BAYESIAN STATISTICS 8, pp. 471–490.
J. M. Bernardo, M. J. Bayarri, J. O. Berger, A. P. Dawid,
D. Heckerman, A. F. M. Smith and M. West (Eds.)

Fast Bayesian Shape Matching Using Geometric Algorithms

Scott C. Schmidler
Duke University, USA
schmidler@stat.duke.edu

Summary

We present a Bayesian approach to comparison of geometric shapes with applications to classification of the molecular structures of proteins. Our approach involves the use of distributions defined on transformation invariant shape spaces and the specification of prior distributions on bipartite matchings. Here we emphasize the computational aspects of posterior inference arising from such models, and explore computationally efficient approximation algorithms based on a geometric hashing algorithm which is suitable for fully Bayesian shape matching against large databases. We demonstrate this approach on the problems of protein structure alignment, structural database searching, and structure classification. We discuss extensions to flexible shape spaces developed in previous work.

Keywords and Phrases: Bioinformatics; Geometric Hashing; Protein Structure; Shape Analysis.

1. INTRODUCTION

Analysis of data obtained as measurements on geometric objects arises in a number of fields, including image processing and computer vision, biological anthropology and anatomy, molecular biology and chemistry, and mechanical engineering to name but a few. Such data differs from standard high dimensional, multivariate data by the necessity of accounting for natural invariances with respect to geometric transformations such as rigid body motions, viewpoint and projection, and scaling. Treatment of such problems falls under the heading of *statistical shape analysis* and a several different directions of theoretical and methodological work have been developed by various authors (Bookstein, 1991; Small, 1996; Dryden and Mardia, 1998; Kendall *et al.*, 1999; Lele and Richtsmeier, 2001). In the current paper we focus on *landmark* methods for shape analysis, which represent objects by a set of n landmark points in d dimensions. An observed configuration is then given by an $n \times d$ matrix X, with $x_i \in \mathbb{R}^d$ the ith row of X representing the coordinates of the ith landmark.

2. SHAPE AND SHAPE SPACE

For simplicity and concreteness we consider in this paper only *rigid-body shape* (or *size-and-shape*) arising from invariance under special Euclidean transformations. The *shape* of X denoted by $[X]$ is then defined as the equivalence class

$$[X] = \{XR + \mathbf{1}\mu' \mid R \in \mathbf{SO}(d),\ \mu \in \mathbb{R}^d\}$$

which is the orbit of X under the special Euclidean group $SE(d)$. In general $SE(d)$ may be replaced with a more general class of transformations; common examples include similarity group (including scaling), the affine group, projective transformations, various smooth nonlinear mappings, and piecewise combinations (Dryden, 1999; Schmidler, 2006).

The set of equivalence classes arising from all configurations of n landmarks in d dimensions is called the *(size-and-)shape space* and denoted by $\mathrm{S}\Sigma_n^d$. Shape analysis involves the analysis of observed configuration data in this space, which is invariant under rigid-body transformations. Distances in shape space for two observed configurations with corresponding landmarks may be obtained via the *Procrustes distance*,

$$d_P^2(X,Y) = \min_{\substack{\mu \in \mathbb{R}^3 \\ R \in \mathbf{SO}(3)}} \|Y - (XR + \mathbf{1}\mu')\|_F^2 = \|X_c\|^2 + \|Y_c\|^2 - 2\mathrm{tr}(D)$$

where $\|X\|_F$ denotes Frobenious norm, X_c denotes centering, and $Y'X = UDV'$ is a singular value decomposition. Procrustes distance and related metrics provide one approach to development of distribution theory and multivariate analysis methodology for statistical analysis of shape data (*e.g.*, Dryden and Mardia, 1998).

To date such analyses have largely been predicated on the existence of a one-to-one landmark correspondence between X and Y. However, an important aspect of shape matching in many applied problems is the need to identify such a correspondence, or to determine if one exists. We have previously developed a Bayesian approach to this problem.

3. BAYESIAN SHAPE MATCHING

We have recently developed a Bayesian approach to shape matching motivated by problems in structural proteomics (Schmidler, 2006; Rodriguez and Schmidler, 2006a; Wang and Schmidler, 2006; Rodriguez and Schmidler, 2006b). Let $X_{n\times 3}$ and $Y_{m\times 3}$ be two configuration matrices. We define an *alignment* between X and Y to be a pair $A = [M, \theta]$ where θ is a transformation (here $\theta = (R, \mu) \in SE(3)$) and M is a bipartite matching. Denote the set of matchings by $\mathcal{M}_{n,m}$; a convenient representation for $M \in \mathcal{M}_{n,m}$ is a match matrix $M_{n\times m} = [m_{ij}]$ such that $m_{ij} = 1$ if landmarks X_i and Y_j are matched, and 0 otherwise. Denote by X_M and Y_M the p non-zero rows of MX and $M'Y$ respectively, giving the coordinates of the matched residues.

We have developed a Bayesian approach to shape alignment which defines a prior distribution on alignments $P(M)$ and obtains the posterior distribution:

$$P(M \mid X, Y) = \frac{P(X, Y \mid M)P(M)}{\sum_M P(X, Y \mid M)P(M)}$$

under a likelihood defined on shape space given by the joint density:

$$\begin{aligned} P(X, Y \mid M) &= P(Y_M \mid X_M) P(X_M) P(Y_{\bar{M}}) P(X_{\bar{M}}) \\ &= (2\pi\sigma^2)^{-\frac{3p}{2}} \exp -\frac{1}{2\sigma^2} d_P^2(X_M, Y_M) \prod_{y_i \in Y_{\bar{M}}} f(y_i; \lambda) \prod_{x_i \in X_{\bar{M}}} f(x_i; \lambda) \quad (1) \end{aligned}$$

where $\|X\|_F = \mathrm{tr}(X'X)$ is the Frobenius norm and $f(\,\cdot\,; \lambda)$ is a one-parameter density for unmatched landmarks with λ having an interpretation as a soft threshold for landmark deviations. This density corresponds to an additive model on shape space

$$[Y_M] = [X_M] + \epsilon \qquad\qquad \epsilon \sim N(\mathbf{0}, \sigma^2 I)$$

with ϵ_p a matrix-normal random perturbation; other forms for the error distribution can also be accommodated (Rodriguez and Schmidler, 2006a).

Strictly speaking the density (1) is defined on a tangent-space approximation to shape space, the difference being in the integral required for normalization (see Small (1996); Dryden and Mardia (1998) and Schmidler (2006)); in our case this normalization cannot be ignored.

An alternative approach is to place prior distributions on transformation parameters θ and obtain a posterior distribution over the alignment pair $A = [M, \theta]$, from which the marginal posterior for M may be obtained by integrating out the 'nuisance' parameters θ (Schmidler, 2006; Wang and Schmidler, 2006); see also Green and Mardia (2006) for a related approach developed independently. In comparison, the likelihood defined directly on shape space (1) identifies transformations θ directly with matchings M, but can also be viewed as a profile likelihood for A. Schmidler (2006) considers both approaches for protein registration, including the marginalization approach under affine transformations where integration may be done analytically. The marginalization approach has the advantage of accounting for uncertainty in θ given M. In addition, in some applications of image analysis (*e.g.*, tracking), θ is the parameter of interest while M may be the nuisance parameter. In this paper, we restrict attention to models of the form (1) for concreteness; the computational approach described in the next sections applies equally to both approaches.

3.1. *Priors on Bipartite Matchings and MCMC*

Priors on matchings $P(M)$ may be specified in a variety of ways. Rodriguez and Schmidler (2006a) use a gap-penalty prior:

$$P(M) \propto \exp -(g\, n_g(M) + h \sum_{i=1}^{n_g(M)} l_i(M)) \tag{2}$$

based on affine gap opening and extension penalties used in biological sequence analysis; this is essentially a Markovian process on M. The prior encourages grouping of matches together but requires pre-specification of a landmark ordering, equivalent to assuming a topological equivalence of two proteins.

The advantage of the class of priors (2) is that it enables exact sampling of $P(M \mid \theta)$ via an efficient dynamic programming algorithm, thus producing an efficient Gibbs sampling algorithm for iteratively drawing $P(M \mid \theta)$ and $P(\theta \mid M)$ (Wang and Schmidler, 2006) or efficient Metropolis proposals for the posterior obtained from (1) (Rodriguez and Schmidler, 2006a). However, this MCMC scheme suffers from the combination of facts that $P(M \mid \theta)$ may be multimodal and that M and θ are strongly coupled in the posterior $P(M, \theta \mid X, Y)$; mixing between multiple modes, when they exist, is therefore unacceptably slow. A solution to this is given by (Rodriguez and Schmidler, 2006a) which constructs a library of θ's to a proposal distribution used during the Metropolis sampling; this approach is effective but somewhat computationally intensive and is less efficient for multiple shape alignment (Wang and Schmidler, 2006). The approach described in this paper is in a sense a generalization of this library approach, which may potentially replace MCMC sampling entirely.

In this paper we avoid the order-preserving assumptions of (2) and consider priors on unrestricted matchings. Thus the approach described here is applicable to general shape matching when such *a priori* assumptions are not acceptable. More general priors are also considered briefly in (Rodriguez and Schmidler, 2006a), where other MCMC moves are mixed with the dynamic programming steps. Further work is needed to explore the impact of informative priors on bipartite matchings.

3.2. *Posterior Summaries*

The model given in the previous section allows for the fully Bayesian comparison of geometric objects. A key advantage of this approach is the ability to accounting for uncertainty in the point correspondences and accurately estimate variability in the resulting matchings, which in many cases may be multimodal. Quantities of interest under such a model include distance measures such as the posterior expectation of Procrustes distance or root-mean-square deviation:

$$E(\text{rmsd} \mid X, Y) = \sum_{M \in \mathcal{M}_{n,m}} |M|^{-\frac{1}{2}} d_P(X_M, Y_M) P(M \mid X, Y)$$

or the posterior probability of a match, which may be calculated, *e.g.*, as the marginal posterior probability $P(|M|^{-\frac{1}{2}} d_P \mid X, Y) \leq \delta$ for some δ or the posterior probability that the fraction of landmarks matched is greater than ϵ. Other important posterior summaries include the posterior mode $\hat{M} = \arg\max_M P(M \mid X, Y)$, or the *marginal alignment matrix* which gives the marginal posterior probability of matching any pair of landmarks:

$$P(m_{ij} \mid X, Y) = \sum_{M \in \mathcal{M}_{n,m}} m_{ij} P(M \mid X, Y) \tag{3}$$

integrating out all other parameters in the model. The space $\mathcal{M}$ grows exponentially in n and m, making exact calculation of these quantities infeasible. Under the gap-penalty based prior, Rodriguez and Schmidler (2006a); Wang and Schmidler (2006) provide Markov chain Monte Carlo (MCMC) sampling algorithms for approximation of such posterior quantities, which take advantage of efficient Gibbs steps involving stochastic dynamic programming recursions. However, the combinatorial nature of the space $\mathcal{M}$ and the existence of multiple posterior modes makes MCMC slow. Thus for applications involving large numbers of shape comparisons such as searching molecular structure or image databases for shapes similar to some query shape, the Bayesian approach remains infeasible. Below we explore deterministic computational approximations based on geometric algorithms to address this issue.

4. BAYESIAN SHAPE CLASSIFICATION

In this paper we also consider the problem of shape *classification*, using a simple Bayesian classifier. Wang and Schmidler (2006) describe Bayesian estimation of *mean shape* under the Bayesian framework above, applied to multiple protein structure alignment and analysis of functional conservation. Given a mean shape Z_c for each class $c \in \mathcal{C}$ of interest, we may classify based on the posterior distribution

$$P(Y \in c \mid Y) = \frac{P(Y \mid Z_c)P(c)}{\sum_{c' \in \mathcal{C}} P(Y \mid Z_{c'})P(c')} = \frac{\sum_{M \in \mathcal{M}} P(Y \mid Z_c, M)P(M)P(c)}{\sum_{c' \in \mathcal{C}} \sum_{M' \in \mathcal{M}} P(Y \mid Z_{c'}, M')P(M)P(c')}.$$

This requires calculation of the marginal likelihood $P(Y, Z_c)$ which, as with the posterior summaries above, involves a sum over the exponential space of all possible alignments. More generally, suppose that for each class c of interest we have k_c example shapes $X_1^c, \ldots, X_{k_c}^c$ from class c, and obtain a posterior distribution over Z_c and associated parameters ζ_c represented by a finite set of samples $(Z_c^{(i)}, \zeta_c^{(i)})$ from $P(Z_c, \zeta_c \mid X_1^c, \ldots, X_{k_c}^c)$. Denote the set of examples by $\mathbf{X} = \cup_{c \in \mathcal{C}}\{X_1^c, \ldots, X_{k_c}^c\}$. Then we may approximate the posterior classification probability of new observation Y by

$$\begin{aligned} P(Y \in c \mid Y, \mathbf{X}) &= \frac{\int\int P(Y \mid Z_c, \zeta_c)P(Z_c, \zeta_c \mid X_1^c, \ldots, X_{k_1}^c)\, dZ_c\, d\zeta_c P(c)}{\sum_{c' \in \mathcal{C}} \int\int P(Y \mid Z_c, \zeta_{c'})P(Z_c, \zeta_{c'} \mid X_1^{c'}, \ldots, X_{k_{c'}}^{c'})\, dZ_{c'}\, d\zeta_{c'} P(c')} \\ &\approx \frac{\sum_i \sum_M P(Y \mid Z_c^{(i)}, \zeta_c^{(i)})P(Z_c^{(i)}, \zeta_c^{(i)} \mid X_1^c, \ldots, X_{k_1}^c)P(c)}{\sum_{c' \in \mathcal{C}} \sum_i \sum_M P(Y \mid Z_c^{(i)}, \zeta_{c'}^{(i)})P(Z_c^{(i)}, \zeta_{c'}^{(i)} \mid X_1^{c'}, \ldots, X_{k_{c'}}^{c'})P(c')}. \end{aligned}$$

In each case we require calculation of the marginal likelihood $P(Y \mid Z_c, \zeta_c)$. In fact this is a closely related problem to that of Bayesian shape matching for database search described above, as can be seen by considering the collected posterior samples $\{(Z_c^{(i)})\}_{c \in \mathcal{C}}$ as a database of shapes. Thus the shared problem is that of marginalizing over the combinatorial space of matchings for a large set of target shapes simultaneously.

5. APPROXIMATE POSTERIORS VIA GEOMETRIC HASHING

As described, computation of posterior quantities by MCMC methods suffers from several drawbacks, including the requirement of order-preserving matchings for efficient Gibbs steps, the difficulty of mixing between multiple posterior modes in the combinatorial space of bipartite matchings, and general infeasibility of Monte Carlo methods for calculations which must be repeated thousands or millions of times to search large databases. In this paper we explore the adaptation of an algorithm from the image processing literature, *geometric hashing*, to compute approximate posterior quantities much more efficiently. Our goal is to approximate the marginal likelihood

$$\begin{aligned} P(Y \mid Z_c, \zeta_c) &= \sum_{M \in \mathcal{M}} P(Y \mid M, Z_c, \zeta_c)P(M) \\ &\approx \sum_{M \in \mathcal{M}^*} P(Y \mid M, Z_c, \zeta_c)P(M) \end{aligned}$$

by consideration of some high posterior density set $\mathcal{M}^*$ which can be calculated efficiently.

5.1. *Geometric Hashing*

Geometric hashing is an algorithmic technique developed in the computer vision literature for object recognition and image analysis to rapidly match scenes to a database of models (Wolfson and Rigoutsos, 1997). It has also been applied effectively to alignment and substructure analysis of protein molecules (Nussinov and Wolfson, 1991; Wallace et al., 1996). A key advantage of the approach is the ability to match against a library of models simultaneously in polynomial time. There are several variations on the algorithm; here we describe and implement one of the simplest although not necessarily the most computationally efficient. Here we briefly introduce the algorithm; see above references for more details. In the next section we show how this algorithm may be adapted to perform the Bayesian shape calculations of Section 3 very efficiently.

Geometric hashing begins by representing each object in the database in a hashtable, to which search objects are then compared. All objects are represented as follows: choose three non-collinear landmarks from X denoted by x_a, x_b, and x_c, and define the unique coordinate frame having $s = x_a, x_b, x_c$ lying in the xy-plane with x_a at the origin and x_b on the positive x-axis, and z-axis given by the right-hand rule. Denote by e_1^s, e_2^s, e_3^s the associated orthogonal basis. All other points $X_{[4:n]}$ are represented in this coordinate frame as $x_i - x_0^s = a_i^s e_1^s + b_i^s e_2^s + c_i^s e_3^s$ where x_0^s is the chosen origin; note that the resulting coordinates (a_i^s, b_i^s, c_i^s) are invariant under $SE(3)$. These coordinates are then used to index a location in a table, where the coordinates are stored along with a pointer to the reference set s. This process is then repeated for (a) every ordered subset s of landmark triplets in X, and (b) for every X in the database. This indexing takes $O(n^4)$ processing time per database object, but can be precomputed once offline and then used repeatedly.

To match a newly observed object Y to the database, a landmark triple s is chosen (perhaps randomly) and used to compute coordinates (a_i^s, b_i^s, c_i^s) for the remaining landmarks. These coordinates again serve as indices into the table, where the associated entry contains a list of (point, reference set, object) triplets for matching points in database objects. Each such element of the list is then assigned a 'vote' for the associated (object,reference set) pair. The database (object, reference set) pair receiving the most such votes is considered the best match. This voting can be given a probabilistic semantics and the resulting best match considered to be a maximum likelihood match among objects in the database (Wolfson and Rigoutsos, 1997). The key to the approach is that comparison of a new object Y to an entire database may be done rapidly in order $O(nc^*)$ where c^* is a constant giving the average entry list length, related to the density of points.

6. FULLY BAYESIAN SHAPE MATCHING BY GEOMETRIC HASHING

The geometric hashing technique allows rapid large-scale parallel comparison of object configurations against a database. In the paper we explore the use of this technique for approximation of marginal posterior quantities under the models described in Sections 3.2 and 4.

We consider approximation of marginal posterior quantities such as (3) or 95% credible sets:

$$C_{95} = \{M \in \mathcal{M} : P(M \mid X, Y) \geq c_\alpha\} \tag{4}$$

where c_α = min s.t. $P(C_{95} \mid X, Y) \geq .95$. As mentioned, the original geometric hashing algorithm may be viewed as an approximation of the MAP estimate

$\arg\max_{M \in \mathcal{M}} P(M \mid X, Y)$ under a uniform prior; however, if the posterior distribution $P(M \mid X, Y)$ is multimodal, the MAP estimate may be a poor summary.

6.1. *Priors*

We consider two simple classes of priors. The first are uniform priors on matchings $P(M) = |\mathcal{M}|^{-1}$, where

$$|\mathcal{M}| = \sum_{k=0}^{\min(n_X, n_Y)} \binom{n_X}{k} \frac{n_Y!}{(n_Y - k)!}.$$

The second are *exchangeable* priors:

$$P(M) = f(|M|) = g(k)$$

which depend only on the number of matches but not the precise pattern.

6.2. *Likelihood Bound*

In order to approximate quantities such as high posterior density credible sets (4) and marginal posterior probabilities (3) to within a given accuracy, we must be able to identify a subset $\mathcal{M}^* \subset \mathcal{M}$ of elements with high posterior probability. Suppose we wish to find all alignments M such that

$$Z(X,Y)^{-1} g(k) (2\pi\sigma^2)^{-\frac{3k}{2}} \mathrm{e}^{-\frac{1}{2\sigma^2} d_P^2(X_M, Y_M)} P(y_i \in Y_{\bar{M}}, x_i \in X_{\bar{M}}) \geq p^*,$$

where $Z(X,Y)$ denotes the normalizing constant or marginal likelihood. For a match of size $|M| = k$ and conditional on σ^2, we then require

$$d_P(X_M, Y_M) \leq \sqrt{2}\sigma \left[\log\left(\frac{p^*}{g(k)\lambda^{(n+m-2k)} Z(X,Y)^{-1}}\right) + \frac{3k}{2} \log(2\pi\sigma^2) \right]^{\frac{1}{2}} = d_k^*.$$

Thus we may construct such a set by

$$\mathcal{M}^* = \bigcup_k \{ M \in \mathcal{M} : \ |M| = k, \ d_P(X_M, Y_M) \leq d_k^* \}.$$

We may adapt the geometric hashing algorithm of Section 5.1 to approximate this set very quickly as described in the following section.

Identification such a set $\mathcal{M}^*$ would allow us to obtain theoretical guarantees on the accuracy of our approximation of the posterior. For example, we would like to guarantee that

$$1 - \alpha \leq \sum_{M \in \mathcal{M}^*} P(M \mid X, Y) = \sum_{M \in \mathcal{M}^*} \frac{L(X, Y \mid M) P(M)}{\sum_{M'} L(X, Y \mid M') P(M')}.$$

However, in practice we do not know the marginal likelihood $Z(X,Y)$ needed to compute d_k^*. In order to bound the contributions of individual alignments to the posterior we therefore require a bound on $Z(X,Y)$ as well, so that

$$\sum_{M \in C_\alpha} L(X, Y \mid M) P(M) \geq \beta \quad \text{and} \quad \sum_{M \notin C_\alpha} L(X, Y \mid M') P(M') \leq \gamma$$

where $\beta\gamma^{-1} \geq \alpha$. If we can find $\mathcal{M}^*$ such that it contains all $M \in \mathcal{M}$ with $P(M)L(X, Y \mid M) \geq \delta^*$, a weak bound on $Z(X, Y)$ may be obtained from

$$P(X,Y) \leq \sum_{M \in \mathcal{M}^*} P(X, Y \mid M)P(M) + (|M| - |M^*|)\delta^*.$$

It is as yet unclear whether this can lead to a practical algorithm with guaranteed bounds. Nevertheless, the above argument shows that there exists such a d_k^* which will give accurate approximations. Even if we cannot choose d_k^* by theoretical arguments with guarantees, we may be able to obtain accurate approximations with good empirical performance by experimentation; we explore this below.

6.2.1. *Bounding Procrustes distance based on geometric hashing*

The above argument shows that there exists a sequence $d_k^*, k = 0, \ldots, \min(n, m)$ such that the set $\mathcal{M}^* = \{M \in \mathcal{M} : d_P(X_M, Y_M) \leq d^*_{|M|}\}$ yields a high posterior credible set and good approximations to posterior quantities. Even if we cannot determine such a d^* directly to obtain theoretical guarantees, this suggests that choosing a d^* large enough may give good empirical performance. Denote by

$$d^2_{GH(a,b,c)}(X, Y) = \|X - Y\hat{R}_{GH(a,b,c)}\|_F^2 = \sum_{i=2}^{p} \|x_i - \hat{y}_j\|^2$$

the sum-of-squared Euclidean distances of matched landmarks under the geometric hashing transform described above using reference set (a, b, c). For all $(a, b, c) \subset X$ we have

$$d_P(X, Y) \leq d_{GH(a,b,c)}(X, Y).$$

Thus a sufficient condition for $d_P(X_M, Y_M) \leq d_k^*$ for $|M| = k$ is that $\max \|x_i - \hat{y}_j\| \leq d_k^*/\sqrt{k}$ where $\hat{y} = y\hat{R}_{GH(a,b,c)}$. Thus we need only find all matchings with maximum distance between matched points less than $d^* = \max_k d_k^*/\sqrt{k}$. (Alternatively we may build n different hashtables for matches of size $k = 1, \ldots, n$, but we have not implemented this here.)

To obtain all such matching points for a given reference set, we need simply adapt the geometric hashing algorithm as follows: choose the resolution of the hash table indexing to be d^*. Now when matching an object to the database, each point used as an index into the table at entry (a, b, c) say, must also check the 26 neighboring entries in the surrounding cube:

$$\{(a-1, b, c), (a, b-1, c), (a, b, c-1), (a-1, b-1, c), \ldots)\}.$$

In this way every point within d^* of the indexed point may be identified with only a constant factor increase in computational time.

We further adapt the geometric hashing algorithm as follows: we use all $\binom{n}{3}$ unordered triplets of the query object. For each reference set we find all landmark matches within d^*, to construct the maximal matching. Using the corresponding matching we then compute the full least-squares Procrustes distance to obtain the associated likelihood. The resulting matching algorithm has increased computational complexity of $O(n^4)$. Note that there are multiple approximations being made here. First, the approximation to the posterior by a high density credible set. Second the set itself is approximated: since the above condition is sufficient but not necessary, the hashing algorithm is not guaranteed to obtain all matches with $d_P \leq d^*$. Finally, each match $M \in \mathcal{M}^*$ is maximal, effectively giving a mode approximation for each region of high posterior probability matches.

Nevertheless, this approach has strong advantages over MCMC-based matching, inheriting the strengths of the original geometric hashing algorithm. First, it is deterministic and non-iterative, making it suitable for rapid, large scale repeated search. Second, it computes the marginal likelihood for the Bayesian match of the query object against an entire set of target objects simultaneously rather than sequentially. Third and related, it is inherently parallelizable. Thus one could imagine a distributed database server which performs shape queries in a fully Bayesian fashion rather than by a standard heuristic optimization criteria.

6.3. *Alternative Priors on* $\mathcal{M}$

It is unclear how general $P(M)$ may be and still be amenable to this rapid hashing approach; this merits further exploration. Priors of the form (2) seem difficult to incorporate into the algorithm, but other types of spatial process priors may prove tractable. In addition, the approximating set $\mathcal{M}^*$ obtained under uniform prior as above may be used to obtain approximations to other priors via importance reweighting:

$$P(Y \mid Z_c, \zeta_c) \approx \sum_{M \in \mathcal{M}^*} P(Y \mid M, Z_c, \zeta_c) w(M) / \sum_{M' \in \mathcal{M}^*} w(M')$$

where $w(M) = P(M)/g(|M|)$. Bounds for this case seem to more difficult and rely on the maximum and minimum of $w(M)$ over $M \in \mathcal{M}$.

7. EXAMPLES

We apply this approach to the problem of pairwise protein structure alignment developed previously (Rodriguez and Schmidler, 2006a).

Fig. 1 shows the results obtained from applying the geometric hashing technique to approximate marginal posteriors as described in Section 6 to match two short protein fragments taken from the N-terminal Helix A region of human deoxyhemoglobin β-chain (4hhb_A) against sperm whale myoglobin (5mbn). The figure shows the marginal posterior match matrices (3) obtained from the hashing approximation, compared with an exact calculation obtained by exhaustive enumeration, which is just feasible for this short $n = 7$ problem.

Fig. 2 shows the approximate marginal match matrix obtained from a larger problem which is chosen to exhibit multimodality, where MCMC performs poorly. An N-terminal fragment of the Helix A region of human deoxyhemoglobin β-chain (4hhb_A) is matched against a stretch of 30 N-terminal residues from sperm whale myoglobin spanning both helices A and B. There are expected to be at least two reasonable matches, to each of the two helices, separated by a break for the loop in between. We see this from the hashing-based posterior, as well as lower posterior modes obtained from reversing the N- to C-terminal orientation of the query helix. Exact calculation for this example is infeasible.

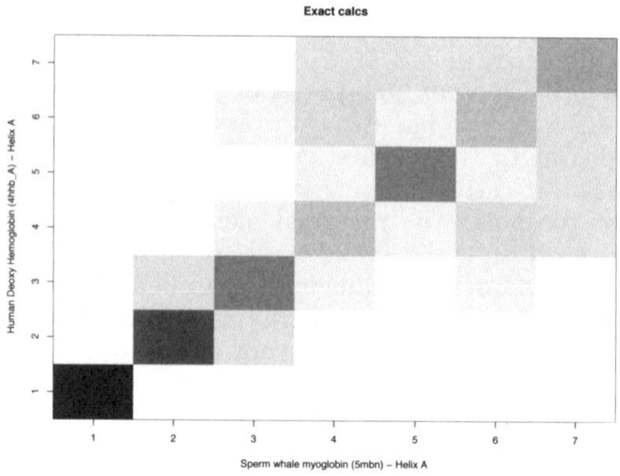

(a)

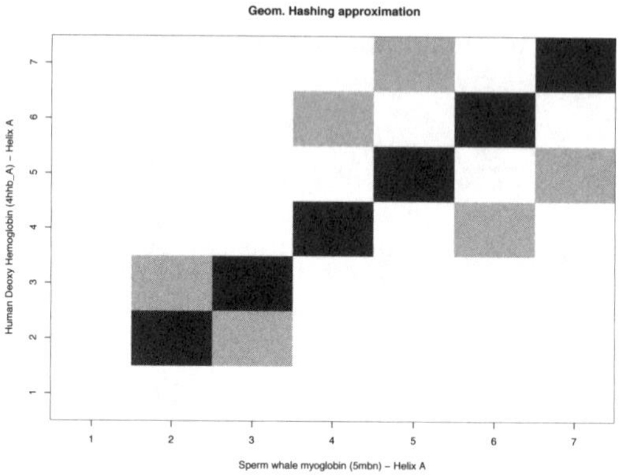

(b)

Figure 1: *Marginal posterior match probability matrix obtained for short 7-residue N-terminal fragment including portion of Helix A for human deoxy-hemoglobin β-chain (4hhb_A) against sperm whale myoglobin (5mbn). Shown are results from (a) exact calculation, and (b) geometric hashing-based approximation as described in text.*

8. CONCLUSIONS AND FURTHER WORK

We have described an approximate calculation technique for Bayesian shape matching under the framework described by (Rodriguez and Schmidler, 2006a) using an extension of the geometric hashing technique developed by (Nussinov and Wolfson, 1991). This technique has the advantage of being highly computationally efficient and inherently parallel, allowing particularly efficient simultaneous calculation when shape matching against a large set of shapes is required. As described, this addresses the fundamental computational problem of Bayesian shape matching of a query object against a large database of possible targets, or for calculation of key quantities in Bayesian shape classification when classes are described by mean shapes or posterior distributions summarized by Monte Carlo samples.

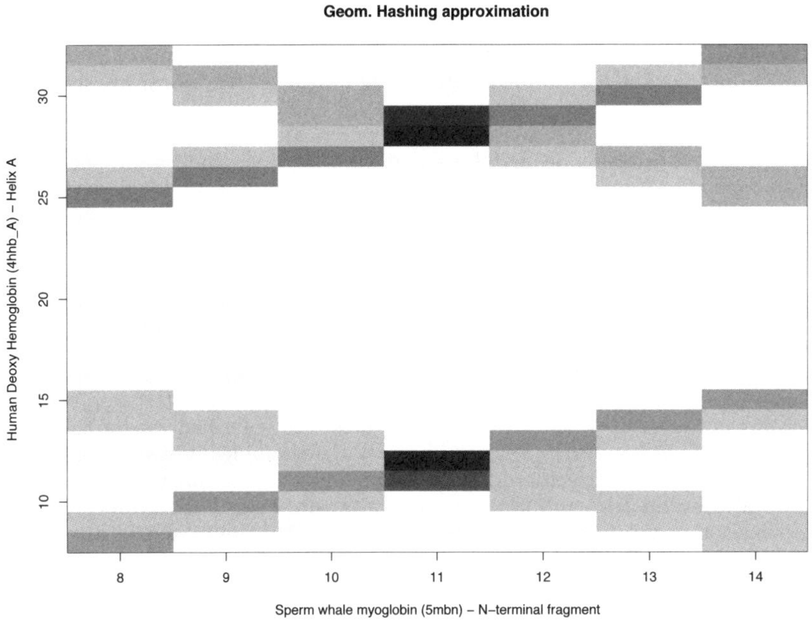

Figure 2: *Application of Bayesian matching algorithm using geometric hashing to larger example exhibiting multimodality: comparison of N-terminal helical fragment of human deoxyhemoglobin β-chain (4hhb_A) against residues 8-32 of sperm whale myoglobin (5mbn) which include helices A and B.*

This basic approach has been demonstrated for test examples in the alignment of protein structures. Further experimentation and optimization of the implementation is needed to explore the large scale applications which motivate this approach; the algorithm is inherently parallelizable and development of a distributed database server which performs shape queries in a fully Bayesian fashion may be feasible.

There are a number of directions for further work along these lines. It is an interesting question whether more sophisticated bounds for the posterior quantities may be obtained to provide theoretical guarantees on the accuracy of approximation. A simple approach is suggested here but more work is needed to determine if this can be useful in practice. However it is worth pointing out that the MCMC approach, which is also in standard usage for an enormous variety of Bayesian computational problems, also comes with few guarantees in practical problems where the mixing time of the Markov chain cannot be bounded polynomially.

It may be fruitful to view the use of maximal matchings to construct the set $\mathcal{M}^*$ as a finite space analog to standard integral approximation algorithms in continuous spaces based on normal approximations at posterior modes (Tierney et al., 1989; Tanner, 1993). In this sense, the 'width' of a mode is clearly missing and it may be helpful to construct a multiplicative factor for the posterior contribution of elements in $\mathcal{M}^*$, *e.g.*, based on the number of subsets of size $k' < k$ for a maximal matching of size k.

The additive error model used here is particularly simple and does not account for spatial covariation between landmark deviations, for example. Addressing this while preserving the efficiency of the hashing calculation requires some care, but increasing d^* and computing the Procrustes distance using appropriate Mahalanobis inner product may be feasible and warrants investigation. Alternatively, as with the discussion of non-factorable priors in Section 6.1, importance reweighting of $\mathcal{M}^*$ may be effective. It remains unclear how practically significant such covariance structure may be for shape matching and classification, and this is likely to be problem specific. It is worth noting that the hashing approach of standardizing the coordinate frame to a single triplet seems reminiscent of Bookstein coordinates (see *e.g.*, Bookstein, 1986; Dryden and Mardia 1998), which is known to induce spurious covariance on the other landmarks. It would be interesting to explore how this affect manifests in the geometric hashing case presented here.

A major advantage of the hashing technique over MCMC sampling is the ability to quickly generate multiple modes which may be well-separated and difficult for a Markov chain algorithm to cross between. One possibility for improving on the hashing approximation described here is to use it to develop a fast approximation which is then refined by MCMC sampling, using the identified modes to generate move proposals in the Markov chain. This has the combined advantage of allowing the MCMC to mix between modes, and of refining the hashing answer in a way which will converge with the usual theoretical guarantees. This is a more powerful generalization of the library-sampling extension to MCMC matching given in Rodriguez and Schmidler (2006a); Wang and Schmidler (2006) without requiring the order-preserving constraints assumed there. This approach would also allow the incorporation of more general covariance structure and more informative priors. While very promising for improving the quality of the approximation, this approach suffers from the fact that any MCMC technique will be significantly slower than the rapid and simultaneous hashing technique described in this paper, and thus infeasible for large-scale comparison against large databases. More experience with this approach is needed in order to understand the impact of trading off these computational and modeling alternatives.

REFERENCES

Bookstein, F. L. (1986). Size and shape spaces for landmark data in two dimensions. *Statist. Science* **1**, 181–242.

Bookstein, F. L. (1991). *Morphometric Tools for Landmark Data: Geometry and Biology.* Cambridge: University Press.

Dryden, I. L. (1999). General shape and registration analysis. *SEMSTAT 3* (W. S. Kendall, O. BarndorffNielsen and M. N. M. van Lieshout, eds.) London: Chapman and Hall, 333–364.

Dryden, I. L. and Mardia, K. V. (1998). *Statistical Shape Analysis.* Chichester: Wiley.

Green, P. J. and Mardia, K. V. (2006) Bayesian alignment using hierarchical models, with applications in protein bioinformatics. *Biometrika* **93**, 235–254.

Kendall, D. G., Barden, D., Carne, T. K., and Le, H. (1999). *Shape and Shape Theory.* Chichester: Wiley.

Lele, S. R. and Richtsmeier, J. T. (2001). *An Invariant Approach to Statistical Analysis of Shapes.* London: Chapman and Hall.

Nussinov, R. and Wolfson, H. J. (1991). Efficient detection of three-dimensional structural motifs in biological macromolecules by computer vision techniques. *Proc. Natl. Acad. Sci. USA* **88**, 10495–10499.

Rodriguez, A. and Schmidler, S. C. (2006a). Bayesian protein structure alignment. *Tech. Rep.*, Duke University, USA.

Rodriguez, A. and Schmidler, S. C. (2006b). Combining sequence and structure information in protein alignments. *Tech. Rep.*, Duke University, USA.

Schmidler, S. C. (2006). Bayesian flexible shape matching with applications to structural bioinformatics. *Tech. Rep.*, Duke University, USA.

Small, C. G. (1996). *The Statistical Theory of Shape.* Berlin: Springer-Verlag.

Tanner, M. A. (1993). *Tools for Statistical Inference.* Berlin: Springer-Verlag.

Tierney, L., Kass, R. E., and Kadane, J. B. (1989). Fully exponential Laplace approximations to expectations and variances of nonpositive functions. *J. Amer. Statist. Assoc.* **84**, 710–716.

Wallace, A., Laskowsi, R., and Thornton, J. (1996). Tess: A geometric hashing algorithm for deriving 3D coordinate templates for searching structural databases: Applications to enzyme active sites. *Prot. Sci.* **6**, 2308–2323.

Wang, R. and Schmidler, S. C. (2006). Bayesian multiple protein structure alignment and analysis of protein families. *Tech. Rep.*, Duke University, USA.

Wolfson, H. J. and Rigoutsos, I. (1997). Geometric hashing: An overview. *IEEE Comp. Sci. and Eng.* **4**, 10–21.

DISCUSSION

DARREN J. WILKINSON (*Newcastle University, UK*)

Matching and alignment. The fundamental problem considered in this paper is one of matching and aligning geometric point configurations. Given two (or more) sets of points (landmarks) in space, the task is to find a subset of those landmarks that occurs in all configurations which *match*, after filtering out noise and a geometrical transformation from some given class. In certain application areas some additional information will be available. For example, the labels attached to each landmark may be ordered, and that ordering may be assumed to be consistent across the point configurations, thus reducing the number of matching configurations that need to be considered (as any matchings that violate the order can be immediately discarded). Alternatively, the labels attached to each landmark may have a *colour*, and here there may be a hard constraint that only labels of the same colour may be matched, or there could be a soft constraint that labels of the same colour are much more likely to be matched.

This generic statistical problem has a range of potential applications in a variety of scientific disciplines. However, the work described here has been motivated by applications to protein bioinformatics, and even within this particular area there are a number of important applications. The example considered in the paper is that of alignment of related proteins, and another example discussed is that of matching a new protein against a library of known proteins. Other examples include identification of protein ligand binding sites and alignment of 2d protein gels, but there are many others.

Computational challenge. The basic problem can be broken down into several distinct components which (conceptually, at least) may be considered separately. The most basic problem is, given a pair of configurations, X and Y, and a proposed matching M, compute the 'optimal' geometric alignment of the configurations, or better, the posterior distribution of the 'correct' alignment. This problem is relatively well studied, and there are a number of reasonable solutions in the literature.

The most commonly adopted solution, and that adopted in this paper, is to use the Procrustes alignment (the alignment minimising Procrustes distance). Based on the alignment, the probability associated with it may be computed. This is computationally fast, and generally leads to a good alignment in practice. However, using this technique has some drawbacks when there is some uncertainty regarding the alignment (and especially when there exists more than one good alignment), as it effectively ignores alignment uncertainty and prevents alignment uncertainty from getting propagated correctly through the rest of the analysis. A fully Bayesian solution would work with the full posterior distribution of the alignment transformation. This is possible but typically computationally demanding. However, if it were possible to work within a fully Bayesian formulation but *integrate out* the geometrical transformation, then alignment uncertainty would be propagated correctly without suffering a significant computation penalty. This problem seems to be the subject of current research activity in this area.

Once a good solution to the alignment problem is in place, attention can turn to the problem of finding a good matching. A matching $M \in \mathcal{M}$ may be represented as a matrix with (i, j)th element equal to one if landmarks X_i and Y_j match, and zero otherwise. The matching matrices therefore consist mainly of zeroes with at most one 1 in each row and column. The number of such matrices (the size of the match space, $\mathcal{M}$) increases in a typical combinatoric way as the number of landmarks increases, making complete enumeration of the space impractical for all but very small problems. However, at least conceptually the problem is straightforward: one simply needs to consider each matching in turn and evaluate the combined matching and alignment probability. This set of probabilities can then be used to find the 'good' matches. Although it is possible to construct MCMC algorithms to simultaneously explore the space of matchings and alignments (see, for example, Green and Mardia, 2006), such algorithms are unlikely to work reliably in the case of multi-modal posteriors, where there are a number of distinct good matches separated by very unlikely regions of match space.

The above is the matching and alignment problem in its most basic form, but there are a number of practically important extensions. The most obvious extension is that of the simultaneous matching of more than two point configurations, and another (considered in the paper), is that of matching a given specific configuration against a library of existing configurations to find the best matching configuration in the library. Again this is conceptually quite straightforward, as all that is needed is a solution to the matching and alignment problem for each configuration in the library. However, computationally this can be a significant challenge if the library being matched against (say, a protein database) is very large.

Let us now discuss the key features of the approach.

The prior model. For a pair of configurations X and Y, a prior model

$$\Pr(M, X, Y) = \Pr(M)\Pr(X, Y|M)$$

is required for a Bayesian analysis. Although several forms for $\Pr(M)$ are mentioned in the paper, the only prior actually used in the context of the proposed algorithm is that which is uniform over all possible matchings. Although this sounds attractive it has a number of consequences that might not be desirable in all cases; in particular, the induced prior on $\Pr(|M|)$, is strongly biased towards larger values of $|M|$.

The prior for $\Pr(X, Y|M)$ uses an additive normal model on size-and-shape space for the matched landmarks (and independent uniform for the unmatched landmarks). This prior effectively assumes that the Procrustes alignment is 'correct', and does not reflect uncertainty in the geometric alignment. As mentioned previously, a fully Bayesian model that allowed the integrating out of the geometrical transformation would clearly be preferable here, but it is hard to quantify the difference it would actually make in practice. More generally, it would be interesting to compare this prior with the Poisson thinning model of Green and Mardia (2006), particularly regarding the degree of analytic tractability that the different prior modelling frameworks offer.

Geometric hashing. The key novel aspect of this work is the use of a geometric hashing technique for the identification of a high posterior probability set of matchings, $\mathcal{M}^\star \subseteq \mathcal{M}$. Geometric hashing is widely used in the computer vision community for the rapid identification of objects in a scene, though in this context it is typically used to find just the best match, and also typically assumes a high signal-to-noise ratio. The idea in this paper is to use a geometric hashing framework to identify a (relatively small) set of matchings, $\mathcal{M}^\star$ which can be used in place of $\mathcal{M}$ for all calculations that require an average over the posterior distribution of the matching. This is a nice idea, but there are several issues associated with using the technique in practice. The issues all relate to the question of how big $\mathcal{M}^\star$ needs to be in order to be confident of not missing some good matches.

The first issue concerns the effect of noise on the algorithm. For a given reference set (a, b, c), it seems clear that if a, b and c are 'close together', then noise is going to have a bigger effect on the resulting hash locations than if they are 'far apart'. It therefore seems clear that matches from some reference sets are going to be more reliable than others. Of course it is suggested in the paper that matching is repeated for all landmark triples, which should help with this problem, as there are presumably more reference sets 'far apart' than 'close together'. It would also be interesting to better understand the effect of spatially correlated noise on the effectiveness of the hashing algorithm.

Another issue which immediately springs to mind is the dependence of the procedure on the true number of matching landmarks. It is clear that if a particular reference set includes an unmatched landmark then a sensible match will not be found. However, if there are relatively few matching landmarks, then the majority of reference sets will contain an unmatched landmark. It is not completely clear what effect this will have on the overall effectiveness of the procedure.

There are also more subtle issues relating to the fact that the true number of matching landmarks is unknown. Arguments are given in section 6.2 for why a critical set of Procrustes distances, $d_k^\star$, should exist. However, these clearly depend on the number of matching landmarks, k. When a hash-table is built, it is done so using a particular resolution, $d^\star$. In the paper it is suggested that $d^\star = \max\{d_k^\star\}$ should be used. One immediate question that arises is whether this should really be $d^\star = \max\{d_k^\star/k\}$, but in any case there is a more general question about how this choice of hash-table resolution might bias the number of matchings. There appear to be at least two different possible errors. First, there are likely to be truly matching landmarks that are missed due to the choice of hash-table resolution. Of course this can always be compensated for by increasing the resolution. Conversely, because a conservative choice of $d^\star$ is used, there are likely to be many false matches. So for a given configuration, even if the hashing procedure correctly identifies a good match, it is quite likely that some of the matching landmarks are spurious false matches. The paper does not explain precisely how the putative set of matches $\mathcal{M}^\star$ is constructed from the geometric hash matches – further details would be very useful here. Again, it is not immediately clear what effect false matches will have on downstream calculations.

Comments and questions.. The fundamental problem being tackled here is the problem of finding a good matching. Since a matching is a graph, this is actually the ubiquitous problem of searching a large graph space for 'good' graphs, accepting that the distribution on graph space may well be multi-modal, with more than one good graph being separated by regions of graph-space with very low probability. In general there are no reliable computationally efficient approaches to this problem, so it seems entirely natural to look for problem-specific solutions. Here, the geometric structure of the spatial point configurations is being used to provide a shortcut to the most promising parts of graph space, and this seems to be a good idea. However, I think that there is still work to be done to demonstrate that this method really works, really scales, and really provides a fast and reliable solution to the matching problem on challenging examples. One way of addressing this would be to get a real handle on $\Pr(\mathcal{M}^\star)$, but I wonder how realistic this is.

The results presented in the paper show fairly substantial differences between the exact method and the proposed algorithm even on a toy problem with seven landmarks on each of two closely related configurations. Are these discrepancies likely to increase with the complexity of the problem? It is encouraging to see a solution for the larger, more complex problem, but of course in that case there is no truth against which one may compare. For this particular problem, since the four dominant modes seem to have roughly equal probability, one could compare the results presented against the combined results from four different fully Bayesian MCMC chains (one for each mode). Of course this isn't the truth, but would be interesting to compare nevertheless.

Finally, I'd be interested to know more about how the techniques are implemented in practice. The approach is described as fully Bayesian, but it is not completely clear from the presentation how tuning parameters such as σ and λ are chosen. Similarly, in the absence of good bounds, how are the $d_k^\star$ actually set? How fast is the whole geometric hashing and matching procedure, in terms of CPU time, for real protein matching problems on a regular PC? What about non-uniform priors on M, and other priors on $X, Y|M$? Are there any practical difficulties (other than CPU time) associated with hashing and matching against a very large library? Is the extension to the simultaneous multiple alignment of more than two configurations straightforward, and if so, how many structures can be aligned before the procedure breaks down? What about hard or soft constraints on landmark colour?

The author is to be congratulated on an interesting and thought-provoking paper, suggesting a novel approach to a difficult and practically important problem. I look forward to reading about further developments of this work in the future.

IAN L. DRYDEN (*University of Nottingham, UK*)

Molecule comparison is a very challenging topic, and this interesting paper provides some fast Bayesian methods towards tackling the problem. The use of statistical shape analysis in molecule matching has seen much recent activity, including the approaches of Kent *et al.* (2004), Green and Mardia (2006), and Dryden *et al.* (2007), as well as Schmidler's work with his colleagues.

The underlying model of Dryden *et al.* (2007) has strong similarities with that of the Procrustes-based model (1). A small difference results from the dimensionality of the tangent space, where the power of σ^{-1} in the density is taken as $Q = pd -$

$d(d-1)/2 - d$ due to the rotation and translation constraints from the Procrustes registration. Another rather cosmetic difference is that X is random and Y is fixed in Dryden *et al.* (2007), although this is just a reparameterisation of the model where both X and Y are random. More substantial differences are in the choices of prior for M, particularly when point ordering (as in sequences) is taken into account in (2).

Two basic approaches for dealing with the nuisance parameters in Bayesian matching are:

(i) marginalization, where a prior is placed on the nuisance parameters θ and then integrated out (Green and Mardia, 2006);

(ii) maximization, where θ is identified from M using Procrustes registration (Schmidler's paper and Dryden *et al.*, 2007).

Which approach is better? From some initial investigations with Kim Evans and Huiling Le, simple Markov chain Monte Carlo algorithms using approach (ii) often get stuck in local modes, although this can be overcome by proposing large moves. Posterior based inference in (i) and (ii) was quite similar in several applications and simulation studies, particularly when variability was small. The MAP estimator under (ii) performed slightly better than that for (i) in some more noisy simulations, although there were other instances when (i) was slightly better. From a shape theory perspective inference should be invariant to translations and rotations of the data, which is the case in (ii) or with a uniform prior for θ in (i). However, from a Bayesian perspective (i) is very natural. Further comparison between the two approaches (i) and (ii) is needed.

The choice of density for the unmatched points and the prior distributions are crucial to the performance of the algorithm. There is very little information in the data to choose between a very close match on a small number of points and a less accurate match on a larger number of points. Specifying the density for the unmatched points , together with other hyper-parameters, enables the user to indicate which scenario is preferred. What form of density is used for $f(\cdot, \lambda)$ in the model?

What is clear with all previous Bayesian approaches is that they are too computationally expensive for searching through large databases of molecules. Hence the work of Schmidler in carrying out fast (approximate) Bayesian inference is very welcome, particularly because uncertainty in the estimates can be quantified.

KANTI V. MARDIA (*University of Leeds, UK*)

I found the paper very stimulating and look forward to also reading the other papers from which this paper has been drawn.

However, the paper leads to certain philosophical issues: What should be the gold standard in unlabelled shape analysis? Though the MCMC maybe too slow in some applications, does it nevertheless have a role to play as a gold standard against fast heuristic approximation? Does the hybrid method of using fast computer algorithm with approximate statistical models have a better future?

I should point out that the Green and Mardia (2006) approach allows also the use of informative priors on θ, for the situation where one does not simply want to integrate θ out, so that size-and-shape space issue is not always relevant. Further, accounting for uncertainty in θ given M is a feature of all methods based on jointly modelling M and θ (not just this 'marginalization approach'). Your method seems to maximise over θ instead of integrating it out – what are the practical consequences of this? Also we have to think of theoretical justification for integrating out for unlabelled shape analysis. Which one is apparently more powerful?

Yes, MCMC methods do require good mixing. This can be a real art! The principle of hybrid methods *e.g.*, refining a fast computational method to bring out statistical summaries is worth exploring. See also Mardia *et al.* (2007) where the graph method is refined through MCMC for active sites (Green and Mardia, 2006). Indeed, in various examples there, the procedure has produced biologically meaningful extensions of the glycine rich motif by this method.

We should distinguish problem related to large databases versus small scale problems. For example, images with few multiple views need a small scale implementation method; the problem of a very efficient algorithm is not as acute. One of our strategies to get candidate matches is to use a hunting mode principle which could incorporate a prior on M based on information from the geometric hashing algorithm Take one motif B (a single unit), to be matched with A (larger having several motifs). Take the best match in A. Then match B with residual from A and so on iteratively (eg, Nyirongo, 2006). Unfortunately, I could not understand the full details of your examples to repeat the calculations. Did you use the C-alpha coordinates, for example?

Yes, Bookstein's size-and-shape coordinates do lead to spurious correlations. The better way of representing isotropic shape variation is through Kendall's shape coordinates. Further the log centroid is often used to represent scale (see also Mardia *et al.*, 2004).

REPLY TO THE DISCUSSION

I would like to thank Professor Wilkinson for his thoughtful discussion of the paper, and I am also especially grateful to professors Dryden and Mardia for taking time to add their own valuable additional insights and suggestions, along with pointers to their own work on related problems. I have found our discussions of our respective work in recent years to be enormously beneficial and am grateful for their contributions.

Dryden and Mardia both note the differences between shape-space models (which maximize over registrations) and marginalization under some prior on θ. (Dryden correctly points out that the shape-space likelihood involves the dimension of the tangent space rather than the Euclidean space.) As referenced in the paper we have developed both approaches in parallel for the protein structure alignment problem, but as yet have no answers as to which is preferable and we look forward to seeing Dryden's work on comparing these approaches more rigorously. The main differences seem likely to arise in data where registrations are poorly determined by matches and where that additional variability is important to take into account. But since the unmatched density effectively provides a soft threshold for inclusion, such variability must be bounded implicitly, and it would be interesting to make this explicit. Of course there are other problems where θ itself is the subject of interest, where the shape-space approach would be less desirable. Mardia asks for the theoretical justification for integrating out θ, but for us marginalization is purely a computational trick for working with the full joint model. Marginalization for computational speed-up is justified for MCMC algorithms by the effect of 'collapsing' on the convergence rate (Liu, 1994); similarly it enables use of other efficient calculations such as the hashing algorithm described here when working with the full joint model. Thus our discussion of marginalization is intended to show how the hashing algorithm of the current paper may be used for both the shape-space and the joint models, without addressing the important issue of which is to be preferred or under what circumstances.

Dryden raises the key and interrelated issues of placing priors on matchings and choosing a density of unmatched points; we have wrestled with the latter for some time and our approach to specifying these densities is described in Rodriguez and Schmidler (2006a), where whole classes of densities give rise to the same single parameter (the 'soft threshold' λ in the current paper), which also seems closely related to the Poisson thinning process of Green and Mardia. By no means is our solution the final say on this issue, which is the real key to the unlabeled shape matching problem. This is effectively a model selection problem however, and as such unlikely to have a unique best solution.

Both Wilkinson and Dryden raise the issue of priors on M, with Wilkinson noting the marginal distribution on $|M|$ arising from a uniform prior on matchings may be undesirable. We agree and Section 6.1 discusses the types of non-uniform priors which may be used immediately with the computational approach given in this paper, essentially those specified in terms of $|M|$. Section 6.3 suggests an importance reweighting approach to deal with even more general priors on M itself, but as mentioned this is subject to the usual potential drawbacks of importance sampling. However the real challenge seems to be how to design appropriate priors on M which yield desirable behavior for matching problems. Again this problem is closely related to model selection and is unlikely to have a unique solution, but there is much room for further research here. The uniform priors used here were chosen only for convenience in demonstrating the algorithm and for comparison with the existing use of geometric hashing for maximization.

Wilkinson asks about reference sets containing landmarks which match poorly; such matches generate small likelihoods and thus contribute negligibly to the posterior. The main disadvantage of choosing a conservative d^* (as opposed to the separate d_k^* suggested in the paper) is computing time; more hashtable entries need be examined in order to find the subset within the threshold. We agree that the examples in the paper are small and serve only as proof-of-principle; significantly more engineering effort will be required to examine how the method scales in practice for large problems. Wilkinson also raises several other good questions which there is insufficient space to address here; many of these questions are good ones and will only be answered by further study.

Finally, Mardia correctly notes that fast calculation is not always required, and that when problems are small enough and time constraints loose enough, MCMC simulation may provide a more accurate approximation to the posterior than that given here (indeed we have used it in our other work). However the focus of this paper is fast computation for the many problems where time constraints dominate, such as online prediction or tracking, or large database search. In such cases, MCMC is far too slow and one must be willing to accept alternative, faster approximations or abandon the Bayesian approach entirely. The current paper provides a possible methodology for producing such fast approximations to the full Bayesian posterior model rather than resorting to a more heuristic criteria as often done in practice. Also, as noted briefly in the paper MCMC is also simply an approximation, for which in this problem (and most others) we have no theoretical guarantees on the accuracy for finite simulations. In fact, as described in Rodriguez and Schmidler (2006a), standard MCMC may be particularly bad in this problem if the multimodality issue is not addressed. Of course there are many methods for addressing multimodality in MCMC, but all require significant computation and alas none provide theoretical guarantees. In Rodriguez and Schmidler (2006a) a method is given for identify potential modes under ordering constraints, in order to improve mixing of MCMC. The hashing method presented here may be viewed as a generalization of this approach which does not require ordering constraints, and indeed this may prove as fruitful a way to utilize the hashing algorithm as direct approximation. Under assumption that any match will have at least three points within δ^* we are guaranteed to find a transformation near any and all modes within $O(n^4)$.

An important related question for applied purposes is whether Bayesian statistical approaches to this problem offer a significant practical improvement over heuristic methods currently used. We believe that it does; however, as is often the case, this is a question that is best answered by empirical comparison. Yet to compare the Bayesian approach on real problems we must be able to compute its solutions on large scales. Thus the ability to do fast computation is a prerequisite, and the algorithms presented in this paper offer one promising approach to doing so.

ADDITIONAL REFERENCES IN THE DISCUSSION

Dryden, I. L., Hirst, J. D., and Melville, J. L. (2007). Statistical analysis of unlabelled point sets: comparing molecules in cheminformatics. *Biometrics* (to appear).

Liu, J. S. (1994). The collapsed Gibbs sampler in Bayesian computations with applications to a gene regulation problem. *J. Amer. Statist. Assoc.* **89**, 958–966.

Kent, J. T., Mardia, K. V., and Taylor, C. C. (2004). Matching problems for unlabelled configurations. *LASR2004 Proceedings: Bioinformatics, images and wavelets* (R. G., Aykroyd, S. Barber and K. V. Mardia, eds.) 33–36.

Mardia, K. V., Nyirongo, V. B., Green, P. J., Gold, N. D. and Westhead, D. R. (2007). Bayesian refinement of protein functional site matching. *BMC Bioinformatics* (to appear).

Mardia, K. V., Kirkbride, J. and Bookstein, F. L. (2004). Statistics of shape, direction and cylindrical variables. *J. Appl. Statist.* **31**, 465–479.

Nyirongo, V. (2006). *Statistical Approaches to Protein Matching in Bioinformatics.* Ph.D. Thesis, University of Leeds, UK.

BAYESIAN STATISTICS 8, pp. 491–524.
J. M. Bernardo, M. J. Bayarri, J. O. Berger, A. P. Dawid,
D. Heckerman, A. F. M. Smith and M. West (Eds.)

Nested Sampling for Bayesian Computations

JOHN SKILLING
Maximum Entropy Data Consultants Ltd, Kenmare, Ireland.
skilling@eircom.net

SUMMARY

Nested sampling is a simple and general algorithm for directly computing the **evidence** Z (*aka* marginal likelihood) in multi-dimensional inference. It yields a Bayesian probability distribution $\Pr(Z)$ from which a central estimate and its uncertainty follow. Samples from the posterior distribution are available as a by-product with no extra computation. The method works by sampling proportionally to the prior, though constrained within a 'nested' sequence of progressively-higher likelihood contours. Exploration depends only on the shape of these contours and not on the associated values, so that nested sampling is invariant to monotonic re-labelling of likelihood. Thus, in multi-phase applications, nested sampling bridges between different phases in a way denied to any thermal method, promising considerable extra scope. Although presented here in a Bayesian context, nested sampling is equally applicable to numerical integration of general positive functions in many dimensions.

Keywords and Phrases: BAYESIAN COMPUTATION, EVIDENCE, MARGINAL LIKELIHOOD, MODEL SELECTION, ALGORITHM, NEST, ANNEALING, NON-THERMAL, PHASE CHANGE.

1. BAYESIAN INFERENCE

You, the reader, seek parameter(s) θ, encapsulated by your model's prior $\pi(\theta)$. Possibly you have several such models, and want to compare them. You also have some data D, encapsulated by your likelihood function $L(\theta)$. Bayesian inference uses this information in the *product rule* of probability calculus:

$$
\begin{array}{ccccc}
\Pr(\theta) \times \Pr(D \mid \theta) & = & \Pr(D) \times \Pr(\theta \mid D) & & \| \text{ model} \\
\pi(\theta) \;\times\; L(\theta) & = & Z \;\times\; P(\theta) & & \\
\text{Prior and Likelihood} & \Rightarrow & \text{Evidence and Posterior.} & &
\end{array}
$$

Dr. John Skilling was previously at the Department of Applied Mathematics and Theoretical Physics, University of Cambridge, UK.

The corresponding *sum rule* expresses the fact that the prior and posterior are probability distributions, summing to unity:

$$\int \pi(\theta)\, d\theta = \int P(\theta)\, d\theta = 1\,.$$

According to these rules, the evidence (*aka* prior predictive, marginal likelihood) is calculated as

$$Z = \int L(\theta)\, \pi(\theta)\, d\theta\,, \tag{1}$$

whilst the posterior is

$$P(\theta) = Z^{-1} L(\theta)\, \pi(\theta)\,. \tag{2}$$

I, the author, propose a 'nested sampling' algorithm for your inferences Z and P.

Equation (1) precedes equation (2), because it defines the Z that appears there. It can be evaded by relaxing (2) to a proportionality $P(\theta) \propto L(\theta)\, \pi(\theta)$ – yet it seems foolish to discard a central quantity so soon. Moreover, the evidence Z dominates the posterior P in application as well as in the algebra, because it assesses the model in question. There's little reason to bother with a model's posterior at all if the evidence for it is out-classed by another. Finally, Z is just a single real number, whereas the posterior is a possibly-complicated function with multi-dimensional argument. For all these reasons, my strategy is to compute the evidence first; the posterior happens to follow as a corollary.

As a matter of historical record, the search took me twenty years, and was circuitous. The idea of nested sampling finally appeared, obliquely, in discussion with David MacKay in the radio astronomy tea room at the University of Cambridge in late 2003. The algorithm works and accounts of it – with example computer code – have appeared elsewhere (Skilling 2004, 2006, Sivia and Skilling 2006), but this account explains the underlying Bayesian rationale more clearly. Application papers (Mukherjee *et al.* 2006, Murray *et al.* 2006, Parkinson *et al.* 2006) are now beginning to appear.

2. HOW I WOULD **LIKE** TO COMPUTE Z

The obvious way of evaluating (1), by rastering θ throughout its multi-dimensional space, is hopelessly slow. However, rastering in some different order might be helpful. Generally, we can re-write (1) as

$$Z = \int L\, dX\,,$$

where $dX = \pi(\theta)\, d\theta$ is the element of prior mass, whose own sum is $\int dX = 1$ so that we can take X to be a scalar coordinate in the range (0,1). Since Z is invariant under re-ordering X, we can take any convenient order. The best order would be one in which the integrand L was as well-behaved as possible, which suggests sorting the elements into either increasing or (as here) decreasing order of likelihood value. We would then have a one-dimensional integral over unit range

$$Z = \int_0^1 L(X)\, dX\,, \tag{3}$$

where the relation between L and X is that

$$X(L') = \int_{L(\theta)>L'} \pi(\theta)\, d\theta \tag{4}$$

is the prior mass enclosed within the contour L', in other words the cumulative distribution function for L exceeding L'. The corresponding inverse $L(X)$ is a positive decreasing function of X, as illustrated in Fig. 1 for a simple 4×4 example.

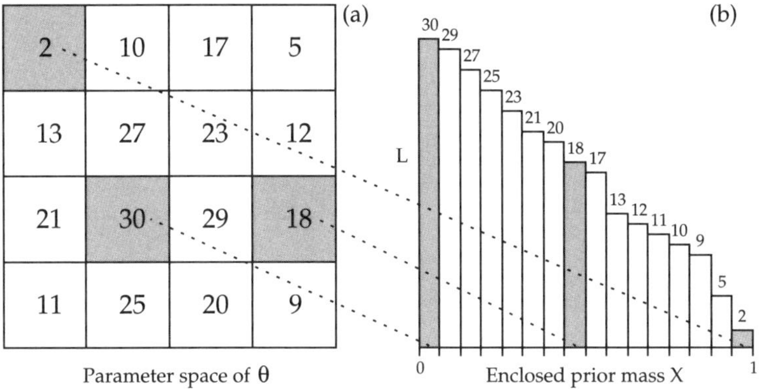

Figure 1: *Likelihood values (a) in parameter space, (b) sorted as $L(X)$.*

All complications of dimensionality, shape, topology and so on are annihilated by such a sorting operation, and the remaining integral (3) would be trivially easy by any numerical method. We would just evaluate L on some grid of X, which could be quite coarse because L is well-behaved, and add up the column areas $L\,\Delta X$.

Some care is needed because the bulk of the posterior, where most of Z is accumulated, is usually found in some tiny patch of the multi-dimensional space of θ. The prior-to-posterior compression ratio is quantifiable as e^H, where H is the 'information';

$$H = \int \frac{P}{\pi} \log\left(\frac{P}{\pi}\right) dX\,. \tag{5}$$

Hence most of the integral for Z is to be found at exponentially small abscissa values, $X \sim e^{-H}$. Though dimensionless, H can be thousands or more in a sizable application: if the likelihood involves R different measurements, H may be roughly $R\log(\text{signal/noise})$ and large. This strongly suggests that X should be sampled geometrically, with successive locations $X_1, X_2, X_3, \ldots$ shrinking according to some ratio $0 < t < 1$. Then, writing $w = \Delta X$ for the column width (also shrinking exponentially), we have

$$X_{i+1} = tX_i\,, \qquad w_{i+1} = X_i - X_{i+1} \tag{6}$$

in terms of which

$$Z = \sum w_i L_i\,, \qquad H = \sum w_i \frac{L_i}{Z} \log\left(\frac{L_i}{Z}\right). \tag{7}$$

This starts at $X_0 = 1$ (where we assign $L_0 = 0$) which covers the entire prior, and would terminate when we deemed the summations to have adequately converged. A refinement could use the trapezoid rule $w_i = (X_{i-1} - X_{i+1})/2$ in (6), but the accuracy gain from $\mathcal{O}(w)$ to $\mathcal{O}(w^2)$ turns out to be insignificant.

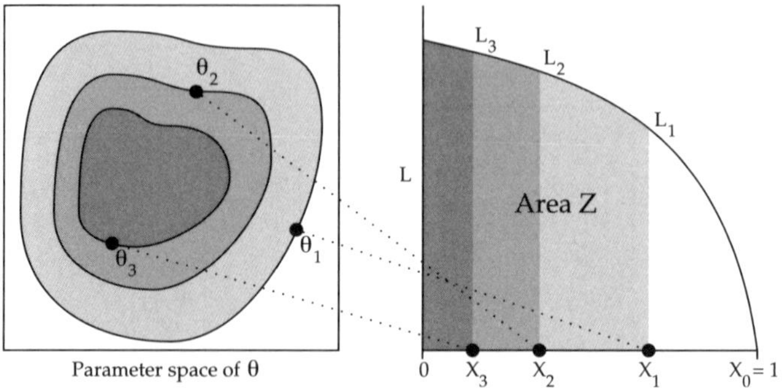

Figure 2: *Nested likelihood contours are sorted to enclosed prior mass X.*

Fig. 2 shows the relationship between contours $L_i = L(\theta_i)$ with their enclosed θ-volumes on the left, and the corresponding X-ranges and column areas on the right. From step i, the next step involves setting $X_{i+1} = tX_i$ within the existing X-range, followed by

$$\theta_{i+1} = \texttt{Sort}^{-1}(X_{i+1}) \tag{8}$$

to find the new point. But, of course, we can't do $\texttt{Sort}^{-1}$ to relate the likelihood-cumulant X back to θ.

So let's be Bayesian. After all, we're doing Bayesian computation.

3. HOW WE **CAN** COMPUTE Z

The trick is to convert incompetence to ignorance, which we can quantify.

You are the application engineer, who knows about L. I am the algorithm designer, trying to compute Z. At step i, I could set, *randomly*,

$$X_{i+1} \text{ uniform in range } X < X_i\,, \quad \theta_{i+1} = \texttt{Sort}^{-1}(X_{i+1}) \tag{9}$$

and give you your next θ. Alternatively, you could get it yourself by sampling

$$\theta_{i+1} \text{ random (over prior) in volume } L > L_i\,, \quad X_{i+1} = \texttt{Sort}(\theta_{i+1}) \tag{10}$$

and give me my next X. The two procedures are equivalent, because the X-range was *defined* to be the prior mass enclosed in the θ-volume, so sampling from one is the same as sampling from the other.

Let us agree that you will adopt the latter procedure (10). At this stage, before you have actually done your work, my belief about the shrunken X that you plan to give me is uniform:

$$\Pr(t_{i+1}) = 1 \text{ in } 0 < t_{i+1} < 1\,, \tag{11}$$

where $t_{i+1} = X_{i+1}/X_i$ is the $(i+1)$'th shrinkage ratio. As far as I am concerned, you are just intending to use a rather complicated but uniform generator. Next, you acquire θ_{i+1} from within $L > L_i$. You could give it to me, and I ought then to know what X_{i+1} actually is, instead of having to guess. But I can't do the `Sort` mapping, so your information would be useless to me. Accordingly, Bayesian methodology requires my belief about the new X to stay uniform and unchanged: if I could use your information I would, but I can't so I don't. Lastly, you evaluate the new likelihood $L(\theta_{i+1})$ and give it to me as L_{i+1}. This is a definitive number that I *can* use. The likelihood constraint now contracts from the contour $L = L_i$ to the contour $L = L_{i+1}$ nested within it – hence the name 'nested sampling' – and the iterative step is complete.

The upshot of this is that I collect a list of increasing likelihoods $L_1, L_2, L_3, \ldots$, accompanied by decreasing enclosed masses $X_1, X_2, X_3, \ldots$ that I do *not* know exactly but can guess *statistically*. These shrinking X-ranges bridge inwards from the prior, to and beyond the bulk of the posterior. To help me, I have the quantified belief that successive shrinkage ratios (6) are each independently chosen from `Uniform`(0,1). This gives me a probability distribution for the X's, and thence for the w's, and lastly (7) for Z. I can't give you a firm result for Z, but I *can* give you the probability distibution $\Pr(Z)$ representing my informed belief about it – computed most simply as a dozen or more Monte Carlo values derived from sampling the underlying t's. A more precise but tedious numerical computation of $\Pr(Z)$ can be performed after the run is complete, by performing the one-dimensional t-integrals in reverse order. In practice, I will then compress $\Pr(Z)$ to a central estimate and uncertainty range, giving you Z and δZ (usually better presented as $\log Z$ and $\delta \log Z$).

Some rough analysis shows what we may expect. The i'th enclosed mass is $X_i = t_1 t_2 \ldots t_i$ (X_0 being 1). The statistics of each t are $\log(t) = -1 \pm 1$ and independent, so X_i behaves as

$$\log X_i = -i \pm \sqrt{i}\,. \tag{12}$$

Though there is individual variability, the enclosed mass shrinks by about a factor of e each step, on average. The bulk of the posterior should be found around $X \sim e^{-H}$, so we expect to need about H steps to reach the dominant domain. If, there, the likelihood involves R useful measurements, it might perhaps be approximated as rank-R multivariate normal. Then the standard folklore '$\chi^2 = R \pm \sqrt{2R}$' suggests a ratio $\exp(\sqrt{2R})$ between the inner and outer enclosed masses. Hence (ignoring $\sqrt{2}$) the algorithm would pass through the dominant domain in about $\sqrt{R}$ steps. Thereafter, the enclosed X would continue to shrink exponentially, but the likelihood would fail to increase fast enough to compensate for ever-diminishing width w, and the summed $Z = \sum w_i L_i$ would approach its final value (exponentially fast if L is bounded above). Now

$$H \approx R \log(\text{signal/noise}) > R > \sqrt{R} > 1 \tag{13}$$

so, according to this analysis, it is usually much more expensive (H steps) to find the posterior than to move through it ($\sqrt{R}$ steps), or to converge (at least apparently) to the final result (about another $\sqrt{R}$ steps).

If we were to ignore uncertainty, we could simply proclaim each t to be e^{-1} and accumulate the corresponding central value of Z. This summation should have about $\sqrt{R}$ significant contributions, reached after about H steps inward. However, these inward steps are themselves uncertain (12) by about $\pm\sqrt{H}$. So all the weights $w = \Delta X$ around that domain are collectively uncertain by correlated $\pm\sqrt{H}$ shifts *in the logarithm.* There will also be some local variability that would be found with a proper Monte Carlo simulation, but it will be less important. The major contribution to uncertainty in Z is this scale factor, potentially many orders of magnitude. Specifically, the standard deviation of $\log Z$ to accompany the estimate (7) is

$$\delta(\log Z) \approx \sqrt{H} \gg 1\,. \tag{14}$$

This does *not* mean that the result (7) is useless. It *does* mean that one needs care when making model comparisons. It's not right to infer that model A is 20 000 ($=e^{10}$) times preferable to model B on the basis of $\log Z_A = 110$ and $\log Z_B = 100$ if, actually, the results were $\log Z_A = 110 \pm 30$ and $\log Z_B = 100 \pm 40$. The difference $\log(Z_A/Z_B)$ is not just 10 but 10 ± 50, for which there is a mere 58% chance that A is better than B. To fix this, the Z's would need to be calculated more accurately, as outlined in Section 7 below.

Taking the example further, if $\log Z_A$ were normally distributed as 110 ± 30, Z_A itself would have a wildly skew distribution with $\texttt{mean}(Z_A) = e^{560}$, a 15σ outlier. This is why results of nested sampling should be quoted in terms of $\log Z$, not Z.

4. THE POSTERIOR $P(\theta)$

This is easy. No more computation need be done. All I need do is pass to you, along with my estimate of $\log Z$ (and its uncertainty), my accompanying list of proportional weights $w_i L_i / Z$. Your computed points θ_i already decompose the total area Z under the $L(X)$ curve into columns of proportion

$$P_i = w_i L_i / Z\,. \tag{15}$$

Samples from the posterior are to be taken randomly from this total area (*i.e.* proportionally to both prior mass and to likelihood, as in Fig. 3), so your θ_i are already samples from the posterior, with those weights. About $\sqrt{R}$ of these points will be significant, which may already be enough for your purposes. You might wish to quantify some property $Q(\theta)$ as

$$\texttt{mean}(Q) = \sum P_i\, Q(\theta_i)\,, \quad \texttt{variance}(Q) = \sum P_i \left(Q(\theta_i) - \texttt{mean}(Q)\right)^2, \tag{16}$$

or you might want equally-weighted posterior samples, to be selected with probability $P_i / \max(P)$. You could then use these to seed conventional Metropolis–Hastings exploration to get further posterior samples, if you did not wish to run nested sampling again to get more.

To be fully professional, the individual probabilities P_i should be averaged over a dozen or more Monte Carlo simulations of the underlying t's. However, the major uncertainty in nested sampling is the $\pm\sqrt{H}$ scaling that affects the important w's and Z equally, so cancels from the P_i, for which the simple proclamation '$t_i = e^{-1}$' works quite well.

5. THE THERMAL ALTERNATIVE

Thermodynamic integration is the standard benchmark method for calculating Z. As the name implies, it is a 'thermal' method that computes with a 'cooled' likelihood L^β instead of L itself. As β increases from 0 to 1 according to some assigned

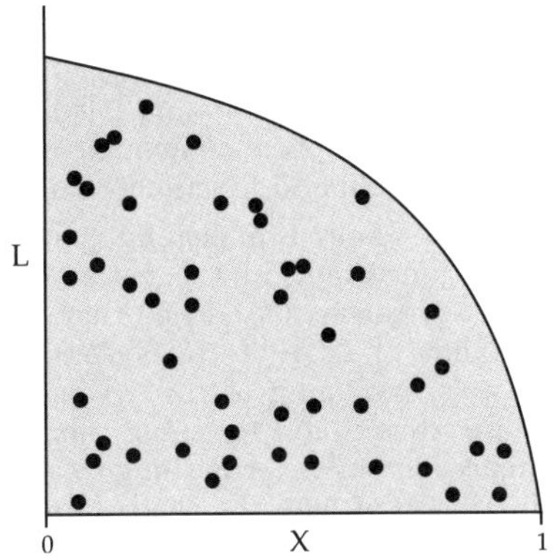

Figure 3: *Posterior samples are scattered randomly over the area Z.*

'schedule', an ensemble of points 'anneals' from an exploration of the prior ($\beta = 0$) to the posterior ($\beta = 1$).

Most of the ensemble points should be found around the maximum of $L^\beta X$, where the annealed likelihood and the prior mass are in balance. This is where the slope $d \log L / d \log X$ on a $\log L / \log X$ plot takes the value $-1/\beta$. Thus (Fig. 4a) annealing tracks the *tangent* of this curve, from arbitrarily steep at the beginning to slope -1 at the end. At each intermediate 'reciprocal temperature' β, the average log-likelihood $\langle \log L \rangle_\beta$ can be read off from the ensemble. This then accumulates the evidence value, according to

$$\log Z = \int_0^1 \langle \log L \rangle_\beta \, d\beta \,. \tag{17}$$

But this assumes that $\log L$ is a *concave* function of $\log X$, so that decreasing slope moves steadily upward in L and inward in X. It may not be.

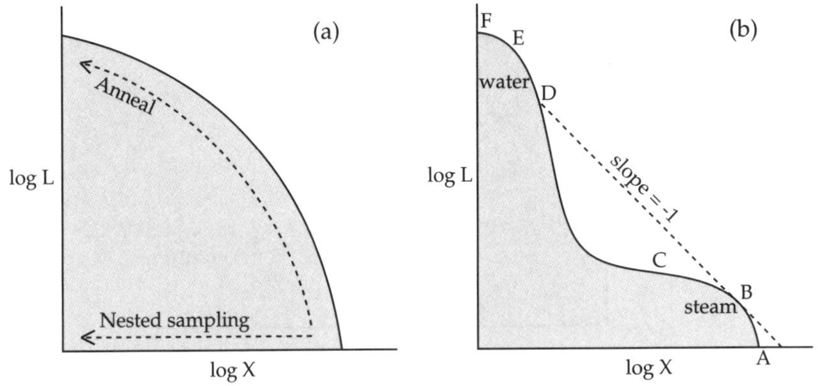

Figure 4: *Nested sampling tracks the abscissa, but annealing tracks the tangent (a), whilst having difficulty with convex likelihood (b).*

Many systems have convex regions in these diagrams, which are never seen because they are intrinsically unstable. Prepare H_2O in the intermediate state between water and steam, if it were possible, and it would immediately split into those two phases. This is a general phenomenon, being a hallmark of collective behaviour. It is reflected in the likelihood functions that model them (Fig. 4b).

No matter what schedule is adopted, annealing is supposed to follow the concave hull of the log-likelihood function as its tangential slope flattens. But this will require jumping right across any convex '$\smile$' region that separates ordinary concave '$\frown$' phases where local maxima of $L^\beta X$ are to be found. At $\beta = 1$, the bulk of the posterior should lie near a maximum of LX, in one or other of two different phases. Call the outer phase 'steam' and the inner phase 'water', as suggested by the supposedly large difference in volume. Annealing to $\beta = 1$ will normally take the ensemble from the neighbourhood of A to the neighbourhood of B, where the slope is $d\log L/d\log X = -1/\beta = -1$. Yet we actually want samples to be found from the more-important inner phase beyond D, finding which will be exponentially improbable unless the intervening convex valley is shallow. Alternatively, annealing could be taken beyond $\beta = 1$ until, when the ensemble is near the point of inflection C, the supercooled steam crashes inward to chilled water, somewhere near F. It might then be possible to anneal back out to unit temperature, reaching the desired water phase near E. However, annealing no longer bridges smoothly during the crash, and the value of the evidence is lost. Lost along with it is the internal Bayes factor Pr(states near E)/Pr(states near B), which should have enabled the program to assess the relative importance of water and steam. If there were three phases instead of just two, annealing might fail even more spectacularly. It is quite possible (Fig. 5) for supercooled steam to condense directly to cold ice, and superheated ice to sublime directly to hot steam, without settling in an intermediate water phase at all. The dominant phase could be lost in the hysteresis, and inaccessible to annealing.

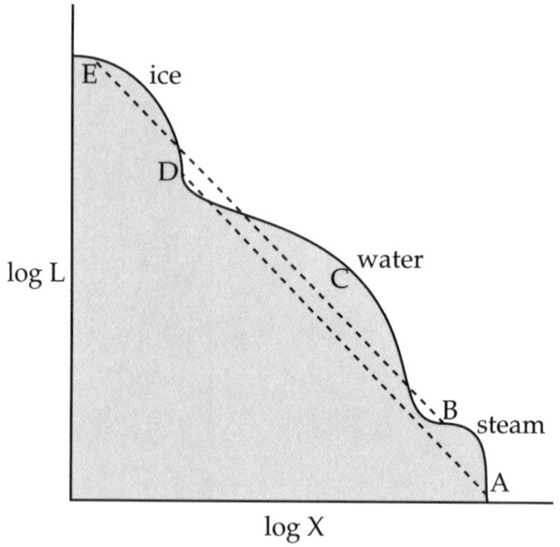

Figure 5: *Dominant 'water' phase is lost in the hysteresis loop* ABED. *The desired region around* C *will be missed by annealing.*

This example shows *an alarming generic failure of thermal methods*: they cannot cope with convex log-likelihoods. There are no convex thermal bridges between different concave phases. At least in retrospect, this is not a surprise. Thermal methods view parameter-space through the lens of a Laplace transform, in which quantities are modulated by and summed over the exponential factor $e^{-\beta E}$ where $E = -\log(L)$ is the thermal 'energy'. Recovering information from a Laplace transform is a notoriously ill-conditioned problem, because the transform destroys all subtlety of detail. Even more seriously, the convex bridge between concave phases is permanently inaccessible no matter how β is manipulated, making it impossible to relate such phases correctly.

Nested sampling, though, marches steadily down the abscissa X, with its hard-edged constraint keeping matters in focus. The likelihood tracks along ABCDEF$\cdots$ regardless of whether the function is concave or convex or even differentiable at all. In fact, it is invariant to *any* monotonic re-scaling of likelihood values. It doesn't matter what the values are, provided they preserve their order, and the contours stay the same shape. In the example, nested points will pass through the steam phase to the supercooled region, then steadily across the convex bridge into superheated water until the ordinary water phase is reached, traversed, and left behind in an optional continued search for ice. All the internal Bayes factors are available, so the dominant phase can be identified and quantified. There is no analogue of temperature, so there can never be any thermal catastrophe. The only danger, in a potentially-convex situation, is of premature termination, because it will seem that Z has converged even though there may actually be an alternative phase of importantly-high likelihood in a small domain yet to be found.

6. HOW TO UPGRADE TO NESTED SAMPLING

Exploration to find new points is conventionally handled by MCMC, using a transition scheme $\theta \leftrightarrow \theta'$ that is faithful to the prior, meaning that forward and backward transitions balance when points θ are populated proportionally to $\pi(\theta)$. Usually, coordinates are chosen in which the prior is flat, though that is not necessary. The possibly-annealed likelihood factor L^β (assuming a concave log-likelihood) is handled by imposing the Metropolis–Hastings rejection rule:

$$\text{Accept proposed } \theta' \text{ if and only if } L(\theta')^\beta \geq L(\theta)^\beta \times \texttt{Uniform}(0,1). \tag{18}$$

Nested sampling can use the same MCMC transition scheme, but rejects transitions (Fig. 6) according to

$$\text{Accept proposed } \theta' \text{ if and only if } L(\theta') \geq L_{\text{constraint}}. \tag{19}$$

This minor simplification is the only change needed to the exploration procedure. Well – not quite. A discrete problem might have finite prior mass associated with a particular likelihood value. This would make the `Sort` operation ambiguous and hence irreversible, to the detriment of nested sampling. To fix this, assign a numerical 'key' to each new trial θ, chosen at random from some library large enough that repeats will not occur. This trick amounts to promoting the dimension of $\theta \in \Theta$ to include this extra 'key' variable $\ell \in \mathbb{R}$, which is initially unknown. Then, in the event that likelihood values coincide, compare the keys to decide which is the greater. `Sort` is then unambiguous, because modified likelihoods will never

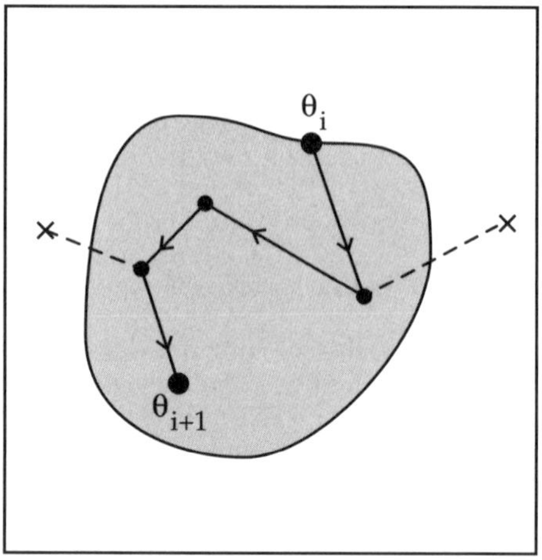

Parameter space of θ

Figure 6: *Transitions to new θ within the likelihood constraint ($\bullet\!\!\longrightarrow\!\!\bullet$) are accepted, those to outside ($\bullet - - - - \times$) are rejected.*

be equal. In effect, the likelihood L is modified to $L + \epsilon\ell\ : \Theta \times \mathbb{R} \to \mathbb{R}$, where ϵ is smaller than any arithmetical precision that will be encountered elsewhere.

Existing refinements (Gibbs sampling, slice sampling, importance sampling and so on) can be used as before. At comparable stages in the calculations, the tolerable step lengths are much the same for (18) and (19), so nested sampling offers no gain or loss of efficiency here. For example, the speed of exploration within a 'hard' sphere of radius r is just the same as the speed within a 'soft' Gaussian sphere of the same variance. In both cases, the best step length for a random exploration direction is $\mathcal{O}(R^{-1/2}r)$, where R is the sphere's dimension, so that $\mathcal{O}(R)$ steps are needed to equilibrate the MCMC chain and reach a 'new' sample. The issue of how much of this equilibration is actually needed per iterate remains open, and presumably is little changed with nested sampling. Since about $\sqrt{R}$ points will contribute significantly to the eventual results, it may be acceptable to relax the equilibration by nearly this factor and still retain enough independence to be useful. No doubt such corner-cutting will depend on the particular application.

7. EXTENSION TO N POINTS

Instead of running with just one point, as above, nested sampling can be operated with N. An iterative step starts (Fig. 7a) with these N points sampled randomly (over the prior) from within the previous iterate's likelihood constraint. The outermost of these points, with smallest likelihood, is identified as θ_i, with likelihood $L_i = L(\theta_i)$, then discarded. The $N-1$ survivors remain random as before, but are confined within the shrunken constraint $L > L_i$ (Fig. 7b). One new point is then randomly generated within this same constraint to recover the full membership N and complete the loop (Fig. 7c). The θ-volume now covers this new point as well as the $N-1$ survivors, so all N are random and on the same footing. After the notional `Sort` operation, the corresponding X-range also holds these N points, still uniformly and independently sampled. The next point $i+1$ is the outermost of these, for which the shrinkage factor is distributed as the largest of N samples from `Uniform`(0,1). The distribution of shrinkage is hence modified from uniform (11) to

$$\Pr(t) = Nt^{N-1} \text{ in } 0 < t < 1 . \tag{20}$$

The corresponding statistics are now $\log(t) = (-1 \pm 1)/N$. By focussing on the outer 1 part in N of the X-range, nested sampling gains extra resolution and, thereby, precision. As might be expected, the algorithm goes N times more slowly, thus generating N times more posterior points, and decreasing the uncertainties by the usual factor $\sqrt{N}$. The algorithm still bridges inward from the prior to and beyond the posterior, but its steps are smaller, with more overlap between successive ranges.

With the original single point ($N = 1$), the natural way of generating the required new point was to explore with MCMC, using the original as the seed. With $N > 1$, it is more natural to copy a randomly chosen survivor (which is already a supposedly-unbiassed sample within the chosen region), and evolve the copy by MCMC.

The standard nested-sampling procedure is now complete:

Start with N points θ from the prior $\pi(\theta)$;
  initialise $X_0 = 1, Z = 0, H = 0$.
Repeat for $i = 1, 2, 3, \ldots$;
  record the lowest of the current likelihood values as L_i,
  proclaim $X_i = \exp(-i/N)$, or, better, use several samples,
  set $w_i = X_{i-1} - X_i$, or, possibly, use trapezoid rule,
  increment Z by $w_i L_i$, and update H similarly,
  replace the point of lowest likelihood (assuming $N > 1$),
    with a copy of a random survivor, then
  evolve this within $L(\theta) > L_i$, proportionally to the prior.
Terminate when Z appears to have converged.
Result is $\log(Z) \pm \sqrt{H/N}$, or, better, get $\Pr(\log Z)$ from samples.

Note that the $\mathcal{O}(N^{-1/2})$ statistical uncertainty $\pm\sqrt{H/N}$ in the logarithm much exceeds any $\mathcal{O}(N^{-1})$ or $\mathcal{O}(N^{-2})$ bias introduced by choice of integration rule, legitimising the un-sophisticated assignment of column width w in (6).

For illustration, consider the limiting case where $L = 1$ within prior mass X, but $L = 0$ outside. The true values are simply $Z = X$ and $H = -\log X$. During nested sampling, the i'th enclosed mass is $X_i = t_1 t_2 \ldots t_i$ (X_0 being 1). Each logarithmic shrinkage $-\log t$ is distributed according to (20) as Beta with mean $1/N$ and variance $1/N^2$. Each is independent, so $-\log X_i$ is distributed as Beta with mean i/N and variance i/N^2. Consequently, the number of steps $k = NH \pm \sqrt{NH}$ needed to reach the likelihood boundary is Poisson with mean NH.

Likelihoods before this are 0, and after it they are 1. Hence, from (6) and (7), the estimate of Z will be X_k (assuming adequate termination). Accordingly, $\log Z = -(k \pm \sqrt{k})/N$ has the well-behaved Beta distribution of X_k. The net effect of randomness during the run, and inferential uncertainty about its track, is that $\log Z$ is recovered with standard error $\sqrt{H/N}$, and that this $\mathcal{O}(N^{-1/2})$ error is faithfully reported as the uncertainty involved. Asymptotically for large N, the correct result is almost certainly recovered. Of course, Z itself is likely to have a wildly skew distribution because of the exponentiation needed to recover it from its logarithm.

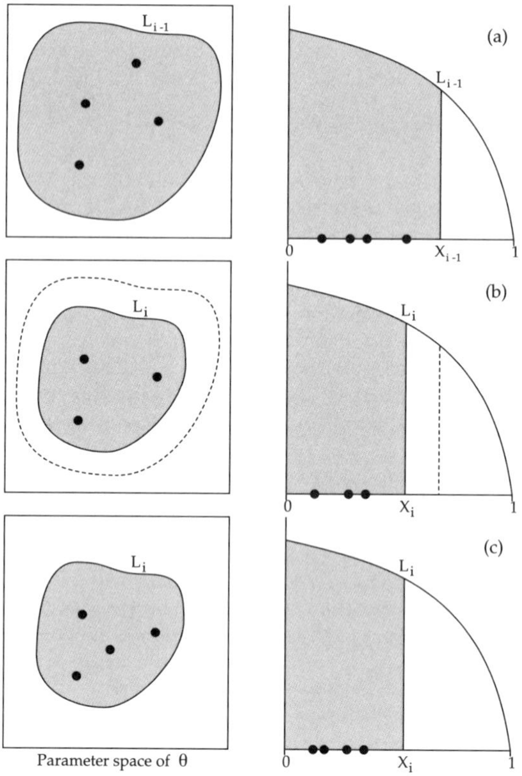

Figure 7: *(a) Nested sampling with $N = 4$ points within the $(i-1)$'th constraint $L > L_{i-1}$; (b) after the outermost point is identified as the i'th and discarded; (c) with one new point within L_i, completing the loop.*

8. COPYING AND MULTI-MODALITY

It is a detail, and it is really only suggested by nested sampling and not required, but the 'copy' operation offers a significant advantage. Suppose the likelihood is multi-modal (Fig. 8), breaking into separate peaks above a critical threshold. Soon after passing the threshold, it becomes practically impossible for MCMC transitions to reach the other peak. Nested sampling requires correct sampling over the whole likelihood-constrained region, whether or not it is simply-connected. Yet MCMC exploration alone is incapable of doing this, and will trap samples in the 'wrong' peak with inferior likelihood bound.

However, the 'copy' operation always replaces the discarded point with one of higher likelihood, regardless of that point's position. Points trapped in the 'wrong' peak will systematically be replaced with copies of points in the 'right' peak, as and when their likelihood advantage becomes apparent. Indeed, when the likelihood bound has climbed above the inferior summit of the 'wrong' peak, that can contain no points at all. All N points will find themselves exploring the 'right' peak, quite

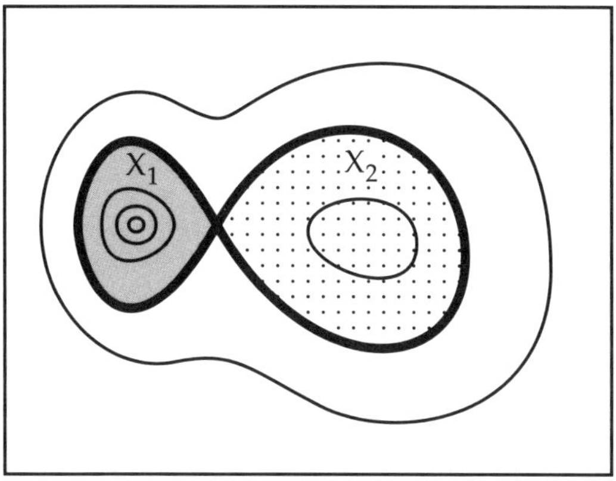

Figure 8: *The threshold likelihood contour encloses a narrow but high 'right' peak X_1, and a wide but low 'wrong' peak X_2.*

automatically and without any specific attention. All that really matters for a good chance of success is that N is large enough for at least one point to find the 'right' peak as the threshold is crossed.

9. EXAMPLES

9.1. *Gaussians*

Let the coordinates θ have uniform prior over the 20-dimensional unit cube $[-\frac{1}{2}, \frac{1}{2}]^{20}$, and let the likelihood be

$$L(\theta) = 100 \prod_{i=1}^{20} \frac{1}{\sqrt{2\pi}u} \exp\left(-\frac{\theta_i^2}{2u^2}\right) + \prod_{i=1}^{20} \frac{1}{\sqrt{2\pi}v} \exp\left(-\frac{\theta_i^2}{2v^2}\right) \tag{21}$$

with $u = 0.01$ and $v = 0.1$. This represents a Gaussian 'spike' of width 0.01 superposed on a Gaussian 'plateau' of width 0.1. The Bayes factor favouring the spike is 100, and the evidence is $Z = 101$. There is only a single maximum, at the origin, and this should surely be an easy problem. Yet $L(X)$ is partly convex (Fig. 9), and an annealing program restricted to $\beta \le 1$ needs roughly a billion (e^{20}) trials to find the spike, and several times e^{25} to equilibrate properly. On the other hand, H is only 63.2, so nested sampling could reach and cross the spike and cover the whole range of Fig. 9 in a mere 100 iterates.

Admittedly, about $N = 16$ objects would be needed if the uncertainty from

$$\log Z \approx \log(101) \pm \sqrt{63.2/N} \tag{22}$$

(and hence in the spike/plateau Bayes-factor logarithm) needs to be reduced to the ± 2 or so required to identify the favoured (spike) mode with reasonable confidence. That multiplies the computational load to something like 1600 evaluations, though this remains comfortably less than e^{25}.

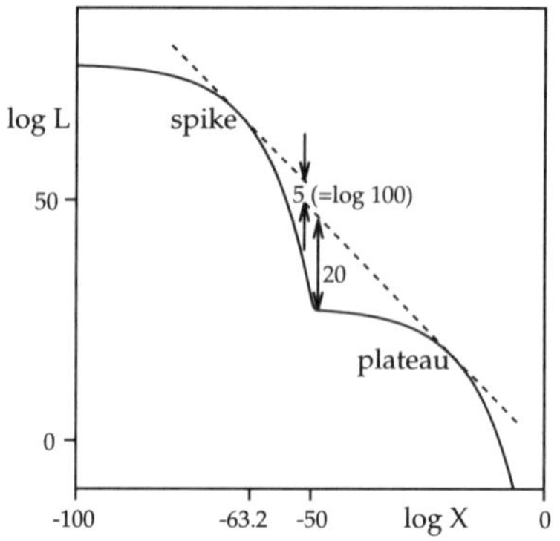

Figure 9: *Gaussian spike on plateau. The spike is favoured by a Bayes factor exp(5), but annealing needs exp(20) trials to find it.*

On the other hand, if the spike was moved off-centre to $(0.2, 0.2, 0.2, \ldots)$, with likelihood

$$L(\theta) = 100 \prod_{i=1}^{20} \frac{1}{\sqrt{2\pi}u} \exp\left(-\frac{(\theta_i - 0.2)^2}{2u^2}\right) + \prod_{i=1}^{20} \frac{1}{\sqrt{2\pi}v} \exp\left(-\frac{\theta_i^2}{2v^2}\right) \tag{23}$$

then nested sampling too would be in difficulty. There are now two maxima over θ and, at the separatrix contour above which the phases separate, the aperture of the plateau is e^{35} times greater than that of the spike. This means that some huge number of trials is needed to have a good chance of finding the spike, even though the $\log L / \log X$ plot is indistinguishable from Fig. 9. That's impossible in practice. General multi-modality remains difficult.

Of course, this particular problem is easily soluble by splitting it into its constituent parts and evaluating the two evidence values separately. With evidence values being so easy to compute, it's better to split problems (where possible) than to attempt joint computation. Splitting avoids the dangerous mismatch between initial gate widths and final Bayes factors: it's better to avoid a problem than to have to solve it. More generally, it's preferable to do separate calculations when trying to compare models of different types (as in cosmology: Mukherjee, Parkinson and Liddle, 2006).

9.2. *Data Analysis*

It is not just large problems with awkward likelihood functions that exhibit phase changes. In data analysis, it frequently happens that there is an initial 'it's all just noise' phase to be overcome before the true interpretation of the signals emerges.

As trivial illustration, consider a small experiment to measure the single coordinate θ, over which the prior $\pi(\theta)$ is flat in (0,1). Its data D yield the likelihood function (Fig. 10)

$$L(\theta) = 0.99 \frac{2q^2}{(\theta + q)^3} + 0.01 \quad \text{with, say, } q = 10^{-9}. \tag{24}$$

This is already a decreasing function, and the sorting operation of nested sampling is just the identity, $X = \theta$.

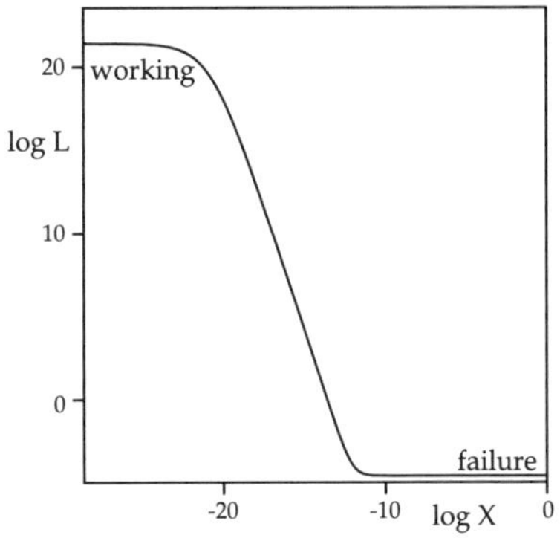

Figure 10: *The 'working' phase on the left is hard to find by annealing.*

An interpretation of (24) is that the experiment was anticipated to work with 99% reliability. If it worked, the likelihood $L = 2q^2/(\theta + q)^3$ would have been appropriate, meaning that $\theta \approx 10^{-9}$ was measured. If it failed, which was anticipated 1% of the time, the likelihood would have been the uninformative $L = 1$, because the equipment would just return a random result. Under annealing, the original hot phase is the failure mode. An annealed ensemble limited to $\beta \leq 1$ is most unlikely to find the 'working' mode unless it is allowed about a billion trials, and will wrongly suggest 'failure', with $Z = 0.01$. Only if β is increased far beyond 1 to a million or so would the ensemble be likely to find the working mode in fewer than thousands of trials. Even then, the samples would crash inward and have to be annealed back out through that factor of a million. And the evidence value would have been lost.

For nested sampling, which steadily tracks $\log X$ instead of trying to use the slope, such problems are easy. All one needs is the determination to keep going for the NH or so shrinkage steps needed to reach and then cross the dominant mode with a collection of N objects. By then, the behaviour of $\log L$ as a function of $\log X$ has been found, so that any distinct phases can be identified along with their Bayes factors, as well as the overall evidence Z.

9.3. *Potts Model*

The Potts (or Ising) model is a two-dimensional rectangular model of atoms of two changeable 'colours' (A and B). The loglikelihood counts the number of bonds between atoms of the same colour (A-A or B-B), with a coefficient set to the critical value:

$$\log L = \log(1 + \sqrt{2}) \times (\text{number of A-A} \ + \ \text{number of B-B}). \tag{25}$$

This is always concave, but a wide swathe of it becomes almost straight as the number of atoms increases. Fig. 11 plots it for a coarse (18×18) grid, and much of the curve is already nearly straight. Supercooling is never needed, but a tiny change in temperature moves the bulk of the posterior from one end of the straight section to the other. Because there is no actual convexity, this is an example of a 'second order' phase transition. It is difficult to anneal it properly, but nested sampling moves steadily inward towards the fully-ordered states without any difficulty, exploring intermediate partially-ordered states on the way.

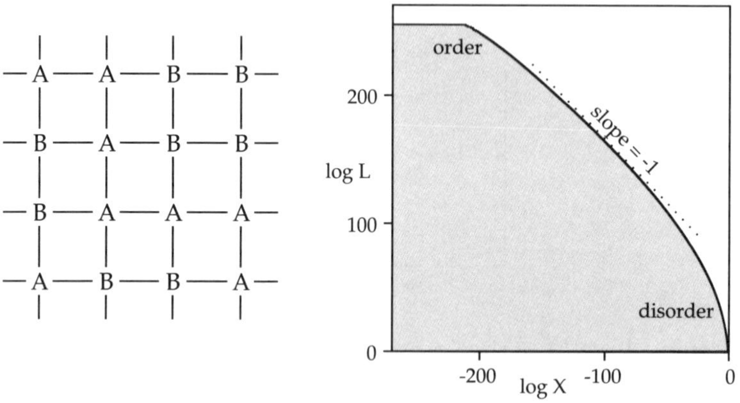

Figure 11: *Potts model (left), on* 18×18 *grid (right). The transition region between disorder and order is almost straight.*

Murray *et al.* (2006) have programmed this on a 256×256 grid using the Fortuin–Kasteleyn–Swendsen–Wang exploration strategy (Edwards and Sokal, 1988), and found the anticipated steady increase in order.

10. REMARKS

Nested sampling reverses the traditional approach by putting equation (1) (the evidence) in its proper place before equation (2) (the posterior). It also incorporates an estimate of uncertainty, arising naturally from straightforward Bayesian reasoning.

The hard constraint of nested sampling is simpler than the soft constraint of thermal methods (which are anyway fallible), and it needs no annealing schedule, so it promises to be more robust.

Nested sampling is invariant under monotonic re-labelling of the likelihood, and behaves sensibly and usefully with multi-modal likelihoods. These properties should give the method increased power in multi-phase applications.

Finally, the idea is extremely simple. It is reminiscent of the fable of how a mathematician finds a needle in a haystack of S straws. The mathematician keeps cutting the haystack in half and discarding the half not containing the needle, thus reaching it in $\log_2(S)$ cuts instead of the naive expectation $S/2$. The Bayesian nested-sampler does much the same, but cutting at random is a little better, and reaches the posterior results in $H = \log_e(S)$ steps.

Remarkably, Bayesian computation *could* have been founded this way in 1953 (by Metropolis *et al.*). All the tools were then in place. Perhaps it should have been, then half a century of diversion into fragile thermal methods would have been avoided.

ACKNOWLEDGMENTS

My primary debt is to the work of Richard Cox (1946) who proved that probability theory was not just a convenient and/or plausible system, but was forced by the basic properties required of rational inference. Hence we can safely ignore alternatives and analogies, and concentrate on the only allowable general calculus. This robustly productive philosophy reached me through my colleague Steve Gull's study of the work of Edwin Jaynes, as collected in the posthumous volume (Jaynes 2003). I was thus able to return to thinking about inference, after having rejected statistics as a student precisely because of the anti-logical frequentist hypothesis-testing that my teachers espoused. More recently, I wish to thank David MacKay and the Inference group journal club at Cambridge for valuable encouragement and advice, and for hosting nested-sampling papers and programs (currently available for download at www.inference.phy.cam.ac.uk/bayesys). Helpful comments from delegates at Valencia8, in particular my discussant Michael Evans, have led to improvements in this final draft.

This work was supported by Maximum Entropy Data Consultants Ltd of Cambridge (England).

REFERENCES

Cox, R. T. (1946). *The Algebra of Probable Inference.* Baltimore: John Hopkins University Press.

Edwards, G. and Sokal, A. (1988). Generalization of the Fortuin-Kasteleyn-Swendsen- Wang representation and Monte Carlo algorithm, *Phys. Rev. D* **38**, 2009–2012.

Jaynes, E. T. (2003). *Probability Theory: the Logic of Science* (L. Bretthorst, ed.) Cambridge: University Press.

Metropolis, M., Rosenbluth, A. W., Rosenbluth, M. N., Teller, A. H. and Teller, E. (1953). Equation of state by fast computing machines, *J. Chem. Phys.* **21**, 1087–1092.

Mukherjee, P., Parkinson, D. and Liddle, A. R. (2006). Nested Sampling Algorithm for Cosmological Model Selection. *Astrophys. J.* **638**, L51–L54.

Murray, I., MacKay, D. J. C., Gharamani, Z. and Skilling, J. (2006). Nested Sampling for Potts Models. *Advances in Neural Information Processing Systems* **18**.

Parkinson, D., Mukherjee, P. and Liddle, A. R. (2006). *Bayesian Model Selection Analysis of WMAP3,* submitted to PRD, arXiv: astro-ph/0605003.

Sivia, D. S. and Skilling, J. (2006). *Data Analysis: a Bayesian Tutorial* (2nd ed., Chapters 9 and 10) Oxford: University Press.

Skilling, J. (2004). Nested sampling. *Amer. Inst. Phys. Conference Proc.* **735**, 395–405.

Skilling, J. (2006). Nested sampling for general Bayesian computation. *Bayesian Analysis* **1**, 833–860.

DISCUSSION

MICHAEL J. EVANS (*University of Toronto, Canada*)

Introduction. The paper contains a number of interesting contributions. We have chosen to comment on two aspects of the paper, namely, the statements concerning the necessity of computing the prior predictive density (the evidence) at the observed data prior to implementing a posterior analysis, and the integration algorithm presented in the paper and referred to as nested sampling.

What is a Bayesian analysis? The ingredients to a Bayesian analysis comprise the sampling model $\{P_\theta : \theta \in \Omega\}$ for the response X, the prior Π for the parameter θ, and the observed value $X = x$. This is equivalent to specifying the joint model $P_\theta \times \Pi$ for (X, θ) and the observed value $X = x$. In many situations we may also have a loss function but we ignore this here as it is not material to our discussion. We refer to $P_\theta \times \Pi$ as the Bayesian model. The Bayesian model and the data comprise the full information available and we ask how this information is to be used in carrying out a statistical analysis.

The joint model can be factored as $P_\theta \times \Pi = M \times \Pi(\cdot \,|\, X)$ where M is the marginal prior predictive for X and $\Pi(\cdot \,|\, X)$ is the posterior for θ. The principle of conditional probability then says that probability statements about θ, that are initially based on the prior Π, should instead be based on the observed posterior $\Pi(\cdot \,|\, X = x)$. What then, is the role of M in a statistical analysis?

If our goal is inference about θ, then it might seem that we can ignore M and proceed directly to work with $\Pi(\cdot \,|\, X = x)$. This may seem even preferable if it is possible to compute or sample from $\Pi(\cdot \,|\, X = x)$ without making any direct reference to M. In fact this is a feature of the MCMC algorithms that have revolutionized Bayesian computing over the past few years. While this may have some appeal, it still leaves one wondering about the role that M might or should have.

A more direct approach to obtaining $\Pi(\cdot \,|\, X = x)$ proceeds as follows. If we denote the densities of P_θ and Π by f_θ and π respectively, then the density of M at x is given by $m(x) = E_\Pi(f_\theta(x))$ and the density of $\Pi(\cdot \,|\, X = x)$ is $\pi(\theta \,|\, x) = f_\theta(x)\pi(\theta)/m(x)$. We can then work with $\pi(\theta \,|\, x)$ to compute various posterior characteristics. This approach makes some reference to M but only through the value of its density at x and we are still left wondering about any role for the full distribution M in the analysis.

Dr. Skilling makes the point that $m(x)$ (denoted as Z and referred to in the paper as *the evidence*) is something of great importance and perhaps even comes logically before the computation of the posterior. The suggestion is made that this value "assesses the model in question' and, if it leads to doubts about the validity of the model, then there is no real need to proceed to the computation of the posterior.

The paper doesn't make clear how one is supposed to use $m(x)$ in this process, but the suggestion is made that, if we had two Bayesian models, leading to $m_1(x)$ and $m_2(x)$ respectively, then we would decide between them using the ratio $m_1(x)/m_2(x)$. This is the Bayes factor in favour of the first Bayesian model and equals $m_1(x)/m_2(x) = (1 - p_1)p_{1|x}/p_1(1 - p_{1|x})$ where p_1 is the prior probability of the first Bayesian model and $p_{1|x}$ is the posterior probability of this model. When more than two models are being considered this ratio is a similar function of the conditional prior and posterior probabilities of the first model given that only these two models are possible.

A point that can be made here, is that calibrating the value of $m_1(x)/m_2(x)$, i.e., saying when this quantity is small or large, depends intrinsically on the assignments of the prior probabilities. When these are known, then the size of the Bayes factor, i.e., is it strong evidence in favour of the first model or otherwise, is interpreted in terms of the posterior probabilities. It may be argued that the Bayes factor does not formally require the specification of the prior probabilities for the models and that this is a virtue of this approach. But then we think it is reasonable to question how we should interpret the value of $m_1(x)/m_2(x)$. So one could argue that in model selection problems the issues are a bit more involved than just the computation of the $m_i(x)$ values. Of course, the computation of the ratios of these values is absolutely necessary.

More generally, the value of $m(x)$ is involved in model criticism. If the value of $m(x)$ is a surprising value, then we have evidence that the Bayesian model is incorrect in some sense. In such circumstances, proceeding to inferences about θ seems at least of questionable validity. It seems essential that checks be made on the ingredients put into an analysis to ensure that these make sense in light of the data obtained. In Box (1980) it was proposed that $m(x)$ be compared with its distribution induced by M to determine if $m(x)$ is surprising and, if it was, conclude that we had evidence against the validity of the Bayesian model.

Actually there are two ways in which a Bayesian model can fail. The sampling model can fail by the data being surprising for every distribution in $\{P_\theta : \theta \in \Omega\}$ or, if the sampling model is correct, the prior may place its mass primarily on values of θ for which the data are surprising. We refer to this second failure as prior-data conflict. The consequences of these two failures are somewhat different. With large amounts of data prior-data conflict can be ignored, as the data swamp the prior, while no amount of data can fix a bad sampling model. Accordingly, as argued in Evans and Moshonov (2006a), it seems sensible to check for these errors separately. First we check the sampling model, if no problems are detected we then check for prior-data conflict, and then depending on this assessment, possibly proceed to inference about θ.

In Evans and Moshonov (2006a) it is argued that the appropriate approach for checking for prior-data conflict is based on comparing the observed value $T(x)$, of a minimal sufficient statistic T, with its conditional prior-predictive distribution $M_T(\cdot \,|\, U(T(x))$ given an ancillary $U(T)$. This leads to the factorization $M = P(\cdot \,|\, T(x)) \times P_{U(T)} \times M_T(\cdot \,|\, U(T(x))$ where $P(\cdot \,|\, T(x))$ is the conditional distribution of the data given $T(x)$ and $P_{U(T)}$ is the marginal distribution of $U(T(X))$. Both $P(\cdot \,|\, T(x))$ and $P_{U(T)}$ are independent of the prior and so are available for checking the sampling model. Interestingly the lack of a unique maximal ancillary does not pose a problem in this context as we simply get different checks with different maximal ancillaries.

We see that this approach leads to using the full information available to the statistician in carrying out a Bayesian analysis. Each step in the analysis is based on a component of a factorization of the joint model. A further factorization of $M_T(\cdot \,|\, U(T(x))$ is sometimes available, as discussed in Evans and Moshonov (2006b), when we want to check for prior-data conflict with specific components of the prior. We also note that this approach avoids the double use of the data problems inherent in posterior model checks, as discussed in Bayarri and Berger (2000).

So we are in agreement with Dr. Skilling's assertion concerning the importance of looking at Z before proceeding to inference about θ and that logically this step cannot be avoided. As just discussed, however, we would go much further than just computing Z.

Nested sampling. Nested sampling appears to be a new integration method that is applicable to the evaluation of any integral. We will first focus on its application to the evaluation of $Z = E_\Pi\,(f_\theta\,(x))$, as is done in the paper, although there does not appear to be anything that makes this problem, as opposed to a general integration problem, particularly suitable for nested sampling. We modify Dr. Skilling's presentation slightly to bring the notation and terminology more in line with what is customary in probability and statistics. One simple approach to estimating Z is to generate a sample $\theta_1, \ldots, \theta_N$ from the prior Π and compute $N^{-1}\sum_{i=1}^{N} f_{\theta_i}\,(x)$. We refer to this as naive importance sampling and note that it is widely recognized as being inadequate for this problem.

To develop nested sampling we put $\psi = \Psi(l) = \Pi(f_\theta(x) > l)$ so that Ψ is the complementary cdf of the random variable $f_\theta(x)$ when $\theta \sim \Pi$ and x is fixed. Now let $\Psi^{-1}(\psi) = \sup\{l : \Psi(l) > \psi\}$. Then, if $\psi \sim U(0,1)$, we have that $\Psi^{-1}(\psi) \sim f_\theta(x)$ and note that this is true whether the distribution of $f_\theta(x)$ is discrete or continuous. Therefore we can write

$$Z = \int_0^1 \Psi^{-1}(\psi)\, d\psi \tag{26}$$

and note that the integrand Ψ^{-1} decreasing. We believe that the representation (26) summarizes what the paper means by ' sorting the likelihood'.

Since the integral in (26) is one-dimensional, we can approximate this by a Riemann sum $Z \approx \sum_{i=1}^m \Psi^{-1}(\psi_i)(\psi_i - \psi_{i+1})$ where $0 = \psi_{m+1} < \psi_m < \cdots < \psi_0 = 1$. The problem with this approximation is the need to calculate $\Psi^{-1}(\psi_i)$, which is the $(1-\psi_i)$-th quantile of the distribution of $f_\theta(x)$ when $\theta \sim \Pi$. Obviously, evaluating this is generally at least as hard as evaluating the original integral.

The paper proposes a novel approach that avoids the need to directly evaluate $\Psi^{-1}(\psi_i)$. There are a number of variants discussed in the paper and we focus on one of these but feel our comments apply generally. The first step in this is to suppose that the ψ_i are randomly generated. These quantities are defined iteratively via $\psi_i = t_1 t_2 \cdots t_i$ where t_i is distributed as the largest order statistic in a sample of N from the $U(0,1)$ distribution. It is then easily shown that $E(\psi_i - \psi_{i+1}) \sim e^{-i/N} - e^{-(i+1)/N}$ as $N \to \infty$. Accordingly we could instead use the approximation $Z \approx \sum_{i=1}^m \Psi^{-1}(\psi_i)(e^{-i/N} - e^{-(i+1)/N})$. This still seems to require the evaluation of $\Psi^{-1}(\psi_i)$, but now suppose we generate $\theta_{i1}, \ldots, \theta_{iN}$ from the conditional distribution $\theta \mid f_\theta(x) > f_{\theta_{i-1}}(x)$ and let $\theta_i \in \{\theta_{i1}, \ldots, \theta_{iN}\}$ be such that $f_{\theta_i}(x) = \min\{f_{\theta_{i1}}(x), \ldots, f_{\theta_{iN}}(x)\}$. We have to be able to implement this sampling but, when we can, then $\Psi^{-1}(\psi_i)$ has the same distribution as $f_{\theta_i}(x)$ and we can approximate (26) via the randomized estimator

$$Z_{m,N} = \sum_{i=1}^m f_{\theta_i}(x)\left(e^{-i/N} - e^{-(i+1)/N}\right). \tag{27}$$

It is of course necessary to show that (27) converges to (26) . For this we note that we can study the more general problem concerning the convergence of $\sum_{i=1}^m g(\psi_i)(\exp\{-i/N\} - \exp\{-(i+1)/N\})$ for a Riemann integrable $g : [0,1] \to R^1$. For the case of a continuous g we have the following result.

Theorem 1 *For continuous $g : [0,1] \to R^1$, and with the ψ_i generated as described, then $\sum_{i=1}^m g(\psi_i)(e^{-i/N} - e^{-(i+1)/N}) \to \int_0^1 g(\psi)\, d\psi$ in probability as $N \to \infty$ and $m/N \to \infty$.*

Proof. Let $\epsilon > 0$ and note that we can find a polynomial h such that $|g(\psi) - h(\psi)| < \epsilon$ for all $\psi \in [0,1]$. Then

$$\left|\sum_{i=1}^m g(\psi_i)\left(e^{-i/N} - e^{-(i+1)/N}\right) - \sum_{i=1}^m h(\psi_i)\left(e^{-i/N} - e^{-(i+1)/N}\right)\right|$$

$$
\begin{aligned}
&\leq \sum_{i=1}^{m} |g(\psi_i) - h(\psi_i)| \left(e^{-i/N} - e^{-(i+1)/N}\right) \leq \epsilon \sum_{i=1}^{m} \left(e^{-i/N} - e^{-(i+1)/N}\right) \\
&= \epsilon \left(e^{-1/N} - e^{-(m+1)/N}\right) \to 0
\end{aligned}
$$

as $N \to \infty$ and $m/N \to \infty$. Therefore, we can prove the result for g a polynomial. Further if we prove the result for all powers $g(\psi) = \psi^k$ with $k > 0$, then the result is established.

For $g(\psi) = \psi^k$ with $k > 0$ we proceed as follows. First we have that, under the sampling scheme specified for the ψ_i, then $E(\psi_i^k) = (E(t_1^k))^i = (1 + k/N)^{-i}$. This leads to

$$
\begin{aligned}
& E\left(\sum_{i=1}^{m} \psi_i^k \left(e^{-i/N} - e^{-(i+1)/N}\right)\right) \\
&= \frac{(1 - e^{-1/N})(1 + k/N)^{-1} e^{-1/N}}{1 - (1 + k/N)^{-1} e^{-1/N}} \left(1 - ((1 + k/N)^{-1} e^{-1/N})^m\right) \qquad (28)
\end{aligned}
$$

and, when $N \to \infty$ and $m/N \to \infty$, this converges to $E(\psi^k) = 1/(k+1)$. Reasoning in a similar fashion we can prove that

$$
Var\left(\sum_{i=1}^{m} \psi_i^k \left(e^{-i/N} - e^{-(i+1)/N}\right)\right) \to 0
$$

as $N \to \infty$ and $m/N \to \infty$. Then, using Markov's inequality, we have that

$$
\begin{aligned}
& P\left(\left|\sum_{i=1}^{m} \psi_i^k \left(e^{-i/N} - e^{-(i+1)/N}\right) - 1/(k+1)|\right| > \eta\right) \\
&\leq \eta^{-2} \left\{E\left(\sum_{i=1}^{m} \psi_i^k \left(e^{-i/N} - e^{-(i+1)/N}\right)\right) - 1/(k+1)\right\}^2 \\
&\qquad + \eta^{-2} Var\left(\sum_{i=1}^{m} \psi_i^k \left(e^{-i/N} - e^{-(i+1)/N}\right)\right)
\end{aligned}
$$

and the right-hand side converges to 0 establishing the result. □

While the proof is for continuous g, it seems that it can be generalized to arbitrary Riemann integrable functions and in particular for step functions. The function Ψ^{-1} will be a step function when $f_\theta(x)$ has a discrete distribution under Π. Applying the theorem when Ψ^{-1} is continuous, establishes that $Z_{m,N} \to Z$ in probability as $N \to \infty$ and $m/N \to \infty$. This also establishes that $\ln Z_{m,N} \to \ln Z$ in probability but doesn't tell us the rate of convergence. For various reasons we may be more interested in $\ln Z$ than in Z. If we can prove that $r_{m,N}(Z_{m,N} - Z) \to V$ in distribution, where V has mean 0 and variance σ^2, then the delta theorem tells us that $r_{m,N}(\ln Z_{m,N} - \ln Z)$ converges in distribution to V/Z and so the rate of convergence does not change by taking the logarithm. The asymptotic coefficient of variation of $Z_{m,N}$ is $\sigma/(r_{m,N}Z)$ while for $\ln Z_{m,N}$ it is $\sigma/(r_{m,N}Z \ln Z)$. Since Z will often be very large, the estimator $\ln Z_{m,N}$ will be a more accurate estimator of $\ln Z$ than $Z_{m,N}$ is of Z. Still, if $Z_{m,N}$ is a terrible estimator of Z, as will be reflected in an extremely large value of σ, then the improvement in accuracy obtained by using $\ln Z_{m,N}$ will be immaterial. To obtain a better understanding of nested sampling, and so determine if it can be used to obtain reliable results, we need to obtain the distribution of V, the rate $r_{m,N}$, and the dependence of σ^2 on Ψ^{-1}.

The theorem requires both $N \to \infty$ and $m/N \to \infty$ and it might be wondered if both conditions are necessary. Consideration of (28) for $g(\psi) = \psi^k$ shows that both of these conditions are necessary even to obtain asymptotic unbiasedness. Restricting nested sampling to decreasing functions, such as Ψ^{-1}, does not remove the need for both conditions, as consideration of the functions $g(\psi) = 1 - \psi^k$ for $k > 1$ shows. There is a variant of nested sampling where N uniformly distributed points are added in the final interval $[0, \exp\{-(m+1)/N\}]$. For this variant, if we fix m and let $N \to \infty$, then we will have $Z_{m,N} \to Z$ in probability. But note that $\exp\{-(m+1)/N\} \to 1$ and so this algorithm is really nothing more than naive importance sampling. When $m/N \to \infty$, then the contribution in $[0, \exp\{-(m+1)/N\}]$ is of no importance as the mass in this interval goes to 0. If we fix N and let $m \to \infty$, then the estimator will be asymptotically biased as (28) shows with $g(\psi) = \psi^k$ for $k > 0$. For example, with $k = 1, N = 1$ the asymptotic bias is 28% of the true value. Basically it seems that we have to choose N large enough so that the bias is small and then choose m, larger than N, to make the variance small. Replicating a biased estimator and averaging, as the paper seems to suggest, will not get rid of the bias. If the bias is large, nothing is really accomplished by this.

We might also consider studying the related quadrature rule $\sum_{i=1}^{m} g(\exp\{-i/N\})$ $(\exp\{-i/N\} - \exp\{-(i+1)/N\})$. This can also be shown to converge to $\int_0^1 g(x)\,dx$ as $N \to \infty$ and $m/N \to \infty$ for continuous g. We might ask for what functions g is this quadrature rule particularly effective and see if these correspond to the kinds of Ψ^{-1} functions encountered in practice.

As previously noted, nested sampling can be applied to any integration problem. When w is a probability density with respect to a support measure μ, we have that

$$\int_{\mathcal{X}} f(x)\,\mu\,(dx) = \int_{\mathcal{X}} \frac{f(x)}{w(x)} w(x)\,\mu\,(dx) = E_w\left(\frac{f(X)}{w(X)}\right)$$

and we can apply the nested sampling approach to $f(X)/w(X)$ with $X \sim w$. From this point-of-view nested sampling reminds us somewhat of randomized quadrature rules as discussed, for example, in Evans and Swartz (2000). Like randomized quadrature, nested sampling could be seen as a potentially useful variance reduction technique in the context of importance sampling.

In a specific problem it seems difficult to determine how to choose m and N, but perhaps study of the asymptotic distribution and the rate of convergence will provide guidance. Certainly it doesn't seem likely that naive choices of these characteristics will result in successful approximations. It seems plausible to us that the curse of dimensionality, that applies to high-dimensional integration generally, is hidden in the values of m and N that are required to produce useful approximations. Many integration algorithms that seem to avoid the curse of dimensionality really do not. For example, importance sampling and Metropolis–Hastings require the choice of good samplers to be effective and there is no easy way to find these for arbitrary high-dimensional integration problems.

Overall we feel that Dr. Skilling has produced an interesting new approach to approximating integrals. Transforming an integration problem to a one-dimensional problem and then using the trick for avoiding the evaluation of the function Ψ^{-1} at a specific point, seems particularly ingenious. Only more study of the algorithm, however, will reveal the extent to which it can be effectively used.

N. FRIEL (*University of Glasgow, UK*) and
A. N. PETTITT (*Queensland University of Technology, Brisbane, Australia*)

This is an interesting paper with a focus on computing the marginal likelihood or evidence. As statisticians we would disagree with the paper's assertion that the evidence should come first. Instead, the informal process of the assessment of prior knowledge,(exploratory) data analysis, simulation and model diagnostics would provide insight into an appropriate model. However, we do agree with the paper giving importance to calculation of the marginal likelihood as, in some cases, it is of foremost importance. For example, in estimation of motor unit numbers (Ridall *et al.*, 2006), in relation to motor neurone disease, the posterior probability of a given value being the number of units is found by calculating a marginal likelihood. Elsewhere, Friel and Pettitt (2005), we have presented a method, called the power posterior, given by $L^{\beta}X$ in the notation of the paper, for calculating the marginal likelihood. This is identical to the paper's annealing method. We have found that the power posterior/annealing method to be a useful method for complex hierarchical models such as the motor unit number estimation where the model is essentially a mixture model with random inclusions of components dependent on a covariate. This model consists of several hierarchies of latent variables so that the powered/annealed complete likelihood is a product of Gaussian densities.

The paper by Skilling suggests that the power posterior/annealing approach would have difficulties if the power posterior $L^{\beta}X$ is not concave. However, we suggest that in many, often used statistical models use of latent variables expresses the complete likelihood so that it is in the exponential family. It is the complete likelihood which is powered/annealed rather than the likelihood derived by marginalising over the latent variables. The powering method is relatively straightforward to implement and monitor given that the original model has been implemented. The three examples presented in the paper appear to be somewhat contrived in nature in order to illustrate problems with the annealing method. The examples are not representative of models found generally in statistical analysis.

NICOLAS CHOPIN (*University of Bristol, UK*) and
CHRISTIAN P. ROBERT (*Université Paris Dauphine and CREST-INSEE, France*)

The approximation of marginal densities is central to the Bayesian approach to testing of hypotheses since ratios $m_1(x)/m_2(x)$ of those marginals provide Bayes factors. It is thus of interest to see the emergence of a novel proposal for the approximative computation, although we are less confident than the author about the applicability of nested sampling in realistic Bayesian problems.

Rewriting Z as an integral over $[0, 1]$. A first difficulty stems from the convoluted presentation of the equivalence between Z in Eqn (1) and z in Eqn (3). We do not see why a discretisation and ordering would be necessary at this stage. Indeed,

$$Z = \mathbb{E}^{\pi}[L(\theta)] = \mathbb{E}^{\tilde{\pi}}[L] = \int_0^{\infty} X(\lambda)\, \mathrm{d}\lambda$$

where $\tilde{\pi}$ denotes the distribution of $L(\theta)$, associated with the cdf $(1 - X(\lambda))$. So it is only under the minimal restriction that X is strictly decreasing that we have the representation of Eqn (3), since

$$Z = \int_0^{L^{\max}} \ell\, \mathrm{d}X(\ell) = \int_0^1 X^{-1}(x)\, \mathrm{d}x\,.$$

We may add at this point that the use of the same notation for the likelihood L and the inverse of the complementary cdf $X(\lambda)$ is unnecessarily confusing. (Another difficulty that is not alluded to in the paper but that we cannot discuss here is the case of an unbounded likelihood, as for instance in the simple case of a two component Gaussian mixture.)

This representation, while not novel, is quite interesting because it looks at a familiar quantity from an unusual perspective.

Constrained sampling . However, this representation involves the implicit function $X(\lambda)$, which is approximated quite crudely in the remainder of the paper, and the corresponding algorithm requires in addition a constrained sampling that is certainly not *'easier than the traditional Metropolis–Hastings sampling involving likelihood-weighting and detailed-balance'.* We also note that the corpus of work on bridge and path sampling for Bayes factor approximation (Chen *et al.*, 2000) *is omitted,* as is the possibility of using reversible jump techniques (Green, 1995) to explore simultaneously a series of models.

The second point is that Eqn (3) is rather useless in justifying the corresponding nested sampling algorithm since it is simply based on the Riemann decomposition (Robert and Casella, 2004, Chapter 2)

$$Z = \int_0^{L^{\max}} \ell \, \mathrm{d}X(\ell) = \sum_{i=1}^{\mathbf{m}} \int_{L_{i-1}}^{L_i} \ell \, \mathrm{d}X(\ell) \approx \sum_{i=1}^{\mathbf{m}} L_i(X_{i-1} - X_i)$$

that works no matter what the values L_i are, as long as the differences $(X_{i-1} - X_i)$ all converge to 0 with m going to infinity (which happens to not be the case here, but hopefully the first differences should have a extremely small contribution in terms of L-values.) The choice made in the paper of simulating the θ_i's from $\pi(\theta)$ is thus pertinent but not necessarily optimal. In particular, vague priors are likely to induce a lack of efficiency when the likelihood function is quite concentrated within the support of π. (In importance sampling, simulation from the prior is only done to ensure finiteness of the variance, as in defensive sampling, Robert and Casella, 2004, Chapter 3). And, obviously, it is impossible to simulate from an improper prior.

From an algorithmic point of view, the experience of perfect sampling (Mira *et al.*, 2001) shows that simulating from $\pi(\theta)$ restricted under the constraint $L(\theta) \geq L(\theta_i)$ is not always possible. In large dimension spaces, simulating from the prior till the constraint is satisfied is unrealistic: simple calculations involving say Gaussian distributions are enough to show that this becomes exponentially difficult in the dimension of the problem. In that respect, Figs 1 and 2 (right side) are quite optimistic, as we would rather expect a very sharp peak on the left, which would fall to zero almost immediately. More fundamentally, sampling from MCMC offers no clear justification for a finite number of simulations. (A single iteration clearly does not work since the chain may remain at the same place and thus repeat the value of $L(\theta_i)$.) Adding a single value to the sample of N points at each iteration of the algorithm also seems quite inefficient, compared with a new generation of a constrained sample of N points, because the chances of getting high values of $L(\theta)$ are then necessarily much lower. We also note that the multimodality issue is not treated in a completely satisfactory manner: if N is too small, the initial sample may miss a narrow but primary mode and it is then quite uncertain whether or not this mode can be recovered at a later stage.

Approximating the X_i's. Our main concern however is with the rather fleeting description of how the X_i's, the X-values attached to the simulated θ_i's, are obtained in Section 7. Sampling X_i (independently?) does not seem to make sense, while approximating it by $\exp(-i/N)$ is rather crude, unless N is very large. And even in that case, to establish if, how quick, and in which sense our approximated integral does converge, in N but also in m, is far from obvious. This is especially true if we follow the author's suggestion to recycle the $N-1$ simulated X_i's that remain after deleting the lowest value.

We propose an alternative description of this particular aspect of the algorithm, which we hope will clarify things: in an initial version, a pair (X_i, L_i) is simulated with respect to an appropriate distribution (constrained prior), in a way that ensures that $L_i = X^{-1}(X_i)$. To improve the algorithm through a form of Rao-Blackwellisation, we replace $\log(X_i)$ by its expectation $-i/N$, but keep L_i as random. This second algorithm is correct provided either (a) the simulated L_i has expectation $\mathbb{E}[L_i] = X^{-1}(\exp(-i/N))$; or (b) it converge to this value in some sense. Condition (a) allows for unbiased estimation (thanks to the linearity of the approximated integral), but should be met only if $X^{-1}(\exp(\cdot))$ is linear, a constrain that never holds. Condition (b) is more reasonable, but should make it difficult to establish convergence results, as the *joint* convergence of all L_i (along iterations) would have to be established.

Moving along X axis at a geometric rate. Because of the curse of dimensionality mentioned above, even geometric steps (along the X-dimension) may not be fast enough to reach the likelihood mode in a reasonable number of iterations. This problem seems to be aggravated by the fact that, in the version of the algorithm with N points, the ratio X_i/X_{i-1} is then the largest of N uniform variates, and therefore should be close to one. This is yet another complication for proving any form of convergence.

More examples. Rather than the toy problems exposed in the paper (which we reproduced to convince ourselves), it would have been nice to have the method illustrated by realistic statistical examples.

REPLY TO THE DISCUSSION

Evans. Prof. Evans has made some interesting comments and developments, in particular usefully translating my 'physics-style' symbols into standard statistical notation, as shown in Table 1.

My symbols are fewer and simpler, mostly because nested sampling *is* simple, and also as a courtesy to the more general readership that I have in mind. I do use one extra symbol, H, for the information that's a central feature of nested sampling. On the other hand, I do not define Ω because it's already implied by θ being a parameter that its dominion set exists, and no other use of Ω is made. Neither do I use the response set $X = \{x\}$ of all possible observed values, on the logical grounds that the observed values are already constrained to the specific data that define the likelihood function, and other data that might have been observed *but were not* should play no part in subsequent analysis. I also overload the likelihood symbol L, which may be a function $L(\theta)$ of θ, or a function $L(X)$ of X, or the i'th element L_i in a list, according to the type of argument. Overloading is standard in computing, and for operators, and can with no ambiguity be applied to functions too. Along

Table 1: *Comparative notation.*

Skilling	'physics'	Evans	'statistics'
θ	parameter	θ	parameter
–		Ω	parameter space
D	data	x	observed value
–		X	response
–		P_θ	sampling model
$\pi(\theta)$	prior	Π	prior with density π
–		$P_\theta \times X$	joint or Bayesian model
$P(\theta)$	posterior	$\Pi(\cdot \mid X)$	posterior
$L(\theta)$	likelihood	$f_\theta(x)$	density of P_θ at x
θ_i	point	$\theta_i(x)$	sample
L_i	likelihood	$f_{\theta_i}(x)$	
X_i	prior mass	ψ_i	
`Sort`		Ψ	complementary cdf
$L(X)$	likelihood	$\Psi^{-1}(\psi)$	
–		M	marginal prior predictive
Z	evidence	$m(x)$	density of M at x
$\int L(\theta)\pi(\theta)\,d\theta$	evidence	$E_\Pi(f_\theta(x))$	density of M at x
H	information	–	

with these simplifications of notation go simplifications of terminology. Thus I use '`Sort`' to suggest the invariance to re-ordering which is the heart of nested sampling – a symmetry that the formal 'complementary cdf' captures less easily. *Keep it simple, keep it easy.*

Evans puts the use of 'evidence' Z for model selection into wider context, correctly reminding readers that the selection of appropriate model(s) also involves one's assessment of their relative prior probabilities. I do not myself go further than this, but that's not the issue here. I'm interested in computing Z.

Preliminaries complete, we move to substantive issues with Evans' equation (1). As correctly stated, this is my equation (3) in alternative notation. Helpfully, Evans then investigates the asymptotic behaviour of nested sampling, a matter that I had not formally addressed. In the passage from his equation (1) to (2), Evans seems to reverse my presentation by taking $\Psi^{-1}(\psi_i)$ (my $L(X_i)$) to be distributed as $f_{\theta_i}(x)$ (suggesting that this is random) and ψ_i to be the proclaimed $e^{-i/N}$ (suggesting that this is known). On my view, $L(X_i)$ is the definitive real number obtained as the particular likelihood L_i obtained at the i'th location θ_i, and it is the X_i's which are the unknown 'random' quantities that may be proclaimed to be approximately $e^{-i/N}$. L_i stopped being distributed when it was computed and our knowledge of it collapsed to a definite number, while our knowledge of X_i remains as vague as it was before the run was started. Apparently changing the view, though, does not change the algebra.

Our task is to estimate the integral

$$Z = \int_0^1 L(X)\,dX$$

of a decreasing function L of X from a list of values $\{L_i, X_i\}, i = 1, \ldots, m$, in which the L_i are known but the X_i are not. All we know about X_i is that it is constructed as $t_1 t_2 \ldots t_i$ where each t_i is distributed as $\Pr(t) = N t^{N-1}$. Leaving the limit $N \to \infty$ until later, the means and covariances are

$$\mathtt{E}(X_i) = \left(\frac{N}{N+1}\right)^i, \qquad \mathtt{E}(X_i X_j) = \left(\frac{N}{N+1}\right)^{j-i} \left(\frac{N}{N+2}\right)^i \qquad (j \geq i).$$

Because L is a decreasing function, the list of values gives lower and upper bounds (conditional on X)

$$Z_- \equiv \sum_{i=1}^{m} \ell_i X_i \;\leq\; Z \;\leq\; Z_+ \equiv \sum_{i=0}^{m} \ell_{i+1} X_i \qquad (\ell_i \equiv L_i - L_{i-1})$$

where $X_0 = 1$ and the likelihood list has been extended to include $L_0 = 0$ and $L_{m+1} = L_{\max}$. For now, assume we have an upper limit $L_{\max}$ on the likelihood: practical data cannot be infinitely likely. These bounds have expectations

$$\mathtt{E}(Z_-) = \sum_{i=1}^{m} \ell_i \left(\frac{N}{N+1}\right)^i, \quad \mathtt{E}(Z_+ - Z_-) = \frac{\mathtt{E}(Z_-)}{N} + \left(\frac{N}{N+1}\right)^m (L_{\max} - L_m)\,.$$

The first task is to assign a termination condition m. For the termination error for $\mathtt{E}(Z)$ to be less than proportion ϵ, it suffices to set m such that

$$\left(\frac{N}{N+1}\right)^m L_{\max} < \epsilon \mathtt{E}(Z_-).$$

Because the left side decreases indefinitely with m and the right side increases, this is bound to hold eventually, when m is some logarithmic multiple of N. The condition ensures that lower and upper bounds are close, with

$$\mathtt{E}(Z_+ - Z_-) < (N^{-1} + \epsilon)\mathtt{E}(Z_-).$$

So, for large N and small ϵ, both bounds must converge to the expectation $\mathtt{E}(Z)$.

As well as expectation over X, there is also an uncertainty, for which it suffices to use the lower-bound variance

$$\mathtt{var}(Z_-) = \sum_{i=1}^{m} \ell_i \left[\left(\frac{N}{N+2}\right)^i - \left(\frac{N}{N+1}\right)^{2i}\right] \left[\ell_i + 2 \sum_{j=i+1}^{m} \ell_j \left(\frac{N}{N+1}\right)^{j-i}\right].$$

The factors in square brackets have the following conservative bounds, giving

$$\mathtt{var}(Z_-) < \sum_{i=1}^{m} \ell_i \left[\frac{1}{2eN}\right] \left[L_m\right] = \frac{L_m^2}{2eN} \leq \frac{L_{\max}^2}{2eN}\,.$$

This shows that as N increases, our knowledge of Z converges. Since there *is* a true value of Z, this convergence can only be to that value. Asymptotics are OK (provided we have the upper limit $L_{\max}$).

A non-Bayesian reader who is uncomfortable with that logic could observe that the number of steps needed to reach a specific X associated with any given L is Poisson with mean and variance $-N \log X$. As N increases, this number becomes proportionally more accurate, so that nested sampling almost certainly tracks the likelihood function ever more closely and ever more finely.

If we are unable to provide the upper bound $L_{\max}$, we are in the ordinary quandary of how many terms to take when summing an infinite series without proof of convergence. In such case, the user must exercise judgment about when Z (which is available as Z_- but not as Z_+) has converged. Nested sampling provides no automatic guarantee of convergence, because some undetected phase of large likelihood might lurk within the small unexplored domain.

Yet, although this asymptotic analysis is of theoretical interest, it is of little practical relevance. Asymptotic behaviour is all very fine, but in practice we want to be able to use conveniently small N, such as 1 or 2. At limited N, there is a big practical difference between the estimates of Z and of $\log Z$. Although moments of both ultimately behave properly as $N \to \infty$, $\log Z$ behaves properly much sooner than Z. If nested sampling relied on the expectation of Z, it would be a terrible method, and quite unusable.

This can be illustrated by the special case of two-valued likelihood, $L = 1$ within prior mass X and $L = 0$ elsewhere, from which all other likelihood functions $L(X)$ can be derived by superposition. Evidence estimates Z are linear in L so follow the same superposition. The true results for this example are $Z = X$ with $H = -\log X$. Think of the logarithmic compression H as a million, as in a biggish problem with a few megabytes of partly-redundant data. Running with N points, nested sampling compresses to prior mass X after k iterates, where k is Poisson with mean and variance NH. The likelihood list is $L_1, \ldots, L_k$ all 0, followed by $L_{k+1}, \ldots, L_m$ all 1 up to termination at iterate m. The corresponding (lower) estimate of Z is

$$Z = X_k - X_m \,.$$

Because the X's almost certainly decrease exponentially fast, the termination error X_m becomes negligible only a few times N iterates beyond the k'th. The major issue is the behaviour with N of the remaining estimate

$$Z = X_k \,.$$

The question is: what is X_k? In nested sampling, $-\log X_k$ is distributed as Beta with mean k/N and variance k/N^2. Thus the result is

$$\log Z = (-k \pm \sqrt{k})/N \,.$$

Since k was Poisson,

$$k = NH \pm \sqrt{NH} \,,$$

and thereby close to NH, nested sampling yields a distribution for Z close to $\log Z = -H \pm \sqrt{H/N}$, whose sampling error away from the true value $-H$ is of order

$N^{-1/2}$. Usefully, nested sampling *also* honestly reports its estimate of $\sqrt{H/N}$ as its uncertainty.

But, if N should be set less than the possibly-large information H, the moments of Z itself come from uselessly far out on the tail of the Beta distribution of its logarithm. With $N = 1$ and $H = 1000000$, the likelihood step might plausibly occur at iterate $k = 1001000$ (a 1-σ sample from Poisson(1000000)). The results (mean $\pm$ std.dev.) from that particular run would be

$$\log Z = -1001000 \pm 1001 \quad \text{(appropriately covering the truth, } -1000000\text{)}$$

from $\Pr(\log Z)$, but the first two moments of $\Pr(Z)$ yield instead

$$Z = 2^{-1001000} \pm (\sqrt{3})^{-1001000} \quad \text{(a hopeless description of our inference).}$$

It's $\log Z$ that we get, not Z. If we want Z to within less than 100% proportional error, we need $\log Z$ to better than ± 1. This needs at least H points, which could be ruinously expensive. In large problems, we need to think logarithmically, and accept that an uncertainty of 1000 powers of e may still be good enough to distinguish between models for millions of data. This is a long way from the asymptotics of $N \to \infty$.

Beyond these statistical sampling errors, there is also a numerical bias $\delta(\log Z)$ of order N^{-1} caused by using the right-hand choice $X_{i-1} - X_i$ of column width in the Riemann sum for Z, instead of the left-hand choice $X_i - X_{i+1}$ or the more symmetrical trapezoid-rule's $(X_{i-1} - X_{i+1})/2$. This is much less than the statistical uncertainty: thus Evans quotes 28% in a specific example, less than a single power of e.

Evans also points out, though, that numerical bias remains under replication of runs with limited N, giving an unpleasant systematic bias to the method that could become apparent in small applications. However, nested sampling has the neat property that different runs with $N_1, N_2, \ldots, N_r$ points can be considered as a *single* run with $N_1 + N_2 + \ldots + N_r$ points simply by merging all the likelihood values into a single, denser, list. The Beta statistics of the X's are just the same as if the list had been computed in one go, but the estimate of $\log Z$ is improved because of the finer sampling of the Riemann sum. Treated thus, the numerical bias *now decreases* proportionally to r^{-1}, so would stay buried well beneath the $r^{-1/2}$ statistical noise, which itself tends to zero as the total number of points increases.

Finally, I should confirm that I do *not* claim to avoid the curse of dimensionality. The curse is relegated to the task of sampling new points θ from the constraint region, which may be arbitrarily difficult. I *do* claim to avoid the curse of thermal methods, which are defeated by convex log-likelihood functions. And I *do* think that nested sampling's automatic estimate of uncertainty is useful.

As for whether the control parameters m (termination) and N (precision) can be set easily, wider experience with applications will doubtless reveal the extent to which nested sampling can be effectively used. I have worked with it for two years, and am optimistic. Try it!

Friel and Pettitt. Friel and Pettitt's 'power posterior', more commonly known as 'simulated annealing', dates back at least 20 years. I completely agree that it often works. Nevertheless, simulated annealing needs a temperature schedule. In their application they choose a power-law schedule, but that's a matter of judgment, and could be inefficient in other problems with different likelihoods. Indeed, I argue that in convex problems, there is *no* workable schedule. Nested sampling has no temperature and needs no schedule. Moreover, nested sampling yields not just the 'evidence' value, but also a realistic estimate of its uncertainty, which is a useful advance.

As for range of application, there is a selection effect. We have become quite good at solving such problems as are amenable to current techniques. We can even judge when a problem will be amenable, as when it is in an exponential family.

We are less good at solving problems that are *not* amenable to current techniques. For some reason such failures attract less publicity, and are 'less representative of models found generally in statistical analysis'. Yet there must be a huge number of unsolved important problems where the 'power posterior' $L^{\beta}X$ is not concave. I surmise that collective behaviour, in which systems can separate into different phases, is the hallmark of such problems. And collective behaviour is ubiquitous in large-scale physics and biology. We wouldn't be here otherwise.

We have solved the simple problems. I want to push the boundaries and solve more difficult problems, and looking to the past may not help. Yes, I chose my examples to illustrate the point, but that doesn't mean they are atypically 'contrived'. I chose them as illustrative of problems that we *want* to solve, not of problems that we have *already* solved.

Chopin and Robert. Sadly, Chopin and Robert (C&R) have missed the point. Instead of indulging in sordid polemical ping-pong by replying in detail to all their 30 or so criticisms, I will comment on the outlook which may have allowed such misconception.

Motivation. One reason we want to calculate Z instead of Bayes ratios Z_1/Z_2 is common courtesy to one's future readers, who will not wish to re-compute one's Z_1 – perhaps making a mistake while tediously re-coding an unfamiliar approach – in order to compare it with their alternatives $Z_2, Z_3, \ldots$ Hence the evidence value Z should *always* accompany publication of a posterior distribution. Human effort should be distributed, not duplicated.

A deeper reason is that we want to deal with today's and tomorrow's large and challenging problems, not just yesterday's toys. Problems with a million data tend to have evidence values of the order of exp(millions). Even quite a minor change in prior model might easily induce a Bayes factor of exp(thousands). In this environment, there is often little or no communication between the models, so that the reversible jump that C&R suggest would not be helpful. The evidence values would have to be computed separately. These calculations may not need to be particularly accurate. An uncertainty of $\pm$hundreds or $\pm$thousands in $\log Z$ may well be all that is needed to discriminate adequately between alternative models, and fortunately nested sampling yields this uncertainty along with $\log Z$ itself.

On the other hand, if models θ and ϕ are close enough to communicate, they can be put together as a joint model ψ to be processed by nested sampling in the usual way. The θ–or–ϕ Bayes factor becomes a property $Q(\psi)$ that can be estimated, with its numerical uncertainty, just like any other property. Reversible jump is then just another exploration technique, along with slice sampling and all the rest, that can be used within nested sampling to scan the parameter space. Those techniques are not what my paper is about.

Practical inference. I see practical inference as a branch of computer science, rather than of pure mathematics. A dataset is something storable in my computer, as a finite list fitting into n 0's and 1's. That storage allows 2^n possible datasets – usually a large number, but still finite.

Analysis of data D will involve a likelihood function $\Pr(D \mid \cdots)$. Whatever this function is, its probability values sum to unity; $\sum_D \Pr(D \mid \cdots) = 1$. Hence individual values are all bounded by 1, regardless of the conditioning. In particular there is *no possibility* of the divergent values that C&R suggest may make difficulty for nested sampling. Indeed, how *could* divergence ever occur in a probabilistic computation?

An answer to this may come when we move from the basics of computation towards traditional mathematics. Often, the bits of data are grouped 32 (say) at a time and held to represent integers or (more generally) real numbers in IEEE or similar format. Instead of needing an individual likelihood function for each of the 2^{32} settings, we can use a simple formula

$$\Pr(x \mid \cdots) = L(x, \cdots) \qquad \text{for real } x$$

and code the function as a short computer procedure $L($`real` $x, \ \cdots)$.

Now, though, we have freedom to make mistakes. For example, we can formally transform $x \in (0,1)$ to $\xi = x^2$, provided we also transform L to

$$\Lambda(\xi, \cdots) = \tfrac{1}{2}\xi^{-1/2} L(\xi^{1/2}, \cdots).$$

The new likelihood function Λ appears to be unbounded near $\xi = 0$, but that is merely an inconsequential and reversible artefact of poor representation. The original individual representation is still valid, and bounded.

Similar remarks apply to the parameters $\theta \in \Theta$ that we wish to infer. Any particular sample θ is storable as a finite list of m 0's and 1's, in a coding sufficiently detailed for our purposes. That storage allows 2^m possible points in Θ – usually a large number, but still finite.

Analysis of parameters θ involves various probabilities $\Pr(\theta \mid \cdots)$ such as prior and posterior, and perhaps intermediate distributions too. Whatever these distributions are, their values always sum to unity; $\sum_\theta \Pr(\theta \mid \cdots) = 1$. Again, all individual values are bounded by 1, regardless of the conditioning. Also, there is *no possibility* of the improper prior that C&R suggest, for where could we put the infinite range? How *could* an improper prior exist?

Again, the answer to this may come when we move from the basics of computation towards traditional mathematics. The bits of θ are usually grouped to represent real numbers with sufficient (and finite) precision for current purposes. It is rather easier to make mistakes now, because it is seductively easy to think of the components of θ as belonging to the continuum, with unbounded range and precision. It even feels more sophisticated and powerful. Unfortunately, in making the grandiose passage

$$\sum\nolimits_\theta \Pr(\text{discrete } \theta \mid \cdots) \ \ldots \quad \longrightarrow \quad \int_{-\infty}^{\infty} d\theta \, \Pr(\text{continuous } \theta \mid \cdots) \ \ldots$$

from computing to mathematics, one risks convincing oneself that one may be dealing with an unbounded likelihood or an improper prior. But an unbounded likelihood is just an artefact of a poor representation, and there is no such thing as an improper prior. Probabilities are bounded and sum to unity. Period. I don't have to protect nested sampling against impossible environments.

From the point of view of basic computation, inference is trivially simple:

One starts with a prior	$\Pr(\theta)$	over (at most) 2^m states
and mutiplies it by a likelihood	$\Pr(D \mid \theta)$	as a $2^n \times 2^m$ matrix
to form a joint distribution	$\Pr(\theta, D)$	over 2^{m+n} states.

We then have the results, namely

the evidence	$\Pr(D) = \sum_\theta \Pr(\theta, D)$	as a list of 2^n possible values
and posterior	$\Pr(\theta \mid D) = \Pr(\theta, D)/\Pr(D)$	as 2^n distributions over 2^m states.

In principle (and, for a sufficiently small problem, in practice) the prior vector and likelihood matrix could be assigned at the outset, and everything else precomputed. Then, when specific data D are observed, the appropriate results could simply be read off. Nothing could ever go wrong.

Computational inference is nothing more than summing the weights assigned to patterns of bits. Regrettably, our technology is presently inadequate to cope directly with the exponentially big sums over $2^{(n \text{ or } m)}$ possibilities. That is why we develop Monte Carlo sampling methods, such as nested sampling. We haven't (yet?) got the gigabit quantum computers which might do what we want directly, so we have to simulate the summations instead. But what we simulate is solidly founded.

Mathematics. Thinking of computational inference as a branch of pure mathematics is a trap. A mathematical integral is an idealisation of a sum. Continuum limits of infinite accuracy and infinite range allow paradoxical behaviour, which never occurs in finite computation. 'In principle, we pass to a limit only after verifying that the limit is well-behaved ... In cases where the limit is well-behaved, it may be possible to get the correct answer by operating directly on the infinite sets, but one cannot count on it. If the limit is not well-behaved, then any attempt to solve the problem directly on the infinite set would have led to nonsense, *the cause of which cannot be seen if one looks only at the limit, and not the limiting process*' (E.T. Jaynes *Probability Theory* Cambridge University Press (2003) p. 663, his italics). A sum is not a mere approximation of an already–dangerous integral – it is the basic and safe quantity.

The danger of thinking in the limit is illustrated by C&R when they criticise my presentation of the likelihood $L(X)$ as a function of enclosed prior mass X. All I do, and it's very simple, is list the 2^m available points θ, sort them by likelihood value, and accumulate prior mass until X is enclosed, at which point the likelihood is $L(X)$. I say it in pictures in Fig. 1, and (as should be obvious) continuum notation is mere convenient shorthand for discrete operation. Yet, by thinking of the likelihood awkwardly as 'the inverse of the complementary cdf [of enclosed prior mass]' without grasping the overall meaning, C&R have confused the issue. They 'do not see why a discretisation and ordering would be necessary'. I reply that discretisation is intrinsic to computation, and that ordering is a sufficient introduction to nested sampling. In the discrete approach, one simply passes over continuum difficulties without noticing them because they aren't there. At worst, some likelihood values coincide, in which case the sorting order is locally ambiguous, but $L(X)$ stays constant and well defined in that locality as elsewhere.

Thinking computationally instead of mathematically extends to notation. Mature programmers are comfortable with overloading function names according to the type of argument, with the obsolete `SQRT` (single precision), `CSQRT` (complex), `DSQRT` (double precision), etc. all subsumed into `sqrt` (generic). There's no ambiguity, simple notation is clearer, and experience demonstrates that mistakes are fewer and comprehension is improved. Likewise in my presentation of nested sampling, it is not necessary to have different symbols for the likelihood L according to whether it is a function $L(\theta)$ of parameter or a function $L(X)$ of enclosed prior mass. C&R say my practice is 'unnecessarily confusing'. I say that simple notation is better.

Objections. I reject C&R's supposed 'improvement' of the algorithm. A run of nested sampling produces an ordered list of explicitly-computed likelihood values L accompanied by enclosed prior masses X which are known through their joint distribution $\Pr(X)$. C&R say the L's are random, but that is wrong. Through numerical integration, the evidence Z is a defined function of the *known* L's and *statistical* X's. The distribution $\Pr(Z)$, or equivalently $\Pr(\log Z)$, is immediately available. This probabilistic inference of Z is straightforward and pure. There is no place in nested sampling for any Rao-Blackwellisation or any non-Bayesian method, replacement, simulation or whatever. Nested sampling is correct as it stands.

C&R say that the hard-edged sampling I need becomes exponentially difficult in the dimension of Gaussian (and other) problems. I stand by my paper (Section 6), where I remark that 'the tolerable step lengths are much the same ... so nested sampling offers no gain or loss of efficiency' of exploration, as compared with traditional Metropolis–Hastings. Anyway, exponential difficulty would have shown up as a million- or billion-fold time penalty when computing my 20-dimensional Gaussian 'toy' problem, which they claim to have reproduced.

Finally, C&R say that nested sampling's constrained sampling is 'certainly not' easier than traditional Metropolis–Hastings, as I claim. That is wrong. Nested–sampling's rigid constraint (Eqn 19) *most certainly <u>is</u>* simpler than Metropolis–Hastings' random constraint (Eqn 18). That minor but critical simplification is the key to nested sampling's greater range of application.

ADDITIONAL REFERENCES IN THE DISCUSSION

Bayarri, M. J. and Berger, J. O. (2000). p-values for composite null models. *J. Amer. Statist. Assoc.* **95:452**, 1127–1142.

Box, G. E. P. (1980) Sampling and Bayes' inference in scientific modelling and robustness. *J. Roy. Statist. Soc. A* **143**, 383–430 (with discussion).

Chen, M. H. Shao, Q. M. and Ibrahim, J. G. (2000). *Monte Carlo Methods in Bayesian Computation.* Berlin: Springer-Verlag.

Evans, M. and Moshonov, H. (2006a). Checking for prior-data conflict. *Bayesian Analysis* **1**, 893–914.

Evans, M. and Moshonov, H. (2006b). Checking for prior-data conflict with hierarchically specified priors. *Bayesian Statistics and its Applications* (S. K. Upadhyay, U. Singh and D. K. Dey, eds.) New Delhi: Anamaya Pub.,

Evans, M. and Swartz, T. (2000). *Approximating Integrals via Monte Carlo and Deterministic Methods.* Oxford: University Press

Friel, N. and Pettitt, A. N. (2005). Marginal likelihood estimation via power posteriors. *Tech. Rep.*, University of Glasgow, UK.

Green, P. J. (1995). Reversible jump MCMC computation and Bayesian model determination. *Biometrika* **82**, 711–732.

Mira, A., Møller, J. and Roberts, G. O. (2001). Perfect slice samplers. *J. Roy. Statist. Soc. B* **63**, 583–606.

Ridall, P. G., Pettitt, A. N., Friel, N., Henderson, R. and McCombe, P. (2006). Motor unit number estimation using reversible jump Markov chain Monte Carlo. *J. Roy. Statist. Soc. C* **56**, 235–269 (with discussion).

Robert, C. P. and Casella, G. (2004). *Monte Carlo Statistical Methods*, 2nd. edn. Berlin: Springer-Verlag

BAYESIAN STATISTICS 8, pp. 525–562.
J. M. Bernardo, M. J. Bayarri, J. O. Berger, A. P. Dawid,
D. Heckerman, A. F. M. Smith and M. West (Eds.)

Objective Bayesian Analysis for the Multivariate Normal Model

DONGCHU SUN
University of Missouri-Columbia and Virginia Tech, USA
sund@missouri.edu

JAMES O. BERGER
Duke University, USA
berger@stat.duke.edu

SUMMARY

Objective Bayesian inference for the multivariate normal distribution is illustrated, using different types of formal objective priors (Jeffreys, invariant, reference and matching), different modes of inference (Bayesian and frequentist), and different criteria involved in selecting optimal objective priors (ease of computation, frequentist performance, marginalization paradoxes, and decision-theoretic evaluation).

In the course of the investigation of the bivariate normal model in Berger and Sun (2006), a variety of surprising results were found, including the availability of objective priors that yield exact frequentist inferences for many functions of the bivariate normal parameters, such as the correlation coefficient. Certain of these results are generalized to the multivariate normal situation.

The prior that most frequently yields exact frequentist inference is the right-Haar prior, which unfortunately is not unique. Two natural proposals are studied for dealing with this non-uniqueness: first, mixing over the right-Haar priors; second, choosing the 'empirical Bayes' right-Haar prior, that which maximizes the marginal likelihood of the data. Quite surprisingly, we show that neither of these possibilities yields a good solution. This is disturbing and sobering. It is yet another indication that improper priors do not behave as do proper priors, and that it can be dangerous to apply 'understandings' from the world of proper priors to the world of improper priors.

Keywords and Phrases: KULLBACK–LEIBLER DIVERGENCE; JEFFREYS PRIOR; MULTIVARIATE NORMAL DISTRIBUTION; MATCHING PRIORS; REFERENCE PRIORS; INVARIANT PRIORS.

This research was supported by the National Science Foundation, under grants DMS-0103265 and SES-0351523, and the National Institute of Health, under grants R01-CA100760 and R01-MH071418.

1. INTRODUCTION

Estimating the mean and covariance matrix of a multivariate normal distribution became of central theoretical interest when Stein (1956, 1972) showed that standard estimators had significant problems, including inadmissibility from a frequentist perspective. Most problematical were standard estimators of the covariance matrix; see Yang and Berger (1994) and the references therein.

In the Bayesian literature, the most commonly used prior for a multivariate normal distribution is a normal prior for the normal mean and an inverse Wishart prior for the covariance matrix. Such priors are conjugate, leading to easy computation, but lack flexibility and also lead to inferences of the same structure as those shown to be inferior by Stein. More flexibile and better performing priors for a covariance matrix were developed by Leonard and Hsu (1992), and Brown (2001) (the generalized inverse Wishart prior). In the more recent Bayesian literature, aggressive shrinkage of eigenvalues, correlations, or other features of the covariance matrix are entertained; see, for example, Daniels (1999, 2002), Liechty (2004) and the references therein. These priors may well be successful in practice, but they do not seem to be formal objective priors according to any of the common definitions.

Recently, Berger and Sun (2006) considered objective inference for parameters of the bivariate normal distribution and functions of these parameters, with special focus on development of objective confidence or credible sets. In the course of the study, many interesting issues were explored involving objective Bayesian inference, including different types of objective priors (Jeffreys, invariant, reference and matching), different modes of inference (Bayesian and frequentist), and different criteria involved in deciding on optimal objective priors (ease of computation, frequentist performance and marginalization paradoxes).

In this paper, we first generalize some of the bivariate results to the multivariate normal distribution; Section 2 presents the generalizations of the various objective priors discussed in Berger and Sun (2006). We particularly focus on reference priors, and show that the right-Haar prior is indeed a one-at-a-time reference prior (Berger and Bernardo, 1992) for many parameters and functions of parameters.

Section 3 gives some basic properties of the resulting posterior distributions and gives *constructive posterior distributions* for many of the priors. Constructive posteriors are expressions for the posterior distribution which allow very simply simulation from the posterior. Constructive posteriors are also very powerful for proving results about exact frequentist matching. (Exact frequentist matching means that $100(1-\alpha)\%$ credible sets arising from the resulting posterior are also exact frequentist confidence sets at the specified level.) Results about matching for the right-Haar prior are given in Section 4 for a variety of parameters.

One of the most interesting features of right-Haar priors is that, while they result in exact frequentist matching, they also seem to yield marginalization paradoxes (Dawid, Stone and Zidek, 1973). Thus one is in the philosophical conundrum of having to choose between frequentist matching and avoidance of the marginalization paradox. This is also discussed in Section 4.

Another interesting feature of the right-Haar priors is that they are not unique; they depend on which triangular decomposition of a covariance matrix is employed. In Section 5, two natural proposals are studied to deal with this non-uniqueness. The first is to simply mix over the right-Haar priors. The second is to choose the 'empirical Bayes' right-Haar prior, namely that which maximizes the marginal likelihood of the data. Quite surprisingly, it is shown that both of these solutions gives inferior answers, a disturbing and sobering phenomenon. It is yet another indication that improper priors do not behave as do proper priors, and that it can be dangerous to apply 'understandings' from the world of proper priors to the world of improper priors.

2. OBJECTIVE PRIORS FOR THE MULTIVARIATE NORMAL DISTRBUTION

Consider the p-dimentional multivariate normal population, $\boldsymbol{x} = (x_1, \cdots, x_p)' \sim N_p(\boldsymbol{\mu}, \boldsymbol{\Sigma})$, whose density is given by

$$f(\boldsymbol{x} \mid \boldsymbol{\mu}, \boldsymbol{\Sigma}) \quad = \quad (2\pi)^{-p/2}|\boldsymbol{\Sigma}|^{-1/2} \exp\Big(-\frac{1}{2}(\boldsymbol{x}-\boldsymbol{\mu})'\boldsymbol{\Sigma}^{-1}(\boldsymbol{x}-\boldsymbol{\mu})\Big). \tag{1}$$

2.1. *Previously Considered Objective Priors*

Perhaps the most popular prior for the multivariate normal distribution is the Jeffreys (rule) prior (Jeffreys, 1961)

$$\pi_J(\boldsymbol{\mu}, \boldsymbol{\Sigma}) = |\boldsymbol{\Sigma}|^{-(p+2)/2}. \tag{2}$$

Another commonly used prior is the independence-Jeffreys prior

$$\pi_{IJ}(\boldsymbol{\mu}, \boldsymbol{\Sigma}) = |\boldsymbol{\Sigma}|^{-(p+1)/2}. \tag{3}$$

It is commonly thought that either the Jeffreys or independence-Jeffreys priors are most natural, and most likely to yield classical inferences. However, Geisser and Cornfield (1963) showed that the prior which is exact frequentist matching for all means and variances (and which also yields Fisher's fiducial distribution for these parameters) is

$$\pi_{GC}(\boldsymbol{\Sigma}) = |\boldsymbol{\Sigma}|^{-p}. \tag{4}$$

It is simple chance that this prior happens to be the Jeffreys prior for $p = 2$ (and perhaps simple chance that it agrees with π_{IJ} for $p = 1$); these coincidences may have contributed significantly to the popular notion that Jeffreys priors are generally successful.

In spite of the frequentist matching success of π_{GC} for means and variances, the prior seems to be quite bad for correlations, predictions, or other inferences involving a multivariate normal distribution. Thus a variety of other objective priors have been proposed in the literature.

Chang and Eaves (1990) ((7) on page 1605) derived the reference prior for the parameter ordering $(\mu_1, \cdots, \mu_p; \sigma_1, \cdots, \sigma_p; \boldsymbol{\Upsilon})$,

$$\begin{aligned} \pi_{CE}(\boldsymbol{\mu}, \boldsymbol{\Sigma})\, d\boldsymbol{\mu}\, d\boldsymbol{\Sigma} &= \frac{1}{|\boldsymbol{\Sigma}|^{(p+1)/2}|\boldsymbol{I}_p + \boldsymbol{\Sigma}^*\boldsymbol{\Sigma}^{-1}|^{1/2}}\, d\boldsymbol{\mu}\, d\boldsymbol{\Sigma} &(5)\\ &= 2^p \left[\prod_{i=1}^{p} \frac{d\mu_i d\sigma_i}{\sigma_i}\right]\left[\frac{1}{|\boldsymbol{\Upsilon}|^{(p+1)/2}|\boldsymbol{I}_p + \boldsymbol{\Upsilon}^*\boldsymbol{\Upsilon}^{-1}|^{1/2}} \prod_{i<j} d\rho_{ij}\right], &(6)\end{aligned}$$

where $\boldsymbol{\Upsilon}$ is the correlation matrix and $\boldsymbol{A}^*\boldsymbol{B}$ denotes the Hadamard product of the squared matrices $\boldsymbol{A} = (a_{ij})$ and $\boldsymbol{B} = (b_{ij})$, whose entries are $c_{ij} = a_{ij}b_{ij}$. In the bivariate normal case, this prior is the same as Lindley's (1965) prior, derived using certain notions of transformation to constant information, and was derived as a reference prior for the correlation coefficient ρ in Bayarri (1981).

Chang and Eaves (1990) also derived the reference prior for the ordering $(\mu_1, \cdots, \mu_p; \lambda_1, \cdots, \lambda_p; \boldsymbol{O})$, where $\lambda_1 > \cdots > \lambda_p$ are the ordered eigenvalues of $\boldsymbol{\Sigma}$, and $\boldsymbol{O}$ is an orthogonal matrix such that $\boldsymbol{\Sigma} = \boldsymbol{O}'\text{diag}(\lambda_1, \cdots, \lambda_p)\boldsymbol{O}$. This reference prior was discussed in detail in Berger and Yang (1994) and has the form,

$$\pi_E(\mu, \boldsymbol{\Sigma})\, d\boldsymbol{\mu}\, d\boldsymbol{\Sigma} = \frac{I_{[\lambda_1 > \cdots > \lambda_p]}}{|\boldsymbol{\Sigma}| \prod_{i<j}(\lambda_i - \lambda_j)}\, d\boldsymbol{\mu}\, d\boldsymbol{\Sigma}\,. \tag{7}$$

Another popular prior is the right Haar prior, which has been extensively studied (see, e.g., Eaton and Sudderth, 2002).

It is most convenient to express this prior in terms of a lower-triangular matrix $\boldsymbol{\Psi}$ with positive diagonal elements and such that

$$\boldsymbol{\Sigma}^{-1} = \boldsymbol{\Psi}'\boldsymbol{\Psi}. \tag{8}$$

(Note that there are many such matrices, so that the right Haar prior is not unique.) The right Haar prior corresponding to this decomposition is given by

$$\pi_H(\mu, \boldsymbol{\Psi})\, d\boldsymbol{\mu}\, d\boldsymbol{\Psi} = \prod_{i=1}^{p} \frac{1}{\psi_{ii}^{i}}\, d\boldsymbol{\mu}\, d\boldsymbol{\Psi}\,. \tag{9}$$

We will see in the next subsection that, this prior is one-at-a-time reference prior for various parameterizations.

Because $d\boldsymbol{\Sigma} = \prod_{i=1}^{p} \psi_{ii}^{i-2(p+1)} d\boldsymbol{\Psi}$, the independence Jeffreys prior $\pi_{IJ}(\boldsymbol{\mu}, \boldsymbol{\Psi})$ corresponding to the left-Haar measure is given by

$$\pi_{IJ}(\boldsymbol{\mu}, \boldsymbol{\Psi}) = \prod_{i=1}^{p} \frac{1}{\psi_{ii}^{p-i+1}}\, d\boldsymbol{\mu}\, d\boldsymbol{\Psi}\,. \tag{10}$$

The right-Haar prior and the independence Jefferys prior are limiting cases of generalized Wishart priors; see Brown (2001) for a review.

We now give the Fisher information matrix and a result about the reference prior for a group ordering related to the right-Haar parameterization; proofs are relegated to the appendix.

Fact 1 *(a) The Fisher information matrix for* $\{\boldsymbol{\mu}, \psi_{11}, (\psi_{21}, \psi_{22}), \cdots, (\psi_{p1}, \cdots, \psi_{pp})\}$ *is*

$$\boldsymbol{J} = -E\left(\frac{\partial^2 \log f}{\partial\boldsymbol{\theta}\partial\boldsymbol{\theta}'}\right) = diag(\boldsymbol{\Sigma}^{-1}, \boldsymbol{\Lambda}_1, \cdots, \boldsymbol{\Lambda}_p), \tag{11}$$

where, for $i = 1, \cdots, p$, $\boldsymbol{\Lambda}_i = \boldsymbol{\Sigma}_i + \psi_{ii}^{-2}\boldsymbol{e}_i\boldsymbol{e}_i'$, *with* $\boldsymbol{e}_i$ *being the* i^{th} *unit column vector.*
(b) The reference prior of $\boldsymbol{\Psi}$ *for the ordered group* $\{\mu_1, \cdots, \mu_p, \psi_{11}, (\psi_{21}, \psi_{22}), \cdots, (\psi_{p1}, \cdots, \psi_{pp})\}$ *is given by*

$$\pi_{R1}(\boldsymbol{\mu}, \boldsymbol{\Psi}) \quad \propto \quad \frac{1}{\prod_{i=1}^{p} \psi_{ii}}\,. \tag{12}$$

Note that the right-Haar prior, the Jeffreys (rule) prior, the independence Jeffreys prior, the Geisser–Cornfield prior and the reference prior π_{R1} have the form

$$\pi_{\boldsymbol{a}}(\boldsymbol{\mu}, \boldsymbol{\Psi}) d\boldsymbol{\mu} d\boldsymbol{\Psi} = \prod_{i=1}^{p} \frac{1}{\psi_{ii}^{a_i}} d\boldsymbol{\mu} d\boldsymbol{\Psi}, \tag{13}$$

where $\boldsymbol{a} = (a_1, \cdots, a_p)$. This class of priors has also received considerable attention in directed acyclic graphical models. See, for example, Roverato and Consonni (2004). Also, Consonni, Gutiérrez-Peña, and Veronese (2004) and their references for reference priors for exponential families with simple quadratic variance function.

Table 1: *Summary of objective priors of $(\boldsymbol{\mu}, \boldsymbol{\Psi})$ as special cases of (13).*

prior	form	$(a_1, \cdots, a_p)$
π_H	$\prod_{i=1}^{p} \psi_{ii}^{-i}$	$a_i = i$
π_J	$\prod_{i=1}^{p} \psi_{ii}^{-(p-i)}$	$a_i = p - i$
π_{IJ}	$\prod_{i=1}^{p} \psi_{ii}^{-(p-i+1)}$	$a_i = p - i + 1$
π_{GC}	$\prod_{i=1}^{p} \psi_{ii}^{-(2-i)}$	$a_i = 2 - i$
π_{R1}	$\prod_{i=1}^{p} \psi_{ii}^{-1}$	$a_i = 1$

2.2. *Reference Priors under Alternative Parameterizations*

Pourahmadi (1999) considered another decomposition of $\boldsymbol{\Sigma}^{-1}$. Let $\boldsymbol{T} = (t_{ij})_{p \times p}$ be the $p \times p$ unit lower triangular matrix, where

$$t_{ij} = \begin{cases} 0 & \text{if } i < j, \\ 1 & \text{if } i = j, \\ \frac{\psi_{ij}}{\psi_{ii}} & \text{if } i > j. \end{cases} \tag{14}$$

Pourahmadi (1999) pointed out the statistical interpretations of the below-diagonal entries of $\boldsymbol{T}$ and the diagonal entries of $\boldsymbol{\Psi}$. In fact, $x_1 \sim N(\mu_1, d_1)$, $x_i \sim N(\mu_i - \sum_{j=1}^{j-1} t_{ij}(x_j - \mu_j), \psi_{ii}^{-2})$, $(j \geq 2)$, so the t_{ij} are the negatives of the coefficients of the best linear predictor of x_i based on $(x_1, \cdots, x_{i-1})$, and ψ_{ii}^2 is the precision of the predictive distribution. Write $\widetilde{\boldsymbol{\Psi}} = diag(\psi_{11}, \cdots, \psi_{pp})$. Clearly

$$\boldsymbol{\Psi} = \widetilde{\boldsymbol{\Psi}} \boldsymbol{T}, \tag{15}$$

$$\boldsymbol{\Sigma} = (\boldsymbol{T}' \widetilde{\boldsymbol{\Psi}}^2 \boldsymbol{T})^{-1}. \tag{16}$$

For $i = 2, \cdots, p$, define $\widetilde{\boldsymbol{\Psi}}_i = diag(\psi_{11}, \cdots, \psi_{ii})$ and denote the upper and left $i \times i$ submatrix of $\boldsymbol{T}$ by $\boldsymbol{T}_i$. Then

$$\boldsymbol{\Psi}_i = \widetilde{\boldsymbol{\Psi}}_i \boldsymbol{T}_i, \tag{17}$$

$$\boldsymbol{\Sigma}_i = (\boldsymbol{T}_i' \widetilde{\boldsymbol{\Psi}}_i^2 \boldsymbol{T}_i)^{-1}, \quad i = 2, \cdots, p. \tag{18}$$

Fact 2 *(a) The Fisher information for* $(\boldsymbol{\mu}, \psi_{11}, (t_{21}, \psi_{22}), (t_{31}, t_{32}, \psi_{33}), \cdots, (t_{p1}, \cdots, t_{p,p-1}, \psi_{ii}))$ *is of the form*

$$\widetilde{\boldsymbol{J}} = diag(\boldsymbol{T}' \widetilde{\boldsymbol{\Psi}}^2 \boldsymbol{T}, \frac{2}{\psi_{11}^2}, \widetilde{\boldsymbol{J}}_2, \cdots, \widetilde{\boldsymbol{J}}_p), \tag{19}$$

where for $i = 2, \cdots, p$,

$$\widetilde{J}_i = diag\left(\psi_{ii}^2 T_{i-1}^{-1} \widetilde{\Psi}_{i-1}^{-2} T_{i-1}'^{-1}, \frac{2}{\psi_{ii}^2} \right). \tag{20}$$

(b) The one-at-a-time reference prior for $\{\mu_1, \cdots, \mu_p, \psi_{11}, \psi_{22}, \cdots, \psi_{ii}, t_{21}, t_{31}, t_{32}, \cdots, t_{p1}, \cdots, t_{p,p-1}\}$, *and with any ordering of parameters, is*

$$\widetilde{\pi}_R(\widetilde{\boldsymbol{\theta}}) = \prod_{i=1}^{p} \frac{1}{\psi_{ii}}. \tag{21}$$

(c) The reference prior in (b) is the same as the right-Haar measure for $\boldsymbol{\Psi}$, *given in (9).*

Consider the parameterization $\boldsymbol{D} = \text{diag}(d_1, \cdots, d_p)$ and $\boldsymbol{T}$, where $d_i = 1/\psi_{ii}^2$. Clearly $\boldsymbol{D} = \widetilde{\boldsymbol{\Psi}}^{-2}$ and $\boldsymbol{\Sigma}^{-1} = \boldsymbol{T}'\boldsymbol{D}^{-1}\boldsymbol{T}$. Also write $\boldsymbol{D}_i = \boldsymbol{\Psi}_i^{*-2}$.

Corollary 1 *(a) The Fisher information for* $(\boldsymbol{\mu}, d_1, \cdots, d_p; t_{21}; t_{31}, t_{32}; \cdots, t_{p1}, \cdots, t_{p,p-1})$ *is of the form*

$$\boldsymbol{J}^{\#} = diag(\boldsymbol{T}'\boldsymbol{D}^{-1}\boldsymbol{T}, \frac{1}{d_1^2}, \cdots, \frac{1}{d_p^2}, \boldsymbol{\Delta}_2, \cdots, \boldsymbol{\Delta}_p), \tag{22}$$

where, for $i = 2, \cdots, p$,

$$\boldsymbol{\Delta}_i = \frac{1}{d_i} \boldsymbol{T}_{i-1}^{-1} \boldsymbol{D}_{i-1} \boldsymbol{T}_{i-1}'^{-1}. \tag{23}$$

(b) The one-at-a-time reference prior for $\{\mu_1, \cdots, \mu_p, d_1, \cdots, d_p, t_{21}, t_{31}, t_{32}, \cdots, t_{p1}, \cdots, t_{p,p-1}\}$, *and with any ordering, is*

$$\widetilde{\pi}_R(\boldsymbol{\theta}) \propto \prod_{i=1}^{p} \frac{1}{d_i}. \tag{24}$$

(c) The reference prior in (b) is the same as the right-Haar measure for $\boldsymbol{\Psi}$, *given in (9).*

Suppose one is interested in the generalized variance $|\boldsymbol{\Sigma}| = \prod_{i=1}^{p} d_i$; the one-at-a-time reference prior is also the right-Haar measure π_H. To see this, define

$$\begin{cases} \xi_1 & = \frac{d_1}{d_2}, \\ \xi_2 & = \frac{(d_1 d_2)^{1/2}}{d_3}, \\ \cdots & \cdots \\ \xi_{p-1} & = \frac{(\prod_{j=1}^{p-1} d_j)^{1/(p-1)}}{d_p}, \\ \xi_p & = \prod_{j=1}^{p} d_j. \end{cases} \tag{25}$$

Fact 3 *(a) The Fisher information matrix for* $(\boldsymbol{\mu}, \xi_1, \cdots, \xi_p; t_{21}, t_{31}, t_{32}, \cdots, t_{p1}, \cdots, t_{p,p-1})$ *is*

$$diag\left(\boldsymbol{\Sigma}^{-1}, \frac{1}{2\xi_1^2}, \frac{2}{3\xi_2^2}, \cdots, \frac{p-1}{p\,\xi_{p-1}^2}, \frac{1}{p\,\xi_p^2}, \boldsymbol{\Delta}_2, \cdots, \boldsymbol{\Delta}_p\right), \tag{26}$$

where $\boldsymbol{\Delta}_i$ *is given by (23).*

(b) The one-at-a-time reference prior of any ordering for $\{\mu_1, \cdots, \mu_p, \xi_1, \cdots, \xi_p; t_{21}, t_{31}, t_{32}, \cdots, t_{p1}, \cdots, t_{p,p-1}\}$ *is*

$$\widetilde{\pi}_R(\boldsymbol{\theta}) \propto \prod_{i=1}^{p} \frac{1}{\xi_i}. \tag{27}$$

(c) The reference prior in (b) is π_H, *given in (9).*

Corollary 2 *Since* $\xi_p = |\boldsymbol{\Sigma}|$, *it is immediate that the one-at-a-time reference prior for* ξ_p, *with nuisance parameters* $(\boldsymbol{\mu}, \xi_1, \cdots \xi_{p-1}, t_{21}, t_{31}, t_{32}, \cdots, t_{p1}, \cdots, t_{p,p-1})$, *is the right-Haar prior* π_H.

Corollary 3 *One might be interested in* $\eta_i \equiv |\boldsymbol{\Sigma}_i| = \prod_{j=1}^{i} d_i$, *the generalized variance of the upper left* $i \times i$ *submatrix of* $\boldsymbol{\Sigma}$. *Using the same arguments as in Fact 3, the Fisher information for* $(\boldsymbol{\mu}, \xi_1, \cdots, \xi_{i-1}, \eta_i, d_{i+1}, \cdots, d_p; t_{21}, t_{31}, t_{32}, \cdots, t_{p1}, \cdots, t_{p,p-1})$ *is*

$$diag\left(\boldsymbol{\Sigma}^{-1}, \frac{1}{2\xi_1^2}, \cdots, \frac{i-1}{i\,\xi_{i-1}^2}, \frac{1}{i\,\eta_i^2}, \frac{1}{d_{i+1}^2}, \cdots, \frac{1}{d_p^2}, \boldsymbol{\Delta}_2, \cdots, \boldsymbol{\Delta}_p\right). \tag{28}$$

The one-at-a-time reference prior for $|\boldsymbol{\Sigma}_i|$, *with nuisance parameters* $\{\mu_1, \cdots, \mu_p, \xi_1, \cdots, \xi_{i-1}, d_{i+1}, \cdots, d_p; t_{21}, t_{31}, t_{32}, \cdots, t_{p1}, \cdots, t_{p,p-1})$ *and any parameter order, is the right-Haar prior* π_H.

3. POSTERIOR DISTRIBUTIONS

Let $\boldsymbol{X}_1, \cdots, \boldsymbol{X}_n$ be a random sample from $N_p(\boldsymbol{\mu}, \boldsymbol{\Sigma})$. The likelihood function of $(\boldsymbol{\mu}, \boldsymbol{\Sigma})$ is given by

$$L(\boldsymbol{\mu}, \boldsymbol{\Sigma}) = (2\pi)^{-np/2}|\boldsymbol{\Sigma}|^{-n/2} \exp\left\{ -\frac{n}{2}(\boldsymbol{X}_n - \boldsymbol{\mu})'\boldsymbol{\Sigma}^{-1}(\boldsymbol{X}_n - \boldsymbol{\mu}) - \frac{1}{2}tr(\boldsymbol{S}\boldsymbol{\Sigma}^{-1})\right\},$$

where

$$\boldsymbol{X}_n = \frac{1}{n}\sum_{i=1}^{n} \boldsymbol{X}_i, \qquad \boldsymbol{S} = \sum_{i=1}^{n}(\boldsymbol{X}_i - \boldsymbol{X}_n)(\boldsymbol{X}_i - \boldsymbol{X}_n)'.$$

Since all the considered priors are constant in $\boldsymbol{\mu}$, the conditional posterior for $\boldsymbol{\mu}$ will be

$$(\boldsymbol{\mu} \mid \boldsymbol{\Sigma}, \boldsymbol{X}) \sim N_p(\boldsymbol{x}, \frac{1}{n}\boldsymbol{\Sigma}). \tag{29}$$

Generation from this is standard, so the challenge of simulation from the posterior distribution requires only sampling from the marginal posterior of $\mathbf{\Sigma}$ given $\boldsymbol{S}$. Note that the marginal likelihood of $\mathbf{\Sigma}$ based on $\boldsymbol{S}$ is

$$L_1(\mathbf{\Sigma}) = \frac{(2\pi)^{-np/2}}{|\mathbf{\Sigma}|^{(n-1)/2}} etr\left(-\frac{1}{2}\mathbf{\Sigma}^{-1}\boldsymbol{S}\right). \tag{30}$$

Throughout the paper, we assume that $\boldsymbol{S}$ is positive definite, as this is true with probability one.

3.1. *Marginal Posteriors of $\mathbf{\Sigma}$ under π_J, π_{IJ}, π_{CE} and π_E*

Marginal Posteriors Under π_J and π_{IJ}: It is immediate that these marginal posteriors for $\mathbf{\Sigma}$ are Inverse Wishart $(\boldsymbol{S}^{-1}, n)$ and Inverse Wishart $(\boldsymbol{S}^{-1}, n-1)$, respectively.

Marginal Posterior Under π_{CE}: This marginal posterior distribution is imposing in its complexity. However, rather remarkably there is a simple rejection algorithm that can be used to generate from it:

Step 1. Generate $\mathbf{\Sigma} \sim$ Inverse Wishart $(\boldsymbol{S}^{-1}, n-1)$.

Step 2. Simulate $u \sim$ Uniform$(0,1)$.
If $u \le 2^{p/2}\,|\boldsymbol{I}_p + \mathbf{\Sigma}^*\mathbf{\Sigma}^{-1}|^{-1/2}$, report $\mathbf{\Sigma}$. Otherwise go back to *Step 1.*

Note that the acceptance probability $2^{p/2}\,|\boldsymbol{I}_p + \mathbf{\Sigma}^*\mathbf{\Sigma}^{-1}|^{-1/2}$ is equal to one if the proposed $\mathbf{\Sigma}$ is diagonal, but is near zero when the proposed $\mathbf{\Sigma}$ is nearly singular. That this algorithm is a valid accept-reject algorithm, based on generation of $\mathbf{\Sigma}$ from the independence Jeffreys posterior, is established in Berger and Sun (2006).

Marginal Posterior Under π_E: It is possible to generate from this posterior using the following Metropolis-Hastings algorithm from Berger et. al. (2005).

Step 1. Generate $\mathbf{\Sigma}^* \sim$ Inverse Wishart $(\boldsymbol{S}^{-1}, n-1)$.

Step 2. Set $\mathbf{\Sigma}' = \begin{cases} \mathbf{\Sigma}^* & \text{with probability } \alpha, \\ \mathbf{\Sigma} & \text{otherwise,} \end{cases}$
where $\alpha = \min\left\{1, \frac{\prod_{i<j}(\lambda_i^* - \lambda_j^*)}{\prod_{i<j}(\lambda_i - \lambda_j)} \cdot \frac{|\mathbf{\Sigma}|^{(p-1)/2}}{|\mathbf{\Sigma}^*|^{(p-1)/2}}\right\}$.

3.2. *Marginal Posterior of $\mathbf{\Sigma}$ under π_a*

In the following, we will write $\mathbf{\Sigma} = \mathbf{\Sigma}_p$, including the dimension in order to dervive some useful recursive formulas. Let $\boldsymbol{\psi}_{p,p-1}$ be the $(p-1)\times 1$ vector of the last column of $\mathbf{\Psi}'_p$, excluding ψ_{pp}. Then

$$\mathbf{\Psi}_1 = \psi_{11},\ \mathbf{\Psi}_2 = \begin{pmatrix} \psi_{11} & 0 \\ \psi_{21} & \psi_{22} \end{pmatrix}, \cdots, \mathbf{\Psi}_p = \begin{pmatrix} \mathbf{\Psi}_{p-1} & \mathbf{0} \\ \boldsymbol{\psi}'_{p,p-1} & \psi_{pp} \end{pmatrix}.$$

We will also write $\boldsymbol{S} = \boldsymbol{S}_p$ and let $\boldsymbol{s}_{p,p-1}$ represent the $(p-1)\times 1$ vector of the last column of $\boldsymbol{S}_p$ excluding s_{pp}. Thus

$$\boldsymbol{S}_1 = s_{11},\ \boldsymbol{S}_2 = \begin{pmatrix} s_{11} & s_{21} \\ s_{21} & s_{22} \end{pmatrix}, \cdots, \boldsymbol{S}_p = \begin{pmatrix} \boldsymbol{S}_{p-1} & \boldsymbol{s}_{p,p-1} \\ \boldsymbol{s}'_{p,p-1} & s_{pp} \end{pmatrix}. \tag{31}$$

Fact 4 *Under the prior $\pi_{\boldsymbol{a}}$ in (13), the marginal posterior of $\boldsymbol{\Psi}$ is given as follows. (a) For given $(\psi_{11}, \cdots, \psi_{pp})$, the conditonal posteriors of the off-diagonal vector $\boldsymbol{\psi}_{i-1,i}$ are independent normal,*

$$(\boldsymbol{\psi}_{i,i-1} \mid \psi_{ii}, \boldsymbol{S}) \sim N(-\psi_{ii}\boldsymbol{S}_{i-1}^{-1}\boldsymbol{s}_{i,i-1},\ \boldsymbol{S}_{i-1}^{-1}). \tag{32}$$

(b) The marginal posteriors of ψ_{ii}^2 $(1 \le i \le p)$ are independent gamma $((n - a_i)/2, w_i/2)$, where

$$w_i = \begin{cases} s_{11}, & \text{if } i = 1, \\ \frac{|\boldsymbol{S}_i|}{|\boldsymbol{S}_{i-1}|} = s_{ii} - \boldsymbol{s}_{i,i-1}'\boldsymbol{S}_{i-1}^{-1}\boldsymbol{s}_{i-1,i}, & \text{if } i = 2, \cdots, p. \end{cases} \tag{33}$$

(c) The marginal likelihood of $\boldsymbol{Y}$ (or the normalizing constant) is

$$\begin{aligned} M &\equiv \int L(\boldsymbol{\mu}, \boldsymbol{\Psi})\pi_{\boldsymbol{a}}(\boldsymbol{\mu}, \boldsymbol{\Psi})d\boldsymbol{\mu}d\boldsymbol{\Psi} \\ &= \frac{\prod_{i=1}^{p}\Gamma(\frac{1}{2}(n-a_i))2^{(n-a_i)/2}}{2^p(2\pi)^{(n-p)p/2}n^{p/2}} \frac{\prod_{i=1}^{p-1}|\boldsymbol{S}_i|^{(a_i-a_{i+1}-1)/2}}{|\boldsymbol{S}|^{(n-a_p)/2}}. \end{aligned} \tag{34}$$

3.2.1. *Constructive Posteriors*

In the remainder of the paper we use, without further comment, the notation that the symbol * appended to a random variable denotes randomness arising from the constructive posterior (i.e., from the random variables used in simulation from the posterior), while a random variable without a * refers to randomness arising from the (frequentist) distribution of a statistic. Also, let Z_{ij} denote standard normal random variables. Whenever several of these occur in an expression, they are all independent (except that random variables of the same type and with the same index refer to the same random variable). Finally, we reserve quantile notation for posterior quantiles, with respect to the * distributions.

Fact 5 *Consider the prior $\pi_{\boldsymbol{a}}$ in (13). Let $\chi_{n-a_i}^{2*}$ denote independent draws from chi-squared distributions with the indicated degree of freedoms, and $\boldsymbol{z}_{i,i-1}^*$ denote independent draws from $N_{i-1}(\boldsymbol{0}, \boldsymbol{I}_{i-1})$. The constructive posterior of $(\psi_{11}, \cdots, \psi_{pp}, \boldsymbol{\psi}_{21}, \cdots, \boldsymbol{\psi}_{p,p-1})$ given $\boldsymbol{X}$ can be expressed as*

$$\psi_{ii}^* = \sqrt{\frac{\chi_{n-a_i}^{2*}}{w_i}}, \quad i = 1, \cdots, p, \tag{35}$$

$$\begin{aligned} \boldsymbol{\psi}_{i,i-1}^* &= \boldsymbol{S}_{i-1}^{-1/2}\boldsymbol{z}_{i,i-1}^* - \psi_{ii}^*\boldsymbol{S}_{i-1}^{-1}\boldsymbol{s}_{i,i-1} \\ &= \boldsymbol{S}_{i-1}^{-1/2}\boldsymbol{z}_{i,i-1}^* - \sqrt{\frac{\chi_{n-a_i}^{2*}}{w_i}}\boldsymbol{S}_{i-1}^{-1}\boldsymbol{s}_{i,i-1}, \quad i = 2, \cdots, p. \end{aligned} \tag{36}$$

Letting $\boldsymbol{\Psi}^* = (\psi_{ij}^*)$, the constructive posterior of $\boldsymbol{\Sigma}$ is simply $\boldsymbol{\Sigma}^* = \boldsymbol{\Psi}^{*-1}(\boldsymbol{\Psi}^{*-1})'$.

Alternatively, let $\boldsymbol{V}$ be the Cholesky decomposition of $\boldsymbol{S}$, i.e., $\boldsymbol{V}$ is the lower-triangular matrix with positive diagonal elements such that $\boldsymbol{S} = \boldsymbol{V}\boldsymbol{V}'$. It is easy to see that

$$w_i = t_{ii}^2, \quad i = 1, \cdots, p. \tag{37}$$

We will also write $\boldsymbol{V} = \boldsymbol{V}_p$ and let $\boldsymbol{v}_{p,p-1}$ represent the $(p-1)\times 1$ vector of the last column of $\boldsymbol{V}_p'$ excluding v_{pp}. We have

$$\boldsymbol{V}_1 = v_{11},\ \boldsymbol{V}_2 = \begin{pmatrix} v_{11} & 0 \\ v_{21} & v_{22} \end{pmatrix},\ \cdots, \boldsymbol{V}_p = \begin{pmatrix} \boldsymbol{V}_{p-1} & 0 \\ \boldsymbol{v}_{p,p-1}' & v_{pp} \end{pmatrix}. \tag{38}$$

Corollary 4 *Under the prior $\pi_{\boldsymbol{a}}$ in (13),*

$$(\boldsymbol{\psi}_{i,i-1} \mid \psi_{ii}, \boldsymbol{V}) \sim N(-\psi_{ii}\boldsymbol{V}_{i-1}^{-1\prime}\boldsymbol{v}_{i,i-1}, (\boldsymbol{V}_{i-1}\boldsymbol{V}_{i-1}')^{-1}), \tag{39}$$

$$(\psi_{ii}^2 \mid \boldsymbol{V}) \sim \textit{inverse gamma}((n-a_i)/2, v_{ii}^2/2). \tag{40}$$

Proof. This follows from Fact 4 and the equality $\boldsymbol{S}_{i-1}^{-1}\boldsymbol{s}_{i,i-1} = \boldsymbol{V}_{i-1}^{-1\prime}\boldsymbol{v}_{i,i-1}$. □

The following fact is immediate.

Fact 6 *Under the assumptions of Fact 5, the constructive posterior of $(\psi_{11}, \cdots, \psi_{pp}, \boldsymbol{\psi}_{21}, \cdots, \boldsymbol{\psi}_{p,p-1})$ given $\boldsymbol{X}$ can be expressed as*

$$\psi_{ii}^* = \frac{\sqrt{\chi_{n-a_i}^{2*}}}{v_{ii}}, \quad i = 1, \cdots, p, \tag{41}$$

$$\boldsymbol{\psi}_{i,i-1}^* = \boldsymbol{V}_{i-1}^{-1\prime}\boldsymbol{z}_{i,i-1}^* - \frac{\sqrt{\chi_{n-a_i}^{2*}}}{v_{ii}}\boldsymbol{V}_{i-1}^{-1\prime}\boldsymbol{v}_{i,i-1}, \quad i = 2, \cdots, p. \tag{42}$$

3.2.2. *Posterior Means of $\boldsymbol{\Sigma}$ and $\boldsymbol{\Sigma}^{-1}$*

Fact 7 *(a) If $n - a_i > 0,\ i = 1, \cdots, p$, then*

$$E(\boldsymbol{\Sigma}^{-1} \mid \boldsymbol{X}) = E(\boldsymbol{\Psi}'\boldsymbol{\Psi} \mid \boldsymbol{X}) = \boldsymbol{V}^{-1\prime}\text{diag}(g_1, \cdots, g_p)\boldsymbol{V}^{-1}, \tag{43}$$

where $g_i = n - a_i + p - i,\ i = 1, \cdots, p$.
(b) If $n - a_i > 2,\ i = 1, \cdots, p$, then

$$E(\boldsymbol{\Sigma} \mid \boldsymbol{X}) = E((\boldsymbol{\Psi}'\boldsymbol{\Psi})^{-1} \mid \boldsymbol{X}) = \boldsymbol{V}\,\text{diag}(h_1, \cdots, h_p)\boldsymbol{V}', \tag{44}$$

where $h_1 = u_1,\ h_j = u_j\prod_{i=1}^{j-1}(1+u_i),\ j = 2, \cdots, p$, with $u_i = 1/(n - a_i - 2)$, $i = 1, \cdots, p$.

Proof. Letting $\boldsymbol{Y} = \boldsymbol{\Psi}\boldsymbol{V}$, then $\boldsymbol{Y} = (y_{ij})_{p\times p}$ is still lower-triangular and

$$[\boldsymbol{Y} \mid \boldsymbol{X}] \propto \prod_{i=1}^p (y_{ii}^2)^{(n-a_i-1)/2} \exp\Big\{-\frac{1}{2}tr(\boldsymbol{Y}\boldsymbol{Y}')\Big\}. \tag{45}$$

From above, we know that all y_{ij}, $1 \le i \le j \le p$, are independent and

$$y_{ij} \sim N(0,1), \quad 1 \le j < i \le p;$$

$$y_{ii} \sim (y_{ii}^2)^{(n-a_i-1)/2} \exp\Big(-\frac{1}{2}y_{ii}^2\Big), \quad 1 \le i \le p.$$

If $n - a_i > 0$, $i = 1, \cdots, p$, then $(y_{ii}^2 \mid \boldsymbol{X}) \sim gamma((n - a_i)/2,\ 1/2)$ and $E(y_{ii}^2 \mid \boldsymbol{X})$ exists. Thus it is straightforward to get (43). For (44), we just need to show $E\{(\boldsymbol{YY}')^{-1} \mid \boldsymbol{X}) = \text{diag}(h_1, \cdots, h_p)$. Under the condition $n - a_i > 1$, $E(y_{ii}^{-2} \mid \boldsymbol{X})$ exists and is equal to u_i, $i = 1, \cdots, p$. Thus we obtain the result using the same procedure as in Eaton and Olkin (1987). □

4. FREQUENTIST COVERAGE AND MARGINALIZATION PARADOXES

4.1. *Frequentist Coverage Probabilities and Exact Matching*

In this subsection we compare the frequentist properties of posterior credible intervals for various quantities under the prior $\pi_{\mathbf{a}}$, given in (13). As is customary in such comparisons, we study one-sided intervals $(\theta_L, q_{1-\alpha}(\mathbf{x}))$ of a parameter θ, where θ_L is the lower bound on the parameter θ (e.g., 0 or $-\infty$) and $q_{1-\alpha}(\mathbf{x})$ is the posterior quantile of θ, defined by

$$P(\theta < q_{1-\alpha}(\boldsymbol{x}) \mid \boldsymbol{x}) = 1 - \alpha.$$

Of interest is the frequentist coverage of the corresponding confidence interval, i.e.,

$$P(\theta < q_{1-\alpha}(\mathbf{X}) \mid \boldsymbol{\mu}, \boldsymbol{\Sigma}).$$

The closer this coverage is to the nominal $1 - \alpha$, the better the procedure (and corresponding objective prior) is judged to be.

Berger and Sun (2006) showed that, when $p = 2$, the right-Haar prior is exact matching prior for many functions of parameters of the bivariate normal distribution. Here we generalize the results to the multivariate normal distribution.

To prove frequentist matching, note first that $(\boldsymbol{S} \mid \boldsymbol{\Sigma}) \sim Wishart(n-1, \boldsymbol{\Sigma})$. It is easy to see that the joint density for $\boldsymbol{V}$ (the Chelosky decomposition of $\boldsymbol{S}$), given $\boldsymbol{\Psi}$, is

$$f(\boldsymbol{V} \mid \boldsymbol{\Psi}) \propto \prod_{i=1}^{p} v_{ii}^{n-i-1} etr\Big(-\frac{1}{2}\boldsymbol{\Psi VV}'\boldsymbol{\Psi}'\Big). \tag{46}$$

The following technical lemmas are also needed. The first lemma follows from the expansion

$$tr(\boldsymbol{\Psi VV}'\boldsymbol{\Psi}') = \sum_{i=1}^{p} \psi^2 v_{ii}^2 + \sum_{i=1}^{p}\sum_{j=1}^{i-1}\Big(\sum_{k=1}^{i} \psi_{ik} v_{kj}\Big)^2. \tag{47}$$

The proofs for both lemmas are straightforwad and are omitted.

Lemma 1 *For $n \ge p$ and given $\boldsymbol{\Sigma}^{-1} = \boldsymbol{\Psi\Psi}'$, the following random variables are independent and have the indicated distributions:*

$$Z_{ij} = \psi_{ii}\Big(v_{ij} + \sum_{k=1}^{i-1} t_{ik} v_{kj}\Big) \sim N(0,1), \tag{48}$$

$$(\psi_{ii} v_{ii})^2 = \chi^2_{n-i}. \tag{49}$$

Lemma 2 *Let $Y_{1-\alpha}$ denote the $1-\alpha$ quantile of any random variable Y.*
(a) If $g(\cdot)$ is a monotonically increasing function, $[g(Y)]_{1-\alpha} = g(Y_{1-\alpha})$ for any $\alpha \in (0,1)$.
(b) If W is a positive random variable, $(WY)_{1-\alpha} \geq 0$ if and only if $Y_{1-\alpha} \geq 0$.

Theorem 1 *(a) For any $\alpha \in (0,1)$ and fixed $i = 1, \cdots, p$, the posterior $1-\alpha$ quantile of ψ_{ii} has the expression*

$$(\psi_{ii}^*)_{1-\alpha} = \frac{\sqrt{(\chi^{2*}_{n-a_i})_{1-\alpha}}}{v_{ii}}. \tag{50}$$

(b) For any $\alpha \in (0,1)$ and any $(\boldsymbol{\mu}, \boldsymbol{\Psi})$, the frequentist coverage probability of the credible interval $(0, (\psi_{ii}^*)_{1-\alpha})$ is

$$P\Big(\psi_{ii} < (\psi_{ii}^*)_{1-\alpha} \mid \boldsymbol{\mu}, \boldsymbol{\Psi}\Big) = P\Big(\chi^2_{n-i} < (\chi^{2*}_{n-a_i})_{1-\alpha}\Big), \tag{51}$$

which does not depend on $(\boldsymbol{\mu}, \boldsymbol{\Psi})$ and equals $1-\alpha$ if and only if $a_i = i$.

Proof. Part (a) follows from (41). Part (b) follows from (49) and Lemma 2 (a) with $g(Y) = Y^2$. □

Corollary 1.1 *For any $\alpha \in (0,1)$, the posterior quantile of $d_i = var(x_i \mid x_1, \cdots, x_{i-1})$ is $(d_i^*)_{1-\alpha} = v_{ii}^2/(\chi^{2*}_{n-a_i})_\alpha$. For any $(\boldsymbol{\mu}, \boldsymbol{\Sigma})$, the frequentist coverage probability of the credible interval $(0, (d_i^*)_{1-\alpha}) = (\chi^2_{n-i} < (\chi^{2*}_{n-a_i})_{1-\alpha})$, is a constant $P(\chi^2_{n-i} > (\chi^{2*}_{n-a_i})_\alpha)$, and equals $1-\alpha$ if and only if $a_i = i$.*

Observing that $|\boldsymbol{\Sigma}_i| = \prod_{j=1}^i d_j$ yields the following results, whose proof is similar to that of Theorem 1 and is omitted.

Theorem 2 *(a) For any α, the posterior $1-\alpha$ quantile of $|\boldsymbol{\Sigma}_i|$ has the expression*

$$(|\boldsymbol{\Sigma}_i|^*)_{1-\alpha} = \frac{\prod_{j=1}^i v_{jj}^2}{\left(\prod_{j=1}^i \chi^{2*}_{n-a_j}\right)_\alpha}. \tag{52}$$

(b) For any $\alpha \in (0,1)$ and any $(\boldsymbol{\mu}, \boldsymbol{\Psi})$, the frequentist coverage probability of the credible interval $(0, (|\boldsymbol{\Sigma}_i|^*)_{1-\alpha})$ is

$$P(|\boldsymbol{\Sigma}_i| < (|\boldsymbol{\Sigma}_i|^*)_{1-\alpha} \mid \boldsymbol{\mu}, \boldsymbol{\Psi}) = P\bigg(\prod_{j=1}^i \chi^2_{n-j} > (\prod_{j=1}^i \chi^{2*}_{n-a_j})_\alpha\bigg), \tag{53}$$

which is a constant and equals $1-\alpha$ if and only if $(a_1, \cdots, a_i)$ is a permutation of $(1, \cdots, i)$.

For the bivariate normal case, Berger and Sun (2006) showed that the right-Haar measure is the exact matching prior for ψ_{21} and t_{21}. We also expect that, for the multivariate normal distribution, the right-Haar prior is exact matching for all ψ_{ij} and t_{ij}.

4.2. *Marginalization Paradoxes*

While the Bayesian credible intervals for many parameters under the right-Haar measure are exact matching priors, it can be seen that the prior can suffer marginalization paradoxes. The basis for such paradoxes (Dawid, Stone, and Zidek, 1973) is that any proper prior has the property: if the marginal posterior distribution for a parameter θ depends only on a statistic T – whose distribution in turn depends only on θ – then the posterior of θ can be derived from the distribution of T together with the marginal prior for θ. While this is a basic property of any proper Bayesian prior, it can be violated for improper priors, with the result then called a marginalization paradox.

In Berger and Sun (2006), it was shown that, when using the right-Haar prior, the posterior distribution of the correlation coefficient ρ for a bivariate normal distribution depends only on the sample correlation coefficient r. Brillinger (1962) showed that there does not exist a prior $\pi(\rho)$ such that the this posterior density equals $f(r \mid \rho)\pi(\rho)$, where $f(r \mid \rho)$ is the density of r given ρ. This thus provides an example of a marginalization paradox.

Here is another marginalization paradox in the bivariate normal case. We know from Berger and Sun (2006) that the right-Haar prior π_H is exact matching prior for ψ_{21}. Note that the constructive posterior of ψ_{21} is

$$\frac{Z^*}{\sqrt{s_{11}}} - \frac{\sqrt{\chi^{2*}_{n-2}}}{\sqrt{s_{11}}} \frac{r}{\sqrt{1-r^2}}, \tag{54}$$

which clearly depends only on (s_{11}, r).

It turns out that the joint density of (s_{11}, r) depends only on (σ_{11}, ρ). Note that the posterior of (σ_{11}, ρ) based on the product of $f(s_{11}, r \mid \sigma_{11}, \rho)$ and the marginal prior for (σ_{11}, ρ) based on π_H is different from the marginal posterior of (σ_{11}, ρ) based on π_H. Consequently, the posterior distribution of ψ_{21} from the right Haar provides another example of the marginalization paradox.

It is somewhat controversial as to whether violation of the marginalization paradox is a serious problem. For instance, in the bivariate normal problem, there is probably no proper prior distribution that yields a marginal posterior distribution of ρ which depends only on r, so the relevance of an unattainable property of proper priors could be questioned.

In any case, this situation provides an interesting philosophical conundrum of a type that we have not previously seen: a complete objective Bayesian and frequentist unification can be obtained for inference about the usual parameters of the bivariate normal distribution, but only if violation of the marginalization paradox is accepted. The prior π_{CE} does avoid the marginalization paradox for ρ_{12}, but is not exact frequentist matching. We, alas, know of no way to adjudicate between the competing goals of exact frequentist matching and avoidance of the marginalization paradox, and so will simply present both as possible objective Bayesian approaches.

5. ON THE NON-UNIQUENESS OF RIGHT-HAAR PRIORS

While the right-Haar priors seem to have some very nice properties, the fact that they depend on the particular lower triangular matrix decomposition of $\mathbf{\Sigma}^{-1}$ that is used is troubling. In the bivariate case, for instance, both

$$\pi_1(\mu_1, \mu_2, \sigma_1, \sigma_2, \rho) = \frac{1}{\sigma_2^2(1-\rho^2)} \quad \text{and} \quad \pi_2(\mu_1, \mu_2, \sigma_1, \sigma_2, \rho) = \frac{1}{\sigma_1^2(1-\rho^2)}$$

are right-Haar priors (expressed with respect to $d\mu_1\, d\mu_2\, d\sigma_1\, d\sigma_2\, d\rho$).

There are several natural proposals for dealing with this non-uniqueness. One is to mix over the right-Haar priors. Another is to choose the 'empirical Bayes' right-Haar prior, that which maximizes the marginal likelihood of the data. These proposals are developed in the next two subsections. The last subsection shows, quite surprisingly, that neither of these solutions works! For simplicity, we restrict attention to the bivariate normal case.

5.1. *Symmetrized Right-Haar Priors*

Consider the symmetrized right-Haar prior

$$\begin{aligned}\tilde{\pi}(\mu_1,\mu_2,\sigma_1,\sigma_2,\rho) &= \pi_1(\mu_1,\mu_2,\sigma_1,\sigma_2,\rho)+\pi_2(\mu_1,\mu_2,\sigma_1,\sigma_2,\rho)\\ &= \frac{1}{\sigma_1^2(1-\rho^2)}+\frac{1}{\sigma_2^2(1-\rho^2)}. \end{aligned} \tag{55}$$

This can be thought of as a 50–50 mixture of the two right-Haar priors.

Fact 8 *The joint posterior of $(\mu_1,\mu_2,\sigma_1,\sigma_2,\rho)$ under the prior $\tilde{\pi}$ is given by*

$$\begin{aligned}\tilde{\pi}(\mu_1,\mu_2,\sigma_1,\sigma_2,\rho\mid \boldsymbol{X}) &= C\pi_1(\mu_1,\mu_2,\sigma_1,\sigma_2,\rho\mid \boldsymbol{X})\\ &\quad +(1-C)\pi_2(\mu_1,\mu_2,\sigma_1,\sigma_2,\rho\mid \boldsymbol{X}),\end{aligned} \tag{56}$$

where

$$C = \frac{s_{11}^{-1}}{s_{11}^{-1}+s_{22}^{-1}}. \tag{57}$$

and $\pi_1(\cdot\mid \boldsymbol{X})$ and $\pi_2(\cdot\mid \boldsymbol{X})$ are the posteriors under the priors π_1 and π_2, respectively.

Proof. Let $p=2$ and $(a_1,a_2)=(1,2)$ in (34). We get

$$\begin{aligned}C_j &= \int L(\mu_1,\mu_2,\sigma_1,\sigma_2,\rho)\pi_j(\mu_1,\mu_2,\sigma_1,\sigma_2,\rho)d\mu_1 d\mu_2 d\sigma_1 d\sigma_2 d\rho\\ &= \frac{\Gamma(\frac{n-1}{2})\Gamma(\frac{n-2}{2})2^{(n-2)/2}s_{jj}^{-1}}{\pi^{(n-3)/2}|\boldsymbol{S}|^{(n-2)/2}},\end{aligned} \tag{58}$$

for $j=1,2$. The result is immediate. □

For later use, note that, under the prior $\tilde{\pi}$, the posterior mean of $\boldsymbol{\Sigma}$ has the form

$$\widehat{\boldsymbol{\Sigma}}_S = \mathrm{E}(\boldsymbol{\Sigma}\mid \boldsymbol{X}) \equiv \mathrm{E}(\boldsymbol{\Sigma}\mid \boldsymbol{X}) = C\,\widehat{\boldsymbol{\Sigma}}_1+(1-C)\,\widehat{\boldsymbol{\Sigma}}_2, \tag{59}$$

where $\widehat{\boldsymbol{\Sigma}}_i$ is the posterior mean under $\pi_i(\mu_1,\mu_2,\sigma_1,\sigma_2,\rho)$, given by

$$\widehat{\boldsymbol{\Sigma}}_i = \frac{1}{n-3}(\boldsymbol{S}+\boldsymbol{G}_i), \tag{60}$$

where

$$\boldsymbol{G}_1 = \begin{pmatrix} 0 & 0\\ 0 & \frac{2}{n-4}\left(s_{22}-\frac{s_{12}^2}{s_{11}}\right)\end{pmatrix}, \quad \boldsymbol{G}_2 = \begin{pmatrix} \frac{2}{n-4}\left(s_{11}-\frac{s_{12}^2}{s_{22}}\right) & 0\\ 0 & 0\end{pmatrix}. \tag{61}$$

Here $\boldsymbol{\Sigma}_1$ is a special case of (44) when $p=2$ and $(a_1,a_2)=(1,2)$.

5.2. *The Empirical Bayes Right-Haar Prior*

The right-Haar priors above were essentially just obtained by coordinate permutation. More generally, one can obtain other right-Haar priors by orthonormal transformations of the data. In particular, define the orthonormal matrix

$$\boldsymbol{\Gamma} = (\gamma_1, \gamma_2) \ ,$$

where the γ_i are orthonormal colomn vectors. Consider the transformation of the data $\boldsymbol{\Gamma x}$, so that the resulting sample covariance matrix is

$$\boldsymbol{S}^* = \boldsymbol{\Gamma S \Gamma}' = \begin{pmatrix} s_{11}^* & s_{12}^* \\ s_{12}^* & s_{22}^* \end{pmatrix} = \begin{pmatrix} \gamma_1' \boldsymbol{S} \gamma_1 & \gamma_1' \boldsymbol{S} \gamma_2 \\ \gamma_2' \boldsymbol{S} \gamma_1 & \gamma_2' \boldsymbol{S} \gamma_2 \end{pmatrix}. \tag{62}$$

The right-Haar prior can be defined in this transformed problem, so that each $\boldsymbol{\Gamma}$ defines a different right-Haar prior.

A commonly employed technique when facing a class of priors, as here, is to choose the 'empirical Bayes' prior, that which maximizes the marginal likelihood of the data. This is given in the following lemma.

Lemma 3 *The empirical Bayes right-Haar prior is given by that* $\boldsymbol{\Gamma}$ *for which*

$$\begin{aligned} s_{11}^* &= \frac{1}{2}(s_{11} + s_{22}) - \frac{1}{2}\sqrt{(s_{11} - s_{22})^2 + 4s_{12}^2}\,, \\ s_{12}^* &= 0\,, \\ s_{22}^* &= \frac{1}{2}(s_{11} + s_{22}) + \frac{1}{2}\sqrt{(s_{11} - s_{22})^2 + 4s_{12}^2}\,. \end{aligned}$$

(Note that the two eigenvalues of $\boldsymbol{S}$ *are* s_{11}^* *and* s_{22}^**. Thus this is the orthonormal transformation such that the sample variance of the first coordinate is the smallest eigenvalue.)*

Proof. Noting that $|\boldsymbol{S}^*| = |\boldsymbol{S}|$, it follows from (58) that the marginal likelihood of $\boldsymbol{\Gamma}$ is proportional to $1/s_{11}^*$. Hence we simply want to find an orthonormal $\boldsymbol{\Gamma}$ to minimize $\gamma_1' \boldsymbol{S} \gamma_1$. It is standard matrix theory that the minimum is the smallest eigenvalue of $\boldsymbol{S}$, with γ_1 being the associated eigenvector. Since $\boldsymbol{\Gamma}$ is orthonormal, the remainder of the lemma also follows directly. □

Lemma 4 *Under the empirical Bayes right-Haar prior, the posterior mean of* $\boldsymbol{\Sigma}$ *is* $\widehat{\boldsymbol{\Sigma}}_E = \mathrm{E}(\boldsymbol{\Sigma} \mid \boldsymbol{X})$ *and given by*

$$\begin{aligned} \widehat{\boldsymbol{\Sigma}}_E &= \frac{1}{n-3}\left(\boldsymbol{S} + \frac{s_{22}^*}{n-4}\left(\boldsymbol{I} + \frac{1}{s_{22}^* - s_{11}^*}\begin{pmatrix} s_{11} - s_{22} & 2s_{12} \\ 2s_{12} & s_{22} - s_{11} \end{pmatrix}\right)\right) \\ &= \frac{1}{n-3}\left(\boldsymbol{S} + \frac{s_{22}^*}{n-4}\left(\boldsymbol{I} + \frac{1}{s_{22}^* - s_{11}^*}\boldsymbol{S} - \frac{1}{\frac{1}{s_{11}^*} - \frac{1}{s_{22}^*}}\boldsymbol{S}^{-1}\right)\right). \end{aligned}$$

Proof. Under the empirical Bayes right-Haar prior, the posterior mean of $\boldsymbol{\Sigma}^* = \boldsymbol{\Gamma \Sigma \Gamma}'$ is

$$\mathrm{E}(\boldsymbol{\Sigma}^* \mid \boldsymbol{X}) \quad = \quad \frac{1}{n-3}(\boldsymbol{S}^* + \boldsymbol{G}^*),$$

where

$$G^* = \begin{pmatrix} 0 & 0 \\ 0 & g^* \end{pmatrix}, \quad g^* = \frac{2}{n-4}\left(s_{22}^* - \frac{s_{12}^{*2}}{s_{11}^*}\right) = \frac{2s_{22}^*}{n-4}.$$

So the corresponding estimate of $\boldsymbol{\Sigma}$ is

$$\mathrm{E}(\boldsymbol{\Sigma} \mid \boldsymbol{X}) = \boldsymbol{\Gamma}' \, \mathrm{E}(\boldsymbol{\Sigma}^* \mid \boldsymbol{X}) \, \boldsymbol{\Gamma} = \frac{1}{n-3}(\boldsymbol{S} + \boldsymbol{\Gamma}' \boldsymbol{G}^* \boldsymbol{\Gamma}).$$

Computation yields that the eigenvector $\boldsymbol{\gamma}_2 = (\gamma_{21}, \gamma_{22})'$ is such that

$$\begin{aligned} \gamma_{21}^2 &= \frac{1}{2} + \frac{1}{2}\frac{s_{11} - s_{22}}{\sqrt{(s_{11} - s_{22})^2 + 4s_{12}^2}}, \\ \gamma_{22}^2 &= \frac{1}{2} - \frac{1}{2}\frac{s_{11} - s_{22}}{\sqrt{(s_{11} - s_{22})^2 + 4s_{12}^2}}, \\ \gamma_{21}\gamma_{22} &= \frac{s_{12}}{\sqrt{(s_{11} - s_{22})^2 + 4s_{12}^2}}. \end{aligned}$$

Thus

$$\begin{aligned} \boldsymbol{\Gamma}' \boldsymbol{G}^* \boldsymbol{\Gamma} &= g^* \boldsymbol{\gamma}_2 \boldsymbol{\gamma}_2' \\ &= \frac{s_{22}^*}{n-4}\left(\boldsymbol{I} + \frac{1}{\sqrt{(s_{11} - s_{22})^2 + 4s_{12}^2}} \begin{pmatrix} s_{11} - s_{22} & 2s_{12} \\ 2s_{12} & s_{22} - s_{11} \end{pmatrix}\right). \\ &= \frac{s_{22}^*}{n-4}\left(\boldsymbol{I} + \frac{1}{s_{22}^* - s_{11}^*} \begin{pmatrix} s_{11} - s_{22} & 2s_{12} \\ 2s_{12} & s_{22} - s_{11} \end{pmatrix}\right). \end{aligned}$$

The last expression in the lemma follows from algebra. □

5.3. *Decision-Theoretic Evaluation*

To study the effectiveness of the symmetrized right-Haar prior and the empirical Bayes right-Haar prior, we turn to a decision theoretic evaluation, utilizing a natural invariant loss function.

For a multivariate normal distribution $N_p(\boldsymbol{\mu}, \boldsymbol{\Sigma})$ with unknown $(\boldsymbol{\mu}, \boldsymbol{\Sigma})$, a natural loss to consider is the entropy loss, defined by

$$\begin{aligned} L(\hat{\boldsymbol{\mu}}, \hat{\boldsymbol{\Sigma}}; \boldsymbol{\mu}, \boldsymbol{\Sigma}) &= 2\int \log\left\{\frac{f(\boldsymbol{X} \mid \boldsymbol{\mu}, \Sigma)}{f(\boldsymbol{X} \mid \hat{\boldsymbol{\mu}}, \hat{\boldsymbol{\Sigma}})}\right\} f(\boldsymbol{X} \mid \boldsymbol{\mu}, \boldsymbol{\Sigma})\, d\boldsymbol{X} \\ &= (\hat{\boldsymbol{\mu}} - \boldsymbol{\mu})' \hat{\boldsymbol{\Sigma}}^{-1} (\hat{\boldsymbol{\mu}} - \boldsymbol{\mu}) + tr(\hat{\boldsymbol{\Sigma}}^{-1}\boldsymbol{\Sigma}) - \log|\hat{\boldsymbol{\Sigma}}^{-1}\boldsymbol{\Sigma}| - p. \quad (63) \end{aligned}$$

Clearly, the entropy loss has two parts, one is related to the means $\hat{\boldsymbol{\mu}}$ and $\boldsymbol{\mu}$ (with $\hat{\boldsymbol{\Sigma}}$ as the weight matrix), and the other is related to $\hat{\boldsymbol{\Sigma}}, \boldsymbol{\Sigma}$. The last three terms of this expression are related to 'Stein's loss,' and is the most commonly used losses for estimation of a covariance matrix (cf. James and Stein,1961; Haff, 1977).

Lemma 5 *Under the loss (63) and for any of the priors considered in this paper, the generalized Bayesian estimator of $(\boldsymbol{\mu}, \boldsymbol{\Sigma})$ is*

$$\hat{\boldsymbol{\mu}}_B = \mathrm{E}(\boldsymbol{\mu} \mid \boldsymbol{X}) = (\bar{x}_1, \bar{x}_2)', \tag{64}$$

$$\widehat{\boldsymbol{\Sigma}}_B = \mathrm{E}(\boldsymbol{\Sigma} \mid \boldsymbol{X}) + \mathrm{E}\{(\widehat{\boldsymbol{\mu}}_B - \boldsymbol{\mu})'(\widehat{\boldsymbol{\mu}}_B - \boldsymbol{\mu}) \mid \boldsymbol{X}\} = \frac{n+1}{n}\mathrm{E}(\boldsymbol{\Sigma} \mid \boldsymbol{X}). \tag{65}$$

Proof. For the priors we consider in the paper,

$$[\boldsymbol{\mu} \mid \boldsymbol{\Sigma}, \boldsymbol{X}] \sim N_2\left((\bar{x}_1, \bar{x}_2)', \frac{1}{n}\boldsymbol{\Sigma}\right), \tag{66}$$

so that (64) is immediate. Furthermore, it follows that

$$\mathrm{E}((\hat{\boldsymbol{\mu}}_B - \boldsymbol{\mu})'\hat{\boldsymbol{\Sigma}}^{-1}(\hat{\boldsymbol{\mu}}_B - \boldsymbol{\mu}) \mid \boldsymbol{X}) = \frac{1}{n} tr(\hat{\boldsymbol{\Sigma}}^{-1}\boldsymbol{\Sigma}) \tag{67}$$

so that the remaining goal is to choose $\hat{\boldsymbol{\Sigma}}$ so as to minimize

$$\begin{aligned} &\mathrm{E}\Big((1+\frac{1}{n})tr(\hat{\boldsymbol{\Sigma}}^{-1}\boldsymbol{\Sigma}) - \log|\hat{\boldsymbol{\Sigma}}^{-1}\boldsymbol{\Sigma}| - p \mid \boldsymbol{X}\Big) \\ &= \mathrm{E}\left(tr(\hat{\boldsymbol{\Sigma}}^{-1}\tilde{\boldsymbol{\Sigma}}) - \log|\hat{\boldsymbol{\Sigma}}^{-1}\tilde{\boldsymbol{\Sigma}}| - p \mid \boldsymbol{X}\right) + \log(1+\frac{1}{n}), \end{aligned} \tag{68}$$

where $\tilde{\boldsymbol{\Sigma}} = (1+\frac{1}{n})\boldsymbol{\Sigma}$. It is standard (see, e.g., Eaton, 1989) that the first term on the right hand side of the last expression is minimized at

$$\widehat{\boldsymbol{\Sigma}} = \mathrm{E}(\tilde{\boldsymbol{\Sigma}} \mid \boldsymbol{X}) = (1+\frac{1}{n})\mathrm{E}(\boldsymbol{\Sigma} \mid \boldsymbol{X}), \tag{69}$$

from which the result is immediate. □

We now turn to frequentist decision-theoretic evaluation of the various posterior estimates that arise from the reference priors considered in the paper. Thus we now change perspective and consider $\boldsymbol{\mu}$ and $\boldsymbol{\Sigma}$ to be given, and consider the frequentist risk of the posterior estimates $\hat{\boldsymbol{\mu}}_B(\boldsymbol{X})$ and $\widehat{\boldsymbol{\Sigma}}_B(\boldsymbol{X})$, now considered as functions of $\boldsymbol{X}$. Thus we evaluate the frequentist risk

$$R(\hat{\boldsymbol{\mu}}_B, \widehat{\boldsymbol{\Sigma}}_B; \boldsymbol{\mu}, \boldsymbol{\Sigma}) = \mathrm{E}L(\hat{\boldsymbol{\mu}}_B(\boldsymbol{X}), \hat{\boldsymbol{\Sigma}}_B(\boldsymbol{X}); \boldsymbol{\mu}, \boldsymbol{\Sigma}), \tag{70}$$

where the expectation is over $\boldsymbol{X}$ given $\boldsymbol{\mu}$ and $\boldsymbol{\Sigma}$. The following lemma states that we can reduce the frequentist risk comparison to a comparison of the frequentist risks of the various posterior means for $\boldsymbol{\Sigma}$ under Stein's loss. It's proof is virtually identical to that of Lemma 5, and is omitted.

Lemma 6 *For frequentist comparison of the various Bayes estimators considered in the paper, it suffices to compare the frequentist risks of the $\widehat{\boldsymbol{\Sigma}}(\boldsymbol{X}) = \mathrm{E}(\boldsymbol{\Sigma} \mid \boldsymbol{X})$, with respect to*

$$R(\widehat{\boldsymbol{\Sigma}}(\boldsymbol{X}); \boldsymbol{\mu}, \boldsymbol{\Sigma}) = \mathrm{E}\left(tr(\widehat{\boldsymbol{\Sigma}}^{-1}(\boldsymbol{X})\boldsymbol{\Sigma}) - \log|\widehat{\boldsymbol{\Sigma}}^{-1}(\boldsymbol{X})\boldsymbol{\Sigma}| - p\right), \tag{71}$$

where the expectation is with respect to $\boldsymbol{X}$.

Lemma 7 *Under the right Haar prior π_H, the risk function (71) is a constant, given by*

$$R(\widehat{\mathbf{\Sigma}}(\boldsymbol{X}); \boldsymbol{\mu}, \mathbf{\Sigma}) = \sum_{j=1}^{p} \log(h_j) + \sum_{j=1}^{p} \mathrm{E} \log(\chi^2_{n-j}). \tag{72}$$

where h_j is given by (44).

The proof of this can be found in Eaton (1989). If $p = 2$, it follows that the risk for the two right-Haar priors is

$$\log(n-2) - 2\log(n-3) - \log(n-4) + \mathrm{E}\log(\chi^2_{n-1}) + \mathrm{E}\log(\chi^2_{n-2}).$$

For instance, when $n = 10$, this risk is approximately 0.4271448.

Table 2 gives the risks for the estimates arising from the two right-Haar priors, $\widehat{\mathbf{\Sigma}}_1$ and $\widehat{\mathbf{\Sigma}}_2$, the estimate $\widehat{\mathbf{\Sigma}}_S$ arising from the symmetrized right-Haar prior, the estimate $\widehat{\mathbf{\Sigma}}_E$ arising from the empirical Bayes right-Haar prior, $\widehat{\mathbf{\Sigma}}_{R\rho}$ arising from the reference prior for ρ, and an interesting estimate $\widehat{\mathbf{\Sigma}}_D$, arising from the spirit of Dey and Srinivasan (1985) and Dey (1988) and defined by

$$\widehat{\mathbf{\Sigma}}_D = (1-C)\,\widehat{\mathbf{\Sigma}}_1 + C\,\widehat{\mathbf{\Sigma}}_2. \tag{73}$$

This is obtainded by switching $\widehat{\mathbf{\Sigma}}_1$ and $\widehat{\mathbf{\Sigma}}_2$ in (59).

Table 2: *Frequentist risks of various estimates of $\mathbf{\Sigma}$ when $n = 10$ and for various choices of $\mathbf{\Sigma}$. These were computed by simulation, using 10,000 generated values of $\boldsymbol{S}$.*

$(\sigma_1, \sigma_2, \rho)$	$R(\widehat{\mathbf{\Sigma}}_1)$	$R(\widehat{\mathbf{\Sigma}}_2)$	$R(\widehat{\mathbf{\Sigma}}_S)$	$R(\widehat{\mathbf{\Sigma}}_E)$	$R(\widehat{\mathbf{\Sigma}}_D)$	$R(\widehat{\mathbf{\Sigma}}_{R\rho})$
$(1,1,0)$	.4287	.4288	.4452	.6052	.3833	.4095
$(1,2,0)$	.4278	.4270	.4424	.5822	.3859	.4174
$(1,5,0)$	.4285	.4287	.4391	.5404	.3989	.4334
$(1,50,0)$	.4254	.4250	.4272	.5100	.4194	.4427
$(1,1,.1)$	.4255	.4266	.4424	.5984	.3810	.4241
$(1,1,.5)$	.4274	.4275	.4403	.5607	.3906	.3936
$(1,1,.9)$	.4260	.4255	.4295	.5159	.4134	.4206
$(1,1,-.9)$	.4242	.4243	.4280	.5119	.4118	.4219

The simulated risks are given in the Table 2 for $\widehat{\mathbf{\Sigma}}_1$ and $\widehat{\mathbf{\Sigma}}_2$ instead of the exact risks, because the comparisons between estimates is then more meaningful (the simulation errors being highly correlated since the estimates were all based on common realizations of sample covariance matrices).

The first surprise here is that the risk of $\widehat{\mathbf{\Sigma}}_S$ is actually worse than the risk of the right-Haar prior estimates. This is in contradiction to the usual belief that, if considering alternate priors, utilization of a mixture of the two priors will give superior performance.

This would also seem to be in contradiction to the known fact for a convex loss function (such as Stein's loss) that, if two estimators $\widehat{\mathbf{\Sigma}}_1$ and $\widehat{\mathbf{\Sigma}}_2$ have equal risk functions, then an average of the two estimators will have lower risk. But this refers to a constant average of the two estimators, not a data-weighted average as in $\widehat{\boldsymbol{\Sigma}}_S$. What is particularly striking is that the data-weighted average arises from the posterior marginal likelihoods corresponding to the two different priors, so the posterior seems to be 'getting it wrong,' weighting the 'bad' prior more than the 'good' prior.

This is indicated in even more dramatic fashion by $\widehat{\boldsymbol{\Sigma}}_E$, the empirical Bayes version, which is based on that right-Haar prior which is 'most likely' for given data. In that the risk of $\widehat{\boldsymbol{\Sigma}}_E$ is much worse than even the risk of $\widehat{\boldsymbol{\Sigma}}_S$, it seems that empirical Bayes has selected the worst of all of the right-Haar priors! Also surprising is that, $\widehat{\boldsymbol{\Sigma}}_D$, the alternative average of $\widehat{\boldsymbol{\Sigma}}_1$ and $\widehat{\boldsymbol{\Sigma}}_2$, has considerably smaller risk in all the cases considered here.

The phenomenon arising here is disturbing and sobering. It is yet another indication that improper priors do not behave as do proper priors, and that it can be dangerous to apply 'understandings' from the world of proper priors to the world of improper priors. (Of course, the same practical problems could arise from use of vague proper priors, so use of such is not a solution to the problem.)

From a formal objective Bayesian position (e.g., the viewpoint from the reference prior perspective), there is no issue here. The various reference priors we considered are (by definition) the correct objective priors for the particular contexts (choice of parameterization and parameter of interest) in which they were derived. It is use of these priors – or modifications of them based on 'standard tricks' – out of context that is being demonstrated to be of concern.

APPENDIX

Proof of Fact 1. The likelihood function of $(\boldsymbol{\mu}, \boldsymbol{\Psi})$ is

$$f(\boldsymbol{x} \mid \boldsymbol{\mu}, \boldsymbol{\Psi}) \quad \propto \quad |\boldsymbol{\Psi}'\boldsymbol{\Psi}|^{\frac{1}{2}} \exp\Big(-\frac{1}{2}(\boldsymbol{x}-\boldsymbol{\mu})'\boldsymbol{\Psi}'\boldsymbol{\Psi}(\boldsymbol{x}-\boldsymbol{\mu})\Big),$$

and the log-likelihood is then

$$\log f \quad = \quad const + \sum_{i=1}^{p} \log(\psi_{ii}) - \frac{1}{2}\sum_{i=1}^{p}\Big(\sum_{j=1}^{i}\psi_{ij}(x_j-\mu_j)\Big)^2.$$

For any fixed $i = 1, \cdots, p$, let $\boldsymbol{\Sigma}_i$ be the variance and covariance matrix of $(x_1, \cdots, x_i)'$. Also, let $\boldsymbol{e}_i$ be the $i \times 1$ vector whose ith element is 1 and 0 otherwise. The Fisher information matrix of $\boldsymbol{\theta}$ is then (11).

Note that $Var(\boldsymbol{x}) = \boldsymbol{\Sigma} = \boldsymbol{\Psi}^{-1}\boldsymbol{\Psi}'^{-1}$. Let $\boldsymbol{\Psi}_i$ be the $i \times i$ left and top sub-matrix of $\boldsymbol{\Psi}$. It is easy to versify that $\boldsymbol{\Sigma}_i = \boldsymbol{\Psi}_i^{-1}\boldsymbol{\Psi}_i'^{-1}$. Using the fact that $|\boldsymbol{B}+\boldsymbol{a}\boldsymbol{a}'| = |\boldsymbol{B}|(1+\boldsymbol{a}'\boldsymbol{B}^{-1}\boldsymbol{a})$ where $\boldsymbol{B}$ is invertible and $\boldsymbol{a}$ is a vector, we can show that

$$|\boldsymbol{\Lambda}_i| \quad = \quad 2\prod_{j=1}^{i}\frac{1}{\psi_{jj}^2}. \tag{74}$$

From (11) and (74), the reference prior of $\boldsymbol{\Psi}$ for the ordered group $\{\boldsymbol{\mu}, \psi_{11}, (\psi_{21}, \psi_{22}), \cdots, (\psi_{p1}, \cdots, \psi_{pp})\}$, is easy to obtain as (12) according to the algorithm in Berger and Bernardo (1992).

Proof of Fact 2. For $i = 2, \cdots, p$, denote $\boldsymbol{t}_{i,i-1} = \boldsymbol{\psi}_{i,i-1}/\psi_{ii}$. Clearly, the Jacobian from $(\boldsymbol{\psi}_{i,i-1}, \psi_{ii})$ to $(\boldsymbol{t}_{i,i-1}, \psi_{ii})$ is

$$\boldsymbol{J}_i = \frac{\partial(\boldsymbol{\psi}_{i,i-1}, \psi_{ii})}{\partial(\boldsymbol{t}_{i,i-1}, \psi_{ii})} = \begin{pmatrix} \psi_{ii}\boldsymbol{I}_{i-1} & \boldsymbol{t}_{i,i-1} \\ \boldsymbol{0}' & 1 \end{pmatrix}. \tag{75}$$

The Fisher information for $\widetilde{\boldsymbol{\theta}}$ has the form (19), where

$$\widetilde{\boldsymbol{\Lambda}}_i = \boldsymbol{J}_i'\boldsymbol{\Lambda}_i\boldsymbol{J}_i = \boldsymbol{J}_i'\left(\boldsymbol{\Psi}_i^{-1}\boldsymbol{\Psi}_i'^{-1} + \frac{1}{\psi_{ii}^2}\boldsymbol{e}_i\boldsymbol{e}_i'\right)\boldsymbol{J}_i. \tag{76}$$

Note that

$$\boldsymbol{\Psi}_i = \begin{pmatrix} \boldsymbol{\Psi}_{i-1} & \boldsymbol{0} \\ \psi_{ii}\boldsymbol{t}_{i,i-1}' & \psi_{ii} \end{pmatrix} \quad \text{and} \quad \boldsymbol{\Psi}_i^{-1} = \begin{pmatrix} \boldsymbol{\Psi}_{i-1}^{-1} & \boldsymbol{0} \\ -\boldsymbol{t}_{i,i-1}'\boldsymbol{\Psi}_{i-1}^{-1} & \frac{1}{\psi_{ii}} \end{pmatrix}.$$

We have that

$$\boldsymbol{J}_i'\boldsymbol{\Psi}_i^{-1} = \begin{pmatrix} \psi_{ii}\boldsymbol{I}_{i-1} & \boldsymbol{t}_{i,i-1} \\ \boldsymbol{0}' & 1 \end{pmatrix}\begin{pmatrix} \boldsymbol{\Psi}_{i-1}^{-1} & \boldsymbol{0} \\ -\boldsymbol{t}_{i,i-1}'\boldsymbol{\Psi}_{i-1}^{-1} & \frac{1}{\psi_{ii}} \end{pmatrix} = \begin{pmatrix} \psi_{ii}\boldsymbol{\Psi}_{i-1}^{-1} & \boldsymbol{0} \\ \boldsymbol{0}' & \frac{1}{\psi_{ii}} \end{pmatrix}. \tag{77}$$

Substituting (77) into (76) and using the fact that $\boldsymbol{J}_i'\boldsymbol{e}_i = \boldsymbol{e}_i$,

$$\widetilde{\boldsymbol{J}}_i = \begin{pmatrix} \psi_{ii}^2\boldsymbol{\Psi}_{i-1}^{-1}\boldsymbol{\Psi}_{i-1}'^{-1} & \boldsymbol{0} \\ \boldsymbol{0}' & \frac{2}{\psi_{ii}^2} \end{pmatrix} = \begin{pmatrix} \psi_{ii}^2\boldsymbol{T}_{i-1}^{-1}\widetilde{\boldsymbol{\Psi}}_{i-1}^{-2}\boldsymbol{T}_{i-1}'^{-1} & \boldsymbol{0} \\ \boldsymbol{0}' & \frac{2}{\psi_{ii}^2} \end{pmatrix}. \tag{78}$$

Part (a) holds. It is easy to see that the upper and left $i \times i$ submatrix of $\boldsymbol{\Lambda}_i^*$ does not depend on $\boldsymbol{t}_{i,i-1}$. Part (b) can be proved using the algorithm of Berger and Bernardo (1992b). Furthermore, part (c) holds because of $|\boldsymbol{J}_i| = \psi_{ii}^{i-1}$.

$$\widetilde{\pi}_R(\widetilde{\boldsymbol{\theta}})\prod_{i=2}^{p}\frac{1}{|\boldsymbol{J}_i|} = \prod_{i=1}^{p}\frac{1}{\psi_{ii}^i} = \pi_H(\boldsymbol{\Psi}).$$

Proof of Fact 3. Note that (25) is equivalent to

$$\begin{cases} d_1 &= \xi_1^{\frac{1}{2}}\,\xi_2^{\frac{1}{3}} \cdots\cdots \xi_{p-2}^{\frac{1}{p-1}}(\xi_{p-1}\xi_p)^{\frac{1}{p}}, \\ d_2 &= \xi_1^{-\frac{1}{2}}\xi_2^{\frac{1}{3}} \cdots\cdots \xi_{p-2}^{\frac{1}{p-1}}(\xi_{p-1}\xi_p)^{\frac{1}{p}}, \\ d_3 &= \xi_2^{-\frac{2}{3}} \cdots\cdots \xi_{p-2}^{\frac{1}{p-1}}(\xi_{p-1}\xi_p)^{\frac{1}{p}}, \\ \cdots & \qquad\qquad \cdots\cdots \\ d_{p-1} &= \xi_{p-2}^{-\frac{p-2}{p-1}}(\xi_{p-1}\xi_p)^{\frac{1}{p}}, \\ d_p &= \xi_{p-1}^{-\frac{p-2}{p}}\,\xi_p^{\frac{1}{p}}. \end{cases}$$

Then, the Hessian is

$$
\begin{aligned}
\boldsymbol{H} &= \frac{\partial(d_1,\cdots,d_p)}{\partial(\xi_1,\cdots,\xi_p)} \\
&= \begin{pmatrix}
\frac{d_1}{2\xi_1} & \frac{d_1}{3\xi_2} & \frac{d_1}{4\xi_3} & \cdots & \frac{d_1}{(p-1)\xi_{p-2}} & \frac{d_1}{p\xi_{p-1}} & \frac{d_1}{p\xi_p} \\
-\frac{d_2}{2\xi_1} & \frac{d_2}{3\xi_2} & \frac{d_2}{4\xi_3} & \cdots & \frac{d_2}{(p-1)\xi_{p-2}} & \frac{d_2}{p\xi_{p-1}} & \frac{d_2}{p\xi_p} \\
0 & \frac{d_3}{3\xi_2} & \frac{d_3}{4\xi_3} & \cdots & \frac{d_3}{(p-1)\xi_{p-2}} & \frac{d_3}{p\xi_{p-1}} & \frac{d_3}{p\xi_p} \\
\vdots & \vdots & \vdots & \vdots & \vdots & \vdots & \vdots \\
0 & 0 & 0 & \cdots & -\frac{(p-2)d_{p-1}}{(p-1)\xi_{p-2}} & \frac{d_{p-1}}{p\xi_{p-1}} & \frac{d_{p-1}}{p\xi_p} \\
0 & 0 & 0 & \cdots & 0 & -\frac{(p-1)d_{p-1}}{p\xi_{p-1}} & \frac{d_{p-1}}{p\xi_p}
\end{pmatrix} \\
&= \boldsymbol{DQ\Xi},
\end{aligned}
$$

where $\boldsymbol{\Xi} = diag(\xi_1,\cdots,\xi_p)$ and

$$
\boldsymbol{Q} = \begin{pmatrix}
\frac{1}{2} & \frac{1}{3} & \frac{1}{4} & \cdots & \frac{1}{p-1} & \frac{1}{p} & \frac{1}{p} \\
-\frac{1}{2} & \frac{1}{3} & \frac{1}{4} & \cdots & \frac{1}{p-1} & \frac{1}{p} & \frac{1}{p} \\
0 & -\frac{2}{3} & \frac{1}{4} & \cdots & \frac{1}{p-1} & \frac{1}{p} & \frac{1}{p} \\
\vdots & \vdots & \vdots & \vdots & \vdots & \vdots & \vdots \\
0 & 0 & 0 & \cdots & -\frac{p-2}{p-1} & \frac{1}{p} & \frac{1}{p} \\
0 & 0 & 0 & \cdots & 0 & -\frac{p-1}{p} & \frac{1}{p}
\end{pmatrix}.
$$

Note that the Fisher information matrix for $(d_1,\cdots,d_p)$ is $(\boldsymbol{D}^2)^{-1}$. The Fisher information matrix for $(\xi\cdots,\xi_p)$ is then

$$
\boldsymbol{H}'\boldsymbol{D}^{-2}\boldsymbol{H} = \boldsymbol{\Xi}'\boldsymbol{Q}'\boldsymbol{D}\boldsymbol{D}^{-1}\boldsymbol{D}^{-1}\boldsymbol{D}\boldsymbol{Q}\boldsymbol{\Xi} = \boldsymbol{\Xi}'\boldsymbol{Q}'\boldsymbol{Q}\boldsymbol{\Xi}.
$$

It is easy to verify that

$$
\boldsymbol{Q}'\boldsymbol{Q} - diag\left(\frac{1}{2},\frac{2}{3},\cdots,\frac{p-1}{p},\frac{1}{p}\right)
$$

We have that

$$
\boldsymbol{H}'\boldsymbol{D}^{-2}\boldsymbol{H} = diag\left(\frac{1}{2\xi_1^2},\frac{2}{3\xi_2^2},\cdots,\frac{p-1}{p\,\xi_{p-1}^2},\frac{1}{p\,\xi_p^2}\right).
$$

This proves part (a). Parts (b) and (c) are immediate.

Proof of Fact 4. We have

$$
\begin{aligned}
M &\equiv \int L(\boldsymbol{\mu},\boldsymbol{\Psi})\pi_{\boldsymbol{a}}(\boldsymbol{\mu},\boldsymbol{\Psi})d\boldsymbol{\mu}d\boldsymbol{\Psi} \\
&= \int \frac{\prod_{i=1}^p(\psi_{ii}^2)^{(n-a_i-1)/2}}{(2\pi)^{(n-1)p/2}n^{p/2}}\exp\left(-\frac{1}{2}tr(\boldsymbol{\Psi S\Psi}')\right)d\boldsymbol{\Psi}. \qquad (79)
\end{aligned}
$$

Note that $w_i = s_{ii} - \boldsymbol{s}'_{i,i-1}\boldsymbol{S}^{-1}_{i-1}\boldsymbol{s}_{i,i-1} > 0$ for $i = 2, \cdots, p$. Also let $\boldsymbol{g}_i = -\psi_{ii}\boldsymbol{S}^{-1}_{i-1}\boldsymbol{s}_{i,i-1}$. We then have a recursive formula,

$$\begin{aligned} tr(\boldsymbol{\Psi}_p\boldsymbol{S}_p\boldsymbol{\Psi}'_p) &= tr(\boldsymbol{\Psi}_{p-1}\boldsymbol{S}_{p-1}\boldsymbol{\Psi}'_{p-1}) + (\boldsymbol{\psi}_{p,p-1} - \boldsymbol{g}_p)'\boldsymbol{S}_{p-1}(\boldsymbol{\psi}_{p,p-1} - \boldsymbol{g}_p) \\ &= \sum_{i=1}^{p}\psi_{ii}^2 w_i + \sum_{i=2}^{p}(\boldsymbol{\psi}_{i,i-1} - \boldsymbol{g}_i)'\boldsymbol{S}_{i-1}(\boldsymbol{\psi}_{i,i-1} - \boldsymbol{g}_i). \end{aligned}$$

Then

$$M = \int \frac{\prod_{i=1}^{p}(\psi_{ii}^2)^{(n-a_i-1)/2}}{(2\pi)^{(n-1)p/2-(p-1)p/2}n^{p/2}\prod_{j=1}^{p-1}|\boldsymbol{S}_j|^{1/2}} \exp\Big(-\frac{1}{2}\sum_{i=1}^{p}\psi_{ii}^2 w_i\Big)\prod_{i=1}^{p}d\psi_{ii}. \quad (80)$$

Let $\delta_i = \psi_{ii}^2$. The conclusion follows from the fact that the right hand side of (80) is equal to

$$\begin{aligned} &\frac{\prod_{j=1}^{p-1}|\boldsymbol{S}_j|^{-1/2}}{2^p(2\pi)^{(n-p)p/2}n^{p/2}}\prod_{i=1}^{p}\int_0^\infty \delta_i^{(n-a_i)/2-1}\exp\Big(-\frac{w_i}{2}\delta_i\Big)d\delta_i \\ =\ &\frac{\prod_{i=1}^{p}\Gamma(\frac{1}{2}(n-a_i))2^{(n-a_i)/2}}{2^p(2\pi)^{(n-p)p/2}n^{p/2}}\prod_{j=1}^{p-1}|\boldsymbol{S}_j|^{-1/2}\frac{1}{s_{11}^{(n-a_1)/2}}\prod_{i=2}^{p}\Big(\frac{|\boldsymbol{S}_{i-1}|}{|\boldsymbol{S}_i|}\Big)^{(n-a_i)/2}. \end{aligned}$$

ACKNOWLEDGMENTS

The authors would like to thank Susie Bayarri and Jose Bernardo for helpful comments and discussions throughout the period when we were working on the project.

REFERENCES

Bayarri, M. J. (1981). Inferencia bayesiana sobre el coeficiente de correlación de una población normal bivariante. *Trab. Estadist.* **32**, 18–31.

Berger, J. O. and Bernardo, J. M. (1992). On the development of reference priors. *Bayesian Statistics 4* (J. M. Bernardo, J. O. Berger, A. P. Dawid and A. F. M. Smith, eds.) Oxford: University Press, 35–60 (with discussion).

Berger, J. O. and Bernardo, J. M. (1992). Reference priors in a variance components problem. *Bayesian Analysis in Statistics and Econometrics* (P. K. Goel and N. S. Iyengar, eds.) Berlin: Springer-Verlag, 177–194.

Berger, J. O., Strawderman, W. and Tang, D. (2005). Posterior propriety and admissibility of hyperpriors in normal hierarchical models, *Ann. Statist.* **33**, 606–646.

Berger, J. O. and Sun, D. (2006). Objective priors for a bivariate normal model. *Tech. Rep.*, ISDS, Duke University, USA.

Brillinger, D. R. (1962). Examples bearing on the definition of fiducial probability with a bibliography. *Ann. Math. Statist.* **33**, 1349–1355.

Brown, P. J. (2001). The generalized inverted Wishart distribution. *Encylopedia of Environmetrics* **2** (A. H. El-Shaarawi and W. W. Piegorsh, eds.), Chichester: Wiley, 1079–1083.

Chang, T. and Eaves, D. (1990). Reference priors for the orbit in a group model, *Ann. Statist.* **18**, 1595–1614.

Consonni, G., Gutiérrez-Peña, E. and Veronese, P. (2004). Reference priors for exponential families with simple quadratic variance function. *J. Multivariate Analysis* **88**, 335–364.

Daniels, M. and Kass, R. (1999). Nonconjugate Bayesian estimation of covariance matrices and its use in hierarchical models, *J. Amer. Statist. Assoc.* **94**, 1254–1263.

Daniels, M. and Pourahmadi, M. (2002). Bayesian analysis of covariance matrices and dynamic models for longitudinal data *Biometrika* **89**, 553–566.

Dawid, A. P., Stone, M. and Zidek, J. V. (1973). Marginalization paradoxes in Bayesian and structural inference. *J. Roy. Statist. Soc. B* **35**, 189–233 (with discussion).

Dey, D. and Srinivasan, C. (1985). Estimation of a covariance matrix under Stein's loss, *Ann. Statist.* **13**, 1581–1591.

Dey, D. (1988). Simultaneous estimation of eigenvalues. *Ann. Inst. Statist. Math.* **40**, 137-147.

Eaton, M. L. (1989). *Group Invariance Applications in Statistics.* Hayward, CA: IMS.

Eaton, M. L. and Olkin, I. (1987). Best equivariant estimators of a Cholesky decomposition. *Ann. Statist.* **15**, 1639–1650.

Eaton, M. L. and Sudderth, W. (2002). Group invariant inference and right Haar measure. *J. Statist. Planning and Inference* **103**, 87–99.

Geisser, S. and Cornfield, J. (1963). Posterior distributions for multivariate normal parameters. *J. Roy. Statist. Soc. B* **25**, 368–376.

Haff, L. (1977). Minimax estimators for a multivariate precision matrix. *J. Multivariate Analysis* **7**, 374–385.

James, W. and Stein, C. (1961). Estimation with quadratic loss.*Proc. First Berkeley Symp.* (J. Neyman ed.) Berkeley: Univ. California Press, 361–380.

Jeffreys, H. (1961). *Theory of Probability.* Oxford: University Press.

Leonard, T. and Hsu, J. S. J. (1992). Bayesian inference for a covariance matrix. *Ann. Statist.* **20**, 1669-1696.

Liechty, J., Liechty, M. and Müller, P. (2004). Bayesian correlation estimation. *Biometrika* **91**, 1–14.

Lindley, D. V. (1965). The use of prior probability distributions in statistical inference and decisions. *Proc. Fourth Berkeley Symp.* **1** (J. Neyman and E. L. Scott, eds.) Berkeley: Univ. California Press, 453–468.

Pourahmadi, M. (1999). Joint mean-covariance models with applications to longitudinal data: Unconstrained parameterisation, *Biometrika* **86**, 677–690.

Pourahmadi, M. (2000). Maximum likelihood estimation of generalised linear models for multivariate normal covariance matrix. *Biometrika* **87**, 425-435.

Roverato, A. and Consonni, G. (2004). Compatible prior distributions for dag models. *J. Roy. Statist. Soc. B* **66**, 47–61.

Stein, C. (1956). Some problems in multivariate analysis. *Tech. Rep.*, Stanford University, USA.

Yang, R. and Berger, J. O. (1994). Estimation of a covariance matrix using the reference prior *Ann. Statist.* **22**, 1195–1211.

DISCUSSION

BRUNERO LISEO (*Università di Roma 'La Sapienza', Italy*)

General comments. I have enjoyed reading this authoritative paper by Jim Berger and Dongchu Sun. The paper is really thought provoking, rich of new ideas, new proposals, and useful technical results about the most used and popular statistical model.

I assume that the role of the discussion leader in a conference is different from that of a journal referee, and his/her main goal should be to single out the particular features of the paper that deserve attention, and to provide a personal perspective and opinion on the subject.

This paper deserves discussion in at least three important aspects: (i) technical results; (ii) philosophical issues; and (iii) guidance on the choice among available 'objective' prior distributions.

Among the technical results I would like to remark first the fact that almost all the proposed 'objective posteriors' are given in a constructive form, and it is usually nearly immediate to obtain a sample from the marginal posterior of the parameter of interest. This simple fact facilitates the reader's task, since it is straightforward to perform other numerical comparisons and explore different aspects of the problem.

There are also many philosophical issues raised by the authors. The paper, as many others written by Berger and Sun, stands on the interface between frequentist and Bayesian statistics. From this perspective, the main contributions of the paper are, in my opinion, the following:

- to provide a classical interpretation of some objective Bayes procedures;
- to provide an objective Bayesian interpretation of some classical procedures;
- to derive optimal and nearly optimal classical/fiducial/objective Bayesian procedures for many inferential problems related to the bivariate and multivariate normal distribution.

Before discussing these issues, I would like to linger over a tentative definition of what is the main goal of an objective Bayesian approach, which seems, in some respects, lacking. I have asked many Bayesian statisticians the question: 'Could you please provide, in a line, a definition of the Objective Bayesian approach?'. I report the most common answers, together with some of the contributions illustrated in Berger (2006) and O'Hagan (2006):

(i) A formal Bayesian analysis using some *conventional* prior information, which is largely 'accepted' by researchers.

(ii) The easiest way to obtain good frequentist procedures.

(iii) A Bayesian analysis where the prior is obtained from the sampling distribution; it is the only feasible approach when there is no chance of getting genuine prior information for our model and we do not want to abandon that model.

(iv) The optimal Bayesian strategy under some specific frequentist criterion (frequentist matching philosophy).

(v) A cancerous oxymoron.

(vi) A convention we should adopt in scenarios in which a subjective analysis is not tenable.

(vii) A collection of convenient and useful techniques for approximating a genuine Bayesian analysis.

While preferring option (iii), I must admit that option (ii) captures important aspects of the problem, in that it provides a way to validate procedures and to facilitate communication among statisticians. On the other hand, option (iv) might be a dangerous goal to pursue, as the authors, perhaps indirectly, illustrate in the paper.

Indeed the authors present many examples where

- If the prior is chosen to be 'exact' in terms of frequentist coverage, then the resulting posterior will suffer from some pathologies like, for example, the marginalization paradox.

- To be 'optimal' one has to pay the price of introducing unusual (albeit, it is the right Haar prior!) default priors. For instance, in the bivariate normal example, when ρ is the parameter of interest, the 'exact' frequentist matching prior is

$$\pi_H(\mu_1, \mu_2, \sigma_1, \sigma_2, \rho) \propto \frac{1}{\sigma_1^2(1-\rho^2)}, \tag{81}$$

  in which, in any possible 'objective sense', the a priori discrimination between the two standard deviations are, at least, disturbing for the practitioner.

I personally do not consider the marginalization paradox so troublesome. After all, it is a potential consequence of using improper priors (Regazzini, 1983). Here the main issue seems to be: once that the myth of optimality has been abandoned, a complete agreement between frequentist, Bayesian and fiducial approaches cannot be achieved. It is time, I believe, to decide whether this is a problem or not. My personal view is that there exist, nowadays, many examples in which frequentist reasonable behaviour of objective Bayesian procedures is simply impossible and 'objective' Bayesians should not consider that as the main 'objective" of a Bayesian analysis. Many examples of irreconcilability between classical and Bayesian analysis arise when the parameter space is constrained, but even in the regular bivariate normal problem there are functions of the parameters for which frequentist matching behaviour cannot be achieved by Bayesian solutions: perhaps the Fieller's problem is the most well known example (Gleser and Hwang, 1987). Another example is the problem of estimating the coefficient of variation of a scalar Normal distribution (Berger et al., 1999).

In the Fieller's problem one has to estimate the ratio of two Normal means, that is $\theta = \mu_1/\mu_2$; Gleser and Hwang (1987) showed that any confidence procedure of level $1-\alpha < 1$ produces infinite sets with positive sampling probability; on the other hand, any 'reasonable' objective prior like Jeffreys' or reference prior always provides finite HPD sets. Also, the frequentist coverage of one side credible sets derived from such objective priors is always far from the nominal level. In cases like this one, the mathematical structure of the model (in which a sort of *local* unidentifiability occurs) simply prevents from deriving an exact and non trivial confidence procedure. What should an objective Bayesian decide to do here? what should his/her goal be in these cases?

There is another compelling reason to be against option (iv) as a possible *manifesto* of objective Bayesian inference: the choice of the optimality criterion is crucial and it sounds like another way to re-introduce subjectivity in the problem. To explore this issue I have performed a small simulation in the bivariate case, when ρ is the parameter of interest. I have compared the three most 'promising' priors in terms of frequentist quadratic loss. Table 3 shows that the results are sensibly different from those obtained from a coverage matching perspective. Notice that in Table 3 π_H is the right Haar prior, the one giving exact frequentist coverage, while

Table 3: *Mean square error for different possible priors. Here we assume that the estimation of ρ is the goal of inference and the three priors are compared in terms of the mean squared errors of the resulting estimators. For each simulation, the bold character indicates the best. The right Haar prior is the winner except when $\mid \rho \mid$ is close to one, and/or σ_1 and σ_2 are both large.*

$(\rho, \sigma_1, \sigma_2)$	π_H	π_{21}	$\pi_{R\rho}$
$(0, 1, 1)$	**0.154**	0.229	0.188
$(0.5, 10, 1)$	**0.143**	0.176	0.180
$(0.5, 1, 10)$	**0.132**	0.170	0.178
$(0.9, 1, 10)$	0.045	**0.034**	0.056
$(-0.9, 10, 1)$	0.046	0.035	**0.028**
$(-0.5, 10, 1)$	**0.142**	0.166	0.148
$(-0.1, 10, 1)$	**0.155**	0.212	0.176
$(0.1, 10, 10)$	0.160	**0.150**	0.182
$(0.8, 10, 10)$	0.084	**0.065**	0.186

π_{IJ} is the Independence-Jeffreys' prior and $\pi_{R\rho}$ is the reference prior when ρ is the parameter of interest.

Improper priors. Another result provided by the authors, that I have found particularly stimulating, is the problem of sub-optimality of the mixed Haar prior. Since, in the multivariate case, the right Haar prior is not unique and a default choice cannot be made, Berger and Sun propose to consider a mixture of the possible priors. Then, in the bivariate case, this procedure produces, for the covariance matrix, an estimator with a frequentist risk higher than those one would have obtained by using a single right Haar prior. I believe that this surprising result can be explained by the fact that a convex combination of two improper priors arbitrarily depends on the weights we tacitly attach to the two priors. In other words, as the authors notice, it seems an example where one extends the use of elementary probability rules to a scenario (the use of improper priors) where they could easily fail. To this end, Heath and Sudderth (1978) state that

> *... many 'obvious' mathematical operations become just formal devices when used in the presence of improper priors...*

This is exactly what happens when the marginalization paradox shows up. There was an interesting debate in the 1980's about the connections between improper priors, non conglomerability and the marginalization paradox: see for example, Akaike (1980), Heath and Sudderth (1978), Jaynes (1980), Sudderth (1980) and Regazzini (1983). In particular, Regazzini (1987) clearly showed that the marginalization paradox is nothing more than a consequence of the fact that the (improper) prior distribution may be non-conglomerable. While this phenomenon never happens with proper priors, its occurrence is theoretically conceivable when σ-additivity is not taken for granted, as is the case with the use of improper priors. Interestingly, in one of the examples discussed by Regazzini (1983) namely the estimation of the ratio of two exponential means (Stone and Dawid, 1972), the marginalization paradox is produced by an improper prior distribution that is obtained as a solution of a Bayesian/Frequentist agreement problem. More precisely, the question was: let $(x_1, \cdots, x_n)$ be an i.i.d. sample from an Exponential distribution with parameter

$\phi\theta$ and let $(y_1, \cdots, y_n)$ be an i.i.d. sample from an Exponential distribution with parameter ϕ and suppose we are interested on θ: does it exist a prior distribution such that the posterior mean of θ is exactly equal to the 'natural' classical point estimate $\bar{y}/\bar{x}$? In that example, it is also possible to show that the only (improper) prior which satisfies the above desideratum, namely $\pi(\theta, \phi) \propto 1/\theta$, produces the marginalization paradox.

Today we all 'live in the sin' of improper priors, but it is still important to discriminate among them. One can achieve this goal in different ways: for example it is possible to check if a given improper prior is coherent (that is, if it has a finitely additive interpretation). However, especially in the applied world, this road might be, admittedly, hard to follow.

Alternatively, one could try to figure out where do these distributions put the prior mass. Sudderth (1980) interestingly stresses the fact that, for example, the uniform improper prior on the real line and the finitely additive limit of uniform priors on a sequence of increasing compact sets show a sensibly different behavior: indeed,

> *... the odds on compact sets versus the whole space are* 0 : 1 *for the finitely additive prior and finite: infinite for the improper prior.*

My personal view is that a reasonable way to discriminate among priors might be to measure, in terms of some Kullback–Leibler type index, the discrepancy of the induced posteriors with respect to some *benchmark* posterior. The reference prior algorithm has its own benchmark in a sort of asymptotic maximization of missing information and it works remarkably well in practice; I believe that other possible benchmarks might be envisaged, perhaps in terms of prediction.

BERTRAND CLARKE (*University of Brithsh Columbia, Canada*)

Sun and Berger (2006) examine a series of priors in various objective senses, but the focus is always on the priors and how well they perform inferentially. While these questions are reasonable, in fact it is the process of obtaining the prior, not the prior so obtained, that makes for objectivity and compels inquiry.

Indeed, the term objective prior is a misnomer. The hope is merely that they do not assume a lot of information that might bias the inferences in misleading directions. However, to be more precise, consider the following which tries to get at the process of obtaining the prior.

Definition. A prior is objective if and only if the information it represents is of specified, known provenance.

This means the prior is objective if its information content can be transmitted to a recipient who would then derive the same prior. The term information is somewhat vague; it is meant to be the propositions that characterize the origin of the prior, in effect systematizing its obtention and specifying its meaning. As a generality, the proliferation of objective priors in the sense of this definition seems to stem from the various ways to express the concept of absence of information. Because absence of information can be formulated in so many ways, choosing one formulation is so informative as to uniquely specify a prior fairly often. Thus it is not the priors themselves that should be compared so much as the assumptions in their formulation.

The information in the prior may stem from a hunch, an expression of conservatism, or of optimism. Or, better, from modeling the physical context of the problem. What is distinctive about this definition of objective priors is that their information content is unambiguously identified. It is therefore ideationally complex enough to admit agreement or disagreement on the basis of rational argument, modeling, or extra-experimental empirical verification while remaining a separate source of information from the likelihood or data.

By this definition, one could specify Jeffreys prior in expression (2) in several equivalent ways. It can be regarded as (1) the asymptotically least favorable prior in an entropy sense, (2) a transformation invariant prior provided an extra condition is invoked so it is uniquely specified, (3) the frequentist matching prior (for $p = 2$). Likewise, the information content of other priors in Sun and Berger (2006) can be specified since they are of known provenance and hence objective.

Jeffreys prior is justly regarded as noninformative in the sense that it changes the most on average upon receipt of the data under a relative entropy criterion. This concept of noninformativitity is appropriate if one is willing to model the data being collected as having been transmitted from a source, and to model the parameter value as a message to be decoded. Jeffreys prior is the logical consequence of this modeling strategy. So, if an experimenter is unhappy with the performance of the Jeffreys prior, the experimenter must model the experiment differently. It is not logically consistent to adopt the data transmission model that leads to Jeffreys prior but then reject the Jeffreys prior.

However, if some information theoretic model is thought appropriate, a natural alternative to the Jeffreys' prior could be Rissanen's (1983) prior which has the plus of being proper on the real line. Rissanen's prior also has an information theoretic interpretation but it's in a signal-to-noise ratio sense rather than a transmission sense. Indeed, there are many ways to motivate physically the choice of prior; several similar information theoretic criteria are used in Clarke and Yuan (2004) resulting in ratios of variances. By contrast, if it is the tail behavior of the prior that matters more than the information interpretation, then the relative entropy might not be the right distance, obviating information theory. Other distances such as the chi-square lead to other powers of the Fisher information, see Clarke and Sun (1997).

Essentially, the argument here is to turn prior selection into an aspect of modeling. Thus, the issue is not whether a given prior gives the best performance, for instance in the decision theoretic sense of Table 2, but whether the information implicit in the prior is appropriate for the problem. In particular, although one might be mathematically surprised that the risk of $\hat{\Sigma}_S$ is higher than the risk of right Haar priors, this is not the point. The point is whether the assumptions undergirding one or another of these priors is justified. For instance, if the model for the experiment includes the belief that minimal risk will be achieved, one would not be led to the mixture of two priors. On the other hand, if the risk is not part of the physical model then one is not precluded from using the mixture of priors if it is justified.

In the same spirit, the issues of the marginalization paradox and frequentist matching can be seen as generally irrelevent. The marginalization paradox does not exist for proper priors and it is relatively straightforward to derive priors that satisfy a noninformativity principle and are proper. Rissanen's prior is only one example. Indeed, the information for prior construction may include 'use the first k data points to upgrade an improper prior chosen in such-and-such a way to propriety and then proceed with $n - k$ data points for inference'. Similarly, matching priors merely represent the desire to replicate frequentist analysis. If the two match closely enough, the models may be undistinguishable, otherwise they are different and can't both be right. Moreover, the fact that matching frequentist analysis and avoiding the marginalization paradox often conflict is just a fact: The models from which the priors derive cannot satisfy both criteria, perhaps for purely mathematical reasons, and little more can be made of it without examining classes of models.

A reasonable implication from this definition is that subjective priors have no place in direct inference because their provenance cannot be evaluated. The main role remaining to subjective priors may be some aspects of robustness analysis. It remains perfectly reasonable to say 'I have consulted my predilections and impressions and think they are well represented if I draw this shape of density, announce that analytic form and then see what the inferential consequences are'. However, since the information in such a prior is not of known provenance, it cannot be subject to inquiry or validation and so without further justification does not provide a firm basis for deriving inferences. Of course, being able to argue that n is large enough that the prior information is irrelevant would be adequate, and in some cases, an extensive robustness analysis around a subjective prior (especially if the robustness included 'objective' priors, like powers of the Fisher information) might lend the subjective approach credence.

GUIDO CONSONNI and PIERO VERONESE
(*University of Pavia, Italy and Bocconi University, Italy*)

The paper by Sun and Berger presents an impressive collection of results on a wide range of objective priors for the multivariate normal model. Some of them are closely related to previous results of ours, which were derived using a unified approach based on the theory for conditionally reducible exponential families. In the sequel we briefly present and discuss some of the main aspects.

In Consonni and Veronese (2001) we discuss Natural Exponential Families (NEFs) having a particular recursive structure, that we called conditional reducibility, and the allied parameterization ϕ of the sampling family. Each component of ϕ is the canonical parameter of the corresponding conditional NEF. One useful feature of ϕ is that it allows a direct construction of "enriched" conjugate families. Furthermore the parameterization ϕ is especially suitable for the construction of reference priors, see Consonni, Gutiérrez-Peña, and Veronese (2004) within the framework of NEFs having a simple quadratic variance function. Interestingly, we show that reference priors for different groupings of ϕ belong to the corresponding enriched conjugate family.

In Consonni and Veronese (2003), henceforth CV03, the notion of conditional reducibility is applied to NEFs having a homogeneous quadratic variance function, which correspond to the Wishart family on symmetric cones, i.e the set of symmetric and positive definite matrices in the real case. In this setting we construct the enriched conjugate family, which is shown to coincide with the Generalized Inverse Wishart (GIW) prior on the expectation Σ of the Wishart, and provide structural distributional properties including expressions for the expectation of Σ and Σ^{-1}. We also obtain a grouped-reference prior for the ϕ-parameterization, as well as for $\Sigma = (\sigma_{ij},\, i,j = 1,\ldots,p)$ according to the grouping $\{\sigma_{11}, (\sigma_{21}, \sigma_{22}), \ldots, (\sigma_{p1}, \ldots, \sigma_{pp})\}$, showing that the reference prior belongs to the enriched conjugate family in this case, too. This in turn allows to prove directly that the reference posterior is always proper and to compute exact expressions for the posterior expectation of Σ and Σ^{-1}.

There is a close connection between our ϕ-parameterization and the Cholesky decomposition of $\Sigma^{-1} = \Psi'\Psi$ in terms of the triangular matrix Ψ of formula (8): in particular ϕ and Ψ are related through a block lower-triangular transformation, (see CV03, where further connections between ϕ and other parameterizations are explored). As a consequence the reference prior on Ψ can be obtained from that of ϕ through a change-of-variable.

Our results on the Wishart family are directly applicable to a random sample of size n from a multivariate normal $N_p(0, \Sigma)$, since the Wishart family corresponds to the distribution of the sample cross-products divided by n (a sufficient statistic).

In the paper by Sun and Berger the multivariate normal model $N_p(\mu, \Sigma)$ is considered, and several objective priors for (μ, Ψ), and other parameterizations, are discussed. Special emphasis is devoted to the class of priors π_a, see (13), that includes a collection of widely used distributions such as the reference prior π_{R1}, see (12), corresponding to the grouping $\{\mu, \psi_{11}, (\psi_{21}, \psi_{22}), \ldots, (\psi_{p1}, \ldots, \psi_{pp})\}$. We remark that the right-hand-side of (12), $\Pi_{i=1}^{p} \psi_{ii}^{-1}$, coincides with the reference on Ψ derived from our reference prior on ϕ. To see why this occurs, notice that the Fisher information matrix for $\{\mu, \psi_{11}, (\psi_{21}, \psi_{22}), \ldots, (\psi_{p1}, \ldots, \psi_{pp})\}$ is block-diagonal, with the first block constant w.r.t. μ, while the remaining blocks do not involve μ and are equal to those that hold under the $N_p(0, \Sigma)$ case. Incidentally, while our reference on Ψ is actually of the GIW type, this is clearly not the case for (12), which is a prior on (μ, Ψ), so that the support is not the space of real and symmetric positive definite matrices.

As the authors point out at the beginning of Section 3, since the conditional posterior for μ given Ψ is normal, interest centers on the marginal posterior of Ψ, which depends on the data only through the sample variance $S/(n-1)$, whose distribution is Wishart, so that our results are still relevant. Specifically the marginal on Ψ under $\pi_a(\mu, \Psi)$, as well as the posterior, belongs to the enriched conjugate/GIW family, whence items (a) and (b) of Fact 4, and Fact 7, are readily available from Corollary 1, Proposition 1 and Proposition 2 of CV03.

Our last point concerns the marginalization paradox in the multivariate setting. Partition Σ into four blocks Σ_{ij}, $i, j = 1, 2$; then the reference prior for ϕ, which corresponds to the prior π_{R1}, does not incur the marginalization paradox for the marginal variance Σ_{11}, the conditional variance $\Sigma_{2|1} = \Sigma_{22} - \Sigma_{21}\Sigma_{11}^{-1}\Sigma_{12}$, and the pair $\{\Sigma_{11}, \Sigma_{2|1}\}$, see CV03.

A. PHILIP DAWID (*University College London, UK*)

By relating the problem to one of fiducial inference, I here present a general argument as to when we can expect a marginal posterior distribution based on a formal right-Haar prior to be frequency calibrated. The following problem formulation and analysis closely follow Dawid, Stone and Zidek (1973) and Dawid and Stone (1982), which should be consulted for full details.

Structural model. Consider a simple 'structural model' (Fraser, 1968):

$$X = \Theta E \tag{82}$$

where the *observable* X, the *parameter* Θ, and the *error variable* E all take values in a group G, and the right-hand side of (82) involves group multiplication. Moreover,

E has known distribution P, independently of the value θ of Θ. Letting P_θ denote the implied distribution of $X = \theta E$, we obtain an induced *statistical model* $\mathcal{P} = \{P_\theta : \theta \in G\}$ for X. (Note however that distinct structural models can induce the same statistical model.)

Now assign to Θ the formal right-Haar prior over G. It then follows (Hora and Buehler, 1966) that the resulting formal posterior Π_x for Θ, based on model $\mathcal{P}$ and data $X = x$, will be identical with the *structural (fiducial)* distribution for Θ based on (82), which is contructively represented by the *fiducial model*:

$$\Theta = xE^{-1} \tag{83}$$

with the distribution of E still taken to be P.

Non-invariant estimation. It easily follows that a suitably invariant level-γ posterior credible set will also be a level-γ confidence interval. Less obviously, such a confidence interpretation of a posterior interval also holds for inference about suitable one-dimensional functions of Θ, even though no invariance properties are retained.

Thus let H be a subgroup of G, and $w : G \to \mathbb{R}$ a maximal invariant function under left-multiplication by H. Suppose further that w is *monotone*:

$$w(\theta_1) \leq w(\theta_2) \Leftrightarrow w(\theta_1 e) \leq w(\theta_2 e) \tag{84}$$

for all $\theta_1, \theta_2, e \in G$. Define $Z := w(X)$ and $\Lambda = w(\Theta)$. It now follows (Dawid and Stone, 1982, §4.2) that, under mild continuity conditions:

- Z is a function of Λ and E: say $Z = f(\Lambda, E)$
- the marginal sampling distribution of Z depends only on Λ;
- the marginal posterior distribution of Λ (which is the same as its marginal fiducial distribution) depends only on Z: say Π_z for $Z = z$;
- Π_z is represented constructively by $\Lambda = g(z, E)$, with E having distribution P; here $g(z, e)$ represents the solution, for λ, of $z = f(\lambda, e)$;
- Π_z agrees with Fisher's fiducial distribution, obtained by differentiating with respect to λ the cumulative distribution function P_λ for Z given $\Lambda = \lambda$;
- if, for each z, λ_z is a posterior level-γ upper credible limit for $\Lambda = w(\Theta)$, *i.e.*, $\Pi_z(\Lambda \leq \lambda_z) = \gamma$, then λ_Z will also be an exact level-γ upper confidence limit for Λ, *i.e.*, $P_\theta(w(\theta) \leq \lambda_Z) = \gamma$, all θ.

The application to the case of the correlation coefficient Λ in a bivariate normal model (albeit with zero means) was treated explicitly in Section 3 of Dawid and Stone (1982), with G the triangular group and H its diagonal subgroup. The results of Berger and Sun (2006) follow directly.

As noted in Section 2.4 of Dawid, Stone and Zidek (1973), use of the right-Haar prior on G will typically entail a marginalization paradox for Λ. For discussion of such logical issues, see Dawid, Stone and Zidek (2007).

Ancillaries. The above theory generalizes readily to problems where E and X live in a general sample space $\mathcal{X}$, and $\Theta \in G$, an exact transformation group on $\mathcal{X}$:

the relevant theory is in Dawid, Stone and Zidek (1973). Let $a(\cdot)$ be a maximal invariant function on $\mathcal{X}$ under the action of G. Then defining $A = a(X)$, it is also the case that $A = a(E)$, and A is thus ancillary. In particular, on observing $X = x$ we learn that $E \in \mathcal{E}_x := \{e : a(e) = a(x)\}$. In the fiducial model (83) we must now restrict E to $\mathcal{E}_x$, defining xe^{-1} as the then unique solution for θ of $x = \theta e$, and assigning to E the distribution over $\mathcal{E}_x$ obtained by conditioning its initial distribution P on $a(E) = a(x)$. We must also evaluate sampling performance conditional on the ancillary statistic $a(X)$. Then all the above results go through (indeed, we obtain the coverage property conditional on A, which is stronger than unconditional coverage).

JAYANTA GHOSH (*Purdue Univ., USA, and Indian Statistical Inst., India*)

I have a couple of general as well as specific comments on objective priors inspired by the paper of Sun and Berger and its discussion by Liseo, both of which are very interesting.

I focus on two major problems. Objective priors for point or interval estimates, generated by standard algorithms, are almost always improper and not unique. They are improper because they are like a uniform distribution on a non-compact parameter space. They are not unique because different algorithms lead to different priors. Typically they may not have all the properties one usually wants, namely, coherence, absence of marginalization paradox, some form of probability matching, and some intuitive motivation. Sun and Berger (2006) show no prior for the bivariate normal can avoid marginalization paradox and also be probability matching for the correlation coefficient. Brunero points out that probability matching or any other Frequentist matching need not be desirable and the marginalization paradox is a consequence of the prior being improper. Incidentally, for one or two dimensions, probability matching priors are like reference priors. For higher dimensions probability matching priors are not well-understood.

Brunero asks why one should try to do Frequentist matching at all to justify an objective prior. He suggests a better way of comparing two objective priors might be to examine where they put most of their mass.

It is worth pointing out that coherence could help us decide which one to choose. A stronger notion, namely, admissibility, has been used by Jim in a number of cases. Probability matching seems to provide a very weak and asymptotic form of coherence. A detailed study of coherence of improper priors is contained in Kadane *et al.* (1999).

Another way of comparing two objective priors, which addresses impropriety directly, could explore how well the posteriors for the improper priors are approximated by the posteriors for the approximating truncated priors on a sequence of compact sets. This is usually not checked but is related to one of the basic requirements for construction of a reference prior, vide Berger and Bernardo (1992). As far as non-uniqueness is concerned, a mitigating factor is that a moderate amount of data lead to very similar posteriors for different objective priors, even though the data are not large enough to wash away most priors and make the posteriors approximately normal. I wonder if this holds for some or all the objective priors for the bivariate normal.

VICTOR RICHMOND R. JOSE, KENNETH C. LICHTENDAHL, JR., ROBERT F. NAU and ROBERT L. WINKLER (*Duke University, USA*)

This paper is a continuation of the work by Berger and Sun to increase our understanding of diffuse priors. Bayesian analyses with diffuse priors can be very useful, although we find the term 'objective' inappropriate and misleading, especially when divergent rules for generating 'objective' priors can yield different results for the same situation. How can a prior be 'objective' when we are given a list of different 'objective' priors, found using different criteria, from which to choose? The term diffuse, which has traditionally been used, is fine, as is weakly informative (O'Hagan, 2006). In the spirit of Savage's (1962) 'precise measurement', diffuse priors can be chosen for convenience as long as they satisfy the goal of providing good approximations to the results that would be obtained with more carefully assessed priors. Issues surrounding so-called 'objective Bayesian analysis' have been discussed at length in papers by Berger (2006) and Goldstein (2006) and the accompanying entertaining commentary. There is neither need nor space to go over all of that ground here, but our views are in line with the comments of Fienberg (2006), Kadane (2006), Lad (2006), and O'Hagan (2006).

Over the years, much time and effort has been spent pointing out basic differences between frequentist and Bayesian methods and indicating why Bayesian methods are fundamentally more sound (they condition appropriately, they address the right questions, they provide fully probabilistic statements about both parameters and observables, etc.). Jim Berger has participated actively in this endeavor. Given this effort, why would we want to unify the Bayesian and frequentist approaches? Why should we be interested in 'the prior that most frequently yields exact frequentist inference', which just leads us to the same place as using frequentist methods, many of which have been found lacking from a Bayesian viewpoint? Why should we care about the frequentist performance of Bayesian methods? Is it not preferable to focus more on how well Bayesian analyses (including those using diffuse priors) perform in a *Bayesian sense*? That means, for example, using scoring rules to evaluate predictive probabilities.

One aim is apparently to market Bayesian methods to non-Bayesians. As Lad (2006) notes, 'The marketing department has taken over from the production department'. Our sense is that, since Bayesian methods are inherently more sound, sensible, and satisfying than frequentist methods, the use of Bayesian methods will continue to increase. Rather than trying to stimulate this by marketing Bayesian procedures as 'objective' (which neither they nor frequentist procedures can be) and meeting frequentist criteria (which is not what they are intended to do), let's invest more effort toward continuing to apply Bayesian methods to important problems and toward making Bayesian methods more accessible. The development of more user-friendly Bayesian software for modeling both prior and likelihood and for handling Bayesian computations would be a big step in the right direction.

Sun and Berger's concern about improper priors is well-founded, since improper diffuse priors are commonly encountered. If we think of the Bayesian framework in terms of a big joint distribution of parameters and observables, improper priors leave us without a proper joint distribution and leave us unable to take advantage of the full Bayesian menu of options. First, it is often argued that if the posteriors following improper diffuse priors are themselves proper, all is well (e.g., Berger, 2006, p. 393). However, posteriors following improper diffuse priors are not always proper. For example, in a normal model with high dimensional data and small sample sizes (large p, small n), the posteriors that follow from many of the priors considered by Sun and Berger are not proper. Second, even though improper priors may (eventually) yield proper posteriors, they do not, in general, provide proper predictive distributions. Although decision-making problems are typically expressed in terms of parameters rather than observables in Bayesian statistics, which means that posterior distributions are of interest, the primary focus in decision analysis is often on observables and predictive distributions. For important preposterior decisions such as the design of optimal sampling plans and for value of information calculations, proper predictive distributions are needed. As a result, many so-called 'objective' priors, including priors proposed by Sun and Berger, leave us unable to make formal preposterior decisions and force us to resort to ad hoc procedures instead.

As an illustration, consider optimal sampling in clinical drug trials. Should we sample at all, and how big should the initial sample be? With improper 'objective' priors, we are unable to provide a formal analysis of such important decisions, which have serious life-and-death implications. As Bayesians, and more generally as scientists, we should actively promote the use of tools that are more conducive to good decisions.

FEDERICO J. O'REILLY (*IIMAS, Mexico*)

The authors ought to be congratulated for providing yet another piece of research where rather than looking for differences in statistical procedures, they try to identify closeness between what do theories provide. This paper is very much in line with the objective Bayesian point of view which as mentioned in Berger (2006), was present in science long before the subjective Bayesian approach made its formal arrival.

One wonders why there has been such an extreme position in stressing differences between Bayesian and classical results. Exact coincidences between reference posterior densities and fiducial densities exist, not too often as shown in Lindley (1958), but in many cases, practical differences are small. And in trying to understand these differences, a compromise between procedures, in this case, with exact coverage probabilities or procedures for which the marginalization paradox does not arise, must be faced. The authors mention this fact very clearly. For some, there is no compromise to be done; they have made their point and should be respected, but for those exploring this compromise, we believe they too, should be respected.

We would like to refer just to one aspect of the paper on the correlation coefficient ρ in the bivariate normal distribution which in our opinion stands out. On the one hand, right Haar priors are elicited following invariance considerations but unfortunately (inherited from the non-uniqueness of the factorization) there are two right Haar measures which appear in a nonsymmetrical fashion despite the symmetrical role that the standard deviations σ_1 and σ_2 'should' have. The authors explore mixing these two Haar measures in an effort, it seems, to get rid of the uncomfortable asymmetry and they do recognize that the effort does not produce a reasonable answer.

On the other hand, the symmetric reference prior $\pi(\rho)$ given in Bayarri (1982), is mentioned as not yielding exact coverage probabilities for the corresponding posterior, but certainly free from the marginalization paradox. Two questions arise at this point. The first one is how different is the coverage probability from the exact one? The second question has to do with the interesting relationship between the two Haar measures and the reference prior, which in this case is the geometric mean.

Do we have to mix with convex linear combinations $(\alpha, 1-\alpha)$? In the discussion, it was mentioned that with improper priors, the weights placed on a convex linear combination have little interpretation, except if one seeks symmetry ($\alpha = 0.5$), as it seems to be the case. A priori weights α and $1 - \alpha$ for the two priors mean a different set of weights for the posteriors when representing the posterior obtained from the mixed prior as a mixture of the posteriors. Why not simply work with the prior obtained by 'mixing' both prior measures using their geometric mean? The result, in general, would provide a proper posterior if the two associated posteriors are proper (Schwarz inequality).

It would have been interesting to see some graphs of the various posteriors discussed for ρ in a few cases with small and medium values for the sample size n when the sampling correlation coefficient r is varied along its range. In some examples we have been doing in non-location scale families, even with very small sample sizes, the graphs of the fiducial and the reference posterior densities are almost indistinguishable. That is the case among others, of the truncated exponential in O'Reilly and Rueda (2006).

REPLY TO THE DISCUSSION

We are grateful to all discussants for their considerable insights; if we do not mention particular points in the discussions, it is because we agree with those points.

Dr. Liseo. We agree with essentially all of Dr. Liseo's illuminating comments, and will mention only two of them.

The survey and comments concerning the definition of the Objective Bayesian approach were quite fun, and we feel that there is some truth in all the definitions (except one), reinforcing the difficulty of making the definition precise.

Dr. Liseo reminds us that there are many situations where objective Bayesians simply cannot reach agreement with frequentists, thus suggesting that frequentist performance should be a secondary criterion for objective Bayesians. He also comments, however, that he does not consider the marginalization paradox to be particularly troublesome, and shows that there is no clear winner when looking at estimation of ρ (which, interestingly, is also the case when entropy loss is used). Hence we are left with the uneasy situation of not having a clear recommendation between the right-Haar and reference prior for ρ.

Dr. Clarke gives a strong argument for the appealing viewpoint that choice of objective priors is simply a choice of the communicable information that is used in its construction, giving examples of priors arising from information transmission arguments. One attractive aspect of this is that it also would apply to a variety of priors that are based on well-accepted understanding, for instance various hierarchical priors or even scientific priors: if all scientists agreed (based on common information they hold) that the uncertainty about μ was reflected by a $N(0, 1)$ distribution, shouldn't that be called an objective prior?

There are two difficulties with implementation of this viewpoint, however. The first is simply a matter of communication; while calling the above $N(0, 1)$ prior objective might well be logical, the name 'objective prior' has come to be understood as something quite different, and there is no use fighting history when it comes to names.

A more troubling difficulty is that this formal view of objective priors does not seem to be practically implementable. For instance, Dr. Clarke suggests that the

Jeffreys prior is always suitable, assuming we choose to view data transmission under relative entropy to be the goal. We know, of course, that the Jeffreys prior can give nonsensical answers for multivariate parameters (e.g., inconsistency in the Neyman-Scott problem). Dr. Clarke's argument is that, if we feel the resulting answer to be nonsensical, then we cannot really have had the data transmission problem as the goal. While this is perhaps a logically sound position, it leaves us in the lurch when it comes to practice; when can we use the Jeffreys prior and when should we not? Until an approach to determination of objective priors provides answers to such questions, we strongly believe in the importance of actually evaluating the performance of the prior that results from applying the approach.

Drs. Consonni and Veronese give a nice discussion of the relationship of some of the priors considered in our paper with their very interesting results on reference priors for natural exponential families with a particular recursive structure, namely conditional reducibility; for these priors, some of the facts listed in our paper are indeed immediate from results of Consonni and Veronese. (Of course, the right Haar prior and those in (5) and (7) are not in this class.) It is also indeed interesting that this class of priors does not lead to a marginalization paradox.

Dr. Dawid's result is quite fascinating, because it shows a wide class of problems in which objective priors can be exact frequentist matching (and fiducial), even though the parameter of interest is not itself fully invariant in the problem. Also, the posterior distribution can be given constructively in these situations. Such results have long been known for suitably invariant parameters, but this is a significant step forward for situations in which full invariance is lacking.

In Berger and Sun (2006), several of the results concerning exact frequentist matching and constructive posteriors could have been obtained using this result of Dr. Dawid. However, some of the results in that paper (e.g. those concerning the parameters $\eta_3 = -\rho/[\sigma_1\sqrt{1-\rho^2}]$ and $\xi_1 = \mu_1/\sigma_1$) required a more difficult analysis; these may suggest further generalizations of Dr. Dawid's result.

Dr. Ghosh: We certainly agree with the insightful comments of Dr. Ghosh. We have not studied the sample size needed for different objective priors to give essentially the same posterior; presumably this would be a modest sample size for the bivariate case, but the multivariate case is less clear.

Drs. Jose, Lichtendahl, Nau, and Winkler give a nice discussion of some of the concerns involving objective priors. We agree with much of what they say. Yet we also are strong believers that objective priors can be of great use in statistics. The discussion paper in Bayesian Analysis covers this 'other side' of the debate.

The property that improper priors need a sufficiently large sample size to yield a proper posterior can be viewed as a strength of objective priors: one realizes whether or not there is enough information in the data to make the unknown parameters identifiable. A proper prior masks the issue by yielding a proper posterior, which can be dangerous in the non-identifiable case: for some of the parameters (or functions thereof) there is then no information provided by the data and the posterior is just the prior – dangerous unless extreme care was taken in developing the prior.

Dr. O'Reilly makes the very interesting suggestion that one should symmetrize the right-Haar priors by geometric mixing, rather than arithmetic mixing, and observes that this indeed yields the reference prior for ρ. This is an appealing suggestion and strengthens the argument for using the reference prior.

Dr. O'Reilly asks if the reference prior for ρ yields credible sets with frequentist coverage that differ much from nominal. We did conduct limited simulations on this and the answer appears to be no; for instance, even with a minimal sample size of five, the coverage of a nominal 95% set appears to be no worse than 93%.

ADDITIONAL REFERENCES IN THE DISCUSSION

Akaike, H. (1980). The interpretation of improper prior distributions as limits of data dependent proper prior distributions. *J. Roy. Statist. Soc. B* **42**, 46–52.

Berger, J. O. (2006). The case for objective Bayesian analysis. *Bayesian Analysis* **1**, 385-402.

Berger, J. O., Liseo, B. and Wolpert, R. L. (1999) Integrated likelihood methods for eliminating nuisance parameters. *Statist. Science* **14**, 1–28.

Clarke, B. and Sun, D. (1997). Reference priors under the Chi-squared distance. *Sankhyā A* **59**, 215–231.

Clarke, B. and Yuan, A., (2004). Partial information reference priors. *J. Statist. Planning and Inference* **123**, 313–345.

Consonni, G. and Veronese, P. (2001). Conditionally reducible natural exponential families and enriched conjugate priors. *Scandinavian J. Statist.* **28**, 377–406.

Consonni, G. and Veronese, P. (2003). Enriched conjugate and reference priors for the Wishart family on symmetric cones. *Ann. Statist.* **31**, 1491–1516.

Dawid, A. P. and Stone, M. (1982). The functional-model basis of fiducial inference. *Ann. Statist.* **10**, 1054–74 (with discussion).

Dawid, A. P., Stone, M. and Zidek, J. V. (2007). The marginalization paradox revisited. In preparation.

Fienberg, S. E. (2006). Does it make sense to be an 'Objective Bayesian'? *Bayesian Analysis* **1**, 429–432.

Fraser, D. A. S. (1968). *The Structure of Inference.* New York: Wiley.

Gleser, L. J. and Hwang, J. T. (1987). The non-existence of $100(1-\alpha)\%$ confidence sets of finite expected diameter in errors-in-variables and related models. *Ann. Statist.* **15**, 1351–1362.

Goldstein, M. (2006). Subjective Bayesian analysis: Principles and practice. *Bayesian Analysis* **1**, 403–420.

Heath, D. and Sudderth, W. (1978). On finitely additive priors, coherence and extended admissibility. *Ann. Statist.* **6**, 333–345.

Hora, R. B. and Buehler, R. J. (1966). Fiducial theory and invariant estimation. *Ann. Math. Statist.* **37**, 643–656.

Kadane, J. B. (2006). Is 'Objective Bayesian Analysis' objective, Bayesian, or wise? *Bayesian Analysis* **1**, 433–436.

Kadane, J., B., Schervish, M. J. and Seidenfeld, T. (1999). *Cambridge Studies in Probability, Induction and Decision Theory.* Cambridge: University Press.

Lad, F. (2006). Objective Bayesian statistics ... Do you buy it? Should we sell it? *Bayesian Analysis* **1**, 441–444.

Lindley, D.V. (1958). Fiducial distributions and Bayes theorem. *J. Roy. Statist. Soc. B* **20**, 102–107.

O'Hagan, A. (2006). Science, subjectivity and software. Comments on the papers by Berger and by Goldstein. *Bayesian Analysis* **1**, 445–450.

O'Reilly, F. and Rueda, R. (2006), Inferences in the truncated exponential distribution. *Tech. Rep.*, IMAS, UNAM, Mexico.

Regazzini, E. (1983). Non conglomerabilità e paradosso di marginalizzazione. *Sulle probabilità coerenti nel senso di de Finetti.* Bologna, Italy: Clueb.

Regazzini, E. (1987). de Finetti coherence and statistical inference. *Ann. Statist.* **15**, 845–864.

Rissanen, J. (1983). A universal prior for integers and estimation by minimum description length. *Ann. Statist.* **15**, 416–431.

Savage, L. J., *et al.* (1962). *The Foundations of Statistical Inference: A Discussion.* London: Methuen.

Stone, M. and Dawid, A. P. (1972). Un-Bayesian implications of improper Bayes inference in routine statistical problems. *Biometrika* **59**, 369–375.

Sudderth, W. (1980). Finitely additive priors, coherence and the marginalization paradox. *J. Roy. Statist. Soc. B* **42**, 339–341.

CONTRIBUTED PAPERS
(synopsis)

BAYESIAN STATISTICS 8, pp. 565–570.
J. M. Bernardo, M. J. Bayarri, J. O. Berger, A. P. Dawid,
D. Heckerman, A. F. M. Smith and M. West (Eds.)

Bayesian Encompassing Specification Test Under Not Completely Known Partial Observability*

CARLOS ALMEIDA and MICHEL MOUCHART
Institut de Statistique, Université Catholique de Louvain, Belgium
almeida@stat.ucl.ac.be mouchart@stat.ucl.ac.be

SUMMARY

A Bayesian specification test based on the encompassing principle for the case of partial observability is proposed. A structural parametric *null model* is compared against a nonparametric *alternative model* at the level of latent variables. A same observability process is introduced in both models. The comparison is made between the posterior measures of the non-Euclidean parameter (of the alternative model) in the extended and in the alternative models. The general development is illustrated with an example where a linear combination of a latent vector is only observed.

Keywords and Phrases: BAYESIAN ENCOMPASSING; BAYESIAN SPECIFICATION TEST; DIRICHLET PRIOR; PARTIAL OBSERVABILITY.

1. INTRODUCTION

When covariance structure models involve ordinal variables one should justify the normality of the latent variables supposedly generating, by discretization, the ordinal variables; *e.g.*, Muthén (1984). As another example of partial observability, consider the observation of a linear combination of latent variables (with unknown coefficients). Formally, models involving partial observability are described by a *structural model*: $\xi \mid \theta \sim P_{\xi|\theta}$, along with an *observability process* defined by $X = g(\xi, \alpha)$ where g is a known function, $\alpha \in \mathbb{R}^q$ is an unknown parameter, ξ is a vector of latent variables and X is a vector of manifest (or observable) variables. The case where g is not a function of α has been treated in Almeida and Mouchart (2005) under the heading of *completely known partial observability*; the case where $\alpha \in \mathbb{R}^q$ is unknown, but the functional form of g is known, is called *not completely*

---

* This is a synopsis. For the full version, see *Bayesian Analysis* **2** (2007), to appear.
Financial support from the IAP research network Nr P5/24 of the Belgian State (Federal Office for Scientific, Technical and Cultural Affairs) is gratefully acknowledged.

known partial observabilitity and is the subject matter of this paper. The Bayesian specification of the structural models (null and alternative) are:

$$\mathcal{E}^0 : (\theta,\xi) \sim Q^0_{\theta,\xi} = M^0_\theta \otimes P^0_{\xi|\theta} \qquad \mathcal{E}^1 : (\psi,\xi) \sim Q^1_{\psi,\xi} = M^1_\psi \otimes P^1_{\xi|\psi} \tag{1}$$

where θ and ψ are the parameters characterizing the respective sampling models for the latent vector ξ and the Markovian product $M^0_\theta \otimes P^0_{\xi|\theta}$, between a marginal (prior) probability M^0_θ and a (sampling) regular conditional probability $P^0_{\xi|\theta}$ defines the joint distribution of (θ,ξ). A general non-parametric alternative is specified by $P^1_{\xi|\psi} = \psi$. In both models, $X = g(\xi,\alpha) \doteq g_\alpha(\xi)$ defines the *partial observability process*, where g is a known bi-measurable function; this implies that:

$$(a)\ X \perp\!\!\!\perp \theta \mid \alpha,\xi; Q^0 \qquad \text{and} \qquad (b)\ X \perp\!\!\!\perp \psi \mid \alpha,\xi; Q^1, \tag{2}$$

and clearly: $P^0_{X|\alpha,\theta,\xi} = P^1_{X|\alpha,\psi,\xi} = \delta_{\{X=g(\xi,\alpha)\}}$, where $\delta_{\{\bullet\}}$ is the *unit mass measure*.

In line with seminal papers by Cox (1961, 1962) about testing non-nested hypotheses, the encompassing principle has been developed for comparing two experiments sharing a same sample space. The main idea is to compare the inference on parameter of the second model using the second one and the inference on the same parameter using the first model extended in order to also include the parameter of the second one. For a detailed study of encompassing testing in a sampling theory approach see Mizon and Richard (1986); in a Bayesian framework see Florens *et al.* (1990, Section 3.5) and Florens and Mouchart (1989, 1993).

In order to include the parameter of the second model in the *extended model*, a *Bayesian Pseudo-True Value* (BPTV) is defined through a probability transition. The use of a conditional independence condition (*BPTV condition:* $\xi \perp\!\!\!\perp \psi \mid \theta; Q^{0,*}$) permits us to interpret the first model as the marginalization by sufficiency of the extended model; the extended model is accordingly written as: $Q^{0,*} = M^0_\theta \otimes P^0_{\xi|\theta} \otimes M_{\psi|\theta}$. A test statistic is constructed through a distance or a divergence between the two posterior distributions $M^{0,*}_{\psi|\xi}$ and $M^1_{\psi|\xi}$; the statistics will be calibrated against the predictive measure in the null model. Florens and Mouchart (1993) suggests a possibility for the specification of the BPTV:

$$M_{\psi|\theta} = \int M^1_{\psi|\xi} dP^0_{\xi|\theta} \qquad (= E^0[M^1_{\psi|\xi} \mid \theta]\ ), \tag{3}$$

this may be viewed as a Bayesian adaptation of the Cox (1961) suggestion.

Recently, Florens *et al.* (2003) considered a non parametric alternative model with a Dirichlet process as prior distribution. Using the BPTV specified as in (3) they show that the posterior measure $M^{0,*}_{\psi|\xi}$ is a mixture of Dirichlet processes and they use direct simulation of Dirichlet process, as developed in Rolin (1992) or in Sethuraman (1994), for computing the test statistic and for its calibration. Later, in Almeida and Mouchart (2005) the encompassing specification test has been extended to the case of partial observability completely known.

2. GENERAL RESULTS

We assume that the sampling distributions of ξ do not depend on α:

$$(a)\ \alpha \perp\!\!\!\perp \xi \mid \theta; Q^0 \qquad \text{and} \qquad (b)\ \alpha \perp\!\!\!\perp \xi \mid \psi; Q^1. \tag{4}$$

In the structural level, θ and ψ are *sufficient* parameters. The *complete models* are:

$$\begin{aligned}\mathcal{E}^0: &\quad (\theta,\alpha,\xi,X) \sim Q^0_{\theta,\alpha,\xi,X} = M^0_{\theta,\alpha} \otimes P^0_{\xi|\theta} \otimes P^0_{X|\xi,\alpha} \\ \mathcal{E}^1: &\quad (\psi,\alpha,\xi,X) \sim Q^1_{\psi,\alpha,\xi,X} = M^1_{\psi,\alpha} \otimes P^1_{\xi|\psi} \otimes P^1_{X|\xi,\alpha}.\end{aligned}$$

The construction of the *extended model* $\mathcal{E}^{0,*}$ uses an *extended Bayesian Pseudo-True Value (BPTV) condition*, namely:

$$\psi \perp\!\!\!\perp \alpha,\xi \mid \theta; Q^{0,*} \Leftrightarrow \begin{cases} (a)\ \psi \perp\!\!\!\perp \alpha \mid \xi,\theta; Q^{0,*} \\ (b)\ \psi \perp\!\!\!\perp \xi \mid \theta; Q^{0,*}. \end{cases} \tag{5}$$

The first one gives a neutrality of the partial observability for interpreting ψ in $Q^{0,*}$ and the second one is the standard BPTV hypothesis within the structural model. Note that (4a) and (5a) are equivalent to:

$$\alpha \perp\!\!\!\perp (\xi,\psi) \mid \theta; Q^{0,*}. \tag{6}$$

Since the structural process acts independently of the partial observability process, we assume that :

$$(a)\ \alpha \perp\!\!\!\perp \theta; Q^0 \qquad \text{and} \qquad (b)\ \alpha \perp\!\!\!\perp \psi; Q^1. \tag{7}$$

The next theorem shows in the extended complete model a complete separation between the structural model and the parameter of the observability process.

Theorem 1 *Using the extended BPTV condition (5), we have under (7a), that:*

$$\alpha \perp\!\!\!\perp (\xi,\psi,\theta); Q^{0,*}. \tag{8}$$

Proof. Clearly (8) is equivalent to (8*a*) $\alpha \perp\!\!\!\perp \xi,\theta$ and (8*b*) $\alpha \perp\!\!\!\perp \psi \mid \xi,\theta$. Property (8*a*) is equivalent to (7a) and $\alpha \perp\!\!\!\perp \xi \mid \theta$, implied by (6), and property (8*b*) is also implied by (6). □

The encompassing test consists in evaluating $d_1(X) = d^*(M^{0,*}_{\lambda|X}, M^1_{\lambda|X})$ as a test statistic, where d^* is a discrepancy or a divergence between probabilities and which must be calibrated against P^0_X and $\lambda = h(\psi)$ is a finite dimensional functional which represents the parameter of actual interest.

The partial observation is not likely to identify the complete parameters α and θ (resp. ψ) in the null (resp. alternative) model, even though we have assumed that θ (resp. ψ) is identified by ξ. This is so unless X were a sufficient statistic in both models, see Mouchart and Oulhaj (2003). Let γ_X (resp. ω_X) be a minimal sufficient parameter for the sampling distribution of X; they verify that:

$$(a)\ X \perp\!\!\!\perp \theta,\alpha \mid \gamma_X; Q^0 \qquad (b)\ X \perp\!\!\!\perp \psi,\alpha \mid \omega_X; Q^1. \tag{9}$$

This raises the question whether $d_1(X)$ should be defined from the distance between the posterior distributions of the complete parameter ψ or of the identified parameter ω_X only. Let us therefore ask how far it is legitimate to concentrate the encompassing test on the identified part of ψ, *i.e.* to evaluate

$$d_2(X) = d^*(M^{0,*}_{\omega_X|X}, M^1_{\omega_X|X})$$

instead of $d_1(X)$. Intuitively, that would be legitimate if in the extended null model and in the alternative model the distributions conditional on the data and the identified parameters would not depend on the data. The next Theorem shows that under the extended BPTV condition, a similar property holds in the extended null model and provides as a corollary, *i.e.*, equation (10), a more operational structure.

Theorem 2 *Under (5), the extended model* $\mathcal{E}^{0,*}$ *satisfies:* $X \perp\!\!\!\perp \psi, \theta, \alpha \mid \gamma_X; Q^{0,*}$

Proof. In $Q^{0,*}$, we have that (5) implies $\psi \perp\!\!\!\perp X \mid \theta, \alpha$, which jointly with (9a) implies the conclusion □

Then, under condition (5) we have $X \perp\!\!\!\perp \omega_X \mid \gamma_X; Q^{0,*}$ and we may write:

$$Q^{0,*}_{\theta,\alpha,\psi,X} = \left[M^0_{\gamma_X} \otimes M^{0,*}_{\omega_X|\gamma_X} \otimes P^0_{X|\gamma_X}\right] \otimes M^{0,*}_{\alpha,\theta,\psi|\gamma_X}. \tag{10}$$

This means that the prior conditional distribution $M^{0,*}_{\alpha,\theta,\psi|\gamma_X}$ needs not be explicitly specified for the construction of the encompassing test.

Finally, the next theorem requires a further condition on the BPTV in order to ensure that ω_X is sufficient, in the extended model, relatively to ψ.

Theorem 3 *Under the extended BPTV condition (5), if*

$$\psi \perp\!\!\!\perp \gamma_X \mid \omega_X; Q^{0,*} \tag{11}$$

then $\psi \perp\!\!\!\perp X \mid \omega_X; Q^{0,*}$

Proof. Clearly the result is implied by $\psi \perp\!\!\!\perp X, \gamma_X \mid \omega_X$ which is equivalent to (11) along with $\psi \perp\!\!\!\perp X \mid \gamma_X, \omega_X$ implied by the conclusion of the Theorem 3. □

Theorem 3 permits us to write: $M^{0,*}_{\psi|X} = \int M^{0,*}_{\psi|\omega_X} dM^{0,*}_{\omega_X|X}$. If, by specification, we impose the coherence condition:

$$M^{0,*}_{\psi|\omega_X} = M^1_{\psi|\omega_X}, \tag{12}$$

a convenient distance for the posterior distributions of ψ can be defined through a distance between posterior distributions of ω_X, as in $d_2(X)$. It should be emphasized that the conditions (11) and (12) are restrictions on the prior specification only and are accordingly not testable as they do not bear on the sampling neither on the structural process. Furthermore, these conditions depend on the specification of $M^{0,*}_{\alpha,\theta,\psi|\gamma_X}$ that leaves, from (10), the testing procedure invariant.

3. AN EXAMPLE

In order to illustrate the computations implied by the test developed, let us consider a bivariate latent vector ξ_i, a linear combination of which is observable, namely: $X_i = g_\alpha(\xi_i) = \alpha'\xi_i$, where $\xi_i = (\xi_{i,1}, \xi_{i,2})'$ and $\alpha = (\alpha_1, \alpha_2)'$. We want to compare a completely normally distributed (with known variances) null model with a non parametric alternative model both satisfying (4) and (7):

$$\mathcal{E}^0 : \begin{cases} \xi \mid \theta, \alpha \sim \mathcal{N}_{(2)}(\theta, A_0) \\ \theta \mid \alpha \sim \mathcal{N}_{(2)}(0, B_0) \\ \alpha \sim M^0_\alpha \end{cases} \qquad \mathcal{E}^1 : \begin{cases} \xi \mid \psi, \alpha \sim \psi \\ \psi \mid \alpha \sim \mathcal{D}i(n^1_0 \mathcal{N}_{(2)}(0, C_0)) \\ \alpha \sim M^1_\alpha, \end{cases} \tag{13}$$

where A_0, B_0 and C_0 are definite positive matrices, $n_0 > 0$ and $\mathcal{D}i$ denotes a Dirichlet process. The corresponding statistical models are:

$$\mathcal{E}^0 : \begin{cases} X \mid \theta, \alpha \sim \mathcal{N}(\alpha'\theta, \alpha' A_0 \alpha) \\ \theta \mid \alpha \sim \mathcal{N}_{(2)}(0, B_0) \\ \alpha \sim M^0_\alpha \end{cases} \qquad \mathcal{E}^1 : \begin{cases} X \mid \psi, \alpha \sim \omega_X \\ \omega_X \mid \alpha \sim \mathcal{D}i(n_0^1 \mathcal{N}(0, \alpha' C_0 \alpha)) \\ \alpha \sim M^1_\alpha. \end{cases} \tag{14}$$

$\omega_X = \psi \circ g_\alpha^{-1}$ is the parameter of $\mathcal{E}^1$ identified by X. Under assumptions (11) and (12) and defining the BPTV as:

$$M_{\omega_X \mid \gamma_X} = \int M^1_{\omega_X \mid \tilde{X}_1^n} dP^0_{\tilde{X}_1^n \mid \gamma_X} = E^0[\, M^1_{\omega_X \mid \tilde{X}_1^n} \mid \gamma_X \,] \tag{15}$$

where $\tilde{X}_1^n$ is a virtual sample from the null sampling process such that $\tilde{X}_1^n \perp\!\!\!\perp X_1^n \mid \gamma_X; Q^{0,*}$, we obtain the posterior distribution of ω_X:

$$M^{0,*}_{\omega_X \mid X_1^n} = E^0[\, M^1_{\omega_X \mid \tilde{X}_1^n} \mid X_1^n \,] \tag{16}$$

Both in the null and in the alternative models the posterior distributions of the parameter ω_X are mixtures of Dirichlet processes. An algorithm based on a direct simulation of trajectories of Dirichlet process as developed in Rolin (1992) and in Sethuraman (1994) and developed in Florens *et al.* (2003) may be applied with a simple adaptation. For the sake of illustration, we have selected for λ the first two moments. The p-value is evaluated by simulation as:

$$\widehat{p\text{-value}} = \frac{1}{N_C} \sum_{1 \le \ell \le N_C} \boldsymbol{1}_{\{d((X_1^n)^*_\ell) \,>\, d(X_1^n)\}}$$

where $(X_1^n)^*_\ell$ are simulated samples from P^0_X, and $d(X_1^n)$ is computed by simulation. *Numerical illustration.* We specify in this exercise the same prior distribution for α *i.e.* $M^0_\alpha = M^1_\alpha$. We use the following reparametrization:

$$\alpha = \sqrt{\alpha_X} R_0^{-1} \begin{pmatrix} \cos\tau \\ \sin\tau \end{pmatrix}; \; \tau \in [0, 2\pi), \quad \alpha_X = \alpha'(A_0 + B_0)\alpha \ge 0, \; (A_0 + B_0) = R_0' R_0$$

The prior distribution is: $\alpha_X \perp\!\!\!\perp \tau$; $\alpha_X \sim Gamma(1,1)$, $\tau \sim \mathcal{U}_{[0,2\pi)}$, with $A_0 = B_0 = \begin{pmatrix} 1 & .5 \\ .5 & 1 \end{pmatrix}$. We want to evaluate how far a sample generated by a distribution in the alternative sampling model would be likely associated to a value of the test statistic relatively far in the tail of the null predictive distribution. Consider accordingly the following specification: $\phi = (\phi_1, \phi_2)'$; $\phi_1 \perp\!\!\!\perp \phi_2$; $\phi_i \sim \chi^2_2$; $\xi = D_0(\tilde{\xi} - (2,2)')$; $\tilde{\xi} = (\tilde{\xi}_1, \tilde{\xi}_2)$; $\tilde{\xi}_1 \perp\!\!\!\perp \tilde{\xi}_2 \mid \phi$; $\tilde{\xi}_i \sim \text{Expo}(\phi_i^{-1})$ where we choose D_0 in such a way that $12\, D_0 D_0' = A_0 + B_0$ in order to ensure that the predictive expectation and variances are the same as in the null model. For the test statistic we consider the Kullback–Leibler divergence.

If one considers the rule 'Reject $\mathcal{E}^0$ if $\widehat{p\text{-value}} \le 0.05$', one may define the empirical coverage as the proportion of the samples in the critical region. The empirical coverage is expected to be higher than 0.05 and to increase with the sample size. Table 1 gives the observed results for four different sample sizes ($n = 10, 50, 100, 200$) and 3 trials. One may notice that, with a slight exception in trial 3, the coverage rate increases monotonically with the sample size and is always higher than 0.05.

Table 1: *Coverage rates.*

n	trial 1	trial 2	trial 3
10	0.14	0.10	0.14
50	0.17	0.13	0.11
100	0.25	0.20	0.18
200	0.28	0.23	0.24

4. CONCLUSIONS

This paper demonstrates the operationality of a Bayesian encompassing test in the framework of partial observability even if numerical issues require powerful computations. We show that the test, from a theoretical point of view, is feasible. The proposed procedure might therefore be adapted to a wider class of problems.

From a numerical point of view, the computations are made easier thanks to the possibility of direct simulations for the trajectories of a Dirichlet process (making use of Rolin–Sethuraman representation). If this were not the case, *e.g.* for other non parametric alternatives, recourse to heavier, and numerically more problematic, procedures, such as those based on MCMC, could probably not be avoided.

REFERENCES

Almeida, C. and Mouchart, M. (2005). Bayesian encompassing test under partial observability. *Tech. Rep.*, Université Catholique de Louvain, Belgium.

Cox, D. R. (1961). Tests of separate families of hypotheses. *Proc. First Berkeley Symp.* (J. Neyman ed.) Berkeley: Univ. California Press 105–123.

Cox, D. R. (1962). Further results on tests of separate families of hypotheses. *J. Roy. Statist. Soc. B* **24**, 406–424.

Florens, J. P. and Mouchart, M. (1993). Bayesian testing and testing Bayesians. *Handbook in Statistics* **11** (G. Maddala, C. Rao, and H. Vinod, eds.) Amsterdam: North-Holland, 303–333.

Florens, J. P., Mouchart, M. and Rolin, J. M. (1990). *Elements of Bayesian Statistics.* New York: Marcel Dekker.

Florens, J.P., Richard, J.F. and Rolin, J.M. (2003). Bayesian encompassing specification test of a parametric model against a nonparametric alternative. *Tech. Rep.*, Université Catholique de Louvain, Belgium.

Florens, J. and Mouchart, M. (1989). Bayesian specification test. *Contributions to Operational Research and Economics* (B. Cornet and H. Tulkens, eds.) Cambridge, Mass: MIT, 467–460.

Mizon, G. E. and Richard, J. F. (1986). The encompassing principle and its application to testing non-nested hypotheses. *Econometrica* **54**, 657–678.

Mouchart, M. and Oulhaj, A. (2003). A Note on partial sufficiency with connection to the identification problem. *Metron* **61**, 267–283.

Muthén, B. (1984). A General structural equation model with dichotomous, ordered categorical variables, and continuous latent variables indicators. *Psychometrika* **49**, 115–132.

Rolin, J.M. (1992). Some useful properties of the Dirichlet process. *Tech. Rep.*, Université Catholique de Louvain, Belgium.

Sethuraman, J. (1994). A constructive definition of Dirichlet priors. *Statist. Sinica* **4**, 639–650.

BAYESIAN STATISTICS 8, pp. 571–576.
J. M. Bernardo, M. J. Bayarri, J. O. Berger, A. P. Dawid,
D. Heckerman, A. F. M. Smith and M. West (Eds.)

Comparing Normal Means: New Methods for an Old Problem*

José M. Bernardo
Universitat de València, Spain
jose.m.bernardo@uv.es

Sergio Pérez
Colegio de Postgraduados, Mexico
sergiop@colpos.mx

Summary

Comparing the means of two normal populations is a very old problem in mathematical statistics, but there is still no consensus about its most appropriate solution. In this paper we treat the problem of comparing two normal means as a Bayesian decision problem with only two alternatives: either to accept the hypothesis that the two means are equal, or to conclude that the observed data are, under the assumed model, incompatible with that hypothesis. The combined use of an information-theory based loss function, the *intrinsic discrepancy* (Bernardo and Rueda, 2002), and an objective prior function, the *reference prior* (Bernardo, 1979; Berger and Bernardo, 1992), produces a new solution to this old problem which, for the first time, has the invariance properties one should presumably require.

Keywords and Phrases: Bayes factor, BRC, comparison of normal means, intrinsic discrepancy, Kullback–Leibler divergence, precise hypothesis testing, reference prior.

1. STRUCTURE OF THE DECISION PROBLEM

Precise hypothesis testing as a decision problem. Assume that available data z have been generated from an unknown element of the family of probability distributions for $z \in \mathcal{Z}$, $\{p_z(\cdot \mid \phi, \omega),\ \phi \in \Phi,\ \omega \in \Omega\}$, and suppose that it is desired to check whether or not these data may be judged to be compatible with the (null) hypothesis $H_0 \equiv \{\phi = \phi_0\}$. This may be treated as a decision problem with only two alternatives; a_0: to accept H_0 (and work *as if* $\phi = \phi_0$) or a_1: to claim that the observed data are incompatible with H_0. Notice that, with this formulation, H_0 is generally a composite hypothesis, described by the family of probability distributions

* This is a synopsis. For the full version, see *Bayesian Analysis* **2** (2007), 571–576.
José M. Bernardo is Professor of Statistics at the Facultad de Matemáticas, Universidad de Valencia, 46100-Burjassot, Valencia, Spain. Dr. Sergio Pérez works at the Colegio de Postgraduados, Montecillo, 56230-Texcoco, Mexico. Work by Professor Bernardo has been partially funded with Grant MTM2006-07801 of the MEC, Spain.

$\mathcal{M}_0 = \{p_{\boldsymbol{z}}(\cdot \mid \boldsymbol{\phi}_0, \boldsymbol{\omega}_0),\ \boldsymbol{\omega}_0 \in \Omega\}$. Simple nulls are included as a particular case where there are no nuisance parameters.

Foundations dictate (see *e.g.*, Bernardo and Smith, 1994, and references therein) that, to solve this decision problem, one must specify utility functions $u\{a_i, (\boldsymbol{\phi}, \boldsymbol{\omega})\}$ for the two alternatives a_0 and a_1, and a joint prior distribution $\pi(\boldsymbol{\phi}, \boldsymbol{\omega})$ for the unknown parameters $(\boldsymbol{\phi}, \boldsymbol{\omega})$; then, H_0 should be rejected if, and only if,

$$\int_\Phi \int_\Omega [u\{a_1, (\boldsymbol{\phi}, \boldsymbol{\omega})\} - u\{a_0, (\boldsymbol{\phi}, \boldsymbol{\omega})\}]\, \pi(\boldsymbol{\phi}, \boldsymbol{\omega} \mid \boldsymbol{z})\, d\boldsymbol{\phi}\, d\boldsymbol{\omega} > 0,$$

where, using Bayes' theorem, $\pi(\boldsymbol{\phi}, \boldsymbol{\omega} \mid \boldsymbol{z}) \propto p(\boldsymbol{z} \mid \boldsymbol{\phi}, \boldsymbol{\omega})\, \pi(\boldsymbol{\phi}, \boldsymbol{\omega})$ is the joint posterior which corresponds to the prior $\pi(\boldsymbol{\phi}, \boldsymbol{\omega})$. Thus, only the utilities difference must be specified, and this may usefully be written as

$$u\{a_1, (\boldsymbol{\phi}, \boldsymbol{\omega})\} - u\{a_0, (\boldsymbol{\phi}, \boldsymbol{\omega})\} = \ell\{\boldsymbol{\phi}_0, (\boldsymbol{\phi}, \boldsymbol{\omega})\} - u_0,$$

where $\ell\{\boldsymbol{\phi}_0, (\boldsymbol{\phi}, \boldsymbol{\omega})\}$ is the non-negative terminal loss suffered by accepting $\boldsymbol{\phi} = \boldsymbol{\phi}_0$ given $(\boldsymbol{\phi}, \boldsymbol{\omega})$, and $u_0 > 0$ is the utility of accepting H_0 when it is true. Hence, H_0 should be rejected if, and only if,

$$t(\boldsymbol{\phi}_0 \mid \boldsymbol{z}) = \int_\Theta \int_\Omega \ell\{\boldsymbol{\phi}_0, (\boldsymbol{\phi}, \boldsymbol{\omega})\}\, \pi(\boldsymbol{\phi}, \boldsymbol{\omega} \mid \boldsymbol{z})\, d\boldsymbol{\phi}\, d\boldsymbol{\omega} > u_0,$$

that is, if the posterior expected loss, the *test statistic* $t(\boldsymbol{\phi}_0 \mid \boldsymbol{z})$ is large enough.

The intrinsic discrepancy loss. As one would expect, the optimal decision depends heavily on the particular loss function $\ell\{\boldsymbol{\phi}_0, (\boldsymbol{\phi}, \boldsymbol{\omega})\}$ used. Specific problems may require specific loss functions, but conventional loss functions may be used to proceed when one does not have any particular application in mind.

A common class of conventional loss function are the *step* loss functions. These *forces* the use of a *non-regular* 'spiked' proper prior which places a lump of probability at $\boldsymbol{\phi} = \boldsymbol{\phi}_0$ and leads to rejecting H_0 if, and only if, its posterior probability is too small or, equivalently, if, and only if, the *Bayes factor* against H_0, is sufficiently large. This will be appropriate wherever preferences are well described by a step loss function, and prior information is available to justify a (*highly informative*), spiked prior. It may be argued that many scientific applications of precise hypothesis testing fail to meet one or both of these conditions.

Another example of a conventional loss function is the ubiquitous *quadratic* loss function. This leads to rejecting the null if, and only if, the posterior expected Euclidean distance of $\boldsymbol{\phi}_0$ from the true value $\boldsymbol{\phi}$ is too large, and may safely be used with (typically improper) 'noninformative' priors. However, as most conventional continuous loss functions, the quadratic loss depends dramatically on the particular parametrization used. But, since the model parametrization is arbitrary, the conditions to reject $\boldsymbol{\phi} = \boldsymbol{\phi}_0$ should be *precisely the same*, for any one-to-one function $\boldsymbol{\psi}(\boldsymbol{\phi})$, as the conditions to reject $\boldsymbol{\psi} = \boldsymbol{\psi}(\boldsymbol{\phi}_0)$. This requires the use of a loss function which is invariant under one-to-one reparametrizations.

It may be argued (Bernardo and Rueda, 2002; Bernardo, 2005b) that a measure of the disparity between two probability distributions which may be appropriate for general use in probability and statistics is the *intrinsic discrepancy*,

$$\delta_{\boldsymbol{z}}\{p_{\boldsymbol{z}}(\cdot), q_{\boldsymbol{z}}(\cdot)\} \equiv \min\left\{\int_{\mathcal{Z}} p_{\boldsymbol{z}}(\boldsymbol{z}) \log \frac{p_{\boldsymbol{z}}(\boldsymbol{z})}{q_{\boldsymbol{z}}(\boldsymbol{z})}\, d\boldsymbol{z},\ \int_{\mathcal{Z}} p_{\boldsymbol{z}}(\boldsymbol{z}) \log \frac{p_{\boldsymbol{z}}(\boldsymbol{z})}{q_{\boldsymbol{z}}(\boldsymbol{z})}\, d\boldsymbol{z}\right\}, \qquad (1)$$

defined as the *minimum* (Kullback-Keibler) logarithmic divergence between them. This is *symmetric*, *non-negative*, and it is zero if, and only if, $p_{\boldsymbol{z}}(\boldsymbol{z}) = q_{\boldsymbol{z}}(\boldsymbol{z})$ a.e. Besides, it is *invariant* under one-to-one transformations of $\boldsymbol{z}$, and it is *additive* under independent observations; thus if $\boldsymbol{z} = \{x_1, \ldots, x_n\}$, $p_{\boldsymbol{z}}(\boldsymbol{z}) = \prod_{i=1}^{n} p_x(x_i)$, and $q_{\boldsymbol{z}}(\boldsymbol{z}) = \prod_{i=1}^{n} q_x(x_i)$, then $\delta_{\boldsymbol{z}}\{p_{\boldsymbol{z}}(\cdot), q_{\boldsymbol{z}}(\cdot)\} = n\, \delta_x\{p_x(\cdot), q_x(\cdot)\}$.

Within a parametric probability model, say $\{p_{\boldsymbol{z}}(\cdot\,|\,\boldsymbol{\theta}), \boldsymbol{\theta} \in \boldsymbol{\Theta}\}$, the intrinsic discrepancy induces a loss function $\delta\{\boldsymbol{\theta}_0, \boldsymbol{\theta}\} = \delta_{\boldsymbol{z}}\{p_{\boldsymbol{z}}(\cdot\,|\,\boldsymbol{\theta}_0), p_{\boldsymbol{z}}(\cdot\,|\,\boldsymbol{\theta})\}$, in which the loss to be suffered if $\boldsymbol{\theta}$ is replaced by $\boldsymbol{\theta}_0$ is not measured terms of the disparity between $\boldsymbol{\theta}$ and $\boldsymbol{\theta}_0$, but in terms of the disparity between the *models* labelled by $\boldsymbol{\theta}$ and $\boldsymbol{\theta}_0$. This provides a loss function which is *invariant* under reparametrization: for any one-to-one function $\boldsymbol{\psi} = \boldsymbol{\psi}(\boldsymbol{\theta})$, $\delta\{\boldsymbol{\psi}_0, \boldsymbol{\psi}\} = \delta\{\boldsymbol{\theta}_0, \boldsymbol{\theta}\}$. Moreover, one may equivalently work with sufficient statistics: if $\boldsymbol{t} = \boldsymbol{t}(\boldsymbol{z})$ is a sufficient statistic for $p_{\boldsymbol{z}}(\cdot\,|\,\boldsymbol{\theta})$, then $\delta_{\boldsymbol{z}}\{\boldsymbol{\theta}_0, \boldsymbol{\theta})\} = \delta_{\boldsymbol{t}}\{\boldsymbol{\theta}_0, \boldsymbol{\theta}\}$. The intrinsic loss may be safely be used with improper priors. In the context of hypothesis testing within the parametric model $p_{\boldsymbol{z}}(\cdot\,|\,\boldsymbol{\phi}, \boldsymbol{\omega})$, the intrinsic loss to be suffered by replacing $\boldsymbol{\phi}$ by $\boldsymbol{\phi}_0$ becomes

$$\delta_{\boldsymbol{z}}\{H_0, (\boldsymbol{\phi}, \boldsymbol{\omega})\} \equiv \inf_{\boldsymbol{\omega}_0 \in \boldsymbol{\Omega}} \delta_{\boldsymbol{z}}\{p_{\boldsymbol{z}}(\cdot\,|\,\boldsymbol{\phi}_0, \boldsymbol{\omega}_0),\, p_{\boldsymbol{z}}(\cdot\,|\,\boldsymbol{\phi}, \boldsymbol{\omega})\}, \tag{2}$$

that is, the intrinsic discrepancy between the distribution $p_{\boldsymbol{z}}(\cdot\,|\,\boldsymbol{\phi}, \boldsymbol{\omega})$ which has generated the data, and the family of distributions $\mathcal{F}_0 \equiv \{p_{\boldsymbol{z}}(\cdot\,|\,\boldsymbol{\phi}_0, \boldsymbol{\omega}_0), \boldsymbol{\omega}_0 \in \Omega\}$ which corresponds to the hypothesis $H_0 \equiv \{\boldsymbol{\phi} = \boldsymbol{\phi}_0\}$ to be tested. If, as it is usually the case, the parameter space $\boldsymbol{\Phi} \times \boldsymbol{\Omega}$ is convex, then the two minimization procedures in (2) and (1) may be interchanged (Juárez, 2005).

As it is apparent from its definition, the intrinsic loss (2) is the *minimum conditional expected log-likelihood ratio* (under repeated sampling) against H_0, what provides a direct *calibration* for its numerical values; thus, intrinsic loss values of about $\log(100)$ would indicate rather strong evidence against H_0.

The Bayesian Reference Criterion (BRC). Any statistical procedure depends on the accepted assumptions, and those typically include many subjective judgements. If has become standard, however, to term 'objective' any statistical procedure whose results only depend on the quantity of interest, the model assumed and the data obtained. The *reference prior* (Bernardo, 1979; Berger and Bernardo, 1992; Bernardo, 2005a), loosely defined as that prior which maximizes the missing information about the quantity of interest, provides a general solution to the problem of specifying an objective prior. See Berger (2006) for a recent analysis of this issue.

The Bayesian reference criterion (Bernardo and Rueda, 2002) is the normative Bayes solution to the decision problem of hypothesis testing described above which corresponds to the use of the *intrinsic loss* and the *reference prior*. Given a parametric model $\{p_{\boldsymbol{z}}(\cdot\,|\,\boldsymbol{\phi}, \boldsymbol{\omega}),\ \boldsymbol{\phi} \in \Phi,\ \boldsymbol{\omega} \in \Omega\}$, this prescribes to reject the hypothesis $H_0 \equiv \{\boldsymbol{\phi} = \boldsymbol{\phi}_0\}$ if, and only if,

$$d(H_0\,|\,\boldsymbol{z}) = \int_0^\infty \delta\, \pi(\delta\,|\,\boldsymbol{z})\, d\delta > \delta_0, \tag{3}$$

where $d(H_0\,|\,\boldsymbol{z})$, termed the *intrinsic (test) statistic*, is the reference posterior expectation of the intrinsic loss $\delta_{\boldsymbol{z}}\{H_0, (\boldsymbol{\phi}, \boldsymbol{\omega})\}$ defined by (2), and where δ_0 is a context dependent positive utility constant, *the largest acceptable average log-likelihood ratio against H_0 under repeated sampling.* For scientific communication, δ_0 could conventionally be set to $\log(10) \approx 2.3$ to indicate some evidence against H_0, and to $\log(100) \approx 4.6$ to indicate strong evidence against H_0.

2. NORMAL MEANS COMPARISON

Problem statement in the common variance case. Let available data $\boldsymbol{z} = \{\boldsymbol{x}, \boldsymbol{y}\}$, $\boldsymbol{x} = \{x_1, \ldots, x_n\}$, $\boldsymbol{y} = \{y_1, \ldots, y_m\}$, consist of two random samples of possibly different sizes n and m, respectively drawn from $\mathrm{N}(x \mid \mu_x, \sigma)$ and $\mathrm{N}(y \mid \mu_y, \sigma)$, so that the assumed model is $p(\boldsymbol{z} \mid \mu_x, \mu_y, \sigma) = \prod_{i=1}^{n} \mathrm{N}(x_i \mid \mu_x, \sigma) \prod_{j=1}^{m} \mathrm{N}(y_j \mid \mu_y, \sigma)$. It is desired to test $H_0 \equiv \{\mu_x = \mu_y\}$, that is, whether or not these data could have been drawn from some element of the family $\mathcal{F}_0 \equiv \{p(\boldsymbol{z} \mid \mu_0, \mu_0, \sigma_0), \mu_0 \in \Re,\ \sigma_0 > 0\}$. To implement the BRC criterion described above one should: (i) compute the *intrinsic discrepancy* $\delta\{H_0, (\mu_x, \mu_y, \sigma)\}$ between the family $\mathcal{F}_0$ which defines the hypothesis H_0 and the assumed model $p(\boldsymbol{z} \mid \mu_x, \mu_y, \sigma)$; (ii) determine the *reference* joint prior $\pi_\delta(\mu_x, \mu_y, \sigma)$ of the three unknown parameters when δ is the quantity of interest; and (iii) derive the relevant *intrinsic statistic*, that is the reference posterior expectation $d(H_0 \mid \boldsymbol{z}) = \int_0^\infty \delta\, \pi_\delta(\delta \mid \boldsymbol{z})\, d\delta$ of the intrinsic discrepancy $\delta\{H_0, (\mu_x, \mu_y, \sigma)\}$.

The intrinsic loss. The (Kullback–Leibler) logarithmic divergence of a normal distribution $\mathrm{N}(x \mid \mu_2, \sigma_2)$ from another normal distribution $\mathrm{N}(x \mid \mu_1, \sigma_1)$ is given by

$$\begin{aligned}
\kappa\{\mu_2, \sigma_2 \mid \mu_1, \sigma_1\} &\equiv \int_{-\infty}^{\infty} \mathrm{N}(x \mid \mu_1, \sigma_1) \log \frac{\mathrm{N}(x \mid \mu_1, \sigma_1)}{\mathrm{N}(x \mid \mu_2, \sigma_2)}\, dx \\
&= \frac{1}{2}\left(\frac{\mu_2 - \mu_1}{\sigma_2}\right)^2 + \frac{1}{2}\left(\frac{\sigma_1^2}{\sigma_2^2} - 1 - \log\frac{\sigma_1^2}{\sigma_2^2}\right). \qquad (4)
\end{aligned}$$

Using its additive property, the (KL) logarithmic divergence of $p(\boldsymbol{z} \mid \mu_0, \mu_0, \sigma_0)$ from $p(\boldsymbol{z} \mid \mu_x, \mu_y, \sigma)$ is $n\,\kappa\{\mu_0, \sigma_0 \mid \mu_x, \sigma\} + m\,\kappa\{\mu_0, \sigma_0 \mid \mu_y, \sigma\}$, which is minimized when $\mu_0 = (n\mu_x + m\mu_y)/(n+m)$ and $\sigma_0 = \sigma$. Substitution yields $h(n,m)\,\theta^2/4$, where $h(n,m) = 2nm/(m+n)$ is the harmonic mean of the two sample sizes, and $\theta = (\mu_x - \mu_y)/\sigma$ is the standardized distance between the two means. Similarly, the logarithmic divergence of $p(\boldsymbol{z} \mid \mu_x, \mu_y, \sigma)$ from $p(\boldsymbol{z} \mid \mu_0, \mu_0, \sigma_0)$ is given by $n\,\kappa\{\mu_x, \sigma \mid \mu_0, \sigma_0\} + m\,\kappa\{\mu_y, \sigma \mid \mu_0, \sigma_0\}$, minimized when $\mu_0 = (n\mu_x + m\mu_y)/(n+m)$ and $\sigma_0^2 = \sigma^2 + (\mu_x - \mu_y)^2\,(mn)/(m+n)^2$. Substitution now yields a minimum divergence $(n+m)/2 \log[1 + h(n,m)/(2(n+m))\,\theta^2$, which is always smaller than the minimum divergence $h(n,m)\,\theta^2/4$ obtained above. Therefore, the required intrinsic loss function is

$$\delta_{\boldsymbol{z}}\{H_0, (\mu_x, \mu_y, \sigma)\} = \frac{n+m}{2} \log\left[1 + \frac{h(n,m)}{2(n+m)}\,\theta^2\right], \qquad (5)$$

a logarithmic transformation of the standardized distance $\theta = (\mu_x - \mu_y)/\sigma$ between the two means. The intrinsic loss (5) increases linearly with the total sample size $n+m$, and it is essentially quadratic in θ in a neighbourhood of zero, but it becomes concave for $|\,\theta\,| > (k+1)/\sqrt{k}$, where $k = n/m$ is the ratio of the two sample sizes, an eminently reasonable behaviour which conventional loss functions do not have. For equal sample sizes, $m = n$, this reduces to $n \log[1 + \theta^2/4]$ a linear function of the sample size n, which behaves as $\theta^2/4$ in a neighbourhood of the origin, but becomes concave for $|\,\theta\,| > 2$.

Reference analysis. The intrinsic loss (5) is a piecewise invertible function of θ, the standardized difference of the means. Consequently, the required objective prior is the joint reference prior function $\pi_\theta(\mu_x, \mu_y, \sigma)$ when the standardized difference of the means, $\theta = (\mu_x - \mu_y)/\sigma$, is the quantity of interest. This may easily be obtained using the orthogonal parametrization $\{\theta, \omega_1, \omega_2\}$, with $\omega_1 = \sigma\sqrt{2(m+n)^2 + m\,n\,\theta^2}$, and $\omega_2 = \mu_y + n\,\sigma\,\theta/(n+m)$. In the original parametrization the required reference prior is found to be

$$\pi_\theta(\mu_x, \mu_y, \sigma) = \frac{1}{\sigma^2}\left(1 + \frac{h(n,m)}{4(m+n)}\left(\frac{\mu_x - \mu_y}{\sigma}\right)^2\right)^{-1/2}. \tag{6}$$

By Bayes theorem, the posterior is $\pi_\theta(\mu_x, \mu_y, \sigma \,|\, \boldsymbol{z}) \propto p(\boldsymbol{z} \,|\, \mu_x, \mu_y, \sigma)\,\pi_\theta(\mu_x, \mu_y, \sigma)$. Changing variables to $\{\theta, \mu_y, \sigma\}$, and integrating out μ_y and σ, produces the (marginal) reference posterior density of the quantity of interest

$$\begin{aligned}\pi(\theta \,|\, \boldsymbol{z}) &= \pi(\theta \,|\, t, m, n)\\ &\propto \left(1 + \frac{h(n,m)}{4(m+n)}\,\theta^2\right)^{-1/2} \mathrm{NcSt}\left(t \,\middle|\, \sqrt{\frac{h(n,m)}{2}}\;\theta,\; n+m-2\right)\end{aligned} \tag{7}$$

where

$$t = \frac{\bar{x} - \bar{y}}{s\sqrt{2/h(n,m)}}\,, \quad s^2 = \frac{n\,s_x^2 + m\,s_y^2}{n+m-2}\,,$$

and $\mathrm{NcSt}(\cdot \,|\, \lambda, \nu)$ is the density of a noncentral Student distribution with noncentrality parameter λ and ν degrees of freedom. The reference posterior (7) is proper provided $n \geq 1$, $m \geq 1$, and $n + m \geq 3$. For further details, see Pérez (2005).

The reference posterior (7) has the form $\pi(\theta \,|\, t, n, m) \propto \pi(\theta)\; p(t \,|\, \theta, n, m)$, where $p(t \,|\, \mu_x, \mu_y, \sigma, m, m) = p(t \,|\, \theta, n, m)$ is the sampling distribution of t. Thus, the reference prior is consistent under marginalization (cf. Dawid, Stone and Zidek, 1973).

The intrinsic statistic. The reference posterior for θ may now be used to obtain the required intrinsic test statistic. Indeed, substituting into (5) yields

$$d(H_0 \,|\, \boldsymbol{z}) = d(H_0 \,|\, t, m, n) = \int_0^\infty \frac{n+m}{2} \log\left[1 + \frac{h(n,m)}{2(m+n)}\,\theta^2\right] \pi(\theta \,|\, t, m, n)\, d\theta, \tag{8}$$

where $\pi(\theta \,|\, t, m, n)$ is given by (7). This has no simple analytical expression but may easily be obtained by one-dimensional numerical integration.

Example. The derivation of the appropriate reference prior allows us to draw precise conclusions even when data are extremely scarce. As an illustration, consider a (minimal) sample of three observations with $\boldsymbol{x} = \{4, 6\}$ and $\boldsymbol{y} = \{0\}$, so that $n = 2$, $m = 1$, $\bar{x} = 5$, $\bar{y} = 0$, $s = \sqrt{2}$, $h(n,m) = 4/3$ and $t = 5/\sqrt{3}$. If may be verified numerically that the reference posterior probability that $\theta < 0$ is

$$\Pr[\theta < 0 \,|\, t, h, n] = \int_{-\infty}^0 \pi(\theta \,|\, t, m, n)\, d\theta = 0.0438,$$

directly suggesting some (mild) evidence against $\theta = 0$ and, hence, against $\mu_x = \mu_y$. On the other hand, using the formal procedure described above, the numerical value of intrinsic statistic to test $H_0 \equiv \{\mu_x = \mu_y\}$ is

$$d(H_0 \,|\, t, m, n) = \int_0^\infty \frac{3}{2} \log\left[1 + \frac{2}{9}\,\theta^2\right] \pi(\theta \,|\, t, m, n)\, d\theta = 1.193 = \log[6.776].$$

Thus, given the available data, the expected value of the average (under repeated sampling) of the log-likelihood ratio against H_0 is 1.193 (so that likelihood ratios may be expected to be about 6.8 against H_0), which provides a precise measure of the available evidence against the hypothesis $H_0 \equiv \{\mu_x = \mu_y\}$.

This (moderate) evidence against H_0 is *not* captured by the conventional frequentist analysis of this problem. Indeed, since the sampling distribution of t under H_0 is a standard Student distribution with $n+m-2$ degrees of freedom, the p-value which corresponds to the two-sided test for H_0 is $2(1 - T_{m+n-2}\{|t|\})$, where T_ν is the cumulative distribution function of an Student distribution with ν degrees of freedom (see, *e.g.*, DeGroot and Schervish, 2002, Section 8.6). In this case, this produces a p-value of 0.21 which, contrary to the preceding analysis, suggests lack of sufficient evidence in the data against H_0.

Further results. The full version of this paper (Bernardo and Pérez, 2007) contains analytic asymptotic approximations to the intrinsic test statistic (8), analyzes the behaviour of the proposed procedure under repeated sampling (both when H_0 is true and when it is false), and discusses its generalization to the case of possibly different variances.

REFERENCES

Berger, J. O. (2006). The case for objective Bayesian analysis. *Bayesian Analysis* **1**, 385–402 and 457–464 (with discussion).

Berger, J. O. and Bernardo, J. M. (1992). On the development of reference priors. *Bayesian Statistics 4* (J. M. Bernardo, J. O. Berger, A. P. Dawid and A. F. M. Smith, eds.) Oxford: University Press, 61–77 (with discussion).

Bernardo, J. M. (1979). Reference posterior distributions for Bayesian inference. *J. Roy. Statist. Soc. B* **41**,113–147.

Bernardo, J. M. (2005a). Reference analysis. *Handbook of Statistics* **25**, D. K. Dey and C. R. Rao, (eds.) Amsterdam: Elsevier17–90.

Bernardo, J. M. (2005b). Intrinsic credible regions: An objective Bayesian approach to interval estimation. *Test* **14**, 317–384.

Bernardo, J. M. and Pérez, S. (2007). Comparing normal means: New methods for an old problem. *Bayesian Analysis* **2**, 45–48.

Bernardo, J. M. and Rueda, R. (2002). Bayesian hypothesis testing: A reference approach. *Internat. Statist. Rev.* **70**, 351–372.

Bernardo, J. M. and Smith, A. F. M. (1994). *Bayesian Theory.* Chichester: Wiley, (2nd edition to appear in 2008).

DeGroot, M. H. and Schervish, M. J. (2002). *Probability and Statistics*, 3rd ed. Reading, MA: Addison-Wesley.

Dawid, A. P., Stone, M. and Zidek, J. V. (1973). Marginalization paradoxes in Bayesian and structural inference *J. Roy. Statist. Soc. B* **35**,189–233 (with discussion).

Juárez, M. A. (2005). Normal correlation: An objective Bayesian approach. *Tech. Rep.*, CRiSM 05-15, University of Warwick, UK.

Pérez, S. (2005). *Objective Bayesian Methods for Mean Comparison.* Ph.D. Thesis, Universidad de Valencia, Spain.

BAYESIAN STATISTICS 8, pp. 577–582.
J. M. Bernardo, M. J. Bayarri, J. O. Berger, A. P. Dawid,
D. Heckerman, A. F. M. Smith and M. West (Eds.)

Integral Priors for the One Way Random Effects Model*

JUAN A. CANO
Universidad de Murcia, Spain
jacano@um.es

MATHIEU KESSLER
Univ. Politécnica de Cartagena, Spain
mathieu.kessler@upct.es

DIEGO SALMERÓN
Consejería de Sanidad de Murcia, Spain
diego.salmeron@carm.es

SUMMARY

The one way random effects model is analyzed from the Bayesian model selection perspective. From this point of view Bayes factors are the key tool to choose between two models. In order to produce objective Bayes factors objective priors should be assigned to each model. However, these priors are usually improper provoking a calibration problem which precludes the comparison of the models. To solve this problem several derivations of automatically calibrated objective priors have been proposed among which we quote the intrinsic priors introduced in Berger and Pericchi (1996) and the integral priors introduced in Cano *et al.* (2007). Here, we focus on the use of integral priors which take advantage of MCMC techniques to produce most of the times unique Bayes factors. Some illustrations are provided.

Keywords and Phrases: BAYESIAN MODEL SELECTION; INTEGRAL PRIORS; INTRINSIC PRIORS; RANDOM EFFECTS MODEL; RECURRENT MARKOV CHAINS.

1. INTRODUCTION

In model selection problems we consider two models M_i, $i = 1, 2$, where the data $\boldsymbol{x}$ are related to the parameter θ_i by a density $f_i(\boldsymbol{x} \mid \theta_i)$, the default (improper) priors are $\pi_i^N(\theta_i) = c_i h_i(\theta_i)$, $i = 1, 2$, and the resulting Bayes factor,

$$B_{21}^N(\boldsymbol{x}) = \frac{m_2^N(\boldsymbol{x})}{m_1^N(\boldsymbol{x})} = \frac{c_2 \int_{\Theta_2} f_2(\boldsymbol{x} \mid \theta_2) h_2(\theta_2)\, d\theta_2}{c_1 \int_{\Theta_1} f_1(\boldsymbol{x} \mid \theta_1) h_1(\theta_1)\, d\theta_1}$$

depends on the arbitrary ratio c_2/c_1. Therefore, we are left with two problems, that is, first the determination of the ratio c_2/c_1 is paramount, second the Bayes factor

* This is a synopsis. For the full version, see *Bayesian Analysis* **2** (2007), 59–68.

using $\pi_i^N(\theta_i)$ is not an actual Bayes factor and inference methods based on proper priors are preferable to those that are not, see Principle 1 in Berger and Pericchi (1996). An attempt for solving these problems (Berger and Pericchi, 1996), consists in using intrinsic priors π_1^I and π_2^I that are solutions to a system of two functional equations. Intrinsic priors provide a Bayes factor free of arbitrary constants but whether or not it is an actual Bayes factor or a limit of actual Bayes factors depends on the model (Berger and Pericchi, 1996). An additional difficulty when considering intrinsic priors is that they might be not unique, see Cano *et al.* (2004) and Cano *et al.* (2007).

On the other hand, in Cano *et al.* (2007) the so-called integral priors for model selection are proposed as solutions to the system of integral equations

$$\pi_1(\theta_1) = \int_{\mathcal{X}} \pi_1^N(\theta_1 \mid x) m_2(x)\, dx \tag{1}$$

and

$$\pi_2(\theta_2) = \int_{\mathcal{X}} \pi_2^N(\theta_2 \mid x) m_1(x)\, dx, \tag{2}$$

where x is an imaginary minimal training sample, see Berger and Pericchi (1996), and $m_i(x) = \int f_i(x|\theta_i)\pi_i(\theta_i)\, d\theta_i,\ i = 1, 2$. The method can be seen as a symmetrization of the equation that defines the expected posterior prior introduced in Pérez and Berger (2002). We emphasize that in this system both priors $\pi_i,\ i = 1, 2$, are the *incognita*.

These integral priors are well behaved, *i.e.* they provide a Bayes factor free of arbitrary constants which in fact is an actual Bayes factor or a limit of actual Bayes factors. Moreover, it turns out to be unique in many situations, and it can be shown that a sufficient condition for the uniqueness is that the Markov chain with transition density

$$Q(\theta_1' \mid \theta_1) = \int g(\theta_1, \theta_1', \theta_2, x, x')\, dx\, dx'\, d\theta_2, \tag{3}$$

where $g(\theta_1, \theta_1', \theta_2, x, x') = \pi_1^N(\theta_1' \mid x) f_2(x \mid \theta_2) \pi_2^N(\theta_2 \mid x') f_1(x' \mid \theta_1)$, is recurrent, see Cano *et al.* (2006).

In this paper, we consider the random effects model

$$M : y_{ij} = \mu + a_i + e_{ij},\ i = 1, ..., k;\ j = 1, ..., n,$$

where the variables $e_{ij} \sim N(0, \sigma^2)$ and $a_i \sim N(0, \sigma_a^2),\ i = 1, ..., k;\ j = 1, ..., n$, are independent. We are interested in the selection problem between models with parameters:

$$M_1 : \theta_1 = (\mu_1, \sigma_1, 0) \text{ and } M_2 : \theta_2 = (\mu_2, \sigma_2, \sigma_a).$$

The default priors we use to derive the integral priors in equations (1) and (2) are the reference priors $\pi_1^N(\theta_1) = c_1/\sigma_1$ and $\pi_2^N(\theta_2) = c_2\sigma_2^{-2}(1 + (\sigma_a/\sigma_2)^2)^{-3/2}$. Note that $\pi_1^N(\theta_1)$ is the reference prior for model M_1 and $\pi_2^N(\theta_2)$ is the reference prior for model M_2 for the ordered group $\{\sigma_a, (\sigma, \mu)\}$ when $n = 1$. We use the prior $\pi_2^N(\theta_2)$ for the sake of simplicity to keep the paper within a methodological level. Under these assumptions the sample densities for the two models considered are

$f_1(\boldsymbol{y} \mid \theta_1) = \prod_{i=1}^k N_n(\boldsymbol{y}_i \mid \mu_1 \mathbf{1}_n, \sigma_1^2 \boldsymbol{I}_n)$, $f_2(\boldsymbol{y} \mid \theta_2) = \prod_{i=1}^k N_n(\boldsymbol{y}_i \mid \mu_2 \mathbf{1}_n, \sigma_2^2 \boldsymbol{I}_n + \sigma_a^2 \boldsymbol{J}_n)$, where $\boldsymbol{y}_i = (y_{i1}, ..., y_{in})'$, $\boldsymbol{y} = (\boldsymbol{y}_1, ..., \boldsymbol{y}_k)'$, $\mathbf{1}_n = (1, ..., 1)'$, $\boldsymbol{I}_n$ is the identity matrix of dimension n and $\boldsymbol{J}_n$ the square matrix of dimension n with all the entries equal to one. Straightforward computations, see Cano *et al.* (2007), yield

$$B_{21}^N(\boldsymbol{y}) = \frac{c_2}{c_1} \int_0^\infty (1+nu^2)^{-\frac{k-1}{2}} \left(1 - \frac{nu^2}{1+nu^2} \frac{S_2}{S}\right)^{-\frac{nk-1}{2}} (1+u^2)^{-\frac{3}{2}} du, \qquad (4)$$

where $S = \sum_{i=1}^k \sum_{j=1}^n (y_{ij} - y)^2 = S_1 + S_2$ and $S_2 = \sum_{i=1}^k n(y_i - y)^2$. Unfortunately, the above Bayes factor depends on the arbitrary ratio c_2/c_1 as we previously stated. To avoid this indeterminacy we will use integral and intrinsic priors instead of the original default priors.

The rest of the paper is organized as follows. Integral and intrinsic priors for the one way random effects model are derived in Section 2, where it is noticed that intrinsic priors are not unique. In Section 3 we address the problem of the uniqueness of integral priors through the consideration of its associated Markov chain. In Section 4 some illustrations are provided and finally, the conclusions are stated in Section 5.

2. INTEGRAL AND INTRINSIC PRIORS

To derive the integral and the intrinsic priors we first need an imaginary minimal training sample. According to the expression (4), when $n = 1$, $k = 2$ and $c_1 = c_2$ the Bayes factor is $B_{21}^N(\boldsymbol{y}(l)) = 1$ and therefore $\boldsymbol{y}(l) = (y_{11}, y_{21})'$ is a minimal training sample. Consequentely, $\{\pi_1^N, \pi_2^N\}$ are integral priors and they are well calibrated when $c_1 = c_2$, see subsection 3.4 in Cano *et al.* (2006).

Likewise, in Moreno *et al.* (1998) the following intrinsic priors $\{\pi_1^I, \pi_2^I\}$ are chosen: $\pi_1^I(\theta_1) = \pi_1^N(\theta_1)$ and $\pi_2^I(\theta_2) = \pi_2^N(\theta_2) E_{\boldsymbol{y}(l)\mid\theta_2}\{B_{12}^N(\boldsymbol{y}(l))\}$, from where we deduce that $\{\pi_1^N, \pi_2^N\}$ are intrinsic priors and they are well calibrated when $c_1 = c_2$ too. Whether or not the integral and the intrinsic priors are unique is a matter of, at least, theoretical interest.

In the case where model M_1 is nested in model M_2, the system of functional equations mentioned in Section 1 reduces to a single equation with two *incognita* and it is a well known result in the literature on intrinsic priors that they are not generally unique. Nevertheless, Moreno *et al.* (1998) provides a limiting procedure for choosing the above pair of sensible intrinsic priors that essentially consists in fixing $\pi_1^N(\theta_1)$ as the intrinsic prior for the simpler model obtaining $\pi_2^I(\theta_2)$ from the above mentioned functional equation as the intrinsic prior for model M_2. In Cano *et al.* (2006), it is shown that integral priors are unique provided the recurrence of an associated Markov Chain, which we now describe for our model.

3. THE MARKOV CHAIN ASSOCIATED WITH THE INTEGRAL PRIORS

The transition density (3) of the associated Markov chain can be described in a simple way. The posterior distribution $\pi_1^N(\theta_1 \mid \boldsymbol{y}(l))$ is obtained as

$$\pi_1^N(\sigma_1 \mid \boldsymbol{y}(l)) \propto \sigma_1^{-2} \exp\left(-\frac{(y_{11} - y_{12})^2}{2\sigma_1^2}\right)$$

and

$$\pi_1^N(\mu_1 \mid \sigma_1, \boldsymbol{y}(l)) = N(\mu_1 \mid \boldsymbol{y}(l), \sigma_1^2/2),$$

where $\overline{\boldsymbol{y}(l)}$ is the average of the components of $\boldsymbol{y}(l)$.

To obtain the posterior distribution $\pi_2^N(\theta_2 \mid \boldsymbol{y}(l))$ first note that

$$\pi_2^N(\mu_2 \mid \sigma_a, \sigma_2, \boldsymbol{y}(l)) = N(\mu_2 \mid \overline{\boldsymbol{y}(l)}, (\sigma_2^2 + \sigma_a^2)/2)$$

and

$$\pi_2^N(\sigma_2, \sigma_a \mid \boldsymbol{y}(l)) \propto \sigma_2^{-2}(1 + (\sigma_a/\sigma_2)^2)^{-3/2}(\sigma_a^2 + \sigma_2^2)^{-1/2} \exp\left(-\frac{(y_{11} - y_{12})^2}{4(\sigma_2^2 + \sigma_a^2)}\right).$$

Then the random variables u, z and ε defined by the equations

$$\frac{\sigma_a}{\sigma_2} = u, \;\; \sigma_2^2 + \sigma_a^2 = \frac{(y_{11} - y_{12})^2}{4} z \text{ and } \mu_2 = \overline{\boldsymbol{y}(l)} + \frac{1}{2}\varepsilon\sqrt{\sigma_2^2 + \sigma_a^2}$$

have densities $q_1(u) = (1+u^2)^{-3/2}$, $q_2(z) \propto z^{-3/2}e^{-1/z}$ and $\varepsilon \sim N(0,1)$. Therefore, the simulation from $\pi_2^N(\theta_2 \mid \boldsymbol{y}(l))$ can be done simulating u, z and ε and solving the above equations. In summary, combining the transitions $\theta_1 \to \boldsymbol{y}(l)' \to \theta_2 \to \boldsymbol{y}(l) \to \theta_1' = (\mu_1', \sigma_1')$ we obtain that

$$\mu_1' = \mu_1 + \sigma_1\alpha \text{ and } \sigma_1' = \sigma_1\beta,$$

where

$$\beta = \frac{\sqrt{w}}{4} \mid \varepsilon_3 - \varepsilon_4 \mid \sqrt{z} \mid \xi_1 - \xi_2 \mid,$$

$$\alpha = \xi + \frac{\varepsilon_2\sqrt{z}}{2\sqrt{2}} \mid \xi_1 - \xi_2 \mid + \frac{\varepsilon_3 + \varepsilon_4}{2}\sqrt{z} \mid \xi_1 - \xi_2 \mid /2 + \beta\varepsilon_1/\sqrt{2}$$

and $\xi_1, \xi_2, \varepsilon_1, \varepsilon_2, \varepsilon_3, \varepsilon_4 \sim N(0,1)$, $u \sim q_1(u) = (1+u^2)^{-3/2}$, $z \sim q_2(z) \propto z^{-3/2}e^{-1/z}$ and $w \sim p(w) \propto w^{-3/2}e^{-1/w}$.

Regarding the autonomous chain (σ_n), $(\log \sigma_n)$ is a recurrent random walk since $E(\log \beta) = 0$ and $E((\log \beta)^2) < +\infty$. Although, for the whole chain (μ_n, σ_n), we have not been able to establish recurrence so far because the second order moments for α and β are not finite. As mentioned previously, if the recurrence of the associated Markov chain is established, we would deduce the uniqueness of the integral priors, see Cano *et al.* (2007) and Cano *et al.* (2006).

4. ILLUSTRATIONS

In the context of variance components problems, two popular data sets can be found in Box and Tiao (1973), pages 246 and 247, respectively, where a parameter estimation approach is used. We compute the Bayes factors for the two sets of data using equation (4) with $c_1 = c_2$, that is, using integral (intrinsic) priors.

The first set of data in Box and Tiao (1973) concerns batch to batch variation in yields of dyestuff. The data arise from a balanced experiment where the total product yield was determined for five samples from each of six randomly chosen batches of raw material. The object of the study was to determine the relative importance of between batch variation versus variation due to sampling and analytic errors. To illustrate the difficulty of the analysis and the need for objective procedures, we have explored the sensitivity of the Bayes factor to the choice of hyperparameters of proper priors. More concretely, assume we choose the following conventional proper priors:

$$\begin{aligned} M_1: &\quad (\mu_1, \sigma_1^2) \sim \mathcal{N}(\bar{y}, \sigma_0^2) \otimes \mathcal{IG}(\alpha, \beta), \\ M_2: &\quad (\mu_2, \sigma_2^2, \sigma_a^2) \sim \mathcal{N}(\bar{y}, \sigma_0^2) \otimes \mathcal{IG}(\alpha, \beta) \otimes \mathcal{IG}(\alpha_a, \beta_a), \end{aligned}$$

where $\mathcal{IG}(\alpha, \beta)$ denotes the Inverse Gamma distribution with mean $\beta/(\alpha-1)$ and variance $\beta^2/((\alpha-1)^2(\alpha-2))$. Fixing $\alpha = 20$, $\alpha_a = 13$, $\beta_a = 20000$ and varying β, numerical values for the associated Bayes factors when $\sigma_0 \to +\infty$ can be found in Table 1. It turns out the Bayes factor is very sensitive to the choice of the hyperparameters and may lead to opposite conclusions.

Table 1: *Associated Bayes factors varying* β

$\beta \times 10^{-3}$	25	30	35	40	45	50
B_{21}	2.99	1.90	1.32	0.98	0.77	0.63

The Bayes factor when the integral priors described in this paper are chosen is found to be $B_{21} = 10.0179$, which means evidence in favor of M_2.

We also computed for this data set the Bayes factor B_{21}^{WG} when the priors suggested in Westfall and Gönen (1996) are chosen. An expression similar to (4) is obtained and the numerical value is found to be $B_{21}^{WG} = 11.9$.

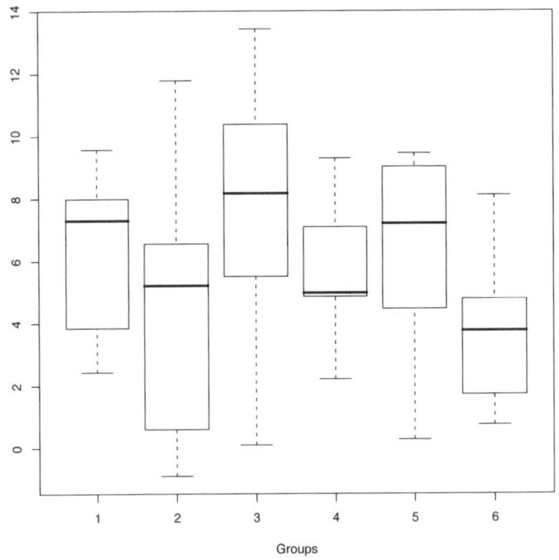

Figure 1: *Boxplot of the simulated data in Box and Tiao (1973), p 247: the intra-group variability overwhelms the known inter-group variability.*

The second set of data are simulated data with $\mu_2 = 5$, $\sigma_2 = 4$, $\sigma_a = 2$, $k = 6$ and $n = 5$. For these data the resulting Bayes factor is $B_{21} = 0.3671$, which means evidence in favor of model M_1 despite the fact that the data were generated from model M_2. Notice that, for these data, we find $B_{21}^{WG} = 0.17$. An explanation for the misbehavior of these Bayes factors is that the standard deviation of the error σ_2 is twice as big as the standard deviation σ_a of the random effects, which, together to the small number of groups, makes difficult the detection of the random effects. A graphical exploration of the data set, see Fig. 1, confirms that even if the groups centers are different the intragroup variation is too big to allow detection. We also have tested the normality (Shapiro-Wilk test) of these data and the resulting *p-value* is 0.735. All these results are in agreement with the results in Box and Tiao (1973), where it is found that the posterior density for σ_a has its mode at the origin and is monotonically decreasing.

5. CONCLUSIONS

The methodology of integral priors introduced in Cano *et al.* (2006) has been applied to solve the one way random effects model from a Bayesian perspective. The main conclusion is that the default priors $\{\pi_1^N, \pi_2^N\}$ are both integral priors and intrinsic priors and they are well *calibrated* when $c_1 = c_2$. This allows to use them to compute the Bayes factor which has been done with two sets of data.

The Markov chain associated when we use this methodology provides some insight on how integral and intrinsic (in this case) priors work out. Finally, we note that intrinsic priors are not unique while under some assumptions the integral priors are unique. However, the question of whether or not the integral priors for this problem are unique is still an open problem.

REFERENCES

Berger, J. O. and Pericchi, L. R. (1996). The intrinsic Bayes factor for model selection and prediction. *J. Amer. Statist. Assoc.* **91**, 109–122.

Box, G. E. P and Tiao, G. C. (1973). *Bayesian Inference in Statistical Analysis.* Reading, MA: Addison-Wesley.

Cano, J. A., Kessler, M. and Moreno, E. (2004). On intrinsic priors for nonnested models. *Test* **13**, 445–465.

Cano, J. A., Kessler, M. and Salmerón, D. (2007). Integral priors for the one way random effects model. *Bayesian Analysis* **2**, 59–68.

Cano, J. A., Salmerón, D. and Robert, C. P. (2007). Integral equation solutions as prior distributions for Bayesian model selection. *Test* (to appear).

Moreno, E., Bertolino, F. and Racugno, W. (1998). An intrinsic limiting procedure for model selection and hypotheses testing. *J. Amer. Statist. Assoc.* **93**, 1451–1460.

Pérez J. M. and Berger J. O. (2002). Expected-posterior prior distributions for model selection. *Biometrika* **89**, 491–511.

Westfall, P. H. and Gönen, M. (1996). Asymptotic properties of ANOVA Bayes factors. *Comm. Statist. Theory and Methods* **25**, 3101–3123.

BAYESIAN STATISTICS 8, pp. 583–588.
J. M. Bernardo, M. J. Bayarri, J. O. Berger, A. P. Dawid,
D. Heckerman, A. F. M. Smith and M. West (Eds.)

Dynamic Matrix-Variate Graphical Models*

CARLOS M. CARVALHO and MIKE WEST
ISDS, Duke University, Durham NC, USA
carlos@stat.duke.edu mw@stat.duke.edu

SUMMARY

This paper introduces a novel class of Bayesian models for multivariate time series analysis based on a synthesis of dynamic linear models and graphical models. The models are then applied in the context of financial time series for predictive portfolio analysis providing a significant improvement in performance of optimal investment decisions.

Keywords and Phrases: DYNAMIC LINEAR MODELS, GAUSSIAN GRAPHICAL MODELS, PORTFOLIO ANALYSIS.

1. INTRODUCTION

Bayesian dynamic linear models (DLMs) (West and Harrison, 1997) are used for analysis and prediction of time series of increasing dimension and complexity in many applied fields. The time-varying regression structure, or state-space structure, and the sequential nature of DLM analysis flexibly allows for the creation and routine use of interpretable models of increasingly realistic complexity. The inherent Bayesian framework naturally allows and encourages the integration of data, expert information and systematic interventions in model fitting and assessment, and thus in forecasting and decision making.

The current work responds to the increasing prevalence of high-dimensional multivariate time series and the consequent needs to scale and more highly structure analysis methods. Contexts of high-dimensional and rapidly sampled financial time series are central examples, though similar needs are emerging in many areas of science, social science and engineering. This paper introduces a broad new class of time series models to address this: the framework synthesises multi- and matrix-variate DLMs with graphical modelling to induce sparsity and structure in the covariance matrices of such models, including time-varying matrices in multivariate time series.

The presentation outlines the framework of matrix-variate DLMs and Gaussian graphical models for structured, parameter constrained covariance matrices based

* This is a synopsis. For the full version, see *Bayesian Analysis* **2** (2007), to appear.

on the use of the family of hyper-inverse Wishart distributions. We then discuss formal model specification and details of the resulting methodology for both constant and, of more practical relevance, time-varying structured covariance matrices in the new models. We summarise the theory that extends DLM sequential updating, forecasting and retrospective analysis to this new model class.

Our applied examples combine these flexible new Bayesian models with Bayesian decision analysis in financial portfolio prediction studies. We discuss theoretical and empirical findings in the context of an initial example using 11 exchange rate time series, and then a more extensive and practical study of 346 securities from the S&P Index. This latter application also develops and applies graphical model search and selection ideas, based on existing MCMC and stochastic search methods now translated to the DLM context, as well as illustrating the real practical utility, and benefits over existing models, of the new methodology.

2. BACKGROUND

2.1. *Matrix-Variate Dynamic Linear Models*

The class of Matrix Normal DLMs (Quintana and West, 1987) represents a general, fully-conjugate framework for multivariate time series analysis and dynamic regression with estimation of cross-sectional covariance structures.

The model for the $p \times 1$ vector $\mathbf{Y}_t$ is defined by

$$\mathbf{Y}_t' = \mathbf{F}_t'\mathbf{\Theta}_t + \boldsymbol{\nu}_t', \qquad \boldsymbol{\nu}_t \sim N(\mathbf{0}, V_t\mathbf{\Sigma}), \tag{1}$$

$$\mathbf{\Theta}_t = \mathbf{G}_t\mathbf{\Theta}_{t-1} + \mathbf{\Omega}_t \qquad \mathbf{\Omega}_t \sim N(\mathbf{0}, \mathbf{W}_t, \mathbf{\Sigma}), \tag{2}$$

where the evolution innovation matrix $\mathbf{\Omega}_t$ follows a *matrix-variate normal* with mean $\mathbf{0}$ (a $n \times p$ matrix), left covariance matrix $\mathbf{W}_t$ and right covariance matrix $\mathbf{\Sigma}$.

2.2. *Gaussian Graphical Models*

Graphical model structuring for multivariate models characterizes conditional independencies via graphs (Dawid and Lauritzen, 1993; Jones *et al*, 2005) and provides methodologically useful decompositions of the sample space into subsets of variables so that complex problems can be handled through the combination of simpler elements. In high-dimensional problems, graphical model structuring is a key approach to parameter dimension reduction and, hence, to scientific parsimony and statistical efficiency when appropriate graphical structures are identified.

In normal distributions, conditional independence restrictions are simply expressed through zeros in the off-diagonal elements of the precision matrix. A p-vector $\mathbf{x}$ with elements x_i has a zero-mean multivariate normal distribution with $p \times p$ variance matrix $\mathbf{\Sigma}$ and precision $\mathbf{\Omega} = \mathbf{\Sigma}^{-1}$ with elements ω_{ij}. Write $G = (V, E)$ for the undirected graph whose vertex set V corresponds to the set of p random variables in $\mathbf{x}$, and whose edge set E contains elements (i,j) for only those pairs of vertices $i, j \in V$ for which $\omega_{ij} \neq 0$. The canonical parameter $\mathbf{\Omega}$ belongs to $M(G)$, the set of all positive-definite symmetric matrices with elements equal to zero for all $(i,j) \notin E$.

The fully conjugate Bayesian analysis of decomposable Gaussian graphical models (David and Lauritzen, 1993) is based on the family of *hyper-inverse Wishart* (HIW) distributions for structured variance matrices. If $\mathbf{\Omega} = \mathbf{\Sigma}^{-1} \in M(G)$, the hyper-inverse Wishart

$$\mathbf{\Sigma} \sim HIW_G(b, \mathrm{D}) \tag{3}$$

has a degree-of-freedom parameter b and location matrix $\mathrm{D} \in M(G)$ implying that for each clique $P \in \mathcal{P}$, $\mathbf{\Sigma}_P \sim IW(b, \mathrm{D}_P)$ where D_P is the positive-definite symmetric diagonal block of D corresponding to $\mathbf{\Sigma}_P$. The full HIW is conjugate to the likelihood from a Gaussian sample with variance $\mathbf{\Sigma}$ on G.

3. SPARSITY IN DLMS: GENERALIZATION TO HIW

Graphical structuring can be incorporated in matrix normal DLMs to provide models for $\mathbf{\Sigma}$ that allow structure, induce parsimony and lead to statistical and computational efficiencies as a result. For a given decomposable graph, take the hyper-inverse Wishart as a conjugate prior for $\mathbf{\Sigma}$; it turns out that the closed-form, sequential updating theory of DLMs can be generalized (Theorem 1 of Carvalho and West, 2007) to this richer model class.

Consider the matrix normal DLM above and suppose $\mathbf{\Omega} = \mathbf{\Sigma}^{-1}$ is constrained by a graph G. If D_t is the data and information set conditioned upon at any time t, assume the NHIW initial prior of the form

$$(\mathbf{\Theta}_0, \mathbf{\Sigma} \mid D_0) \sim NHIW_G(\mathbf{m}_0, \mathbf{C}_0, b_0, \mathbf{S}_0). \tag{4}$$

In components,

$$(\mathbf{\Theta}_0 \mid \mathbf{\Sigma}, D_0) \sim N(\mathbf{m}_0, \mathbf{C}_0, \mathbf{\Sigma}) \qquad \text{and} \qquad (\mathbf{\Sigma} \mid D_0) \sim HIW_G(b_0, \mathbf{S}_0), \tag{5}$$

which incorporates the conditional independence relationships from G into the prior. This is in fact the form of the conjugate prior for sequential updating at all times t, (detailed in Carvalho and West, 2007), generating a sequence of NHIW priors and posteriors for $(\mathbf{\Theta}, \mathbf{\Sigma})$.

4. TIME-VARYING $\mathbf{\Sigma}_T$

Importantly, the above development extends to the practically critical context of time-varying $\mathbf{\Sigma} \rightarrow \mathbf{\Sigma}$ to induce a novel class of *graphical multivariate volatility models*. Models of $\mathbf{\Sigma}_t$ varying stochastically over time are key in areas such as finance, where univariate and multivariate volatility models have been center-stage in both research and front-line financial applications for over two decades, as well as areas in engineering and the natural sciences.

Based on a specified discount factor δ, $(0 < \delta \leq 1)$, beginning at $t-1$ with current posterior

$$(\mathbf{\Sigma}_{t-1} \mid D_{t-1}) \sim HIW_G(b_{t-1}, \mathbf{S}_{t-1}),$$

the Beta-Bartlett stochastic evolution of $\mathbf{\Sigma}_{t-1}$ to $\mathbf{\Sigma}_t$ implies the following prior at time t

$$(\mathbf{\Sigma}_t \mid D_{t-1}) \sim HIW_G(\delta b_{t-1}, \delta \mathbf{S}_{t-1}). \tag{6}$$

The time-evolution maintains the inverse-Wishart form for the prior of $\mathbf{\Sigma}_t$, while increasing the spread of the HIW distribution by reducing the degrees-of-freedom and maintaining the location at $\mathbf{S}_{t-1}/b_{t-1}$. Observing $\mathbf{Y}_t$ generates the realized forecast error $\mathbf{e}_t$ and the time t prior is updated as before, with the discount factor modification implied by the modified time t prior; that is,

$$(\mathbf{\Sigma}_t \mid D_t) \sim HIW_G(b_t, \mathbf{S}_t)$$

with $b_t = \delta b_{t-1} + 1$ and $\mathbf{S}_t = \delta \mathbf{S}_{t-1} + \mathbf{e}_t \mathbf{e}_t'$.

5. LARGE-SCALE DYNAMIC PORTFOLIO ALLOCATION

Bayesian forecasting models and Bayesian decision analysis in asset allocation problems has been routine for a number of years, from the seminal paper of Quintana (1992) to more recent work (e.g., Aguilar and West, 2000, Quintana *et al.*, 2003) with increasingly large problems. Forecast future returns are the key components of mean-variance portfolio optimization methods that allow for parameter change and uncertainty to be taken into account in sequential investment decisions.

At time t, given the first two moments $(\mathbf{f}_t, \mathbf{Q}_t)$ of the predictive distribution of a vector of next-period returns and a fixed scalar return target m, the investor decision problem reduces to choosing the vector of portfolio weights $\mathbf{w}_t$ to minimize the one-step ahead portfolio variance $\mathbf{w}_t'\mathbf{Q}_t\mathbf{w}_t$ subject to constraints $\mathbf{w}_t'\mathbf{f}_t = m$ and $\mathbf{w}_t'\mathbf{1} = 1$. The optimal portfolio weights can be expressed in terms of the precision matrix $\mathbf{K}_t = \mathbf{Q}_t^{-1}$, via

$$w_{ti}^{(m)} = \lambda \frac{f_{ti} - \sum_{j \neq i} (k_{tij}/k_{tii}) f_{tj}}{k_{tii}^{-1}} \tag{7}$$

(Stevens, 1998), where λ is a Lagrange multiplier. In normal models, the weight assigned to asset i depends on the ratio of the intercept of its regression on all other assets relative to the conditional variance of the regression. Hence the investment in asset i depends on the ratio of the expected return that cannot be explained by the linear combination of assets to the *unhedgeable* risk.

In higher-dimensional portfolios, the optimal weights can be very volatile over time due to the uncertainty in the estimation of covariance matrices. Structured variance models should help: the above equations suggest that conditional independence assumptions can directly influence the uncertainty about $\mathbf{w}_t$. If, in fact, the unhedgeable risk can be obtained by a regression involving a smaller number of regressors (i.e. having some of the k_{tij}'s equal to zero) this has to be taken into account; failing to do so implies that unnecessary parameters are being estimated and nothing but noise is added to the problem.

Two examples bear this out: 'sparse' (with graphical modelling constraints) models lead to portfolios that are both less risky and more profitable than under the standard construction with 'full' (unconstrained) variance matrices.

5.1. *Example: International Exchange Rates*

This first application is a dynamic version of the example in Carvalho, Massam and West (2007) where portfolios for $p = 11$ international currency exchange rates relative to the US dollar were compared. The study here uses the graph displayed in Fig. 1.

For each model at each time point, mean-variance and minimum-variance portfolios were computed based on the one-step ahead forecasts. In comparing the impact on portfolio predictions and decisions of the proposed structured model *vis-a-vis* the unconstrained DLM, the overall conclusion is that the DLM graphical model uniformly outperforms the unconstrained, full variance matrix DLM across the full time period of portfolio decisions. The uniform dominance is reflected in higher realized cumulative returns, lower risk portfolios in terms of one-step ahead predictive variances and lower volatility of the optimal portfolio weights as they are sequentially revised, consistent with the idea of more stable portfolios.

This example demonstrates the relevance of appropriate model structuring: the graphical model DLM generates more accurate predictions and optimal portfolio

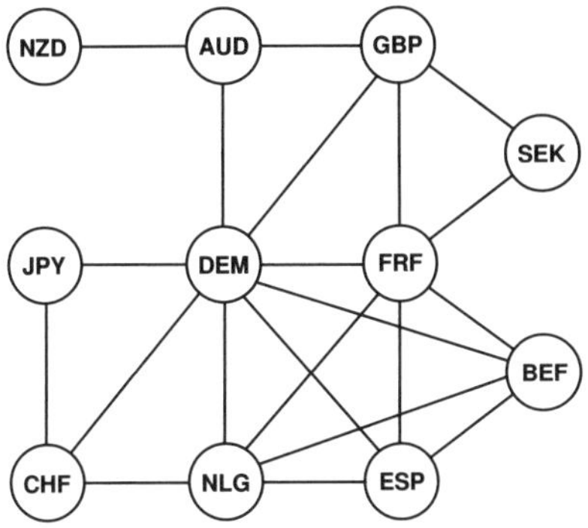

Figure 1: *Graph determining the conditional independence structure in the exchange rate/portfolio investment example.*

decisions, with lower risk in terms of both the nominal predicted portfolio risk and in terms of realized outcomes. In addition, the more stable portfolios add to the practical benefits since they would imply, in a live context, reduced costs in terms of transaction fees for moving money between currencies period-to-period.

5.2. *Example: Portfolio Allocation in the S&P 500*

A higher-dimensional application involves $p = 346$ securities forming part of the S&P500 stock index. In this higher-dimensional setting, graphical models to induce structure are particularly key.

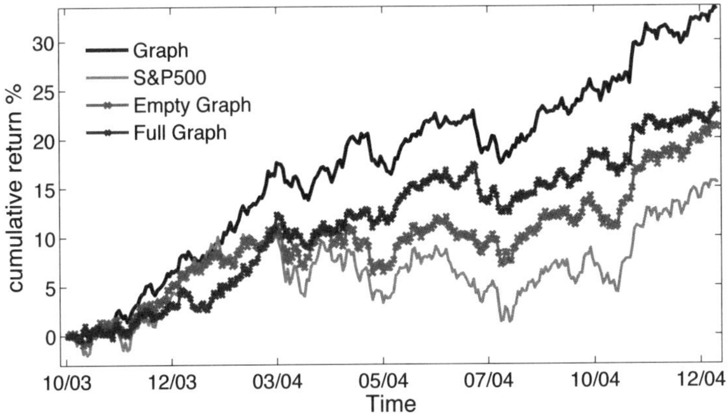

Figure 2: *S&P 500 portfolios: comparison of cumulative returns under different models* ($\delta = 0.98$).

The example also addresses graphical model structure selection, evaluating models using Metropolis stochastic search (Jones *et al.*, 2005) to explore the full space of graphical DLMs using only the first 1 200 observations in the data set. On the remaining data, we sequentially updated portfolios use a few graphical DLMs selected based on the posterior at $t = 1200$. The key conclusions (see Figure 2 here, and details in Carvalho and West, 2007) are that (a) selection of graphs G according to high posterior probability on the training data of 1 200 observations leads to graphical model DLMs that generate substantial improvements in realized portfolio applications; and that (b) higher realized returns are coupled with lower risk and lower volatility of time trajectories of portfolio weights.

6. SUMMARY COMMENTS

The marriage of DLMs with graphical models defines a new, rich class of matrix DLMs that allow for the incorporation of conditional independence structure in the stochastically varying, cross-sectional structure of a set of time series. Our examples focus on Bayesian decision analysis for sequential portfolio allocation, where the utility and benefits of the new models are sharply illuminated. The value of data-consistent structuring and constraints on variance matrices across series is evident; the implied parsimony in parametrization, statistical efficiency in estimation and reduced uncertainty translates into more accurate predictions and decisions. Our second example explores aspects of graphical model uncertainty and model selection, evaluating posterior distributions over graphical structures G as well as time-varying state and variance parameters on a given graph. The models open up a rich new area for application of Bayesian modelling, as well as research directions in the class of dynamic graphical models.

REFERENCES

Aguilar, O. and West, M. (2000). Bayesian dynamic factor models and variance matrix discounting for portfolio allocation. *J. Business Econ. Statist.* **18**, 338–357.

Carvalho, C.M. and West, M. (2007). Dynamic matrix-variate graphical models. *Bayesian Analysis* **2**, 69–98.

Carvalho, C.M., Massam, H. and West, M. (2007) Simulation of hyper inverse-Wishart distributions in graphical models. *Biometrika* (to appear).

Dawid, A.P. and Lauritzen, S.L. (1993). Hyper-Markov laws in the statistical analysis of decomposable graphical models. *Ann. Statist.* **3**, 1272–1317.

Jones, B., Carvalho, C., Dobra, A., Hans, C., Carter, C. and West, M. (2005). Experiments in stochastic computation for high-dimensional graphical models. *Statist. Science* **20**, 388–400.

Quintana, J.M. and West, M. (1987). Multivariate time series analysis: New techniques applied to international exchange rate data. *The Statistician* **36**, 275–281.

Quintana, J.M. (1992). Optimal portfolios of forward currency contracts. *Bayesian Statistics 4* (J. M. Bernardo, J. O. Berger, A. P. Dawid and A. F. M. Smith, eds.) Oxford: University Press, 753–762.

Quintana, J.M., Lourdes, V., Aguilar, O. and Liu, J. (2003). Global gambling (with discussion). *Bayesian Statistics 7* (J. M. Bernardo, M. J. Bayarri, J. O. Berger, A. P. Dawid, D. Heckerman, A. F. M. Smith and M. West, eds.) Oxford: University Press, 349-368.

Stevens, G. (1998). On the inverse of the covariance matrix in portfolio analysis. *J. Finance* **53**, 1821–1827.

West, M. and Harrison, J. (1997). *Bayesian Forecasting and Dynamic Models* (2nd Edn.), New York: Springer-Verlag.

BAYESIAN STATISTICS 8, pp. 589–594.
J. M. Bernardo, M. J. Bayarri, J. O. Berger, A. P. Dawid,
D. Heckerman, A. F. M. Smith and M. West (Eds.)

A Gamma Model for DNA Mixture Analyses*

ROBERT G. COWELL
Cass Business School, London, UK
rgc@city.ac.uk

STEFFEN L. LAURITZEN
University of Oxford, UK
steffen@stats.ox.ac.uk

JULIA MORTERA
Università Roma Tre, Italy
mortera@uniroma3.it

SUMMARY

We present a new methodology for analysing forensic identification problems involving DNA mixture traces where several individuals may have contributed to the trace. The model used for identification and separation of DNA mixtures is based on a gamma distribution for peak area values. In this paper we illustrate the gamma model and apply it on a real example from forensic casework.

Keywords and Phrases: DNA MIXTURE, FORENSIC IDENTIFICATION, MIXTURE SEPARATION, PEAK AREA.

1. INTRODUCTION

We present a methodology for analysing the important problem of identification and separation or deconvolution of *mixed DNA traces*, where several individuals may have contributed to a DNA sample. Here we illustrate its use for analysing DNA mixtures arising in forensic casework. However, the problem of identifying components of mixtures has potential applications outside this area.

A *mixed DNA profile* is typically obtained from an unidentified biological stain or other trace found at a scene of crime. This occurs in rape cases, in burglaries where objects might have been handled by more than one individual, and also in a scuffle or brawl.

The analysis of DNA mixtures, for calculating the likelihoods of all hypotheses involving a specified set of known and unknown contributors to the mixture, based solely on the qualitative information on which alleles were present in the mixture, is illustrated in Evett *et al.* (1991) and Weir *et al.* (1997). In Mortera *et al.* (2003) Bayesian networks have also been constructed to address some of the challenging

* This is a synopsis. For the full version, see *Bayesian Analysis* **2** (2007), to appear.

problems that arise in the interpretation of mixed trace evidence based on qualitative allele information.

However, the analysis becomes relatively complex when we want to use quantitative peak area values. These contain important additional information about the composition of the mixture–see Fig. 1, and Butler (2005) for further background information about STR markers and DNA profiles. To handle such cases more sophisticated probabilistic modelling is required. This paper is concerned with the quantitative modelling of the peak areas to analyse DNA mixed profiles in order, for example, to infer the genotypes of individuals contributing to the mixture.

The outline of the paper is as follows. In Section 2 we summarize our model for peak areas. In Section 3 we illustrate the model on a real forensic casework example and show how well it predicts the genotypes of the individuals who have contributed to the mixture. Finally, in Section 4 we discuss further work required to bring the methodology to a point where it could be applied to the routine analysis of casework.

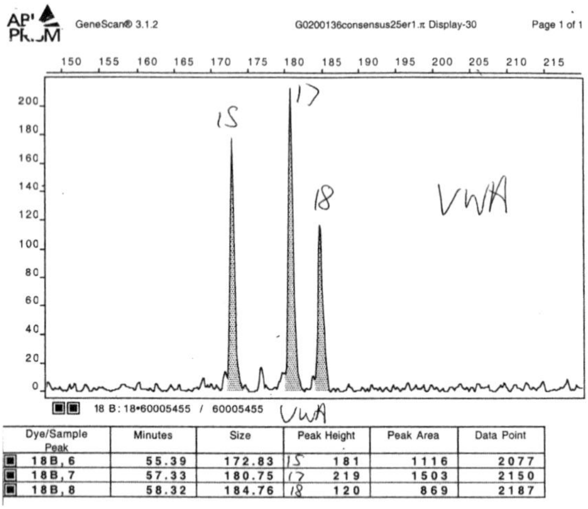

Figure 1: *An electropherogram (EPG) of marker VWA from a mixture. Peaks represent alleles at 15, 17 and 18 and the areas and height of the peaks express the quantities of each. Since the peak around allelic position 17 is the highest this indicates that the 17 allele is likely to be a homozygote or a shared allele between two heterozygotes. This image is supplied courtesy of LGC Limited, 2004.*

2. THE MODEL

2.1. *Notation and Setup*

We consider I potential contributors to a DNA mixture, assume that M markers are to be used in the analysis of the mixture, and that marker m has A_m allelic types, $m = 1, \ldots, M$. Let θ_i denote the proportion of DNA from individual i.

For a specific marker m, the *peak weight* W_a is *e.g.*, *peak area* at allele a multiplied by *allele number*. This is a simple, but adequate correction for preferential amplification of alleles. Our model is idealized in that it ignores complicating artifacts such as stutter, drop-out alleles and so on, and makes the following further assumptions:

- The (pre-amplification) proportion of DNA θ is constant across markers.
- W_a is approximately proportional to the amount of DNA of type a.
- W_+ is the sum of the allele a weights of all contributors.

Our key modelling assumptions is that for each W_{ia}, the contribution of individual i to peak weight at allele a, is assumed to have a *Gamma distribution*:

$$W_{ia} \sim \Gamma(\rho\gamma_i n_{ia}, \eta)$$

where

- $\gamma_i = \gamma\theta_i$ is the amount of DNA from individual i in mixture;
- n_{ia} is the number of alleles of type a carried by individual i;
- η determines scale and ρ is the amplification factor. Both may be marker dependent.

By scaling the weight of each allele by the total for the marker we obtain *relative weights* $R_a = W_{+a}/W_{++}$, where it follows that

$$R_a \sim Dir(\rho B_a),$$

where $B_a = \sum_i \gamma_i n_{ia}$ is the weighted allele number, and $B_+ = 2\gamma$ is twice the total amount of DNA γ and is marker independent. Note that

$$\begin{aligned} \mathbf{E}(R_a) &= \mu_a = B_a/B_+ = \sum_i \theta_i n_{ia}/2, \\ \mathbf{V}(R_a) &= \mu_a(1-\mu_a)/(2\rho\gamma+1) = \sigma^2\mu_a(1-\mu_a). \end{aligned}$$

where we define $\sigma^2 = 1/(2\rho\gamma + 1)$. In any given mixture case, γ is unknown, so sensitivity analysis is desirable.

2.2. *The Network Model*

The diagram in Fig. 2 indicates the model structure for each marker in the case of two contributors where four distinct alleles are present in the mixture. The central layer of the diagram represents genotypes of the two contributors each of which could originate from a specific individual, here named *suspect* and *victim* respectively, or from an unknown contributor. The central layer also includes the joint genotype of the two contributors. The genotypes of suspect, victim and unknown contributors are represented in the top layer. The nodes in the bottom layer represent expected values of each of four possible peak areas as determined by the number of alleles in the genotypes of the contributors and the fraction θ contributed by the the first of these. Similar diagrams represent other markers and these all form elements in an object-oriented Bayesian network used to perform an exact calculation of relevant quantities using efficient local computation algorithms (Lauritzen and Spiegelhalter,

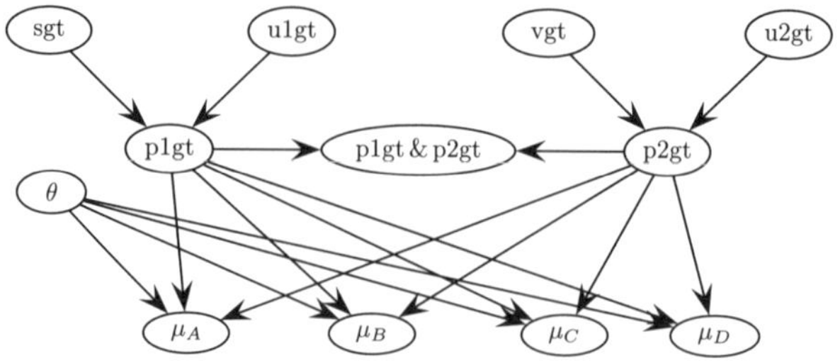

Figure 2: *Network model for a specific marker describing a mixture with two contributors.*

1998), thus avoiding the use MCMC methods. For details on this implementation see Cowell *et al.* (2006b).

Conditional on $\mu^m = (\mu_a^m, a = 1, \ldots, A_m)$, the total peak weight W_{++}^m and relative peak weights are independent and independent of everything else in the network model. Thus information about mixture composition from the peak areas about μ^m is contained in the relative weights R_a^m and the corresponding likelihood factorizes as

$$\begin{aligned} L(\mu \mid W) &= f(W \mid \mu) = L(\mu \mid R, W_{++}) \\ &\propto L(\mu \mid R) = \prod_a \frac{r_a{}^{2\rho\gamma\mu_a - 1}}{\Gamma(2\rho\gamma\mu_a)} = \prod_a \frac{r_a{}^{\mu_a(\frac{1}{\sigma^2} - 1)}}{\Gamma\left(\mu_a(\frac{1}{\sigma^2} - 1)\right)}, \end{aligned}$$

where we have suppressed the dependence on the marker m. This factorization now enables efficient computation of quantities of interest using a probabilistic expert system (Cowell *et al.*, 2006b).

3. APPLICATION OF OUR MODEL TO FORENSIC CASEWORK

An analysis of a mixed trace can have different purposes, several of which can be relevant simultaneously, making a unified approach particularly suitable. In an *evidential calculation*, a *suspect* with known genotype is held and we want to determine the likelihood ratio for the hypothesis that the suspect has contributed to the mixture vs. the hypothesis that the contributor is a randomly chosen individual. We may distinguish two cases: the other contributor could be (i) a *victim* with a known genotype; or (ii) a *contaminator* with an unknown genotype possibly without a direct relation to the crime. This could be a laboratory contamination or any other source of contamination from an unknown contributor.

Another use of our network is the *separation of profiles*, i.e. identifying the genotype of each of the possibly unknown contributors to the mixture, the evidential calculation playing a secondary role. We illustrate this type of calculation here.

Deconvolution of mixtures or separating a mixed DNA profile into its components has been studied by Perlin and Szabady (2001); Wang *et al.* (2002); Bill *et al.* (2005), among others. A mixed DNA profile has been collected and the genotypes of one or more unknown individuals who have contributed to the mixture is desired, for example with the purpose of searching for a potential perpetrator among an existing database of DNA profiles.

For a two-person mixture, the easiest case to consider is clearly that of separation of a single unknown profile, i.e. when the genotype of one of the contributors to the mixture is known. The case when both contributors are unknown is more difficult. In the latter situation this is only possible to a reasonable accuracy when the contributions to the DNA mixture has taken place in quite different proportions.

We concentrate on the problem of separating a mixture into two components, using peak area and allele repeat number information but no information regarding the two contributors to the mixture. In Table 1 we show the results from separating a two person mixture reported by Clayton *et al.* (1998). Our system predicts the profiles of both contributors correctly on every marker.

Table 1: *Predicted genotypes of both contributors for Clayton data, with victim (p1) and male suspect (p2) correct on every marker. The number in brackets is the product of individual marker probabilities. We have used* $\sigma^2 = 0.01$ *on every marker.*

Marker	Genotype p1	Genotype p2	Probability
Amelogenin	X X	X Y	0.9979
D8	13 13	14 15	0.9661
D18	16 18	14 15	0.9996
D21	30 32.2	28 36	0.9844
FGA	23 23	22 23	0.9994
THO	5 7	7 7	0.9019
vWA	17 19	15 16	0.9995
joint			0.8797 (0.8785)

4. DISCUSSION

We have presented a probabilistic expert system based on a gamma model for the peak areas can be used for analysing DNA mixtures using peak area information, yielding a coherent way of predicting genotypes of unknown contributors and assessing evidence for particular individuals having contributed to the mixture. The gamma model appears to perform well, and is suitable for elaboration to include other artifacts such as silent alleles, dropout and stutter phenomena. Further details of our model and its application to other examples may be found in the extended version of this paper (Cowell *et al.*, 2007).

Methods for diagnostic checking and validation of the model should be developed based upon comparing observed weights to those predicted when genotypes are assumed correct. Such methods could also be useful for calibrating the variance parameter σ^2. Cowell *et al.* (2006a) uses a model which also includes measurement error. This is particularly simple to incorporate in the conditional Gaussian model, but appears to have only a minor influence on the results.

In a general review of the analysis of DNA evidence, Foreman *et al.* (2003) include several applications of PES and emphasize their potential by predicting that this methodology 'will offer solutions to DNA mixtures and many more complex problems in the future'. We hope that the model presented in this paper shall indeed be part of the solutions anticipated by Foreman *et al.* (2003).

ACKNOWLEDGEMENT

This research was supported by a PRIN05 grant from MIUR. We thank Caryn Saunders for supplying the EPG image used in Fig. 1.

REFERENCES

Bill, M., Gill, P., Curran, J., Clayton, T., Pinchin, R., Healy, M., and Buckleton, J. (2005). PENDULUM – a guideline-based approach to the interpretation of STR mixtures. *Forensic Science International* **148**, 181–189.

Butler, J. M. (2005). *Forensic DNA Typing.* (2nd edn) Amsterdam: Elsevier.

Clayton, T. M., Whitaker, J. P., Sparkes, R. and Gill, P. (1998). Analysis and interpretation of mixed forensic stains using DNA STR profiling. *Forensic Science International* **91**, 55–70.

Cowell, R. G., Lauritzen, S. L., and Mortera, J. (2007). Identification and separation of DNA mixtures using peak area information. *Forensic Science International* **166**, 28-34.

Cowell, R. G., Lauritzen, S. L. and Mortera, J. (2006b). Object-oriented Bayesian networks for DNA mixture analyses. Manuscript available at `http://www.staff.city.ac.uk/~rgc`

Cowell, R. G., Lauritzen, S. L., and Mortera, J. (2007). A gamma model for DNA mixture analyses. *Bayesian Analysis* **2** (to appear).

Evett, I. W., Buffery, C., Wilcott, G., and Stoney, D. (1991). A guide to interpreting single locus profiles of DNA mixtures in forensic cases. *J. Forensic Science Soc.* **31**, 41–47.

Foreman, L., Champod, C., Evett, I., Lambert, J. and Pope, S. (2003). Interpreting DNA Evidence: A Review. *Internat. Statist. Rev.* **71**, 473–495.

Lauritzen, S. L. and Spiegelhalter, D. J. (1998). Local computations with probabilities on graphical structures and their application to expert systems. *J. Roy. Statist. Soc. B* **50**, 157–224, (with discussion).

Mortera, J., Dawid, A. P., and Lauritzen, S. L. (2003). Probabilistic expert systems for DNA mixture profiling. *Theoretical Population Biology* **63**, 191–205.

Perlin, M. and Szabady, B. (2001). Linear mixture analysis: a mathematical approach to resolving mixed DNA samples. *J. Forensic Sciences* **46**, 1372–1378.

Wang, T., Xue, N., and Wickenheiser, R. (2002). Least square deconvolution (LSD): A new way of resolving STR/DNA mixture samples. *13th International Symposium on Human Identification*, Phoenix, AZ.

Weir, B. S., Triggs, C. M., Starling, L., Stowell, L. I., Walsh, K. A. J. and Buckleton, J. S. (1997). Interpreting DNA mixtures. *J. Forensic Sciences* **42**, 213–222.

BAYESIAN STATISTICS 8, pp. 595–600.
J. M. Bernardo, M. J. Bayarri, J. O. Berger, A. P. Dawid,
D. Heckerman, A. F. M. Smith and M. West (Eds.)

Geographically Assisted Elicitation of Expert Opinion for Regression Models*

ROBERT J. DENHAM
Queensland Department of Natural Resources and Water, Australia
robert.denham@nrm.qld.gov.au

KERRIE MENGERSEN
Queensland University of Technology, Australia
k.mengersen@qut.edu.au

SUMMARY

One of the perceived strengths of Bayesian modelling is the ability to include prior information. Although objective or noninformative priors might be preferred in some situations, in many other applications the Bayesian framework offers a real opportunity to formally combine data with information available from experts. The question addressed in this paper is how to elicit this information in a form suitable for prior modelling. Particular attention is paid to geographic data for which maps might be used to assist in the elicitation. A case study is described in which the method was used to elicit information for the prediction of the distribution of the brush tailed rock wallaby *(Petrogale penicillata)*. This paper provides a synopsis of an approach described by Denham and Mengersen (2007).

1. INTRODUCTION

The value of expert opinion, particularly when data are scarce, has been recognised by the ecology community, although usually not formally or explicitly incorporated into data analyses. Pearce, Cherry, Drielsms, Ferrier and Whish (2001) used expert opinion to modify a generalised linear model (GLM) during three stages of their analysis: the pre-modelling stage involved creation of a new explanatory variable based on fine-scale vegetation and growth-stage information; the model-fitting stage involved expert-based choice of a subset of explanatory variables, and the post-modelling stage included expert-based modification of maps produced from the GLM. The use of expert opinion in Bayesian logistic regression appears to have been first documented by Bedrick, Christensen and Johnson (1996,1997), who describe a conditional means prior in which an expert provides their estimate of the probability of a success at k carefully chosen points in the explanatory variable space.

---

* This is a synopsis. For the full version, see *Bayesian Analysis* **2** (2007), to appear.
This research was supported by Australian Research Council Discovery and Linkage grants.

One challenge, then, is how to elicit information from an expert in a manner that allows appropriate construction of prior distributions. We agree with the statement of Kadane and Wolfson (1998) that *the goal of elicitation ... is to make it as easy as possible for subject-matter experts to tell us what they believe, in probabilistic terms, while reducing how much they need to know about probability theory to do so.* With this in mind, careful consideration must be given to the tools used to elicit expert opinion.

The approach proposed here capitalises on the spatial nature of a problem such as species distribution modelling, as well as the geographic inclination of the experts. It aims to support different types of experts, from those who are able to describe the change in response of the species as the explanatory variables themselves change, to those who have a good knowledge of where the species might occur but not why it occurs there. The prototype software includes a number of interactive tools that allow the expert to provide opinion in a spatial context, and provide information back to the expert.

We illustrate this approach by using expert opinion to help model the distribution of the brush-tailed rock wallaby (*Petrogale penicillata*). Although the brush-tailed rock wallaby has a coastal to sub-coastal distribution, its range has declined substantially in recent years. Distribution mapping for this species will assist in learning about its ecology and habitat requirements, and thus contribute to its conservation and management (Carter and Goldizen 2003). This paper is a synopsis of work described more fully in Denham and Mengersen (2007).

2. METHODS

ArcGIS is a widely used geographic information system (GIS) among the ecology and natural resource science communities. We customised this environment by scripting with Visual Basic for Applications (VBA) and communicating with other applications written for Windows. We wrote additional functions to provide simple buttons and menus on the ArcGIS interface for the expert to access the elicitation functions We linked this to the software package R (Ihaka and Gentleman 1996) using the R(D)Com package to provide sophisticated statistical capability. We used the third party graphing software TeeChart to allow interactive graphical presentation of the derived statistical information.

Two experts were invited to provide assistance for the case study. They first suggested appropriate variables which comprised *aspect, geology,* terrain represented by *slope* and moisture, represented by the annual mean moisture index, *mind.* Whereas the other variables were continuous, *geology* was classified into four groups based on dominant rock type. The variables *aspect* and *mind* were considered to have non-linear (quadratic) effects on response. The variable *mind* illustrates a variable for which it is more difficult to elicit information.

Letting y_i represent observed presence or absence of the species at the ith site and X the matrix of explanatory variables, the model in which we are interested is

$$\begin{aligned} y_i &\sim \text{Bernoulli}(p_i) \\ \text{logit}(p_i) &\sim N((X\beta)_i, \sigma^2) \\ \beta &\sim \text{MVN}(b, \Sigma). \end{aligned} \tag{1}$$

A dataset of 38 sites was available for this study. We wish to elicit from the expert their prior for β, that is a multivariate normal distribution with mean vector b and variance-covariance matrix Σ.

Each expert was questioned separately for this activity. Calibration and quantification were assisted by a short discussion on the concepts of medians and quantiles and a reference or double lottery system. A number of additional layers were available in the GIS to help the expert obtain contextual information, locate themselves and visualise the terrain.

The expert was asked to choose a set of design points that covered the range of probabilities of species occurrence. Experience has shown us that this is preferable when working with rare or uncommon species, since the majority of sites will have a low probability of presence and even careful preselection can provide a poor range of probabilities at this stage of the analysis.

The software provides the user with a map of the area of interest. The expert chooses a design point or a *virtual field site* by clicking on a point in the map, upon which the values of the explanatory variables at that site are displayed and an interactive dialog pops up (Fig. 1). The dialog includes a plot of a beta distribution with three adjustable points located at the median and at the 0.05 and 0.95 quantiles. The expert was asked to provide their best estimate of the probability of presence at that site by clicking on the median point and sliding it left or right until its position matches their belief. They were then questioned about the possible range of values of the probability of presence. To help estimate the quantiles, the expert was asked to consider placing 100 field sites at locations like the one chosen, in the same vicinity. The median was the proportion of sites in which the expert believed that the species would be found, given an exhaustive search each time. The upper and lower 0.05 quantiles were chosen by asking the higher and lower number which the expert thought was believable, but surprising, as based on the lottery exercise.

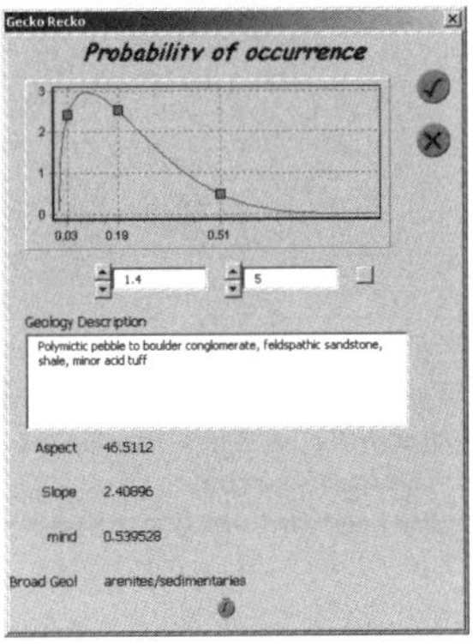

Figure 1: *Beta distribution elicitation dialog. This dialog allows the expert to provide their estimate of the probability of occurrence of the species at a given X.*

The software also provides the expert with the option to adjust the parameters for the beta distribution directly, which automatically updates the plot of the beta distribution. This was found to be convenient for either fine tuning their chosen distribution, or for easily choosing specific distributions such as the uniform if they had no prior opinion for the point (which never happened) or for the extremes (almost no chance of occurrence at that point, or almost definitely present).

This activity provides an expert prior for the probability of presence of the species, $p(p_i) = \text{Beta}(\alpha_i, \beta_i)$ at the ith site. Based on a (specified) minimum number of such sites, the software creates univariate response curves that depict the change in probability of presence with changes in each of the explanatory variables (Fig. 2). This was achieved as follows. Using the mean for the estimate of the probability of presence, $\hat{p}_i = \alpha_i/(\alpha_i + \beta_i)$, we determine the pseudo-observations and binomial sample size such that the binomial variance is approximately equivalent to the variance of $p(p_i)$, so that $\hat{n}_i = \alpha_i + \beta_i + 1$ and $\hat{y}_i = \hat{n}_i\hat{p}_i$. We then fit a logistic regression to these pseudo-observations based on the explanatory variables of interest at each site which are derived from ArcGIS.

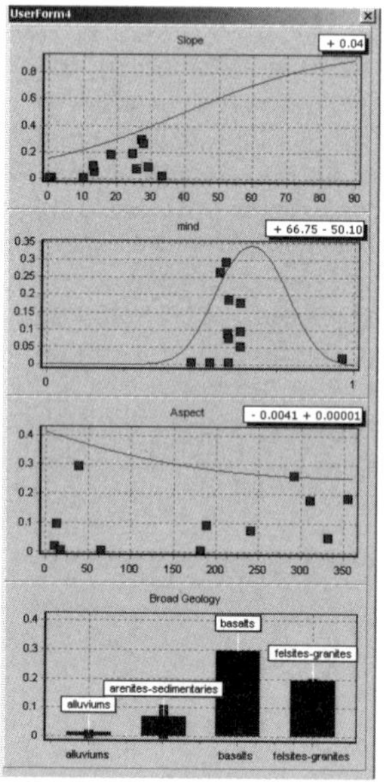

Figure 2: *Interactive response curves. The expert is able to review the relationship between each variable and the probability of occurrence. For each explanatory variable X_i, all other variables $X_{1,\cdots i-1,i+1,\cdots k}$ are fixed at their mean for continuous variables, or the most common value for factor variables. Each displayed point is one of the sites for which the expert has previously provided information. They may click on any of these points to adjust the values and see how it affects the response curve.*

The response curves displayed in the software are interactive, in that the user can click on any point, which will zoom the map to that point and display the beta fitting form. The user can then modify $p(p_i)$ and immediately see the effect the adjustment has on the response curves. The expert also has the option of reviewing those points which are influential, or appear to be outliers. An interactive graph shows a plot of the residuals, and a plot of Cook's distances. Again the expert can click on a point and review the site. The graphs are linked, so that a selected point in one graph will cause the corresponding point in each of the other graphs to be highlighted. The expert continues this procedure, gaining feedback from the response curves until the combined set of results from both map and response curves match their belief about the relationship between the species and its environment.

Although a weighted regression approach similar to that described above could have been used to generate final priors for the set of regression parameters $\{\beta_1, .., \beta_p\}$, we used instead a simulation-based approach. While this takes a little more computational time, it makes no assumptions to derive pseudo sample sizes, nor asymptotic normality assumptions in the GLM estimation stage. At each iteration j and for each of the expert's design points, we sample a probability of occurrence $\tilde{p}_{i,j}$ from the distribution chosen by the expert, that is $\text{Beta}(\alpha_i, \beta_j)$. We then fit the nominated linear regression model to $\text{logit}(\tilde{p}_{i,j})$, thus obtaining one draw of $\{\beta_{1,j}, ..., \beta_{p,j}\}$. These values can then be used directly in the MCMC algorithm for the Bayesian regression (1).

In the rock wallaby exercise, each elicitation procedure took approximately one hour and resulted in 10 design points from one expert and 13 from the other. The priors elicited from the experts were reasonably informative, with the posteriors using these priors clearly different from that obtained using a non-informative prior (Fig. 3).

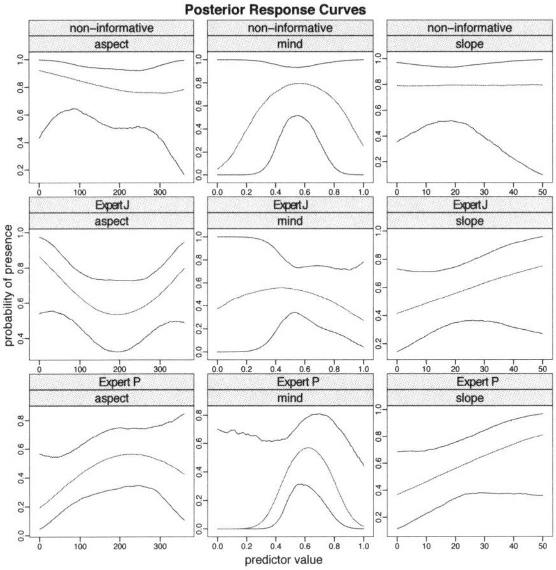

Figure 3: *Posterior response curves for the variables* aspect, mind *and* slope. *The posterior using informative priors elicited from expert J (row 2) and expert P (row 3) are different from that using the non-informative prior (row 1). Note in particular that the models from both experts suggest an increase in probability as slope increases, which is not supported by the model using a non-informative prior.*

3. DISCUSSION

The aim of this paper is to provide an example of how expert opinion can be combined with data to produce a statistical model that balances our prior beliefs with what the data tell us. It was not to create an accurate prediction of the distribution of the brush-tailed rock wallaby, for which much more detailed attention to the priors, model, selected variables and data would have been required. For example, we have made no mention of the how the data were sampled, nor how representative they are of the range of possible habitats for the species. That said, this dataset is not atypical of that faced by ecologists and statisticians working in ecology. The challenge, then, is not to reject the data as inadequate, but to make the best of a bad situation. When the data suggest a relationship between the response and the environmental variables which does not make physical sense, the expert opinion can be used to force a compromise between a believable model and a purely data driven model.

We believe that our proposed approach for elicitation of expert information in a geographic context has a number of key advantages. It allows the expert to make use of contextual, spatial information when asking questions about the response at various levels of the explanatory variables. The structural feedback with the predictive questions suits a range of experts: an expert unfamiliar with the current region of study will use the points as a means to get to the curves they want, whereas someone with good local knowledge but limited understanding of the requirements of the species may ignore the curves and focus on mapping where it lives. The approach is better than simply adjusting response curves, since it facilitates a better assessment of the expert's prior uncertainty. It is also better than eliciting predictions alone, since it capitalises on the spatial context of the problem, avoids requiring the expert to answer questions on esoteric explanatory variables. Finally, this approach allows the analyst to alter the model and corresponding variables without further reference to the expert; this is not possible in more traditional elicitation contexts that ask about responses given other variables in the model, since changing the variables will necessarily change the response.

REFERENCES

Bedrick, E. J., R. Christensen, and W. Johnson (1996). A new perspective on priors for generalized linear models. *J. Amer. Statist. Assoc.* **91**, 1450–1460.

Carter, K. and A. W. Goldizen (2003). Habitat choice and vigilance behaviour of brush-tailed rock-wallabies (*Petrogale penicillata*) within their nocturnal foraging ranges. *Wildlife Research* **30**, 355–364.

Denham, R. and K. Mengersen (2007). Geographically assisted elicitation of expert opinion for regression models. *Bayesian Analysis* **2** (to appear).

Ihaka, R. and R. Gentleman (1996). R: A language for data analysis and graphics. *J. Comp. Graphical Statist.* **5**, 299–314.

Kadane, J. B. and L. J. Wolfson (1998). Experiences in elicitation. *The Statistician* **47**, 3–19.

Pearce, J. L., K. Cherry, M. Drielsma, S. Ferrier, and G. Whish (2001). Incorporating expert opinion and fine-scale vegetation mapping into statistical models of faunal distribution. *J. Appl. Ecology* **38**, 412–424.

BAYESIAN STATISTICS 8, pp. 601–606.
J. M. Bernardo, M. J. Bayarri, J. O. Berger, A. P. Dawid,
D. Heckerman, A. F. M. Smith and M. West (Eds.)

Hierarchical Multiresolution Hazard Model for Breast Cancer Recurrence*

VANJA DUKIĆ and JAMES DIGNAM
University of Chicago, USA
vdukic@health.bsd.uchicago.edu jdignam@health.bsd.uchicago.edu

SUMMARY

The Bayesian multiresolution hazard estimator (MRH) has been recently adapted for estimation of hazard functions (Bouman *et al.*, 2007). In this paper, we extend the previously proposed MRH methods into the hierarchical multiresolution hazard setting (HMRH), to accommodate separate hazard functions within each of several patient strata and some common covariate effects across all strata, while accounting for within-stratum correlation. We apply this method to examine patterns of recurrence after treatment for early stage breast cancer, using data from two large-scale randomized clinical trials.

Keywords and Phrases: MULTIRESOLUTION, HAZARD, BREAST CANCER

1. INTRODUCTION

In survival analysis, because the hazard function $h(t)$ often exhibits unstable behavior making it difficult to reliably discern patterns of change or make comparisons between groups, aggregates of the hazard such as the cumulative hazard or survival function are used for summary and inference on failure risk. While useful for most purposes, these summaries can partially obscure important patterns in the hazard of failure over time. Alternatively, the hazard function itself can reveal important properties of the failure process (Aalen and Gjessing 2001). Generally, some type of smoothed estimate of the hazard function is used to characterize its shape, which is indicative of how failure risk (in the population) changes over time. While a variety of approaches have been proposed, methodological challenges remain for both estimation and associated inferential procedures. In particular, flexible modeling approaches are needed, because in contrast to the functional form in most

* This is a synopsis. For the full version, see *Bayesian Analysis* **2** (2007), to appear.
Vanja Dukić and James Dignam are assistant professors in the Department of Health Studies, The University of Chicago, Chicago, USA. Support for this research was partially provided by the Susan G. Koman Breast Cancer Foundation and Public Health Service grant NCI-U10-CA-69651 from the National Cancer Institute, National Institutes of Health, Department of Health and Human Services.

parametric survival models, the hazard may exhibit complex non-unimodal shape with 'change-points' that may reveal important information about the process under study.

In this article, we investigate the hazard of disease recurrence among women treated for breast cancer with hormonal or cytotoxic chemotherapy agents administered after surgery (referred to as adjuvant therapy) and followed over several years. There is considerable interest in the patterns of recurrence hazards after diagnosis and initial treatment, both to gain biological insights and to better manage the disease clinically. To model the recurrence hazards, we apply the semiparametric multi-resolution hazard (MRH) estimator recently presented in work by Bouman and colleagues (2005, 2007). Employing a piece-wise constant prior for the hazard rate which is constructed in a tree-based and self-consistent manner, the MRH approach permits flexible modeling with the ability to incorporate a variety of *a priori* assumptions about the shape and smoothness of hazard functions.

2. RECURRENCE RISK AFTER EARLY STAGE BREAST CANCER

Despite appreciable progress in therapeutic strategies for early stage (i.e., localized and operable, as opposed to metastatic) breast cancer, the clinical course after diagnosis remains heterogeneous and thus highly unpredictable for individuals. Characteristics that identify which women are at greater or lesser risk of treatment failure are needed to aid in individually tailoring therapy for optimal disease management. Answers may also lie partly in gaining a better understanding of the intermediate and long-term clinical course of the disease, identifying patterns that can portend periods of increased recurrence risk. It is well known that breast cancer recurrence risk remains elevated for a long period after initial diagnosis and tumor removal, and there is longstanding interest in the the prospect of 'cure' after sufficient tumor-free time has been achieved (Berg and Robbins 1966). Some long-term follow-up studies have suggested that a lengthy 'dormancy period' exists (possibly over 20 years), when tumor recurrences may still appear (Gordon 1990; Demicheli et al. 1996; Karrison et al. 1999). Studies examining the shape of the recurrence hazard show a peak 12-24 months after diagnosis, followed by a decline over time, although remaining persistently elevated relative to individuals never having had breast cancer (Saphner et al. 1996; Hess et al. 2003).

The National Surgical Adjuvant Breast and Bowel Project (NSABP) is a U.S. National Cancer Institute sponsored multi-center cancer clinical trials group that has investigated a spectrum of treatments for breast and colorectal cancers. These studies provide an ideal data source for examining factors influencing the hazard of disease recurrence. One key determinant of both expected prognosis and choice of adjuvant therapy type is the presence and quantity of estrogen receptors (ER) on the tumor. From 1982–1988, patients were accrued into one of two trials conducted in parallel based on ER status: NSABP Protocol B-13 randomized 760 patients with ER−negative (ER−) tumors to no further treatment after surgery or to cytotoxic chemotherapy. Protocol B-14 randomized 2 892 patients with ER−positive (ER+) tumors to placebo or the estrogen antagonist drug tamoxifen after surgery. Primary findings first obtained in 1989 showed a significant reduction in breast cancer recurrence risk for patients receiving the adjuvant therapy regimens, and longer follow-up eventually revealed a survival advantage (Fisher et al. 1996a; Fisher et al. 1996b). Further details of the trial designs and findings can be found in the published primary reports. Follow-up continues to date, with mean follow-up of over 15 years and over 900 recurrence events observed. In this paper, we model the cause-specific hazard for breast recurrence in relation to treatment, ER, and other important covariates.

3. THE HIERARCHICAL MULTIRESOLUTION HAZARD MODEL

In breast cancer, it is well-known that women with ER− and ER+ tumors have different expected prognosis due to association of ER with both tumor pathology and clinical characteristics (Hess et al. 2003). Thus, we would like to avoid imposing any proportionality constraint on ER in the model. While proportionality appears to hold better between treatment groups within ER categories, we also wish to permit a different hazard shape by treatment type, as biologic hypotheses concerning the action of adjuvant therapy would suggest this possibility (Skipper 1971). Thus, we allow separate hazard rates for each of the four strata defined by the ER−by-treatment-group combinations (ER− surgery only, ER− chemotherapy, ER+ surgery only (placebo), ER+ tamoxifen), while simultaneously estimating other covariate effects that could be reasonably assumed common over strata.

Multiresolution models for estimation of a discretized intensity function were developed for astrophysics applications by Kolaczyk (1999) and adapted to hazard estimation by Bouman *et al.*, (2005, 2007). The MRH approach consists of first choosing the 'time resolution' – a set of time points $\{t_0, t_1, ..., t_J\}$ – and then estimating the underlying baseline cumulative hazard $H_{\text{b}}(t)$ and its discrete increments $d_j \equiv H_{\text{b}}(t_j) - H_{\text{b}}(t_{j-1}) = \int_{t_{j-1}}^{t_j} h_{\text{b}}(s)ds$, where $h_{\text{b}}(t)$ represents the baseline hazard rate at time t. For those times t such that $t_{j-1} < t < t_j, j = 1, \ldots, J$, a piecewise-constant hazard rate is assumed, where the value of the hazard rate in the jth interval is the hazard increment d_j.

Denoting the total cumulative baseline hazard $H(t_J)$ as $H_{0,0}$, and the hazard increments d_1 as $H_{M,0}$, d_2 as $H_{M,1}, \ldots$, and d_J as $H_{M,2^M-1}$, the multiresolution hazard tree prior is built recursively by defining $H_{m-1,p} = H_{m,2p} + H_{m,2p+1}$, for $m = 1, \ldots, M$, and $p = 0, \ldots, 2^{m-1} - 1$. We refer to m as the level of resolution and p as the position within that level. If we further define $R_{m,p} \equiv H_{m,2p}/H_{m-1,p}$, we can parametrize the hazard increments by $H_{0,0}$ and the 'splits' $R_{1,0}, \ldots, R_{M,2^{M-1}-1}$. For example, when $M = 3$ $(J = 8)$ we have $d_1 = H_{0,0}R_{1,0}R_{2,0}R_{3,0}$, $\ldots$, $d_8 = H_{0,0}(1 - R_{1,0})(1 - R_{2,1})(1 - R_{3,3})$. Though numerous piece-wise constant hazard models have been considered before, the uniqueness of the MRH model lies in its efficient and clever construction of the tree-based prior for a piece-wise constant function, so that the prior is self-consistent, as it does not depend on the final resolution level M (i.e., it is invariant to the height of the tree).

The hierarchical MRH (HMRH) model considers the hazard rate for each stratum $s, s \in \{1, 2, 3, 4\}$, as a random MRH tree, parametrized by the stratum cumulative hazard $H_s(t_J)$ and the stratum set of splits $\mathbf{R}_{m,p,s}$ as follows: $H_s(t_J) \sim \mathcal{G}a(a_s, \lambda_s)$, $R_{1,0,s} \sim \mathcal{B}e(2\gamma_{1,0,s}k_s a_s, 2(1 - \gamma_{1,0,s})k_s a_s)$, $R_{2,p,s} \sim \mathcal{B}e(2\gamma_{2,p,s}k_s^2 a_s, 2(1 - \gamma_{2,p,s})k_s^2 a_s), p = 0, 1$, and so on. The shape parameters of each of these beta priors and the hyperparameters for $H_s(t_J)$ determine the prior expectations of the hazard increments $d_j, j = 1, \ldots, J$. To allow for smoothing in the multiresolution prior, we multiply the shape parameter of the beta priors at each additional level of the hierarchy by a hyperparameter k. Specifically, choosing k less or greater than 0.5 yields, respectively, negative (rougher hazard) or positive (smoother hazard) prior correlation among the d_j's (Bouman 2005, 2007). Smoother hazard may in particular be desirable in the presence of heavy censoring. On each a_s we place a zero-truncated Poisson (ZTP) hyperprior with mean μ_a; on each λ_s, we use an exponential prior with mean μ_λ; and k_s can be also given exponential priors with mean μ_k, or be fixed if specific smoothness (positive or negative correlation) is desired *a priori*.

If for patient i in stratum s, we observe the failure or censoring time $T_{i,s}$, a censoring indicator $\delta_{i,s}$, and covariates $\mathbf{X}_{i,s}$, her likelihood contribution is:

$$L_{i,s} = \left[\exp(\mathbf{X}'_{i,s}\boldsymbol{\beta})h_{\mathrm{b,s}}(T_{i,s})\right]^{\delta_{i,s}} \exp(-\exp(\mathbf{X}'_{i,s}\boldsymbol{\beta})H_b(\min(T_{i,s}, t_J))). \quad (1)$$

Failure times $T_{i,s} > t_J$ are administratively censored ($\delta_{i,s} = 0$) and contribute $\exp(\mathbf{X}'_{i,s}\boldsymbol{\beta})\exp(-H_b(t_J))$ to the likelihood.

The posterior is obtained by combining the priors with the likelihood for all patients from all strata, and is simulated by Gibbs Sampling. Similarly to Bouman et al. (2005, 2007) the full conditional posterior distributions for H_s are gamma, and the full conditional distributions for each $R_{m,p,s}$ and $\boldsymbol{\beta}$ are log-concave and therefore easy to sample from. However, the full conditionals for λ_s, a_s, and k_s are in general difficult to sample from as they are not log-concave; they were simulated via rejection Metropolis sampling (see Gilks, Best, and Tan, 1995).

4. ANALYSIS OF THE RECURRENCE HAZARD AFTER BREAST CANCER

The 16-bin multiresolution model with the 'flat' prior hazard rate for each stratum (with all $\gamma_{m,p,s}$ and all k_s set to 0.5), were fitted. Larger tumor size and younger age at diagnosis were found associated with increased recurrence hazard: within each stratum, an increase of 10 years in age resulted in approximate 20% reduction, while an increase of 1mm in tumor size resulted in approximately 2% increase in recurrence hazard. A higher concentration of tumor progesterone receptors was very weakly indicative of lesser failure risk. While menopausal status is generally an important factor in breast cancer prognosis, here it was only marginally associated with recurrence hazard after stratification and inclusion of age at diagnosis.

Fig. 1 displays by treatment/ER strata the median posterior estimates for the 16-bin baseline hazard increments (corresponding to constant 11.3-month hazard function values), with the post-analysis smoothing of piece-wise constant rate estimates performed via the median-spline method. Several notable features are seen. First, all four groups have the distinctive hazard peak around 12–24 months after surgery, with the ER− groups experiencing the peak a bit earlier. This pattern is similar to that noted by Hess et al. (2003) in their study of recurrence hazards by ER status. Second, the peak is greatest for the ER− patients receiving surgery only, and is substantially reduced by chemotherapy, being lower than untreated patients from the more favorable ER+ group. The lowest peak is among ER+ patients randomized to tamoxifen. Interestingly however, at longer follow-up times the hazard in this group is no lower than that of chemotherapy treated and even untreated ER− patient groups, both of which have smaller hazard than untreated ER+ patients. Ultimately, the ER− chemotherapy treated group has the lowest recurrence hazard.

However, with the exception of time points around the hazard peaks, credible intervals for the four hazard estimates tend to overlap, particularly at longer follow-up times. Thus, we currently cannot reliably conclude whether there is a crossover of failure hazard among the groups at later time points. We note that in other analyses collapsing across one or the other stratification factors (ER or treatment) and treating ER or treatment as covariates in modeling, large treatment effects within ER groups were apparent, while differences between ER groups within treatment modalities (surgery, adjuvant) were large initially but attenuated over time, substantively violating the proportional hazards assumption.

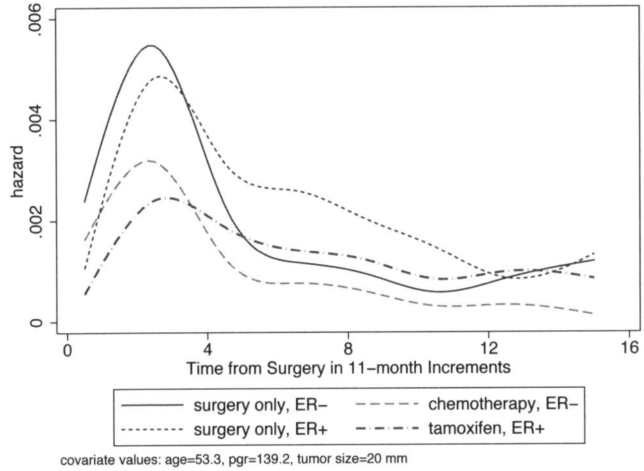

Figure 1: *Smoothed recurrence hazard increments for the 16-bin HMRH model.*

An alternate model with 32 time intervals of length just under 6 months each was fitted to compare the hazard estimates under this alternative resolution and to assess the impact of additional resolution level on covariate effect estimates. Based on the invariance properties of the MRH prior, one can expect that some of that invariance is preserved in the posterior as well; and we observe this to a large degree: the 16-bin hazard estimate obtained by fitting the 32-bin model and then aggregating the neighboring hazard increments is almost indistinguishable from the original 16-bin model hazard estimate. With respect to covariate effects, the effects of increasing the resolution were minimal as well.

5. DISCUSSION

We have illustrated the application of a flexible extension of the familiar Cox proportional hazards model to jointly estimate covariate effects and separate hazard rate functions for several patient strata. This approach allows us to incorporate covariate effects and perform inference related to shapes and change-points in the hazard over time, our primary interest here. The direct examination of the hazard functions reveals important patterns not readily apparent from quantities such as the survival functions. However, the hazard function remains a difficult quantity to draw robust inference from, as even in this large dataset, estimates suggest potentially important differences in shape, but variability estimates preclude any definitive conclusions pending additional analyses.

The observation that among those patients with initially higher risk disease (ER− tumors) those escaping the early failure risk go on to have substantially lower long-term failure risk than those with initially more favorable prognosis (ER+ tumors), has significant implications in both clinical management and further developments of adjuvant therapies. In fact, there has been much recent interest in the development of 'switching' strategies whereby women with ER+ tumors discontinue tamoxifen and begin use of other hormonal treatments, in order to improve on the benefit of this treatment modality. Currently, little is known about what factors might be key to optimizing the switching strategy. For ER− patients, newer chemotherapy and molecularly targeted agents that act on specific tumor vulnerabilities may offer the best opportunity for cure once early failure is avoided.

REFERENCES

Aalen, O. and Gjessing, H. (2001). Understanding the shape of the hazard rate: A process point of view. *Statist. Science* **16**, 1–22.

Berg, J. W. and Robbins, G. F. (1966). Factors influencing short and long term survival of breast cancer patients. *Surgical and Gynecologic Obstetrics* **122**, 1311–1316.

Bouman, P., Dukić, V. and Meng, X.L. (2005). A Bayesian multiresolution hazard model with application to an AIDS reporting delay study. *Statistica Sinica* **15**, 325–357.

Bouman, P., Meng, X. L., Dignam, J. and Dukić, V. A. (2007). Multiresolution hazard model for multi-center survival studies: Application to tamoxifen treatment in early stage breast cancer. *J. Amer. Statist. Assoc.* (to appear).

Cox, D. R. (1972). Regression models and life tables. *J. Roy. Statist. Soc. B* **34**, 187–220.

Demicheli, R., Abbattista, A., Miceli, R., Valagussa, P. and Bonadonna, G. (1996). Time distribution of the recurrence risk for breast cancer patients undergoing mastectomy: further support about the concept of tumor dormancy. *Breast Cancer Research and Treatment* **41**, 177–185.

Dukić, V. and Dignam, J. (2007). Bayesian hierarchical multiresolution hazard model for the study of time-dependent failure patterns in early stage breast cancer. *Bayesian Analysis* **2** (to appear).

Fisher, B., Dignam, J., Mamounas, E. P., Costantino, J., Wickerham, D. L., Redmond, C., Wolmark, N., Dimitrov, N. V., Bowman, D., Glass, A. G., Atkins, J.N., Abramson, N, Sutherland, C., Aron, B. and Margolese R. G. (1996). Sequential methotrexate and fluorouracil for the treatment of node-negative breast cancer patients with estrogen receptor-negative tumors: eight-year results from National Surgical Adjuvant Breast and Bowel Project B-13 and first report of findings from NSABP B-19 comparing methotrexate and fluorouracil with conventional CMF. *J. of Clinical Oncology* **14**, 1982–1992.

Fisher, B., Dignam, J,, Bryant, J., DeCillis. A., Wickerham, D. L., Wolmark, N., Costantino. J., Redmond, C, Fisher, E. R., Bowman, D. M., Deschenes, L., Dimitrov, N. V., Margolese, R. G., Robidoux, A., Shibata, H., Terz, J., Paterson, A. H, Feldman, M. I., Farrar, W., Evans, J. and Lickley, H. L. (1996). Five versus more than five years of tamoxifen therapy for breast cancer patients with negative lymph nodes and estrogen receptor positive tumors. *J. of the National Cancer Institute* **88**, 1529–1542.

Gilks, W., Best, N. and Tan, K. (1995). Adaptive rejection Metropolis sampling. *Appl. Statistics* **44**, 455–472.

Gordon, N. (1990). Application of the theory of finite mixtures for the estimation of 'cure'. *Statistics in Medicine* **9**, 397–407.

Hess, K. R., Pusztai, L., Buzdar, A. and Hortobagyi, G. N. (2003). Estrogen receptors and distinct patterns of breast cancer relapse. *Breast Cancer Research and Treatment* **78**, 105–118.

Karrison, T., Ferguson, D. and Meier, P. (1999). Dormancy of mammary carcinoma after mastectomy. *J. National Cancer Institute* **91**, 80–85.

Kolaczyk, E. D. (1999). Bayesian multiscale models for Poisson processes. *J. Amer. Statist. Assoc.* **94**, 920–933.

Saphner, T., Tormey, D. and Gray, R. (1996). Annual hazard rates of recurrence for breast cancer after primary therapy. *J. Clinical Oncology* **14**, 2738–2746.

Skipper, H. E. (1971) Kinetics of mammary tumor cell growth and implications for therapy. *Cancer* **28**, 1479–1499.

BAYESIAN STATISTICS 8, pp. 607–612.
J. M. Bernardo, M. J. Bayarri, J. O. Berger, A. P. Dawid,
D. Heckerman, A. F. M. Smith and M. West (Eds.)

Bayesian Regression of Piecewise Constant Functions*

MARCUS HUTTER
RSISE / ANU / NICTA, Canberra, Australia
marcus@hutter1.net

SUMMARY

We derive an exact and efficient Bayesian regression algorithm for piecewise constant functions of unknown segment number, boundary location, and levels. It works for any noise and segment level prior, e.g. Cauchy which can handle outliers. We derive simple but good estimates for the in-segment variance. We also propose a Bayesian regression curve as a better way of smoothing data without blurring boundaries. The Bayesian approach also allows straightforward determination of the evidence, break probabilities and error estimates, useful for model selection and significance and robustness studies. We briefly mention the performance on synthetic and real-world examples. The full version of the paper contains detailed derivations, more motivation and discussion, the complete algorithm, the experiments, and various extensions.

Keywords and Phrases: BAYESIAN REGRESSION, EXACT POLYNOMIAL ALGORITHM, NON-PARAMETRIC INFERENCE, PIECEWISE CONSTANT FUNCTION, DYNAMIC PROGRAMMING, CHANGE POINT PROBLEM.

1. INTRODUCTION

We consider the problem of fitting a piecewise constant function through noisy one-dimensional data, where the segment number, boundaries and levels are unknown. Regression with piecewise constant (PC) functions, a special case of change point detection, has many applications. For instance, determining DNA copy numbers in cancer cells from micro-array data, to mention just one recent.

Bayesian piecewise constant regression (BPCR). We provide a full Bayesian analysis of PC-regression. For a fixed number of segments we choose a uniform prior over all possible segment boundary locations. Some prior on the segment levels and data noise within each segment is assumed. Finally a prior over the number of segments is chosen. From this we obtain the posterior segmentation probability distribution

---

* This is a synopsis. For the full version, see *Bayesian Analysis* **2** (2007), to appear.

(Section 2). In practice we need summaries of this complicated distribution. A simple maximum (MAP) approximation or mean does not work here. The right way is to proceed in stages from determining the most critical segment number, to the boundary location, and finally to the then trivial segment levels. We also extract the evidence, the boundary probability distribution, and an interesting non-PC regression curve including error estimate (Section 2). We derive an exact polynomial-time dynamic-programming-type algorithm for all quantities of interest. Our algorithm works for any noise and level prior. We consider more closely the Gaussian 'standard' prior and heavy-tailed robust-to-outliers distributions like the Cauchy (Sections 3 and 5). Finally, some hyper-parameters like the global data average and variability and local within-level noise have to be determined. We introduce and discuss novel, efficient semi-principled estimators, thereby avoiding problematic or expensive numerical EM or Monte-Carlo estimates (Section 4). See Hutter (2006) for detailed derivations, more motivation and discussion, the complete algorithm, experimental evaluation on synthetic and real data, and various extensions (Section 7). The simulations show that our method handles difficult data with high noise and outliers well.

Comparison to other work. Sen and Srivastava (1975) developed a frequentist solution to the problem of detecting a single (the most prominent) segment boundary (called change or break point). Olshen *et al.* (2004) generalize this method to detect pairs of break points, which improves recognition of short segments. Both methods are then (heuristically) used to recursively determine further change points. Another approach is penalized Maximum Likelihood (ML). For a fixed number of segments, ML chooses the boundary locations that maximize the data likelihood (minimize the mean square data deviation). Jong et al. (2003) use a population based algorithm as minimizer, while Picard et al. (2005) use dynamic programming, which is structurally very close to our core recursion, to find the exact solution in polynomial time. An additional penalty term has to be added to the likelihood in order to determine the correct number of segments. The most principled penalty is the Bayesian Information Criterion (Schwarz, 1978; Kass and Wasserman, 1995). Since it can be biased towards too simple (Weakliem, 1999) or too complex (Picard et al., 2005) models, in practice often a heuristic penalty is used. An interesting heuristic, based on the curvature of the log-likelihood as a function of the number of segments, has been used by Picard et al. (2005). Our Bayesian regressor is a natural response to penalized ML. Yao (1984); Barry and Hartigan (1992); Fearnhead (2006) develop a different exact Bayesian PCR algorithm based on a renewal processes that unfortunately has a highly informed prior on the number of segments k. Another related work to ours is Bayesian bin density estimation by Endres and Földiák (2005), who also average over all boundary locations, but in the context of density estimation. Many other approximate, heuristic or numerical approaches to PCR exist; too many to list them all.

Advantages of Bayesian regression. A full Bayesian approach (when computationally feasible) has various advantages over others: A generic advantage is that it is more principled and hence involves fewer heuristic design choices. This is very important for estimating the number of segments. Further, one can decide among competing models solely on evidence, Bayes often works well in theory and practice, and provides probability estimates and variances for all quantities of interest. Note that we are not claiming here that BPCR works better than the other mentioned approaches. In a certain sense Bayes is optimal if the prior is 'true'. Practical superiority likely depends on the type of application. A comparison for microarray data is in progress. The major aim of this paper is to introduce the BPCR model, to derive an efficient algorithm, and demonstrate the gains of BPCR beyond bare PC-regression, like the (predictive) regression *curve* (which is better than local smoothing which wiggles more and blurs jumps).

2. THE GENERAL MODEL

Setup, likelihood, prior. We are given a sequence $\boldsymbol{y} = (y_1, ..., y_n)$, e.g. times-series data or measurements of some function at locations $1...n$, where each $y_i \in I\!R$ resulted from a noisy 'measurement', i.e. we assume that the y_i are independently (e.g. Gaussian or Cauchy) distributed with means (or locations) μ'_i and variances (or scales) $\sigma_i'^2$, so the data *likelihood* is $P(\boldsymbol{y}|\boldsymbol{\mu}', \boldsymbol{\sigma}') := \prod_{i=1}^n P(y_i|\mu'_i, \sigma'_i)$. The estimation of the true underlying function $f = (f_1, ..., f_n)$ is called regression. We assume or model f as *piecewise constant* with k segments with boundaries $0 = t_0 < t_1 < ... < t_{k-1} < t_k = n$. If the noise is i.i.d., we have $\mu'_i = \mu_q$ for $t_{q-1} < i \leq t_q$ $\forall q$ and $\sigma'_i = \sigma$ $\forall i$. Our goal is to estimate the segment levels $\boldsymbol{\mu} = (\mu_1, ..., \mu_k)$, boundaries $\boldsymbol{t} = (t_0, ..., t_k)$, and their number k. Bayesian regression proceeds in assuming a prior for these *quantities of interest.* We model the segment levels by a broad (e.g. Gaussian or Cauchy) distribution with mean ν and variance ρ^2. For the segment boundaries we take a uniform distribution among all segmentations into k segments. Finally we take some prior (e.g. uniform) over the number of segments k. So our *prior* $P(\boldsymbol{\mu}, \boldsymbol{t}, k)$ is the product of $P(\mu_q) = P(\mu_q|\nu, \rho)$ $\forall q$ and $P(\boldsymbol{t}|k) = \binom{n-1}{k-1}^{-1}$ and $P(k) = 1/n$.

Evidence, posterior, MAP, mean. Given the prior and likelihood we can compute the data *evidence* $P(\boldsymbol{y}) = \sum_{k,\boldsymbol{t}} \int P(\boldsymbol{y}|\boldsymbol{\mu}, \boldsymbol{t}, k) P(\boldsymbol{\mu}, \boldsymbol{t}, k)\, d\boldsymbol{\mu}$ and *posterior* $P(\boldsymbol{\mu}, \boldsymbol{t}, k|\boldsymbol{y}) = P(\boldsymbol{y}|\boldsymbol{\mu}, \boldsymbol{t}, k) P(\boldsymbol{\mu}, \boldsymbol{t}, k)/P(\boldsymbol{y})$. The posterior contains all information of interest, but is a complex object for practical use. So we need summaries like the maximum (MAP) or mean and variances. MAP over continuous parameters ($\boldsymbol{\mu}$) is problematic, since it is not reparametrization invariant. This is particularly dangerous if MAP is across different dimensions (k), since then even a linear transformation ($\boldsymbol{\mu} \rightsquigarrow \alpha\boldsymbol{\mu}$) scales the posterior (density) exponentially in k (by α^k). This severely influences the maximum over k, i.e. the estimated number of segments. The mean of $\boldsymbol{\mu}$ does not have this problem. On the other hand, the mean of $\boldsymbol{t}$ makes only sense for fixed (e.g. MAP) k. The most natural solution is to proceed in stages similar to as the prior has been formed.

Quantities of interest. We first determine the posterior probability of k segments $P(k|\boldsymbol{y})$ and the MAP segment number $\hat{k} = \arg\max_k P(k|\boldsymbol{y})$, then, for each boundary t_q its posterior $P(t_q|\boldsymbol{y}, \hat{k})$ and MAP estimate $\hat{t}_q = \arg\max_{t_q} P(t_q|\boldsymbol{y}, \hat{k})$ (or mean) based on $\hat{k}$, and finally the segment level means $\hat{\mu}_q = \int P(\mu_q|\boldsymbol{y}, \hat{\boldsymbol{t}}, \hat{k}) \mu_q d\mu_q$ for this MAP segmentation. The estimate $(\hat{\boldsymbol{\mu}}, \hat{\boldsymbol{t}}, \hat{k})$ defines a (single) piecewise constant function $\hat{f}$, which is our estimate of f. A (very) different quantity is to Bayes-average over all piecewise constant functions and to ask for the mean $\hat{\mu}'_i = \int P(\mu'_i|\boldsymbol{y}) \mu'_i d\mu'_i$ (or $\hat{k}$ fixed) at location i as an estimate for f_i. This 'predictive' Bayesian regression *curve* $\boldsymbol{\mu}'$ behaves similar to a local smoothing of $\boldsymbol{y}$, but without blurring true jumps. Standard deviations of all estimates may also be reported.

3. EFFICIENT SOLUTION

Dynamic Programming can be used to (exactly) compute all quantities of interest in time $O(n^3)$, since fixing one break makes data left and right of the break independent. The only essential assumption is the uniform prior on $\boldsymbol{t}$.

Single segment distribution. The evidence and moments

$$A_{ij}^r := \int P(\mu_q) \prod_{t=i+1}^{j} P(y_t|\mu_q)\mu_q^r d\mu_q$$

of a single segment from $i+1$ to j can be computed analytically for the exponential family with conjugate prior like Gauss or numerically (one-dimensional integral) for others like Cauchy.

Gaussian model. The standard assumption on the noise is independent Gauss $P(y_i|\mu_i',\sigma_i') = \frac{1}{\sqrt{2\pi}\sigma_i'}\exp(-\frac{(y_t-\mu_i')^2}{2\sigma_i'^2})$. The corresponding standard 'conjugate' prior on the means μ_q for each segment q is also Gauss $P(\mu_q|\nu,\rho) = \frac{1}{\sqrt{2\pi}\rho}\exp(-\frac{(\mu_q-\nu)^2}{2\rho^2})$. This makes A_{ij}^0 an (unnormalized) Gaussian integral, which can be computed in closed form, and similarly for A_{ij}^r.

Cauchy model. The standard problem with Gauss is that it does not handle outliers well. If we do not want to or cannot remove outliers by hand, we have to properly model them as a prior with heavier tails. This can be achieved by a mixture of Gaussians or by a Cauchy distribution $P(y_i|\mu_i',\sigma_i') = \frac{1}{\pi}\frac{\sigma_i'}{\sigma_i'^2+(y_i-\mu_i')^2}$. Note that μ_i' and σ_i' determine the location and scale of Cauchy but are not its mean and variance (which do not exist). The prior on the levels μ_q may as well be modeled as Cauchy $P(\mu_q|\nu,\rho) = \frac{1}{\pi}\frac{\rho}{\rho^2+(\mu_q-\nu)^2}$. Actually, the Gaussian noise model may well be combined with a non-Gaussian prior and vice versa if appropriate.

Dynamic programming. Now consider the probability $P(y_1..y_j|k) =: L_{kj}/\binom{j-1}{k-1}$ of first j data given k segments, and $P(y_{i+1}..y_n|k) =: R_{ki}/\binom{n-i-1}{k-1}$ of last $n-i$ data given k segments. The evidence of $y_1..y_j$ with $k+1$ segments = evidence of $y_1..y_h$ with k segments $\times$ single-segment evidence of $y_{h+1}..y_j$, summed over all locations h of boundary k, i.e. we get the *left recursion* $L_{k+1,j} = \sum_{h=k}^{j-1} L_{kh}A_{hj}^0$ with $L_{1j} = A_{0j}^0$. Similarly, we get the *right recursion* $R_{k+1,i} = \sum_{h=i+1}^{n-k} A_{ih}^0 R_{kh}$ with $R_{1n} = A_{in}^0$.

Quantities of interest. From L and R we get the evidence $E := P(\boldsymbol{y}) = \frac{1}{n}\sum_{k=1}^{n} L_{kn}/\binom{n-1}{k-1}$, the k-posterior $C_k := P(k|\boldsymbol{y}) = L_{kn}/\binom{n-1}{k-1}nE$ and its MAP estimate $\hat{k} = \arg\max_k C_k$, the probability that boundary q is located at h, $B_{qh} := P(t_q = h|\boldsymbol{y},\hat{k}) = L_{qh}R_{\hat{k}-q,h}/L_{\hat{k}n}$, the MAP segment boundary $\hat{t}_q := \arg\max_h B_{qh}$, the segment level moments $\widehat{\mu_q^r} = A_{\hat{t}_{q-1}\hat{t}_q}^r/A_{\hat{t}_{q-1}\hat{t}_q}^0$, and the regression curve $\widehat{\mu_t'^r} = \sum_{i=0}^{t-1}\sum_{j=t}^{n} F_{ij}^r$ with $F_{ij}^r = \sum_{q=1}^{\hat{k}} L_{q-1,i}A_{ij}^r R_{\hat{k}-q,j}/L_{\hat{k}n}$.

4. DETERMINATION OF THE HYPER-PARAMETERS

Hyper-Bayes and Hyper-ML. The developed regression model still contains three (hyper)parameters, the global variance ρ^2 and mean ν of $\boldsymbol{\mu}$, and the in-segment variance σ^2. If they are not known, one may choose the parameters such that the evidence $P(\boldsymbol{y}) = P(\boldsymbol{y}|\sigma,\nu,\rho)$ is maximized (empirical Bayes, hyper-ML or ML-II estimate), or simpler the following approximations:

Estimate of global mean and variance ν and ρ. A reasonable choice for ν and ρ are the empirical global mean $\hat{\nu} := \frac{1}{n}\sum_{t=1}^{n} y_t$ and variance $\hat{\rho}^2 := \frac{1}{n-1}\sum_{t=1}^{n}(y_t-\hat{\nu})^2$ of the data $\boldsymbol{y}$. This overestimates the variance ρ^2 of the segment levels, since the expression also includes the in-segment variance σ^2.

Estimate of in-segment variance σ^2. At first there seems little hope of estimating the in-segment variance σ^2 from $\boldsymbol{y}$ without knowing the segmentation, but actually we can use a simple trick. If $\boldsymbol{y}$ would belong to a single segment, i.e. the y_t were i.i.d. with variance σ^2, then $\mathbf{E}[\frac{1}{n}\sum_{t=1}^{n}(y_t-\mu_1)^2] =: \sigma^2 = \frac{1}{2(n-1)}\mathbf{E}[\sum_{t=1}^{n-1}\Delta_t^2]$ would hold, where $\Delta_t = y_{t+1} - y_t$. That is, instead of estimating σ^2 by the squared deviation of the y_t from their mean, we can also estimate σ^2 from the average squared difference of successive y_t. This remains true even for multiple segments if we exclude the segment boundaries in the sum. On the other hand, if the number of segment boundaries is small (more precisely $k\rho^2/n\sigma^2$ small), the error from including the boundaries will be small, i.e. the expression remains approximately valid. Hence we may estimate σ^2 by the upper bound $\hat{\sigma}^2 := \frac{1}{2(n-1)}\sum_{t=1}^{n-1}\Delta_t^2$.

Estimates by quantiles. If mean and variance do not exist or the distribution is quite heavy-tailed, we need other estimates. We use median and quartiles of $\boldsymbol{y}$ for estimating ν and ρ, and quartiles of $\boldsymbol{\Delta}$ for estimating σ. More precisely, $\hat{\nu} = [\boldsymbol{y}]_{n/2}$ and $\hat{\rho} = ([\boldsymbol{y}]_{3n/4} - [\boldsymbol{y}]_{n/4})/2\alpha$, where $[\boldsymbol{y}]$ denotes the sorted $\boldsymbol{y}$ array, and similarly $\hat{\sigma} = ([\boldsymbol{\Delta}]_{3n/4} - [\boldsymbol{\Delta}]_{n/4})/2\beta$. α and β depend on the shape of the noise distribution, e.g. $\alpha = 1$ and $\beta = 2$ for Cauchy and $\alpha \doteq 0.6744$ and $\beta \doteq 0.6744\sqrt{2}$ for Gauss. Use of quartiles for estimating σ is also robust to the 'outliers' caused by the segment boundaries, so yields better estimates than mean-square $\boldsymbol{\Delta}$ if noise is low.

5. THE ALGORITHM

The computation of A, L, R, E, C, B, $\hat{t}_p$, $\widehat{\mu_m^r}$, F, and $\widehat{\mu_t'^r}$ by the formulas/recursions derived in Section 3, are straightforward. For efficiency reasons one should compute the products/sums incrementally from $j \rightsquigarrow j+1$. Similarly $\widehat{\mu_t'^r}$ should be computed incrementally. Typically $r = 0, 1, 2$. In this way, all quantities can be computed in time $O(k_{max}n^2)$ and space $O(n^2)$, where $k_{max} \leq n$ is an upper bound on the number of segments. Space can be reduced to $O(k_{max}n)$ by computing A on-the-fly in the various expressions at the cost of a slowdown by a constant factor. The complete algorithm can be found in (Hutter, 2006). Actually the ratios A_{ij}^r/A_{ij}^0 and the logarithms of A^0, L, R, and E have to be computed and stored to avoid over/underflow.

6. EXPERIMENTS

We tested our BPCR algorithm on various synthetic data sets, sampled from piecewise constant functions with Gaussian and Cauchy noise of varying magnitude. Our Bayesian regressor (piecewise constant $\hat{f}$ and predictive curve $\boldsymbol{\mu}'$) performed very well even for data with 'Cauchy outliers' or so noisy that it was impossible to detect any segments by eye. The other claimed properties, e.g. closeness of $\hat{\sigma}$ to its hyper-ML estimate, have also been confirmed. We also tested robustness of BPCR under misspecification. As expected, Cauchy BPCR with Gaussian data is more robust than reversely, but for high noise BPCR fails in both misspecification directions. Comparing evidences clearly indicated the misspecification. We also used BPCR on DNA micro-array data, which are *very* noisy piecewise constant functions (DNA copy numbers in cancer cells) (Pinkel et al., 1998). Without tuning, the results matched the best available other methods. Deeper biological evaluation is necessary to determine whether it even outperforms the other methods (Hutter, 2006). In any case, the additional variance and accuracy information BPCR provides is useful extra information in this application. Finally, the Bayesian regression curve behaves very nicely (and uniquely). It is very flat, *i.e.*, smoothes the data in long and clear segments, wiggles in less clear segments, and has jumps at the segment boundaries. One of the most critical steps for a good segmentation is determining the right number of segments $\hat{k}$. The distribution $P(k|\boldsymbol{y})$ gives interesting additional insight.

7. CONCLUSION

We considered Bayesian regression of piecewise constant functions with unknown segment number, location and level. We derived an efficient algorithm that works for any noise and segment level prior, e.g. Cauchy which can handle outliers. We derived simple but good estimates for the in-segment variance. We also proposed a Bayesian regression curve as a better way of smoothing data without blurring boundaries. The Bayesian approach also allowed us to straightforwardly determine the global evidence, break probabilities and error estimates, useful for model selection and significance and robustness studies. We discussed the performance on synthetic and real-world examples. BPCR can (easily) be modified and extended in a variety of ways: For discrete segment levels, segment dependent variance, piecewise linear and non-linear regression, non-parametric noise prior, and others. See Hutter (2006) for detailed derivations, more motivation and discussion, the complete algorithm plus examples and data, the experiments, extensions, etc.

REFERENCES

Barry, D. and Hartigan, J. A. (1992). Product partition models for change point problems. *Ann. Statist.* **20**, 260–279.

Endres, D. and Földiák, P. (2005). Bayesian bin distribution inference and mutual information. *IEEE Transactions on Information Theory* **51**, 3766–3779.

Fearnhead, P. (2006). Exact and efficient Bayesian inference for multiple changepoint problems. *Statist. Computing* **16**, 203–213.

Hutter, M. (2006). Bayesian regression of piecewise constant functions. *Tech. Rep.*, IDSIA-14-05, Switzerland. http://arxiv.org/abs/math.ST/0606315.

Jong, K. *et al.* (2003). Chromosomal breakpoint detection in human cancer. *Applications of Evolutionary Computing: EvoWorkshops'03*, Berlin: Springer-Verlag, 54–65.

Kass R. E. and Wasserman, L. (1995). A reference Bayesian test for nested hypotheses with large samples. *Journal of the ACM* **90**, 773–795.

Olshen, A. B., Venkatraman, E. S., Lucito R. and Wigler, M. (2004) Circular binary segmentation for the analysis of array-based DNA copy number data. *Biostatistics* **5**, 557–572.

Picard, F., *et al.* (2005). A statistical approach for array CGH data analysis. *BMC Bioinformatics* **6**, 27.

Pinkel, D., *et al.* (1998). High resolution analysis of DNA copy number variation using comparative genomic hybridization to microarrays. *Nature Genetics* **20**, 207–211.

Schwarz, G. (1978). Estimating the dimension of a model. *Ann. Statist.* **6**, 461–464.

Sen, A. and Srivastava, M. S. (1975). On tests for detecting a change in mean. *Ann. Statist.* **3**, 98–108,.

Weakliem, D. L. (1999). A critique of the Bayesian information criterion for model selection. *Sociol. Meth. Research* **27**, 359–397.

Yao, Y.-C. (1984). Estimation of a noisy discrete-time step function: Bayes and empirical Bayes approaches. *Ann. Statist.* **12**, 1434–1447.

BAYESIAN STATISTICS 8, pp. 613–618.
J. M. Bernardo, M. J. Bayarri, J. O. Berger, A. P. Dawid,
D. Heckerman, A. F. M. Smith and M. West (Eds.)

Identification of Thyroid Gland Activity in Radiotherapy*

LADISLAV JIRSA
Acad. Sciences, Czech Republic
jirsa@utia.cas.cz

ANTHONY QUINN
Trinity College, Dublin, Ireland
aquinn@tcd.ie

FERDINAND VARGA
Charles University, Prague, Czech Republic
varga@utia.cas.cz

SUMMARY

The Bayesian identification of a linear regression model for time dependence of thyroid gland activity in ^{131}I radiotherapy is presented. Prior knowledge is elicited via hard parameter constraints and via the merging of information from an archive of patient records. This prior regularization is shown to be crucial in the reported context, where data typically comprise only 2-3 high-noise measurements. The posterior distribution is simulated via a Langevin diffusion algorithm, whose optimization for the thyroid activity application is explained. Improved patient-specific predictions of thyroid activity are reported. The posterior inference of the total patient radiation dose is presented, allowing the uncertainty of the dose to be quantified in a consistent manner.

Keywords and Phrases: EXTERNAL INFORMATION; INFORMATION MATRIX; PRIOR CONSTRAINTS; NORMAL-INVERSE-GAMMA; LANGEVIN DIFFUSION; NONPARAMETRIC STOPPING RULE.

1. INTRODUCTION

Radioactive ^{131}I is used for radiotherapy in thyroid gland carcinoma, since anorganic iodine is selectively accumulated in the thyroid gland. The activity of the organ, A_t (MBq), at time t (days) is related to the total *dose* (energy absorbed per unit mass), $\mathcal{D} = \mathcal{S}\xi$, via $\xi = \int_0^{+\infty} A_t \; dt$, where $\mathcal{S}$ is a known organ- and isotope-specific constant. ξ is therefore a key quantity in the design of an effective and safe administration of ^{131}I. For practical reasons, measurement pairs, $\{(t_i, d_{t_i})\}_{i=1}^{n} \equiv D$, are available at only a small number, $2 \leq n \lesssim 9$, of non-uniform sampling times, t_i, and are subject to considerable measurement noise. The principal task of this paper is to infer A_t, and hence ξ, in this difficult measurement regime.

---

* This is a synopsis. For the full version, see *Bayesian Analysis* **2** (2007), to appear.
Grants AV ČR 1ET 1007 50404 and MŠMT ČR 1M0572 partially supported this work.

2. A REGRESSION MODEL FOR THYROID ACTIVITY

A_t should observe the following prior constraints: (i) $A_t \geq 0$, $= 0$ at $t = 0$ and as $t \to +\infty$; (ii) A_t is unimodal, achieving a maximum at t_m, where $t_m \in (t_l, t_u)$; (iii) from medical experience $t_l = 4$ hours (0.167 days) and $t_u = 72$ hours (3 days); and (iv) for some $t_h > t_m$, the decrease of A_t, $t > t_h$, is faster than the physical decay of ^{131}I. The following three-parameter model was proposed in (Heřmanská, 2001):

$$\ln A_t = a_1 + a_2 \ln(ct) + a_3 (ct)^{\frac{2}{3}} \ln(ct) - \frac{t}{T_p} \ln 2, \tag{1}$$

Here, $a = [a_1, a_2, a_3]'$ is the *patient-specific* vector of unknown linear regression coefficients, T_p is the physical half-life of ^{131}I (8.04 days) and c is a known time-scale coefficient, the value of which will be discussed shortly. The positive measurement process, d_t, is taken as conditionally log-normal. Since $A_t \gg 0$, it follows that

$$x_t \equiv \ln d_t + \frac{t}{T_p} \ln 2 = \psi_t' a + e_t, \tag{2}$$

where $e_t \sim \mathcal{N}(0, r)$ is uncorrelated (white) noise, with variance r. In (2), $\psi_t = [1,\ \ln(ct),\ (ct)^{2/3} \ln(ct)]'$ is the regressor at time t. A sample activity curve, A_t, is illustrated in Fig. 1.

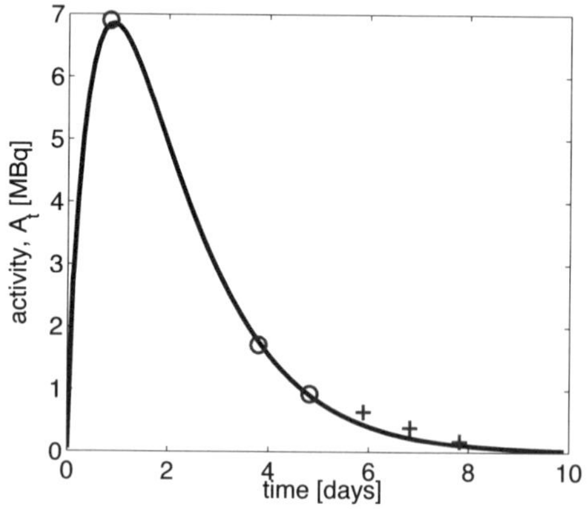

Figure 1: *Sample activity curve, A_t, identified on three data pairs.*

3. BAYESIAN CONJUGATE INFERENCE OF THYROID ACTIVITY

The conjugate parameter prior for $f(x_t|a, r)$ in (2) is $\mathcal{N}i\mathcal{G}(V_0, \nu_0)$, *i.e. Normal--inverse-Gamma*, with prior parameters V_0 and ν_0. However, a hard constraint, $a \in \mathbb{A} \subset \mathbb{R}^3$, must be respected in order to satisfy (i)–(iv) (Section 2). This is expressed via the uniform prior, $\chi_{\mathbb{A}}(a)$, with support $\mathbb{A}$. The marginal posterior distribution of a is then (Kárný, 2005)

$$f(a|L, \Lambda, \nu) \propto \tilde{f}(a|L, \Lambda, \nu)\chi_{\mathbb{A}}(a), \tag{3}$$

$$\tilde{f}(a|L, \Lambda, \nu) = \mathcal{I}^{-1}(L, \Lambda, \nu)\left[1 + \lambda_{11}^{-1}\left(a - \tilde{\mathsf{E}}a\right)' L_{aa}' \Lambda_{aa} L_{aa}\left(a - \tilde{\mathsf{E}}a\right)\right]^{-\frac{1}{2}(\nu-2)} \tag{4}$$

Here, the n measurements, D (Section 1), have entered the posterior inference via the LD-decomposition (Kárný, 2005) $V = L'\Lambda L$, of the *extended information matrix*,

$$V = V_0 + \sum_{i=1}^{n} \Psi_{t_i}\Psi'_{t_i}, \tag{5}$$

where $\Psi_{t_i} = (x_{t_i}, \psi'_{t_i})'$, from (2), is the *extended regressor* at sampling time t_i. The lower-diagonal matrix, L, has been partitioned as

$$L = \begin{pmatrix} l_{11} & 0 \\ l_{a1} & L_{aa} \end{pmatrix},$$

where $l_{11} = 1$ is the $(1,1)$ element. Similarly, λ_{11} is the $(1,1)$ element of diagonal matrix, Λ. In (4), $\nu = \nu_0 + n$ is the *degrees-of-freedom* parameter. The normalizing constant, $\mathcal{I}(L, \Lambda, \nu)$, of the unconstrained distribution (4) is available in closed form, but that of the complete inference (3) must be evaluated numerically, if needed. Finally, the posterior moments of a with respect to the *unconstrained* inference, $\tilde{f}(a|L, \Lambda, \nu)$ (4), are, respectively,

$$\tilde{\mathsf{E}}[a|L,\Lambda,\nu] \equiv \tilde{\mathsf{E}}a = L_{aa}^{-1}L_{a1}, \qquad \tilde{\mathsf{cov}}[a|L,\Lambda,\nu] = \frac{\lambda_{11}}{\nu - 7}\, L_{aa}^{-1}\Lambda_{aa}^{-1}\left(L'_{aa}\right)^{-1}. \tag{6}$$

3.1. *Prior Support, $\mathbb{A}$*

Constraints (i) and (ii) in Section 2 are satisfied if $a_3 < 0 < a_2$. Examining the function $g_t = \psi'_t a$ (2), its unique maximizer, t_{mg}, solves $g'_t = 0$. Then, the unique maximizer, t_m, of A_t (see (ii)) exists and $t_{mg} > t_m$ since $\ln 2/T_p > 0$. For $t > t_{mg}$, $g'_t < 0$, satisfying (iv) for $t_h = t_{mg}$. Imposing $t_l < t_m < t_u$ (see (iii)) via the derivative $(\ln A_t)'$, we get

$$-a_3(ct_l)^{\frac{2}{3}}\left(\frac{2}{3}\ln(ct_l)+1\right) + \frac{t_l}{T_p}\ln 2 < a_2 < -a_3(ct_u)^{\frac{2}{3}}\left(\frac{2}{3}\ln(ct_u)+1\right) + \frac{t_u}{T_p}\ln 2.$$

The first inequality here can be written as $a_2 + ka_3 > q$, where $q > 0$. For $t_m = t_l$, it becomes $a_2 + ka_3 = q$. If $k < 0$, then it is necessary to have $a_2 < 0$ for some values of a_3, in order that t_m can reach its lower limit. This violates the condition $a_2 > 0$. Hence, to guarantee $a_2 > 0$, c can be chosen to ensure that $k \geq 0$. In particular, $k = 0$ if $c = \frac{1}{t_l}\exp\left(-\frac{2}{3}\right) \equiv 1.3388$. The first inequality above becomes $a_2 > t_l \ln 2/T_p$, $\forall a_3$. Then, $a_2 > 0$ is redundant.

These bounds may be combined into the following linear matrix element-wise inequality, whose solution space, $\mathbb{A}$, provides the support for (3):

$$M\,a < b, \qquad M = \begin{pmatrix} 0 & 0 & 1 \\ 0 & 1 & 4.8687 \\ 0 & -1 & 0 \end{pmatrix}, \quad b = \begin{pmatrix} 0 \\ 0.2586 \\ -0.0144 \end{pmatrix}. \tag{7}$$

3.2. *Prior Parameters, V_0 and ν_0*

External information is merged from a database of 3 876 sampled patient activity curves, collected at the Clinic of Nuclear Medicine at Motol Hospital in Prague. V_0

is constructed using these data sequences $D_j, j = 1, \ldots, 3\,876$, each scaled to $(0, 1]$. They form m clusters, one for each of the standard measurement times in the Hospital, and so can be described by a mixture, $M(x_t, t)$, of components in (x_t, t)-space (2). The full-rank prior information matrix is then $V_0 = w \int M(x_t, t)\Psi\Psi' \,\mathrm{d}\Psi$, using the information merging technique in (Kracík, 2005). The weight, w, is assigned the value 5.3×10^{-5}, which optimizes the prediction of the fourth measurement (Section 4). This value of V_0 in (5) ensures that $\tilde{\mathsf{E}}[a|V_0, \nu_0] \in \mathbb{A}$ (Jirsa, 2007).

ν_0 was chosen as 7.05 so that $\tilde{\mathsf{cov}}[a|L_0, \Lambda_0, \nu_0]$ exists. In the radiotherapy context, the minimal number of observations is $n = 2$, giving $\nu = 9.05$, ensuring finiteness of the posterior variance of r.

3.3. *Posterior Inference Methodology*

The hard constraint, $\chi_{\mathbb{A}}(a)$ (3), obviates analytical evaluation of the posterior moments of a. Moreover, the posterior inference of ξ (Section 1) is unavailable since the mapping $a \rightarrow \xi(a)$ cannot be expressed in closed form. Hence, a simulation-based method is required.

The distribution of the transformed variables, $a^* = T(a - \tilde{\mathsf{E}}a)$, which uses (6) with $T = \sqrt{\frac{\nu-7}{\lambda_{11}}\Lambda_{aa}}L_{aa}$, was sampled. Here, $\sqrt{}\Lambda_{aa}$ denotes the element-wise square-root. The transformed support, $\mathbb{A}^*$ (7), is the solution space of $M^*\ a^* < b^*$, with $M^* = MT^{-1}$ and $b^* = b - M\tilde{\mathsf{E}}a$. From (6), the unconstrained posterior distribution of a^* is therefore zero-mean with identity covariance matrix. Markov chain realizations were generated efficiently in these transformed variables, via the Langevin diffusion algorithm (Roberts, 1998).

The chain was initialized at the MAP estimate, found by constrained optimization of the quadratic denominator (4). Within a burn-in period (200–400 samples), the initial step-size of the Markov chain was trimmed to reach its near-optimum value (Roberts, 1998), ensuring good mixing of the chain.

After the burn-in, the samples were each inverted via the linear mapping above, generating approximate realizations from (3). Finally, each realization of a was substituted into (1), and the equivalent realizations from $f(\xi|L, \Lambda, \nu)$ were obtained by numerical integration (Section 1).

Using a nonparametric Bayesian stopping rule (Quinn, 2007), sampling from $f(\xi|L, \Lambda, \nu)$ in a specific patient case is stopped when the uncertainty in the 95% confidence interval bounds and width falls below 3 %. The average number of samples at stopping, for 700 archive data sequences, is 4 529. Bayesian confidence intervals for ξ can be evaluated under this procedure in fractions of a second, when implemented in C++ on a standard PC. Hence, it is suitable for on-line use in clinical practice.

4. PREDICTION OF THYROID ACTIVITY AND INFERENCE OF DOSE

The influence of the priors above was examined in the context of a database of 2355 records, each containing at least four measurement pairs, D. For each patient case, A_t (1) was inferred from the first three measurements, and used to predict the fourth, which typically followed after 1–3 days (see Fig. 1). This reflects the usual practice of taking no more than $n = 3$ measurements per patient. Predictions were generated for each patient as follows: (a) via (4), with $V_0 = \epsilon I_4$, $\epsilon = 0^+$, and $\nu_0 = 7.05$ (in which case about 40 % had to be rejected, as the inferred A_t was physically impossible); (b) via (3), initialized as in (a) above; and (c) via (3), with

V_0 assigned as in Section 3.2. The major prediction advantage in case (c) is evident in Table 1. The statistics describe the distribution of relative prediction errors over the set of 2355 records. The 'data' column gives the number of data sequences yielding valid predictions. The percentage of outliers in relative prediction error (being more than three standard deviations from the mean), for valid predictions, is in the last column.

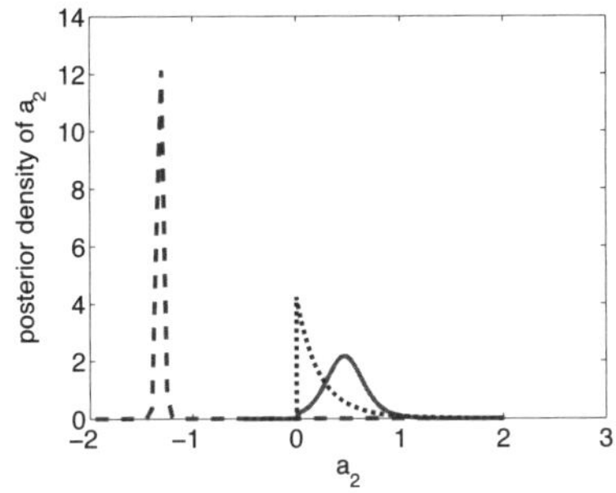

Figure 2: *Marginal posterior inference of a_2 for a specific patient. The* dashed *curve corresponds to case (a) in Table 1, the* dotted *curve to case (b), and the* solid *curve to case (c).*

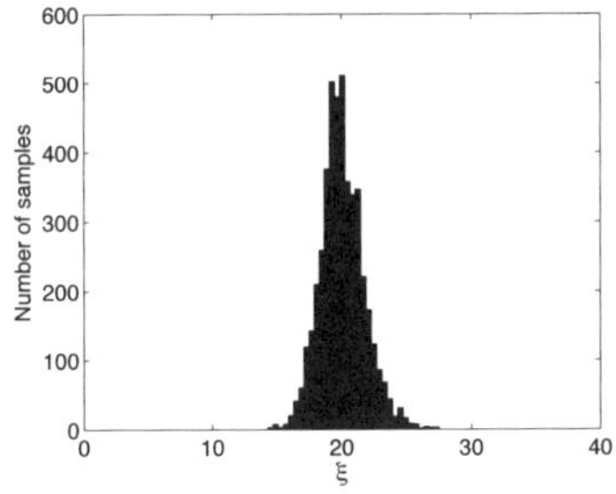

Figure 3: *Empirical distribution of ξ estimated using the data in Fig. 1.*

The influence of externally-informed V_0 (Section 3.2) on the inference of a_2 is illustrated in Fig. 2 for a specific patient case.

Table 1: *Relative prediction error for three prior knowledge structures: (a) without hard parameter constraints or externally informed V_0; (b) with hard parameter constraints but without externally informed V_0; and (c) with both hard parameter constraints and side-informed V_0 (see Section 3.2).*

Prior	mean	median	st.dev.	data	outliers
(a)	0.0576	−0.0066	0.475	1 403	2.28 %
(b)	−0.0968	−0.1456	0.431	2 355	0.85 %
(c)	−0.0004	−0.0544	0.416	2 355	0.81 %

The empirical distribution of ξ, corresponding to the data in Fig. 1, is shown in Fig. 3. For data from 700 patients, the mean skewness of $f(\xi|L, \Lambda, \nu)$ is 1.66, whereas that of $f(\ln \xi | L, \Lambda, \nu)$ is 0.29, supporting the hypothesis of a nonsymmetric pdf such as the log-normal.

5. DISCUSSION AND CONCLUSION

The hard constraint (7) has been effective in ensuring physically realizable inferences of thyroid activity in ^{131}I radiotherapy (Table 1). This, and the merging of external information from a large database of patient data, via prior parameter V_0 (Section 3.2), provides reliable activity predictions in this data-parsimonious, high-noise context. The influence of this prior information on identification of parameter a_2 is illustrated in Fig. 2. Although the parametric form of $f(\xi | L, \Lambda, \nu)$ (see Fig. 3) is not known, we found that it satisfied several log-normal properties, notably positive support and skewness. This is in accordance with (Hamby, 1999), where $f(\xi)$ is found to be analytically log-normal, although their inference is not patient-specific. In future work, the fixed time-scale coefficient, c (1), could be introduced via another hard-constrained patient-specific parameter, resulting in a model with 4 regression coefficients. Other methods of exploiting data-based prior information can be tested, although the one presented in Section 3.2 provides significant improvement in identification, as reported in Section 4, as well as improved chain convergence and computational speed. The identified model will be used clinically in the design of patient-specific optimized administrations of ^{131}I, contributing to treatment quality and radiation protection.

REFERENCES

Hamby, D. M. and Benke, R. R. (1999). Uncertainty of the iodine-131 ingestion dose conversion factor. *Radiation Protection Dosimetry*, **82**, 245–256.

Heřmanská, J., Kárný, M., Zimák, J., Jirsa, L., Šámal, M. and Vlček, P. (2001). Improved prediction of therapeutic absorbed doses of radioiodine in the treatment of thyroid carcinoma. *J. Nuclear Medicine*, **42**, 1084–1090.

Jirsa, L., Quinn, A. and Varga, F. (2007). Identification of Thyroid Gland Activity in Radiotherapy. *Bayesian Analysis* **2** (to appear).

Kárný, M., Böhm, J., Guy, T. V., Jirsa, L., Nagy, I., Nedoma, P. and Tesař, L. (2005). *Optimized Bayesian Dynamic Advising: Theory and Algorithms*. Berlin: Springer-Verlag

Kracík, J. and Kárný, M. (2005). Merging of data knowledge in Bayesian estimation. *Proc. 2nd International Conference on Informatics in Control, Automation and Robotics* (J. Filipe, J. A. Cetto and J. L. Ferrier, eds.) Barcelona, 229–232.

Quinn, A. and Kárný, M. (2007). Learning for nonstationary Dirichlet processes. *Internat. J. Adaptive Control and Signal Processing* **21** (to appear).

Roberts, G. O. and Rosenthal, J. S. (1998). Optimal scaling of discrete approximation to Langevin diffusions. *J. Roy. Statist. Soc. B* **60**, 255–268.

BAYESIAN STATISTICS 8, pp. 619–624.
J. M. Bernardo, M. J. Bayarri, J. O. Berger, A. P. Dawid,
D. Heckerman, A. F. M. Smith and M. West (Eds.)

Partial Convexification of Random Probability Measures*

GEORGE KOKOLAKIS and GEORGE KOUVARAS
National Technical University of Athens, Greece
kokolakis@math.ntua.gr gkouv@math.ntua.gr

SUMMARY

The construction of an absolutely continuous random probability measure is examined here. To achieve this we propose a 'partial convexification' procedure of a random process, such as the Dirichlet, resulting in a multimodal distribution function.

Keywords and Phrases: CONVEXITY; DIRICHLET PROCESS; MULTIMODAL DISTRIBUTIONS; POLYA TREES; RANDOM PROBABILITY MEASURES.

1. INTRODUCTION

The Dirichlet process, introduced in Ferguson (1973, 1974), is usually chosen to represent nonparametric prior information on a space of probability measures. The major drawback of a Dirichlet process is that it selects discrete distributions with probability one. Alternative methods such as mixtures of Dirichlet processes (Antoniak, 1974, Escobar and West, 1995 and Walker *et al.*, 1999) and Polya trees (Lavine, 1992, 1994) have been proposed to overcome this problem. In this paper, we present a Bayesian procedure that results to absolutely continuous random probability measures that have finite expected number of modes (Kokolakis and Kouvaras, 2007). For the construction of random Dirichlet probability measures, a variant of Polya trees (Kokolakis, 1983, Kokolakis and Dellaportas, 1996) is used.

In Section 2, we review some concepts on unimodality in $\mathbb{R}$ and provide the theoretical background on 'partial convexification' which is needed to implement our methodology to get multimodal distributions. In Section 3, we present some illustrative results and possible extensions are discussed.

---

* This is a synopsis. For the full version, see *Bayesian Analysis* **2** (2007), to appear.
George Kokolakis is Professor of Statistics at the National Technical University of Athens.
George Kouvaras is PhD student at the National Technical University of Athens.

2. FROM UNIMODALITY TO PARTIAL CONVEXIFICATION

Much of the work with unimodal distribution functions is based on a representation theorem due to Khincin and Shepp (see Feller, Vol. 2, p. 158). In particular, we have:

Theorem 1 *The c.d.f. F is unimodal with mode at zero if and only if it is of the form:*

$$F(t) = \int_0^1 G(t/u)\,du, \qquad t \in \mathbb{R}.$$

This means that F is the distribution of the product of two independent random variables U and Y, with U uniformly distributed on $(0,1)$ and Y having an arbitrary c.d.f. G.

It is easy to realize that Theorem 1 takes the following equivalent form:

Corollary 1 *The c.d.f. F is convex on the negative real line and concave on the positive, if and only if there exists a distribution function G on $\mathbb{R}$ such that F admits the representation:*

$$F(x) = G(0-) + \int_{\mathbb{R}} H_y(x) G(\,dy), \qquad x \in \mathbb{R}, \tag{1}$$

where $H_y(\cdot)$ is, for $y \neq 0$,

$$H_y(x) = \begin{cases} 0, & \dfrac{x}{y} \leq 0, \\[2ex] \dfrac{x}{|y|}, & 0 < \dfrac{x}{y} < 1, \\[2ex] \dfrac{y}{|y|}, & \dfrac{x}{y} \geq 1, \end{cases}$$

and for $y = 0$, $H_0(\cdot)$ is degenerate at zero.

We notice that with $y = 0$ we have $p \equiv Pr[X = 0] = Pr[Y = 0]$. With $y \neq 0$, $H_y(x)$ is a bounded function of x with bounded left and right derivatives. Applying the bounded convergence theorem we conclude that the c.d.f. F is differentiable a.e. in $\mathbb{R}$, and its derivative, wherever it exists, is

$$f(x) \equiv DF(x) = \int_{\mathbb{R}\setminus\{0\}} DH_y(x) G(\,dy) = \begin{cases} \displaystyle\int_{(-\infty,x)} \frac{1}{|y|} G(\,dy), & x < 0, \\[3ex] \displaystyle\int_{(x,+\infty)} \frac{1}{|y|} G(\,dy), & x > 0. \end{cases} \tag{2}$$

In this case $\int f(x)\,dx = 1 - p$. When $p = 0$, f is a density function.

Introducing the above result into (1) we get:

$$G(x-) + Pr[Y = x] \cdot I_{(x \geq 0)} = F(x) - x f(x), \qquad x \in \mathbb{R}, \tag{3}$$

and, under differentiability conditions,

$$g(x) = -xf'(x), \qquad x \in \mathbb{R}.$$

Special cases of (2) and (3) can be found in Brunner (1992) and in Hansen and Lauritzen (2002).

According to the above procedure we always get a c.d.f. F with single mode at zero, no matter what the distribution G, we start with, is. To overcome this limitation we propose a 'partial convexification' procedure of a c.d.f. G that produces a c.d.f. F with a finite expected number of modes.

'Partial convexification' procedure relies on using a Un(α, 1) distribution, with $0 < \alpha < 1$, instead of Un(0,1). The parameter α can be fixed, or random with a prior distribution $p(\alpha)$, on the interval (0, 1).

In our Bayesian model specification we assume the following:

(i) $Y \sim G$, where G is a random c.d.f. distributed as a Dirichlet process $\mathrm{Di}(\beta(\cdot))$ and $\beta(\cdot)$ a σ–additive nonnull finite measure on $(\mathbb{R}_+, \mathcal{B}_+)$,

(ii) $U \sim \mathrm{Un}(\alpha, 1)$ with α fixed on the interval (0, 1).

(iii) Y and U are independent.

Then

$$F(x) = \int_{(0,+\infty)} H_y(x)G(dy), \quad x \in \mathbb{R}_+, \tag{4}$$

where $H_y \sim \mathrm{Un}(\alpha y, y)$, *i.e.*,

$$H_y(x) = \begin{cases} 0, & x \leq \alpha y, \\ \dfrac{x - \alpha y}{(1-\alpha)y}, & \alpha y < x < y, \\ 1, & x \geq y, \end{cases} \tag{5}$$

and the c.d.f. F of the product $X = UY$ admits a.e. a derivative f. Specifically, the following equation is satisfied:

$$F(x) = G(x) + xf(x) - \frac{\alpha}{1-\alpha}\left\{G\left(\frac{x}{\alpha}\right) - G(x)\right\}, \qquad x \in \mathbb{R}_+. \tag{6}$$

It is interesting to notice that when Y has a discrete probability function $\{q_k = Pr[Y = k], \quad k = 1, 2, ...\}$, equation (4) becomes

$$F(x) = \sum_{k=1}^{[x]} q_k + \frac{x}{1-\alpha} \sum_{k=[x]+1}^{[\frac{x}{\alpha}]} \left(\frac{1}{k} - \frac{\alpha}{x}\right) q_k, \qquad x \in \mathbb{R}_+, \tag{7}$$

where $[\cdot]$ stands for the integer part, and

$$f(x) = \frac{1}{1-\alpha} \sum_{k=[x]+1}^{[\frac{x}{\alpha}]} \frac{1}{k} q_k, \qquad x \in \mathbb{R}_+. \tag{8}$$

According to (4) we obtain a prior distribution on the subspace of multimodal c.d.f.'s. The expected number of modes of F increases from one, when $\alpha = 0$, to infinity, when $\alpha = 1$, having a finite number of modes when $0 < \alpha < 1$. This means that when $0 < \alpha < 1$, the c.d.f. $F(x)$ alternates between local concavities and local convexities, *i.e.*, a 'partial convexification' of F is produced.

3. APPLICATION

For demonstration purposes we have simulated ten datasets from a mixture of two normal distributions, specifically $N(15, 3^2)$ and $N(30, 4^2)$, with weights $w_1/w_2 = 2/3$.

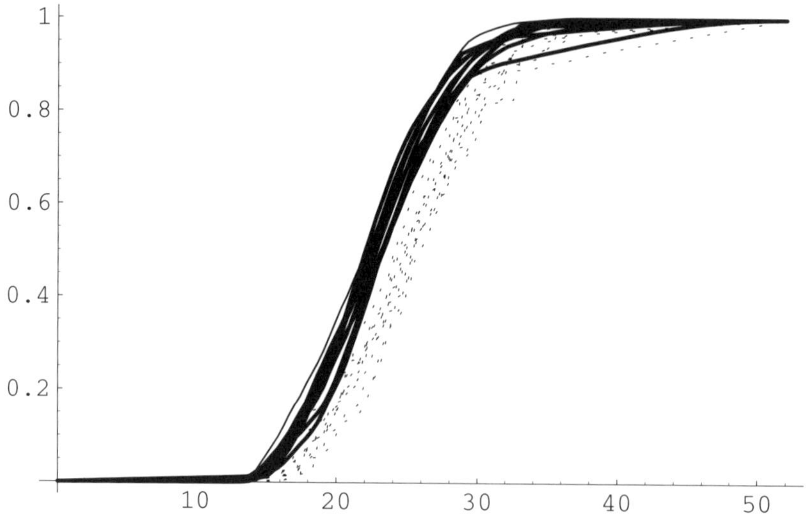

Figure 1: *Locally convexified random Dirichlet probability distributions (continuous lines) and preconvexified random Dirichlet probability distributions (dotted lines).*

The sample sizes have all been taken equal to 250. Dirichlet processes have been produced with parameter $50 \times \text{Ga}$, where Ga stands for the Gamma distribution function with mean $\mu = 25$ and standard deviation $\sigma = 5$. The 'partial convexification' procedure has been applied with the parameter $\alpha = 0.80$.

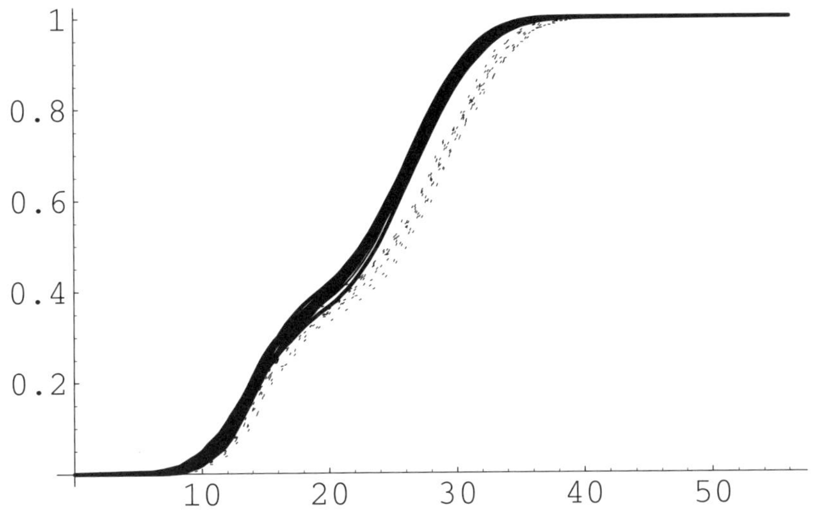

Figure 2: *Locally convexified posterior Dirichlet probability distributions (continuous lines) and posterior Dirichlet probability distributions (dotted lines).*

REFERENCES

Antoniak, C.E. (1974). Mixtures of Dirichlet processes with applications to Bayesian nonparametric problems. *Ann. Statist.* **2**, 1152–1174.

Brunner, L.J. (1992). Bayesian nonparametric methods for data from a unimodal density. *Statistics and Probability Lett.* **14**, 195–199.

Escobar, M.D. and West, M. (1995). Bayesian density estimation and inference using mixtures. *J. Amer. Statist. Assoc.* **90**, 577–588.

Ferguson, Th. S. (1973). A Bayesian analysis of some nonparametric problems. *Ann. Statist.* **1**, 209–230.

Ferguson, Th. S. (1974). Prior distributions on spaces of probability measures. *Ann. Statist.* **2**, 615–629.

Feller, W. (1971). *An Introduction to Probability Theory and its Applications.* Vol. 2. (2nd ed.) New York: Wiley.

Hansen, M. and Lauritzen, S. (2002). Nonparametric Bayes inference for concave distribution functions. *Statistica Neerlandica* **56**, 110–127.

Kokolakis, G. (1983). A new look at the problem of classification with binary data. *The Statistician* **32**, 144–152.

Kokolakis, G. and Dellaportas, P. (1996). Hierarchical modelling for classifying binary data. *Bayesian Statistics 5* (J. M. Bernardo, J. O. Berger, A. P. Dawid and A. F. M. Smith, eds.) Oxford: University Press, 647–652.

Kokolakis, G. and Kouvaras, G. (2007). On the multimodality of random probability measures. *Bayesian Analysis* **2**, 213–220.

Lavine, M. (1992). Some aspects of Polya tree distributions for statistical modelling. *Ann. Statist.* **20**, 1222–1235.

Lavine, M. (1994). More aspects of Polya tree distributions for statistical modelling. *Ann. Statist.* **22**, 1161–1176.

Walker, S.G., Damien, P., Laud, P.W. and Smith, A.F.M. (1999). Bayesian nonparametric inference for random distributions and related functions. *J. Roy. Statist. Soc. B* **61**, 485–528 (with discussion).

BAYESIAN STATISTICS 8, pp. 625–630.
J. M. Bernardo, M. J. Bayarri, J. O. Berger, A. P. Dawid,
D. Heckerman, A. F. M. Smith and M. West (Eds.)

Bayesian Multivariate Areal Wombling*

Haijun Ma
Amgen Inc., USA
hma@amgen.com

Bradley P. Carlin
University of Minnesota, USA
brad@biostat.umn.edu

Summary

Multivariate data summarized over areal units (counties, zip codes, etc.) are common in the field of public health. Statistical methods for areal wombling (boundary analysis) must account for correlations across both diseases and locations. Utilizing recent developments in multivariate conditionally autoregressive (MCAR) distributions and spatial structural equation modeling, we suggest a variety of Bayesian hierarchical models for multivariate areal boundary analysis, including some that incorporate random neighborhood structure. We illustrate our methods using Minnesota county-level esophagus, larynx, and lung cancer data, using methods that permit identification of both composite and cancer-specific boundaries. Our results indicate primary boundaries in both the composite and cancer-specific response surface separating the mining- and tourism-oriented northeast counties from the remainder of the state.

Keywords and Phrases: Areal data; Cancer; Multivariate conditionally autoregressive (MCAR) model.

1. INTRODUCTION

Recently, there has been increasing interest in the spatial problem of detecting barriers separating regions of high and low response for certain quantities of interest. This area is often referred to as *boundary analysis* or *wombling*, the latter name paying tribute to an early important paper in the area (Womble, 1951). In public health, wombling is useful for detecting regions of significantly different disease mortality or incidence, thus improving decision-making regarding disease prevention and control, allocation of resources, and so on.

Multivariate areal data are common in public health studies. Much of this data is *areal* (aggregated at a certain regional level) to protect subjects' privacy. Observations collected over a map are often spatially correlated. At the same time, the

* This is a synopsis. For the full version, see *Bayesian Analysis* **2** (2007), to appear.
Haijun Ma is is Senior Biostatistician with Amgen Inc., Thousand Oaks, CA, USA. Bradley P. Carlin is Professor of Biostatistics in the School of Public Health at the University of Minnesota, USA.

multiple variables under study (e.g., diseases sharing etiologic or other risk factors) are often correlated too. As an example, consider the Surveillance, Epidemiology and End Results (SEER; `http://seer.cancer.gov`) database, which provides the numbers of deaths and corresponding numbers of person-years at risk in quinquennial age brackets for each county in various U.S. states and each of several cancer sites. Here we will study SEER data on the numbers of deaths due to cancers of the lung, larynx and esophagus in the years from 1990 to 2000 at the county level in the US state of Minnesota. These three cancer sites are all part of the upper aerodigestive tract, and hence closely related anatomically.

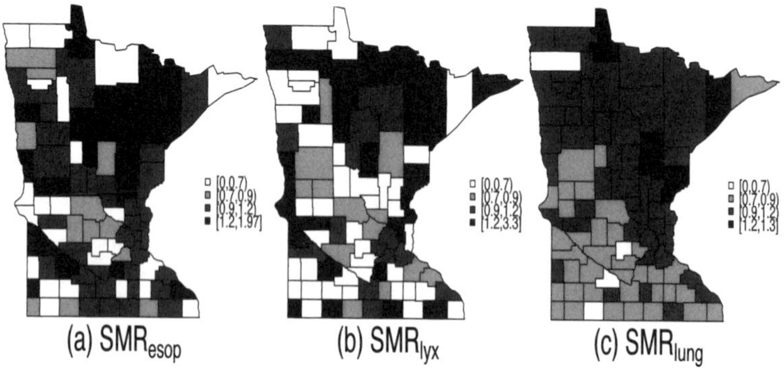

Figure 1: *Maps of age-adjusted standardized mortality ratios: (a) esophagus cancer; (b) larynx cancer; (c) lung cancer.*

Fig. 1 gives the raw county-level standardized mortality ratios (SMRs) based on our data, calculated as Y_{ik}/E_{ik} for $i = 1, \ldots, n$ and $k = 1, 2, 3$, where Y_{ik} is the cancer death count and E_{ik} is the age-adjusted expected count for cancer k in county i. The SMR maps for all three cancers indicate a pattern of decrease from northeast to southwest, suggesting positive association among them. The pattern is strongest for lung cancer, which also has less variable SMRs due to the higher case counts. Our primary substantive goal is to identify significant boundaries on these maps for the cancers individually, as well as for any underlying spatial common factor, all while accounting for correlation across both counties and cancers. Boundaries are important here for public health professionals tasked with identifying geographic regions in need of certain cancer-related intervention efforts, e.g., a cancer education or screening campaign focused at a few shopping malls or other public locations. They are also useful for identifying regions of rapid change in the fitted cancer surface, so that these areas can be studied in more detail for clues (say, missing covariates) that might explain why cancer mortality differs across the identified boundaries.

For correlated areal variables, the most popular modeling approach has been through the conditionally autoregressive (CAR) distribution (Besag, 1974) and its variants. Mardia (1988) described the theoretical background for multivariate CAR (MCAR) specifications using Gaussian Markov random fields (MRFs). Many generalizations of MCAR allow more flexible modeling of associations between different variables and areal units; see e.g. Kim et al. (2001), Carlin and Banerjee (2003), and Gelfand and Vounatsou (2003).

The rest of this article is organized as follows. Section 2 is devoted to providing background information on univariate areal wombling and spatial factor analysis. Section 3 then describes our proposed multivariate areal wombling techniques. The methods are illustrated using our SEER cancer data in Section 4, where disease-specific boundaries are constructed and mapped. Finally, Section 5 concludes and mentions directions for further work in this area. Readers interested in more information may wish to consult the full version of this paper (Ma and Carlin, 2007).

2. METHODOLOGICAL BACKGROUND

2.1. *Existing Methods for Univariate Areal Boundary Analysis*

Boundary analysis typically involves estimation or testing of 'lines" on a continuous surface. In the areal case, however, defensible boundaries can only be a subset of the borders that determine the areal units, since we lack within-unit information.

Ma et al. (2006) embedded areal boundary analysis in a Bayesian hierarchical modeling framework. For a single disease having observed counts Y_i, expected counts E_i, and covariate vector $\mathbf{x}_i$ for region i, they assume $Y_i \mid \boldsymbol{\beta}, \phi_i^S \sim Poisson(\mu_i)$, with

$$\log \mu_i = \log E_i + \mathbf{x}_i' \boldsymbol{\beta} + \phi_i^S, \quad i = 1, \ldots, n \,. \tag{1}$$

The paper then proposed a 'site-edge" (SE) approach that jointly models two types of random effects. First, the set of site-level (areal) random effects $\boldsymbol{\phi}^S = \{\phi_i^S\}$ are given a CAR prior to account for spatial association among the areas. Then a second set of *edge-level* random effects $\boldsymbol{\phi}^E = \{\phi_i^E\}$ are included and their distribution modeled using edge adjacency information from the map, enabling smooth, connected boundaries. More specifically, the SE model assumes

$$p(\boldsymbol{\phi}^S \mid \boldsymbol{\phi}^E, \tau_\phi) \quad \propto \quad \exp\left\{ -\frac{\tau_\phi}{2} \sum_{i \sim j} (1 - \phi_{ij}^E)(\phi_i^S - \phi_j^S)^2 \right\} , \tag{2}$$

$$\text{and} \quad p(\boldsymbol{\phi}^E) \quad \propto \quad \exp\left\{ -\nu \sum_{ij \sim kl} \phi_{ij}^E \phi_{kl}^E \right\} , \tag{3}$$

where $\phi_i^S \in \Re$ and $\phi_{ij}^E \in \{0, 1\}$. That is, $\phi_{ij}^E = 1$ if edge (i, j) is a boundary element, and 0 otherwise, so smoothing of neighboring site effects ϕ_i^S and ϕ_j^S is only encouraged if there is no boundary between them. The prior for $\boldsymbol{\phi}^E$ in (3) is an *Ising* model with 'binding strength" parameter ν (Geman and Geman, 1984, p. 725). Smaller (or even negative) values of ν lead to more connected boundary elements, hence more separated areal units. Finally, edges ij and kl are considered adjacent ($ij \sim kl$) if they are connected on the map. Ma et al. (2006) use this approach to find Medicare service area boundaries for two competing hospice systems headquartered in Duluth, Minnesota, but their work did not address potential dependence between the two hospices.

2.2. *Spatial Factor Analysis*

Often it is scientifically interesting to find whether multiple diseases share common underlying factors, which are often understood as a mixture of shared risk factors, socioeconomic status, and so on. An example is the shared component model of

Knorr-Held and Best (2001), illustrated in the WinBUGS user manual. The model partitions the geographical variation in two diseases into a common (shared) component $\boldsymbol{\theta}$, and two disease-specific (residual) components $\boldsymbol{\psi}_1$ and $\boldsymbol{\psi}_2$. Assuming the death counts Y_{ik} for disease k in area i to be independent Poisson variables with mean μ_{ik}, $k = 1, 2$, they model

$$\log(\mu_{ik}) = \log E_{ik} + \mathbf{x}'_{ik}\boldsymbol{\beta}_k + \delta_k\theta_i + \psi_{ik} \;, \tag{4}$$

where E_{ik} are the expected counts, and the scaling parameters δ_k allow different 'risk gradients" for different diseases. The covariates $\mathbf{x}_{ik}$ are the multiple disease analog of those in (1); if unavailable, the $\boldsymbol{\beta}_k$ become univariate intercept parameters. Convolution priors (Besag et al., 1991) may be given to the three components $\boldsymbol{\theta}, \boldsymbol{\psi}_1$, and $\boldsymbol{\psi}_2$, expressing each areal random effect as a sum of a spatially structured (usually CAR) random effect and a non-spatial *i.i.d.* white noise random effect.

3. MULTIVARIATE BOUNDARY ANALYSIS

In this section, we develop several multivariate boundary analysis methods using the building blocks discussed in the previous section. We use the above Poisson likelihood and the link function given in (4), but slightly different priors than in Subsection 2.2, namely

$$\boldsymbol{\theta}|\tau_s \;\sim\; CAR(\tau_s, W) \;, \tag{5}$$

$$\text{and} \;\; \psi_{ik}|\tau_{\psi_k} \;\stackrel{iid}{\sim}\; N(0, 1/\tau_{\psi_k}) \;. \tag{6}$$

Boundaries for the latent factor $\boldsymbol{\theta}$ identify edges of abrupt change that are common to all of the diseases. Note that (6) assumes the residuals ψ_{ik} are independent across both regions and diseases. We could easily generalize to independent CAR models (i.e., $\boldsymbol{\psi}_k|\tau_{\psi_k}, W \stackrel{ind}{\sim} CAR(\tau_{\psi_k}, W)$) or even use an MCAR for $\boldsymbol{\psi} = (\boldsymbol{\psi}'_1, \ldots, \boldsymbol{\psi}'_k)'$, although our inclusion of the shared component $\boldsymbol{\theta}$ in (5) may well make such complexity unnecessary. Defining the 'shared" BLV for edge ij as $\Delta_{\theta,ij} = \theta_i - \theta_j$, boundary elements can be identified based on posterior summaries of $\Delta_{\theta,ij}$ (say, $E(\Delta_{\theta,ij}|\mathbf{y})$ or their magnitudes). Disease-specific residual boundaries (separating areas that differ due to factors other than the spatial common factor) can analogously be based on $\Delta_{\psi,ij,k} = \psi_{ik} - \psi_{jk}$ for each disease k. We hasten to add that, as in all factor analysis modeling, constraints need to be imposed on δ to avoid identifiability problems. For $p = 2$, Knorr-Held and Best (2001) set $\delta_1 = 1/\delta_2$ and place a lognormal prior on δ_1; Held et al. (2005) generalize this approach for $p > 2$. We instead follow Liu et al. (2005) and fix $\delta_p = 1$, but leave the remaining δ_k unconstrained.

To avoid the oversmoothing often associated with the CAR model in boundary analysis, we can allow the areal adjacency matrix for the common factor to be random. Boundary analysis is straightforward using this approach: we simply replace the prior in (5) with the random adjacency CAR given in (2) and (3). That is, we assume $\boldsymbol{\theta}|\boldsymbol{\phi}^E, \tau_s \sim CAR(\tau_s, W)$, where the $\phi^E_{ij} \equiv 1 - w_{ij}$ have the Ising prior given in (3). In other words,

$$p(\boldsymbol{\phi}^E) \propto \exp\{-\nu \boldsymbol{\phi}^{E\prime} W^* \boldsymbol{\phi}^E\} \;, \tag{7}$$

where W^* is the edge-space adjacency matrix (i.e., $W^*_{mm'} = 1$ if edge $m \equiv ij$ is connected to edge $m' \equiv kl$). We refer to this model as SESHARED, and adopt it for our wombled boundary maps in the next section. The reader is referred to the full paper (Ma and Carlin, 2007) for details regarding selection among a variety of alternate multivariate areal wombling models using the Deviance Information Criterion (DIC; Spiegelhalter et al., 2002).

4. DATA ANALYSIS

Previous epidemiological studies and basic anatomical relations suggest that esophagus, larynx and lung cancer are related. As we noticed in Fig. 1, the raw SMRs for the three cancers show similar patterns of decrease from northeast to southwest. Thus it seems plausible to model an underlying common factor connecting the three diseases.

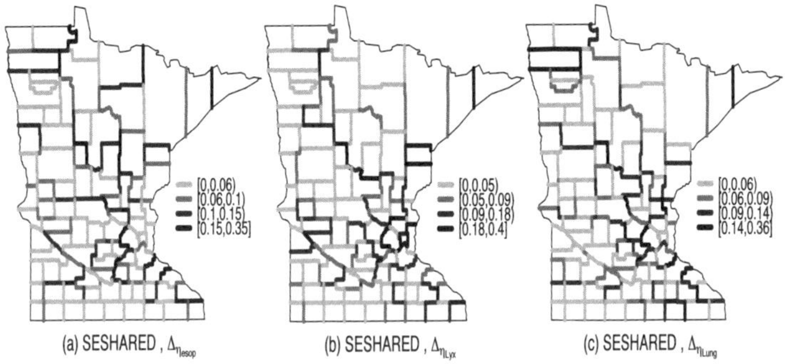

Figure 2: *Disease-specific boundary maps based on posterior medians of $\Delta_{\eta,ij,k}$, Minnesota SEER data: (a) esophagus cancer; (b) larynx cancer; (c) lung cancer.*

Using the usual BLV idea, disease-specific boundary maps can be constructed based on $\Delta_{\eta,ij,k}$ posterior summaries, where $\eta_{ik} = \mu_{ik}/E_{ik}$, the fitted relative risks for the three cancers. These boundaries are shown in Fig. 2. The boundary maps indicate similar patterns across disease, but subtle differences can be appreciated; e.g., east of the Twin Cities metro area (about one third of the way up the map on the right hand side) in panel (b), the larynx map.

5. DISCUSSION AND FUTURE WORK

In this paper, we have proposed an approach for multivariate areal boundary analysis. Accounting for correlations across both space and variables can improve the modeling, both in terms of improved DIC scores and more meaningful and easily interpretable maps and other summaries of the corresponding enhanced model parameters. The SESHARED model emerged as best for our Minnesota cancer dataset, but many other models improve on the IndCAR and IndSE, two models that carefully account for spatial association but ignore correlation among the cancers.

The southwest-to-northeast pattern evident in Fig. 1 motivates a search for a suitable spatially-oriented covariate to include in our models. Sadly, while a thorough search of the US Census Bureau site `quickfacts.census.gov` yielded several county-level income, poverty, and occupational covariates, none emerged as worthy of inclusion in any of our spatial models. The proportion of each county's business establishments classifed as forestry, fishing, hunting, mining, or agriculture support was significantly associated with lung cancer for some areal wombling models, but even this did not lead to worthwhile improvements in DIC score.

Finally, multivariate areal wombling as described herein leads naturally to methods of *spatiotemporal* areal wombling, as needed to track changes in areal boundaries over time (say, for annual cancer surveillance purposes). Here the temporal units play the role of the p variables above. Space-time separability would be a natural assumption to ease conceptual and computational difficulties, but may or may not be justified for any given dataset.

REFERENCES

Besag, J. (1974). Spatial interaction and the statistical analysis of lattice systems. *J. Roy. Statist. Soc. B* **36**, 192–236, (with discussion).

Besag, J., York, J.C., and Mollié, A. (1991). Bayesian image restoration, with two applications in spatial statistics. *Ann. Inst. Statist. Math.* **43**, 1–59, (with discussion).

Carlin, B.P. and Banerjee, S. (2003). Hierarchical multivariate CAR models for spatio-temporally correlated survival data. *Bayesian Statistics 7* (J. M. Bernardo, M. J. Bayarri, J. O. Berger, A. P. Dawid, D. Heckerman, A. F. M. Smith and M. West, eds.) Oxford: University Press, 45–63, (with discussion).

Gelfand, A.E. and Vounatsou, P. (2003). Proper multivariate conditional autoregressive models for spatial data analysis. *Biostatistics* **4**, 11–25.

Geman, S. and Geman, D. (1984). Stochastic relaxation, Gibbs distributions and the Bayesian restoration of images. *IEEE Trans. Patt. Anal. Mach. Intelligence* **6**, 721–742.

Held, L., Natário, I., Fenton, S.E., Rue, H., and Becker, N. (2005). Towards joint disease mapping. *Statistical Methods in Medical Research* **14**, 61–82.

Kim, H., Sun, D. and Tsutakawa, R.K. (2001). A bivariate Bayes method for improving the estimates of mortality rates with a twofold conditional autoregressive model. *J. Amer. Statist. Assoc.* **96**, 1506–1521.

Knorr-Held, L. and Best, N.G. (2001). A shared component model for detecting joint and selective clustering of two diseases. *J. Roy. Statist. Soc. A* **164**, 73–85.

Liu, X., Wall, M. and Hodges, J. (2005). Generalized spatial structural equation models. *Biostatistics* **6**, 539–557.

Ma, H. and Carlin, B. (2007). Bayesian multivariate areal wombling for multiple disease boundary analysis. *Bayesian Analysis* **2** (to appear).

Ma, H., Carlin, B.P., and Banerjee, S. (2006). Hierarchical and joint site-edge methods for Medicare hospice service region boundary analysis. *Tech. Rep.*, University of Minnesota, USA.

Mardia, K.V. (1988). Multi-dimensional multivariate Gaussian Markov random fields with application to image processing. *J. Multivariate Analysis* **24**, 265–284.

Spiegelhalter, D.J., Best, N., Carlin, B.P., and van der Linde, A. (2002). Bayesian measures of model complexity and fit. *J. Roy. Statist. Soc. B* **64**, 583–639, (with discussion).

Womble, W.H. (1951). Differential systematics. *Science* **114**, 315–322.

BAYESIAN STATISTICS 8, pp. 631–636.
J. M. Bernardo, M. J. Bayarri, J. O. Berger, A. P. Dawid,
D. Heckerman, A. F. M. Smith and M. West (Eds.)

Cluster Allocation Design Networks*

ANA MARIA MADRIGAL
University of Warwick and Unilever Colworth, UK
am.madrigal@warwickgrad.net

SUMMARY

Policy makers usually define a strategy that involves policy assignment and recording mechanisms. Causal inference is sensitive to the specification of these mechanisms. Influence diagrams have been used for causal reasoning within a Bayesian decision-theoretic framework (Dawid, 2002). Design Networks (DNs) expand this framework by including experimental design decision nodes. DNs provide semantics to discuss how a design decision strategy might assist the identification of intervention causal effects. The DN framework is extended to Cluster Allocation. Cases of 'pure' cluster (all individuals in a cluster receiving the same intervention) and 'non-pure' cluster (only a subset receiving the policy) are discussed in terms of causal effects.

Keywords and Phrases: INFLUENCE DIAGRAMS, CAUSAL INFERENCE, DAGS, IDENTIFICATION OF POLICY EFFECTS, CLUSTER ALLOCATION.

1. INTRODUCTION

Different data sets provide different types of information. The distributions that can be learnt (or not) might vary among apparently similar data sets. This is an important consideration to the analyst before learning model parameters. Discussions of causal reasoning have assumed that the graph representing the system implicitly includes the underlying (experimental) mechanism that is generating the data (see Pearl, 2000). In this fixed 'natural' or 'idle' system, whether the future policy intervention F_T effect is identifiable is addressed. Knowing the details of the policy assignment mechanism and a well-planned recording of the data become very relevant issues to measure the right 'causal' effects (see Rubin, 1978). Influence Diagrams (IDs) are used to represent the system dynamics and interventions graphically. Our interpretation of causal effects for interventions is Bayesian decision-theoretic, where an intervention on a system is regarded as a decision. Dawid's (2002) extended influence diagrams are augmented by including 'experimental design' decisions nodes to create what we call a Design Network (DN), to provide semantics to discuss how a 'design' strategy (such as clustering) might assist the systematic identification of intervention causal effects. Design Networks were introduced in Madrigal and Smith (2004). Discussions about the identifiability of causal effects have been

* This is a synopsis. For the full version, see *Bayesian Analysis* **2** (2007), to appear.

usually phrased as 'Is the causal effect of T on Y identifiable?'. The role of data structures is made explicit by phrasing the identifiability question as 'Is the causal effect of T on Y identifiable *from the data available*?'.

The simplest form of external intervention is when a single variable X is forced to take on some fixed value x'. This is known as an 'atomic intervention' and, following Pearl (2000), it is denoted by $do(X = x')$. The atomic intervention replaces the original mechanism: $p(x \mid pa(x))$ by $p(x \mid pa(x); do(X = x')) = 1$ **if** $X = x'$ where $pa(x)$ denotes the parent nodes of X. Pearl's $do(\cdot)$ corresponds to an external intervention. By recognising interventions as decisions, the Bayesian decision-theoretical framework embeds Pearl's *doing* operation and provides a stronger framework for causal inference. Dawid (2002) incorporates intervention nodes F where $F_X = do(X = x)$ corresponds to 'setting' the value of node X to x, and $F_X = \emptyset$, corresponds to X having its 'natural' distribution. There is interest in establishing if the causal effect of F_T on Y can be identified from the available data. The structure of the data available is defined by the set of conditional independencies that are derived from the graph. The discussion is conducted in terms of the relevance of learning the strategy that gave the value t' to T, namely whether it arose from the original experimental setting ($F_T = \emptyset$) or whether it was set externally ($F_T = do(T = t')$). Conditional independencies of the form $(Y \perp\!\!\!\perp F_T \mid T, \cdot)_{d_E}$ are used for this.

Policy makers have to define a strategy that involves 'policy intervention' actions ($D_T = \{d'\}$) and 'experimental design' actions ($D_E = \{d^*\}$). The former includes decisions related to how the policy is implemented. The latter includes experimental design decisions that define the conditions under which the study is carried out and the data (Δ) recorded. If $\mathcal{D} = \{d'_1, .., d'_{\mathcal{D}_T}, d^*_1, .., d^*_{\mathcal{D}_\mathcal{E}}\}$ are the components of a particular decision strategy $\pi_\mathcal{D}$, the interest lies in describing $\pi_D(\mathcal{D} \mid E)$. We say that policy intervention actions (D_T) are concerned with intervening 'the world', while experimental design actions (D_E) relate to intervening the statistician's 'view of the world'. It is important to distinguish between 'choosing a policy' and 'choosing a design'. The success of a policy intervention D_T is measured in terms of its efficacy to provoke 'better' values on the response variable Y through its overall effects reflected by $p(Y \mid F_T = do(T = t'); D_E)$. The efficacy of an experimental intervention, D_E, is measured in terms of its ability to isolate the policy effect as much as possible. Making an explicit representation of both types of interventions will assist decisions of the *experimenter* and considerations of the *analyst*, when the aim is to evaluate the causal effect of F_T . Thus the need to represent in the ID and formulae both: the atomic (future) policy intervention $F_T \in D_T$ (allocating treatment $T = t'$ with probability one) and the (contingent or randomised) experimental allocation strategy $A_T \in D_E$ (allocating treatment $T = t^*$ according to θ).

2. CAUSAL GRAPHICAL ANALYSIS FOR CLUSTER DESIGN NETWORKS

Most of the literature in Cluster Randomised Trials (CRTs) has emphasised the fact that Fisher's principle is violated, as the experimental unit does not coincide with the analysis unit. When introducing the need for the evaluation of a future intervention F_T it is important to acknowledge the fact that the future intervention ($F_T = do(T = t')$), the experimental allocation ($A_T = do(T = t^*)$) and the response variable (Y) could each be at cluster/individual level and not necessarily coincide.

In its simplified version, let $D_E = \{A, R(B)\}$ where A contains the policy assignment mechanisms and $R(B)$ contains the recording mechanism, such that $R(B^q) = 1$, for $q = 1, 2...Q$, if variable B^q is recorded and $R(B^q) = 0$ if B^q is either unobservable or not recorded. *Assignment nodes* A and *recording nodes* R

can be included in the DAG as decision nodes to create a *design network (DN)*. The design network shows the mechanisms that generate the data available Δ_{d_E}. By representing simultaneously the design nodes $D_E = \{A, R(B)\}$ and the future intervention node F_T, the augmented design network is useful to make conclusions about two different tasks involved in policy evaluation and design: (1) the identifiability of the causal effect of T on Y, given a design $d_E \in D_E$; and (2) the choice of a design strategy d_E to collect data when the aim is to evaluate the effect of intervention F_T. To discuss the identifiability of the causal effect of T on Y, we are interested in comparing the relevance of the choice of F_T given particular experimental conditions d_E^*, namely comparing $p(y \mid t', F_T = do(T = t'); D_E = d_E^*)$ and $p(y \mid t', F_T = \emptyset; D_E = d_E^*)$. The identifiability of intervention F_T depends on the data (determined by mechanisms d_E); and the efficacy of experimental design (d_E) is always to be determined with respect to the effects it tries to isolate (here F_T). Thus, D_E and F_T are and should always be read in the light of each other.

A cluster-randomised study implicitly involves two design decisions: (1) the decision of clustering (i.e. to allocate treatments to clusters of individuals) and (2) the decision of randomising (i.e. to allocate using a random procedure). For a cluster-level intervention we distinguish two cases. In a 'pure-cluster' intervention (denoted by $d_C = 1$), the intervention affects all individuals in the cluster and individuals within the cluster cannot choose not to be affected by the policy intervention. In a 'non-pure' cluster allocation (denoted by $d_C = 0$), not all individuals, but only a subset of them (e.g. eligible individuals) within a cluster, will be subject to the intervention: 'all eligible individuals k in cluster j will receive policy t' ' via $A_{Tj} = do(T_j = t')$. Two eligible individuals in the same cluster cannot be allocated different policies (contrary to what would happen if the policy allocation was done at individual level) and the treatment of different units within a cluster is dependent on certain individual-level background variables.

If the interest lies on an individual level response, we recognise two possible (causal) intervention effects of interest: the distribution of the individual response given a cluster intervention, $P(Y_{jk} \mid F_{Tj} = do(T_j = t'))$, and the distribution of the individual response given an individual intervention, $P(Y_{jk} \mid F_{Tjk} = do(T_{jk} = t'))$. The former would try to estimate the effect of a cluster-level intervention (usually the interest in social policy) and the latter would try to estimate the effect of an individual-level intervention (as could be the goal of many medical trials).

2.1. *Effects of Cluster Allocation Design Decisions*

A design network for a cluster-level future intervention F_{Tj} is presented in Fig. 1(a). The levels are represented in the graph by squares, following Spiegelhalter's notation (see WinBUGS), meaning that the same graphical structure applies for each of the observations at the same level. Design decisions can then be taken at both individual and cluster level. Let cluster j (for $j = 1, 2, \ldots, J$) have K_j units and let T_j and T_{jk} be variables for intervention status (Treatment/Control) at cluster and unit level respectively. Let B_j and B_{jk} represent the background variables at cluster and unit level and Z_j some recorded cluster-level covariates. Nodes A_j and A_{jk} will correspond to the assignment mechanisms to allocate policy at cluster and individual levels respectively. The recording mechanisms could be defined over the set of cluster and individual background variables, $R(B_j)$ and $R(B_{jk})$, respectively. The response Y_{jk} correspond to individual k in cluster j.

The individual assignment mechanism A_{jk} is considered to be dependent on the policy that was allocated to cluster j, T_j, and (possibly) on some individual back-

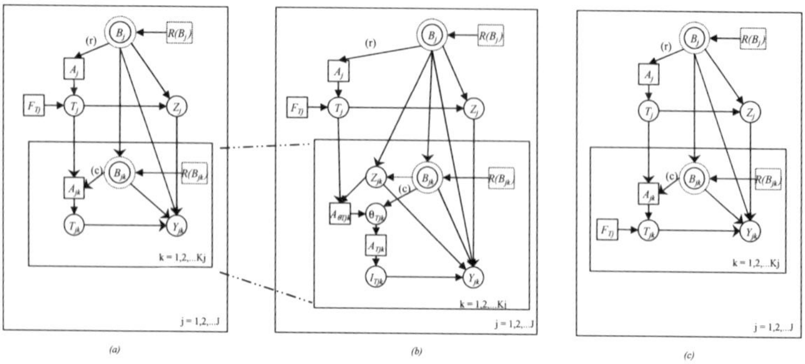

Figure 1: *(a) Design network for cluster allocation and cluster-level future intervention* F_{Tj}. *(b) Close-up of individual-level plateau. (c) Design Network for cluster allocation and individual-level future intervention* F_{Tjk}

ground variables. Thus, the action assigning policy t^* to individual k $A_{Tjk} = do(T_{jk} = t^*)$ is considered to be dependent on T_j and B_{jk} such that $\theta_{Tjk} = p\,(A_{Tjk} = do(T_{jk} = t^*) \mid T_j, B_{jk}) = q(T_j, B_{jk})$. The decision of running a 'pure' cluster allocation experiment ($d_C = 1$) will imply that the intervention is done equally to all members in the cluster. Once the treatment for cluster j T_j is fixed by action $A_{Tj} = do(T_j = t^*)$, this fully implies actions $A_{Tjk} = do(T_{jk} = t^*)$, and so the values of the treatments T_{jk} for all K_j individuals in cluster j. Then $t_j = t_{jk} = t_{jk'}$ for all individuals $k, k' = 1, 2, \ldots, K_j$ in cluster j. The effect of pure-clustering prohibits individual covariates from influencing the choice of treatment, breaking any links from B_{jk} (background individual-level covariates) to T_{jk} in the graph. Arrow (c) dissapears and $\theta_{Tjk} = p\,(A_{Tjk} = do(T_{jk} = t^*) \mid A_{Tj} = do(T_j = t^*), d_C = 1) = 1$. The following proposition is thus established.

Theorem 1 (Pure-Clustering). *If the experimental design strategy* d_E *includes the action of performing a 'pure cluster' experiment such that* $\{d_C = 1\} \in d_E$, *then the 'structural' effect of* $d_C = 1$ *on the 'original' set of conditional independencies, is to introduce the set of conditional independencies* $(T_{jk} \perp\!\!\!\perp B_{jk})_{d_C = 1}$ *that will hold on the data* Δ_{d_E} *generated by* d_E.

In 'non-pure' cluster allocation ($d_C = 0$), although the experimental allocation is made at cluster level, individual k within cluster j might receive treatment or not depending on some individual-level covariates, and thus t_{jk} might differ from $t_{jk'}$ for $k \neq k'$. Figure 1*(b)* includes a close-up of the individual-level plateau in Fig 1(a), in which the design network has been extended for node A_{jk} and variables Z_{jk} introduced. Consider a non-pure cluster experiment such that, within each cluster j, individual policy allocation follows a deterministic rule based on individual-level observed covariates Z_{jk}. The final policy allocated to an individual through A_{jk} will be a function of Z_{jk}, and any other possible influences on T_{jk} from background variables B_{jk} (other than Z_{jk}) are eliminated. The prevalences of T_{jk} in the experimental data available Δ will depend on the policy allocated to the cluster T_j and Z_{jk} but not on B_{jk} (e.g. $\theta_{Tjk} = p\,(A_{Tjk} = do(T_{jk} = t^*) \mid T_j, Z_{jk}) = q(T_j, Z_{jk})$) Arrow (c) in Figure 1*(b)* is deleted when this 'deterministic' allocation takes place.

Theorem 2 *If a 'non-pure cluster' experiment includes a design strategy in which policies at individual level are allocated following a 'deterministic' rule defined by the experimenter based on observed covariates* Z_{jk} *, then conditional independencies* $(T_{jk} \perp\!\!\!\perp B_{jk} \mid Z_{jk})_{d_E}$ *are introduced and will hold on the data* Δ_{d_E} *generated by* d_E*.*

A design decision strategy d_E including random allocation of policies (i.e. $A_{\theta_T} = do(\theta_T = \theta_T^*) \in d_E$) might modify the structure of the data we are to collect. By allocating the treatments completely at random (i.e. $A_{\theta_{T_j}} = do(\theta_{Tj} = \theta_{Tj}^*)$) we ensure that the treatment received is independent of any background variables B_j that, otherwise, might have an influence on the policy assignment mechanism. Then, when random allocation takes place, arrow (r) from B_j to A_j in Figure 1*(a)* disappears and the conditional independence statement $(T_j \perp\!\!\!\perp B_j)_{d_E}$ holds. When randomising at cluster level we ensure that the level of treatment that is received by cluster j is independent of the level received by cluster j' (i.e. knowing that cluster j was assigned to t^* does not give any information about the allocation at cluster j').

2.2. *Identifying Cluster-Level Interventions*

When the interest lies in an individual-level response Y_{jk}, it can be seen in Fig.1(a) that, if d_E^* includes a random cluster allocation procedure and arrow (r) is not present, the conditional independencies $(Y_{jk} \perp\!\!\!\perp F_{Tj} \mid T_j)_{d_E^*}$ hold for all j, k. Thus, once the value of the policy assigned to the cluster T_j is known, learning whether the policy status T_j arose from the future policy implemented $F_{Tj} = do(T_j = t')$ or from the 'original' experimental allocation $A_{Tj} = do(T_j = t^*)$ when $F_{Tj} = \emptyset$, is irrelevant. Therefore, direct identifiability of the effect $F_{Tj} = do(T_j = t')$ on Y_{jk} holds and $p(y_{jk} \mid F_{Tj} = do(T_j = t'))$ can be directly obtained from data $\Delta_{d_E^*}$ available as long as $t' \in \{t^*\}$ such that $p(y_{jk} \mid F_{Tj} = do(T_j = t'); \Delta_{d_E^*}) = p(y_{jk} \mid T_j = t'; \Delta_{d_E^*})$. This does not disregard the fact that individuals belonging to the same cluster will have a positive correlation, which must be taken into account in any model used for the analysis and estimation of the effect on an individual-level response.

When non-random cluster allocation is performed as part of the experimental design d_E^*, $(Y_{jk} \perp\!\!\!\perp F_{Tj} \mid T_j)_{d_E^*}$ does not hold anymore, but the conditional independencies $(Y_{jk} \perp\!\!\!\perp F_{Tj} \mid T_j, B_j)_{d_E^*}$ hold for all j,k. Thus, the identifiability of the effect of a future policy $F_{Tj} = do(T_j = t')$ on Y_{jk} will depend on the recordability of cluster-background variables $R(B_j)$ and the causal effect will need to be obtained through an 'adjustment' procedure such that, as before, using the back-door criteria$p(y_{jk} \mid F_{Tj} = do(T_j = t')) = \int p(y_{jk} \mid T_j = t', B_j)p(B_j)dB_j$. Unless we are ready to assume some prior distribution for $p(B_j)$, the recording of variables B_j as part of the design ($\{R(B_j) = 1\} \in d_E^*$) are needed to achieve an 'adjusted identification' of the causal effect. Different recording mechanisms might assist identification. For instance, if we are ready to assume that cluster background variables do not have a direct effect on the individual response, such that arrow from B_j to Y_{jk} is deleted, then conditioning on T_j, B_{jk} and Z_j would be enough to provide identifiability. Thus, if cluster background variables B_j were not accessible to the experiment, this new set of covariates $\{B_{jk}, Z_j\}$ could assist identification.

2.3. *Identifying Individual Interventions from Clustered Data*

Consider the case where the main interest is in obtaining the individual-level causal effect, namely $P(Y_{jk} \mid F_{Tjk} = do(T_{jk} = t'))$ from data that is clustered. Suppose that it is not feasible to randomise at individual level, but to intervene clusters is possible. From the design network in Fig. 1(c) it can be seen that $(Y_{jk} \perp\!\!\!\perp F_{Tjk} \mid T_{jk})$ does not hold even if arrows (r) and (c) are deleted from the graph. So, the effect of policy intervention $F_{Tjk} = do(T_{jk} = t')$ on Y_{jk} cannot be identified directly from the data and some adjustment will be needed.

When a 'non-pure' cluster allocation takes place in the experiment, individual policy assignment will depend on both the policy allocated T_j and individual background variables B_{jk}. Conditioning on this set of variables, the irrelevance of F_{Tjk} is gained such that $(Y_{jk} \perp\!\!\!\perp F_{Tjk} \mid T_{jk}, T_j, B_{jk})$ Thus the recording of B_{jk} is needed in order to obtain adjusted identifiability through $p(y_{jk} \mid F_{Tjk} = do(T_{jk} = t'); d_C = 0) = \int p(y_{jk} \mid T_{jk} = t', T_j = t^*, B_{jk}) p(T_j = t^*, B_{jk}) dB_{jk} dT_j$. If recording at individual level for B_{jk} is not undertaken, the causal effect will be unidentifiable.

When 'pure' cluster allocation is done, and as a result arrow (c) is deleted, then there are no individual level confounders and all possible confounders will be at cluster-level. So, 'pure cluster' assignment might improve identification of individual intervention effects. In particular, when randomisation at cluster level is feasible or in the case when cluster-level confounders (B_j) are easier to observe and/or control than individual-level confounders (B_{jk}).

3. CONCLUSION

The primary contribution of this paper is to expand on Dawid's (2002) model for causality reasoning within the Bayesian decision-theoretic framework: to 'adapt' it to policy analysis, to include experimental nodes allowing intervention nodes 'do' parameters nodes, to discuss the relevance (or irrelevance) of experimental design and to include interventions at different levels (clusters) of units. Observational data is considered a degenerate type of experimental data. The choices of the experimenter affect the characteristics (units, variables and distributions) of the database. The inspection of influence diagrams has been shown to be useful to decide if the data available is sufficient for obtaining consistent estimates of the target causal effect of policy intervention $F_T = do(T = t')$. If so, we can derive a closed-form expression for the target quantity in terms of distributions of available quantities. If it is not sufficient, this framework can help suggest a set of observations and experiments that, if performed, would render a consistent estimate feasible. Certain policy assignment mechanisms, such as randomised cluster allocation, add 'extra' independencies to the ID. Design networks were introduced for cluster allocation. Spiegelhalter's (2001) Bayesian hierarchical model has been used to illustrate a causal analysis of a simplified version of a mexican social programme; due to space constraints, it is not presented in this synopsis.

REFERENCES

Dawid, A. P. (2002). Influence diagrams for causal modelling and inference *Internat. Statist. Rev.* **70**, 2, 161–189.

Madrigal, A. M. and Smith J. Q. (2004) Causal identification in design networks. *MICAI 2004: Advances in Artificial Intelligence* (R. Monroy *et al.* , eds.) Berlin: Springer-Verlag, 517–526.

Pearl, J. (2000). *Causality.* Cambridge: University Press.

Rubin, D. (1978). Bayesian inference for causal effects: The role of randomization. *Ann. Statist.* **6**, 34–58.

Spiegelhalter, D. J. (2001). Bayesian methods for cluster randomised trials with continuous responses. *Statistics in Medicine* **20**, 435–452.

BAYESIAN STATISTICS 8, pp. 637–642.
J. M. Bernardo, M. J. Bayarri, J. O. Berger, A. P. Dawid,
D. Heckerman, A. F. M. Smith and M. West (Eds.)

Logistic Regression Modelling of Proteomic Mass Spectra in a Case-Control Study on Diagnosis for Colon Cancer*

Bart J. A. Mertens

Leiden University Medical Center, The Netherlands
b.mertens@lumc.nl

Summary

We adapt logistic regression modelling for the evaluation of mass spectroscopic data in proteomic case-control studies. Instead of a direct attempt to model the observed case-control data as regression on peaks, we parameterize the predictor as a linear combination of Gaussian basis functions along the mass/charge axis. The location of these basis functions is treated as a random variable and must be estimated from the data. A fully Bayesian implementation is pursued, which allows the number of functional components within the regression parameter vector to be specified as a random variable. Calculations are implemented through birth–death process modelling.

Application of the model is presented on data from a randomized blocked case-control designed experiment, which was carried out recently at the Leiden University Medical Centre and has motivated this research. The experiment compares mass spectra of serum samples of 63 colon cancer patients with spectra from 50 control patients. We present a-posteriori analyses of the fitted model which allows researchers to select specific spectral regions for further investigation and identification of the associated differentially expressed peptides.

Keywords and Phrases: Mass spectrometry proteomic profiling; diagnosis; logistic regression; classification

1. INTRODUCTION

Mass spectrometry is currently attracting much interest for the simultaneous profiling of hundreds of proteins in tissue, urine or serum samples. It holds great potential for the construction of new diagnostic rules for the detection of pathological states.

* This is a synopsis. For the full version, see *Bayesian Analysis* **2** (2007), to appear.

This applies particularly for diseases where either no diagnostic methods are available or when the reliability of existing methods is poor.

A good example where new diagnostic procedures are urgently needed is in the early detection of colon cancer. To evaluate potential of mass spectrometry diagnosis in this context, a case-control experiment was set up at the Leiden University Medical Center, which contrasts serum samples of 50 healthy controls with those from 63 colon cancer patients. For each sample, the corresponding spectrum is measured (MALDI-TOF), each of which consists of hundreds of peaks along the mass charge axis (Dalton)which ranges from 1000 up to 5000 Dalton. The entire signal is discretized on a pre-defined and fixed grid of contiguous small bins of about 1 Dalton wide, along the mass/charge axis. We must refer to de Noo *et al.* (2006) or the supplementary materials for full details on the design, measurement protocols and spectral data. Fig. 1 shows a plot contrasting the sample mean spectra.

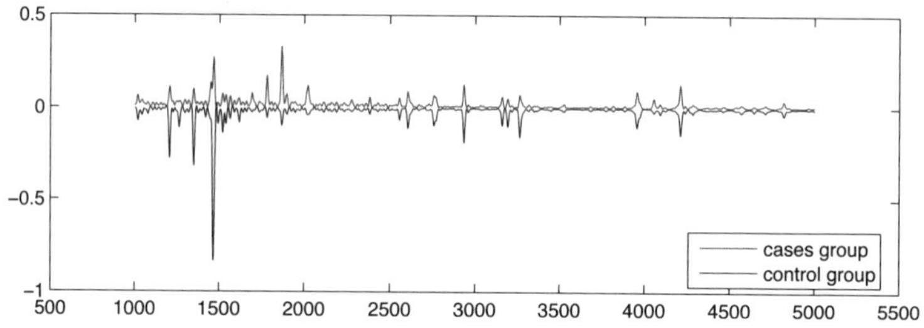

Figure 1: *Mean spectra for each group separately, after preprocessing. We plot negative intensity value for the control group (bottom mean spectrum).*

Adaptations of classical procedures, such as linear discriminant or logistic regression, based on either dimension reduction, penalized estimation or even fully Bayesian shrinkage, could be applied on the discretized spectral data, but tend to suffer from extreme correlations in the spectral data. They are also sensitive to the dominant sources of variation which are typically associated with the major spectral peaks. This paper presents an approach to calibration of the logistic regression model, based on parametrization of the vector of regression coefficients, which leads to much more parsimonious models which are also more easily interpretable. At the time of writing and based on a number of experiments, besides the one presented in this paper, we have found the model to be highly competitive in comparison to the 'standard' approaches available. It is currently used in all new case-control evaluations for mass spectral data at the Leiden Medical Center. In the following, we outline the major ideas. We must refer to the full paper for details as well as additional results on new data for further validation of the methodology.

2. CONDITIONAL MODELLING OF CLASS OUTCOME USING MASS SPECTRA

Let $y_i \in \{0,1\}$, $i = 1, ..., n$ be the binary class indicators denoting presence or absence of the clinical condition of interest and n the sample size. We will write $\boldsymbol{x}_i = (x_{i1}, \ldots, x_{iq})$ for the associated mass spectrum from each ith sample (after preprocessing), which is an ordered sequence of intensity values on the grid selected to record the spectrum. Similarly, we write $\mathbf{m} = \{m_1, \ldots, m_q\}$ for the ordered set of ordinates of the bins along the mass/charge axis corresponding to the sequence of q measured intensity values from each sample. We consider the binary regression model

$$y_i \sim Bernoulli(p_i),$$

with $g(p_i) = \eta_i$, where $\eta_i = \boldsymbol{x}_i\boldsymbol{\beta}$ is a linear predictor $g(p) = \log(p/(1-p))$ and $\boldsymbol{\beta}$ a vector of regression parameters. The vector of regression parameters $\boldsymbol{\beta}$ is parametrized in terms of a set of Gaussian basis functions along the mass/charge axis

$$\psi_j(x, \mu_j, \sigma_j) = \exp[-\frac{1}{2}\frac{(x-\mu_j)^2}{\sigma_j^2}],$$

for $j = 1, ..., k$, with k a finite dimension and the parameters μ_j and σ_j denoting the locations and widths (expressed as standard deviation). We now write the regression parameter vector $\boldsymbol{\beta} = (\beta_1, ..., \beta_q)$ as a linear combination

$$\boldsymbol{\beta} = \sum_{j=1}^{k} \alpha_j \Psi_j$$

with $\Psi_j = (\psi_j(m_1), ..., \psi_j(m_q))'$, for $j = 1, ..., k$. The latter functions may be interpreted as operators on $\boldsymbol{x}$ and hence, we will write $\Psi_j(\boldsymbol{x}) = \boldsymbol{x}\Psi_j$ for $j = 1, ..., k$.

With this formulation, the linear predictor equation may be rewritten as

$$\eta_i = \boldsymbol{x}_i\boldsymbol{\beta} = \boldsymbol{x}_i \sum_{j=1}^{k} \alpha_j \Psi_j = \sum_{j=1}^{k} \alpha_j \Psi_j(\boldsymbol{x}_i)$$

which reveals the linear predictor as a linear combination of basis functions in $\boldsymbol{x}$. For full generality, we will also include an intercept term into the model by introducing the constant term $\Psi_0(\boldsymbol{x}_i) = 1$, such that the linear predictor becomes $\eta_i = \alpha_0 + \sum_{j=1}^{k} \alpha_j \Psi_j(\boldsymbol{x}_i) = \sum_{j=0}^{k} \alpha_j \Psi_j(\boldsymbol{x}_i)$ where α_0 is the intercept term which augments the vector of regression parameters $\boldsymbol{\alpha} = (\alpha_0, \alpha_1, ..., \alpha_k)'$. In the remainder of this paper and in order to simplify terminology, we will also refer to the basis functions as 'basis components', though it must be clearly understood this is shorthand terminology at all times for 'basis function components within the regression coefficients' and that these constitute an assumption at the level of the regression parameters.

2.1. *Hierarchical Specification of Logistic Models*

For reasons of computational efficiency, we employ a 'linearizing' representation of the above model based on auxiliary variables, as described in Holmes and Held (2006) and Albert and Chib (1993). This is achieved by rewriting the logistic model as

$$y_i = 1 \quad \text{if} \quad z_i > 0, \quad \text{and} \quad 0 \quad \text{otherwise},$$

for an auxiliary variable $z_i = \boldsymbol{x}_i\boldsymbol{\beta} + \epsilon_i$ and $\epsilon_i \sim N(0, (2\omega_i)^2)$, where the ω_i are Kolmogorov-Smirnow random variables (Devroye, 1986). Within a fully Bayesian paradigm, we consider k, $\boldsymbol{\theta} = \{\mu_j, \sigma_j; j = 1, ..., k\}$ and $\boldsymbol{\alpha}$ as random variables.

2.2. *Prior Structure*

To complete the hierarchical model, we specify a normal prior

$$(\alpha_1, ..., \alpha_k) \sim N_k(0, \tau v^2 \boldsymbol{I}_k)$$

on the regression parameters with v a known scale factor and $\tau = 1/\varsigma$ a randomly distributed re-scaling factor, such that ς has a $Gamma(a, b)$ distribution, with a and b positive real numbers. The intercept is given a weakly informative normal prior $\alpha_0 \sim N(0, 10^2)$. For the dimensionality parameter k, we specify a discrete uniform hyperprior distribution on the set of integers $\{0, 1, 2, .., k_{max}\}$ with k_{max} a large positive integer. The prior distribution of location parameters μ_j, $j = 1, ..., k$ is given a uniform discrete distribution on the grid of location values $\boldsymbol{m}$. For the prior distribution on the width parameters σ_j, $j = 1, ..., k$, we postulate a truncated normal distribution $p(\sigma_j) \sim \phi(\sigma_j - \gamma)I(\sigma_j)$ for all $j = 1, ..., k$ with $I()$ the heaveside function and $\phi()$ the standard normal density. γ is a positive parameter which denotes the prior mode of the width of basis components within the regression parameter vector. We return to aspects of prior choice in the data analysis section.

2.3. *Variable Dimension MCMC Sampling for Logistic Regression*

Estimation proceeds via birth–death sampling moves on the location grid $\boldsymbol{m}$. Our approach essentially follows methodology outlined by Holmes and Held (2006) and Denison *et al.*(2002) for linear and probit models. Methodological details of hybrid MCMC sampling methods for variable dimension models have been discussed by a number of publications over the past few years (Green (1995), Richardson and Green (1997) and Denison *et al.*(2002), among others). Hence, to save space in this synopsis, we refer to the full paper or to previously mentioned literature for details.

3. DATA ANALYSIS AND APPLICATION TO THE COLON CANCER DATA

We applied the model to analysis of the colon cancer mass spectral data, using $k_{max} = 150$. Prior choice of γ is based on the observation that peaks within the spectral data itself will have a typical width of about 3 Dalton. It is important to acknowledge that γ itself represents a prior assumption on the regression coefficients instead. However, it is reasonable to assume that within the confines of our model and if there is any differential expression of peaks within the data and between groups this should give rise to components of differential expression within the regression coefficient vector which are of similar width. Hence we use $\gamma = 3$ as prior choice, irrespective of the location of any regression coefficient component proposed. Our prior choice of v originates from the knowledge that approximately the variance-covariance structure of $(\alpha_1, ..., \alpha_k)$ is $n{*}(\Psi'\Psi)^{-1}$, with $\Psi = [\Psi_1, ..., \Psi_k]$. This leads to a prior guess of $v^2 = 0.1$. We set $a = b = 1$ for the analyses presented.

3.1. *Identification of Differentially Expressed Mass Spectral Regions*

Fig. 2 shows for each potential basis function location within the set $\mathbf{m}$, the number of times that location was selected into the model, across all models simulated and expressed as a percentage of the total number of models considered. The plot identifies two major discriminating sources of variation in the data, the first centered at 1352.4 Dalton (corresponding basis function selected to nearly 25% of all models) and the second at 1867.2 Dalton (almost 15% of all models). The corresponding

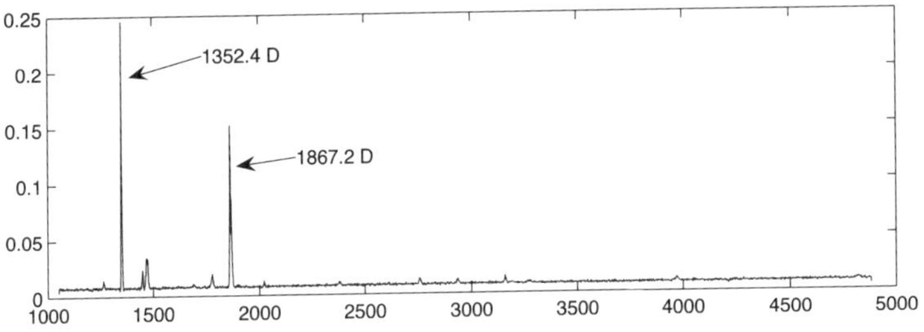

Figure 2: *Marginal posterior density of peak component locations.*

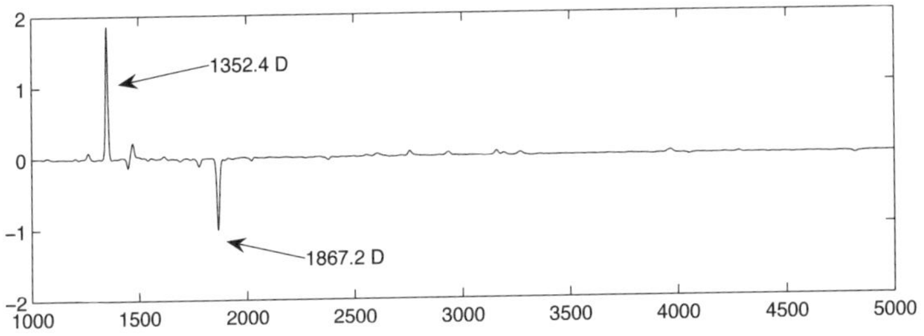

Figure 3: *Plot of marginal mean discriminant coefficients versus mass-charge value.*

Fig. 3 shows the marginal mean of logistic regression coefficients

$$\sum_{m \in \mathsf{M}} \boldsymbol{\beta}_m / M = \sum_{m \in \mathsf{M}} \sum_{j=1}^{k} \alpha_{mj} \Psi_{mj} / M$$

across all simulated models, where m is an indicator identifying the model within the set of all simulated models M and M is the total number of models simulated. The plot reveals that the discriminating information corresponds to a contrast between bin intensities centered at 1352.4 and 1867.2 Dalton. This interpretation can be easily confirmed through a scatter plot of bin intensities at these locations (Fig. 4). As can be seen, large intensities at 1352.4 Dalton for one group correspond to small intensities at 1867.3 Dalton for the other group, and vice versa. We note Figs 2 and 3 also identify a number of other interesting locations in the spectra, most notably a small contrasting between locations 1448.6 and 1471.5 Dalton. There is some indication of other separating regions in the spectra at 1265.7 Dalton and at 1778.9 Dalton, though the evidence is not as strong. The corresponding bin locations are selected to less than 5% of all simulated models.

3.2. *Evaluation of Classification Performance*

To assign observations, we calculate for each observation the mean a-posteriori class probabilities $P(y_i = 1 \mid \boldsymbol{x}_i) = \sum_{m \in \mathsf{M}} P_m(y_i = 1 \mid \boldsymbol{x}_i)/M$, of group-membership,

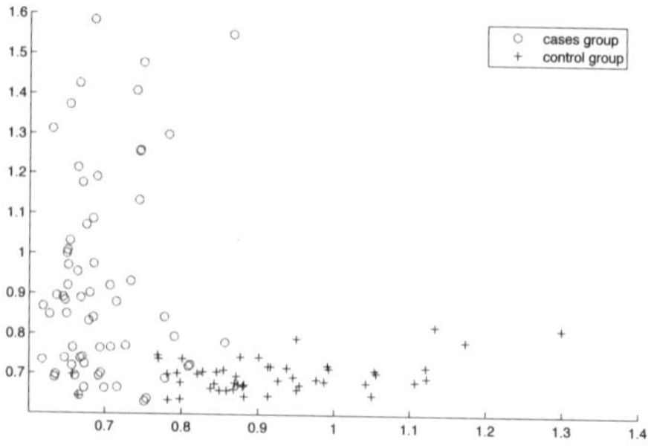

Figure 4: *Scatter plot of spectral intensity values at 1352. Dalton versus those at 1867.2 Dalton, using distinct plotting symbols to distinguish cases (o) from controls (+).*

with $i = 1, \ldots, n$, where P_m denotes the a-posteriori class probability calculated from the mth model simulated within the MCMC chain and the sum is across all models simulated. Using a cut-off value of 0.5 (*e.g.*) to assign observations based on these mean a-posteriori class probabilities, we find a sensitivity and specificity of both 0.92, such that the global misclassification error rate equals 0.08. The Brier distance, defined as

$$B = \frac{1}{n} \sum_i [1 - P(c(i) \mid \mathbf{x}_i)]^2,$$

equals 0.066, where $c(i)$ denotes the true class label of the ith observation with $P(c(i) \mid \mathbf{x}_i)$ the posterior mean class probability for the ith observation for the true class. Receiver Operating Curve (ROC) calculations based on the marginal mean a-posteriori class probability give and Area Under the Curve (AUC) of 0.978.

The full paper provides more details on results, as well as comparison with classical analysis approaches and a sensitivity study on prior choice. Comparative results on new and more complex spectral data is also presented, which further demonstrates the performance of the method as compared to existing approaches.

REFERENCES

Albert, J. H. and Chib, S. (1993) Bayesian analysis of binary and polychotomous response data. *J. Amer. Statist. Assoc.* **88**, 669–679.

Denison, D. G. T., Holmes, C. C., Mallick, B. K. and Smith, A. F. M. (2002) *Bayesian Methods for Nonlinear Classification and Regression.* New York: Wiley.

Devroye, L. (1986). *Non-uniform Random Variate Generation.* Berlin: Springer-Verlag.

Green, P. (1995) Reversible jump Markov chain Monte Carlo computation and Bayesian model determination. *Biometrika* **82**, 711–732.

Holmes, C. C. and Held, L. (2006) Bayesian auxiliary variable models for binary and multinomial regression. *Bayesian Analysis* **1**, 145–168.

de Noo, M. E., Mertens, B. J., Ozalp, A., Bladergroen, M. R., van der Werff, M. P. J., van de Velde, C. J., Deelder, A. M., and Tollenaar, R. A. E. M. (2006). Detection of colorectal cancer using MALDI-TOF serum protein profiling. *Eur. J. Cancer* **42**, 1068–1076.

Richardson, S. and Green, P. (1997) On Bayesian analysis of mixtures with an unknown number of components. *J. Roy. Statist. Soc. B* **59**, 731–792, (with discussion).

BAYESIAN STATISTICS 8, pp. 643–648.
J. M. Bernardo, M. J. Bayarri, J. O. Berger, A. P. Dawid,
D. Heckerman, A. F. M. Smith and M. West (Eds.)

Ergodic Averages Via Dominating Processes*

JESPER MØLLER
Aalborg University, Denmark
jm@math.aau.dk

KERRIE MENGERSEN
Queensland Univ. Tech., Australia
k.mengersen@qut.edu.au

SUMMARY

We show how the mean of a monotone function (defined on a state space equipped with a partial ordering) can be estimated, using ergodic averages calculated from upper and lower dominating processes of a stationary irreducible Markov chain. In particular, we do not need to simulate the stationary Markov chain and we eliminate the problem of whether an appropriate burn-in is determined or not. Moreover, when a central limit theorem applies, we show how confidence intervals for the mean can be estimated by bounding the asymptotic variance of the ergodic average based on the equilibrium chain.

Keywords and Phrases: ASYMPTOTIC VARIANCE; BAYESIAN MODELS; BURN-IN; ERGODIC AVERAGE; ISING MODEL; MARKOV CHAIN MONTE CARLO; MIXTURE MODEL; MONOTONOCITY; PERFECT SIMULATION; RANDOM WALK; SPATIAL MODELS; UPPER AND LOWER DOMINATING PROCESSES

1. INTRODUCTION

Suppose that π is a given target distribution on a state space Ω and we wish to estimate the mean $\mu = \int \phi(x)\pi(\mathrm{d}x)$ for a given real function ϕ. In many applications it is not known or at least not straightforward to generate a stationary chain, so instead a non-stationary chain $Y_1, Y_2 \ldots$ is generated by Markov chain Monte Carlo (MCMC) and μ is estimated by the ergodic average $\sum_{t=M+1}^{N} \phi(Y_t)/(N-M)$, where $M \geq 0$ is an 'appropriate' burn-in and $N \gg M$ is 'sufficiently" large, (see, for example, Robert and Casella 2004). This estimator is consistent provided the chain is irreducible and M is independent of the Y chain. The problem is to determine M and N so that the estimator is close to μ with a high degree of confidence.

Propp and Wilson (1996) showed how upper and lower dominating processes can be used for generating a perfect (or exact) simulation of a stationary Markov chain at a fixed time, provided the chain is monotone with respect to a partial ordering

---

* This is a synopsis. For the full version, see *Bayesian Analysis* **2** (2007), to appear.
This research was supported by The Danish Natural Science Research Council and the Australian Research Centre for Dynamic Systems and Control.

on Ω for which there exists unique maximal and minimal states. In this paper we introduce similar ideas but our aim is to obtain reliable estimates of mean values rather than perfect simulations.

More specifically, we consider irreducible Markov chains with π as the invariant distribution and make the following additional assumptions. Let $X = (X_t; t = 1, 2, \ldots)$ denote the possibly unknown equilibrium chain, i.e. $X_1 \sim \pi$ and hence $X_t \sim \pi$ for all $t \geq 1$, and let $\bar{\phi}_t = \sum_{s=1}^{t} \phi(X_s)/t$ denote the ergodic average estimating μ. Assume there exist stochastic processes $U = (U_t; t = 1, 2, \ldots)$ and $L = (L_t; t = 1, 2, \ldots)$ so that

$$\bar{\phi}_t^L \leq \bar{\phi}_t \leq \bar{\phi}_t^U, \quad t = 1, 2, \ldots, \tag{1}$$

where the ergodic averages $\bar{\phi}_t^L = \sum_{s=1}^{t} \phi(L_s)/t$ and $\bar{\phi}_t^U = \sum_{s=1}^{t} \phi(U_s)/t$ are consistent estimators of μ. To ensure (1) we assume that with respect to a partial ordering $\prec$ on Ω, U and L are bounding X, i.e.

$$L_t \prec X_t \prec U_t, \quad t = 1, 2, \ldots, \tag{2}$$

and ϕ is monotone

$$x \prec y \text{ implies } \phi(x) \leq \phi(y) \tag{3}$$

(or, as discussed later, ϕ is a linear combination of monotone functions). Then (1) holds, so it suffices to consider the processes $(\bar{\phi}_t^L; t = 1, 2, \ldots)$ and $(\bar{\phi}_t^U; t = 1, 2, \ldots)$. Consequently, we do not need to simulate the equilibrium chain and we eliminate the problem of whether an appropriate burn-in is determined or not. Assuming a central limit theorem applies, confidence intervals for the mean can be estimated by bounding the asymptotic variance of $\bar{\phi}_t$. Note also that to assess if the process $(\phi(X_t); t = 1, 2, \ldots)$ has stabilised into equilibrium, it suffices to consider the processes $(\phi(L_t); t = 1, 2, \ldots)$ and $(\phi(U_t); t = 1, 2, \ldots)$. In contrast to the Propp-Wilson algorithm we do not require that U_t and L_t coalesce for all sufficiently large t. Equivalently, we do not require that X is uniformly ergodic (Foss and Tweedie, 1998).

This synopsis presents our ideas in the general setting and discussed applications and extensions. The full paper, to appear in the *Bayesian Analysis Journal*, provides a more complete description of the methods and their application.

2. GENERAL SETTING AND METHODS

Assume that (2)–(3) and hence (1) are satisfied, where the equilibrium chain X is irreducible and $\bar{\phi}_t^L$ and $\bar{\phi}_t^U$ are consistent estimators of μ. Moreover, assume that a central limit theorem (CLT) applies, so that $\sqrt{t}(\bar{\phi}_t - \mu)$ converges in distribution to $N(0, \sigma^2)$ as $t \to \infty$, where

$$\sigma^2 = \sum_{t=-\infty}^{\infty} \gamma_{|t|} < \infty, \quad \gamma_t = \text{Cov}(\phi(X_1), \phi(X_{t+1})), \, t \geq 0 \, . \tag{4}$$

Sufficient conditions for the CLT to hold can be found in Meyn and Tweedie (1993), Geyer (1996), Chan and Geyer (1994) and Roberts and Rosenthal (1998). For instance, it suffices to establish that X is geometrically ergodic and, if X is reversible, that $\text{E}\phi(X_t)^2 < \infty$.

Assuming that X is reversible, we estimate σ^2 using, for example, a window type estimator (Geyer, 1992) or batch means Ripley, 1987). For specificity, we consider here a window type estimator

$$\hat{\sigma}_N^2 = \sum_{t=-m}^{m} \hat{\gamma}_{|t|,N} \ ; \ \hat{\gamma}_{t,N} = \frac{1}{N} \sum_{s=1}^{N-t} (\phi(X_{s+t}) - \bar{\phi}_N)(\phi(X_s) - \bar{\phi}_N) \ ;$$

see, for example, Priestly (1981, pp. 323–324). Geyer's initial series estimator is given by letting $m = 2l + 1$ where l is the largest integer so that the sequence $\hat{\gamma}_{2t,N} + \hat{\gamma}_{2t+1,N}$, $t = 0, \ldots, l$, is strictly positive, strictly decreasing and strictly convex. For an irreducible and reversible Markov chain this provides a consistent conservative estimator of σ^2, cf. Geyer (1992).

For a real number or function f, write $f_+ = \max\{0, f\}$ for its positive part and $f_- = \max\{0, -f\}$ for its negative part, so $f = f_+ - f_-$. By (2)-(3) we have that

$$0 \le \phi_+(L_t) \le \phi_+(X_t) \le \phi_+(U_t), \quad 0 \le \phi_-(U_t) \le \phi_-(X_t) \le \phi_-(L_t),$$

$$0 \le \bar{\phi}_{N+}^L \le \bar{\phi}_{N+} \le \bar{\phi}_N^U, \quad 0 \le \bar{\phi}_{N-}^U \le \bar{\phi}_{N-} \le \bar{\phi}_{N-}^L.$$

Hence $\hat{\sigma}_N^2$ is bounded from above and below by

$$\hat{\sigma}_{\max,N}^2 = \sum_{t=-m}^{m} a_{|t|,N}, \quad \hat{\sigma}_{\min,N}^2 = \sum_{t=-m}^{m} b_{|t|,N},$$

where for $t \ge 0$,

$$\begin{aligned} a_{t,N} = \ & \tfrac{1}{N} \textstyle\sum_{s=1}^{N-t} \Big\{ \phi_+(U_{s+t})\phi_+(U_s) - \phi_-(U_{s+t})\phi_+(L_s) \\ & -\phi_+(L_{s+t})\phi_-(U_s) + \phi_-(L_{s+t})\phi_-(L_s) - \phi_+(L_{s+t})\bar{\phi}_{N+}^L \\ & +\phi_+(U_{s+t})\bar{\phi}_{N-}^L + \phi_-(L_{s+t})\bar{\phi}_{N+}^U - \phi_-(U_{s+t})\bar{\phi}_{N-}^U \\ & -\phi_+(L_s)\bar{\phi}_{N+}^L + \phi_+(U_s)\bar{\phi}_{N-}^L + \phi_-(L_s)\bar{\phi}_{N+}^U \\ & -\phi_-(U_s)\bar{\phi}_{N-}^U + \bar{\phi}_N^{U\,2} - 2\bar{\phi}_{N+}^L\bar{\phi}_{N-}^U + \bar{\phi}_N^{L\,2} \Big\} \end{aligned}$$

and

$$\begin{aligned} b_{t,N} = \ & \tfrac{1}{N} \textstyle\sum_{s=1}^{N-t} \Big\{ \phi_+(L_{s+t})\phi_+(L_s) - \phi_-(L_{s+t})\phi_+(U_s) \\ & -\phi_+(U_{s+t})\phi_-(L_s) + \phi_-(U_{s+t})\phi_-(U_s) - \phi_+(U_{s+t})\bar{\phi}_{N+}^U \\ & +\phi_+(L_{s+t})\bar{\phi}_{N-}^U + \phi_-(U_{s+t})\bar{\phi}_{N+}^L - \phi_-(L_{s+t})\bar{\phi}_{N-}^L \\ & -\phi_+(U_s)\bar{\phi}_{N+}^U + \phi_+(L_s)\bar{\phi}_{N-}^U + \phi_-(U_s)\bar{\phi}_{N+}^L \\ & -\phi_-(L_s)\bar{\phi}_{N-}^L + \bar{\phi}_N^{L\,2} - 2\bar{\phi}_{N+}^U\bar{\phi}_{N-}^L + \bar{\phi}_N^{U\,2} \Big\}. \end{aligned}$$

These bounds depend entirely on the upper and lower processes and not on the equilibrium chain.

2.1. *Method 1*

Our first method is based on combining (1), the CLT and the upper bound on $\hat{\sigma}_N^2$ to obtain a conservative confidence interval for μ: letting $\hat{\sigma}_{\max,N} = \sqrt{\hat{\sigma}^2_{\max,N}}$, then asymptotically with at least probability $2(1-\alpha)$,

$$\bar{\phi}_N^L - q_\alpha \hat{\sigma}_{\max,N} \le \mu \le \bar{\phi}_N^U + q_\alpha \hat{\sigma}_{\max,N} \ .$$

2.2. *Method 2*

One potential problem with Method 1 is meta-stability: the processes $\bar{\phi}_N^L$ and $\bar{\phi}_N^U$ may appear to have converged at time N, but they have not yet done so. A more conservative alternative is to use i.i.d. blocks of upper and lower processes. The relative merit of one method over the other depends on the particular model.

Assume that there exist unique elements $\hat{0}, \hat{1} \in \Omega$ so that $\hat{0} \prec x \prec \hat{1}$ for all $x \in \Omega$. Suppose that $((U_t^{(1)}, L_t^{(1)}),_{t=1,\ldots,T_1}, T_1)$, $((U_t^{(2)}, L_t^{(2)}),_{t=1,\ldots,T_2}, T_2), \ldots$ are i.i.d. 'blocks", where $T_1, T_2, \ldots$ are either equal fixed times or i.i.d. random times so that

$$U_1^{(i)} = \hat{1}, \quad L_1^{(i)} = \hat{0}, \quad i = 1, 2, \ldots,$$

$$U_t^{(1)} = U_t, \quad L_t^{(1)} = L_t, \quad t = 1, \ldots, T_1,$$

$$L_t^{(2)} \prec L_{t+T_1} \prec U_{t+T_1} \prec U_t^{(2)}, \quad t = 1, \ldots, T_2,$$

$$L_t^{(3)} \prec L_{t+T_1+T_2} \prec U_{t+T_1+T_2}, \prec U_t^{(3)}, \quad t = 1, \ldots, T_3,$$

and so on. We may, for example, choose T_i as the first time n_i at which

$$\frac{1}{n_i} \sum_{t=1}^{n_i} \left(\phi(U_t^{(i)}) - \phi(L_t^{(i)}) \right) \le \epsilon,$$

where $\epsilon > 0$ is a user-specified parameter.

By (2)–(3), for $N = T_1 + \ldots + T_m$ and $m = 1, 2, \ldots$, $\tilde{\phi}_N^L \le \bar{\phi}_N^L \le \bar{\phi}_N \le \bar{\phi}_N^U \le \tilde{\phi}_N^U$, where we set

$$\tilde{\phi}_N^U = \frac{1}{N} \sum_{i=1}^{m} W_i^U, \quad W_i^U = \sum_{s=1}^{T_i} \phi(U^{(i)}_{s+T_0+\ldots+T_{i-1}}),$$

$$\tilde{\phi}_N^L = \frac{1}{N} \sum_{i=1}^{m} W_i^L, \quad W_i^L = \sum_{s=1}^{T_i} \phi(L^{(i)}_{s+T_0+\ldots+T_{i-1}}),$$

and $T_0 = 0$. On one hand these new bounds are more conservative: in Method 1, $\bar{\phi}_N^U$ and $\bar{\phi}_N^L$ are consistent estimators of μ, whereas $\tilde{\phi}_N^U$ and $\tilde{\phi}_N^L$ almost surely converge to $\mathrm{E}W_1^U/\mathrm{E}T_1$ and $\mathrm{E}W_1^L/\mathrm{E}T_1$, respectively, which in general are different from μ. On the other hand, since the blocks are i.i.d., we may better 'trust" the bounds $\tilde{\phi}_N^U$ and $\tilde{\phi}_N^L$: if these bounds are close, we may expect that $\bar{\phi}_N^U$ and $\bar{\phi}_N^L$ have been stabilised. If $T_i = n_i$ then of course $\tilde{\phi}_N^U - \tilde{\phi}_N^L \le \epsilon$.

Finally, the classical CLT and strong law of large numbers apply for the i.i.d. blocks so that as $m \to \infty$, $\tilde{\phi}_N^U$ and $\tilde{\phi}_N^L$ are approximately normally distributed with variances $(\text{Var}W_1^U)/(m(\text{E}T_1)^2)$ and $(\text{Var}W_1^L)/(m(\text{E}T_1)^2)$ provided the moments exist. It is straightforward to estimate these moments from the i.i.d.. blocks and thereby obtain consistent estimates $\tilde{\sigma}_{\max,N}$ and $\tilde{\sigma}_{\min,N}$ for the standard deviations. Thus asymptotically with at least probability $2(1-\alpha)$,

$$\tilde{\phi}_N^L - q_\alpha \tilde{\sigma}_{\min,N} \leq \mu \leq \tilde{\phi}_U^L + q_\alpha \tilde{\sigma}_{\max,N}.$$

3. APPLICATIONS AND EXTENSIONS

The full paper (Møller and Mengersen, to appear in *Bayesian Analysis Journal*) provides details of our methods applied to a simple random walk, an Ising model and a mixture model for which the weights are unknown. Our methods also apply when ϕ is anti-monotone, i.e. when $x \prec y$ implies $\phi(x) \geq \phi(y)$, and when ϕ is a linear combination of monotone functions. Many lattice and point process models are of an exponential family type where case each coordinate function of the canonical sufficient statistic is a linear combination of monotone functions.

Method 1 easily extends to a time-continuous setting. For example, spatial birth-death processes have been successfully used for perfect simulation of spatial point processes (Kendall and Møller, 2000), and Method 1 can straightforwardly be modified to this case. However, Method 2 does not easily apply in that case, since there is no maximal element.

In particular, our methods apply for many stochastic models used in statistical physics and spatial statistics. Examples include Ising and hard-core models, many of Besag's auto-models (Besag, 1974) and for coupling constructions (Møller, 1999). Moreover, many spatial point process models, including the Strauss process and other repulsive pairwise interaction point process models can be handled, using the modification of Method 1 discussed above.

Our methods are suitable for posterior distributions associated with mixtures of exponential families and conjugate priors (Casella *et al.*, 2002) using the upper and lower chains introduced in Mira *et al.* (2001), where other examples of applications are also given. They also apply when using the upper and lower processes for the perfect simulated tempering algorithms and the Bayesian models considered in Møller and Nicholls (1999) and Brooks *et al.* (2002). However, our methods so far are only applicable to general Bayesian problems for which it is possible to construct upper and lower dominating processes and for which the functions ϕ of interest are linear combinations of monotone functions.

REFERENCES

Besag, J. (1974). Spatial interaction and the statistical analysis of lattice systems. *J. Roy. Statist. Soc. B* **36**, 192–326 (with discussion).

Brooks, S., Fan, Y. and Rosenthal, J. (2002). Perfect forward simulation via simulated tempering. *Tech. Rep.*, University of Cambridge, UK.

Casella, G., Mengersen, K. , Robert, C. P. and Titterington, D. (2002) Perfect slice samplers for mixtures of distributions. *J. Roy. Statist. Soc. B* **64**, 777–790.

Chan, K. and Geyer, C. (1994). Discussion of 'Markov chains for exploring posterior distributions". *Ann. Statist.* **22**, 1747–1758.

Foss, S. and Tweedie, R. (1998). Perfect simulation and backward coupling. *Stochastic Models* **14**, 187–203.

Geyer, C. (1992). Practical Monte Carlo Markov chain *Statist. Science* **7**, 473–511 (with discussion).

Geyer, C. (1996). Estimation and optimization of functions. *Markov Chain Monte Carlo in Practice* (W. Gilks, S. Richardson and D. Spiegelhalter, eds.) London: Chapman and Hall, 241–258.

Kendall, W. and Møller, J. (2000). Perfect simulation using dominating processes on ordered spaces, with application to locally stable point processes. *Adv. Appl. Probability* **32**, 844–885.

Mira, A., Møller, J. and Roberts, G. (2001). Perfect slice samplers. *J. Roy. Statist. Soc. B* **63**, 583–606.

Møller, J. (1999). Perfect simulation of conditionally specified models. *J. Roy. Statist. Soc. B* **61**, 251–264.

Møller, J. and Nicholls, G. (1999). Perfect simulation for sample-based inference. *Tech. Rep.*, R-99-2011, Aalborg University, Denmark.

Priestly, M. B. (1981). *Spectral Analysis and Time Series.* New York: Academic Press.

Propp, J. and Wilson, D. (1996). Exact sampling with coupled Markov chains and applications to statistical mechanics. *Random Structures and Algorithms* **9**, 223–252.

Ripley, B. (1987). *Stochastic Simulation.* New York: Wiley.

Robert, C. P. and Casella, G. (2004). *Monte Carlo Statistical Methods* (2nd Edition). Berlin: Springer-Verlag.

Roberts, G. and Rosenthal, J. (1998). Markov chain Monte Carlo: Some practical implications of theoretical results. *Can. J. Statist.* **26**, 5–32.

BAYESIAN STATISTICS 8, pp. 649–654.
J. M. Bernardo, M. J. Bayarri, J. O. Berger, A. P. Dawid,
D. Heckerman, A. F. M. Smith and M. West (Eds.)

Bayesian Model Diagnostics Based on Artificial Autoregressive Errors*

MARIO PERUGGIA
The Ohio State University, USA
peruggia@stat.ohio-state.edu

SUMMARY

This article provides a synopsis of the material presented in Peruggia (2007), where I develop a simple and effective device for ascertaining the quality of the modeling choices and detecting lack-of-fit in hierarchical Bayes models. These models will typically specify regression equations for one or more parameters, with errors that are often assumed to be i.i.d. Gaussian. To build the diagnostic tool, I specify an artificial autoregressive structure (AAR) in the probability model for the errors that incorporates the i.i.d. model as a special case and assess lack-of-fit by examining the posterior distribution of the AAR parameters.

Keywords and Phrases: ALLOMETRY; ASYMPTOTIC NORMALITY; AUTOCORRELATION; HIERARCHICAL MODELS; RESPONSE TIMES.

1. INTRODUCTION

It is common statistical practice to evaluate the fit of a model by examining the behavior of the realized residuals. Realized residuals incompatible with i.i.d. errors may indicate inadequacies in the model and suggest avenues to improve the fit. The use of residual plots and related diagnostics is well established in the frequentist arena and virtually all statistical software packages provide easy access to computational and graphical tools to perform residual-based model checks.

Hierarchical Bayes models provide a natural way of incorporating covariate information into the inferential process through the elaboration of regression equations for one or more of the model parameters, with errors that are often assumed to be i.i.d. Gaussian. The approach is conceptually simple and the development of efficient

---

* This is a synopsis. For the full version, see *Bayesian Analysis* **2** (2007), to appear.
Mario Peruggia is a Professor in the Department of Statistics at The Ohio State University. This material is based upon work supported by the National Science Foundation under Awards No. SES-0214574, SES-0437251, and DMS-0605052. The author thanks Peter Craigmile, Ilenia Epifani, Steve MacEachern and Trisha Van Zandt for their helpful comments.

Markov Chain Monte Carlo (MCMC) computational techniques has made its implementation feasible even in rather complicated settings. Unfortunately, building adequate regression models is a complicated art form that requires the practitioner to make numerous hard decisions along the way. By comparison to its frequentist counterpart, the contents of the Bayesian model-building toolbox look quite scanty. All the more so when the regression model to be constructed is for a parameter at a higher stage in the hierarchy, a quantity for which, typically, the intuition is not as well developed as the intuition for quantities that are directly observable.

In Peruggia (2007) I demonstrate that residual analysis is also a powerful diagnostic tool for Bayesian model building. Rather than examining the realized residuals directly (a difficult task, especially when the regression models are for parameters that appear at higher levels of the model hierarchy), I introduce an artificial autoregressive (AAR) structure in the probability model for the errors that incorporates the i.i.d. model as a special case. Lack-of-fit can be detected by examining the posterior distribution of the AAR parameters. In general, posterior distributions that assign considerable mass to a region of the AAR parameter space away from zero provide evidence that apparent dependencies in the errors are compensating for misspecifications of some other model aspects (typically conditional means).

2. AAR ERRORS IN A SIMPLE LINEAR REGRESSION MODEL

The basic premise underlying the proposed diagnostic device can be easily and clearly illustrated in the context of a linear regression model. In Peruggia (2007) I consider a data set comprised of $n = 101$ pairs of observations (x_i, y_i) simulated from the model $y_i = 1.0 + (0.1)x_i + u_i$, where, for $i = 0, \ldots, n$, $x_i = i$ and the u_i are independent standard normal errors.

To model these data, I assume that, for $i = 0, \ldots, n$, $y_i = \alpha + bx_i + \eta_i$, where b is a fixed constant, possibly mispecified, and $\alpha \sim N(0, 10^3)$. The errors η_i have an AAR structure specified as follows: $\eta_1 \sim N(0, \sigma_1^2)$, and, for $i = 1, \ldots, n$, $\eta_i = \phi\eta_{i-1} + \epsilon_i$, with $\epsilon_i \sim N(0, \sigma_\epsilon^2)$, $1/\sigma_1^2 \sim Gamma(2.5, 10.0)$, $1/\sigma_\epsilon^2 \sim Gamma(0.1, 0.1)$, and $\phi \sim N(0, 1.0)$. Using BUGS, I fit five regression models with an AAR error structure and a possibly misspecified slope. The five values of b under consideration, $b_1 = 0.10$, $b_2 = 0.09$, $b_3 = 0.07$, $b_4 = 0.05$, and $b_5 = 0.00$, move progressively away from the slope of 0.1 used to simulate the data.

For $1 \leq j \leq 5$, consider the fitted regression line given by $y = \hat{\alpha}_j + b_j x$, where $\hat{\alpha}_j$ is an estimate of the posterior mean of α obtained by averaging $1,000$ posterior draws generated by BUGS based on the model with slope b_j. The residuals from the fitted line determine the behavior of the diagnostic device.

The residuals from the horizontal line, corresponding to slope $b_5 = 0.00$, exhibit the strongest systematic trend. As x increases, they progressively shift from having large negative values to having large positive values. This increasing trend is due to the erroneous specification of the conditional mean for the observations. Failure to remove the trend induces strong positive autocorrelations in the residuals that are captured by the AAR model. As a consequence, the posterior distribution of the AAR parameter ϕ is concentrated near one. Based on the 1000 draws from the posterior distribution of ϕ used to construct the histogram, the posterior mean is estimated at 0.93 and the 95% equal-tailed posterior probability interval is $[0.83, 1.01]$ (see Table 1).

By contrast, the residuals from the fitted line corresponding to slope $b_1 = 0.1$ are well behaved and manifest no obvious of trends. This, of course, is not surprising because the data were simulated from a regression model with slope equal to 0.1,

Table 1: *ARR Parameter Estimates for the SLR Example*

slope b_j	0.10	0.09	0.07	0.05	0.00
post. mean of ϕ	0.02	0.09	0.46	0.71	0.93
post. 2.5 percentile of ϕ	-0.19	-0.12	0.27	0.56	0.83
post. 97.5 percentile of ϕ	0.24	0.31	0.65	0.87	1.01

and, in this case, a model of independence for the AAR error structure ($\phi = 0$) would be appropriate. Accordingly, the posterior distribution of the AAR parameter ϕ is centered in the vicinity of zero. Based on the 1,000 draws from the posterior distribution of ϕ used to construct the histogram, the posterior mean is estimated at 0.02 and the 95% equal-tailed posterior probability interval is $[-0.19, 0.24]$ (see Table 1). The intermediate cases corresponding to slopes b_2, b_3, and b_4, are characterized by correspondingly worsening fits and posterior distributions of the AAR parameter ϕ that become progressively more concentrated about a central location that shifts from zero towards one (see Table 1).

3. ASYMPTOTIC CONSIDERATIONS

The following considerations based on asymptotic arguments help to formalize the intuition developed in the example presented in Section 2.

Let f and g be smooth functions on a finite closed interval, that, without loss of generality, can be taken to be $[0, 1]$. I will consider regular infill asymptotics, assuming that the true data generating mechanism is given by $y_i = f(i/n) + u_i$, where, for $i = 0, \ldots, n$, the u_i are independent normal errors with mean zero and variance σ_u^2. The fitted model with AAR errors is given by $y_i = g(i/n) + \eta_i$, with $\eta_i = \phi\eta_{i-1} + \epsilon_i$, where the i.i.d. zero-mean normal noise process $\{\epsilon_i\}$ has variance σ_ϵ^2.

A full Bayesian model would typically specify a parametric family $g(t \mid \boldsymbol{\theta})$ to which $g(t)$ belongs as well as a prior distribution for the model parameters $\boldsymbol{\theta}$, ϕ, and σ_ϵ^2. For all fixed n, the model will define a likelihood $p_n(\boldsymbol{y}_n \mid (\boldsymbol{\theta}, \phi, \sigma_\epsilon^2))$ for the data $\boldsymbol{y}_n = (y_0, \ldots, y_n)$ collected under the regular infill scheme. Let $(\boldsymbol{\theta}^*, \phi^*, (\sigma_\epsilon^*)^2)_n$ be the value of $(\boldsymbol{\theta}, \phi, \sigma_\epsilon^2))$ that minimizes the Kullback–Leibler divergence of the distribution$p_n(\boldsymbol{y}_n \mid (\boldsymbol{\theta}, \phi, \sigma_\epsilon^2))$ from the true distribution of the data, $p_n(\boldsymbol{y}_n)$, let $(\boldsymbol{\theta}^*, \phi^*, (\sigma_\epsilon^*)^2) = \lim_{n\to\infty}(\boldsymbol{\theta}^*, \phi^*, (\sigma_\epsilon^*)^2)_n$, and let

$$V(\boldsymbol{\theta}, \phi, \sigma_\epsilon^2) = \lim_{n\to\infty} (n+1)[I_n(\boldsymbol{\theta}, \phi, \sigma_\epsilon^2)]^{-1},$$

where $I_n(\boldsymbol{\theta}, \phi, \sigma_\epsilon^2)$ denotes the Fisher information matrix based on a sample of size $n + 1$. Under suitable regularity conditions (see Gelman et al. (2004) for the case of i.i.d. observations), the joint posterior distribution of $\sqrt{n+1}\,[(\boldsymbol{\theta}, \phi, \sigma_\epsilon^2) - (\boldsymbol{\theta}^*, \phi^*, (\sigma_\epsilon^*)^2)]$ will converge to a normal distribution with mean $\mathbf{0}$ and covariance matrix $V[(\boldsymbol{\theta}^*, \phi^*, (\sigma_\epsilon^*)^2)]$. The member of the family $g(t \mid \boldsymbol{\theta})$ that best approximates $f(t)$ in the sense of minimizing $\int_0^1 (f(t) - g(t \mid \boldsymbol{\theta}))^2\, dt$ is $g(t \mid \boldsymbol{\theta}^*)$. Let $r^*(t) = f(t) - g(t \mid \boldsymbol{\theta}^*)$. Then, marginally, the posterior of $\sqrt{n+1}\,(\phi - \phi^*)$ will converge to a univariate normal distribution with mean 0 and variance $(\sigma_\epsilon^*)^2/(\int_0^1 [r^*(t)]^2\, dt + \sigma_u^2)$, where

$$\phi^* = \frac{\int_0^1 [r^*(t)]^2\, dt}{\int_0^1 [r^*(t)]^2\, dt + \sigma_u^2}. \tag{1}$$

The expression in Equation (1) can be interpreted as a *signal to noise ratio* for detecting unmodeled trend: the integrated squared bias in the numerator quantifies the amount of unmodeled trend and the denominator quantifies the amount of unmodel trend plus the amount of noise in the data. Peruggia (2007) illustrates an application of Equation (1) to the regression example of Section 2.

4. AAR ERRORS IN DATA ANALYSIS APPLICATIONS

In Peruggia (2007) I present several examples of the application of the diagnostic device in several data analysis settings.

Example 1 (Random Effects Models). In one example I demonstrate that the AAR method can also be employed when the independent variable x does not play the role of time and does not take on equally spaced values. For illustration, I consider the 100 complete pairs of brain and body weights for placental mammalian species published in an allometry study by Sacher and Staffeldt (1974). In addition to the weights, the data set also records the order and sub-order to which each species belongs. This data set was used in MacEachern and Peruggia (2002) to illustrate the performance of some numerical and graphical diagnostic tools in detecting the shortcomings of a simple linear regression (SLR) model compared to a variance components (VC) model.

In Peruggia (2007) I fit both a SLR model with AAR errors and a Bayesian VC model with AAR errors which incorporates random effect terms for the 13 orders and 19 sub-orders in the taxonomy. Based on 1,000 draws generated using BUGS with a burn-in of 10,000 and a thinning rate of 1 in 10, I computed summaries of the posterior distributions of the AAR parameter ϕ for the SLR and VC models. For the SLR model, the posterior expectation of ϕ is estimated at 0.61 with a 95% equal-tailed posterior probability interval given by $[0.45, 0.78]$. For the VC model, the posterior expectation of ϕ is estimated at 0.28 with a 95% equal-tailed posterior probability interval given by $[-0.07, 0.59]$.

There is a clear indication that the SLR model is not adequately capturing the variability in the data despite the strong linear trend evidenced by the scatterplot. The reasons for this are carefully outlined in MacEachern and Peruggia (2002) Essentially, the SLR model neglects to account for dependencies of species within orders and sub-orders. The VC model ameliorates the situation by inducing positive correlations within these groups. In this example, the AAR diagnostic device exploits the existence of a trade-off between the residual autocorrelation structure and the presence of random effects in the model (cf. Pinheiro and Bates, 2000, p. 398). Specifically, when the correlations within groups are not taken into consideration by the SLR model, the residuals within each order and sub-order tend to behave similarly and the AAR error structure is sensitive to this grouping because the observation are arranged in the data file according to orders and sub-orders.

Example 2 (Model Comparison). In this example I look at a data set of Williams (1959). This data set was analyzed by Carlin and Chib (1995) and Spiegelhalter et al. (1996) to illustrate the use of Bayes factors to compare non-nested regression models. The response variable y is the maximum compressive strength parallel to the grain measured on 42 specimens of radiata pine. There are two possible predictors: x, denoting the density of the specimens, and z, denoting the density adjusted for resin content. Corresponding to the two predictors, there are two competing SLR models (Model A and B) with AAR errors.

In Peruggia (2007) I show how the AAR error diagnostic device is able to detect the superiority of Model B over Model A. In this example there is no natural ordering of the observations, not even a partial ordering, such as that induced by the taxonomy of the mammals data set of Example 1. If one fits Models A or B using the original ordering in the data file, the posterior distribution of the AAR parameter ϕ turns out to put considerable mass around zero and no evidence of lack-of-fit is uncovered in either case. Here, however, the main concern is a comparison of the fit provided by the two models and the ordering of the observations should relate to this comparison. The models differ in the predictor they use, so the relative quality of the fit will be determined by differences in the two predictors. This suggests simply ordering the data according to the increasing values of the differences $x_i - z_i$ (alternatively, an ordering based on relative differences, normalized for the size of the predictors, could also be used). The posterior distributions of ϕ corresponding to the two fits of the ordered data exhibit the following behavior. While the posterior distribution of ϕ in Model B continues to put considerable mass around zero, the posterior distribution of ϕ in Model A is shifted to the right of zero, thus providing evidence of lack-of-fit. Model A fits poorly systematically at those points where the differences between the x and y predictors are large, a feature that AAR diagnostic can readily detect.

Example 3 (A Complex Hierarchical Model). In this example I examine the ability of the AAR diagnostic device to uncover aspects of model inadequacy in the context of a complex hierarchical model for a large set of response time data. The data were collected in a series of recognition memory trials conducted on four subjects over ten non-consecutive days.

For each trial, a subject was initially asked to study a list of 40 words randomly selected from a database of 2337 common English words. The subject was then presented with a sequence of 40 words, 20 selected from the study list and 20 selected from words in the database not included in the study list. The words were displayed sequentially on a computer monitor and the subject was asked to strike one of two keys on the keyboard depending on whether she thought the word was an 'old' word contained in the study list or a 'new' word not contained in the study list. The times in milliseconds elapsed between the appearances on the words on the screen and the keystrokes were recorded, along with indicators of the accuracy of the responses. Each subject participated in two trials on each day. There were thus a total of $4 \times 2 \times 10 = 80$ trials, each contributing 40 response times.

The hierarchical model considered here is a refinement of the model presented in Peruggia et al. (2002). For the current analysis I used a two parameter Weibull likelihood to model the shifted response times obtained by subtracting a value equal to 95% of the minimum response time for each list from all of the 40 response times for that list. The logarithm of the scale parameter for each of the 80 trials is endowed with a regression structure that includes random effects primarily intended to model different levels of subject specific learning as days go by. The regression also includes fixed effects for the nature of the words ('old' vs. 'new'), the accuracy of the responses ('right' vs. 'wrong'), and their interaction. The prior specifications for the full Bayesian hierarchy are detailed in Peruggia (2007).

The error terms $\eta_{i,d,l,w}$ in the regressions for the log-scale parameters act as diagnostic devices for the stated model of conditional independence of the 40 response times within each list, following the basic AAR structure used throughout the article with AAR coefficients $\phi_{i,d,l}$ that depend on subject i, day d, and list l. The AAR structure is specified with respect to a given ordering of the response times corresponding to the words w within a list. In the diagnostic analysis, I considered two separate orderings, the one corresponding to the sequence in which the words were presented to the subjects and the one corresponding to a rearrangement in which the 'new' words are made to precede the 'old' words, while preserving the original ordering within each group.

For a basic diagnostic analysis, I quickly scanned the histograms of the 80 sets of 1,000 realizations from the marginal posterior distributions of the AAR parameters $\phi_{i,d,l}$ generated using WinBUGS with a burn-in period of 20,000 and a subsequent thinning rate of 1 in 500. Most distributions put considerable mass around zero but there are several interesting cases in which the shapes of the posterior distributions of the AAR parameters suggest inadequacies in the fit for certain word lists. For these cases I was then able to understand the source of the lack-of-fit by a direct examination of the residuals for the corresponding regression models.

For example, the AAR parameter diagnostic flags subject 3, day 9, list 1, uncovering the fact that, after adjusting for the values of the covariates, the earlier response times appear to be slower than the model would predict and the later response times appear to be faster. Another interesting case deals with subject 3, list 6, day 1. The AAR diagnostic based on the original ordering of the RTs provides no indication of lack-of-fit. However, when the RTs are rearranged so that those corresponding to the 'new' words are made to precede those corresponding to the 'old' words (while preserving the original ordering within each group) the AAR diagnostic detects lack-of-fit, due to the fact that that responses to the new words were systematically faster than the model would predict and responses to the old words were systematically slower. Several other interesting cases are examined in Peruggia (2007).

5. DISCUSSION

In Peruggia (2007) I proposed a general procedure based on AAR errors for detecting lack-of-fit in hierarchical Bayes model. The principal appeal of the procedure lies in its flexibility and its ease of implementation. The behavior of the residuals can shed much light on the fit of a model and suggest directions for refinement if lack-of-fit is detected. The strength of the proposed method is that the AAR device can be applied to assess the behavior of residuals at different levels of the hierarchy (without computing them explicitly), probing different aspects of the fit, without much conceptual or practical difficulty.

The potential benefits of the method when applied to complex hierarchical models is clearly illustrated in Example 3. There, I introduced, at a higher level of the hierarchy, a separate AAR error device for each of 80 sets of multidimensional observations (the 80 sets of RTs for each of the 80 word lists). The posterior distribution of each AAR parameter is a low dimensional summary that can quickly detect if the model of conditional independence provides an inadequate fit to the RTs for the corresponding list. Scanning such summaries to assess and quantify lack-of-fit is much easier than scanning observation-level residual plots. Finally, The AAR device is based on orderings of the observations. To detect lack-of-fit it is important to find a *meaningful* ordering for which the AAR parameter is significant.

REFERENCES

Carlin, B. P. and Chib, S. (1995). Bayesian model choice via Markov chain Monte Carlo methods. *J. Roy. Statist. Soc. B* **57**, 473–484.

Gelman, A., Carlin, J. B., Stern, H. S. and Rubin D. B. (2004). *Bayesian Data Analysis* (2nd ed.) London: Chapman and Hall.

MacEachern, S. N. and Peruggia, M. (2002). Bayesian tools for EDA and model building: A brainy study. *Case Studies in Bayesian Statistics V* (C. Gatsonis, R. E. Kass, B. Carlin, A. Carriquiry, A. Gelman, I. Verdinelli and M. West, eds.) New York: Springer-Verlag, 345–362.

Peruggia, M. (2007). Bayesian model diagnostics based on artificial autoregressive errors. *Bayesian Analysis* **2** (to appear).

Peruggia, M., Van Zandt, T., and Chen, M. (2002). Was it a Car or a Cat I Saw? An analysis of response times for word recognition. *Case Studies in Bayesian Statistics VI* (C. Gatsonis, R. E. Kass, A. Carriquiry, A. Gelman, D. Higdon, D. K. Pauler and I. Verdinelli, eds.) New York: Springer-Verlag, 319–334.

Pinheiro, J. C. and Bates, D. M. (2000). *Mixed-effects Models in S and S-PLUS*. Berlin: Springer-Verlag.

Sacher, G. A. and Staffeldt, E. F. (1974). Relation of gestation time to body weight for placental mammals: Implications for the theory of vertebrate growth. *Amer. Naturalist* **108**, 593–615.

Spiegelhalter, D. J., Thomas, A., Best, N. G., and Gilks, W. R. (1996). *BUGS Examples* **2**, Version 0.5(ii). Cambridge, UK: MRC Biostatistics Unit.

Williams, E. (1959). *Regression Analysis*. New York: Wiley.

BAYESIAN STATISTICS 8, pp. 655–660.
J. M. Bernardo, M. J. Bayarri, J. O. Berger, A. P. Dawid,
D. Heckerman, A. F. M. Smith and M. West (Eds.)

Estimation of Faraday Rotation Measures of the Near Galactic Sky, Using Gaussian Process Models*

MARGARET B. SHORT
University of Alaska at Fairbanks, USA
ffmbs1@uaf.edu

DAVID M. HIGDON and PHILIPP P. KRONBERG
Los Alamos National Laboratory, USA
dhigdon@lanl.gov kronberg@lanl.gov

SUMMARY

Our primary goal is to obtain a smoothed summary estimate of the magnetic field generated in and near to the Milky Way by using Faraday rotation measures (RM's). Each RM in our data set provides an integrated measure of the effect of the magnetic field along the entire line of sight to an extragalactic radio source. The ability to estimate the magnetic field generated locally by our galaxy and its environs will help astronomers distinguish local versus distant properties of the universe. The RM data are collected over locations in the galactic sky that correspond to locations on the unit sphere. In order to model such data, we employ a Bayesian process convolution approach which uses Markov chain Monte Carlo (MCMC) for estimation and prediction. Complications arise due to contamination in the RM measurements, and we resolve these by means of a mixture prior on the errors.

Keywords and Phrases: PROCESS CONVOLUTION, SMOOTHING SPLINE, OUTLIER MODELING

1. INTRODUCTION

There is widespread interest in understanding the role of magnetic fields on all scales and at all epochs in the history of the Universe. The strength and structure of large scale magnetic fields in astrophysics is just beginning to be probed with meaningful precision now that Faraday rotation measurements (RM), the prime detector of such fields, are available in sufficiently large numbers. This now justifies

---

* This is a synopsis. For the full version, see *Bayesian Analysis* **2** (2007), to appear.

appropriate statistical methods to facilitate interpretation of the RM values in terms of cosmic magnetic fields.

These 1566 RM measurements, spread over the spherical sky, contain multiple causes for their RM values. Most of the measurements are influenced only by our Milky Way galaxy; others also include the influence of distant galaxies and other distant sources (see Fig. 1). For this application, we are only concerned with the portion of the RM signal that is due to our local galaxy. It is not trivial to *a priori* separate out different contributions to a given RM measurement.

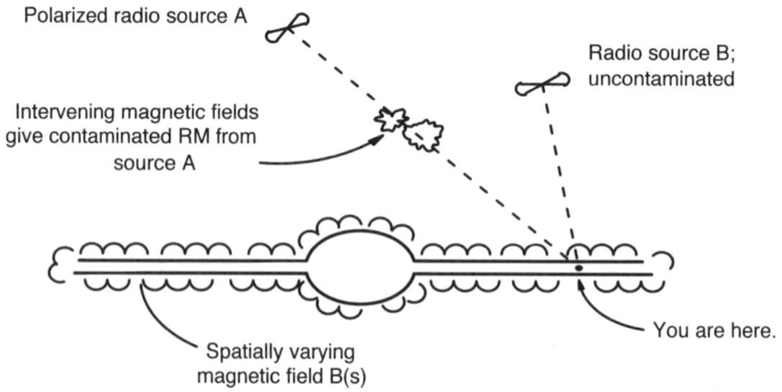

Figure 1: *Polarized radio sources are altered as they pass through ionized gases in the presence of a magnetic field, giving rise to observed rotation measures (RM's). The RM obtained from radio source B is determined only by the ionized gases and magnetic field from the Milky Way, whereas the RM from radio source A is contaminated by extragalactic interveners along the line of sight*

To date, a number of estimates of the RM sky have been produced (Simard-Normandin and Kronberg, 1980; Frick etal., 2001). This analysis is unique in that it uses the most extensive collection of RM measurements to date, uses a principled approach for accounting for outliers, accounts for uncertainties in the spatial modeling as well as the outlier modeling, and can accommodate the next generation of RM measurements, which will likely be in the tens of thousands.

2. FARADAY ROTATION MEASURES

Faraday rotation measure is defined as the line integral over a path of length L (in parsecs, with 1 pc = 3×10^{18} cm)

$$RM = k\int_0^L n_e \boldsymbol{B}\cdot ds, \tag{1}$$

where $k = 8.1 \times 10^5$, n_e is the electron density (measured in cm^{-3}), and $\boldsymbol{B}$ is magnetic field (in units of gauss); it is measured in units of radians per square meter of wavelength. Earth corresponds to one end point of the integral. Looking outward in any radial direction to a given radio source gives us another line integral at some position (l, b) on the galactic sky. Thus we may think of RM as a scalar quantity (positive or negative) which is a function of location on a sphere.

The locations on the sphere at which we are able to directly estimate RM are only in the selected directions where extragalactic radio sources such as radio galaxies and quasars happen to lie. These sources give off highly polarized signals from which it is possible to estimate the RM. Fig. 2 shows RM values, which correspond to 1566 distant radio sources. Red circles indicate positive RM's, blue circles negative RM's. More intense reds and blues correspond to RM's having larger magnitude. In this figure, the equator corresponds to the plane of the Milky Way. The center point $(l, b) = (0^o, 0^o)$ on the figure points directly at the center of the Milky Way. This is an Aitoff projection map, showing the entire sky. By convention, galactic longitudes l increase as one moves left.

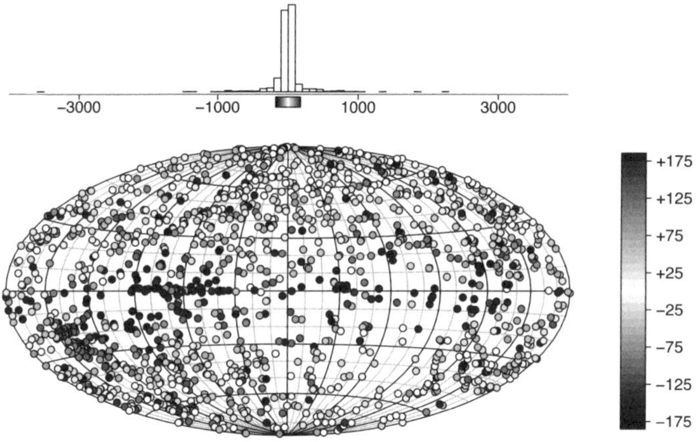

Figure 2: *1566 RM observations; units are rad/m^2. Large RM's (intense red and blue circles) tend to cluster along the plane of the Milky Way; higher latitudes exhibit smaller RM's. The histogram shows the extremely long tails, corresponding to "contaminated" observations*

Our goal is to estimate a smooth RM field due to the effect of the Milky Way galaxy. However, as shown in Fig. 1, some of the RM's are contaminated by intervening galaxies (or clusters of galaxies) that have substantial magnetic fields. These interveners can produce outliers that are apparent in Fig. 2. Our analysis accounts for the possibility that a substantial fraction of the RM's have some sort of contamination which is the sum total of the magnetic fields the signal has passed through on its way to Earth.

3. A PROCESS CONVOLUTION MODEL ON THE SPHERE

A fairly general Gaussian process (GP) model for an arbitrary space $\mathcal{S}$ with distance metric $d(\cdot, \cdot)$ can be created by taking a collection of uniform, regularly spaced knot locations $\boldsymbol{w}_1, \ldots, \boldsymbol{w}_J$, and assigning to each of these locations a knot value $x_1, \ldots, x_J$

which are assumed to have iid $N(0, [J\lambda_x]^{-1})$ distributions. Convolving these knot values with a simple smoothing kernel $k(\cdot)$ then results in the GP model

$$z(\boldsymbol{s}) = \sum_{j=1}^{J} x_j k(d(\boldsymbol{s}, \boldsymbol{w}_j)), \qquad \boldsymbol{s} \in \mathcal{S}. \tag{2}$$

Fig. 3 shows an example where $\mathcal{S}$ is the unit circle.

The covariance between any two locations is given by the formula

$$\mathrm{Cov}(z(\boldsymbol{s}_1), z(\boldsymbol{s}_2)) = [J\lambda_x]^{-1} \sum_{j=1}^{J} k(d(\boldsymbol{s}_1, \boldsymbol{w}_j))k(d(\boldsymbol{s}_2, \boldsymbol{w}_j)). \tag{3}$$

For bounded $\mathcal{S}$, as $J \to \infty$, the covariance function (3) becomes a function only of the distance $d(\boldsymbol{s}_1, \boldsymbol{s}_2)$ and λ_x, so that $z(\boldsymbol{s})$ is a stationary GP in the limit. For a finite J, (2) is only approximately stationary. Through empirical studies, we have found that if $k(\cdot)$ is normal with SD σ_k, a knot-to-knot spacing of no more than σ_k leads to a very good approximation to the limiting stationary model when $\mathcal{S}$ is the surface of the sphere in 3-d.

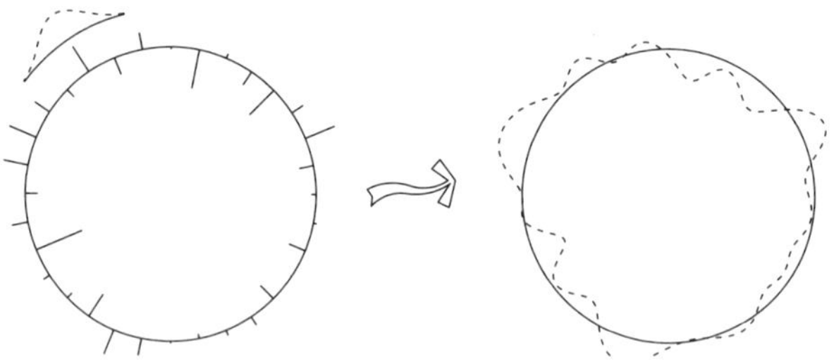

Figure 3: *A realization of a process convolution model on the unit circle. Left panel shows knot locations and values (positive and negative), along with a typical smoothing kernel. Right panel shows the resulting realization obtained by a convolution of the knot values with the smoothing kernel.*

4. MODEL AND ANALYSIS

For this application, we take $\mathcal{S}$ to be the surface of the unit sphere, $d(\cdot, \cdot)$ to be great circle distance, and $k(\cdot; \sigma_k)$ to be a normal density with standard deviation σ_k. A recursive tessellation algorithm is used to distribute $J = 2562$ knots over the unit sphere, giving a neighboring knot-to-knot distance of $2\pi/80$. The prior for the spatial RM field $z(\boldsymbol{s})$ is induced by the above specification. The N observations taken at locations $\boldsymbol{s}_1, \ldots, \boldsymbol{s}_N$ are modeled as

$$\boldsymbol{Y}(\boldsymbol{s}_i) = z(\boldsymbol{s}_i) + \epsilon_i, \; i = 1, \ldots, N$$

where the errors are independent with $N(0, [\omega_i \lambda_\epsilon]^{-1})$ distributions. The ω_i's, which modify the error precision, account for potentially contaminated observations. Priors for the remaining parameters complete the specification:

$$\boldsymbol{x}|\lambda_x \quad \sim \quad MVN(0, {\lambda_x}^{-1} I_{J\times J})$$

$$
\begin{aligned}
\sigma_x &\sim U(2\pi/80, 5\cdot 2\pi/80) \\
\lambda_{\mathrm{x}} &\sim \Gamma(1, .001) \\
\lambda_\epsilon &\sim \Gamma(1, .001) \\
\omega_i &\overset{\text{iid}}{\sim} \begin{cases} \delta_1 & \text{with probability 0.8} \\ \text{Uniform}(0,1) & \text{with probability 0.2.} \end{cases}
\end{aligned}
$$

Here the gamma priors for λ_{x} and λ_ϵ have a prior mean of 1000; δ_1 denotes a point mass at 1. Note that any overall mean for the RM field is to be absorbed in $z(\boldsymbol{x})$. Also, the prior probability of 0.8 that a given RM measurement is contaminated is set from expert judgement. Sensitivity studies have shown the posterior for $z(\boldsymbol{s})$ is rather insensitive to this value over a range from .5 to .9. Finally, the prior for the kernel width σ_x is constrained to be no smaller than the knot-to-knot distance from the 2562 knot locations. This ensures that $z(\boldsymbol{s})$ contains no artifacts due to overly dispersed knot locations.

The posterior mean for $z(\boldsymbol{s})$ is shown in Fig. 4. A large portion of the signal in the posterior mean field is from lines of sight along the equator which go through a substantial portion of the disk shaped Milky Way galaxy. Although this posterior map shows 5120 locations, predictions and their uncertainties can be quickly obtained for any location in the sky because of the constructive process convolution approach used here.

The average proportion of the observations identified as contaminated ($\omega_i < 1$) is 28%. The posterior distribution for the ω_i's, conditional on $\omega_i < 1$, is U-shaped, with additional mass near 0 and near 1.

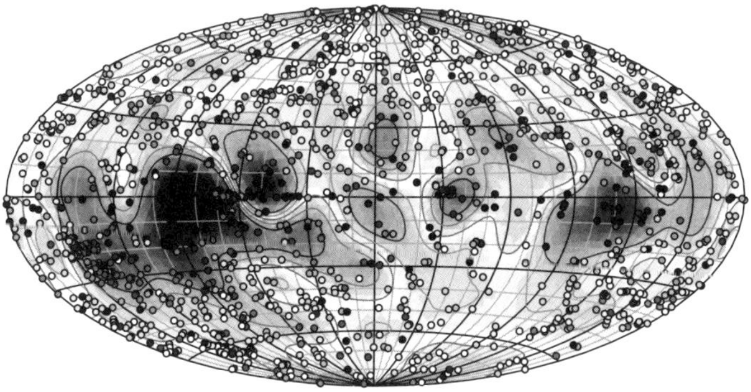

Figure 4: *Posterior plot for final model. Contour lines appear at levels corresponding to RM values of* $-50, -25, 0, 25$ *and* 50 *rad/m*^{2}. *The most extreme data are depicted by circles that are larger than those used for the remainder of the data.*

In conclusion we emphasize that this analysis gives a principled, statistical estimate of the RM field in the near galactic sky while accounting for contamination from extra-galactic sources, using the most complete set of RM measurements to date. The large number of observations in this current collection of measurements has forced us to make modeling choices that adequately address the specific features of this application while still making a fully Bayesian estimation analysis feasible via MCMC. The process convolution model for the RM field allows the MCMC scheme to avoid computationally costly inverses of large matrices. It also facilitates the inclusion of non-standard error models in the analysis.

REFERENCES

Besag, J. Green, P. J. Higdon, D. M. and Mengersen, K. (1995). Bayesian computation and stochastic systems. *Statist. Science* **10**, 3–66 (with discussion).

Calder, C. A. Holloman, C. and Higdon, D. M. (2002). Exploring space-time structure in ozone concentration using a dynamic process convolution model. *Case Studies in Bayesian Statistics VI* (C. Gatsonis, R. E. Kass, A. Carriquiry, A. Gelman, D. Higdon, D. K. Pauler and I. Verdinelli, eds.) New York: Springer-Verlag, 165–176.

Frick, P. Stepanov, R. Shukurov, A. and Sokoloff, D. (2001). Structures in the rotation measure sky. *Monthly Not. Royal Astronomical Soc.* **325**, 649–664.

Higdon, D. M., Swall, J. and Kern, J. (1999). Non-stationary spatial modeling. *Bayesian Statistics 6* (J. M. Bernardo, J. O. Berger, A. P. Dawid and A. F. M. Smith, eds.) Oxford: University Press, 761–768.

Higdon, D. M. (2002). Space and space-time modeling using process convolutions. *Quantitative methods for current environmental issues*, (C. Anderson, V. Barnett, P. C. Chatwin and A. H. El-Shaarawi, eds.) Berlin: Springer-Verlag, 37–56.

Lee, H. K. H. Higdon, D. M., Calder, C. A. and Holloman, C. H. (2005). Efficient models for correlated data via convolutions of intrinsic processes. *Statistical Modelling* **5** 53–74.

Oh, H. and Li, T. (2004). Estimation of global temperature fields from scattered observations by a spherical-wavelet-based spatially adaptive method. *J. Roy. Statist. Soc. B* **66**, 221–238.

Simard-Normandin, M. and Kronberg, P. (1980). Rotation measures and the galactic magnetic field. *The Astrophysical J.* **242**, 74–94.

Simard-Normandin, M. Kronberg, P. and Button, S. (1981). The Faraday rotation measures of extragalactic radio sources. *The Astrophysical J. Supplement series* **45**, 97–111.

Stepanov, R. Frick, P. Shukurov, A. and Sokoloff, D. (2002). Wavelet tomography of the Galactic magnetic field I. The method. *Astronomy & Astrophysics* **391**, 361–368.

BAYESIAN STATISTICS 8, pp. 661–666.
J. M. Bernardo, M. J. Bayarri, J. O. Berger, A. P. Dawid,
D. Heckerman, A. F. M. Smith and M. West (Eds.)

An Asymptotic Viewpoint on High-Dimensional Bayesian Testing*

DAN J. SPITZNER
Virginia Tech, Blacksburg, Virginia, USA
dan.spitzner@vt.edu

SUMMARY

The Bayesian point-null testing problem is studied asymptotically under a high-dimensional normal means model. Associated procedures are intended for application in finite dimensional settings, but their behavior is studied as the dimensionality grows arbitrarily. Thus, 'high-dimensionality' is formulated as an asymptotic concept, and interest is in the behavior of the procedures 'in the limit'. The approach is to allow the prior null probability to vary with dimension and with prior dispersion parameters, and to guide its parameterization so that the posterior null probability behaves in accordance with Bayesian consistency concepts. Among issues studied are the objectivity of setting the prior null probability to one-half, the Jeffreys–Lindley paradox, and the influence of smoothness constraints. Relevance to applications in functional data analysis and goodness-of-fit testing is also discussed.

Keywords and Phrases: BAYESIAN TESTING; HIGH-DIMENSIONAL TESTING; RATES OF TESTING; GOODNESS-OF-FIT TESTING

1. INTRODUCTION

This article discusses various issues in Bayesian testing of a point null hypothesis in high dimensions. The framework is much in the spirit of Robert (1993), in the sense that it considers a prior structure in which the prior null probability, a point mass on H_0, is connected to the dispersion parameters of the prior on H_1. The present investigation extends this dependency to account for the dimensionality of the model. It is shown that in high dimensions many difficulties associated with the construction of 'objective' Bayesian tests are naturally circumvented.

In particular, it is shown that the problem of arbitrary normalizing constants associated with improper priors may be largely avoided in high dimensions, and

* This is a synopsis. For the full version, see *Bayesian Analysis* **2** (2007), to appear.
Dan Spitzner is Assistant Professor, Department of Statistics (0439), Virginia Tech, Blacksburg, VA 24061, USA. The author is grateful for the comments of Judith Rousseau and Dongchu Sun.

Jeffreys–Lindley paradox largely resolved (*cf.* Spiegelhalter and Smith, 1982; Aitkin, 1991). In addition, it is argued that high-dimensional solutions that specify a prior null probability of one-half necessarily lack objectivity, but an alternative 'noninformative' solution is proposed. The second part of the article is concerned with the behavior of high-dimensional tests under geometric constraints intended to reflect a 'smooth' functional model, such as might be used in goodness-of-fit testing (*cf.*, Rayner and Best, 1989; Verdinelli and Wasserman, 1998) or functional data analysis (*cf.* Fan and Lin, 1998; Spitzner, 2006a). A connection is made to frequentist results on high-dimensional testing, which consequently identifies a favorable prior specification that is 'weighted' to account for smoothness. Finally, the issue of whether to condition on the geometric restrictions is considered, and it is concluded that such conditioning is ill-advised.

2. HIGH-DIMENSIONAL ANALYSIS OF THE BASIC MODEL

The data $Y = [Y_1, \ldots, Y_p]'$ are modelled according to

$$Y_j = \theta_j + \sigma e_j, \tag{1}$$

for mean parameter $\theta = [\theta_1, \ldots, \theta_p]'$, scale parameter σ, and independent standard normal errors e_j. Here, σ is assumed known, but might in more general contexts be treated with its own prior specification. The objective is to test $\mathrm{H}_0 : \theta_j = 0$ for all $j = 1, \ldots, p$ against a general alternative H_1.

Following a standard Bayesian setup for hypothesis testing (*cf.* Robert, 2001), a prior mass $\rho_0 \in (0,1)$ is placed on the null hypothesis and a continuous distribution $(1-\rho_0)\pi(\theta|\mathrm{H}_1)$ on the alternative, where $\pi(\theta|\mathrm{H}_1)$ is a specified density. The evidence provided by Y about H_0 is reported as $\mathrm{P}[\mathrm{H}_0|Y]$, the posterior probability of H_0. There is also the possibility of reporting a Bayes factor (*cf.* Kass and Raftery, 1995) $B_{01} = \{\mathrm{P}[\mathrm{H}_0|Y]/\rho_0\}/\{(1-\mathrm{P}[\mathrm{H}_0|Y])/(1-\rho_0)\}$, but it is simpler, for present purposes, to focus on $\mathrm{P}[\mathrm{H}_0|Y]$.

The present analysis takes $\theta|\mathrm{H}_1 \sim N(0, \tau^2 W)$, for $W = \mathrm{diag}(w_1 \ldots, w_p)$, and considers asymptotics as $p \to \infty$. A key parameterization of the prior is

$$\log\left(\rho_0^{-1} - 1\right) = -\frac{1}{2}\left\{\sum_{j=1}^{p} \log\left(1 - v_{j,1}\right) + \|v_{1/2}\|^2 + r_p\|v_1\|\right\}, \tag{2}$$

where where v_k denotes the p-dimensional vector with j'th entry $v_{j,k} = \{1 + \sigma^2(\tau^2 w_j)^{-1}\}^{-k}$, and r_p is some sequence that does not depend on Y or θ. The posterior null probability under (2) becomes

$$\mathrm{P}[\mathrm{H}_0|Y, \mathcal{B}_{s,M}] = \left[1 + \exp\left\{\frac{1}{2}\left(n\|v_{1/2}Y/\sigma\|^2 - \|v_{1/2}\|^2 - r_p\|v_1\|\right)\right\}\right]^{-1}. \tag{3}$$

This parameterization is completely general, but suggests a connection between ρ_0 and the prior dispersion parameters τ and $w_1, \ldots, w_p$ through the vector v.

To make sense of (2), Diaconis and Freedman's (1986) 'what if' principle is used to set up ground rules for favorable behavior of our Bayesian test. The question is 'What if the data came out that way?', and is intended to scrutinize whether the posterior distribution 'updates' the prior distribution in a meaningful way. It avoids the pitfalls of the frequentist consistency concept by focussing on hypothetical *data*, rather than *probabilities*. It is particularly easy to apply in the high-dimensional context, since each particular θ may be associated with a with a set of data that occurs almost surely.

With the 'what if' principle in mind, it will be taken as minimal requirement that $\mathrm{P}[\mathrm{H}_0|Y] \to 0$ for data associated with H_0. An analysis of (3) shows this is identical to the requirement $\limsup_p r_p > -\infty$. The following considerations are also relevant: (i) One would want $\mathrm{P}[\mathrm{H}_0|Y] \to 1$ for data associated with H_0; In terms of (2) this is identical to $r_p \to \infty$. (ii) One would want $\mathrm{P}[\mathrm{H}_0|Y] \to 0$ for data associated with H_1. This latter consideration is complicated by the presence of an 'indistinguishable region' of parameters values θ in a region around H_0 for which $\limsup_p \mathrm{P}[\mathrm{H}_0|Y] > 0$. Nevertheless, this region is smaller for r_p diverging to infinity at slower rates, suggesting one would want $r_p \to \infty$ but at the slowest possible rate.

There remains a fair bit of ambiguity in how to specify such r_p in practice, since for each sequence there is always another that diverges at a slower rate (*e.g.*, $\log r_p$). Practical-minded resolutions are discussed, but for purposes of mathematical analysis the basic rule '$r_p \to \infty$ as slowly as possible' is interpreted as follows: whenever another sequence diverging to infinity is encountered, it will be assumed to diverge at a rate faster than r_p. With this convention, it is possible to deduce a number of interesting observations about the high-dimensional testing problem.

2.1. *Objectivity of* $\rho_0 = 1/2$

In low-dimensional settings it is often argued that $\rho_0 = 1/2$ serves as a 'noninformative' setting for Bayesian testing. Analysis under our high-dimensional framework calls this label into question. Specifically, it is seen from (2) that the property $\lim_p \rho_0 = 1/2$ (or any other value strictly between zero and one) might be possible by solving for the correct τ. However, it is shown that a *necessary* condition for $\lim_p \rho_0 = 1/2$ is that $\lim_j w_j = 0$. The implication for prior selection is that to achieve $\rho_0 = 1/2$ in a high-dimensional context one must specify a suitable ordering of the dimension indices, then specify relative weightings such that there is greater certainty of $\theta_j \approx 0$ in the higher-indexed dimensions (*i.e.*, for large j). Clearly, such considerations would compromise the 'noninformative' character of the prior.

2.2. *Diffuse Priors*

Rather than treating $\rho_0 = 1/2$ as 'noninformative', an alternative is to allow $\rho_0 \to 0$, then set $w_j = 1$ for each j and take $\tau/\sigma \to \infty$ at a very fast rate. An accurate approximation to (3) is

$$\mathrm{P}[\mathrm{H}_0|Y] \quad \approx \quad \left[1 + \exp\left\{\frac{1}{2}\left(\|Y/\sigma\|^2 - p - r_p p^{1/2}\right)\right\}\right]^{-1}, \tag{4}$$

provided each r_p has a finite limit as $\tau/\sigma \to \infty$. Given a suitable specification of r_p, this is proposed as an extension of Robert's (1993) noninformative solution to high-dimensions.

Robert (1993) provides a basic argument for the appropriateness of allowing $\rho_0 \to 0$, which is that the essential size of H_1 grows as τ increases, and so it makes sense to shift prior mass away from H_0 in order to better reflect comparisons against a larger space of alternatives. In the high-dimensional context, H_1 grows even more dramatically since it is not only τ but p that is taken to increase.

Although this approach avoids use of an improper prior at each fixed p, the approximation (4) nevertheless suggests a normalizing constant for a parallel solution derived from a flat prior. If one fixes $\rho_0 = 1/2$, then the flat prior $\pi(\theta|\mathrm{H}_1) = c$ on H_1 leads to exact agreement with the right side of (4) upon setting $c = (2\pi\sigma^2)^{-p/2}\exp\{-(p + r_p p^{1/2})/2\}$.

2.3. *Jeffreys–Lindley Paradox in High Dimensions*

The Jeffreys–Lindley 'paradox' refers to the observation that a Bayesian's evidence assessment may be very strongly in favor of H_0 while at the same time the frequentist observes a very small p-value. Details of this conflict are investigated by directly comparing $\mathrm{P}[\mathrm{H}_0|Y]$ with the frequentist p-value, $\hat{p}$, calculated by treating $\mathrm{P}[\mathrm{H}_0|Y]$ as a 'test statistic' *à la* Neyman–Pearson testing.

Generally, when $\theta|\mathrm{H}_1 \sim N(0, \tau^2 W)$ and taking $r_p \to \infty$ as slowly as possible, it is shown that $\mathrm{P}[\mathrm{H}_0|Y]/\hat{p} \to \infty$ for data such that $\hat{p} \to 1$. Thus, for p-values near one, both the Bayesian and frequentist would conclude very little evidence against H_0, but $\mathrm{P}[\mathrm{H}_0|Y]$ would be closer to one than $\hat{p}$. A similar though slightly more complex comparison is made for small p-values. For data such that $\hat{p} \to 0$, both $\mathrm{P}[\mathrm{H}_0|Y] \to 0$ and $\hat{p} \to 0$, but the comparison is different depending on the rate at which $\hat{p}$ grows. Writing $\hat{z}$ for the normalized test statistic, it is shown that $\mathrm{P}[\mathrm{H}_0|Y]/\hat{p} \to \infty$ if $\lim_p \|v_1\|/\hat{z} \leq 1/\sqrt{2}$ and $\mathrm{P}[\mathrm{H}_0|Y]/\hat{p} \to 0$ otherwise. That there is a cutoff point suggests a region of data, each data point exhibiting some evidence against H_0, for which $\mathrm{P}[\mathrm{H}_0|Y]$ and $\hat{p}$ take similar values. These observations indicate a change in the nature of the Jeffreys–Lindley paradox in high-dimensions, alleviating much of what has been argued to make it paradoxical.

3. HIGH-DIMENSIONAL ANALYSIS OF THE 'SMOOTH' MODEL

High-dimensional Bayesian testing is explored in contexts where the parameter space is constrained to reflect a 'smoothness' assumption of an underlying functional model. For this purpose, a 'sample size' parameter, n, is introduced with which the model is reparameterized so that $\sigma = \tilde{\sigma}/\sqrt{n}$. Updated asymptotics take $n \to \infty$ as $p \to \infty$, reflecting a situation where the challenge of handling many dimensions at once is alleviated somewhat by correspondingly small errors. Such ideas are especially relevant for applications in functional data analysis and goodness-of-fit testing.

A suitable means of incorporating a smoothness assumption is to restrict the model to a 'Sobolev space',

$$\mathcal{B}_{s,M} = \left\{ \tilde{\theta} : \sum_{j=1}^{\infty} j^{2s}\tilde{\theta}_j^2 \leq M \right\}, \tag{5}$$

where $\tilde{\theta} = \theta/\tilde{\sigma}$, $s > 1/2$, and $M > 0$. When the $\tilde{\theta}_j$ are derived as coefficients of an orthogonal basis decomposition it is sensible to interpret (5) as an assumption of a certain number of derivatives (more derivatives for larger s) in an underlying functional representation. The interest now is to ascertain how the prior might be modified appropriately to to reflect and exploit these assumptions.

3.1. *Rates of Testing*

Recall from Section 2 the interest in formulating a prior so as to shrink the 'indistinguishable' region the greatest extent possible. Under (5) it becomes possible to use the same consideration to deduce favorable settings for the prior parameters $w_1, \ldots, w_p$.

A precise criterion for measuring the 'size' of the indistinguishable region arises from consideration of sequences $\delta_n \to 0$ for which the criterion $\sum_{j=1}^p \tilde{\theta}_j^2 = O(\delta_n^2)$ characterizes the alternatives $\tilde{\theta}$ that are indistinguishable. The rate at which $\delta_n \to 0$ is a 'rate of testing', for which faster rates are associated with smaller indistinguishable regions. Our interest is to determine which $w_1, \ldots, w_p$ lead to the fastest such rates. It is fortunate that a direct mathematical connection can be made with known performance bounds deduced under this criterion in the frequentist context. For instance, general performance bounds are implied from the results of Ingster (1993) and Spokoiny (1996). Most relevant are bounds deduced in Spitzner (2006b) for tests derived from quadratic forms. One distinction within the literature on this topic is whether s, a parameter of the geometry, is known precisely or known only up to fixed bounds. The latter sets up the 'adaptive' framework for rates of testing.

In the present context, and that of Spitzner (2006b), the relative rate of p to n plays a critical role in determining rates of testing. Set $L = \lim_p\{\tilde{s}\log p/(2\log n)\}$, which shall be assumed to exist. If $L < 1$ than all of H_0 is indistinguishable. The setting $L = 1$ with $w_1, \ldots, w_p$ such that $v_{1,j} = j^{-(1-r)/2}$ for any $0 < r \leq 1$ leads to Ingster's optimal rate of testing given by $\delta_n = n^{-2s/\tilde{s}}$, where $\tilde{s} = 4s + 1$. Observe that to specify $L = 1$ requires s, so this is not an adaptive rate. However, one might instead take $L \geq 1$, which requires only the lower bound $s > 1/2$. This is now an adaptive context, for which the results of Spitzner (2006b) imply that the best adaptive rates of testing, among priors $\theta|\mathrm{H}_1 \sim N(0, \tau^2 W)$, are given by $w_1, \ldots, w_p$ such that $v_{1,j} = (j\log j)^{-1/2}$. The corresponding rate of testing is given by $\delta_n = \{n^2(\log n)^{-1}\}^{-s/\tilde{s}}$ and it is interesting that this is faster than (hence falls short of) Spokoiny's minimax adaptive rate given by $\delta_n = \{n^2(\log\log n)^{-1}\}^{-s/\tilde{s}}$.

3.2. *Prior Conditioning on Smoothness*

It would be reasonable to argue that smoothness assumptions represented by (5) are most appropriately treated as prior information and should be incorporated into the prior distribution by conditioning. That is, $\theta|\mathrm{H}_1 \sim N(0, \tilde{\tau}W/n)$ is to be reformulated as $\theta|\mathrm{H}_1 \sim N(0, \tilde{\tau}W/n|\mathcal{B}_{s,M})$. Subsequently, it is shown that the distinction between $\mathrm{P}[\mathrm{H}_0|Y]$ and $\mathrm{P}[\mathrm{H}_0|Y, \mathcal{B}_{s,M}]$ is completely determined by a 'smoothness factor', given by

$$SF(Y) \quad = \quad \frac{\mathrm{P}[Q_2 \leq M|Y]}{\mathrm{P}[Q_1 \leq M]} \tag{6}$$

where

$$Q_1 = \frac{\tilde{\tau}^2}{n\tilde{\sigma}^2}\sum_j j^{2s} w_j Z_j^2 \quad \text{and} \quad Q_2 = \frac{1}{n}\sum_{j=1}^p j^{2s} v_{j,1}\left\{Z_j + v_{j,1/2}\sqrt{n}Y_j/\tilde{\sigma}\right\}^2, \tag{7}$$

for independent standard normal random variables $Z_1, \ldots, Z_p$. It is of interest to determine whether such intuition conflicts with the guidelines offered by 'what if' criteria. By evaluating (6) under a saddlepoint approximation (Reid, 1988), it is deduced that there is, in fact, a conflict, by which the indistinguishable region is made larger under the prior formulated conditionally on $\mathcal{B}_{s,M}$. When the relative rate of increase of p to n is fast, it may even be the case that $\mathrm{P}[\mathrm{H}_0|Y] \to 1$ almost surely, even for data associated with H_1.

4. CONCLUDING DISCUSSION

A discussion is given on issues raised by the consequences of (2), and issues yet to be resolved. It is suggested the results of Section 2.1 point to a need for probabilistic interpretation of ρ_0 to take dimension and geometry into consideration. The problem of how to elicit r_p in practice is discussed, and various 'objective' arguments are given to simplify the issue. A discussion is also given on the relevance of results deduced in the high-dimensional setting to low-dimensional problems.

REFERENCES

Aitkin, M. (1991). Posterior Bayes factors. *J. Roy. Statist. Soc. B* **53**, 111-142.

Diaconis, P. and Freedman, D. (1986). On the consistency of Bayes estimates. *Ann. Statist.* **14**, 1–26.

Fan, J. and Lin, S.-K. (1998). Test of significance when data are curves. *J. Amer. Statist. Assoc.* **93**, 1007–1021.

Ingster, Y. I. (1993). Asymptotically minimax hypothesis testing for nonparametric alternatives I-III. *Mathematical Methods of Statistics* **2**, 85–114; **3**, 171–189; **4**, 249–268.

Kass, R. E. and Raftery, A. E. (1995). Bayes factors. *J. Amer. Statist. Assoc.* **90**, 773–795.

Rayner, J. C. W. and Best, D. J. (1989). *Smooth Tests of Goodness-of-Fit.* Oxford: University Press.

Reid, N. (1988). Saddlepoint methods and statistical inference. *Statist. Science* **3**, 213-238, (with discussion).

Robert, C. P. (1993). A note on Jeffreys–Lindley paradox. *Statistica Sinica* **3**, 603–608.

Robert, C. P. (2001). *The Bayesian Choice.* Berlin: Springer-Verlag.

Spiegelhalter, D. J. and Smith, A. F. M. (1982). Bayes factors for linear and log-linear models with vague prior information. *J. Roy. Statist. Soc. B* **44**, 377–387.

Spitzner, D. J. (2006a). Use of goodness-of-fit procedures in high-dimensional testing. *J. Statist. Computation and Simulation* **76**, 447–457.

Spitzner, D. J. (2006b). Testing in functional data analysis using quadratic forms. *Tech. Rep.*, Virginia Tech, USA.

Spokoiny, V. G. (1996). Adaptive hypothesis testing using wavelets. *Ann. Statist.* **24**, 2477–2498.

Verdinelli, I. and Wasserman, L. (1998). Bayesian goodness-of-fit testing using infinite-dimensional exponential families. *Ann. Statist.* **26**, 1215–1241.

BAYESIAN STATISTICS 8, pp. 667–672.
J. M. Bernardo, M. J. Bayarri, J. O. Berger, A. P. Dawid,
D. Heckerman, A. F. M. Smith and M. West (Eds.)

The Marginalization Paradox and Probability Limits*

TIMOTHY C. WALLSTROM
Los Alamos National Laboratory, USA
tcw@lanl.gov

SUMMARY

Dawid, Stone, and Zidek have shown that when improper priors are used in a particular type of problem, it is often impossible to obtain a marginal posterior from an associated likelihood through formal application of Bayes' law. This is the marginalization paradox (MP). Sudderth has shown that the MP can be resolved in the setting of finitely-additive probability. In this paper, I show that it can also be resolved in the usual countably-additive setting.

Keywords and Phrases: MARGINALIZATION PARADOX; IMPROPER PRIORS; COHERENCE; FORMAL POSTERIOR; PROBABILITY LIMIT.

1. INTRODUCTION

The marginalization paradox (MP) concerns the following situation. Let $p(x|\theta)$ be the density function for some statistical model, $\pi(\theta)$ a possibly improper prior, and $\pi(\theta|x)$ the corresponding posterior, as computed from Bayes' law. If $\pi(\theta)$ is improper, i.e., if its integral is infinite, we call $\pi(\theta|x)$ a *formal* posterior. Now suppose that $x = (y, z)$ and $\xi = (\eta, \zeta)$, that the marginal density $p(z|\theta)$ depends on θ only through ζ, and that the marginal posterior $\pi(\zeta|x)$ depends on x only through z. We denote these functions by $\tilde{p}(z|\zeta)$ and $\tilde{\pi}(\zeta|z)$, respectively.

Intuition would now suggest that $\tilde{\pi}(\zeta|z) \propto \tilde{p}(z|\zeta)\,\omega(\zeta)$, for some function $\omega(\zeta)$. That is, the marginalized quantities should satisfy Bayes' law. However, it was discovered by Stone and Dawid (1972), and by Dawid, Stone and Zidek (1973), that when $\pi(\theta)$ is improper, this is often impossible. This result is known as the 'marginalization paradox'.

The MP shows that formal posteriors can lead to inconsistencies, and any theory of improper inference needs to come to terms with this fact. Responses have included the rejection of improper inference (Lindley, 1973), allowing the improper prior to depend on the estimand (reference posteriors (Bernardo, 1979), objective Bayes), allowing the improper prior to depend on the data (Akaike, 1980), and the use of finitely-additive probability (Sudderth, 1980). Jaynes (2003) has argued that $\tilde{\pi}(\zeta|z)$ and $\tilde{p}(z|\zeta)\,\omega(\zeta)$ implicitly use different prior information, so there is no

---

* This is a synopsis. For the full version, see *Bayesian Analysis* **2** (2007), to appear. This paper is dedicated to Professor Mervyn Stone.

need for agreement and no paradox. In this paper, I argue for a different response. Specifically, I suggest that we regard probability limits, rather than formal posteriors, as the proper generalization of the notion of 'Bayesian inference' to the setting of improper priors.

The issue raised by the MP is whether an improper Bayesian inference remains a Bayesian inference under marginalization. If it does not, then improper Bayesian inference is internally inconsistent. This raises the question, however, as to what 'Bayesian inference' means in the improper context. There are foundational arguments for considering the probability limit, rather than the formal posterior, as the appropriate generalization (Section 5.1). Dawid, Stone and Zidek (1973) show that if interpret 'Bayesian inference' as 'formal posterior', marginalization produces an inconsistency. The present paper shows that if interpret it as 'probability limit', marginalization will not lead to an inconsistency.

Specifically, I consider the MP for inferences $q(\theta|x)$ that are probability limits, as defined by Stone (1970). $q(\theta|x)$ may or may not be a formal posterior. Assume that the marginal $q(\zeta|x)$ depends on x only through z; we denote the resulting function $\tilde{q}(\zeta|z)$. I show that if $q(\theta|x)$ is a probability limit for $p(x|\theta)$, then $\tilde{q}(\zeta|z)$ is a probability limit for $\tilde{p}(z|\zeta)$. Thus, if we regard a Bayesian inference as a probability limit, the marginal of a Bayesian inference is still a Bayesian inference.

This result is perhaps surprising, because $\tilde{q}(\zeta|z)$ will often be incompatible with $\tilde{p}(z|\zeta)$, in the sense that the two quantities cannot be related formally by Bayes' law. However, for an inference to be a probability limit, it is only necessary that it obey Bayes' law locally, in the region where data are expected (Eq. 4). Thus, an inference can be a probability limit even if the inference and likelihood are not related by Bayes' law globally. Marginalization, in fact, enables us to construct probability limits that are not formal posteriors (Example 1).

The present result is closely related to that of Sudderth (1980), who showed that the MP could be resolved in the setting of finitely additive probability theory. The structure of his argument is the same as mine, with the difference that he replaces the class of formal posteriors with the class of posteriors based on proper, finitely additive priors. Equivalently, this class may be characterized as the set of 'coherent' inferences (Heath and Sudderth, 1978). He shows that this class is closed under marginalization. As with probability limits, there are foundational arguments supporting the use of coherent inferences (Sudderth, 1980). Some, however, may find this resolution inadequate, because it uses finitely-additive probabilities, which are not in general use, have many nonintuitive properties, and are afflicted with their own paradoxes, such as nonconglomerability (Kass and Wasserman, 1996).

The current paper shows that the MP may be resolved in the setting of ordinary countably-additive probability theory. This result is new, and is not implied by previous work. Probability limits are coherent (Section 5.2), so Sudderth's theorem implies that the marginal of a probability limit will be coherent. It does not, however, show that the marginal is a probability limit, nor does it produce a sequence of ordinary posteriors that the marginal will approximate.

In Section 2, we define probability limits, coherent inferences, and formal posteriors. In Section 3, we prove Theorem 1, which is our central result, and in Section 4, we provide an example. An extensive discussion may be found in the full paper.

2. IMPROPER INFERENCE

We define three different concepts that are used to make inferences from countably-additive improper priors.

Let $(\mathscr{X}, \mathscr{M})$ and $(\Theta, \mathscr{N})$ be measurable spaces, to be identified as the data and parameter spaces, respectively, where $\mathscr{M}$ and $\mathscr{N}$ are σ-algebras. All measures are assumed to be countably additive, unless stated otherwise. Let $p(A|\theta)$ be a *Markov kernel* on $\mathscr{M} \times \Theta$, by which we mean that, for each $\theta \in \Theta$, $p(\cdot|\theta)$ is a probability measure on $\mathscr{X}$, and for each $A \in \mathscr{M}$, $p(A|\cdot)$ is measurable on Θ. We use $\|\cdot\|$ to denote the total variation norm, $\pi(d\theta)$ to denote a *prior*, which is a probability measure on Θ, and $m_\pi(dx) = \int p(dx|\theta)\, \pi(d\theta)$ to denote the marginal data distribution.

Definition 1 *Let $p(dx|\theta)$ and $q(d\theta|x)$ be Markov kernels, and let $\{\pi_n(d\theta)\}$ be a sequence of probability distributions. Then $q(d\theta|x)$ is a probability limit for $p(dx|\theta)$ and $\{\pi_n(d\theta)\}$ if*

$$\|p(dx|\theta)\, \pi_n(d\theta) - q(d\theta|x)\, m_n(dx)\| \to 0. \tag{1}$$

If $\{\pi_n(d\theta)\}$ is omitted, it means that $q(d\theta|x)$ is a probability limit for $p(dx|\theta)$ and some sequence $\{\pi_n(d\theta)\}$, and similarly if $p(dx|\theta)$, or both $p(dx|\theta)$ and $\{\pi_n(d\theta)\}$, are omitted. We do not require that q be unique, nor that it be a formal posterior.

The notion of probability limit can be interpreted geometrically. Let $\mathscr{P}$ be the set of probability measures on Θ, and let $\mathscr{S}$ be the set of signed measures on $\mathscr{X} \times \Theta$, which is a Banach space in the total variation norm. Given $p(dx|\theta)$ and $q(d\theta|x)$, define a mapping $r_{p,q} : \mathscr{P} \to \mathscr{S}$, by $r_{p,q}(\pi) = p(dx|\theta)\, \pi(d\theta) - q(d\theta|x)\, m_\pi(dx)$. Let $\mathscr{R}_{p,q} = r_{p,q}(\mathscr{P})$. Then q is a probability limit for p if and only if the measure $\mu \equiv 0$ is in the closure of the convex set $\mathscr{R}_{p,q}$. This interpretation was suggested by ideas of Eaton and Sudderth (1998), p. 368.

In connection with this interpretation, it is worth noting that marginalization is a contraction in the space $\mathscr{S}$. Let μ be a measure on some space (S, Σ). From an abstract perspective, a marginal of μ is a restriction of μ to some subalgebra $\Sigma_1 \subset \Sigma$. If $\mu_1 \equiv \mu|\Sigma_1$, it follows trivially from the definition of total variation that $\|\mu_1\| \leq \|\mu\|$. (Cf., e.g., Dunford and Schwartz (1966), III.9.14.)

If we require only that the measures to be finitely-additive, Definition 1 is essentially the same as that used by Heath and Sudderth (1989); they call $q(d\theta|x)$ 'approximable by proper priors'. It is convenient to reserve the term 'probability limit' for the countably-additive case.

The terminology 'probability limit' comes from the fact that if the measures have densities, then $q(\theta|x)$ is a probability limit for $p(x|\theta)$ and $\{\pi_n(\theta)\}$ if and only if, for any $\epsilon > 0$, $m_n\{\|\pi_n(\theta|x)\, d\theta - q(\theta|x)\, d\theta\| > \epsilon\} \to 0$. The notion of probability limit is due to Stone (Stone 1965; Stone, 1970).

Definition 2 *Let $p(dx|\theta)$ and $q(d\theta|x)$ be Markov kernels, and let $\pi(d\theta)$ be a probability measure. Then $q(d\theta|x)$ is a posterior for $p(dx|\theta)$ and $\pi(d\theta)$ if*

$$\|p(dx|\theta)\, \pi(d\theta) - q(d\theta|x)\, m_\pi(dx)\| = 0. \tag{2}$$

If Definition 2 is used with finitely-additive measures, we use the term 'coherent inference', because q is then a coherent inference in the sense of Heath and Sudderth (Heath and Sudderth, 1978; Lane and Sudderth, 1984). Coherent inferences are not actually based on improper priors. However, a finitely-additive proper prior can sometimes be viewed as the analog of a countably-additive improper prior (Sudderth, 1980). Therefore, we view coherent inferences as another method for associating inferences with countably-additive improper priors. The notion of coherence defined here is sometimes called 'HLS-coherence' (Kass and Wasserman, 1996), to distinguish it from an alternate notion of coherence due to Regazzini (1987).

Assume now that all measures on $\mathscr{X}$ and Θ have densities relative to otherwise arbitrary measures dx and $d\theta$. We can then define a third concept:

Definition 3 *Let $p(x|\theta)\,dx$ be a Markov kernel, and $\pi(\theta)$ a nonintegrable density on Θ. The formal posterior, for $p(x|\theta)$ and $\pi(\theta)$, is the density*

$$\pi(\theta|x) = p(x|\theta)\pi(\theta)/m_\pi(x), \tag{3}$$

if $m_\pi(x) > 0$, and zero elsewhere.

Definition 3 requires inferences to obey a global version of Bayes' law. Probability limits require that inferences obey a local and approximate version. If $q(\theta|x)$ is a probability limit for $p(x|\theta)$ (with the obvious extension of terminology), then for any $\epsilon > 0$, there exists a probability density $\pi(\theta)$ such that

$$\iint \left| \frac{p(x|\theta)\,\pi(\theta)}{m_\pi(x)} - q(\theta|x) \right| m_\pi(x)\,d\theta\,dx < \epsilon. \tag{4}$$

Note that, with Definition 2, the two inferences need only agree in the region where $m_\pi(x)$ is large.

3. THE MARGINALIZATION PARADOX

Assume now that $\mathscr{X} = \mathscr{Y} \times \mathscr{Z}$ and $\Theta = \mathscr{Y} \times \mathscr{Z}$, and denote the measure algebras of the component spaces by $\mathscr{M}(\mathscr{Y})$, etc. Let $p(dx|\theta)$ be the sampling distribution for some statistical model and let $q(d\theta|x)$ be an inference, i.e., a Markov kernel. The MP is concerned with situations in which the following assumption holds.

Assumption 1 *The marginal distribution, $p(dz|\eta,\zeta)$, is independent of η, and the marginal inference $q(d\zeta|y,z)$, is independent of y.*

When Assumption 1 is valid, define

$$\tilde{p}(dz|\zeta) \equiv p(dz|\eta,\zeta) \qquad \text{and} \qquad \tilde{q}(d\zeta|z) \equiv q(d\zeta|y,z). \tag{5}$$

These quantities are well-defined by virtue of Assumption 1. If the measures have densities, and if there is no $\omega(\zeta)$ such that $\tilde{q}(\zeta|z) \propto \tilde{p}(z|\zeta)\,\omega(\zeta)$, then we say that there is a *marginalization paradox.* For numerous examples, cf. Stone and Dawid (1972) and Dawid, Stone and Zidek (1973). Our main result is the following:

Theorem 1 *If Assumption 1 is valid, and if $q(d\theta|x)$ is a probability limit for $p(dx|\theta)$, then $\tilde{q}(d\zeta|z)$ is a probability limit for $\tilde{p}(dz|\zeta)$.*

Proof. Let $\pi(d\theta)$ be a probability measure, let

$$\mu(dx, d\theta) = p(dx|\theta)\,\pi(d\theta) - q(d\theta|x)\,m_\pi(dx),$$

and let μ_1 be the restriction of μ to the algebra of sets of the form $\mathscr{Y} \times A \times \mathscr{Y} \times A$, where $A \in \mathscr{M}(\mathscr{Z})$ and $A \in \mathscr{M}(\mathscr{Z})$. We have

$$\begin{aligned} \mu_1(\mathscr{Y}, dz, \mathscr{Y}, d\zeta) &= \int_{\mathscr{Y}} \int_{\mathscr{Y}} p(dy, dz|\eta,\zeta)\,\pi(d\eta, d\zeta) - q(d\eta, d\zeta|y,z) m_\pi(dy,dz) \\ &= \tilde{p}(dz|\zeta)\,\pi(d\zeta) - \tilde{q}(d\zeta|z)\,m_\pi(dz), \end{aligned}$$

where we have used Assumption 1. Then $\|\mu_1\| \leq \|\mu\|$, as in Section 2, and the theorem follows trivially. □

Corollary 1 *(Sudderth.) Let $p(dx|\theta)$ and $q(d\theta|x)$ be finitely-additive Markov kernels, and let $\pi(d\theta)$ be a finitely additive probability distribution. If Assumption 1 is valid, and if $q(d\theta|x)$ is a coherent inference for $p(dx|\theta)$ and $\pi(d\theta)$, then $\tilde{q}(d\zeta|z)$ is a coherent inference for $\tilde{p}(dz|\zeta)$ and $\pi(d\zeta)$.*

Proof. The result is a Corollary of the proof, which does not use countable additivity. □

Comments. 1. The theorem provides a new way of establishing the existence of a probability limit. 2. Definitions 1 and 2 assume that $\pi(d\theta)$ is a probability measure. Theorem 1 can fail without this assumption, because the value of μ_1 can become $\infty - \infty$.

4. EXAMPLE

We give an example to show that the hypothesis in Theorem 1 does not preclude the existence of a MP. This proves, in particular, that there are probability limits that are not formal posteriors.

We assume a basic familiarity with Haar measure (Bourbaki, 2004; Nachbin, 1965). We also need the following result. Let G be a connected locally compact topological group. Let $\mathscr{X} = \Theta = G$, and assume that we have a *group model*:

$$p(dx|\theta) = f(\theta^{-1}x)\,\mu(dx), \tag{6}$$

where μ is left Haar measure. Stone (1970) has shown that if $\pi(d\theta) = \nu(d\theta)$, where ν is right Haar measure, and if G is *amenable* (Bondar and Milnes, 1981; Reiter and Stegeman, 2000), then $\pi(d\theta|x)$ is a probability limit.

Our example will involve the affine group in one dimension, which is solvable and thus amenable. Let $G = \{(h,k) : h > 0, k \in \mathbb{R}\}$, and define group multiplication as

$$(h_1, k_1) \circ (h_2, k_2) = (h_1 h_2, h_1 k_2 + k_1).$$

Left and right Haar measures are given by $\mu(dg) = dhdk/h^2$ and $\nu(dg) = dhdk/h$. (Cf. Lehmann and Casella, 1998, Eq. 4.4.15).

Example 1 (Coefficient of Variation. Stone and Dawid (1972)).
Consider $n \geq 2$ samples from $X \sim N(\mu, \sigma^2)$. $x = (s, \overline{x})$ is a sufficient statistic for $\theta = (\sigma, \mu)$, where $x = n^{-1}\sum x_i$ and $s^2 = \nu^{-1}\sum(x_i - x)^2$, where $\nu = n - 1$. The sampling distribution is

$$p(dx|\theta) = C\,\frac{s^n}{\sigma^n}\exp\left[-\frac{n(x-\mu)^2}{2\sigma^2} - \frac{\nu s^2}{2\sigma^2}\right]\mu(dx),$$

where C is a constant, and $\mu(dx) = dx\,ds/s^2$. Identify x and θ as elements of G, the affine group in one dimension. $p(dx|\theta)$ is then of the form of Eq. (6), as follows from the invariance of the model: $p(gA|g\theta) = p(A|\theta)$, where $A \in \mathscr{M}$.

Let the prior be right Haar. By Stone's theorem (Stone, 1970), $\pi(\theta|x)$ is a probability limit. However, as shown in Stone and Dawid (1972) (and can readily be verified), there is also a MP. This example also shows that the MP can arise for coherent inferences (Heath and Sudderth, 1978, Example 4.2).

ACKNOWLEDGMENTS

This paper is dedicated to Professor Mervyn Stone, in admiration for his work in laying the foundations of probability limits. I thank Professors A. P. Dawid, M. Stone, W. D. Sudderth and J. V. Zidek for helpful comments and encouragement; Dr. David Higdon, for suggesting that I present this work at the Valencia conference; and the referees, for many insightful comments. I acknowledge support from the Department of Energy under contract DE-AC52-06NA25396.

REFERENCES

Akaike, H. (1980). The interpretation of improper prior distributions as limits of data dependent proper prior distributions. *J. Roy. Statist. Soc. B* **42**, 46–52.

Bernardo, J. M. (1979). Reference posterior distributions for Bayesian inference. *J. Roy. Statist. Soc. B* **41**, 113-147, (with discussion).

Bondar, J. V. and Milnes, P. (1981). Amenability: A survey for statistical applications of Hunt-Stein and related conditions on groups. *Z. Wahrsch. verw. Gebiete* **57**, 103–128.

Bourbaki, N. (2004). *Elements of Mathematics, Integration II.* Berlin: Springer-Verlag.

Dawid, A. P., Stone, M. and J. V. Zidek (1973). Marginalization paradoxes in Bayesian and structural inference. *J. Roy. Statist. Soc. B* **35**, 189–233 (with discussion).

Dunford, N. and Schwartz, J. T. (1966). *Linears Operators. Part I: General Theory.* Chichester: Wiley.

Eaton, M. L. and Sudderth, W. D. (1998). A new predictive distribution for normal multivariate linear models. *Sankhyā A* **60**, 363–382.

Heath, D. and Sudderth, W. D. (1978). On finitely additive priors, coherence, and extended admissibility. *Ann. Statist.* **6**, 333–345.

Heath, D. and Sudderth, W. D. (1989). Coherent inference from improper priors and from finitely additive priors. *Ann. Statist.* **17**, 907–919.

Jaynes, E. T. (2003). *Probability Theory: The Logic of Science.* Cambridge: University Press.

Kass, R. E. and Wasserman, L. (1996). The selection of prior distributions by formal rules. *J. Amer. Statist. Assoc.* **91**, 1343–1370.

Lane, D. A. and Sudderth, W. D. (1984). Coherent predictive inference. *Sankhya A* **46**, 166–185.

Lehmann, E. L. and Casella, G. (1998). *Theory of Point Estimation* (2nd edn) Berlin: Springer-Verlag

Lindley, D. V. (1973). Discussion of Dawid, Stone and Zidek (1973). *J. Roy. Statist. Soc. B* **35**, 218–219.

Nachbin, L. (1965). *The Haar Integral.* New York: Van Nostrand.

Regazzini, E. (1987). de Finetti's coherence and statistical inference. *Ann. Statist.* **15**, 845–864.

Reiter, H. and Stegeman, J. D. (2000). *Classical Harmonic Analysis and Locally Compact Groups.* Oxford: University Press.

Stone, M. (1965). Right Haar measure for convergence in probability to quasi posterior distributions. *Ann. Math. Statist.* **36**, 440–453.

Stone, M. (1970). Necessary and sufficient condition for convergence in probability to invariant posterior distributions. *Ann. Math. Statist.* **41**, 1349 – 1353.

Stone, M. and Dawid, A. P. (1972). Un-Bayesian implications of improper Bayes inference in routine statistical problems. *Biometrika* **59**, 369–375.

Sudderth, W. D. (1980). Finitely additive priors, coherence, and the marginalization paradox. *J. Roy. Statist. Soc. B* **42**, 339–341.

BAYESIAN STATISTICS 8, pp. 673–678.
J. M. Bernardo, M. J. Bayarri, J. O. Berger, A. P. Dawid,
D. Heckerman, A. F. M. Smith and M. West (Eds.)

A Hidden Markov Dirichlet Process Model for Genetic Recombination in Open Ancestral Space*

E. P. Xing and K-A. Sohn
Carnegie Mellon University, Pittsburgh, USA
epxing @cs.cmu.edu ksohn @cs.cmu.edu

Summary

We present a new statistical framework called hidden Markov Dirichlet process (HMDP) to jointly model the genetic recombinations among possibly infinite number of founders and the coalescence-with-mutation events in the resulting genealogies. The HMDP posits that a haplotype of genetic markers is generated by a sequence of recombination events that select an ancestor for each locus from an unbounded set of founders according to a 1st-order Markov transition process. Conjoining this process with a mutation model, our method accommodates both between-lineage recombination and within-lineage sequence variations, and leads to a compact and natural interpretation of the population structure and inheritance process. An efficient sampling algorithm based on a two-level nested Pólya urn scheme was also developed.

Keywords and Phrases: Dirichlet Process; HMM; MCMC; Statistical Genetics; Recombination; Population Structure; SNP.

1. INTRODUCTION

Recombinations between ancestral chromosomes during meiosis play a key role in shaping the patterns of linkage disequilibrium (LD)–the non-random association of alleles at different loci–in a population. Uneven occurrence of recombination events along chromosomal regions during genetic history can lead to 'block structures' in molecular genetic polymorphisms such that within each block only low level of diversities are present in a population. The problem of inferring recombination hotspots is essential for understanding the origin and characteristics of genome variations; several combinatorial and statistical approaches have been developed for uncovering optimum block boundaries from single nucleotide polymorphism (SNP) haplotypes (Daly *et al.*, 2001; Anderson and Novembre, 2003; Patil *et al.*, 2001; Zhang *et al.*,

---

* This is a synopsis. For the full version, see *Bayesian Analysis* **2** (2007), to appear.
Eric P. Xing is an Assistant Professor of Computer Science at CMU, and Kyung-Ah Sohn is a CS Ph.D. student at CMU. Work supported by the NSF Grant No. 0523757, and by the Pennsylvania Health Research Program under Grant ME-01-739. E.P.X. is also supported by an NSF CAREER Award under Grant No. DBI-0546594.

2002). The deluge of SNP data also fuels the long-standing interest of analyzing patterns of genetic variations to reconstruct the evolutionary history and ancestral structures of human populations, using, for example, variants of admixture models on genetic polymorphisms (Rosenberg *et al.*, 2002). These progress notwithstanding, the statistical methodologies developed so far mostly deal with LD analysis and ancestral inference separately, using specialized models that do not capture the close statistical and genetic relationships of these two problems. Moreover, most of these approaches ignore the inherent uncertainty in the genetic complexity (*e.g.*, the number of genetic founders of a population) of the data and rely on inflexible models built on a pre-fixed, closed genetic space. Recently, Xing *et al.* (2004) have developed a nonparametric Bayesian framework for modeling genetic polymorphisms based on the Dirichlet process mixtures and extensions, which attempts to allow more flexible control over the number of genetic founders In this paper, we leverage this approach and present a unified framework to model complex genetic inheritance process that allows recombinations among possibly infinite founding alleles and coalescence-with-mutation events in the resulting genealogies.

We assume that individual chromosomes in a modern population are originated from an unknown number of ancestral haplotypes via biased random recombinations and mutations (Fig. 1). The recombinations between the ancestors follow a a state-transition process we refer to as hidden Markov Dirichlet process (originated from the infinite HMM by Beal *et al.* (2001), which travels in an open ancestor space. Our model draws inspiration from the HMM proposed in Greenspan and Geiger (2003), but we employ a two-level Pólya urn scheme akin to the hierarchical DP (Teh *et al.*, 2006) to accommodate an open ancestor space, and allow full posterior inference of the recombination sites, mutation rates, haplotype origin, ancestor patterns, etc., conditioning on phased SNP data, rather than estimating them using information theoretic or maximum likelihood principles.

2. HIDDEN MARKOV DIRICHLET PROCESS FOR RECOMBINATION

Sequentially choosing recombination targets from a set of ancestral chromosomes can be modeled as a hidden Markov process (Niu *et al.*, 2002; Greenspan and Geiger, 2003), in which the hidden states correspond to the index of the candidate chromosomes, the transition probabilities to the recombination rates between the recombining chromosome pairs, and the emission model to a mutation process that passes the chosen chromosome region in the ancestors to the descents. When the number of ancestral chromosomes is not known, it is natural to consider an HMM whose state space is countably infinite (Beal *et al.*, 2001; Teh *et al.*, 2006). In this section, we describe such an infinite HMM formalism, which we would like to call *hidden Markov Dirichlet process*, for modeling recombination in an open ancestral space.

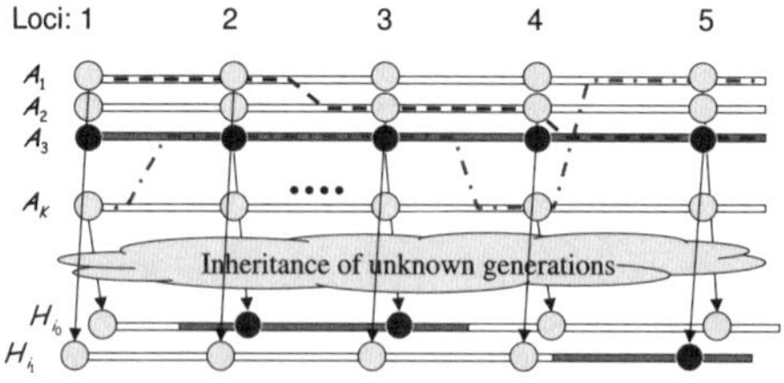

Figure 1: *An illustration of a hidden Markov Dirichlet process for haplotype recombination and inheritance.*

2.1. *Dirichlet Process Mixtures*

Under a well-known genetic model known as *coalescence-with-mutation* (but without recombination), one can treat a haplotype from a modern individual–the joint allele configuration of a contiguous list of SNPs located on one of his/her chromosome (Fig. 1)–as a descendent of an unknown ancestor haplotype (i.e., a founder) via random mutations. It can be shown that such a coalescent process in an infinite population leads to a partition of the population that can be succinctly captured by the following Pólya urn scheme. Consider an urn that at the outset contains a ball of a single color. At each step we either draw a ball from the urn and replace it with two balls of the same color, or we are given a ball of a new color which we place in the urn. One can see that such a scheme leads to a partition of the balls according to their color. Letting parameter τ define the probabilities of the two types of draws, and viewing each (distinct) color as a sample from Q_0, and each ball as a sample from Q, Blackwell and MacQueen (1973) showed that this Pólya urn model yields samples whose distributions are those of the marginal probabilities under the *Dirichlet process*. One can associate mixture component with colors in the Pólya urn model, and thereby define a 'clustering' of the data. The resulting model is known as a *DP mixture*. Note that a DP mixture requires no prior specification of the number of components. Back to haplotype modeling, following Xing *et al.* (2004), let $H_i = [H_{i,1}, \ldots, H_{i,T}]$ denote a haplotype over T SNPs from chromosome i; let $A_k = [A_{k,1}, \ldots, A_{k,T}]$ denote an ancestor haplotype (indexed by k) and θ_k denote the *mutation rate* of ancestor k; and let C_i denote an *inheritance variable* that specifies the ancestor of haplotype H_i. Then, under a DP mixture, we have the following Pólya urn scheme for sampling modern haplotypes:

- Draw first haplotype:

$a_1 \mid \mathrm{DP}(\tau, Q_0) \sim Q_0(\cdot)$, sample the first founder;

$h_1 \sim P_h(\cdot|a_1, \theta_1)$, sample the first haplotype from an inheritance model defined on the 1st founder;

- For subsequent haplotypes:

– sample the founder indicator for the ith haplotype:

$$c_i|\mathrm{DP}(\tau, Q_0) \sim \begin{cases} p(c_i = c_j \text{ for some } j < i|c_1, \ldots, c_{i-1}) = \frac{n_{c_j}}{i-1+\tau} \\ p(c_i \neq c_j \text{ for all } j < i|c_1, \ldots, c_{i-1}) = \frac{\tau}{i-1+\tau} \end{cases}$$

where n_{c_i} is the *occupancy number* of class c_i, the number of previous samples in c_i.

– sample the founder of haplotype i (indexed by c_i):

$$\phi_{c_i}|\mathrm{DP}(\tau, Q_0) \begin{cases} = \{a_{c_j}, \theta_{c_j}\} \text{ if } c_i = c_j \text{ for some } j < i \text{ (i.e., } c_i \text{ refers to an inherited founder)} \\ \sim Q_0(a, \theta) \text{ if } c_i \neq c_j \text{ for all } j < i \text{ (i.e., } c_i \text{ refers to a new founder)} \end{cases}$$

– sample the haplotype according to its founder:

$h_i \mid c_i \sim P_h(\cdot|a_{c_i}, \theta_{c_i})$.

Notice that the above generative process assumes each modern haplotype to be originated from a single ancestor, this is only plausible for haplotypes spanning a short region on a chromosome. Now we consider long haplotypes possibly bearing multiple ancestors due to recombinations between an unknown number of founders.

2.2. *Hidden Markov Dirichlet Process (HMDP)*

In a standard HMM, state-transitions across a discrete time- or space-interval take place in a fixed-dimensional state space, thus it can be fully parameterized by, say, a K-dimensional initial-state probability vector and a $K \times K$ state-transition probability matrix. As first proposed in Beal *et al.* (2001), and later discussed in Teh *et al.* (2006), one can 'open' the state space of an HMM by treating the now infinite number of discrete states of the HMM as the support of a DP, and the transition probabilities to these states from some source as the masses associated with these states. In particular, for each source state, the possible transitions to the target states need to be modeled by a unique DP. Since all possible source states and target states are taken from the same infinite state space, overall we need an open set of DPs with different mass distributions on the SAME support. In the sequel, we describe such a nonparametric Bayesian HMM using an intuitive hierarchical Pólya urn construction. We call this model a *hidden Markov Dirichlet process.*

We set up a single 'stock' urn at the top level, which contains balls of colors that are represented by at least one ball in one or multiple urns at the bottom level. At the bottom level, we have a set of *distinct* urns (say, HMM-urns) which are used to define the initial and transition probabilities of the HMDP model. Specifically, one of HMM urns, u_0, is set aside to hold colored balls to be drawn at the onset of the HMM state-transition sequence. Each of the remaining HMM urns is used to hold balls to be drawn during the execution of a Markov chain of state-transitions. Now let's suppose that at time t the stock urn contains n balls of K distinct colors; the number of balls of color k in this urn is denoted by n_k. For urn u_0 and urns $u_1, \ldots, u_K$, let $m_{j,k}$ denote the number of balls of color k in urn u_j, and $m_j = \sum_{k \in C} m_{j,k}$ denote the total number of balls in urn u_j. Suppose that at time $t-1$, we had drawn a ball with color k'. Then at time t, we either draw a ball randomly from urn $u_{k'}$, and place back two balls both of that color; or with probability $\tau/(m_j + \tau)$ we turn to the top level. From the stock urn, we can either draw a ball randomly and put back two balls of that color to the stock urn and one to $u_{k'}$, or obtain a ball of a new color $K+1$ with probability $\gamma/(n+\gamma)$ and put back a ball of this color to both the stock urn and urn $u_{k'}$ of the lower level. Essentially, we have a master DP (the stock urn) that serves as a base measure for infinite number of child DPs (HMM-urns).

2.3. *HMDP Model for Recombination and Inheritance*

For each modern chromosome i, let $C_i = [C_{i,1}, \ldots, C_{i,T}]$ denote the sequence of inheritance variables specifying the index of the ancestral chromosome at each SNP locus. When a recombination occurs, say, between loci t and $t+1$, we have $C_{i,t} \neq C_{i,t+1}$. We can introduce a Poisson point process to control the duration of non-recombinant inheritance. That is, given that $C_{i,t} = k$, then with probability $e^{-dr} + (1 - e^{-dr})\pi_{kk}$, where d is the physical distance between two loci, r reflects the rate of recombination per unit distance, and π_{kk} is the self-transition probability of ancestor k defined by HMDP, we have $C_{i,t+1} = C_{i,t}$; otherwise, the source state (i.e., ancestor chromosome k) pairs with a target state (e.g., ancestor chromosome k') between loci t and $t+1$, with probability $(1-e^{-dr})\pi_{kk'}$. Hence, each haplotype H_i is a mosaic of segments of multiple ancestral chromosomes from the ancestral pool $\{A_{k,\cdot}\}_{k=1}^{\infty}$. Essentially, the model we described so far is a time-inhomogeneous infinite HMM. The emission process of the HMDP corresponds to an inheritance model from an ancestor to the matching descendent. For simplicity, we adopt the *single-locus mutation model* in Xing *et al.* (2004), which is widely used in statistical genetics as an approximation to a full coalescent genealogy (Liu *et al.*, 2001).

The two-level nested Pólya urn schemes described in Section 2.2 motivates an efficient and easy-to-implement MCMC algorithm to sample from the posterior associated with HMDP. Details of this algorithms is available in Xing and Sohn (2007).

3. EXPERIMENTS

We have applied the HMDP model to both simulated and real haplotype data. Our analyses focus on the three popular problems in statistical genetics of ancestral inference, LD-block analysis and population structural analysis.

3.1. *Analyzing Simulated Haplotype Population*

We simulated a population of 200 individual haplotypes from $K_s = 5$ (unknown to the HMDP model) ancestor haplotypes, via a $K_s = 5$-dimensional HMM.

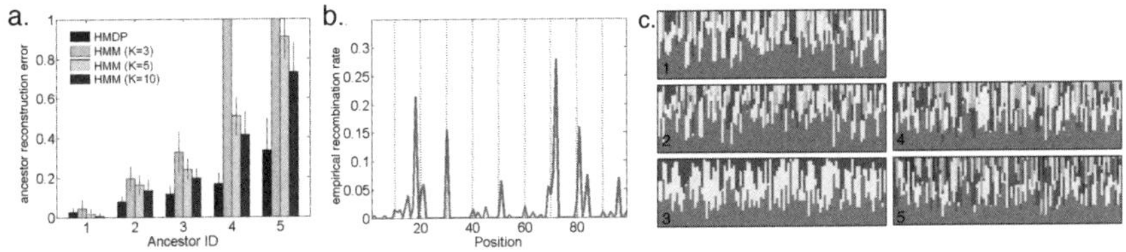

Figure 2: *(a) Ancestor reconstruction errors (the ratio of incorrectly recovered loci over all the loci). (b) The empirical recombination rates along 100 SNP loci with the pre-specified recombination hotspots (dotted lines). (c) The true (panel 1) and estimated (panel 2 for HMDP, and panels 3–5 for the HMMs with 3, 5, 10 states, repsectively.) population maps of ancestral compositions.*

Ancestral inference. Using HMDP, we successfully recovered the correct number (i.e., $K = 5$) of ancestors in 21 out of 30 simulated populations; for the remaining 9 populations, we inferred 6 ancestors. From samples of ancestor states $\{a_{k,t}\}$, we reconstructed the ancestral haplotypes under the HMDP model. For comparison, we also inferred the ancestors under the 3 standard HMMs using EM (Fig. 2a).

LD-block analysis. From samples of the inheritance variables $\{c_{i,t}\}$ under HMDP, we can infer the recombination status of each locus of each haplotype. We define the empirical recombination rates λ_e at each locus to be the ratio of individuals who had recombinations at that locus over the total number of haploids in the population. Fig. 2b shows a plot of the λ_e in one simulated population. We can identify the recombination *hotspots* directly from such a plot based on an empirical threshold λ_t (i.e., $\lambda_t = 0.05$). The inferred hotspots (i.e., the λ_e peaks) show reasonable agreement with the true hotspots shown as vertical dotted lines.

Population structural analysis. Finally, from samples of the inheritance variables $\{c_{i,t}\}$, we can also uncover the genetic origins of all loci of each individual haplotype in a population. For each individual, we define an empirical *ancestor composition vector* η_e, which records the fractions of every ancestor in all the $c_{i,t}$'s of that individual. Fig. 2c displays a *population map* constructed from the η_e's (the thin vertical lines) of all individuals. Five population maps, corresponding to (1) true ancestor compositions, (2) ancestor compositions inferred by HMDP, and (3–5) ancestor compositions inferred by HMMs with 3, 5, 10 states, respectively, are shown in Fig. 2c. The L_1 distance between the HMDP-derived population map and the true map is 0.190, whereas the distance between HMM-map and true map is 0.319.

3.2. *Analyzing Two Real Haplotype Datasets*

We also applied HMDP to two real haplotype datasets, the single-population Daly data (Daly *et al.*, 2001), and the two-population HapMap data (Thorisson *et al.*,

2005); our method was able to uncover known recombination hotspots and population structures underlying these data. For details, see Xing and Sohn (2007).

4. CONCLUSION

We have proposed a new Bayesian approach for joint modeling genetic recombinations among possibly infinite founding alleles and coalescence-with-mutation events in the resulting genealogies. By incorporating a hierarchical DP prior to the stochastic matrix underlying an HMM, our method can efficiently infer a number of important genetic variables, such as recombination hotspot, mutation rates, haplotype origin, and ancestor patterns, jointly under a unified statistical framework. HMDP can also be easily adapted to more complicated genetics problems (e.g., analyzing unphased genotype data) and many engineering and information retrieval contexts such as object and theme tracking in open space. Due to space limit, we leave out some details of the algorithms and more results of our experiments, which are available in the full version of this paper (Xing and Sohn, 2007).

REFERENCES

Anderson, E. C. and Novembre, J. (2003). Finding haplotype block boundaries by using the minimum- description-length principle. *Am. J. Hum. Genet.* **73**, 336–354.

Beal, M. J., Ghahramani, Z. and Rasmussen, C. E. (2001). The infinite hidden Markov model. *Advances in Neural Information Processing Systems* **14**, 577–584.

Blackwell, D. and MacQueen, J. B. (1973). The infinite hidden Markov model *Ann. Statist.* **1**, 353–355.

Daly, M. J., Rioux, J. D., Schaffner, S. F., Hudson, T. J. and Lander, E. S. (2001). High-resolution haplotype structure in the human genome. *Nature Genetics* **29**, 229–232.

Greenspan, D. and Geiger, D. (2003). Model-based inference of haplotype block variation. *Proceedings of RECOMB* **7** 131-137

Liu, J. S., Sabatti, C., Teng, J., Keats, B. and Risch, N. (2001). Bayesian analysis of haplotypes for linkage disequilibrium mapping. *Genome Res.* **11**, 1716–1724.

Niu, T., Qin, S., Xu, X., and Liu, J. (2002). Bayesian haplotype inference for multiple linked single nucleotide polymorphisms. *Am J Hum Genet* **70**, 157–169.

Patil, N., Berno, A. J., *et al.* (2001), Blocks of limited haplotype diversity revealed by high-resolution scanning of human chromosome 21. *Science* **294**, 1719–1723.

Rosenberg, N. A., Pritchard, J. K., Weber, J. L., Cann, H. M., Kidd, K. K., Zhivotovsky, L. A. and Feldman, M. W. (2002). Genetic structure of human populations. *Science* **298**, 2381–2385.

Teh, Y., Jordan, M. I., Beal, M., and Blei, D. (2006). Hierarchical Dirichlet processes. *J. Amer. Statist. Assoc.* **101**, 1566–1581.

Thorisson, G.A., Smith, A.V., Krishnan, L. and Stein, L. D. (2005). The international hapmap project web site. *Genome Research* **15**, 1591–1593.

Xing, E. P., Sharan, R., and Jordan, M. (2004). Bayesian haplotype inference via the Dirichlet process. *Proc. of the 21st International Conference on Machine Learning*, New York: ACM Press. 879–886.

Xing, E. P. and Sohn, K.-A. (2007). Hidden Markov Dirichlet process: Modeling genetic inference in open ancestral space. *Bayesian Analysis* **2** (to appear).

Zhang, K., Deng, M., Chen, T., Waterman, M., and Sun, F. (2002). A dynamic programming algorithm for haplotype block partitioning. *Proc. Natl. Acad. Sci. USA* **99**, 7335–7339.